Modern Business Process Automation

Arthur H. M. ter Hofstede • Wil M. P. van der Aalst
Michael Adams • Nick Russell

Editors

Modern Business Process Automation

YAWL and its Support Environment

Editors

Arthur H. M. ter Hofstede
Queensland University of
Technology
School of Information Systems
Fac. Information Technology
GPO Box 2434, level 5,
126 Margaret Street,
Brisbane QLD 4001
Australia
a.terhofstede@qut.edu.au

Michael Adams
Queensland University of
Technology
School of Information Systems
Fac. Information Technology
GPO Box 2434, level 5,
126 Margaret Street,
Brisbane QLD 4001
Australia
m3.adams@qut.edu.au

Wil M. P. van der Aalst
Eindhoven University of
Technology
Dept. Mathematics &
Computer Science
Den Dolech 2
5600 MB Eindhoven
Netherlands
w.m.p.v.d.aalst@tue.nl

Nick Russell
Eindhoven University of
Technology
Fac. Technology Management
Den Dolech 2
5600 MB Eindhoven
Netherlands
n.c.russell@tue.nl

ISBN 978-3-642-03120-5 e-ISBN 978-3-642-03121-2
DOI 10.1007/978-3-642-03121-2
Springer Heidelberg Dordrecht London New York

Library of Congress Control Number: 2009931714

ACM Computing Classification (1998): J.1, H.4

Cover design: KuenkelLopka GmbH

Printed on acid-free paper

Springer is part of Springer Science+Business Media (www.springer.com)

Preface

The field of Business Process Management (BPM) is marred by a seemingly endless sequence of (proposed) industry standards. Contrary to other fields (e.g., civil or electronic engineering), these standards are not the result of a widely supported consolidation of well-understood and well-established concepts and practices. In the BPM domain, it is frequently the case that BPM vendors opportunistically become involved in the creation of proposed standards to exert or maintain their influence and interests in the field. Despite the initial fervor associated with such standardization activities, it is no less frequent that vendors either choose to drop their support for standards that they earlier championed on an opportunistic basis or elect only to partially support them in their commercial offerings.

Moreover, the results of the standardization processes themselves are a concern. BPM standards tend to deal with complex concepts, yet they are never properly defined and all-too-often not informed by established research. The result is a plethora of languages and tools, with no consensus on concepts and their implementation. They also fail to provide clear direction in the way in which BPM standards should evolve.

One can also observe a dichotomy between the "business" side of BPM and its "technical" side. While it is clear that the application of BPM will fail if not placed in a proper business context, it is equally clear that its application will go nowhere if it remains merely a motivational exercise with schemas of business processes hanging on the wall gathering dust.

An important observation that can be made about the state-of-the-art in BPM relates to tool support. Tool support has evolved considerably in the past decade, both in terms of the breadth of functionality that is provided and in terms of the range and capabilities of the vendors who are involved. However, because of the lack of effective standardization and direction in the field, BPM technology is not widely used. Commercial BPM tools are rarely used in small and medium-sized enterprises because of their prohibitive total cost of ownership. Acquisition costs tend to be high, and ongoing operational support and maintenance commitments can be even higher. One factor in this is that the closed nature of these products means that their customization to specific client requirements may be difficult or even impossible to achieve, and only the biggest users may be able to influence their future feature set. Another issue is the scarcity of knowledge about individual

BPM offerings and the fact that product knowledge does not generalize. This means that it remains difficult for end users to really leverage the capabilities of their BPM investment.

So where does this leave those interested in the field of BPM? First, it is unwise to become too aligned with or invest too heavily in particular standards, tools, or technologies. These all come and go. It is imperative that the conceptual, formal, and technological *foundations* of BPM are thoroughly understood. Only this way can one survive the onslaught of "the latest and greatest" in BPM and not have the wool pulled over one's eyes by vendors, training organizations, and other interested parties. Second, it is important to not only develop an understanding of the business context of BPM and its main drivers but also of the fundamentals of *business process automation*. To automate processes, it is vital to have a correct understanding of the operation of the business processes. This can be obtained by process mining and advanced simulation techniques. Measurable benefits can be realized by business process automation if the right set of patterns is supported. Finally, the (increased) uptake of Linux and other open-source products in the past decade has provided real momentum for the open-source movement. This has also influenced the BPM landscape, and where once there was no alternative to commercial, closed-source, and expensive solutions, such alternatives have now become viable in many instances. The markedly lower acquisition costs and the ability to modify the software to suit one's own needs cannot be ignored as factors in software selection processes and make the use of BPM technology an attractive proposition for a much wider range of potential users, including those in small and medium-sized enterprises for whom it was previously cost-prohibitive.

The book in front of you is the culmination of more than ten years of research and development conducted at universities in different parts of the world. It aims to provide the reader with a deep insight into *fundamental* concepts and techniques that are core to understanding BPM and its application. The focus is not on business/motivational aspects, though these aspects are not ignored.

The book is primarily intended as a textbook for undergraduate and postgraduate students in the field of workflow, or BPM more broadly. However, it is also eminently suitable as a reference for BPM researchers and professionals as it covers a wide range of BPM-related topics in a highly accessible yet thorough way. Exercises to deepen the reader's knowledge are provided throughout and the chapter notes at the end of many chapters provide the reader with references to further work in the area. The book uses *YAWL (Yet Another Workflow Language)*, a powerful language based on the well-known workflow patterns, and its open-source support environment, to explain advanced concepts and their realization.

This book is an edited work and would not have been possible without the contributions of a wide range of experts in the field. They take the reader through the fundamentals of business process modeling and automation, various aspects of flexibility, a number of current industry standards, and advanced topics such as integration, verification, mining, and configuration. In addition, implementation aspects relevant to modern BPM environments are addressed in depth and applications of BPM in the domains of health and screen business are discussed.

We feel privileged to have worked with many enthusiastic, committed, and knowledgeable people over the years. In addition to the authors of the various chapters in this book, we wish to thank the following people for their contributions to the YAWL initiative: Lindsay Bradford, Carmen Bratosin, Ross Brown, Francesco Cardi, Evan Chen, David Edmond, Tore Fjellheim, Matt Flaherty, Mike Fowler, Andrew Hastie, Saphira Heijens, Sean Kneipp, Jan-Christian Kuhr, Giancarlo La Medica, Massimiliano de Leoni, Alfredo Nantes, Ignatius Ong, Helen Paik, Jessica Prestedge, Guy Redding, Dan Simkins, Alex Streit, David Truffet, Sébastien Vicente, and Kenneth Wang.

We also gratefully acknowledge support from the Australian Research Council, the ARC Centre of Excellence for Creative Industries and Innovation, and the Nederlandse Organisatie voor Wetenschappelijk Onderzoek (NWO).

We sincerely hope that this book provides a valuable resource to the reader in the years to come.

Brisbane, Australia	Arthur ter Hofstede
Eindhoven, the Netherlands	Wil van der Aalst
Brisbane, Australia	Michael Adams
Eindhoven, the Netherlands	Nick Russell

Contents

Contributors

W.M.P. (Wil) van der Aalst Eindhoven University of Technology, Eindhoven, the Netherlands

M. (Michael) Adams Queensland University of Technology, Brisbane, Australia

L. (Lachlan) Aldred Queensland University of Technology, Brisbane, Australia

B. (Birger) Andersson Stockholm University and The Royal Institute of Technology, Stockholm, Sweden

P.J.M. (Piet) Bakker Academic Medical Center, Amsterdam, the Netherlands

S. (Stephan) Clemens Queensland University of Technology, Brisbane, Australia

G. (Gero) Decker University of Potsdam, Berlin, Germany

R.M. (Remco) Dijkman Eindhoven University of Technology, Eindhoven, the Netherlands

M. (Marlon) Dumas University of Tartu, Tartu, Estonia

C.J. (Colin) Fidge Queensland University of Technology, Brisbane, Australia

L. (Luciano) García-Bañuelos Universidad Autónoma de Tlaxcala, Tlaxcala, Mexico

F. (Florian) Gottschalk Eindhoven University of Technology, Eindhoven, the Netherlands

A.H.M. (Arthur) ter Hofstede Queensland University of Technology, Brisbane, Australia

M.W.M. (Monique) Jaspers University of Amsterdam, Amsterdam, the Netherlands

P. (Paul) Johannesson Stockholm University and The Royal Institute of Technology, Stockholm, Sweden

M. (Marcello) La Rosa Queensland University of Technology, Brisbane, Australia

R.S. (Ronny) Mans Eindhoven University of Technology, Eindhoven, the Netherlands

J. (Jan) Mendling Humboldt-Universität zu Berlin, Berlin, Germany

A.J. (Arnold) Moleman Academic Medical Center, Amsterdam, the Netherlands

C. (Chun) Ouyang Queensland University of Technology, Brisbane, Australia

M. (Maja) Pesic Eindhoven University of Technology, Eindhoven, the Netherlands

A. (Anne) Rozinat Eindhoven University of Technology, Eindhoven, the Netherlands

M.H. (Helen) Schonenberg Eindhoven University of Technology, Eindhoven, the Netherlands

N.C. (Nick) Russell Eindhoven University of Technology, Eindhoven, the Netherlands

H.M.W. (Eric) Verbeek Eindhoven University of Technology, Eindhoven, the Netherlands

P. (Petia) Wohed Stockholm University and The Royal Institute of Technology, Stockholm, Sweden

M.T. (Moe) Wynn Queensland University of Technology, Brisbane, Australia

Part I
Introduction

Chapter 1
Introduction

Wil van der Aalst, Michael Adams, Arthur ter Hofstede, and Nick Russell

1.1 Overview

The area of Business Process Management (BPM) has received considerable attention in recent years due to its potential for significantly increasing productivity and saving cost. In BPM, the concept of a *process* is fundamental and serves as a starting point for understanding how a business operates and what opportunities exist for streamlining its constituent activities. It is therefore not surprising that the potential impact of BPM is wide-ranging and that its introduction has both managerial as well as technical ramifications.

While benefits can be derived from BPM even when its application is restricted to what can be described as "pen-and-paper" exercises, such as the visualization of business process models in order to discuss opportunities for change and improvement, there is potentially much more to be gained if such an analysis serves as the blueprint for subsequent automation of key business processes. In the area of *Business Process Automation* (BPA), sometimes referred to as workflow management, precise business process descriptions are used to guide the performance of business activities. Work is delivered to selected resources, which can be either humans or software applications, when it needs to be executed. Progress can be monitored and may give rise to the escalation of certain tasks where their deadline has passed or is not likely to be met. Events, such as the completion of a certain task by a certain resource, are logged and the resulting log files can be exploited for analysis purposes, an area of interest in its own right typically referred to as *process mining*.

Substantial cost and time savings can be achieved through the use of workflow technology. When describing a workflow, which is an executable process, one has to capture all aspects relevant to automation, such as the individual activities (or *tasks*) and their execution order, data that is to be entered and passed on, and the way resources are involved. By taking this holistic view of a business process and capturing both the tasks and the data involved, it is less likely that inconsistencies arise from data not being entered or updated during the execution of a certain

W. van der Aalst (✉)
Eindhoven University of Technology, Eindhoven, the Netherlands
e-mail: w.m.p.v.d.aalst@tue.nl

A.H.M. ter Hofstede et al. (eds.), *Modern Business Process Automation*,
DOI 10.1007/978-3-642-03121-2_1, © Springer-Verlag Berlin Heidelberg 2010

process. Consider, for example, the case of an employee being granted permission for leave; this should not only result in an email notification to the applicant, but also in an update to the Human Resources records. By capturing both tasks and resources, it is possible to expedite processes using information regarding the current availability and workload of resources. Or, consider the case of a travel application delayed for 5 days, because it ended up in the in-tray of the absent director, while it should have been automatically rerouted to the acting director. Or the case of the sales inquiry that was left unanswered as it was directly addressed to a sales representative who had since left the company. A task allocation on the basis of roles rather than individuals would have avoided this problem. As a consequence of the explicit representation of tasks and their chronological dependencies, as well as the involvement of resources in the execution of these tasks, it is easier to adapt business processes in order to react in a timely manner to environmental changes, for example, market fluctuations or legislative adaptations. Instead of having to make changes somewhere deep in application code, these changes can be made at the specification level. Analysis and simulation support may help decide whether these changes satisfy certain correctness criteria or are likely to have their intended effect before they are actually deployed. Monitoring capabilities provide scope for rapid detection of problems and subsequent escalation, while post-execution log analysis (i.e., process mining) can provide a solid basis for process improvement.

In the field of BPM, it is recognized that business processes go through various stages of the so-called *BPM life-cycle*, cf. Fig. 1.1. Business processes start this life-cycle when they are created, either from scratch or through configuration of an existing model. This corresponds to the *process (re)design* phase in Fig. 1.1. The business process is subsequently implemented by configuring the corresponding information system. This *system configuration* phase may require substantial implementation efforts or may be a simple selection step. This all depends on the underlying technology and on how much customization is needed. Then the process can be executed and monitored in the *process enactment and monitoring* phase. Finally, in the *diagnosis* phase one can learn from the running process and use this as input for business process improvement. Diagnosis of the actual process

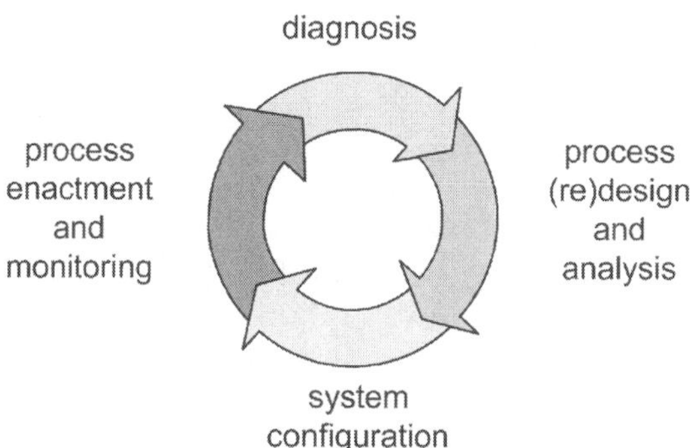

Fig. 1.1 The BPM life-cycle

execution may result in its adaptation. Adaptation may involve some modeling, which then may lead to a new or revised version of the business process for which the life-cycle starts again. Because of the automated support for managing the business process life-cycle, businesses can rapidly adapt to their ever-changing environment.

Much has been written about business processes; however, there is a lack of consensus about how they are best described for the purposes of analysis and subsequent automation. This has resulted in a plethora of approaches for capturing business processes, though not all are intended to support direct automation. There are two main reasons to which this situation can be attributed:

- Business processes can be complex. Their specification may involve capturing complex ordering dependencies between tasks and complex resourcing strategies. Process modeling languages tend to lack the concepts to be able to deal with the broad range of requirements one may encounter when trying to precisely capture business scenarios.
- Standardization efforts in the field have essentially failed. One may argue that this is the result of the standardization processes being partly driven by vested business interests. Whatever the reason, it is clear that today's standards lack widespread adoption and suffer from all kinds of technical problems.

The inherent complexity of business processes and the question of what fundamental concepts are necessary for business process modeling gave rise to the development of a collection of *workflow patterns*. These patterns describe process modeling requirements in an implementation independent manner.

In this chapter, it is shown how the workflow language YAWL and its corresponding system emerged from the Workflow Patterns Initiative. However, before doing so, the role of models in BPM is discussed and some of the standard approaches are reviewed.

1.2 On the Role of Models in BPM

Models can serve different purposes in BPM.

First of all, models may aim at providing *insight*. When developing or improving an information system, it is important that the different stakeholders get insight into the processes at hand and the way that these processes can or should be supported. Models can be used to discuss requirements, to support design decisions, and to validate assumptions. Moreover, the modeling process itself typically provides new and valuable insights, because the modeler is triggered to make things explicit.

Second, models may be used to *analyze* the system and/or its processes. Depending on the type of model, particular types of analysis are possible or not. In the context of BPM, analysis may focus on the business processes or on the information system itself. For example, the performance of a system (e.g., response times) is not the same as the performance of the processes it supports. Traditionally, most techniques used for the analysis of business processes originate from operations research. Students taking courses in operations management will learn to apply techniques such as simulation, queueing theory, and Markovian analysis. The focus mainly is on *performance analysis* and less attention is paid to the correctness of models. However, *verification* is needed to check whether the resulting system is free of logical errors. Many process designs suffer from deadlocks and livelocks that could have been detected using verification techniques. Notions such as soundness can be used to verify the correctness of the systems.

Finally, models can be used to *enact* processes. In the context of a BPM system, models are often used for enactment, that is, based on a model of the process, the corresponding runtime support is generated. In a workflow management system, a model of a process suffices to generate the corresponding system support. In other environments, the set of processes is often hard-coded. For example, although ERP systems like SAP have a workflow engine, most processes are hard-coded into the system and can only be changed by programming or changing configuration parameters. As a result, modifications are either time-consuming (because of substantial programming efforts) or restricted by the set of predefined configuration parameters.

Figure 1.2 shows another view on the role of (process) models. Models can be used to model and analyze operational processes without explicitly considering the information system. Consider, for example, the use of business process simulation in environments where Information Technology (IT) plays only a minor role. However, models can also be used to configure and/or implement systems. The basic idea of workflow technology is to generate systems on the basis of models. Figure 1.2 emphasizes the role of event logs. Information systems record events and this information can be used when designing or analyzing processes. In fact, workflow management systems provide excellent facilities for logging events at

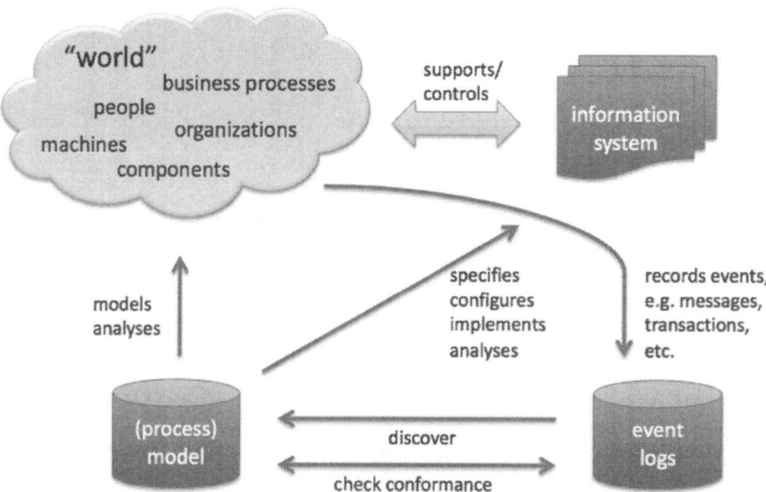

Fig. 1.2 On the role of (process) models in a BPM setting

the business process level. This information can be used to automatically discover models based on frequent patterns. Moreover, if a model is given, then the conformance of the process with respect to this model can be measured. The latter is interesting when analyzing compliance and process flexibility. Both conformance checking and process discovery are part of the process mining domain.

1.3 BPM Standard Approaches

Over the years, there have been many approaches toward the specification of business processes. Many BPM tools supported their own languages and it was often unclear how these languages compared. Over time a number of standards and/or widely used approaches emerged, and we will briefly look at some of the more important ones in this section.

One of the first attempts to define a standard approach to the specification of executable business processes was the XML Process Definition Language (XPDL) 1.0, defined in the nineties by the Workflow Management Coalition (WfMC), an industry body promoting the spread and further development of workflow technology. XPDL was intended to facilitate interoperability between workflow environments. The language offered a minimal set of generally occurring routing constructs such as various splits and joins, and these were defined in natural language. Due to this minimalist approach and to the fact that various interpretations of even these basic constructs was possible, the goal of interoperability was not achieved and gradually XPDL 1.0 became irrelevant.

In 2003, the Business Process Execution Language (BPEL) was proposed. This language combined Microsoft's XLANG and IBM's Web Services Flow Language (WSFL) and is therefore a language that marries two fundamentally different approaches to the specification of executable business processes. Generally speaking, BPEL is a block-structured language where business processes are specified in terms of self-contained blocks that are composed to form larger, more complex, blocks. However, BPEL is not fully block-structured as it supports the specification of dependencies that cross block boundaries through the use of so-called control links. While BPEL was a clear step forward in terms of its support for the specification of control-flow dependencies, the language provided no support for the involvement of human participants in the execution of business activities. In addition, the language has no graphical representation; specifications have an XML-based depiction.

The Business Process Modeling Notation (BPMN) was introduced to provide an easily understood graphical notation that could serve as a front end to various approaches for the execution of business processes. The language itself is not intended to be directly executable, rather specifications are expected to be transformed to an executable language to achieve their enactment. BPMN provides fairly strong support for the specification of control-flow dependencies and is graph-structured rather than block-structured. Contrary to BPEL, BPMN imposes no restrictions on the specification of loops, and loops are allowed to have multiple entry and/or exit points. A consequence of this is that mapping a BPMN specification to a corresponding BPEL specification can be a less than trivial matter, and contemporary support tools typically impose restrictions on BPMN diagrams for the purpose of subsequent transformation to BPEL. Similar to BPEL, though slightly better, BPMN does not make much provision for the various ways in which human participants can be involved in the execution of a business process, and given that BPMN does not have a formalization accepted by a standards organization, the interpretation of some of its concepts may vary. Nonetheless, BPMN can be seen as a move in the direction of more expressive languages, and its continued evolution and increased adoption makes it likely to have some longevity. In recognition of this, XPDL has been reinvented and its 2.0 incarnation is an XML serialization of BPMN.

Although not a formal standard, Event-driven Process Chains (EPCs) are a well-known approach to process specification and the notation has been around for over 15 years. EPCs are supported by the ARIS environment, where they are used for business process modeling and simulation. EPCs are not directly executable and they provide a fairly minimal set of control-flow constructs. Extended EPCs augment EPCs with notations for the involvement of participants and the use of data elements. EPCs do not have a formal foundation, though one has been defined by Wil van der Aalst in the late nineties. As he argued that the semantics of the OR join connector are "not clear" and "subject to multiple interpretations," this formalization does not incorporate this particular connector. It has since been argued that there are inherent semantical problems with this concept in the presence of so-called "vicious circles."

Another well-known approach to process specification are the Activity Diagrams of the Unified Modeling Language (UML). In their 1.4 incarnation, these were based on statecharts, while in their 2.0 incarnation, their semantics was more inspired by Petri nets. Because of the fact that UML 2.0 activity diagrams do not have a notion that corresponds to the concept of a place in Petri nets, the link between UML 2.0 activity diagrams and Petri nets is rather complicated. This is relevant because of two reasons. First of all, certain business scenarios mixing concurrency and choice cannot be expressed easily. Second, there is no simple and clear semantics. UML activity diagrams are not intended for direct execution and, although a formal semantics for them has been defined, no formalization has been officially endorsed by the Object Management Group (OMG), which is the standardization body behind UML. Furthermore, it seems that in recent years UML Activity Diagrams have been eclipsed by BPMN in the context of the specification of business processes.

1.4 The Workflow Patterns Initiative

The concept of workflow has been around for decades and can be traced back to early work on office automation. Despite its early origins, widespread uptake of workflow management systems in practice and the integration of this technology with other types of systems, generally described as process-aware information systems, did not occur until the mid to late nineties. Although there are technological considerations involved, the lack of a commonly accepted foundation for the specification of executable business processes played a major role in the slow progression of workflow to broad adoption. This resulted in a plethora of approaches, where concepts with similar names could have significant semantic differences. The Workflow Management Coalition (WfMC) failed to provide a standard that was (1) sufficiently precise, and (2) sufficiently expressive. As part of its definition of "interface 1," it provided natural language definitions of a number of commonly occurring workflow concepts. These definitions led to a situation where vendors could legitimately claim to abide by the WfMC standard, even though their interpretation of these concepts was fundamentally different and could lead to the same workflow model being executed in different ways. In addition, the concepts that were described by the WfMC captured only simple control-flow dependencies. This meant that workflow migration was not only hampered by different interpretations of similarly named concepts, but also by the fact that some concepts were supported in one environment but did not have a counterpart in another.

These were the circumstances that gave rise to the *Workflow Patterns Initiative* in the second half of 1999. The original founders recognized that there was a need to distill the essential features of the many workflow management systems that existed. This would allow an unbiased comparison of different approaches to the specification of executable business processes and provide the basis for the adaptation and refinement of existing techniques as well as supporting the development of new approaches.

The approach chosen focussed on identifying the constructs required for the specification of control-flow dependencies between tasks. The ability to explicitly capture tasks and their chronological dependencies is an essential capability of workflow management systems. Initially, 13 commercial workflow management systems and two research prototypes were examined with respect to the constructs they offered for control-flow specification. This resulted in a collection of 20 control-flow *patterns*. Following the book by Gamma et al., which provided patterns for object-oriented design, patterns have become a popular way of identifying recurring problems and corresponding solutions in various areas of computer science. The Workflow Patterns consisted of a description of desired control-flow functionality, problems with realizing this functionality and implementation approaches. Workflow management systems were rated on a three point scale against these patterns; the highest score was given where there was direct support for a pattern, while the intermediate score indicated there were some restrictions in terms of direct pattern support. The lowest score did not mean that the pattern could not be realized from a theoretical point of view, as scripting languages are all Turing-complete, but simply that the system involved did not provide direct support. Therefore, the patterns are not concerned with *expressive power* but with *suitability*, a notion that refers to the alignment between a modeling language and a problem domain.

Over time, the patterns-based evaluations were extended to business process modeling languages (e.g., UML Activity Diagrams, BPMN), standards for web service composition (e.g., BPML, BPEL), and open-source workflow management systems (e.g., jBPM, OpenWFE). The patterns collection itself was revised and significantly extended with data patterns, resource patterns, exception handling patterns, service interaction patterns, flexibility patterns, etc.

1.5 Petri Nets and Workflow Nets

While there has not been a commonly accepted formal foundation for workflow management, in the mid nineties, Wil van der Aalst articulated three reasons why *Petri nets* would make a good candidate. The theory of Petri nets, developed by Carl Adam Petri, provides an elegant approach to describing and solving concurrency related problems.

A Petri net is a bipartite graph where the nodes are either places or transitions. A Petri net has a graphical representation where places are represented by circles, transitions by squares, and their connections by directed arcs. A Petri net may have an associated *marking*, which is an assignment of tokens to places. A marking represents the state of a system. A transition is said to be *enabled* in a certain marking when each of its input places contains at least one token. An enabled transition can fire by taking a token from each of its input places and producing a token in each of its output places. Transitions thus correspond to allowed stated changes. A vast body of theory exists for the formal analysis of Petri nets.

As van der Aalst pointed out, the graphical nature of Petri nets, their explicit representation of the notion of state, and the existence of analysis techniques made them eminently suitable for workflow specification. To increase their suitability for workflow specification, he introduced *workflow nets*. A workflow net is a Petri net with one start place and one end place. All tasks have to be on a path from the start place to the end place. To simplify the representation of typical workflow routing constructs, such as AND-splits and XOR-splits, a number of graphical abbreviations were introduced. Workflow nets form a subclass of Petri nets for which the analysis of desirable properties, for example, whether process instances of a workflow can always terminate, is feasible. Tool support for this type of analysis was developed in the form of tools such as Woflan (Workflow Analyzer), ProM, WoPeD, etc.

1.6 The Emergence of YAWL

After the development of the initial collection of workflow patterns, Petri nets were revisited in terms of their suitability for the specification of control-flow dependencies in workflows. While it turned out that many of the patterns could be expressed in a straightforward manner, some patterns were not so easily captured. This observation led to the development of YAWL (Yet Another Workflow Language). In YAWL, Petri nets were taken as a starting point and extended with dedicated constructs to deal with patterns that Petri nets have difficulty expressing, in particular patterns dealing with cancelation, synchronization of active branches only, and multiple concurrently executing instances of the same task.

YAWL was given a formal semantics based on a state transition system and therefore its nets cannot be simply mapped to Petri nets. Hence, while YAWL is inspired by Petri nets, it cannot be seen as a set of notational abbreviations on top of this formalism. Specific verification approaches needed to be developed to deal with the new constructs offered by YAWL. One formalism that turned out to be particularly useful for reasoning about YAWL nets were *reset nets*. These are Petri nets extended with the concept of a reset arc. When a transition is executed, all tokens are removed from places that are connected with a reset arc to this transition. Reset arcs allow the concept of cancelation to be directly expressed. In workflows cancelation implies that the execution of a certain task should lead to other nominated tasks being terminated or not being available for further execution. YAWL offers direct support for cancelation, while its predecessor, workflow nets, does not.

After YAWL was defined, work started on the implementation of a support environment. This effort intended to demonstrate that it was possible to provide comprehensive support for the (original) workflow control-flow patterns. As such the YAWL environment can be seen as a reference implementation that supports the Workflow Patterns. The term YAWL became synonymous with both the language and the support environment. Over time as the environment evolved, the ambitions increased and YAWL grew into a full-fledged workflow management system, which is used in a wide variety of academic and industrial settings.

1.7 A Brief Overview of YAWL

To illustrate and further concretize some of the concepts we have mentioned thus far, let us consider a simplified scenario involving a car accident. In this scenario, the first step that needs to be taken is to obtain a quote for the costs involved in dealing with the damage. After this quote has been received, a preliminary insurance claim is lodged, and a choice needs to be made whether the car is going to be fixed or whether buying a new car is more cost-effective. After this latter decision has been made the bill can be settled, and when this has happened and the preliminary insurance claim has been lodged, the final insurance claim can be lodged.

In Fig. 1.3, a Petri net is shown that captures the flow of control of this simple example. The rectangles in this net are the transitions, which correspond to tasks that need to be performed, while the places correspond to moments in-between processing. The place that is input for the transitions *Have Car Fixed* and *Buy New Car* captures a choice made by the environment. When a token has been produced by the transition *Obtain Quote* for this place, both transitions *Have Car Fixed* and *Buy New Car* are enabled. Once one of these has been chosen, the token is removed from this place and the other transition is no longer enabled. When a transition has multiple output places, this represents parallelism as all these places receive a token upon

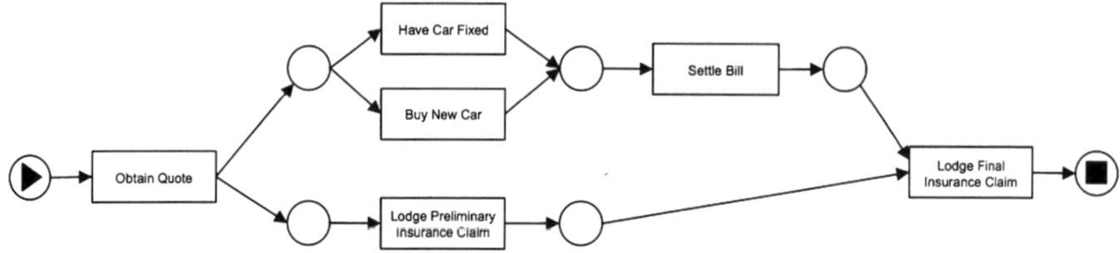

Fig. 1.3 A Petri net for the Accident workflow

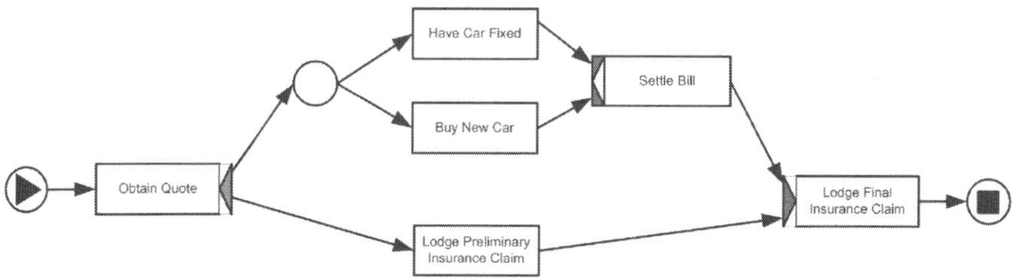

Fig. 1.4 A YAWL net for the Accident workflow

completion of this transition. Similarly, if a transition has multiple input places, this represents a synchronization point as at least one token is required in each input place for this transition to be enabled.

Figure 1.4 shows the YAWL representation of the Petri net of Fig. 1.3. As can be seen, in a YAWL net, there is a more explicit representation of the join and split behavior of the various tasks. In addition, there is no need to explicitly represent all places; those that simply connect two tasks can be omitted. The YAWL model captures the order in which the tasks need to be presented at runtime. To make the specification available to the runtime environment, it needs to be specified in the Editor, see Fig. 1.5, which can save the model in an XML format that can be interpreted by the Engine. In addition, one needs to specify how work is assigned to participants. To keep this example simple, let us assume that all tasks are to be executed by the *Claimant*, which we have captured as a role with one participant. Also, tasks are offered to the *Claimant* and he/she can choose when to start working on them.

At runtime, the Engine uses the YAWL model to determine when certain tasks are to be offered to the *Claimant*. In Fig. 1.6, one can see the worklist containing the work items (i.e. task instances) offered to the *Claimant* after the task *Obtain Quote* has been performed. Note that both options *Buy New Car* and *Have Car Fixed* are offered and the *Claimant* needs to make a choice.

Assuming that the *Claimant* completed the choice for having their car fixed or having the current one repaired, the task *Settle Bill* is offered. In Fig. 1.7, the data input form for the corresponding work item is shown. One can see that performing this work item involves filling in a number of data fields. The presentation of these fields is governed by the type of the data involved (which in this example are all simple types) and whether the fields contain editable values or are for presentation

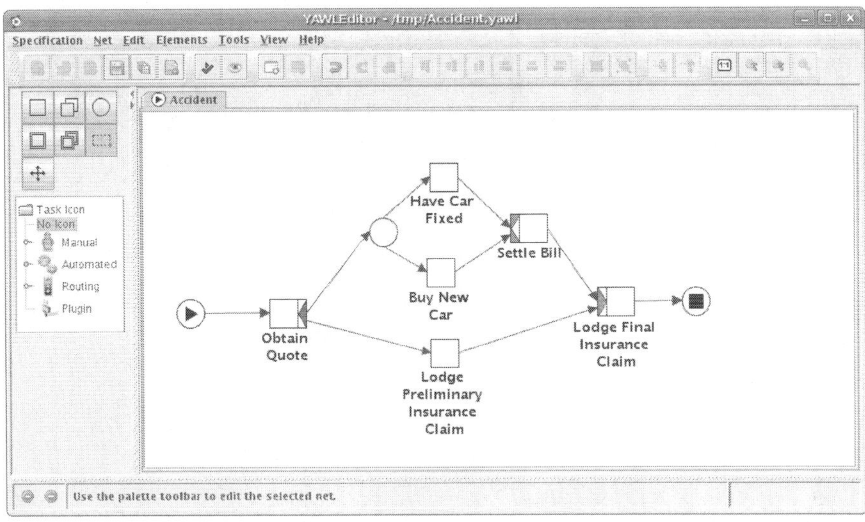

Fig. 1.5 A YAWL net for the Accident workflow in the Editor

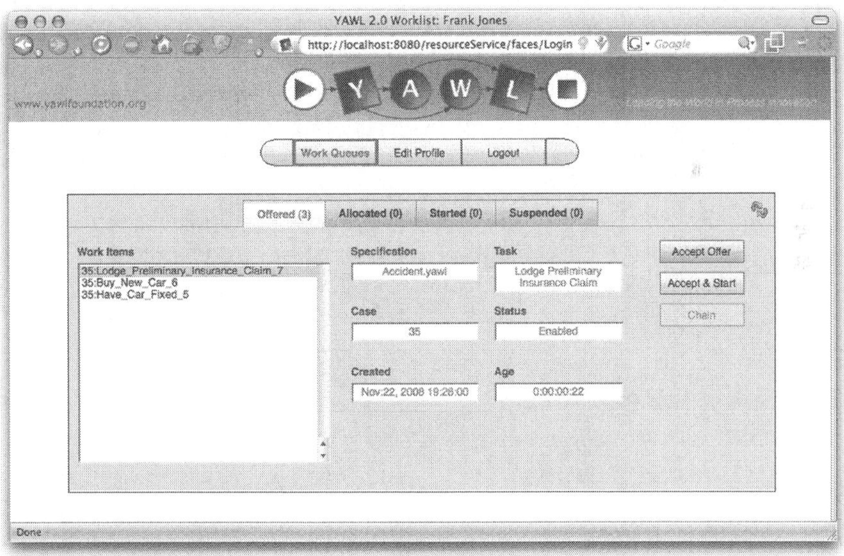

Fig. 1.6 Sample worklist for the Accident workflow

purposes only. Upon completion of the work item, this information is sent back to the Engine, which, in general, may pass it on to other tasks or use it to determine which tasks should be performed next.

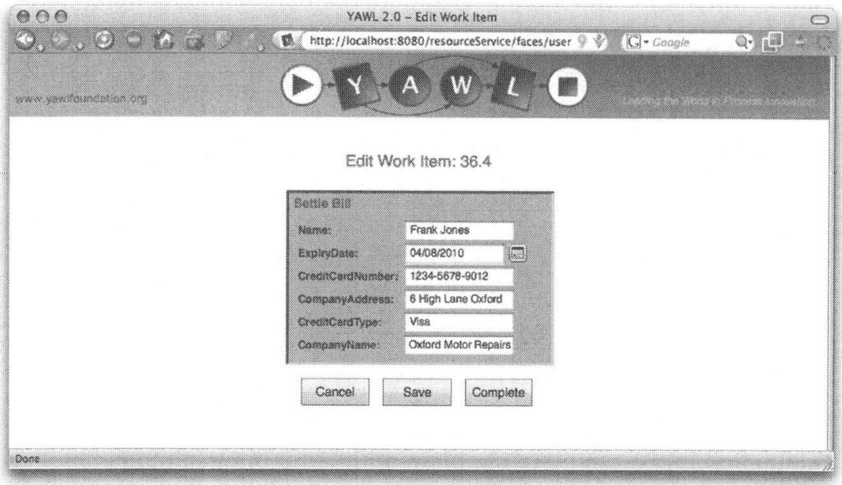

Fig. 1.7 Performing the task *Settle Bill*

The simple example illustrates that in the YAWL environment, as in virtually all BPM environments, there is a distinction between *build time* and *runtime*. Models are constructed in the build time component, the Editor (discussed in detail in Chap. 8), and subsequently deployed in the runtime environment. The runtime environment itself can consist of a number of components, but at its core are the Engine (discussed in detail in Chap. 9) and the Resource Service (discussed in detail in Chap. 10). The Engine deals with the control-flow logic and workflow data, while the Resource Service is concerned with the routing of work items to appropriate resources. A simplified overview of the YAWL architecture is presented in Fig. 1.8 (a detailed discussion of the YAWL architecture can be found in Chap. 7).

1.8 Positioning of YAWL

YAWL has a number of features that position it uniquely in the crowded field of BPM. The language could be developed without the pressures of vested interests and a sole focus on providing *comprehensive support for the Workflow Patterns* was possible. In contrast to BPMN, comprehensive control-flow patterns support was achieved through the introduction of a minimal set of constructs, rather than a construct-per-pattern approach. More recently, comprehensive support for the resource patterns was realized in YAWL, both the language and the support environment.

Another distinguishing feature of YAWL is the fact that it has a *formal foundation*, that is, both a precisely defined syntax and a precisely defined semantics. In the latest version of YAWL, the (abstract) syntax has been defined through the use of set

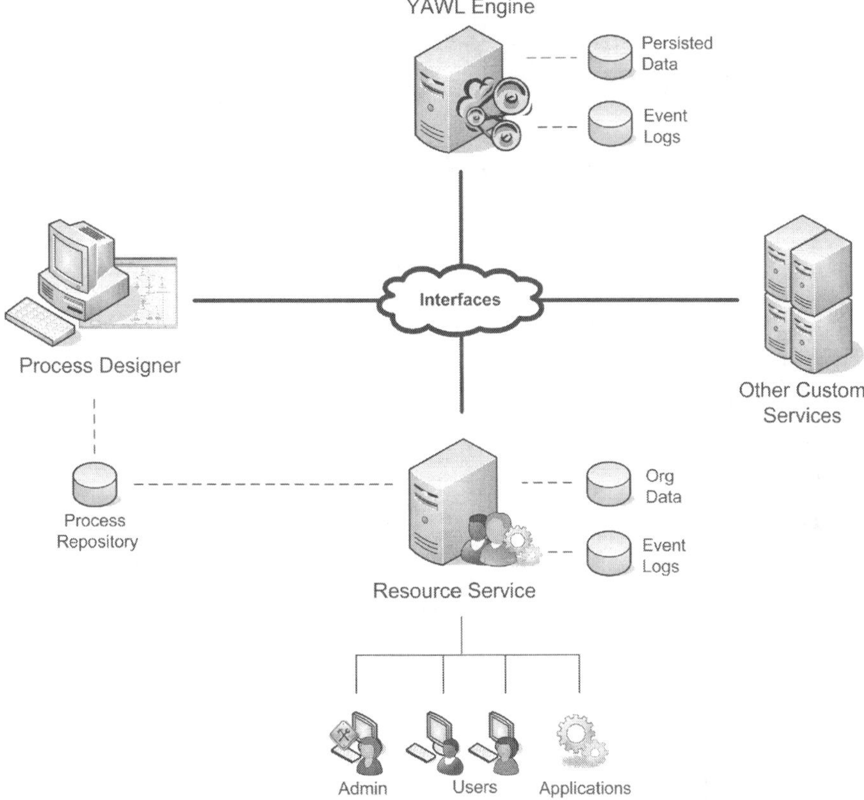

Fig. 1.8 Simplified architectural overview of the YAWL environment

theory and predicate logic, while the semantics has been defined in terms of a large Colored Petri net (CPN), which can interpret YAWL specifications. This formal foundation removes any ambiguity associated with the interpretation of complex constructs and their interplay and also allows for the development of sophisticated verification techniques that allow the detection of inherent flaws in an executable process model before it is deployed (for a detailed treatment of verification in YAWL see Chap. 20). While it is sometimes claimed that certain standards or oft-used approaches have a formal foundation, the problem is usually that either (1) the connection between the language and the formal theory remains unclear, or (2) the formalization is not generally accepted, and certainly not by a standard body.

The importance of *sophisticated flexibility support* in BPM systems has long been recognized. Because of the complexity of providing this support, it has taken quite a long time for satisfactory solutions to become available. Flexibility requirements may take different forms. For example, exceptions may arise that were not anticipated in advance, processes may be more easily captured by rules rather than through an explicit representation of all possible execution paths, and processes may evolve over time due to changes in a business and/or its environment. As in different contexts different solutions may be desirable, YAWL not only provides services supporting various flexibility requirements but also allows these services to interoperate. This leads to a powerful approach in dealing with such requirements. For example, it is possible to combine the procedural YAWL language with worklets (cf. Chap. 4) and Declare (cf. Chap. 6). Worklets can be used to select process fragments at runtime based on rules. Declare can be used to enact processes based on (temporal) constraints.

The YAWL support environment is *open source* and is therefore not only freely available but can also be extended as desired, thus avoiding vendor lock-in. Its *service-oriented* architecture, with a rich set of interfaces, allows the system to interact with other systems and to be extended in a variety of ways.

Through a *link with the ProM environment*, YAWL logs can be analyzed in a number of ways and YAWL specifications can be simulated. For example, it is possible to do a simulation that starts with the current state and that looks into the near future, for example, to decide whether it is beneficial to hire more resources or not. The log analysis can reveal, among others, what the typical walk-throughs are through a process or what the probabilities are of the process choosing one alternative path over another.

Finally, one can *leverage the benefits of BPMN* through the BPMN2YAWL plug-in, where BPMN models can be mapped onto YAWL specifications. Contrary to BPEL, the resulting YAWL models remain readable and maintain a close link with the original BPMN models. This is a consequence of the fact that in YAWL arbitrary cycles can be specified and that YAWL offers comprehensive support for the control-flow patterns in general. The BPMN2YAWL plug-in is discussed in detail in Chap. 13.

1.9 Overview of the Book

This book is about YAWL, both the language and the open source support environment. It discusses the foundations of the language, its unique support for flexibility, the support environment, services that can interoperate with this environment, its relationship to other well-established approaches in the field of BPM, and several applications. There are many technical papers that focus on various aspects of YAWL. This book aims to make this work accessible to a wider audience and is therefore less technical in nature. Where possible, references are provided so that the interested reader can deepen her understanding of specific topics. Another objective of this book is to bring the main material concerning YAWL together in one place and to properly integrate this material. To facilitate the understanding of the reader, a running example is provided in the area of supply chain management.

The book is divided into nine parts. Part II provides the conceptual foundation of YAWL. It explains the concepts required for workflow specification and introduces the running example.

Workflows may need to evolve over time, have to deal with unforeseen exceptions, and sometimes their specification is more easily achieved by specifying constraints that need to be satisfied at runtime rather than by providing an explicit road map for all possible execution paths. In Part III, various aspects of flexibility are examined.

Part IV discusses the core YAWL environment, its architecture, its design component, and its main runtime component. YAWL's architecture is service-oriented, which means that its various components communicate through well-defined interfaces. The design component allows the creation and verification of workflows. Once completed, they can be passed to the runtime component, the Engine, for execution.

In Part V, a number of services are presented, which can interact with the core YAWL environment and provide additional functionality. The services discussed provide concrete ways of achieving flexibility in workflow specification and execution (the *Worklet Service* and the *Declare Service*) or deal with work distribution, allowing the Engine to remain resource-agnostic (the *Resource Service*).

Part VI positions YAWL with respect to a number of well-known approaches to business process specification and/or execution, the Business Process Modeling Notation (BPMN), Event-driven Process Chains (EPCs), the Business Process Execution Language (BPEL), and a number of other well-known open-source workflow management systems. It is shown how BPMN models can be mapped to YAWL nets, how EPCs compare to YAWL and how they can be mapped to YAWL nets and vice versa, how BPEL relates to YAWL, and how open-source workflow management systems such as jBPM and OpenWFE compare with YAWL.

In Part VII a number of advanced topics in the field of BPM are examined. The connection with the Process Mining (ProM) framework is discussed and it is demonstrated how this framework can be used to mine valuable information from execution logs that have been generated by the YAWL environment. Correctness notions for YAWL specifications are introduced and precisely defined and it is shown how, and to what extent, these notions can be automatically verified. Sometimes new process models can be derived from existing process models and the notion of process configuration is explored in more depth through the C-YAWL (Configurable YAWL) approach. Finally, workflows may be running in different settings (e.g. different organizations or departments), but there may be a need for them to exchange information. Ideally such needs are expressed at the specification level, not at the implementation level.

Part VIII discusses two applications in which the YAWL environment has been used. One of these applications is in the healthcare domain, and the other is in the domain of film production.

Part IX concludes the book and provides an outlook for future developments.

Exercises

Exercise 1. Why is it important that a workflow specification language has a formal syntax and semantics? Provide at least three reasons.

Exercise 2. Explain the difference between the notions of expressive power and suitability in the context of a workflow specification language.

Exercise 3. Explain the value proposition of patterns.

Chapter Notes

The original founders of the Workflow Patterns Initiative were Wil van der Aalst, Alistair Barros, Arthur ter Hofstede, and Bartek Kiepuszewski. The first paper appeared in the CoopIS conference in Eilat in 2000 [7] and the main reference appeared in the Distributed and Parallel Databases journal in 2003 [16]. Since then there have been many additional contributors. Since 2005 Nick Russell has been the main driver in developing workflow patterns for the data [225] and the resource perspectives [222] as well as for exception handling [221]. He also led the revision effort of the original control-flow patterns [224]. Marlon Dumas and Petia Wohed have been involved in many patterns-based evaluations (see, e.g., [84, 223, 262–265]). Nataliya Mulyar devoted her PhD work [178] to identifying additional patterns. The main reference site for the Workflow Patterns is www.workflowpatterns.com. A book on the Workflow Patterns is expected to be published by MIT Press in the near future.

An early influential book in the area of workflow management was the book by Stefan Jablonski and Christoph Bussler [124]. Among others, a number of perspectives were described for workflow management, some of which are referred to in this chapter, though under different names. The book by Wil van der Aalst and Kees van Hee [11] is a more recent book on workflow management and, among other topics, discusses in depth the application of Petri nets in this area. Mathias Weske's book [260] discusses the original control-flow patterns, but also workflow nets and YAWL, and, more generally, provides a treatment to the field of business process management from a computer science perspective. Although it is impossible to provide an overview of workflow literature here, we would also like to mention the workflow books by Leymann and Roller [150], Marinescu [157], zur Muehlen [174], and Lawrence [146]. A recent overview of important topics can be found in the book on process-aware information systems [82].

Patterns development became popular in computer science due to the work of Erich Gamma, Richard Helm, Ralph Johnson, and John Vlissides [100] in the area of software design patterns. Other examples of well-known patterns collections in software analysis and design were developed by Gregor Hohpe and Bobby Woolf [120] and by Martin Fowler [98].

Kiepuszewski et al. [131] examined the consequences of various different interpretations of workflow routing constructs defined by the Workflow Management Coalition from an expressiveness point of view.

There are several comprehensive treatments of Petri nets (e.g., [179, 196]). Their use for workflow specification was advocated by Wil van der Aalst [1]. Dufourd et al. [80, 81] provide an introduction to reset nets.

In 2002, van der Aalst and ter Hofstede defined the YAWL language [15]. The development of the support environment started the following year [6]. The first open-source release of this environment occurred in the same year. In subsequent releases, the functionality was extended considerably. In [219], YAWL and some of its extensions were modeled using CPN Tools.

Part II
Concepts

Chapter 2
The Language: Rationale and Fundamentals

Nick Russell and Arthur ter Hofstede

2.1 Introduction

The Business Process Management domain has evolved at a dramatic pace over the past two decades and the notion of the business process has become a ubiquitous part of the modern business enterprise. Most organizations now view their operations in terms of business processes and manage these business processes in the same way as other corporate assets. In recent years, an increasingly broad range of generic technology has become available for automating business processes. This is part of a growing trend in the software engineering field throughout the past 40 years, where aspects of functionality that are potentially reusable on a widespread basis have coalesced into generic software components. Figure 2.1 illustrates this trend and shows how software systems have evolved from the monolithic applications of the 1960s developed in their entirety often by a single development team to today's offerings that are based on the integration of a range of generic technologies with only a small component of the application actually being developed from scratch.

In the 1990s, generic functionality for the automation of business processes first became commercially available in the form of workflow technology and subsequently evolved in the broader field of business process management systems (BPMS). This technology alleviated the necessity to develop process support within applications from scratch and provided a variety of off-the-shelf options on which these requirements could be based. The demand for this technology was significant and it is estimated that by 2000 there were well over 200 distinct workflow offerings in the market, each with a distinct conceptual foundation. Anticipating the difficulties that would be experienced by organizations seeking to utilize and integrate distinct workflow offerings, the Workflow Management Coalition (WfMC), an industry group formed to advance technology in this area, proposed a standard reference model for workflow technology with an express desire to seek a common platform for achieving workflow interoperation.

Although a worthy aim, the proposal met with conceptual difficulties when it came to specifying the details associated with workflow operation and potential

N. Russell (✉)
Eindhoven University of Technology, Eindhoven, the Netherlands
e-mail: nickr@computer.org

A.H.M. ter Hofstede et al. (eds.), *Modern Business Process Automation*,
DOI 10.1007/978-3-642-03121-2_2, © Springer-Verlag Berlin Heidelberg 2010

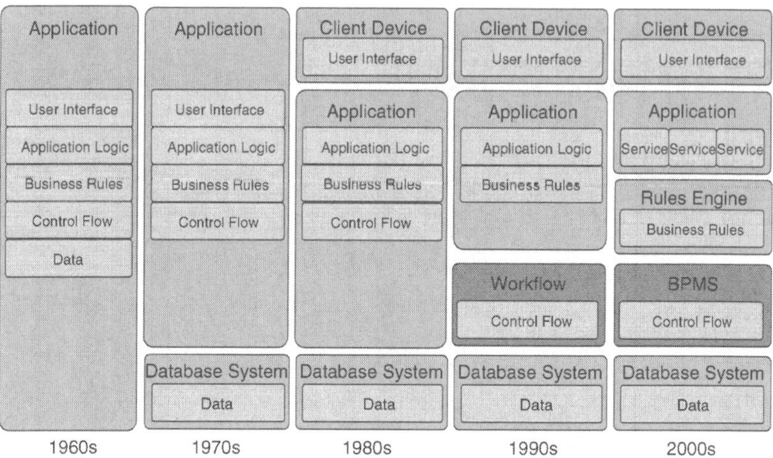

Fig. 2.1 Evolution of BPM technology

interaction schemes. Moreover, individual workflow vendors were reluctant to commit to a common operational platform that would leave them with minimal opportunity for product differentiation. The net result was that the Workflow Reference Model and the associated standards proffered by the WfMC essentially constituted the lowest common denominator of workflow concepts acceptable to all parties rather than laying a foundation for the workflow domain more generally.

Nonetheless, the issues identified remain unaddressed and there is a marked absence of a common conceptual foundation for workflow technology or for the area of business process management more generally. Furthermore, there are a plethora of competing approaches to business process modeling and enactment, and the lack of an agreed set of fundamentals in the domain means that direct comparisons between them and integration of their functionality is extremely difficult. In light of these issues, in 1999, the Workflow Pattern Initiative was conceived as an empirical means of identifying the core functionality required for workflow systems.

During the past 10 years, over 100 patterns have been identified that are relevant to workflow technology and to the various perspectives of business processes more generally. One of the criticisms that the patterns faced early on was that they represented isolated process concepts and did not give a guide as to the form that a process language should take. In response to this, YAWL (Yet Another Workflow Language) was developed. Initially, it sought to show the manner in which the original 20 control-flow patterns should be operationalized in a workflow language. More recently, it has been expanded to encompass a broader range of the overall set of workflow patterns. In tandem with the language effort, the YAWL System has also been developed with the aim of providing a reference implementation for the YAWL language and workflow technology.

In this chapter, we will explore the fundamental underpinnings of the YAWL language, looking at the precursing workflow patterns, then examining the formal foundations on which the language is based and finally reviewing the language constructs of which it is comprised.

2.2 Workflow Patterns

In 1999, the Workflow Patterns Initiative commenced with the aim of identifying the generic, recurrent constructs relevant to business processes. In the 10 years since the initial proposal, the patterns have been extended to provide comprehensive coverage of the desirable constructs in the control-flow, data and resource perspectives of business processes and now delineate 126 distinct patterns. This section will present an overview of the main pattern groupings in each of these perspectives and discuss the more significant patterns in further detail.

2.2.1 Control-Flow Patterns

The control-flow patterns describe language features for managing the flow of control amongst the various tasks that make up a business process. These patterns can be divided into eight distinct groups based on their area of focus:

- **Branching** patterns capture scenarios where the thread of control in a process divides into two or more independent execution threads, possibly in distinct branches.
- **Synchronization** patterns describe scenarios where several independent threads of control in a process (possibly in distinct branches) need to be synchronized into a single branch.
- **Repetition** patterns describe various ways in which repetitive tasks or subprocesses can be specified in a process.
- **Multiple instance (MI)** patterns characterize situations where multiple concurrent instances of a task or subprocess execute simultaneously and may need to be synchronized upon completion.
- **Concurrency** patterns reflect situations where restrictions are imposed on the extent of concurrent execution within a process.
- **Trigger** patterns identify constructs that allow the execution in a process past a specified point to be made contingent on the receipt of a signal from the operating environment.
- **Cancelation** and *completion* patterns categorize the various cancelation scenarios that may arise in a business process.
- **Termination** patterns which address the issue of when the execution of a process can be considered to be finished.

Each of these groups contains a number of patterns that describe desirable behaviors in a business process and we will now consider these in more detail on a group-by-group basis.

Branching Patterns

Branching patterns identify various types of branching behaviors that are generally associated with split operators in a process. Four distinct types of split are commonly delineated:

- The AND-split, which diverges the thread of control in a given branch into multiple concurrent execution threads in several branches.
- The XOR-split (or the eXclusive OR-split), which routes the thread of control in a given branch into one of several possible outgoing branches on the basis of an execution decision made at runtime.
- The OR-split, which routes the thread of control in a given branch into one, several or all outgoing branches on the basis of an execution decision(s) made at runtime.
- The Thread-split, which diverges the thread of control in a given branch into multiple concurrent execution threads in the same branch.

This distinction between each of these operators is illustrated in Fig. 2.2a–d, which provide examples of the various split constructs in the context of making a cup of tea. Figure 2.2a shows an AND-split, where after the *pour tea* task, both the *add milk* and *add sugar* tasks must always occur. In contrast, in Fig. 2.2b, based on an XOR-split, after the *pour tea* task, only one of the *add milk* or *add lemon* tasks executes. Figure 2.2c illustrates a more flexible scenario based on an OR-split, where after the *pour tea* task, one or both of the *add milk* and *add sugar* tasks executes depending on desired drinking preferences. Finally, Fig. 2.2d illustrates the Thread-split operator,

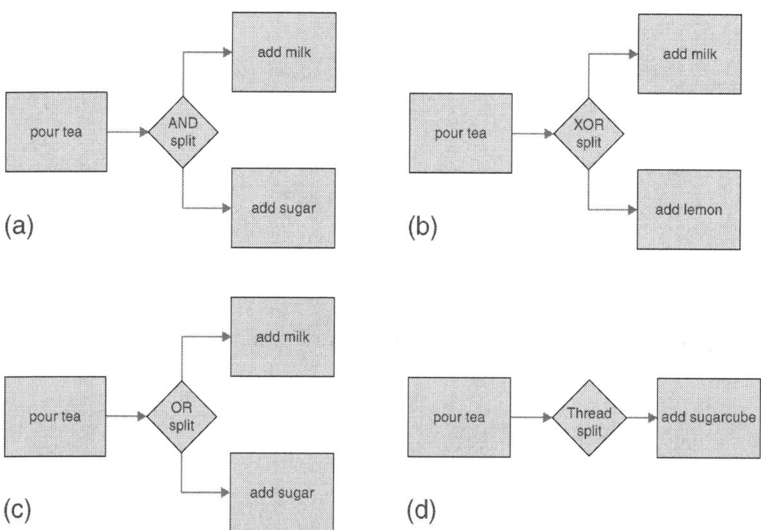

Fig. 2.2 Branching constructs

Table 2.1 Split construct–pattern mapping

Split construct	Pattern
AND-split	Parallel split
XOR-split	Exclusive choice
	Deferred choice
OR-split	Multichoice
Thread-split	Thread split

where after the *pour tea* task, the execution thread diverges into multiple concurrent threads causing the *add sugarcube* task to execute several times.

Each of these situations can be mapped to a specific pattern that precisely describes the desired split behavior as shown in Table 2.1. In the case of the XOR-split, there are two possible ways in which it can be implemented and consequently there are two patterns that describe these alternatives.

The operation of the XOR-split can be realized in two distinct ways depending on whether the process itself contains sufficient information to make the required routing decision when the split construct is reached or whether additional external input is required to make the decision. If we consider the example shown in Fig. 2.2b, two possible courses of action are possible when the XOR-split construct is reached and these are recognized by two specific patterns:

- *Exclusive Choice*, where the process contains sufficient information to determine where to route the thread of control when the decision construct is reached and does so automatically. In the example above, this would immediately result in either the *add milk* or *add lemon* task being initiated when the XOR-split was reached.
- *Deferred Choice*, where the decision is not made automatically from within the context of the process but is deferred to an entity in the operational environment. In the context of the example above, the process would halt at the XOR-split pending this decision, which would be signaled by either the *add milk* or the *add lemon* task being initiated.

The branching patterns provide various mechanisms for diverging the thread of control within a process. The following group of control-flow patterns provide various mechanisms for synchronizing threads of control within a process.

Synchronization Patterns

Synchronization patterns are the logical counterpart to the branching patterns. They identify various types of synchronization and merging behaviors that are generally associated with join constructs in a process. Four distinct types of join are commonly utilized:

- The AND-join, which synchronizes the thread of control in multiple branches and merges them into a single execution thread in an outgoing branch.

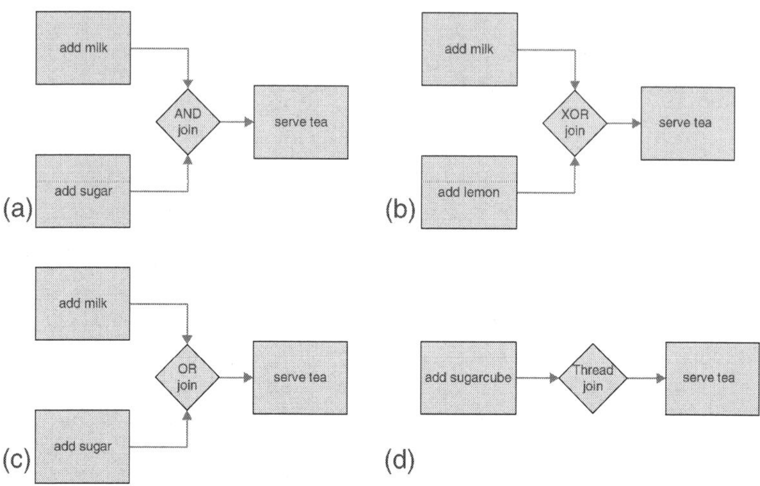

Fig. 2.3 Synchronization constructs

- The XOR-join (or the eXclusive OR-join), which merges the thread of control from one of the several incoming branches into an outgoing branch.
- The OR-join, which synchronizes the various execution threads in those incoming branches that are currently active and merges them into a single execution thread in an outgoing branch.
- The Thread-join, which synchronizes and merges multiple concurrent execution threads in the same branch into a single execution thread.

This distinction between each of these operators is illustrated in Fig. 2.3a–d, which continues the previous example. Figure 2.3a shows an AND-join, which only allows the *serve tea* task to commence once both the *add milk* and *add sugar* tasks have been completed. In contrast, in Fig. 2.3b, based on an XOR-join, once one of the *add milk* or *add lemon* tasks has completed, the *serve tea* task can be initiated. Figure 2.3c utilizes an OR-join and allows the *serve tea* task to commence once all of the preceding *add milk* and *add sugar* tasks that have been initiated have completed. Finally, Fig. 2.3d illustrates the Thread-join construct where multiple concurrent instances of the *add sugarcube* task must complete and the associated threads of control must be synchronized and merged into a single thread before the *serve tea* task can commence.

AND-Join Variants

Although the behavior of each of these join constructs appears to be well defined and predictable, upon closer examination it becomes clear that there are a variety of ways in which they can be actually implemented. If we look at the AND-join, for example, we can identify several possible variations in the way in which it can be operationalized when we consider factors such as the following:

1. *Extent of synchronization required* – do all incoming branches need to be triggered before the join can fire? For example, it might be desirable for the join to be able to fire once a specified threshold of the incoming branches have been enabled, provided it is certain that all of the incoming branches will ultimately be triggered. This quality is particularly attractive in time-critical processes.
2. *Untriggered branch handling* – in the event of partial synchronization, how should any remaining (i.e., untriggered) branches be handled? Where this situation could possibly occur, there are a number of ways in which it can be dealt with.
3. *Likelihood of concurrency* – is it possible that concurrent execution threads may occur in an incoming branch and, if so, how should this be handled.

A number of combination of these factors are permissible during process execution and constitute desirable behaviors, thus they form the basis for a number of distinct patterns as illustrated in Table 2.2.

The distinction between these patterns is as follows:

- The *Synchronization pattern* provides a means of synchronizing all of the execution threads reaching the incoming branches of an AND-join, where there is no likelihood of more than one triggering being received on each incoming branch.
- The *Generalized AND-join*, similar to the *Synchronization* pattern, provides a means of synchronizing all incoming branches; however, it is able to cater for situations where more than one incoming trigger is received on a branch.
- The *Structured Partial Join* is a pattern describing an approach to AND-join implementation that allows the join to fire once a threshold of the incoming branches have been enabled, provided there is certainty that all incoming branches will ultimately be triggered. After the join has fired, it cannot do so again until a trigger has been received on each of the remaining (untriggered) branches. The *Structured Partial Join* can only operate in a process that is structured in form (i.e., has "balanced" splits and joins). The rationale of this is that the structured nature of the process simplifies the implementation of the partial

Table 2.2 AND-join pattern variants

Pattern	Extent of synchronization	Untriggered branch handling	Likelihood of concurrency
Synchronization	All branches	Not required	Not possible
Generalized AND-join	All branches	Not required	Allowed
Structured partial join	Some branches	Do nothing	Not possible
Blocking partial join	Some branches	Block them	Allowed
Canceling partial join	Some branches	Cancel them	Not possible

join; however, this has the consequence that the partial join can only be used in a limited range of situations.
- The *Blocking Partial Join* is an alternative approach to partial join implementation, which guarantees that each incoming branch receives only a single triggering by utilizing the notion of a "blocking region," which allows only one execution thread to travel on each incoming branch and prevents any additional execution threads from flowing down each branch after until the join has reset. Once the partial join has reset, it is once again re-enabled and can fire when it receives the required threshold of triggers on any combination of its incoming branches. The *Blocking Partial Join* has the advantage that it can be used in a broader range of situations than the *Structured Partial Join* such as processes that contain loops.
- The *Canceling Partial Join* provides a more expeditious approach to partial join implementation by canceling any remaining execution threads in incoming branches after the join has fired.

The AND-join variants of the synchronization patterns deal with synchronizing control-flow in multiple branches within a process. In contrast, the XOR-join variants focus more on merging control-flow from one of the several incoming branches into an outgoing branch.

XOR-Join Variants

There are two possible operational forms that the XOR-join can take. These are recognized by two distinct patterns:

- The *Simple Merge*, which supports the merging of execution threads from multiple incoming branches to the outgoing branch where only one of the incoming branches is active at any given time.
- The *Multiple Merge*, which operates in the same way as the *Simple Merge* but is able to handle multiple concurrent execution threads in the incoming branches, each of which leads to a distinct execution thread in the outgoing branch.

OR-Join Variants

The OR-join construct is analogous to the AND-join, in that it synchronizes the execution threads in multiple incoming branches and merges them into a single execution thread in an outgoing branch; however, it has the important distinction that not all of the incoming branches may necessarily be active. In general, an OR-join is preceded by a corresponding OR-split earlier in a process that makes the decision about which branches will be activated. There are three distinct patterns that provide alternate approaches to implementing the OR-join:

- The *Structured Synchronizing Merge*, which requires that the process fragment between the OR-join and the corresponding preceding OR-split be structured in format (i.e., has "balanced" splits and joins). Once the preceding OR-split has fired, it provides information to the OR-join to indicate how many incoming branches are active and will require synchronization. The restriction of the pattern to a structured process simplifies the implementation of the OR-join; however, this has the consequence that it can only be used in a limited range of situations.

- The *Local Synchronizing Merge*, which requires that the OR-join be able to make a decision about when it should fire based on information available locally to it (a notion that is often termed "local semantics"). The approach allows the OR-join to operate in a wider range of processes, although it cannot be utilized in scenarios involving unstructured loops.
- The *General Synchronizing Merge*, where the OR-join can use any information available about the current process state, including an analysis of possible future states, to determine whether it should fire based on currently enabled branches or whether it should wait for any additional branches to complete before firing.

Figure 2.4a–c illustrates the main differences between these OR-join variants. In Fig. 2.4a, the structured nature of the process in which the *Structured Synchronizing Merge* operates is evident, as are the "bypass" branches corresponding to each branch. Figure 2.4b illustrates the *Local Synchronizing Merge*, which is able to function in unstructured processes, including those involving loops, but relies on the use of local semantics and cannot be used on a general basis unlike the *General*

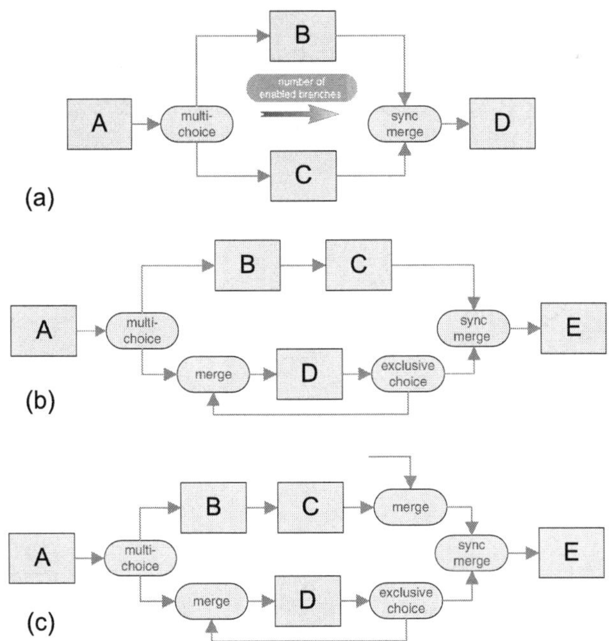

Fig. 2.4 Synchronizing Merge pattern examples

Synchronizing Merge shown in Fig. 2.4c, implementing an unstructured process involving an OR-join that cannot be accommodated by either of the two preceding patterns.

Thread-Join

The Thread-join construct is directly recognized by the *Thread Join* pattern that coalesces a specified number of execution threads in a branch into a single execution thread.

Repetition Patterns

Repetition patterns identify three alternative mechanisms for a task or group of tasks to be executed repeatedly. These are illustrated in Fig. 2.5 and described below:

- *Structured Loop*, where a process contains a dedicated construct that allows an individual task or a set of tasks to be repeated in a structured way. A structured loop has either a pretest associated with it that decides at the beginning of each interaction whether execution of the loop should continue, or a posttest that makes the decision at the end. Combination loops involve both pre- and posttests. Figure 2.5c–e illustrates pretested, posttested, and combination loops, respectively.

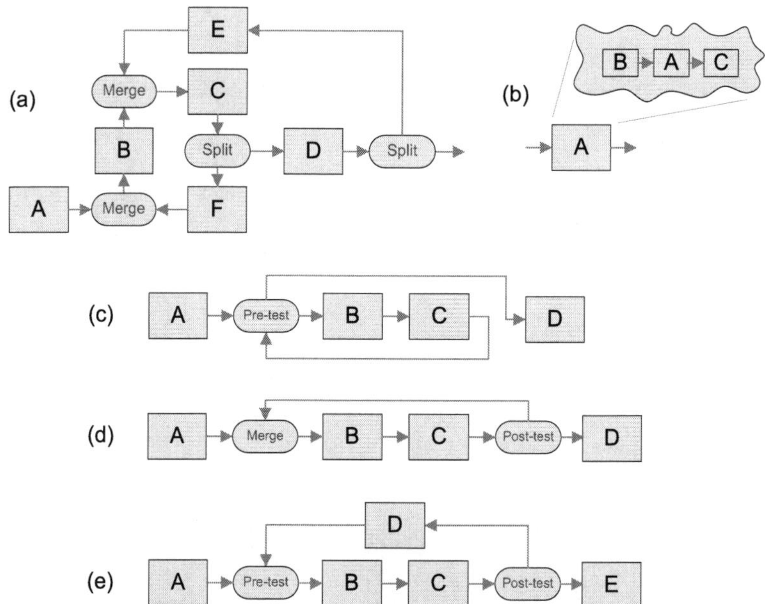

Fig. 2.5 Repetition patterns

Table 2.3 Multiple instance patterns

Pattern	Instance determination	Extent of synchronization	Post-synchronization task handling
Multiple instances without synchronization	Runtime before task initiation	None	No action
Multiple instances with *a priori* design time knowledge	Design time	All instances	No action
Multiple instances with *a priori* runtime knowledge	Runtime before task initiation	All instances	No action
Multiple instances without *a priori* runtime knowledge	Runtime before task completion	All instances	No action
Static partial join for multiple instances	Runtime before task initiation	Some instances	No action
Canceling partial join for multiple instances	Runtime before task initiation	Some instances	Cancel incomplete instances
Dynamic partial join for multiple instances	Runtime during task completion	Some instances	No action

- *Arbitrary Cycles*, where a process contains a path in the form of a cycle that connects a set of tasks, thus allowing them to be executed multiple times. An arbitrary cycle may have more than one entry or exit point. Figure 2.5a illustrates two overlapping cycles involving tasks DEC and BCF.
- *Recursion* provides a means of repeated execution of a task through the use of self-invocation. Figure 2.5b provides an example of this pattern, where the implementation of task *A* is made up of a set of tasks that also include task *A*, hence during its execution, task *A* may invoke subsequent instances of itself in order to fulfill its operational requirements.

Repetition patterns involve the iterative execution of one or more tasks. Another desirable property of processes is the ability for multiple instances of a task to execute concurrently, possibly with some form of synchronization. The multiple instance patterns document the various options in this area.

Multiple Instance (MI) Patterns

Multiple instance patterns identify scenarios where multiple instances of a task are required to execute concurrently. There are various options for the controlled concurrency of a task depending on the time that the required number of instances is determined, on what basis the various instances need to be synchronized, and what happens to any remaining instances after this synchronization has occurred. Various combinations of these factors lead to distinct patterns as illustrated in Table 2.3.

Each of these patterns corresponds to a distinct scenario involving concurrent execution of the same task. Figure 2.6a–e illustrate the operation of the patterns.

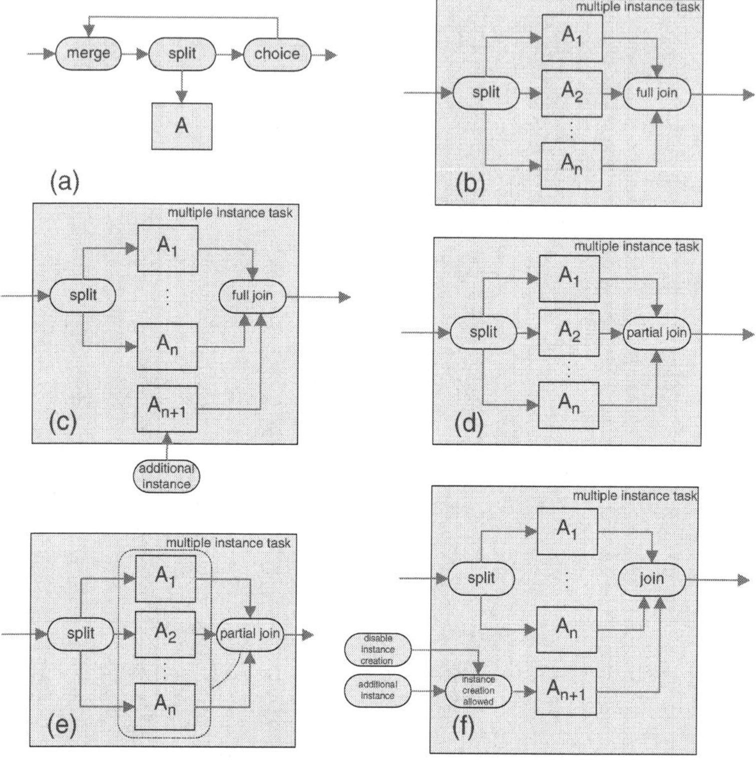

Fig. 2.6 Examples of multiple instance patterns

Figure 2.6a demonstrates the *Multiple Instances without Synchronization* pattern where a loop basically spawns off the required number of task instances and there is no thought of subsequent synchronization. Figure 2.6b illustrates two of the patterns that essentially have the same form requiring a specific number of task instances to be started and then synchronized before the thread of control can be passed to subsequent tasks. The *Multiple Instances with a priori Design Time Knowledge* pattern requires that the number of instances (n) be specified at design time, the *Multiple Instances with a priori Runtime Knowledge* requires that it be determined before the task is initiated at runtime. The *Multiple Instances without a priori Runtime Knowledge* illustrated in Fig. 2.6c is similar to these two patterns, except that it allows additional instances to be added at runtime. The remaining three patterns are illustrated in Fig. 2.6d–f, respectively. The *Static Partial Join for Multiple Instances* involves a partial join that only requires that a predetermined subset of the task instances be completed before synchronization can occur and the thread of control be passed on to subsequent tasks. Any remaining tasks are ignored. The *Canceling Partial Join for Multiple Instances* pattern operates in a similar way, except that when the partial join fires, it cancels the remaining instances. In both cases, the number of task instances is known at design time. Finally, the *Dynamic Partial Join for*

Multiple Instances allows additional instances to be added at runtime and supports a different types of synchronization condition that does not necessarily need to relate to the number of instances that have completed.

Multiple instance patterns address concurrency for a single task. The concurrency patterns extend this notion to a group of tasks.

Concurrency Patterns

Concurrency patterns denote situations involving restrictions on the extent of concurrency that is permissible between a group of tasks in a process. The most restrictive of these patterns is the *Sequence* which requires that a group of tasks cannot execute concurrently and furthermore that they must execute in a specified order. The *Interleaved Parallel Routing* pattern relaxes the total ordering requirement, allowing the overall set of tasks to be broken into subsets of tasks where each group has a specific ordering. Tasks from these subgroups can then execute (potentially on an interleaved basis) until all of them have executed, provided no tasks execute concurrently at any time. The *Interleaved Routing* pattern further relaxes the concurrency constraints for the group of tasks requiring only that none of the tasks execute concurrently. They may execute in any order and all of them must execute exactly once. Analogous to this pattern is the *Critical Section* pattern, which requires that two or more regions of a process model are identified such that the thread of control can only be active in one of these regions at any time. The final pattern in this group is the *Milestone*, which requires that the execution of a nominated task only be allowed to proceed when the process has reached a specific state (typically denoted by the thread of control in a distinct branch being at a specific point).

The concurrency patterns focus on restrictions on concurrent execution between tasks or groups of tasks. The next set of patterns denote mechanisms for actively intervening in the execution of a task or process instance to bring it to a final state either through termination or completion.

Cancelation and Completion Patterns

This group of patterns identifies various mechanisms for triggering the cancelation or completion of tasks. For each of them, the cancelation/completion action must be triggered from within the process. Five distinct patterns of this form are denoted:

- *Cancel Task* identifies the cancelation of a specific (atomic) task. If it is active when the cancelation is effected, its execution is terminated and recorded as having completed unsuccessfully. The canceled task does not trigger any subsequent tasks.
- *Cancel Case* indicates the cancelation of all active task instances in a case. Once triggered, these tasks are terminated and are recorded as having terminated

unsuccessfully. There is no further execution of the canceled case and the case is marked as canceled.

- *Cancel Multiple Instance Task* identifies the cancelation of a specific multiple instance task. Since a multiple instance task may potentially have multiple concurrent execution instances either active or pending at the time of cancelation, the cancelation action extends to all of these instances and requires that their execution be terminated and recorded as having completed unsuccessfully. None of the canceled task instances nor the canceled task itself can trigger any subsequent tasks.
- *Complete Multiple Instance Task* denotes the forced completion of a multiple instance task. Unlike the preceding pattern, it seeks to bring the various instances of a multiple instance to a conclusion by removing any currently executing instances and marking them as complete and by withdrawing any pending instances that have not yet started executing. The multiple instance task is deemed to have completed successfully and any subsequent tasks are triggered.
- *Cancel Region*, which provides a general form of cancelation that is able to terminate a group of tasks.

The cancelation and completion patterns deal with termination of executing task instances from within a process. In contrast, the trigger patterns provide a means of initiating task execution from outside a process.

Trigger Patterns

The trigger patterns provide a means for the execution of a process to be synchronized with its broader operational environment. Triggers can take various forms, examples of which are shown in Fig. 2.7a, which illustrates a message-based trigger, and Fig. 2.7b, which denotes a time-based trigger. Where a trigger is associated with a task, the task can only commence execution when it has (1) been enabled and (2) received an instance of the required trigger. Two types of trigger are recognized by distinct patterns: the *Persistent Trigger* denotes triggers that are durable in form and are retained if the task instance to which they are directed has not yet received the execution thread, whereas the *Transient Trigger* denotes triggers that are ephemeral in form and if they are not immediately consumed by the task to which they are directed (i.e., if the task does not yet have the thread of control), then they are discarded.

Fig. 2.7 Examples of trigger patterns

In contrast to these two patterns that deal with task initiation, the final group of control-flow patterns denote schemes for determining when a process instance is complete.

Termination Patterns

The termination patterns identify two distinct schemes for determining when a process instance is complete:

- *Implicit Termination* considers a process instance to be complete when there is no remaining work left to do now or at any future time. Moreover, the process instance must not be in deadlock or livelock.
- *Explicit Termination* requires a process to have a dedicated endpoint, which signifies the point of completion. When the thread of control reaches this point in the process, it is considered to be complete and any remaining work is discarded.

The preceding control-flow patterns delineate a series of desirable characteristics relating to control-flow issues within a business process. Historically, the control-flow perspective has received a significant amount of attention both in terms of modeling and enactment of business processes and has tended to be the central metaphor for processes against which most other aspects are described. In spite of this bias, the data perspective is an equally important aspect of a business process, and in many cases, the aim of processes is to manage the effective capture and distribution of information to process participants. In the following section, we identify a series of data patterns that describe recurrent data-related constructs and usage scenarios in business processes.

2.2.2 Data Patterns

The data patterns describe language features for defining and managing data resources during business process execution. They can be divided into four groups:

- **Data visibility**, patterns which relate to the scope and visibility of data elements in a process.
- **Data interaction**, patterns which focus on the way in which data is communicated between active elements within a process.
- **Data transfer**, patterns which consider the means by which the actual transfer of data elements takes place between process elements and describe the various mechanisms by which data elements can be passed across the interface of a specific process element.
- **Data-based routing**, patterns which characterize the manner in which data elements can influence the operation of other aspects of a process, particularly the control-flow perspective.

Each of these groups is discussed in detail in the following sections, starting with data visibility patterns.

Data Visibility Patterns

Data visibility patterns define the binding of a data element and its region of visibility. Data elements are usually defined in the context of a particular component of a process and this binding usually defines the scope in which the data element is both visible to and capable of being used by other process elements. There are eight distinct patterns in this group, which describe different degrees of data scoping as follows:

- *Task Data*, which corresponds to a data element defined in the context of a particular task. It is typically only accessible within a given task instance and has the same life-span as the task instance to which it is bound. The *Task Data* pattern is illustrated by the variable *tvar* that is associated with task *A* in Fig. 2.8 and is accessible only within the context of this task.
- *Block Data*, which corresponds to a data element defined in the context of a particular block or net within a process that is hierarchical in form. The data element is visible throughout the block and it has a life-span corresponding to the life-span of the block. The *Block Data* pattern is illustrated by the variables *bvar* and *bvar′* that are associated with the two process block depicted in Fig. 2.8. *bvar* is accessible only by tasks in the upper block, while *bvar′* is accessible only to tasks within the lower block that corresponds to the subprocess decomposition of task *C* in the upper block.
- *Scope Data*, which corresponds to a data element bound to a set of tasks in a process. If the process is hierarchical in form, the tasks are assumed to be in the same block although they need not be directly connected. A scope data element is visible only to the tasks of which the scope is comprised. It has the same life-span as the block in which it resides. The *Scope Data* pattern is illustrated by the variable *svar* shown in Fig. 2.8, which is accessible only by tasks *B* and *C*.

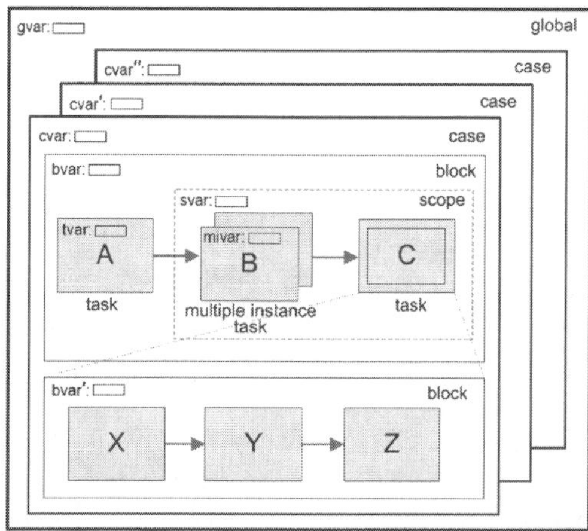

Fig. 2.8 Examples of data visibility patterns

- *Multiple Instance Data* corresponds to a data element in a specific execution instance of a task, where the task may execute multiple times, possibly concurrently. Examples of this are multiple instance tasks, tasks that are triggered multiple times within a process (e.g., that are part of loop), and tasks in a subprocess decomposition shared by several tasks in a process. It is defined in the context of a given task and is accessible only within a single execution instance of the task. It shares the same life-span as the execution instance. The *Multiple Instance Data* pattern is illustrated by the variable *mivar* that is associated with the multiple instance task *B* in Fig. 2.8. Each instance of a multiple instance variable (e.g., *mivar*) is accessible only within the context of a specific instance of a multiple instance task.
- *Case Data*, which corresponds to a data element that is accessible to all tasks within a process instance. It has a life-span that is the same as the process instance. Figure 2.8 illustrates *Case Data* Pattern via the data elements *cvar*, $cvar'$ and $cvar''$.
- *Folder Data*, which corresponds to a data element that is defined outside the context of the process that is able to be coupled with particular process instances during their execution. Once married up with a process instance, the data element is accessible to all tasks in the process instance.
- *Global Data*, which corresponds to a data element that is accessible throughout all tasks in cases for a given process. Figure 2.8 illustrates *Global Data* Pattern via the data element *gvar*, which is accessible throughout all tasks in all process instances.
- *Environment Data*, which correspond to a data element that is defined outside the context of the process but accessible within it during execution. It is possible for an environment data element to be accessed by different cases of different processes.

The data visibility patterns describe the types of data elements that are permissible within the context of a process. The next group of patterns describe various ways in which those data elements are passed between the main components within a process and also between a process and the operating environment.

Data Interaction Patterns

Data interaction patterns divide into two main groups: *internal data interaction* patterns, which describe the various ways in which data elements are communicated between the main components within a process, and *external data interaction* patterns, which describe the various ways in which data elements are communicated between the main components in a process and the operating environment.

There are four internal data interaction patterns describing scenarios for data transfer between active components within a process instance or between process instances:

- *Data Interaction between Tasks* distinguishes the way in which data elements are passed between tasks in a process instance. There are three distinct means by which this can be achieved as illustrated in the figures below. Figure 2.9a

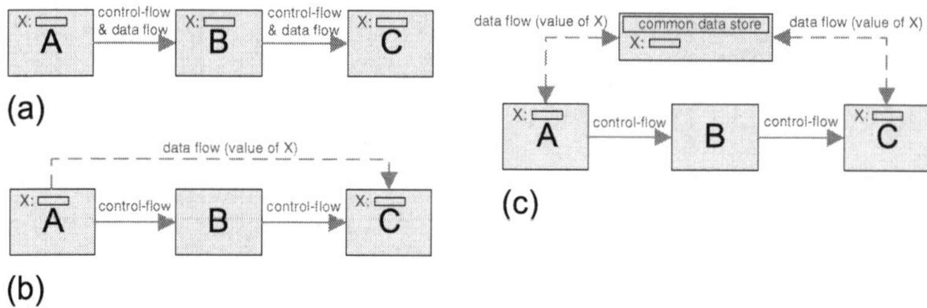

Fig. 2.9 Data interaction between tasks

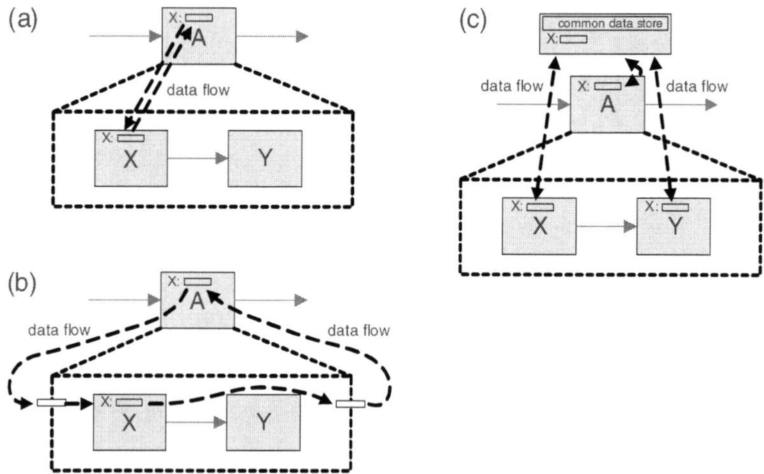

Fig. 2.10 Data interaction between block task and subprocess decomposition

shows the situation where control-flow and data passing occurring between tasks are integrated. Figure 2.9b denotes the situation where data-flow occurs separately from control-flow, reducing the need to distribute data elements to tasks other than those that require them. Figure 2.9c illustrates the scenario where data passing occurs implicitly via a global data store common to all tasks in a process.

- *Data Interaction between Block Task and Subprocess Decomposition* describes the manner in which data elements are passed to and from a composite task, which has an associated subprocess decomposition. This can be done in one of the three ways. Figure 2.10a–c illustrate data interaction with a subprocess using data channels, formal parameters, and implicit shared data stores, respectively.
- *Data Interaction with Multiple Instance Tasks* describes the manner in which data is passed to and from a multiple instance task. As there are multiple instances of the same task, each of these instances replicates the same data elements. Two possible scenarios arise. Either, as illustrated in Fig. 2.11a, the data element that is passed to the instances is composite in form and each of them receives a distinct part of it or, as shown in Fig. 2.11b, each of them replicates the same input

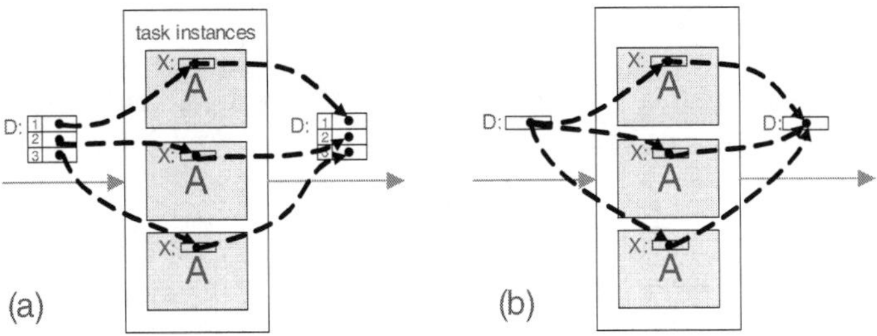

Fig. 2.11 Data interaction with multiple instance tasks

Table 2.4 External data interaction patterns

Pattern	Initiator	Data source	Data target
Data interaction – Process to environment – Push-oriented	Process	Process	Environment
Data interaction – Environment to process – Pull-oriented	Process	Environment	Process
Data interaction – Environment to process – Push-oriented	Environment	Environment	Process
Data interaction – Process to environment – Pull-oriented	Environment	Process	Environment

element. Similar to the factors associated with passing the data elements to the task instances, there is also the consideration of how to pass them to subsequent tasks. In the former scenario, the data elements are coalesced and the composite data element is passed on; in the latter, a specific instance is selected for subsequent usage.

- *Data Interaction between Cases* denotes the communication of a data element from one case to another. Typically this is based on communication between a task instance in one case and a target task instance in another, although other approaches are possible.

There are four external data interaction patterns describing scenarios for data transfer between a process instance and the operating environment. These are identified in Table 2.4. These patterns are differentiated on the basis of two factors: the initiating party of the data interaction and the direction of the data transfer.

Although the data interaction patterns describe the passing of data elements at a macro level, the data transfer patterns focus on the specific details associated with the trafficking of a data element across the interface of the process component that is receiving or emitting it.

Data Transfer Patterns

Data transfer patterns deal with the mechanics of how a data element is actually transported to or from a task instance in an executing process. The actual hand-over of a data element can occur in four distinct ways, each of which is recognized by an individual pattern.

- *Data Transfer by Value* recognizes the situation where the transport of a data element is actually based on copying its value between one task and another. This pattern is utilized in data passing approaches that rely on the actual evaluation of an expression (such as an XQuery statement or database query) prior to the assignment of the value to its target location.
- *Data Transfer by Reference – Unlocked* corresponds to the scenario where data elements are passed by reference rather than by value. Implicit in this approach is the notion that the data value can be seamlessly transferred from the source to the target location.
- *Data Transfer by Reference – With Lock* is an extension of the preceding pattern that also provides a measure of concurrency control by registering a lock over the transferred data element to ensure that the receiving task retains complete control over it and can rely on its integrity. The lock is relinquished when the data element is passed to a subsequent task.
- *Data Transfer – Copy In/Copy Out* corresponds to the situation where the data element being sourced resides in a distinct location (typically outside of the process) such as a database that normally is considered to retain ownership of the data, and when the data element is transferred to the process or task instance, it is done so with the proviso that it must be transferred back (i.e., synchronized) when the process or task instance completes.

The remaining two data transfer patterns deal with the transformation of data elements when they are transferred to and from a task instance. These are the *Data Transformation – Input* and *Data Transformation – Output* patterns.

While this group of patterns has focussed on the mechanics of data transfer at the micro level, the final set of data patterns examines how data can affect other aspects of process execution.

Data-Based Routing Patterns

Data routing patterns focus on the various ways in which the data perspective can influence other aspects of process execution. The first four of these patterns deal with task pre- and post-conditions that cause task commencement or completion to be delayed until specified conditions based either on existence of data elements or nominated data values are satisfied. These patterns are the following:

- *Task Precondition – Data Existence.*
- *Task Precondition – Data Value.*
- *Task Post-condition – Data Existence.*
- *Task Post-condition – Data Value.*

The next two patterns *Event-based Task Trigger* and *Data-based Task Trigger* also provide mechanisms for synchronizing task commencement; however, they are based on the occurrence of specified events such as the receipt of anticipated data elements from the operating environment or from elsewhere in the process instance. From both patterns, either a delayed task instance can be allowed to commence or a completely new task instance can be triggered.

Finally, the *Data-based Routing* pattern denotes the situation where the value of one or more data elements can affect the flow of control in a process. It aggregates two control-flow patterns – *Exclusive Choice* and *Multichoice* – where data elements are used as the selection mechanism for determining whether the thread of control should be routed down a given branch in the process.

While the data patterns focus on identifying the various ways in which data impacts on and is distributed by a business process, the resource perspective delineates the ways in which resources, both human and non-human, are involved in the conduct of work associated with business processes.

2.2.3 Resource Patterns

Resource patterns characterize the way in which work is distributed to the resources associated with a process and managed through to completion. There are seven distinct groups of patterns that describe the various aspects associated with the involvement of resources in automated business processes. These are listed below.

- **Creation** patterns describe design time work distribution directives that are nominated for tasks in a process model.
- **Push** patterns describe situations where the system proactively distributes work items to resources.
- **Pull** patterns relate to situations where individual resources take the initiative in identifying and committing to undertake work items.
- **Detour** patterns describe various ways in which the distribution and life-cycle of work items can deviate from the directives specified for them at design time.
- **Auto-start** patterns identify alternate ways in which work items can be automatically started.
- **Visibility** patterns indicate the extent to which resources can observe pending and executing work items.
- **Multiple-resource** patterns identify situations where the correspondence between work items and resources is not one-to-one.

One of the major determinants of the way in which individual resource patterns apply is related to the state of a given task instance that a resource may encounter at runtime. Figure 2.12 illustrates the life-cycle for a task instance or work item as they are often known in a workflow context. The diagram indicates the various states through which a work item passes during execution and the arcs indicate the

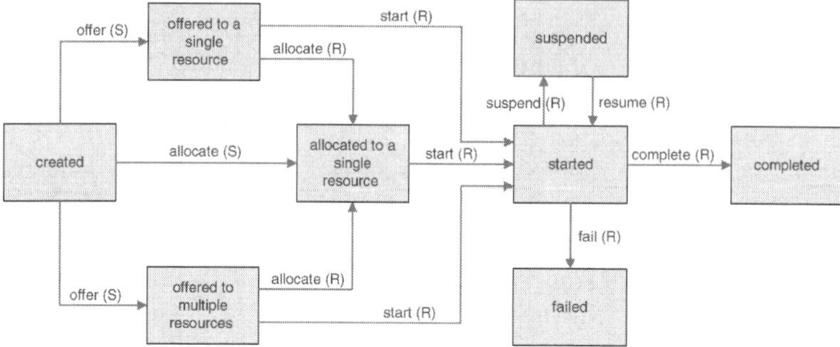

Fig. 2.12 Task life-cycle

permissible paths between states. The label on each arc indicate the common name for the state change and the R or S tag indicates whether the state change is initiated by a resource or the system responsible for automating the process.

Creation Patterns

The first set of patterns describe various ways in which work items can be allocated to one or several resources at runtime. Typically these allocation notions correspond to design time directives that are captured in the underlying process model, which indicate the way in which various tasks in the process are intended to be undertaken by available resources. Consequently, they come into play at the moment that an instance of a task is triggered. There are eleven creation patterns:

- *Direct Distribution* corresponds to the situation where a work item is directly offered or allocated to one or more specifically named resources, for example, allocate the *prepare statement* task to user *jsmith*.
- *Role-based Distribution* corresponds to the situation where a work item is directly offered or allocated to one or more specifically named roles, each of which contain one or more users, for example, offer the *referee match* task to the *northern-referee*. and *southern-referee* roles.
- *Deferred Distribution* corresponds to the situation where the identification of the resource that a work item will be offered or allocated to is delayed until runtime, typically by nominating a data resource from which it can be obtained, for example, at runtime, allocate the *organize meeting* work item to the resource identified in the *daily-coordinator* variable.
- *Authorization* identifies privileges that can be assigned to specific resources during the execution of a process. These privileges define the range of actions that the resource may undertake during the execution of the process, for example, allow the user *jsmith* to skip the *check-pressure* task if it is not deemed necessary.
- *Separation of Duties* (also known as the "four eyes principle") corresponds to a constraint that exists between two tasks, requiring that they not be executed by

the same user within a given process instance, for example, the *sign-cheque* task cannot be undertaken by the same user who completed the *lodge-purchase-order* task.

- *Case Handling* corresponds to the ability to allocate all work items in a case to the same resource at the time of commencement, for example, all work items in an instance of the insurance claim process are allocated to user *jsmith*.
- *Retain Familiar* corresponds to a constraint that exists between two tasks, requiring that where possible they be executed by the same user within a given process instance, for example, the *order-materials* task should be undertaken by the same user who completed the *calculate-materials* task.
- *Capability-based Distribution* corresponds to the situation where a work item is offered or allocated to one or more resources based on capabilities that they possess, for example, allocate the *audit account* task to a user having a *CPA* qualification.
- *Organization-based Distribution* corresponds to the situation where a work item is offered or allocated to one or more resources based on their position or other responsibilities within the organization, for example, allocate the *authorize order* task to a user at the level of *manager* or above.
- *History-based Distribution* corresponds to the situation where a work item is offered or allocated to one or more resources based on their preceding execution history, for example, allocate the *audit major customer* task to the user who has completed most audits in the past three months.
- *Automatic Execution* corresponds to the situation where a work item can be executed without needing to be distributed to a resource, for example, the *send-order* work item can execute automatically without requiring a user to complete it.

While this set of patterns focus on establishing the identity of the resources to which work items should be distributed, the next group – push patterns – deal with the manner in which the system having identified the resources can distribute work items to them.

Push Patterns

Push patterns apply in situations where work items are forwarded to resources by the system that is automating a business process. These patterns correspond to the offer and allocate arcs initiated by the system (i.e., tagged (S)) in Fig. 2.12. The various push patterns are as follows.

- *Distribution by Offer – Single Resource* corresponds to the situation where a work item is offered to a specific resource on a nonbinding basis, for example, the *review-feedback* work item is offered to user *jsmith* to complete, although they are not obliged to do so.
- *Distribution by Offer – Multiple Resources* corresponds to the situation where a work item is offered to several resources on a nonbinding basis with the expectation that one of them might commit to undertaking it, for example, the

process-application work item is offered to members of the *administrator* role to complete, although they are not committed to do so.

- *Distribution by Allocation – Single Resource* corresponds to the situation where a work item is allocated to a specific resource on a binding basis and they are expected to complete it at some future time, for example, the *process-rejection* work item is allocated to the user *jsmith*.
- *Random Allocation* corresponds to the allocation of a work item to a specific resource selected from a group of resources on a random basis, for example, allocate the *process-claim* work item to a member of the *manager* role selected on a random basis.
- *Round Robin Allocation* corresponds to the allocation of a work item to a specific resource selected from a group of resources on a round robin basis, for example, allocate the *process-support-request* work item to the member of the *sysadmin* role who did it least recently.
- *Shortest Queue* corresponds to the allocation of a work item to a specific resource selected from a group of resources based on who has the least work pending (i.e., the shortest work queue), for example, allocate the *annual-review* work item to the member of the *sysadmin* role who has the least pending work.
- *Early Distribution* corresponds to the situation where a work item can be offered or allocated to a resource ahead of the time that it is actually enabled and can be completed, for example, allocate the *annual-review* work item to the member of the *sysadmin* role with the requirement that they do not start it until they are told to.
- *Distribution on Enablement* corresponds to the situation where a work item is offered or allocated to a resource at the same time that it is enabled and can be completed, for example, offer the *finalize-quote* work item to members of the *sales-staff* role as soon as the work item is triggered.
- *Late Distribution* corresponds to the situation where a work item is offered or allocated to a resource at some time after the time at which it is enabled, for example, allocate the *process-defect* work item to the user *jsmith* only when they do not have any other work items queued for them (this is often termed “heads down processing”).

This group of patterns focuses on the distribution of work items by the system, and the next group – pull patterns – deals with work distribution that is enabled by the resources that are doing the work.

Pull Patterns

Pull patterns correspond to work distribution actions that are initiated by the actual resources undertaking them. As such, they provide a means of empowering users in the conduct of their work activities. There are six pull patterns as follows:

- *Resource-Initiated Allocation* corresponds to the situation where a resource commits to undertaking a work item that has been offered to them.

- *Resource-Initiated Execution – Allocated Work Item* corresponds to the situation where a resource starts a work item that has been offered to them.
- *Resource-Initiated Execution – Offered Work Item* corresponds to the situation where a resource commits to undertake and immediately starts a work item that has been offered to them.
- *System-Determined Work Queue Content* corresponds to the ability of the system to impose an ordering on the sequence in which resources see and/or can undertake their work items, for example, display all work items in order of priority.
- *Resource-Determined Work Queue Content* corresponds to the ability for the resource to impose an ordering/reordering on the sequence in which they see and/or can undertake their work items, for example, reorder work items in date received sequence.
- *Selection Autonomy* corresponds to the ability for a resource to choose which work item they undertake next.

All of the patterns discussed to date consider the "normal" state of affairs where the work distribution approach captured in the underlying process model is essentially followed directly. While desirable, this approach does not always allow for actual events that occur in practice, for example, resources are unexpectedly absent, become overloaded, etc. The following set of patterns – detour patterns – recognize approaches for dealing with these unexpected situations.

Detour Patterns

Detour patterns identify approaches for deviating from the work distribution strategy implied by the process model and provide various means for resources, as well as the enabling system itself, to deal with unexpected or undesirable workload situations that may arise. There are nine detour patterns as described below.

- *Delegation* corresponds to the situation where a resource allocates a work item that is currently allocated to them to another resource, for example, user *jsmith* delegates the *pick-order* work item to user *fbrown*.
- *Escalation* corresponds to the situation where the system changes an existing work item offer or allocation and redistributes the work item to another user with the goal of expediting a work item, for example, the system identifies that the *finalize-audit* work item currently allocated to user *jsmith* has exceeded its intended completion deadline and chooses to remove it from *jsmith's* worklist and allocate it to *fbrown*.
- *Deallocation* corresponds to the situation where a resource (or group of resources) chooses to make a work item previously offered or allocated to them available for redistribution to other resources, for example, user *jsmith* recognizes that the *run-trial-balance* work item allocated to them will not be completed in time and chooses to make it available to other resources who may be able to complete it sooner.

- *Stateful Reallocation* corresponds to the situation where a resource chooses to allocate a work item that they have already started (but not completed) executing to another resource and to retain any associated state information, for example, the user *jsmith* chooses to reallocate the *conduct-audit* task to user *fbrown* with full retention of state such that any results from the partial execution of the task are retained and it does not need to be restarted.
- *Stateless Reallocation* corresponds to the situation where a resource chooses to allocate a work item that they have already started (but not completed) executing to another resource without retaining any associated state information, for example, the user *jsmith* chooses to reallocate the *quality-check* task to user *fbrown* without retaining any results from the partial execution of the task, thus requiring that it commence execution from the start.
- *Suspension/Resumption* corresponds to the ability of a resource to suspend and resume execution of a work item, for example, user *jsmith* suspends execution of the *conduct-audit* work item and switches their attention to other work priorities.
- *Skip* corresponds to the ability of a resource to skip a work item allocated to it and mark the work item as complete, for example, user *jsmith* skips the *check-news* work item such that they can commence the next work item in the process.
- *Redo* corresponds to the ability of a resource to redo a work item that has previously been completed in a case. This may necessitate that work items subsequent to it (and hence dependent on its results) also be repeated, for example, user *jsmith* repeats the *enter order* work item and all subsequent work items for the current case.
- *Pre-Do* corresponds to the ability of a resource to execute a work item in the current case ahead of the time that it has been offered or allocated to any resources, for example, user *jsmith* executes the *finalize ticketing* work item even though preceding work items are not complete on the basis that they are sure that the details recorded for earlier work items will not change.

The detour patterns provide various mechanisms for ensuring that a process delivers acceptable outcomes even in the face of unanticipated resource variations or difficulties. In a similar vein, the next set of patterns seek to expedite the execution of a process.

Auto-Start Patterns

Auto-start patterns identify different ways of expediting process execution by automating various aspects of work item handling. There are four such patterns.

- *Commencement on Creation* corresponds to the ability of a resource to commence execution on a work item as soon as it is created, for example, as soon as an instance of the *emergency-shutdown* work item is created, allocate it to the user *jsmith* in a started state.
- *Commencement on Allocation* corresponds to the ability of a resource to commence execution on a work item as soon as it is allocated to them, for example,

once the *emergency-shutdown* work item is allocated to user *jsmith*, change its state to started.

- *Chained Execution* corresponds the ability to automatically start the next work item in a case once the previous one has completed. Typically, the use of this pattern falls under the auspices of an individual resource for work items associated with a specific case, for example, for the current case, user *jsmith* indicates that as soon as they have completed a work item, the following work item should be automatically allocated to them in a started state.
- *Piled Execution* corresponds to the ability to initiate the next instance of a work item corresponding to a given task (perhaps in a different case) once the previous one has completed, such that all work items are allocated to the same resource. Typically, the use of this pattern falls under the auspices of an individual resource for work items associated with a specific task. Only one resource can be in *Piled Execution* mode for a given task at any time, for example, user *jsmith* indicates they wish to undertake any instances of the *finalize-audit* work item regardless of the case in which they arise. In the interest of expediting the completion of these work items, they are to be automatically allocated to *jsmith's* worklist in a *started* state.

The various patterns discussed thus far focus on various qualities associated with the distribution of work items. In contrast, the next group – visibility patterns – deal with the external observability of work items in a process.

Visibility Patterns

Visibility patterns delineate the ability to configure the extent of disclosure about the state of progress on particular work items. There are two such patterns.

- *Configurable Unallocated Work Item Visibility* corresponds to the ability to configure the visibility of unallocated work items by process participants, for example, restrict the reporting of work items not yet allocated or started to users to whom they might be offered
- *Configurable Allocated Work Item Visibility* corresponds to the ability to configure the visibility of allocated or executing work items by process participants, for example, allow any users to see the state of allocated or started instances of the *adjudicate-case* work item.

The final group of patterns – multiple resource patterns – deal with variations in the cardinality between work items and resources.

Multiple Resource Patterns

Multiple resource patterns identify work situations where the correspondence between work items and resources is not one-to-one. There are two of these patterns.

- *Simultaneous Execution* corresponds to the ability for a resource to execute more than one work item simultaneously, for example, user *jsmith* can work on multiple work items that have been assigned to them at the same time.
- *Additional Resources* corresponds to the ability for a given resource to request additional resources to assist in the execution of a work item that it is currently undertaking, for example, user *jsmith* requests the allocation of more resources to the *review-findings* work item in order to speed its execution.

Although the patterns presented thus far have been described in an informal manner, each of them has a precise underlying description. The notion of formality is of critical importance if a business process is to be described and interpreted in an unambiguous way. The next section examines the importance of formality in a business process and presents various approaches to describing them in a complete and precise manner.

2.3 Formal Foundation

One of the frequent criticisms of modeling notations is that they are imprecise and, as a consequence, subject to varying interpretations by different parties. Describing a candidate modeling notation in terms of a formal technique provides an effective means of minimizing the potential for ambiguity. To do so, it is necessary to describe both the syntax and semantics of the modeling formalism using a well-founded technique. Suitable techniques for doing so generally stem from mathematical foundations and include general-purpose modeling approaches such as Petri nets and process algebras together with techniques more specifically focussed on software specification such as the formal specification languages Z, SSADM, and VDM.

Petri nets have proven to be a particularly effective mechanism for modeling the dynamic aspects of processes. As indicated by van der Aalst, they have three specific advantages:

- Formal semantics despite the graphical nature.
- State-based instead of event-based.
- Abundance of analysis techniques.

For these reasons, Petri nets and two specific variants – workflow nets and reset nets – have been chosen as the formal underpinning for the YAWL language. In this section, we will examine the operation of these techniques and their use for describing various operational aspects of the YAWL language. We start with Petri nets.

2.3.1 Petri Nets

Petri nets were originally conceived in the early 1960s as a way of characterizing concurrent behavior in distributed systems. Since that time, Petri nets (which are also known as place/transition nets or P/T nets) have proven useful as a means of

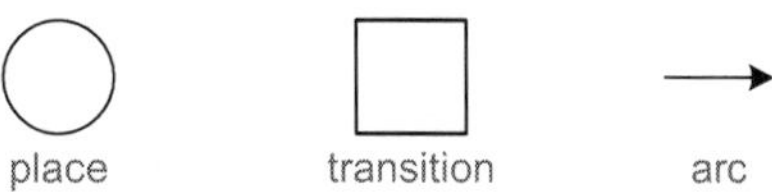

Fig. 2.13 Petri net constructs

describing the dynamics of processes and are widely utilized. Although they have a simple graphical format, they also have a precise operational semantics that makes them an attractive option for modeling the static and dynamic aspects of processes.

A Petri net takes the form of a directed bipartite graph where the nodes are either transitions or places. Figure 2.13 illustrates the elements appearing in a Petri net. Places represent intermediate states that may exist during the operation of a process and transitions correspond to the activities or events of which the process is made up. Arcs connect places and transitions in such a way that places can only be connected to transitions and vice versa. It is not permissible for places to be connected to places or for transitions to be connected to other transitions. A directed arc from a place to a transition indicates an input place to a transition and a directed arc from a transition to a place indicates an output place of a transition. These places are significant when discussing the operational semantics of Petri nets. A Petri net can be characterized formally as follows.

Definition 1. (Petri net). A Petri net is a triple (P, T, F):

- P is a finite set of places,
- T is a finite set of transitions where $P \cap T = \emptyset$,
- $F \subseteq (P \times T) \cup (T \times P)$ is a set of arcs known as the flow relation.

A place p is called an *input place* of a transition t if and only if there is a directed arc from p to t. Similarly, a place p is called an *output place* of transition t if and only if there is a directed arc from t to p. The set of input and output places for a transition t are denoted $\bullet t$ and $t\bullet$, respectively. The notations $\bullet p$ and $p\bullet$ have analogous meanings: for example, $p\bullet$ is the set of transitions that have p as an input place.

The operational semantics of a Petri net are described in terms of *tokens*, which signify a thread of control flowing through a process. Places in a Petri net can contain any number of tokens. The distribution of tokens across all of the places in a net is called a *marking*. It precisely describes the current state of an operational instance of a process and can be characterized by the function $M : P \rightarrow \mathbb{N}$. A state can be compactly described as shown in the following example: $2p_1 + 1p_2 + 0p_3 + 1p_4$ is the state with two tokens in place p_1, one token in p_2, no tokens in p_3, and one token in p_4. We can also represent this state in the following (equivalent) way: $2p_1 + p_2 + p_4$.

We can usefully describe an ordering function $\geq$ over the set of possible states such that for any two states M_1 and M_2, $M_1 \geq M_2$ if and only if for each $p \in P$: $M_1(p) \leq M_2(p)$. $M_1 > M_2$ iff $M_1 \geq M_2$ and $M_1 \neq M_2$.

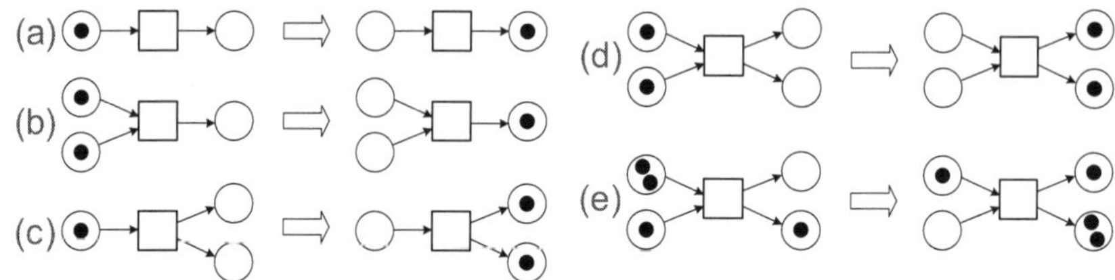

Fig. 2.14 Petri net operational semantics

The operational semantics of Petri nets are characterized by the notion of a transition executing or "firing". A transition in a Petri net can "fire" whenever there are one or more tokens in each of its input places. Such an action causes the state of a Petri net to change. The execution of a transition occurs in accordance with the following firing rules:

1. A transition t is said to be enabled if and only if each input place p of t contains at least one token.
2. An enabled transition may fire. If transition t fires, then t consumes one token from each input place p of t and produces one token for each output place p of t.

Figures 2.14a–e illustrate the act of firing for various Petri net configurations. It is assumed that the firing of a transition is an atomic action that occurs instantaneously and cannot be interrupted. The manner in which a Petri executes is deliberately intended to be nondeterministic. Hence when there are multiple transitions in a Petri net which are enabled, any one of them may fire; however, for execution purposes, it is assumed that they cannot fire simultaneously. Furthermore, a distinction is drawn between a transition being able to fire and it actually firing, and one of the salient features of Petri nets is the fact that an enabled transition is not obliged to fire immediately but can do so at a time of its choosing. These features make Petri nets particularly suitable for modeling concurrent process executions such as those that occur in business processes.

The specific details associated with the enablement and firing of transitions can be summarized as follows.

Definition 2. (Petri net enabling and firing rules). Given a Petri net (P, T, F) and an initial state M_1, we have the following notations that characterize the firing of a given transition (or sequence of transitions) and the resultant state change:

- $M_1 \xrightarrow{t} M_2$: indicates that if transition t is enabled in state M_1, then firing t in M_1 results in state M_2.
- $M_1 \rightarrow M_2$: indicates that there is a transition t such that $M_1 \xrightarrow{t} M_2$.
- $M_1 \xrightarrow{\sigma} M_2$: denotes the firing sequence $\sigma = t_1 t_2 t_3 \ldots t_{n-1}$ that leads from state M_1 to state M_n, i.e. $M_1 \xrightarrow{t_1} M_2 \xrightarrow{t_2} \ldots \xrightarrow{t_{n-1}} M_n$.

- A state M_n is called *reachable* from state M_1 (denoted $M_1 \stackrel{*}{\rightarrow} M_n$) if and only if there is a firing sequence $\sigma = t_1 t_2 t_3 \ldots t_{n-1}$ such that $M_1 \stackrel{t_1}{\rightarrow} M_2 \stackrel{t_2}{\rightarrow} \ldots \stackrel{t_{n-1}}{\rightarrow} M_n$.

We use (PN, M) to denote a Petri net PN with an initial state M. On the basis of this, we can define some additional properties for Petri nets.

Definition 3. (Live). A Petri net (PN, M) is live if and only if, for every reachable state $M'$and every transition t, there is a state M'' reachable from M', which enables t.

The notion of *liveness* is important as a Petri net which demonstrates that at least one transition can fire in every reachable state is deadlock free. Being deadlock free is a desirable property for a Petri net. Another important property is *boundedness*, which ensures that the number of tokens in a net cannot grow arbitrarily.

Definition 4. (Bounded). A Petri net (PN, M) is bounded if and only if, for every reachable state and every place p, the number of tokens in p is bounded.

Another useful and related property is *coverability*, which allows two states to be compared.

Definition 5. (Coverable): A marking M' in Petri net (PN, M) is said to be coverable if there exists a reachable marking M'' such that $M'' \geq M'$.

A final property of Petri nets that proves to be useful when reasoning about them is that of *strong connectivity*, defined as follows.

Definition 6. (Strongly connected). A Petri net is strongly connected if and only if, for every pair of nodes (i.e., places and transitions) x and y, there is a directed path leading from x to y.

Figure 2.15 provides an example of a Petri net used for modeling a business process. In this case it shows an order fulfillment process. This proceeds first with a *take order* task and then the *pack order* and *check account* tasks execute in parallel. When they have both completed, the *credit check* task executes, and if the customer has sufficient credit remaining, the order is despatched. If not, the *decline order* task runs and the customer is advised and finally the *return stock* task ensures that the items from the order are returned to the warehouse.

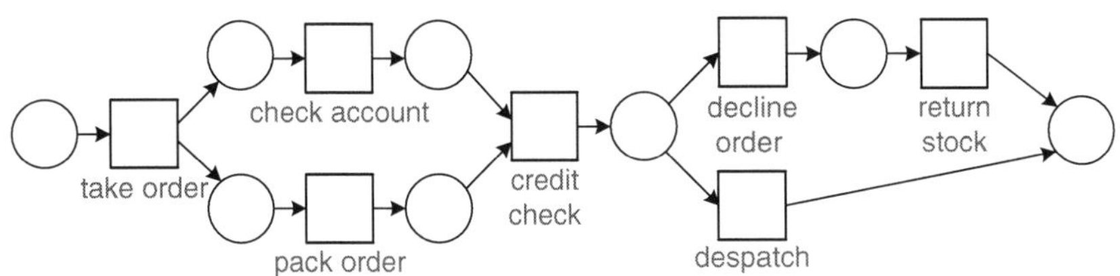

Fig. 2.15 Petri net example: order fulfillment process

Over the years since their initial definition, Petri nets have proven to be an extremely useful mechanism for capturing the details associated with concurrent systems and reasoning about them. To increase their applicability to the business process domain, the workflow nets were devised.

2.3.2 *Workflow Nets*

Workflow nets were developed by van der Aalst as a means of describing workflow processes in a precise way that makes them amenable to subsequent analysis both at design time and runtime. Workflow nets are directly based on Petri nets and consequently share all of their advantages in terms of determinism and the range of available analysis techniques. In a workflow net, transitions represent the tasks that comprise a business process and places represent the conditions preceding and following the tasks. Although they are based on Petri nets, workflow nets are subject to some additional constraints:

1. A workflow net has a single start place and a single end place. Aside from analytic considerations, this means that workflow nets closely correspond to real-life processes that tend to have a specific starting point and a specific end point.
2. Every transition in the workflow net is on a path from the start to the end place. This ensures that each transition in a workflow net contributes to the progression of an executing instance towards its end state.

We can formalize the notion of a workflow net and its correspondence with a Petri net in the following way:

Definition 7. (WF-net). A Petri net $PN = (P, T, F)$ is a WF-net (workflow net) if and only if:

- PN has two special places: i and o. Place i is a source place: $\bullet i = \varnothing$. Place o is a sink place: $o\bullet = \varnothing$.
- If we add a transition $t*$ to PN, which connects place o with i (i.e. $\bullet(t^*) = o$ and $(t^*)\bullet = i$), then the resulting Petri net is strongly connected.

It is important to note that these are the minimal requirements for a workflow net. By themselves they do not guarantee that a candidate workflow net will not potentially be subject to deadlock or livelock. To ensure that this is the case, there are two additional requirements that apply to a workflow net:

- Any executing instance of a workflow net must eventually terminate, and at the moment of termination there must be precisely one token in the end place o and all other places in the workflow net must be empty.
- There should be no dead tasks, that is, it should be possible to execute an arbitrary task by following the appropriate route through the WF-net.

The application of these two additional constraints corresponds to the *soundness* property, which ensures that any given process instance behaves in a predictable way.

Definition 8. (Sound). A procedure modeled in the form of a WF-net $PN = (P, T, F)$ is sound if and only if:

- For every state M reachable from state i, there exists a firing sequence leading from state M to state o. Formally:

$$\forall_M \left[(i \xrightarrow{*} M) \Rightarrow (M \xrightarrow{*} o) \right]$$

- State o is the only state reachable from state i with at least one token in place o. Formally:

$$\forall_M \left[(i \xrightarrow{*} M \wedge M \geq o) \Rightarrow (M = o) \right]$$

- There are no dead transitions in (PN, i). Formally:

$$\forall_{t \in T} \exists_{M,M'} \left[i \xrightarrow{*} M \xrightarrow{t} M' \right]$$

The format of workflow nets is essentially the same as for Petri nets, although there are some notational enhancements (often termed "syntactic sugar") for split and join constructs that simplify the specification of a workflow net. Figure 2.16 illustrates the range of workflow-net constructs.

A key aspect of a workflow net is that it identifies the way in which a task is initiated. This can occur in one of the four ways: (1) *automatic* tasks execute as soon as they are enabled, (2) *user* tasks are passed to human resources for execution once enabled, (3) *external* tasks only proceed once they are enabled and a required message or signal is received from the operating environment, and (4) *time* tasks only proceed once they are enabled and a specified (time-based) deadline occurs.

Figure 2.17 illustrates the order fulfillment process shown earlier in the form of a workflow net. In this model, the split and join constructs are explicitly identified. Also the manner in which tasks are triggered is shown. Most are undertaken by human resources (i.e., staff); however, the *take order* task is externally triggered when an order request is received and the *decline order* task runs automatically with the customer receiving a notification either by email or fax.

While Petri nets and workflow nets provide an effective means of modeling a business process, they are not able to capture the notion of cancelation. This is a

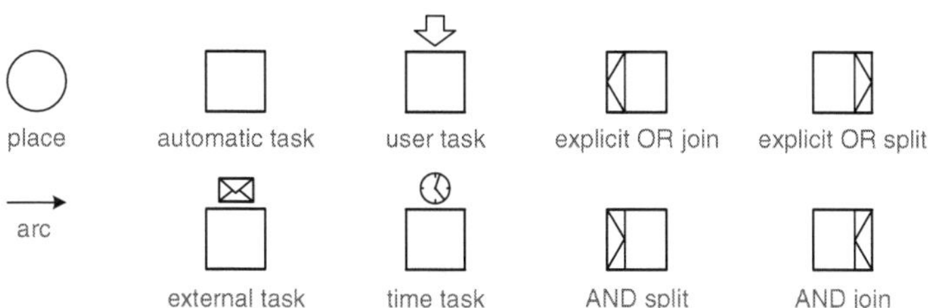

Fig. 2.16 Workflow net constructs

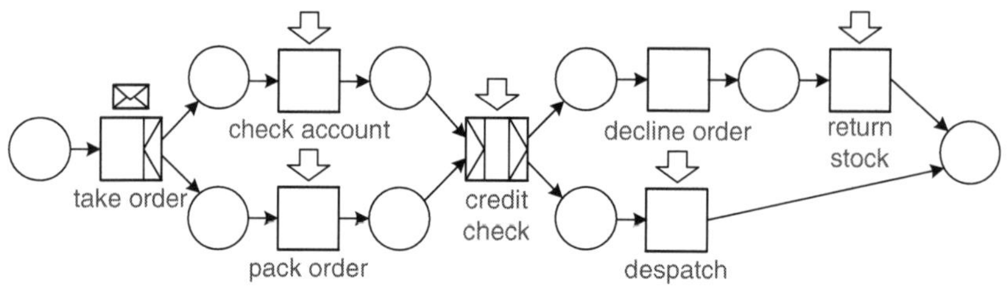

Fig. 2.17 Workflow net example: order Fulfillment process

Fig. 2.18 Reset net constructs

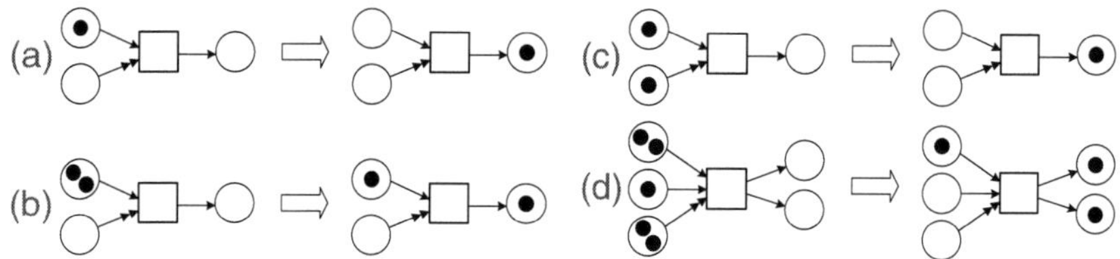

Fig. 2.19 Reset net operational semantics

significant omission, given the relative frequency of cancelation in real-life processes.

2.3.3 Reset Nets

The inability of standard Petri nets and workflow nets to directly capture the notion of cancelation within a business process spurred the use of *reset nets* for this purpose. Reset nets are able to explicitly depict notions of cancelation within a process definition.

In Fig. 2.18, it can be seen that reset nets are made up essentially of the same constructs as standard Petri nets with the addition of a reset arc. Like normal input arcs, reset arcs connect places to transitions; however, they operate in a different way. When the transition to which a reset arc is connected fires, any places connected to the transition by reset arcs are emptied of any tokens that they may contain. Figure 2.19a–d illustrate various scenarios of reset net operation. Note that the tokens in places attached to a transition by reset arcs play no part in the enablement of the transition.

As with other Petri net-based modeling formalisms for business processes, one of the advantages offered by reset nets is that they can be defined in a formal way.

Definition 9. (Reset net). A reset net is a tuple (P, T, F, R), where

- (P, T, F) is a classical Petri net with a finite set of places P, a finite set of transitions T, and a flow relation $F \subseteq (P \times T) \cup (T \times P)$, and
- $R \in T \rightarrow 2^P$ is a function defining the reset arcs associated with each transition t.

A reachable marking M' is defined by first removing the tokens needed to enable t from its input places $\bullet t$, then removing all tokens from reset places associated with t (i.e., $R(t)$), and finally by adding tokens to its output places $t\bullet$. The notions of enablement and firing in a reset net can be formalized as follows.

Definition 10. (Reset net enabling and firing rules). Let $N = (P, T, F, R)$ be a reset net and M be a marking of the net.

- A transition $t \in T$ is *enabled* if and only if $\bullet t \leq M$.
- An enabled transition t in state M can *fire* changing the state to M', denoted $M \xrightarrow{t} M'$, if and only if $\bullet t \leq M$ and $M' = (M - \bullet t)[P \setminus R(t)] + t\bullet$.

Definition 11. (Occurrence sequence). Let $N = (P, T, F, R)$ be a reset net with marking M_0. Let $M_1, \ldots, M_n$ be markings of the reset net and $t_0, t_1, \ldots, t_{n-1}$ be transitions in T. Sequence $\sigma = M_0 t_0 M_{1 \ldots} t_{n-1} M_n$ is an occurrence sequence if and only if $M_i \xrightarrow{t_i} M_{i+1}$ for all $i, 0 \leq i \leq n-1$. A marking M' is reachable from a marking M, written $M \xrightarrow{*} M'$, if and only if there is an occurrence sequence with initial marking M and final/last marking M'.

The three formalisms described above provide the formal basis for the YAWL language and allow business process models developed using the YAWL language to be enacted in an unambiguous way. In the next three sections, we discuss the control-flow, data, and resource perspectives of the YAWL language in detail and relate them to the range of control-flow, data, and resource patterns introduced earlier.

2.4 Control-flow

The control-flow patterns provide a guide to the range of features that are desirable in a workflow language. YAWL provides an intuitive range of constructs for describing the control-flow aspects of a business process and in doing so directly supports 31 control-flow patterns. In this section, we discuss the range of control-flow constructs that underpin YAWL and provide both examples of their usage and a precise operational definition for each of them.

2.4.1 Constructs

Figure 2.20 identifies the complete set of language elements that make up the control-flow perspective of YAWL. A YAWL model or specification is composed

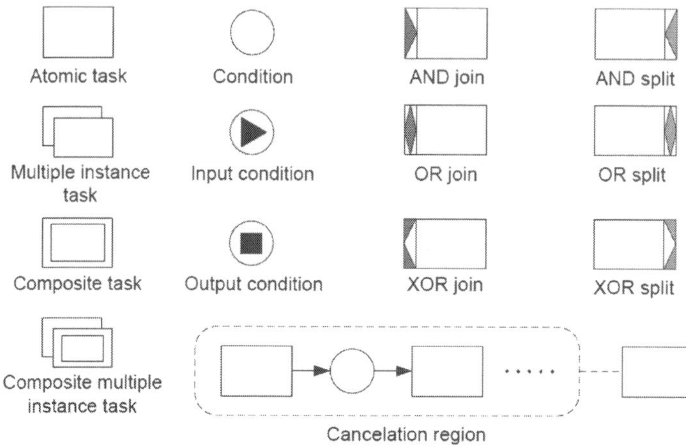

Fig. 2.20 YAWL control-flow symbols

of a set of YAWL nets in the form of a rooted graph structure. Each YAWL net is composed of a series of tasks and conditions. Tasks and conditions in YAWL nets play a similar role to transitions and places in Petri nets. Atomic tasks have a corresponding implementation that underpins them. Composite tasks refer to a unique YAWL net at a lower level in the hierarchy that describes the way in which the composite task is implemented. One YAWL net, referred to as the *top level process* or *top level net*, does not have a composite task referring to it and it forms the root of the graph.

Each YAWL net has one unique input and output condition. The input and output conditions of the top level net serve to signify the start and endpoint for a process instance.

Placing a token in the input condition for a YAWL net initiates a new instance of a process, also known as a *case*. The token corresponds to the thread of control in the process instance and it flows through the YAWL net in accordance with the rules specified in Sect. 2.3.1, triggering tasks as it does so. An enabled instance of a task corresponds to a new unit of work that needs to be completed and is referred to as a *task instance* or *work item*.

Similar to Petri nets, conditions and tasks are connected in the form of a directed graph; however, there is one distinction in that YAWL allows for tasks to be directly connected to each other. In this situation, it is assumed that an implicit condition exists between them.

It is possible for tasks (both atomic and composite) to be specified as having multiple instances (as indicated in Fig. 2.20). Multiple instance tasks (abbreviated hereafter as MI tasks) can have both lower and upper bounds on the number of instances created after initiating the task. It is also possible to specify that the task completes once a certain threshold of instances have completed. If no threshold is specified, the task completes once all instances have completed. If a threshold is specified, the task completes when the threshold is reached, but any remaining instances continue to execute until they complete normally. However, their completion is inconsequential and does not result in any other side-effects. Should the task commence with the required number of minimum instances and all of these complete but the required threshold of instance completions not be reached, the multiple instance task is deemed complete once there are no further instances being executed (either from the original set of instances when the task was triggered or additional instances that were started subsequently). Finally, it is possible to specify whether the number of task instances is fixed after creating the initial instances (i.e., the task is *static*) or whether further instances can be added while the multiple instance task has not yet completed execution (i.e., the task is *dynamic*).

Tasks in a YAWL net can have specific join and split behaviors associated with them. The supported join and split constructs are the AND-join, OR-join, XOR-join, AND-split, OR-split, and XOR-split. The operation of each of the joins and splits in YAWL is as follows:

- *AND-join* – the branch following the AND-join receives the thread of control when all of the incoming branches to the AND-join in a given case have been enabled.
- *OR-join* – the branch following the OR-join receives the thread of control when either (1) each active incoming branch has been enabled in a given case or (2) it is not possible that any branch that has not yet been enabled in a given case will be enabled at any future time with the currently enabled branches continuing to be marked with at least one token.
- *XOR-join* – the branch following the XOR-join receives the thread of control when one of the incoming branches to the XOR-join in a given case has been enabled.
- *AND-split* – when the incoming branch to the AND-split is enabled, the thread of control is passed to all of the branches following the AND-split.
- *OR-split* – when the incoming branch to the OR-split is enabled, the thread of control is passed to one or more of the branches following the OR-split, based on the evaluation of conditions associated with each of the outgoing branches.
- *XOR-split* – when the incoming branch to the XOR-split is enabled, the thread of control is passed to precisely one of the branches following the XOR-split, based on the evaluation of conditions associated with each of the outgoing branches.

Finally, YAWL supports the notion of a *cancelation region*, which encompasses a group of conditions and tasks in a YAWL net. It is linked to a specific task in the same YAWL net. At runtime, when an instance of the task to which the cancelation region is connected completes executing, all of the tasks in the associated cancelation region that are currently executing for the same case are withdrawn. Similarly any tokens that reside in conditions in the cancelation region that correspond to the same case are also withdrawn. We now review the constructs for the control-flow perspective via some illustrative examples.

Figure 2.21 illustrates a review process comprising three main tasks: *schedule review*, *conduct review*, and *report findings*. The *conduct review* task has a *timeout* associated with it that is triggered by the AND-split associated with the *schedule*

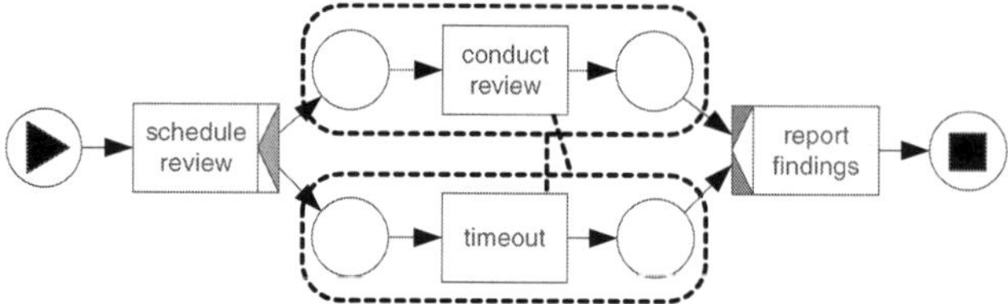

Fig. 2.21 YAWL control-flow example: 1

review task. This ensures that the countdown to the timeout commences as soon as the *schedule review* task completes. (Note that YAWL includes timer facilities for tasks that can be used for implementing timeouts. By using these, this model could be simplified by including a deadline timer in the *conduct review* task, hence removing the requirement for the preceding AND-split. However, for the purposes of illustration, in this example, we explicitly model the timeout as a distinct task).

Both the *conduct review* and *timeout* tasks have cancelation regions associated with them. In effect, a race condition exists between the two tasks and the first to complete cancels the other task. Thus one of two scenarios is possible: the *conduct review* task competes before the deadline is reached or if the deadline is reached, the task is canceled. Whichever scenario occurs, the *report findings* task will be triggered exactly once before the process completes as a consequence of the XOR-join associated with the task. Note that the conditions preceding and following the *conduct review* and *timeout* tasks are explicitly shown in the model and are included in the associated cancelation regions, as the states prior to and immediately after the execution of these two tasks are also subject to cancelation when the other task completes. For example, if the *timeout* completes first then the *conduct review* task is canceled, and if it is waiting to commence or has just completed (i.e., there is a token in the condition before or after the task) then the token is also removed.

Figure 2.22 shows a process for organizing travel. It commences when a travel request arrives triggering the *lodge travel request* task. Once this has completed, one or both of the *organize visa* and *estimate cost* tasks is initiated depending on whether just travel arrangements, visa procurement, or both need to be organized. The process of organizing a visa is fairly straightforward; however, the organization of the actual travel is more complex and involves several tasks. First, the likely costs need to be determined via the *estimate cost* task. Then one or both of the *book accommodation* and *book travel* tasks is initiated, and after those have been completed, it is determined if the bookings are complete, if not, there are further iterations of *estimate costs* and then *book accommodation* and/or *book travel* tasks as required. Once it is determined that the bookings are complete, the final costings are checked. If they are incomplete or vary markedly from the original request, the thread of control is returned to the *estimate cost* task and another iteration of the travel arrangement tasks occur. Once any necessary visas have been organized and the costings are complete, the *archive paperwork* task can execute and the process is complete. The use of the OR-splits associated with the *lodge travel request* and *estimate cost* tasks allows one or both outgoing branches to be selected and the

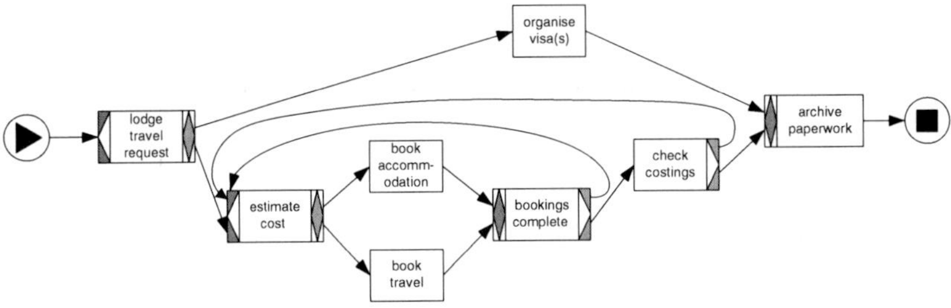

Fig. 2.22 YAWL control-flow example: 2

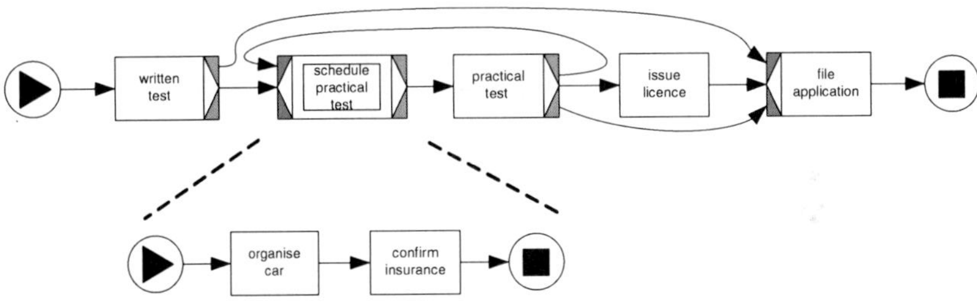

Fig. 2.23 YAWL control-flow example: 3

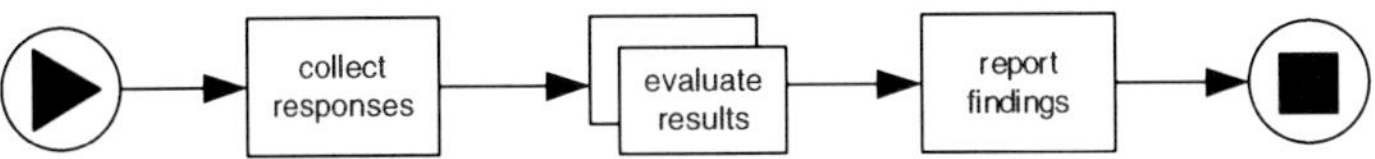

Fig. 2.24 YAWL control-flow example: 4

OR-joins associated with the *bookings complete* and *archive paperwork* tasks ensure that they do not commence until all active preceding tasks have been completed.

Figure 2.23 depicts a driving license examination process. First there is a *written test* task. If the applicant passes this, they are then able to progress to the *schedule practical test* task. If not, the application is deemed to be unsuccessful and it is filed and the test process for the applicant concludes. The *schedule practical test* task is composite in form and consists of a subprocess with two tasks: *organize car* and *confirm insurance*. Once it is complete, the *practical test* task can occur. Depending on its outcome, if it is successful, the *issue license* task is triggered, and if it is unsuccessful, the applicant can attempt a subsequent practical test (on up to two subsequent occasions, after this the application is deemed unsuccessful and is finalized) and the thread of execution is routed back to the *schedule practical test* task again. After the license is issued, the *file application* task completes the process.

Figure 2.24 provides a simple example of the use of the multiple instance task. It shows a medical review process consisting of three tasks. First the *collect responses* task gathers the details of the medical cases being investigated from patients. Then a distinct *evaluate results* task executes concurrently for each patient case being examined. The number of these tasks depends on the data provided by the *collect*

Table 2.5 Control-flow pattern support

Pattern group	Supported patterns
Branching patterns	All except the *Thread Split*
Synchronization patterns	All except the *Thread Merge*, *Partial Join* variants, and the *Blocking Discriminator*
Repetition patterns	*Arbitrary Cycles*, *Recursion*. Partial support for the *Structured Loop* via task pre/post-conditions facilitated via the Exception Service
Multiple instance patterns	All except for the *Dynamic Partial Join for Multiple Instances*, which is only partially supported as a consequence of the fact that multiple instance tasks can only have threshold-based completion conditions and they cannot be specified in a more general manner
Concurrency patterns	All
Trigger patterns	No support as task execution cannot be externally synchronized
Cancelation and completion patterns	All except for *Complete Multiple Instance Task*, which can only be indirectly achieved through the use of a timer to force complete remaining instances
Termination patterns	*Explicit Termination*

responses task. Finally, once all evaluations have been completed, the *report findings* task is completed.

These diagrams serve to give an indication of the control-flow aspects of a YAWL process model. In the next section, we examine the formal representation of the control-flow perspective in a YAWL model.

2.4.2 Control-Flow Patterns Coverage

The various constructs that comprise the control-flow perspective of YAWL are based upon the control-flow patterns discussed in Sect. 2.2.1. It provides an interesting insight into the conceptual power of YAWL to examine the range of control-flow patterns that it provides support for. These are identified in Table 2.5

In total, YAWL directly supports 31 of the 43 control-flow patterns and provides partial solutions for a further three patterns.

2.4.3 Object Role Model

This section provides a description of the main elements that make up the control-flow perspective of a YAWL model using the object-role modeling notation as shown in Fig. 2.25. In summary, a YAWL workflow specification is comprised of a number of YAWL nets. Each net provides a description of a portion of an overall YAWL process. One of them is designated as the root net (i.e., the top-level net in a process definition), while the others correspond to decompositions of composite tasks. Each net comprises a number of net elements, which can be conditions or tasks. Two conditions are specially designated in each net as the start and end condition, which signify the initiation and completion of an instance of the net.

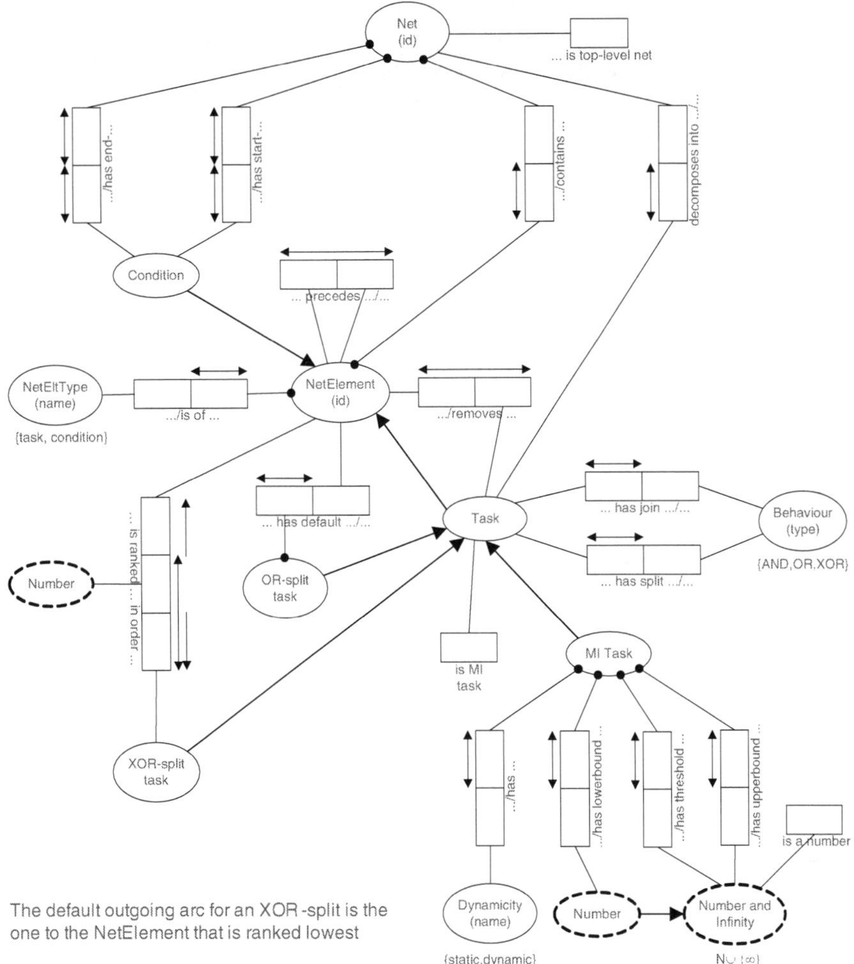

Fig. 2.25 Object role model: YAWL control-flow perspective

There is a precedence relationship that provides an ordering between all net elements such that every net element is on a path from the start to the end condition. Net elements are connected using directed arcs that indicate the direction in which the thread of execution flows. The set of directed arcs for a net is termed the flow relation.

Where a net element is a task, it may demonstrate split or join behavior of AND, OR, or XOR type. Also, a task may be atomic in form or it may have multiple instances. Where it is a multiple instance task, it has a lower and upper bound specified, which indicate the minimum and maximum number of instances that may run. It also has a threshold, which indicates how many instances must be complete before the thread of control can be passed on, and a dynamicity, which indicates whether the task is static, that is, all instances are started at commencement or dynamic, that is, additional instances can be initiated during execution. Where a task has an OR-split associated with it, it has a branch nominated as its default, should all conditions associated with branches evaluate to false. Where the task has an XOR-split associated with it, an ordering is specified, indicating the sequence in which conditions associated with output branches are evaluated. In the event that none of the conditions evaluate positively, the lowest ranked branch is selected as the default.

2.5 Data

The data perspective of YAWL encompasses the definition of a range of data elements, each with a distinct scoping. These data elements are used for managing data with a YAWL process instance and are passed between process components using query-based parameters. To integrate the data and control-flow perspectives, support is provided in YAWL for specifying logical conditions (known as link conditions) based on data elements that define whether the thread of control can be passed to a given branch in a process. All of these YAWL data constructs are discussed subsequently.

2.5.1 Data Element Support

YAWL incorporates a wide range of facilities for data representation and handling. Variables of three distinct scopings are recognized as follows:

- *Net* variables are bound to a specific instance of a *YAWL net*. At runtime, a new instance of the net variable is created for every instance of the *YAWL net* that is initiated. This variable instance is accessible throughout the *YAWL net* at runtime.
- *Task* variables are bound to a specific task. At runtime, a new instance of the task variable is created for every instance of the task that is initiated. This variable instance is accessible only to the corresponding task instance at runtime.
- *Multiple instance* variables are bound to a specific instance of a multiple instance task. At runtime, a new instance of the multiple instance variable is created for every instance of the multiple instance task that is initiated. This variable instance is accessible only to the corresponding task instance at runtime.

Figure 2.26 illustrates the configuration of task and multiple instance variables for the *Evaluate Results* task in the Medical Review process shown in Fig. 2.24. It also shows how the values of these variables are populated from net variables using XQuery-based parameters.

2.5.2 Data Interaction Support

There are two facilities for transferring data elements to and from tasks in YAWL. *Simple parameters* can be used to transfer the value of a data element between a net

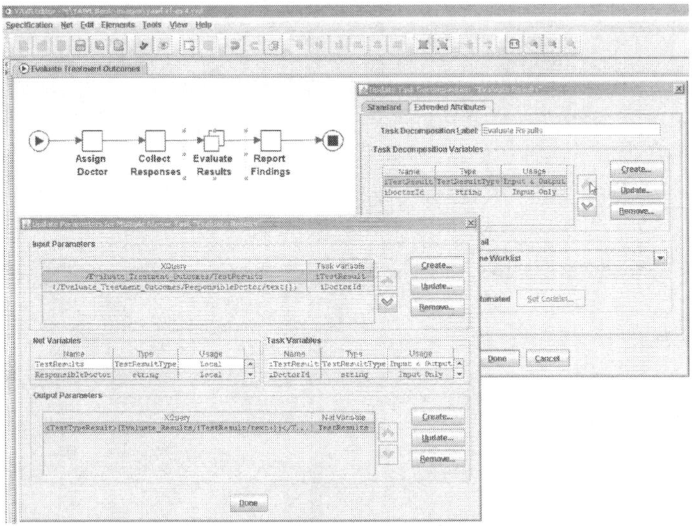

Fig. 2.26 Variable specification and parameter passing in YAWL

variable and a given task. The task can be of any form: atomic, composite, multiple instance, or composite multiple instance; however, in all cases, the parameter operates on every instance of the task to which it applies. Each task may have multiple simple parameters associated with it. There are two types of simple parameter:

- *Input parameters* involve the application of an XQuery function to a net variable in the block surrounding a task and placing the resultant value in a nominated task variable.
- *Output parameters* involve the application of an XQuery function to a task variable and placing the resultant value in a nominated net variable in the surrounding block.

Figure 2.27 illustrates the use of simple parameters in the Driving Test Process shown earlier. In this example, they are used to map data values between the net variable *ApplicantStatus* and the task variable *iApplicantStatus*. In each case, the parameter is specified in the form of an XQuery expression. The usage descriptor for the *iApplicantStatus task* variable is Input & Output, denoting that the value of the variable is passed in at task commencement and passed out at its conclusion.

The situation is essentially the same whether the parameter is transferring a data element to/from an atomic task or a composite task, except that the target variable is identified as *both* a task variable in the composite task and as a net variable in the associated decomposition. Figure 2.28 illustrates the situation where the task being targeted is composite in form and has an associated decomposition specified in the form of a subprocess. In this situation, the net variable *ApplicantStatus* is passed to the variable *iApplicantStatus* that appears as both a task variable for the *Schedule Practical Test* task and is also a net variable in the *Practical Test Scheduling* subpro-

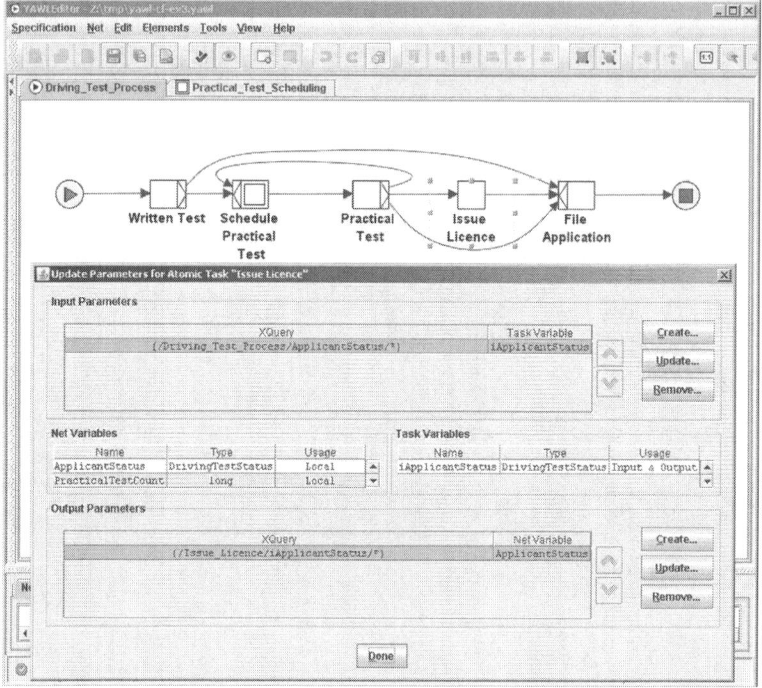

Fig. 2.27 Data passing using simple parameters in YAWL

cess. Both variables share the same name: *iApplicantStatus*. The XQuery expression that identifies the parameter mapping is also shown in this figure.

Multiple instance parameters operate on multiple instance tasks. Unlike simple parameters that result in the application of the XQuery to every instance of the task with which they are associated, multiple instance parameters are used to partition net variables containing record-based data across individual instances of a multiple instance task or to coalesce variables in instances of a given multiple instance task into a nominated net variable. Only one set of multiple instance parameters can be associated with a specific multiple instance task. The multiple instance parameter set is comprised of four distinct components as illustrated in Fig. 2.29.

For data input to a multiple instance task, there are two queries:

- The *accessor query* gathers the data that is to be partitioned across various task instances.
- The *splitter query*, as its name implies, splits the data element derived by the accessor query across each task instance. The number of segments into which the data element is split determines the number of task instances that will be started. Each segment of data is mapped to a distinct task instance. All task instances share the same task variable names (although each task instance has a distinct instance of a given variable). In the example shown in Fig. 2.29, separate variables are created for *Test* and *Nr* in each task instance.

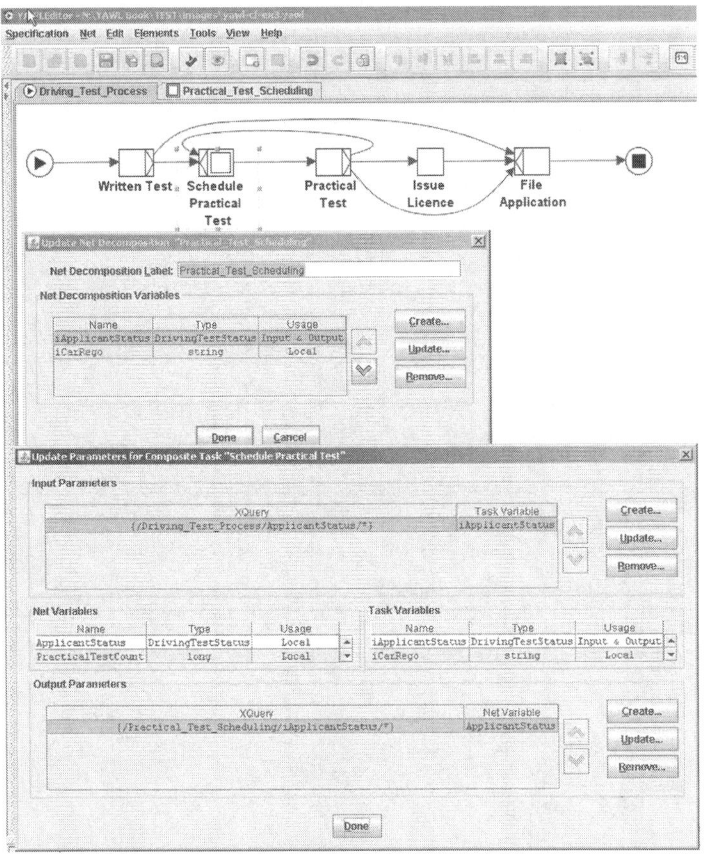

Fig. 2.28 Data passing to composite task in YAWL

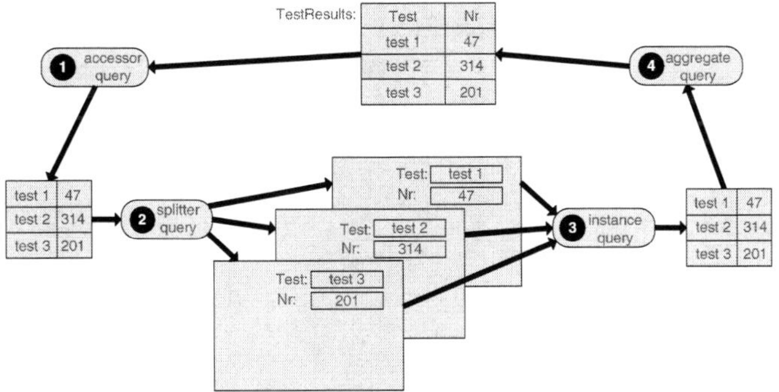

Fig. 2.29 Multiple instance input parameter handling

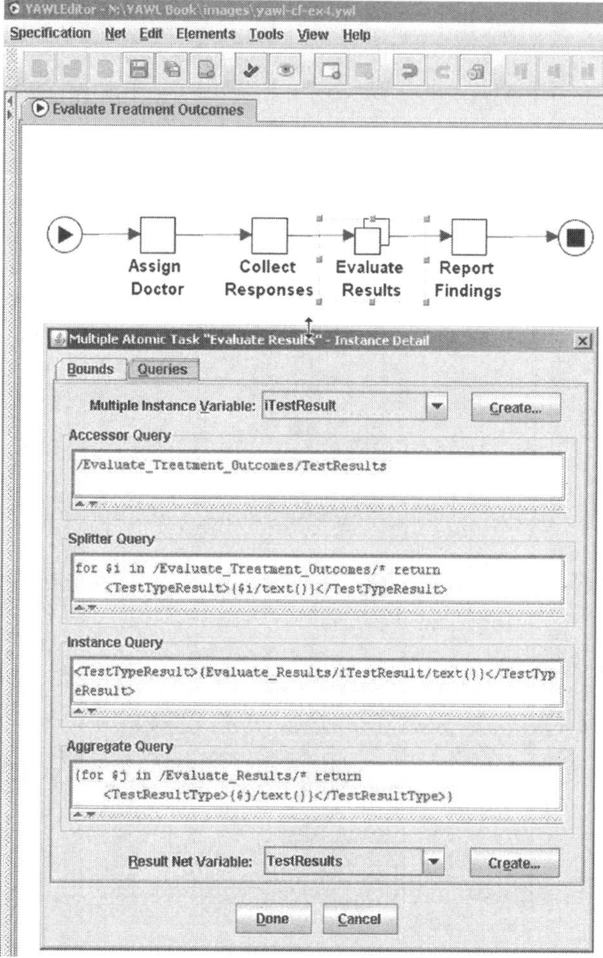

Fig. 2.30 Data passing to/from a multiple instance task in YAWL

Similarly for data output from a multiple instance task, there are also two queries:

- The *instance query* gathers the instances of the task variables from each task instance at their completion.
- The *aggregate query* coalesces the resultant data elements into a single record-based net variable.

Figure 2.30 illustrates how individual instances of the iTestResult variable are populated in the *Evaluate Results* multiple instance task from the TestResults net variable and also how the results are coalesced at the conclusion of the task. The XQuery expression required for each of the queries is also shown.

2.5.3 Link Conditions

As a consequence of a fully fledged data perspective, YAWL is able to support conditions (known as link or flow conditions) on outgoing arcs from OR-splits and XOR-splits. These conditions take the form of XPath expressions, which evaluate to a Boolean result, indicating whether the thread of control can be passed to this link or not. For OR-splits, all outgoing link conditions are evaluated and the thread of control is passed to all that evaluate to true. A default link is nominated to which the thread of control is passed if none of the link conditions evaluate to true. For XOR-splits, the link conditions are evaluated in a specified evaluation sequence and once the first link condition evaluates to true, then the thread of control is passed to that link and any further evaluation of link conditions ceases. If none of the link conditions evaluate to true, then the thread of control is passed to the link that is last in the evaluation sequence. Figure 2.31 illustrates the configuration of the flow conditions for the XOR-split associated with the *Practical Test* task in the Driving Test Process shown earlier in Fig. 2.23.

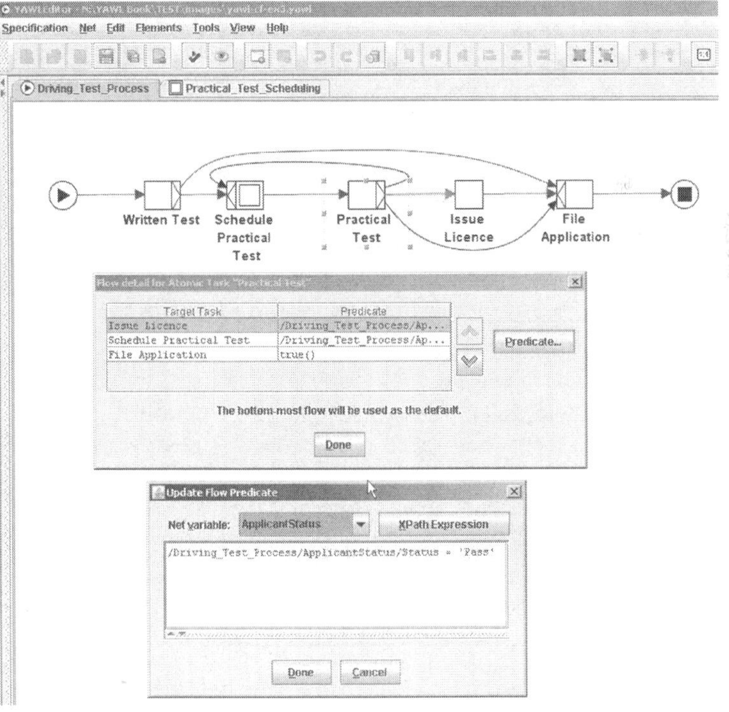

Fig. 2.31 Link condition specification in YAWL

Table 2.6 Data pattern support

Pattern Group	Supported Patterns
Data visibility patterns	*Task Data*, *Block Data*, and *Multiple Instance Data*
Data interaction patterns – internal	All except *Data Interaction – Case to Case*
Data interaction patterns – external	None are directly supported
Data transfer patterns	*Data Transfer by Value*, *Data Transformation – Input*, *Data Transformation – Output*
Data routing patterns	*Data-based Routing*

2.5.4 Data Patterns Coverage

The various facilities that underpin the data perspective in YAWL are influenced by the data patterns presented in Sect. 2.2.2. The extent of data pattern support is summarized in Table 2.6.

In total, YAWL directly supports 10 of the data patterns. There are partial solutions for a further five patterns: *Data Interaction – Process to Environment – Push-Oriented* is indirectly supported through the use of codelets via the Resource Service (cf. Chap. 10) and the various *Task Pre-* and *Post-condition* patterns can be facilitated via task configurations supported by the Exception Service.

2.5.5 Object Role Model

This section provides a conceptual model for the data perspective of a YAWL model using the object-role modeling notation as shown in Fig. 2.32. The data perspective centers around the concepts of nets, tasks, variables, and parameters. Variables are used to represent a data element and have a specific name together with a data type specified using XMLSchema. Variables either correspond to a net or a task. Parameters are used to map data elements from one variable to another at the commencement or conclusion of a task. Parameters may be simple (which can be used by any task) or multiple instance (which only apply to multiple instance tasks). In either case, however, they are specified in terms of an XQuery expression. Parameters always have an input and output data element. Simple parameters may be either input or output parameters, indicating the direction in which they operate. In the case of multiple instance parameters, the situation is more complicated and the parameter is made up of four distinct queries: the accessor, splitter, instance, and aggregate query. Tasks have a precedence relationship between them (termed a *flow*), which indicates the direction in which the thread of control flows between them in a given process. In the case where a task has an OR-split or XOR-split associated with it, its outgoing flows are conditional and there is a Boolean condition, specified in terms of an XPath expression, for each flow. This condition is evaluated at runtime to determine whether the thread of control is passed down the flow when the task completes execution.

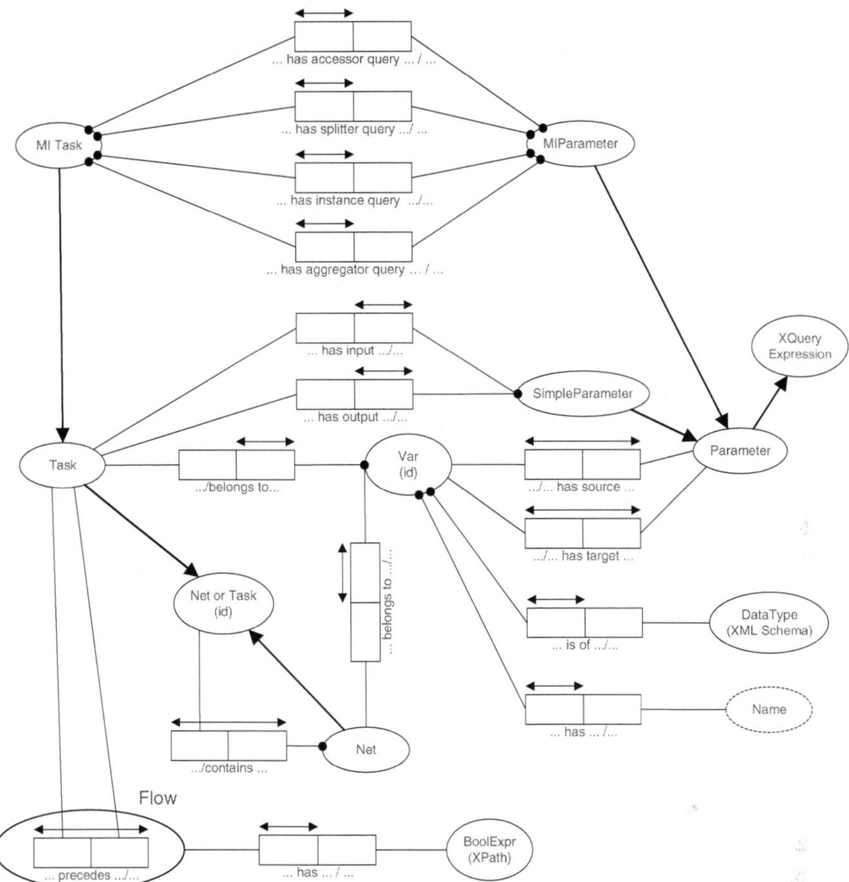

Fig. 2.32 Object role model: YAWL data perspective

2.6 Resources

YAWL provides support for a broad range of work distribution facilities, inspired by the *Resource Patterns* that have not been previously embodied in other PAISs. Traditional approaches to work item routing based on itemization of specific users and roles are augmented with a sophisticated array of new features. There are a variety of differing ways in which work items may be distributed to users. Typically these requirements are specified on a task-by-task basis and have two main components:

1. The *interaction strategy* by which the work item will be communicated to the user, their commitment to executing it will be established and the time of its commencement will be determined.
2. The *routing strategy*, which determines the range of potential users that can undertake the work item.

YAWL also provides additional routing constraints that operate with a given case to restrict the users a given work item is distributed to based on the routing decisions associated with previous work items in the case. There is also the ability to specify privileges for each user in order to define the range of operations that they can perform when undertaking work items. Finally, there are two advanced operating modes that can be utilized in order to expedite work throughput for a given user. Each of these features is discussed in detail in the following sections.

2.6.1 Organizational Model

YAWL assumes that a process operates in an organizational context, hence there is an organizational model underpinning the resource and work distribution aspects of each process. Figure 2.33 identifies the main organizational and resource-related concepts that are applicable to a YAWL process and dictate the context in which it operates. While relatively simple, the YAWL organizational model is deliberately intended to demonstrate general applicability to a broad range of organizational situations in which it might be utilized.

Work in a YAWL process may be completed by a human resource who in turn is assumed to be a member of an organization. An organization is a formal grouping of resources that undertake work items pertaining to a common set of business objectives. They usually have a specific position within that organization and, in

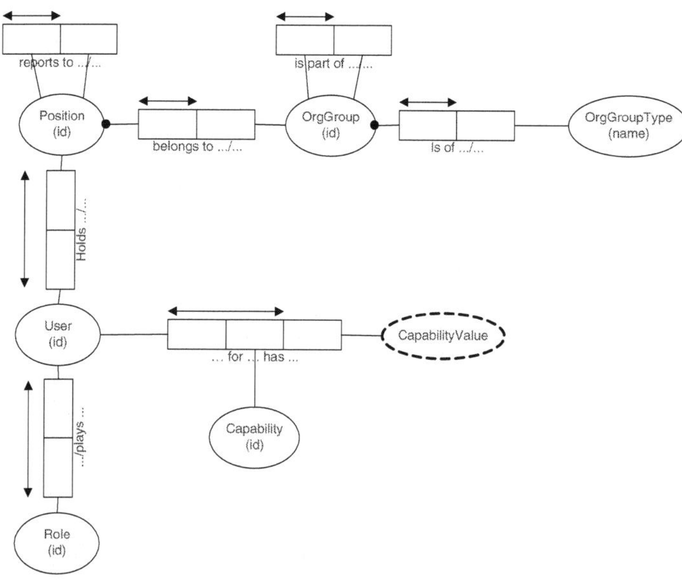

Fig. 2.33 Object role model: YAWL organizational model

general, most organizational characteristics that resources possess relate to the position(s) that they occupy rather than directly to the resource themselves. There are two sets of characteristics that are exceptions to this rule, however: roles and capabilities. Roles serve as another grouping mechanism for human resources with similar positions or responsibility levels, for example, managers, union delegates, etc. Each resource may have one or more corresponding roles. Individual resources may also possess capabilities or attributes that further clarify their suitability for various kinds of work. These may include qualifications and skills as well as other position-related or personal attributes such as specific responsibilities held or previous work experience. Each position is attached to an organizational group, which are groupings of human resources within the organization that undertake work items relating to a common set of business objectives. These may be permanent or temporary in nature (in which case they are formed for a specific purpose and are often known as teams). Each position is generally associated with a specific branch, which defines a grouping of resources within the organization at a specific physical location. It may also belong to a division, which defines a large scale grouping of resources within an organization either along regional geographic or business purpose lines. In terms of the organizational hierarchy, each position may have a number of specific relationships with other positions. Their direct report is the resource to whom they are responsible for their work. Generally, this is a more senior resource at a higher organizational level. Similarly, a position may also have a number of subordinates for whom they are responsible and to who each of them report.

2.6.2 Life-Cycle in YAWL

Once a task is triggered in a YAWL process, a work item is created that can be distributed to these resources. There is a specific life-cycle associated with this work item as it is distributed to these resources and is ultimately managed through to completion. Figure 2.34 illustrates this life-cycle.

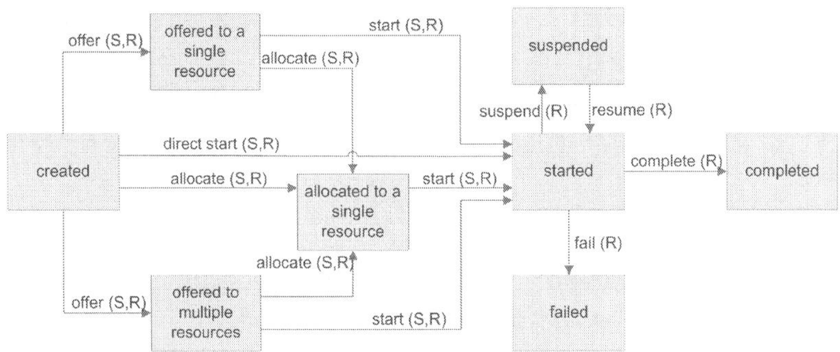

Fig. 2.34 Work item life-cycle in YAWL from a resource perspective

A work item comes into existence in the *created* state. This indicates that the preconditions required for its enablement have been satisfied and it is capable of being executed. At this point, however, the work item has not been allocated to a resource for execution and there are a number of possible paths through these states that individual work items may take. Each edge within this diagram is prefixed with either an S or an R, indicating that the transition is initiated by the system (i.e., the software environment in which instances of the process execute) or resource (i.e., an actual user), respectively. Transitions between all states once the work item has been *created* up until when it is *started* can be initiated either by the system or by a resource depending on how the specific work distribution directives for the associated have been configured. Once a work item is *started*, any further state transitions are triggered by the resource responsible for the work item up until the time that it is *completed* or *failed*. Each state has a specific meaning in terms of how the work item is handled:

- *Created* means that the work item has come into existence but no resources are yet aware of it.
- *Offered to a single resource* indicates that the opportunity to undertake the work item rests with a single resource and that the resource has not yet committed to undertaking it.
- *Offered to multiple resources* indicates that several resources have been offered the opportunity to undertake the work item but none have elected to do so as yet. Once one of them does commit to executing the work item, the offers received by other resources are withdrawn.
- *Allocated* indicates that a specific resource has elected (or been requested) to undertake the work item but has not yet started working on it.
- *Started* indicates that the responsible resource has commenced executing the work item.
- *Suspended* indicates that the responsible resource has elected to cease execution of the work item for a period, but does intend to continue working on it at a later time.
- *Failed* indicates that the work item cannot be completed and that the resource will not work on it any further.
- *Completed* identifies that a work item that has been successfully executed to completion.

2.6.3 Interaction Strategies

The potential range of interaction strategies that can be specified for tasks in YAWL are listed in Table 2.7. They are based on the specification at three main interaction points – offer, allocation, and start – of the identity of the party that will be responsible for determining when the interaction will occur. This can be a resource (i.e., an actual user) or the system. Depending on the combination of parties specified for each interaction, a range of possible distributions are possible as described. From

Table 2.7 Work item interaction strategies supported in YAWL

	Offer	Allocation	Start	Effect
SSS	System	System	System	The system directly allocates work to a resource and it is automatically started.
SSR	System	System	Resource	The system directly allocates work to a resource. It is started when the user selects the *start* option.
SRS	System	Resource	System	The system offers work to one or more users. The first user to choose the *select* option for the work item has the work item allocated to them and it is automatically started. It is withdrawn from other users' worklists.
SRR	System	Resource	Resource	The system offers work to one or more users. The first user to choose the *select* option for the work item has the work item allocated to them. It is withdrawn from other users' worklists. The user can choose when to start the work item via the *start* option.
RSS	Resource	System	System	The work item is passed to a manager who decides which resource the work item should be allocated to. The work item is then directly allocated to that user and is automatically started.
RSR	Resource	System	Resource	The work item is passed to a manager who decides which resource the work item should be allocated to. The work item is then directly allocated to that user. The user can choose when to start the work item via the *start* option.
RRS	Resource	Resource	System	The work item is passed to a manager who decides which resource(s) the work item should be offered to. The work item is then offered to those user(s). The first user to choose the *select* option for the work item has the work item allocated to them and it is automatically started. It is withdrawn from all other user's worklists.
RRR	Resource	Resource	Resource	The work item is passed to a manager who decides which resource(s) the work item should be offered to. The work item is then offered to those user(s). The first user to choose the *select* option for the work item has the work item allocated to them. It is withdrawn from all other users' worklists. The user can choose when to start the work item via the *start* option.

the perspective of the resource, each interaction strategy results in a distinct experience in terms of the way in which the work item is distributed to them. The range of strategies supported range from highly regimented schemes (e.g., SSS), where the work item is directly allocated to the resource and started for them, and the resource has no involvement in the distribution process through to approaches that empower the resource with significant autonomy (e.g., RRR), where the act of committing to undertake a work item and deciding when to start it are completely at the resource's discretion.

As an aid to understanding the distinctions between the various interactions described in Table 2.7, it is possible to illustrate them quite effectively using UML *Sequence Diagrams* as depicted in Fig. 2.35. These show the range of interactions between the system and resources that can occur when distributing work items. An arrow from one object to another indicates that the first party sends a request to the second, for example, in the RRR interaction strategy, the first request is a *manual_offer* from the system to the workflow administrator.

Figure 2.36 illustrates how the interaction strategy is configured for a task in YAWL. In this case, the *Book Accommodation* task in the Process Travel Request process is set to an SRR strategy. This means any associated work items will be automatically offered to prospective users by the system when the task is enabled; however, users may choose to commit to undertaking the task (i.e., allocation) and starting it at a time of their own discretion.

2.6.4 Routing Strategies

The second component of the work distribution process concerns the routing strategy employed for a given task. This specifies the potential user or a group of users from which the actual user will be selected, who will ultimately execute a work item associated with the task. There are a variety of means by which the task routing may be specified as well a series of additional constraints that may be brought into use at runtime. These are summarized below. Combinations of these strategies and constraints are also permissible.

Task Routing Strategies

There are a variety of routing strategies that can be defined at the level of an individual task as described below.

- *Direct user distribution*: This approach involves routing to a specified user or group of users.
- *Role-based distribution*: This approach involves routing to one or more roles. A role is a "handle" for a group of users who allows the group population to be changed without the necessity to change all the task routing directives. The population of the role is determined at runtime at the time of the routing activity. Figure 2.37 continues the configuration of the *Book Accommodation* task and illustrates the use of a role-based distribution strategy for it. In this case, combined with the interaction approach specified in Fig. 2.36, the task would be offered to members of the Travel Consultant role. The same configuration screen also allows for the specification of direct user and deferred distribution approaches, though they are not utilized in this example.
- *Deferred distribution*: This approach allows a net variable to be specified that is accessed at runtime to determine the user or role that the work item associated with the task should be routed to.

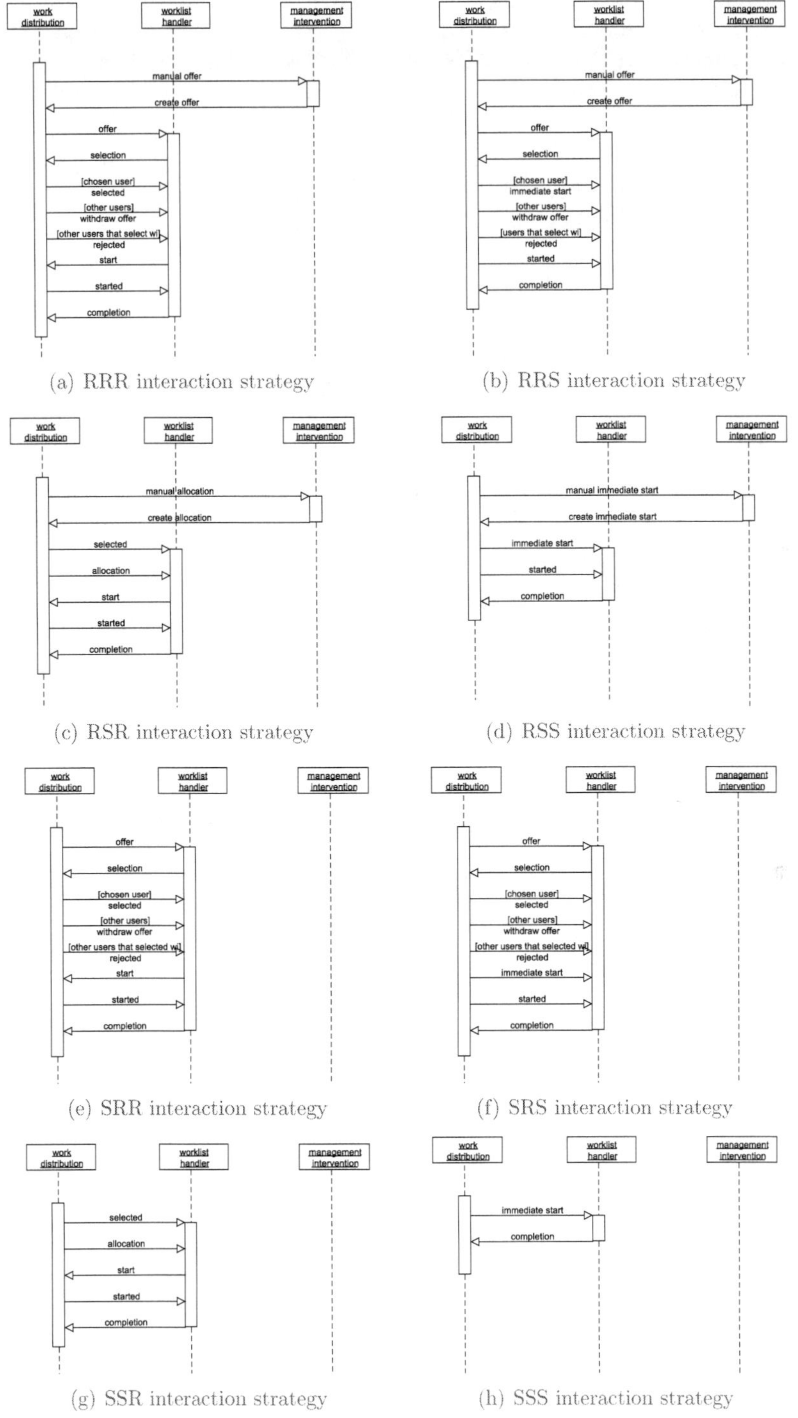

(a) RRR interaction strategy

(b) RRS interaction strategy

(c) RSR interaction strategy

(d) RSS interaction strategy

(e) SRR interaction strategy

(f) SRS interaction strategy

(g) SSR interaction strategy

(h) SSS interaction strategy

Fig. 2.35 Work item interaction strategies in YAWL

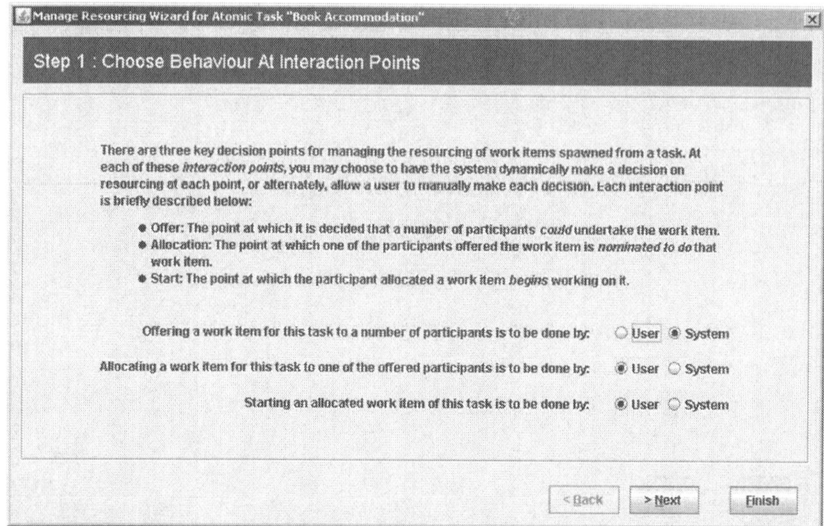

Fig. 2.36 Interaction strategy for YAWL task

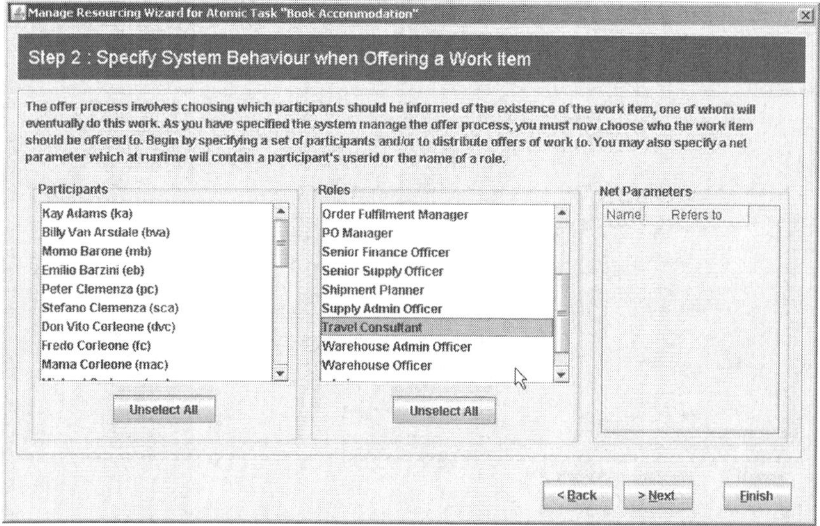

Fig. 2.37 Work distribution specification (users and roles)

- *Organizational distribution*: This approach allows an organizational distribution approach to be specified for a task that utilizes organizational data to make a routing decision. In doing so, the distribution of a task can be limited to users in a specified position or organizational group.
- *Capability-based distribution*: Capabilities can be specified for each user, which describe the qualities that they possess, which may be of relevance in making

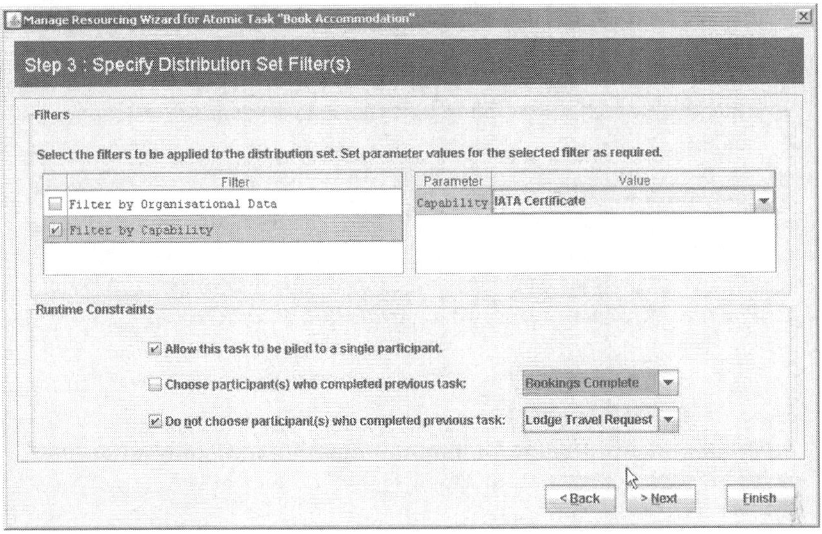

Fig. 2.38 Work distribution specification (other constraints)

routing decisions. A capability-based distribution filter can be specified for each task, which allows user capabilities to be used in making work distribution decisions. Figure 2.38 continues the specification of the work distribution approach for the *Book Accommodation* task specifying a capability distribution filter, which requires that the task only be distributed to users with an IATA Certificate (a standard travel industry qualification). In combination with the previous configurations, this would result in work items corresponding to the task being offered to resources in the *Travel Consultant* role with an *IATA Certificate*. Thus, it can be seen that relatively fine-grained work distribution strategies can be specified for YAWL tasks. This figure also shows some other configuration items that will be discussed shortly.

- *Historical distribution*: The use of pluggable allocators allows historical data – essentially the content of the execution log – to be used in making work distribution decisions. There are a range of standard predefined allocators that can be utilized to implement history-based task distribution strategies such as distribute it to the user who completed it least recently, most recently, has the highest success rate, and completes it the quickest.

Additional Routing Constraints

There are two additional constraints supported by YAWL that can be used to further refine the manner in which work items are routed to users. They are used in conjunction with the routing and interaction strategies described above.

- *Retain familiar*: This constraint on a task overrides any other routing strategies and allows a work item associated with it to be routed to the same user who undertook a work item associated with a specified preceding task in the same process instance. Where the preceding task has been executed several times within the same process instance (e.g., as part of a loop or it is a multiple instance task), it is routed to one of the users who undertook a preceding instance of the task.
- *Four eyes principle*: This constraint on a task is essentially the converse of the *Retain familiar* constraint. It ensures that the potential users to whom a work item associated with a task is routed does *not* include the user who undertook a work item associated with a nominated preceding task in the same process instance. Where the preceding task has been executed several times within the same process instance or is a multiple instance task, it cannot be routed to any of the users who undertook a preceding instance of the task. Figure 2.38 illustrates a four eyes constraint for the *Book Accommodation* task, requiring that it not be distributed in a given case to the same user who undertook the *Lodge Travel Request* task.

Allocation Strategies

Allocation strategies provide a means of selecting one specific user from a range of users identified by the routing strategy specified for a task. The task is then allocated to the selected user. There are various means by which this can be done as indicated below.

- *Random allocation*: This filter ensures that any work items associated with a task are only ever routed to a single user, where the user is selected from a group of potential users on a random basis.
- *Round robin (by time) allocation*: This filter ensures that any work items associated with a task are only ever routed to a single user, where the user is selected from a group of potential users on a cyclic basis such that each of them execute work items associated with the task the same number of times (i.e., the distribution is intended to be equitable). The actual means of selecting the user is based on the user in the nominated group that executed the task least recently.
- *Round robin (by least frequency) allocation*: This filter operates in a similar way to the *Round robin (by time) allocation* described above; however, rather than simply selecting the user who executed the task least recently, it actually involves keeping track of how many times each eligible user has undertaken it and allocating the task to the person who has executed it least times.
- *Round robin (by experience) allocation*: This filter operates in a similar way to the *Round robin (by least frequency) allocation* described above, except that it selects the user who executed the task most times.

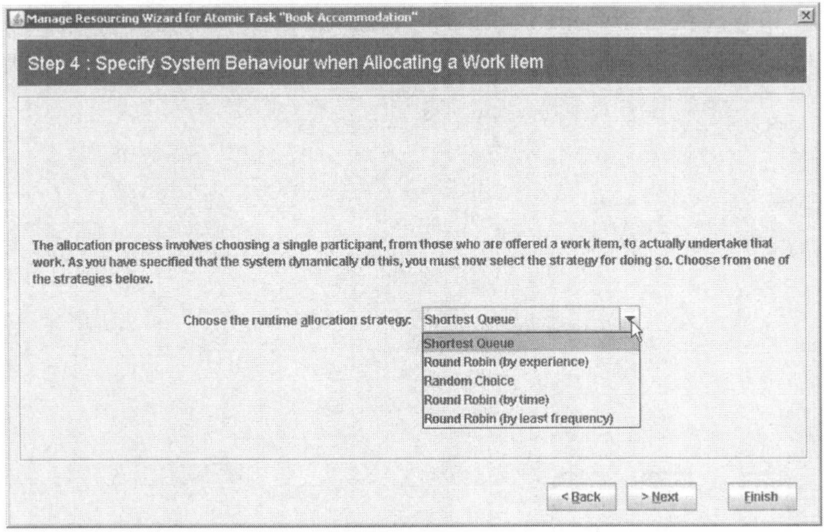

Fig. 2.39 Allocation strategies for work distribution

- *Shortest queue allocation*: This filter ensures that any work items associated with a task are only routed to a single user where the user is selected from the group of potential users on the basis of which of them has the shortest work queue. In YAWL this means that the selected user has the least number of incomplete work items in their worklist. In the event that two user have the same number of incomplete work items, one of them is selected at random.

Figure 2.39 illustrates the configuration of allocation strategies in YAWL. It is only possible to configure these strategies for tasks that are allocated by the system, that is, in the screen shown in Fig. 2.36 where the interaction strategy is specified for a task, and the second option must be set such that the allocation is done by the system.

Advanced User Operating Modes

YAWL supports two advanced operating modes for user interaction with the system. These modes are intended to expedite the throughput of work by imposing a defined protocol on the way in which the user interacts with the system and work items are allocated to them. These modes are described below.

- *Chained execution*: Chained execution is essentially an operating mode that a given user can choose to enable. Once they do this, upon the completion of a given work item in a process, should any of the immediately following tasks in the process instance have potential to be routed to the same user (or to a group of users that include the user), then these routing directives are overridden and

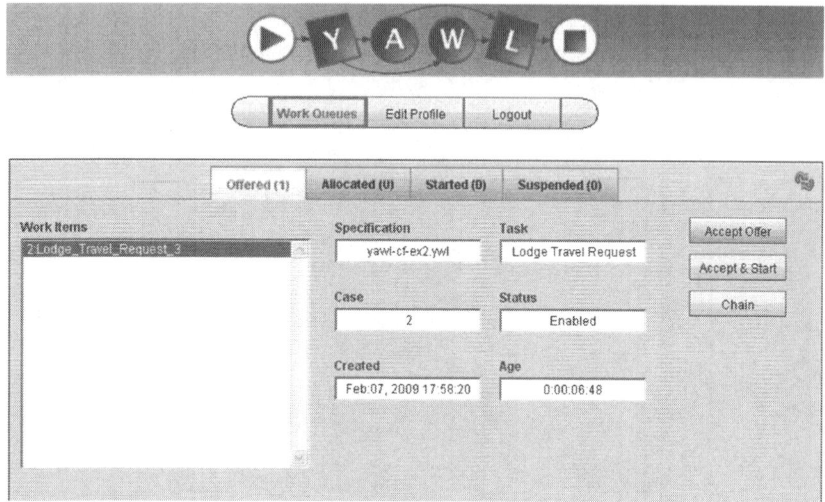

Fig. 2.40 Chained execution initiation in the Resource Service's worklist

the associated work items are placed in the user's worklist with a *started* status. There are two parts to the configuration and use of chained execution: (1) the user must have the chained execution privilege (see Fig. 2.43 for details of how this occurs) and (2) the user must choose to enter the chained execution mode via their worklist handler (the option for this is shown in Fig. 2.40). Once initiated, chained execution continues until the user disables it or the case completes.

- *Piled execution*: Piled execution is another operating mode; however, it operates across multiple process instances. It is enabled for a specified user-task combination, and once initiated, it overrides any routing directive for the nominated task and ensures that any work items associated with the task in any process instance are routed to the nominated user. There are two steps in enabling piled execution for a task: (1) it needs to be specified in the YAWL model that the task can be subject to execution in piled mode, as is the case for the *Book Accommodation* task in Fig. 2.38 and (2) piled execution mode needs to be initiated by a user for a given task using the option for this available in their worklist handler as shown in Fig. 2.41. Once initiated, piled execution continues until the user disables it.

2.6.5 Privileges

YAWL provides support for a number of privileges that can be enabled on a per-task or per-user basis that affect the way in which work items are distributed and the various interactions that the user can initiate to otherwise change the normal manner in which the work item is handled. Table 2.8 summarizes the privileges that can be set for individual users.

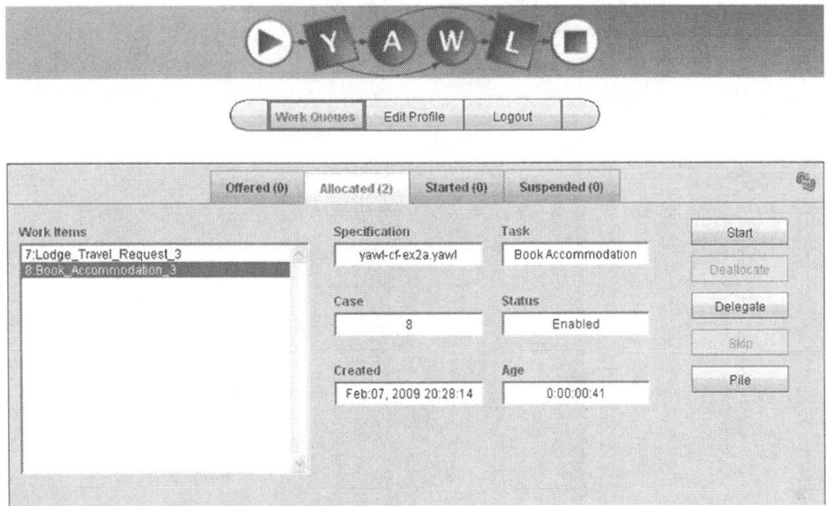

Fig. 2.41 Piled execution initiation in the Resource Service's worklist

Table 2.8 User privileges supported in YAWL

Privilege	Explanation
Choose	The ability to select the next work item to start execution on
Concurrent execution	The ability to execute more than one work item simultaneously
Reorder	The ability to reorder items in the worklist
View team	The ability to view work items offered to other users in the same team(s) as a given user
View group	The ability to view work items allocated to other users in the same group(s) as a given user
Chained execution	The ability to enter the *chained execution* operating mode
Manage cases	The ability to administer cases and redistribute work items to users

The actual screen for doing this forms part of the runtime environment and is shown in Fig. 2.43.

Additionally, there are also privileges that can be set for individual tasks. These are summarized in Table 2.9.

The facility for specifying these options forms part of the YAWL editor. The actual screen for specifying task privileges is shown in Fig. 2.42.

2.6.6 Resource Patterns Coverage

The resource perspective in YAWL has been heavily influenced by the resource patterns, and it provides comprehensive support for them as indicated in Table 2.10. In total, it directly implements 37 of the 43 resource patterns. As YAWL assumes

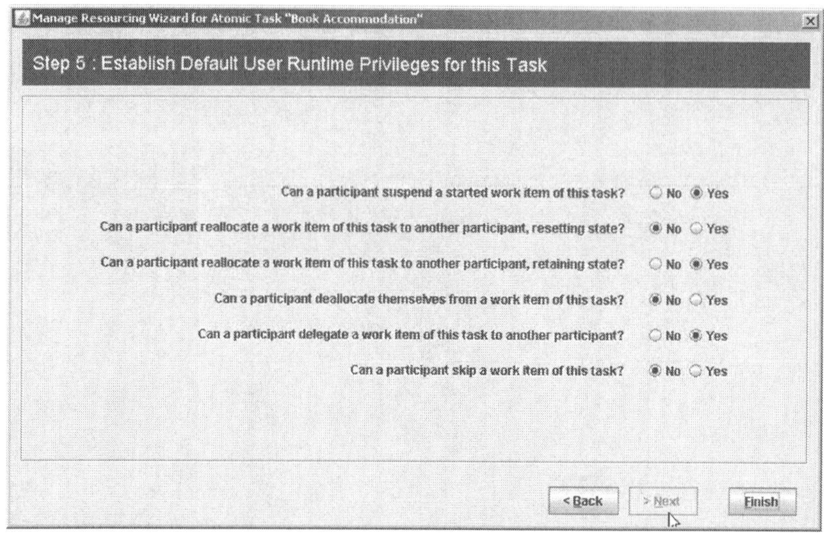

Fig. 2.42 Privilege specification for a YAWL task

Table 2.9 Task privileges supported in YAWL

Privilege	Explanation
Suspend	The ability for a user to suspend execution of work items corresponding to this task
Reallocate without state retention	The ability for the user to reallocate work items corresponding to this task (which have been commenced) to other users without any implied retention of state
Reallocate with state retention	The ability for the user to reallocate work items corresponding to this task (which have been commenced) to another user and retain the state of the work item
Deallocate	The ability for the user to deallocate work items corresponding to this task (which have not yet been commenced) and cause them to be *reallocated*
Delegate	The ability for the user to delegate work items corresponding to this task (which have not yet been commenced) to another user
Skip	The ability for the user to skip work items corresponding to this task

a deterministic process execution environment, the main omission from the broad range of functionality implied by the patterns is in the area of case handling, which requires that an offering be able to (temporarily) deviate from the underpinning process model and allow tasks to be executed out of sequence and for them to be initiated at times other than that would normally be implied by the flow of control within a case. The specific patterns supported by YAWL are as follows.

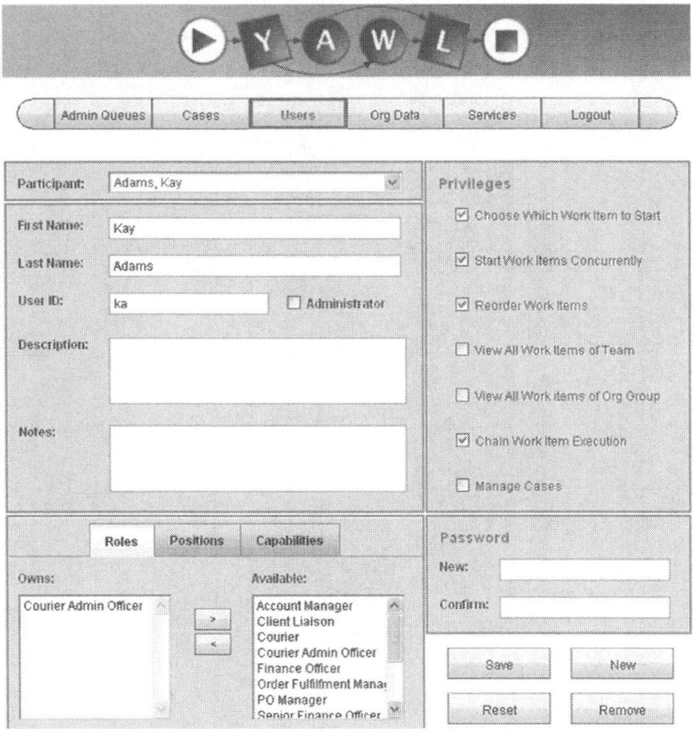

Fig. 2.43 Privilege specification for a YAWL user

Table 2.10 Resource pattern support in YAWL

Pattern Group	Supported Patterns
Creation patterns	All except for *Case Handling*
Push patterns	All except for *Early Distribution* and *Late Distribution*
Pull patterns	All
Detour patterns	All except for *Redo* and *Pre-Do*
Auto-start patterns	All
Visibility patterns	All
Additional resource patterns	*Simultaneous Execution*

2.6.7 Object Role Model

This section provides a conceptual model for work distribution in a YAWL model using the object-role notation as shown in Fig. 2.44. These aspects of the resource perspective center around the notion of a *user task*, which corresponds to an individual task in a YAWL model that will be distributed to a user for execution. This contrasts with a *system task*, which is performed automatically by a software

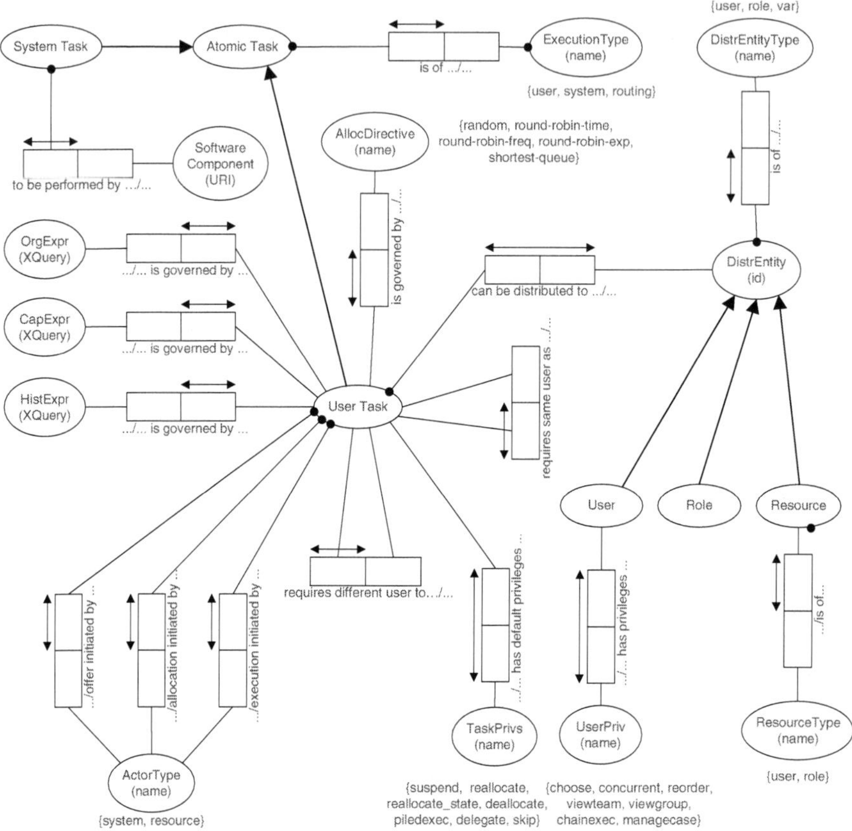

Fig. 2.44 Object role model: YAWL work distribution perspective

component. Capturing the interaction strategy for a task requires recording the type of actor (system or resource) responsible for initiating the offer, allocation, and execution of a work item corresponding to the task. A work item corresponding to a user task is distributed to a distribution entity, which can be a user, a role, or a resource variable (stored in a net variable) which identifies a user or a role. Distribution filters can be specified for a task based on organizational, capability, or historical expressions that restrict the potential range of user to whom a task may be distributed. There is the option to specify an allocation directive for a task. There is also the option to specify that a user task be distributed to the same user who executed another task in the same case or to explicitly specify that it not be distributed to the same user who executed another task. Finally, a range of user and task privileges can be specified for a YAWL model that delineate the range of actions permissible by users with respect to task interactions at runtime.

2.7 Syntax

This section presents a complete abstract syntax for YAWL. A YAWL *specification* is a set of *YAWL nets* that form a rooted graph structure. Each *YAWL net* is composed of a series of tasks. All of these notions are now formalized, starting with the YAWL *specification*.

Definition 12. (YAWL Specification) A *YAWL specification* is a tuple = *(NetID, ProcessID, TaskID, MITaskID, VarID, ParamID, TNmap, NYmap, VarName, DataType, VName, DType, VarType, VNmap, VTmap, VMmap)* such that:

(* global objects *)

- *NetID* is the set of net identifiers (i.e., the top-level process together with all subprocesses);
- *ProcessID* ∈ *NetID* is the process identifier (i.e., the top-level net);
- *TaskID* is the set of task identifiers in nets;
- *MITaskID* ⊆ *TaskID* is the set of identifiers of multiple instance tasks;
- *VarID* is the set of variable identifiers used in nets;
- *ParamID* is the set of parameter identifiers used in nets;

(* decomposition *)

- $TNmap$: $TaskID \nrightarrow NetID$ defines the mapping between composite tasks and their corresponding subprocess decompositions, which are specified in the form of a YAWL net, such that for all t, $TNmap(t)$ yields the *NetID* of the corresponding YAWL-net, if it exists;
- $NYmap$: $NetID \rightarrow YAWLnet$, that is, each net has a complete description of its contents such that for all $n \in NetID$, *NYmap(n)* is governed by Definition 13, where the notation T_n denotes the set of tasks that appear in a net n. Tasks are not shared between nets, hence $\forall_{m,n \in NetID}[T_m \cap T_n \neq \varnothing \Rightarrow m = n]$. *TaskID* is the set of tasks used in all nets and is defined as $TaskID = \bigcup_{n \in NetID} T_n$;
- In the directed graph defined by $G = (NetID, \{(x, y) \in NetID \times NetID \mid \exists_{t \in T_x}[t \in dom(TNmap) \wedge TNmap(t) = y]\})$, there is a path from *ProcessID* to any node $n \in NetID$;

(* variables *)

- *VarName* is the set of variable names used in all nets;
- *DataType* is the set of data types;
- $VName$: $VarID \rightarrow VarName$ identifies the name for a given variable;
- $DType$: $VarID \rightarrow DataType$ identifies the underlying data type for a variable;
- $VarType$: $VarID \rightarrow \{Net, Task, MI\}$ describes the various variable scopings that are supported. The notation $VarID^x = \{v \in VarID \mid VarType(v) = x\}$ identifies variables of a given type;
- $VNmap$: $VarID^{Net} \rightarrow NetID$ identifies the specific net to which each net variable corresponds, such that $dom(VNmap) = VarID^{Net}$;

- *VTmap*: $VarID^{Task} \rightarrow TaskID$ identifies the specific task to which a task variable corresponds, such that $dom(VTmap) = VarID^{Task}$;
- *VMmap*: $VarID^{MI} \rightarrow MITaskID$ identifies the specific task to which each multiple instance variable corresponds, such that $dom(VMmap) = VarID^{MI}$.

Having described the global characteristics of a *YAWL specification*, we can now proceed to the definition of a *YAWL net*. Note that *YAWL nets* is the set of all instances governed by Definition 13. Each *YAWL net* is identified by a unique *nid* in the form of a tuple (C, T, F), which takes its form from classical Petri nets where C corresponds to the set of conditions, T to the set of tasks, and F to the flow relation (i.e., the directed arcs between conditions and tasks). However, there are two distinctions: (1) i and o describe specific conditions that denote the start and end condition for a net and (2) the flow relation allows for direct connections between tasks in addition to links from conditions to tasks and tasks to conditions.

Expressions are denoted informally via *Expr*, which identifies the set of expressions relevant to a *YAWL net*. It may be divided into a number of disjoint subsets including *BoolExpr*, *IntExpr*, *NatExpr*, *StrExpr*, and *RecExpr*, these being the sets of expressions that yield Boolean, integer, natural number, string, and record-based results when evaluated. There is also recognition for work distribution purposes of capability-based, historical, and organizational distribution functions that are denoted by the *CapExpr*, *HistExpr*, and *OrgExpr* subsets of *Expr*, respectively. Note that in actuality, all forms of *Expr* are realized using XQuery and XPath functionality.

Definition 13. (YAWL net) A YAWL net is a tuple *(nid, C, i, o, T,* T_A, T_C, *M, F, Split, Join, Default,* $<_{XOR}$, *Rem, Nofi, ArcCond)* such that:

(* basic control-flow elements *)

- $nid \in NetID$ is the identity of the *YAWL net*;
- C is a set of conditions;
- $i \in C$ is the input condition;
- $o \in C$ is the output condition;
- T is the set of tasks;
- $T_A \subseteq T$ is the set of atomic tasks;
- $T_C \subseteq T$ is the set of composite tasks;
- T_A and T_C form a partition over T;
- $M \subseteq T$ is the set of multiple instance tasks;
- $F \subseteq (C \setminus \{o\} \times T) \cup (T \times C \setminus \{i\}) \cup (T \times T)$ is the flow relation, such that every node in the graph $(C \cup T, F)$ is on a directed path from i to o;
- $Split : T \nrightarrow \{AND, XOR, OR\}$ specifies the split behavior of each task;
- $Join : T \nrightarrow \{AND, XOR, OR\}$ specifies the join behavior of each task;
- $Default \subseteq F, Default : dom(Split \rhd \{OR\}) \rightarrow T \cup C$ denotes the default arc for each OR-split. If none of the outgoing arc expressions evaluate to true, the default arc indicated by $Default(t)$ is selected, thus ensuring that at least one outgoing arc is enabled;

- $<_{XOR} \subseteq \{t \in T \mid Split(t) = XOR\} \times \mathbb{P}((T \cup C) \times (T \cup C))$ describes the evaluation sequence of outgoing arcs from an XOR-split such that for any $(t, V) \in <_{XOR}$ we write $<^t_{XOR} = V$ and V is a strict total order over $t\bullet = \{x \in T \cup C \mid (t, x) \in F\}$. Link conditions associated with each arc are evaluated in this sequence until the first evaluates to true. If none evaluate to true, the minimum element (which corresponds to the default path and is denoted as $\perp_t$) is selected;
- $Rem : T \twoheadrightarrow \mathbb{P}^+(T \cup C \backslash \{i, o\})$ specifies the additional tokens to be removed by emptying a part of the net and tasks that should be canceled as a consequence of an instance of this task completing execution;
- $Nofi$: $M \rightarrow \mathbb{N} \times \mathbb{N}^{inf} \times \mathbb{N}^{inf} \times \{dynamic, static\}$ specifies the multiplicity of each task – in particular, the lower and upper bound of instances to be created at task initiation, the threshold for continuation indicating how many instances must complete for the thread of control to be passed to subsequent tasks and whether additional instances can be created "on the fly" once the task has commenced;

(* conditions on arcs *)

- $ArcCond$: $F \cap (dom(Split \rhd \{XOR, OR\}) \times (T \cup C)) \rightarrow BoolExpr$ identifies the specific condition associated with each branch of an OR or XOR split.

Each *YAWL net* has a *Data passing model* associated with it that describes how data is passed between tasks in a YAWL specification. The abstract syntax for this model is given below.

Definition 14. (Data passing model) Within the context of a *YAWL net nid*, there is a data passing model *(ParamVar, InPar, OutPar, Accessor, Splitter, Instance, Aggregate)* with the following components:

(* parameter variable definition *)

- $ParamVar$: $ParamID \rightarrow VarID \times VarID$ is a function identifying the source and target variables for a given parameter;

(* data passing to/from atomic tasks *)

- $InPar$: $ParamID \times T \rightarrow Expr$ is a function identifying the input parameter mappings to a task at initiation;
- $OutPar$: $ParamID \times T \rightarrow Expr$ is a function identifying the output parameter mappings from a task at completion;

(* data passing to/from multiple instance tasks *)

- $Accessor$: $T_M \rightarrow RecExpr$ is a function identifying the accessor query for a multiple instance task;
- $Splitter$: $T_M \rightarrow Expr$ is a function identifying the splitter query for a multiple instance task;
- $Instance$: $T_M \rightarrow Expr$ is a function identifying the instance query for a multiple instance task;

- *Aggregate*: $T_M \rightarrow RecExpr$ is a function identifying the aggregate query for a multiple instance task.

We now proceed to the abstract syntax for the resource perspective in YAWL. This comprises two distinct components: the *Organizational model*, which provides a description of the overall structure of the organization in terms of users, roles, organizational groups, positions, and reporting lines, and the *Work distribution model*, which defines the manner in which work items are distributed to users at runtime for execution as well as identifying the interactions that individual users are able to invoke to influence the way in which this distribution occurs. These two models are specified in more detail below.

Definition 15. (Organizational model) Within the context of a YAWL specification *ProcessID*, there is an organizational model described by the tuple *(UserID, RoleID, CapabilityID, OrgGroupID, PositionID, CapVal, RoleUser, OrgGroupType, GroupType, PositionGroup, OrgStruct, Superior, UserQual, UserPosition)* as follows:

(* basic definitions *)

- *UserID* is the set of all individuals to whom work items can be distributed;
- *RoleID* is the set of designated groupings of those users;
- *CapabilityID* is the set of qualities that a user may possess that are useful when making work distribution decisions;
- *OrgGroupID* is the set of groups within the organization;
- *PositionID* is the set of all positions within the organization;
- *CapVal* is the set of values that a capability can have;

(* organizational definition *)

- *RoleUser*: $RoleID \rightarrow \mathbb{P}(RoleUser)$ indicates the set of users in a given role;
- $OrgGroupType = \{team, group, department, branch, division, organization\}$ identifies the type of a given organizational group;
- *GroupType*: $OrgGroupID \rightarrow OrgGroupType$;
- *PositionGroup*: $PositionID \rightarrow OrgGroupID$ indicates which group a position belongs to;
- *OrgStruct*: $OrgGroupID \nrightarrow OrgGroupID$ forms an acyclic intransitive graph with a unique root, which identifies a composition hierarchy for groups;
- *Superior*: $PositionID \nrightarrow PositionID$ forms an acyclic intransitive graph, which identifies the reporting lines between positions;

(* user definition *)

- *UserQual*: $UserID \times CapabilityID \rightarrow CapVal \cup \{Undefined\}$ identifies the capabilities that a user possesses;
- *UserPosition*: $UserID \rightarrow \mathbb{P}(PositionID)$ maps a user to the positions that they hold.

The *YAWL Organizational model* takes the form of a tree, based on the reporting relationships between groups, where the most senior group within the organization is the root node of the tree. This model is deliberately chosen to be simple and generic so that it applies to a relatively broad range of situations in which YAWL may be used. Finally, the *Work distribution model* is presented that captures the various ways in which work items are distributed to users and any constraints that need to be taken into account when doing so.

Definition 16. (Work distribution model) Within the context of a *YAWL net nid*, it is possible to describe the manner in which work items are distributed to users for execution. A work distribution model is a tuple *(ResourceVarID, Auto, T_M, Initiator, DistUser, DistRole, DistVar, OrgDist, CapDist, SameUser, FourEyes, UserSel, UserPriv, UserTaskPriv)* as follows:

(* work allocation *)

- $ResourceVarID \subseteq VarID$ is the set of variables that identify resources or roles;
- $Auto \subseteq T_A$ is the set of tasks that execute automatically without user intervention, where T_A is the set of atomic tasks;
- $T_M \subseteq T_A \backslash Auto$ is the set of atomic tasks that must be allocated to users for execution;
- $Initiator\colon T_M \rightarrow \{system, resource\} \times \{system, resource\} \times \{system, resource\}$ indicates who initiates the offer, allocate, and commence actions;
- $DistUser\colon T_M \nrightarrow \mathbb{P}(User)$ identifies the users to whom a task should potentially be distributed;
- $DistRole\colon T_M \nrightarrow \mathbb{P}(Role)$ identifies the roles to whom a task should potentially be distributed;
- $DistVar\colon T_M \nrightarrow \mathbb{P}(ResourceVarID)$ identifies a set of variables holding either user or roles to whom a task should potentially be distributed;
- dom(*DistUser*), dom(*DistRole*) and dom(*DistVar*) form a partition over T_M;
- $OrgDist\colon T_M \nrightarrow OrgExpr$ identifies the organizational criterion that users who execute the task must satisfy;
- $CapDist\colon T_M \nrightarrow CapExpr$ identifies the capability that users who execute the task must possess;
- $SameUser\colon T_M \nrightarrow T_M$ is an irreflexive function that identifies that a task should be executed by one of the same users who undertook another specified task in the same case;
- $FourEyes\colon T_M \nrightarrow T_M$ is an irreflexive function that identifies a task that should be executed by a different user to the one(s) who executed another specified task in the same case;
- $UserSel\colon T_M \nrightarrow$ *{random, round-robin-time, round-robin-freq, round-robin -exp, shortest-queue}* indicates how a specific user who will execute a task should be selected from a group of possible users;

(* user privilege definition *)

- *UserPriv*: *UserID* $\rightarrow$ $\mathbb{P}$(*UserAuthKind*) indicates the privileges that an individual user possesses, where *UserAuthKind* = {*choose, concurrent, reorder, viewteam, viewgroup, chainedexec, managecase*};
- *UserTaskPriv*: *UserID* $\times$ *TaskID* $\rightarrow$ $\mathbb{P}$(*UserTaskAuthKind*) indicates the privileges that an individual user possesses in relation to a specific task, where *UserTaskAuthKind* = {*suspend, reallocate, reallocate_state, deallocate, piledexec, delegate, skip*}.

2.8 Working Example

Throughout this book for illustrative purposes we will use a common working example that depicts the Order Fulfillment process followed by the fictitious Genko Oil company. In this section, we will give a brief introduction to the content and operation of the process from a control-flow, data, and resource perspective. The complete process model is presented in Appendix A.

2.8.1 Control-Flow Perspective

The top level of the order fulfillment process is shown in Fig. 2.45. It comprises five main composite tasks, each of which has an associated subprocess that describes the details of its implementation. We will only discuss the main intentions of each of these tasks here. Full details of their implementation can be found in Appendix A.

The first step in the order fulfillment process is the *Ordering* task, which involves the receipt and processing of a purchase order. As part of this activity, it may require various approvals and modifications before it can finally be confirmed and readied for supply. There is a 3 day time limit for the confirmation of a newly received purchase order. If this requirement is not met or the supply of the purchase order is not fully approved, it may be discarded.

The next step in the process is the *Carrier Appointment* task where the transportation arrangements for shipping the order are determined and any necessary documentation is prepared. As the size of an order may vary from an individual package right up to one requiring a truck (and possibly several trailers) for its transportation, the actual process for organizing the shipment may vary markedly. At its conclusion, however, all required arrangements with shippers have been made and required documentation is available for despatch with the goods ordered. If the task does not complete within 5 days, it is canceled and the process terminates.

The next two steps in the process occur in parallel. The *Payment* task involves two main activities: (1) securing payment for the goods supplied from the customer and (2) organization of payment for the freight company that transported the order. The

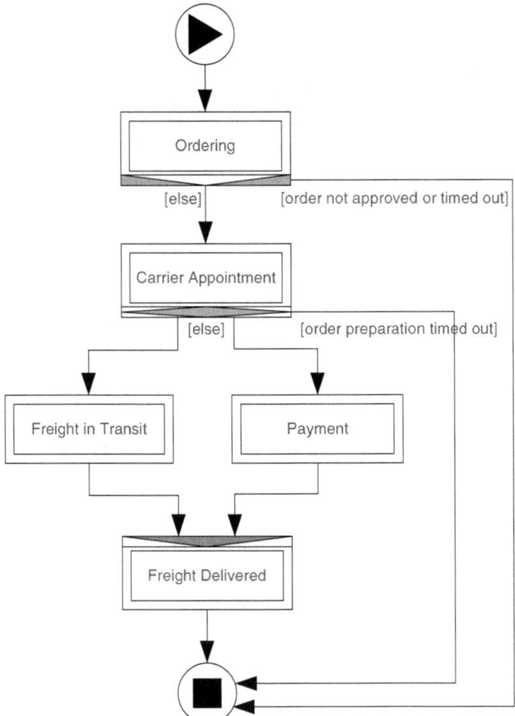

Fig. 2.45 Working example: control-flow

Freight in Transit task involves tracking an order while it is shipped to a customer and handling any enquiries they may have about its status.

Once these two tasks have completed, the last task – *Freight Delivered* – can execute. This involves handling any requests from customers to return goods or any claims for loss or damage during its transportation that may be received. These must be received within a specified timeframe, otherwise the order is deemed to be complete and the process completes.

2.8.2 Data Perspective

The data perspective of the order fulfillment process is relatively straightforward and centers around the notion of a purchase order. Figure 2.46 illustrates the data elements passed between the various nets in the order fulfillment process. In the main, the process-relevant data is retained in the Overall (i.e., the top-level) process. It also illustrates the net variables in each of the nets.

The user-defined data types underpinning the net variables depicted in Fig. 2.46 are summarized in Listing 2.1.

```
<xs:schema xmlns:xs="http://www.w3.org/2001/XMLSchema">
  <xs:complexType name="PurchaseOrderType">
    <xs:sequence>
      <xs:element name="Company" type="CompanyType" />
      <xs:element name="Order" type="OrderType" />
      <xs:element name="FreightCost" type="xs:double" />
      <xs:element name="DeliveryLocation" type="xs:string" />
      <xs:element name="InvoiceRequired" type="xs:boolean" />
      <xs:element name="PrePaid" type="xs:boolean" />
    </xs:sequence>
  </xs:complexType>
  <xs:complexType name="CompanyType">
    <xs:sequence>
      <xs:element name="Name" type="xs:string" />
      <xs:element name="Address" type="xs:string" />
      <xs:element name="City" type="xs:string" />
      <xs:element name="State" type="xs:string" />
      <xs:element name="PostCode" type="xs:string" />
      <xs:element name="Phone" type="xs:string" />
      <xs:element name="Fax" type="xs:string" />
      <xs:element name="BusinessNumber" type="xs:string" />
    </xs:sequence>
  </xs:complexType>
  <xs:complexType name="OrderType">
    <xs:sequence>
      <xs:element name="OrderNumber" type="xs:string" />
      <xs:element name="OrderDate" type="xs:date" />
      <xs:element name="Currency" type="xs:string" />
      <xs:element name="OrderTerms" type="xs:string" />
      <xs:element name="RevisionNumber" type="xs:integer" />
      <xs:element name="Remarks" type="xs:string" />
      <xs:element name="OrderLines" type="OrderLinesType" />
    </xs:sequence>
  </xs:complexType>
  <xs:complexType name="OrderLinesType">
    <xs:sequence>
      <xs:element maxOccurs="unbounded" name="Line" type="LineType" />
    </xs:sequence>
  </xs:complexType>
  <xs:complexType name="LineType">
    <xs:sequence>
      <xs:element name="LineNumber" type="xs:integer" />
      <xs:element name="UnitCode" type="xs:string" />
      <xs:element name="UnitDescription" type="xs:string" />
      <xs:element name="UnitQuantity" type="xs:integer" />
      <xs:element name="Action" type="xs:string" />
    </xs:sequence>
  </xs:complexType>
  <xs:complexType name="RouteGuideType">
    <xs:sequence>
      <xs:element name="OrderNumber" type="xs:string" />
      <xs:element name="DeliveryLocation" type="xs:string" />
      <xs:element name="Trackpoints" type="TrackpointsType" />
    </xs:sequence>
  </xs:complexType>
  <xs:complexType name="TrackpointsType">
    <xs:sequence>
      <xs:element maxOccurs="unbounded" name="Trackpoint" type="xs:string" />
    </xs:sequence>
  </xs:complexType>
  <xs:complexType name="TrailerUsageType">
    <xs:sequence>
      <xs:element name="OrderNumber" type="xs:string" />
      <xs:element name="OrderLines" type="OrderLinesType" />
      <xs:element name="Packages" type="PackagesType" />
    </xs:sequence>
  </xs:complexType>
  ...
</xs:schema>
```

Listing 2.1 Working example: user-defined data type definition using XML schema

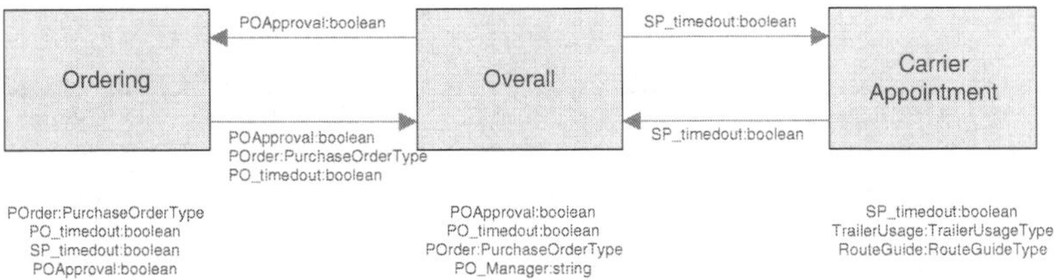

Fig. 2.46 Working example: data passing between nets and net variables within nets

2.8.3 Resource Perspective

The order fulfillment process is based on a simple organizational model that delineates a number of users and the roles that they play in an organizational context. It also identified capabilities of individual users and the privileges that they possess. Table 2.11 summarizes the main elements of this model.

In terms of other aspects of the resource perspective:

- Most tasks in the process utilize an SRR or SRS interaction strategy such that the system offers tasks to potential users, but the selection and commencement of the tasks is under the auspices of the individual user. On occasions where a distribution strategy is used that requires direct allocation of a task to a specific user, a specific allocation strategy is used to select a single user for the task, such as for the following tasks
  - *Carrier Appointment* (SSR): random allocation
  - *Create Carrier Manifest* (SSR): round robin
  - *Prepare Quote Guide* (SSR): shortest queue
  - *Estimate Trailer Usage* (SSR): random allocation
  - *Issue Debit Adjustment* (SSS): shortest queue
  - *Issue Credit Adjustment* (SSS): shortest queue
- A minimalistic set of privileges are specified for tasks, with the main privileges specified in this area being as follows:
  - *Prepare Transport Quote*: suspend, reallocate (with and without state)
  - *Create Purchase Order*: suspend, reallocate (with and without state), deallocate
  - *Approve Purchase Order*: delegate
  - *Confirm Purchase Order*: deallocate
  - *Modify Purchase Order*: skip
  - *Authorize Return Merchandize*: delegate
  - *Authorize Loss or Damage Claim*: delegate

Table 2.11 Working example: organizational model

User	Role(s)	Position	Capabilities	Privileges
Kay Adams	Carrrier Admin Officer	Head of CD		choose, start, reorder
Billy Van Arsdale	Junior Supply Officer	SD clerk		choose, start, reorder
Momo Barone	Courier	CD clerk		choose, start, reorder
Emilio Barzini	Finance Officer	FD clerk		choose, start, reorder
Peter Clemenza	Courier	CD clerk		choose, start, reorder
Stefano Clemenza	Shipment Planner, Courier	CD clerk		choose, start, reorder
Don Vito Corleone	Order Fulfillment Manager	CEO		manage cases
Fredo Corleone	PO Manager	Head of OD		choose, start, reorder
Mama Corleone	Shipment Planner	CD clerk		choose, start, reorder
Michael Corleone	PO Manager	OD clerk		choose, start, reorder
Sonny Corleone	PO Manager	OD clerk		choose, start, reorder
Carmine Cuneo	Shipment Planner	CD clerk		choose, start, reorder
Don Carmine Cuneo	Account Manager	FD clerk		choose, start, reorder, chained execution
Johnny Fontaine	Shipment Planner	CD clerk		choose, start, reorder
Tom Hagen	Senior Supply Officer, Supply Admin Officer	Head of SD, Assistant head of OD	Master in SCLM	choose, start, reorder, view group
Jaggy Jovino	Senior Supply Officer	SD clerk	Master in SCLM	choose, start, reorder
Jo Luccadello	Senior Supply Officer	Warehouse clerk, SD clerk	Bachelor in SCLM	choose, start, reorder
Vincent 'Vinnie' Mancini-Corleone	PO Manager	OD clerk		choose, start, reorder
Carmine Marino	PO Manager, Client Liaison	OD clerk		choose, start, reorder
Capt. McCluskey	Finance Officer	FD clerk		choose, start, reorder
Arturo d'Ofstede	Finance Officer	FD clerk		choose, start, reorder
Carlo Rizzi	Junior Supply Officer	SD clerk	Bachelor in SCLM	choose, start, reorder
Connie Corleone Rizzi	Warehouse Administration Officer	Head of warehouse		choose, start, reorder
Marcello La Rosa	Courier	CD clerk		choose, start, reorder
Virgil 'The Turk' Sollozzo	Senior Finance Officer	Head of FD		choose, start, reorder, view group
Sal Tessio	Client Liaison	OD clerk		choose, start, reorder
Jack Woltz	Warehouse Officer	Warehouse clerk		choose, start, reorder

- Retain familiar and four eyes constraints are specified between the following tasks when distributing work:
  - *Update Shipping Purchase Order* and *Issue Shipment Purchase Order* (retain familiar)
  - *Arrange Delivery Appointment* (SP) and *Arrange Pickup Appointment* (SP) (retain familiar)
  - *Arrange Delivery Appointment* (FTL) and *Arrange Pickup Appointment* (FTL) (four eyes)

2.9 Conclusion

In this chapter, we have introduced the YAWL language in both conceptual and operational terms, describing its major features in the control-flow, data, and resource perspectives. The derivation of these language features from the Workflow Patterns is described in detail as are the formal modeling foundations on which YAWL is based. In doing so, we trust that we have laid the groundwork for the reader to tackle subsequent chapters of this book that assume some knowledge of the material discussed in this section.

Exercises

Exercise 1. Identify the main YAWL split and join operators. Model each of these in terms of Petri nets.

Exercise 2. Model the following process as a Petri net and then as a YAWL model. Highlight the differences between the two models.

- A travel agency makes travel arrangements for people
- A travel arrangement may include one or more of the following activities: book flight, book car, and book hotel
- These activities occur in parallel
- As they involve dealing with third parties, each of the booking activities may be either successful or unsuccessful
- When all booking activities requested by a customer have completed successfully, the payment activity is triggered
- If any of the booking activities are unsuccessful, any further progress on the booking activities (if any) is canceled
- The process completes after either a payment or cancelation activity

Exercise 3. Model the following process as a Petri net and then as a YAWL model. Highlight the differences between the two models. Identify the patterns that exist in each model.

- A quality management process exists within a company to receive improvement recommendations from staff
- An instance of the process is triggered when an improvement suggestion is received
- The first step in the process is to *register* the suggestion
- Two subsequent activities are then triggered in parallel: (1) a *request for further information* is sent to the staff member and (2) the *initial suggestion evaluation* occurs
- After the *request for further information* has been sent, there is a *14 day waiting period* for a response. If further details are received within this time frame, then the *process response* task executes. If not, then the *process response* task is skipped. In either case, the state after the *process response* task has executed or been skipped is known as the *response handling complete* state
- After the *initial suggestion evaluation* has completed, depending on the outcome, one of two possible states is entered: either (1) the decision is made to *accept the suggestion* for action or (2) the decision is made to reject the suggestion and the *processing complete* state is entered
- If the decision is made to accept the suggestion, and the *response handling complete* state has been reached, then a *detailed examination* of the suggestion occurs to determine what resolution action to take
- After the *detailed examination* has completed, the proposed response to the improvement suggestion is forwarded to a manager for *confirmation*
- If the *confirmation* outcome is positive, then the *resolution* task executes
- If the *confirmation* outcome is negative, the process reverts to the *accept the suggestion* state
- After the *resolution* task has completed, the *processing complete* state is entered
- The *archive* task can execute once the process is in the *response handling complete* and *processing complete* state. The process then terminates.

Exercise 4. Model the following process as a Petri net and then as a YAWL model. Highlight the differences between the two models.

- An accident investigation process exists within a farm machinery manufacturing organization to assess whether any reported farm accidents require modifications to machinery currently being made
- An instance of the process is reported each time an accident is reported
- The first task to execute is the *travel to accident location* task
- After this task, two subsequent tasks are initiated in parallel: (1) the *review situation* task and (2) the *determine if investigation required* task
- Once the *review situation* task has completed, multiple concurrent instances of the *interview witness* task are executed concurrently depending on the number of witnesses who saw the accident
- The *determine if investigation required* task allows a summary decision to be arrived at as to whether to complete a full investigation of the accident or not. If

it is decided not to investigate the accident further, any instances of the *review situation* and *determine if investigation required* tasks currently running are canceled. Regardless of the decision, the process enters the *assessment complete* state

- Once the process has reached the *assessment complete* state and any remaining instances of the *interview witness* task have completed, the *report findings* task can execute and the process can complete execution

Exercise 5. Identify the possible execution traces for each of the following YAWL processes.

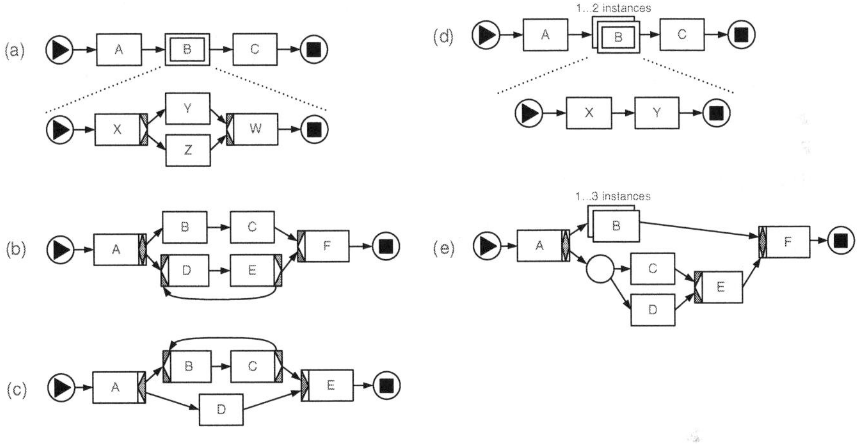

Exercise 6. Please provide answers for the following questions.

(a) Identify four distinct ways of passing data between a task instance and the environment. What are the distinctions between them?
(b) Data passing by reference and value are two common schemes for data handling in workflows. What are the implications of each strategy?
(c) Identify possible uses for data transformation patterns.

Exercise 7. In a workflow system such as YAWL, work items can be in an offered, allocated, or started state. As such, there are a number of distribution options for work items depending on the timing of the various distribution options, for example, offer, allocate, start results in a work item first being offered to several users and appearing in their worklist handler in an offered state as shown in the first line (a) of the diagram below. When one user selects it for allocation, its state changes to allocated. When (at a later time) the user chooses to start it, its state changes to started. Which interaction sequences do the other strategies in lines (b), (c), and (d) correspond to?

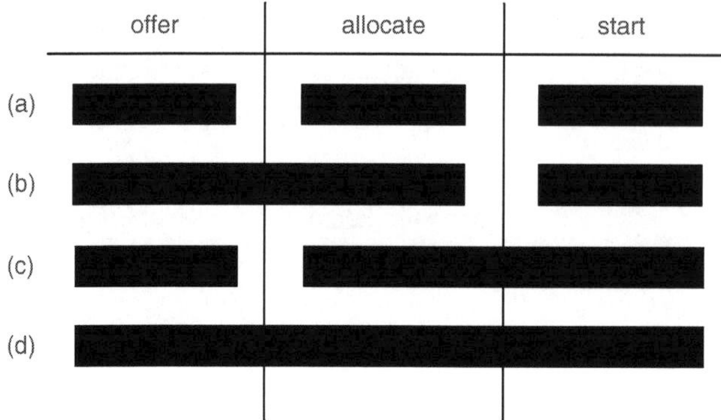

Exercise 8. Which patterns do the following approaches to work distribution correspond to?

(a) Allocate the *batch rivet* task to a manager
(b) Allocate the *batch rivet* task to the resource with the least work
(c) Distribute the *deliver invoice* task to the resource identified in the *responsible operator* field for the contract record corresponding to this customer
(d) Offer the *dangerous mission* task to all *junior soldiers*
(e) Allocate the *dangerous mission* task to the soldier who least recently undertook a mission
(f) Allocate the *dangerous mission* task to a soldier who serves in the *marine* or *infantry division*
(g) Allocate the *dangerous mission* task to a soldier with one or more bravery citations
(h) Allocate the *batch rivet* task to the resource who undertook the *prepare solder* task
(i) Allocate the *count cash* task to a resource who has undertaken the *prepare reconciliation* task for more than 2 weeks
(j) Distribute the *coordinate journey* task to the least busy colonel in the *marine corps* who has undertaken the training course

Chapter Notes

Readers who are interested in pursuing some of the topics introduced in this chapter in more depth may find the following publications of interest.

Evolution of BPM Technology

Figure 2.1 is derived from teaching material developed by van der Aalst and is loosely based on Fig. 1.11 in [11].

Petri Nets

The original definition of Petri nets is contained in the PhD thesis [197] of Carl Adam Petri. Comprehensive introductions to the field of Petri nets can be found in the work of Peterson [196], Reisig [209], and Desel and Esparza [77]. A comprehensive introduction to Colored Petri nets by Jensen can be found in [128]. The treatment of Petri nets is taken from [2].

Workflow Nets

The standard reference on workflow nets and their application in the context of workflow technology is from van der Aalst [2]. The three advantages of using Petri nets for process modeling identified in Sect. 2.3 are taken from this work.

Reset nets

Comprehensive introductions to reset nets and their use can be found in work by Dufourd et al. [80] and Finkel et al. [95,96]. The definitions presented in this chapter are based on these works.

Workflow Patterns

The seminal Workflow Patterns paper is published in [16]. This work has recently been revised in [224], leading to an augmented set of workflow control-flow patterns. Initial work in the area focussed on the control-flow perspective. Subsequent publications extended it to the data [225], resource [222], and exception [221] perspectives. A complete description of the Workflow Patterns is contained in Russell's PhD thesis [219].

YAWL

The initial proposal for YAWL, a workflow language based on the workflow patterns, was presented at the CPN Workshop in 2002 [14]. This was shortly followed by a complete description of the language and its semantics in [15]. The language proposal soon spurred the development of an actual implementation of the YAWL language, which is described in detail in [6]. Initial versions of both the YAWL language and system focussed on control-flow aspects. In light of new insights into the full range of patterns that are relevant to a process, particularly in the data and resource perspectives, a complete revision of the YAWL language, termed *new*YAWL, has recently been proposed [219].

Object-Role Modeling

The standard reference on Object-Role Modeling (commonly referred to as ORM) is [115].

Chapter 3
Advanced Synchronization

Moe Wynn, Wil van der Aalst, and Arthur ter Hofstede

3.1 Introduction

The OR-join is one of the three synchronization constructs supported in YAWL and the only one that depends on non-local semantics, that is, not only the current state but also possible future states need to be considered. The other two synchronization constructs are the XOR-join and the AND-join. Both have local semantics, that is, it suffices to consider the current state without extrapolating to possible future states when making a firing decision. An XOR-join requires no synchronization, that is, as soon as there is a token in one of its input conditions, an XOR-join is enabled. An AND-join requires full synchronization, that is, it is enabled when there is at least one token each in all of its input conditions. The drawback of using an AND-join construct is that a workflow can deadlock when all paths leading to the input conditions of the AND-join are not active. On the other hand, an OR-join construct allows more flexibility as it supports only the synchronization of *active paths*. Hence, the use of an OR-join construct in process models is desirable and *necessary* in situations where it is not possible to know in advance which paths will be active in a particular workflow instance (e.g., paths coming out of a multi-choice construct).

To determine whether an OR-join should be enabled at a particular state of a workflow, we need to look ahead to see if there are other active paths for which we should wait in future states of the workflow. Hence, the OR-join semantics are non-local and the analysis required to decide whether an OR-join should be enabled at a particular workflow state is non-trivial. The decision requires an awareness of the current state as well as possible future states of the workflow. Defining the non-local semantics of an OR-join is difficult even when a workflow language does not support complex constructs (e.g., cancelation) and/or puts certain restrictions on the models (e.g., no loops or only allow structures where an OR-join is preceded by an OR-split). This analysis becomes even more complicated when there are multiple

M. Wynn (✉)
Queensland University of Technology, Brisbane, Australia
e-mail: m.wynn@qut.edu.au

A.H.M. ter Hofstede et al. (eds.), *Modern Business Process Automation*,
DOI 10.1007/978-3-642-03121-2_3,

OR-joins in the workflow or when other complex constructs such as cancelation and loops are present in the workflow.

In this chapter, we explain the OR-join semantics in YAWL as well as the motivation behind these semantics and how it has been operationalized in the YAWL framework. Please note that this chapter aims to provide an overview of the OR-join semantics and does not detail the formal foundations behind the semantics. The rest of the chapter is organized as follows. Section 3.2 describes the informal and the formal semantics of the OR-join. Section 3.3 highlights the different interpretations of the construct in different process modeling languages and motivates the reasoning behind the semantics chosen for YAWL. Section 3.4 discusses the implementation in supporting the semantics and demonstrates how the OR-join is supported in the YAWL framework using two optimization techniques. Section 3.5 provides concluding remarks.

3.2 The OR-Join Semantics

First, we approach the OR-join from an informal and intuitive point of view using a series of YAWL nets. This is followed by a brief discussion of the formal semantics of an OR-join.

3.2.1 Informal Semantics

In general, an OR-join task is enabled only if there is at least one token in one of its input conditions, and it is not possible for more tokens to arrive in other (currently empty) input conditions in future states (i.e., there is no need/possibility to synchronize other paths before proceeding). If it is possible for tokens to arrive in currently empty input conditions in future states, then the OR-join should wait before proceeding. This is the desired behavior of an OR-join and we will refer to this as the *informal semantics* of an OR-join.

Next, a series of YAWL nets are presented to explain the semantics of an OR-join in more detail. For simplicity, the YAWL nets use short labels (e.g., A, $c1$, etc.) to identify tasks and conditions. This allows us to focus on the control-flow requirements for a particular net. The reader is reminded of the notions of marking and reachability as defined for Petri nets in Chap. 2. As with Petri nets, the term *marking* is used to describe the state of a YAWL net and is represented as the number of tokens in certain conditions of a net (e.g., $M = c1 + c5$ represents a state of the workflow where there is one token each in conditions $c1$ and $c5$). A marking is *reachable* from another marking if there is a sequence of tasks that can fire from the first marking to arrive at the second marking.

Using this Petri net terminology, a more precise (but still informal) explanation of the OR-join semantics is given here. An OR-join is enabled at a marking if and only if at least one of its input conditions is marked and it is not possible to reach a

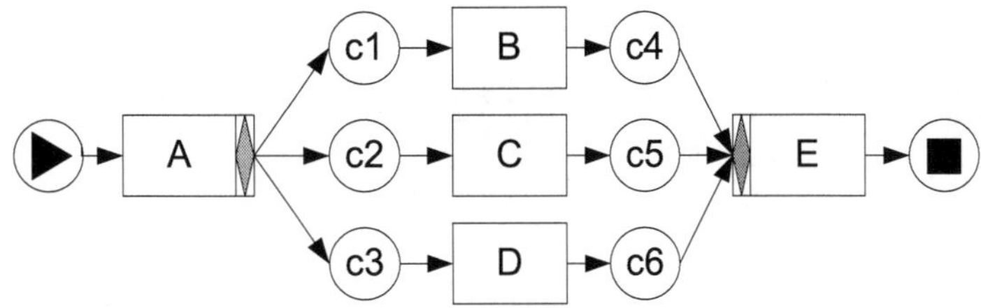

Fig. 3.1 A structured YAWL net with an OR-split task A and an OR-join task E

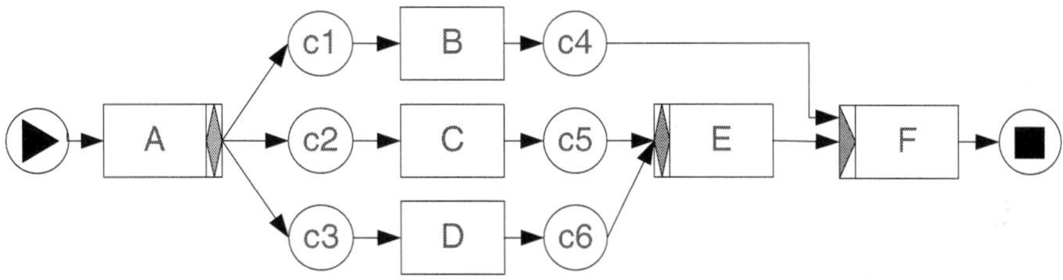

Fig. 3.2 A YAWL net modified from Fig. 3.1

marking that still marks all currently marked input conditions (possibly with fewer tokens) and at least one that is currently unmarked.

Figure 3.1 is a net where A is an OR-split task and E is an OR-join task. The initial marking for the net has exactly one token in the start condition. At the initial marking, task A is enabled and can be fired. After A has executed, tokens are put into conditions $c1$, $c2$, and $c3$ according to the OR-split behavior. Note that the OR-split allows one or more paths to be selected after executing the task (e.g., $c1 + c2$, $c3$, and $c1+c2+c3$). Consider a marking $M = c1+c5$, which results from the scenario where two outgoing paths leading to B and to C were selected after completing task A, and where task C has been executed. At M, there is a token in the input condition $c5$ of the OR-join task E. To determine whether task E should be enabled at M, we need to find out whether tokens could be put into $c4$ or $c6$ in the reachable markings of M. It is possible to reach a new marking $M' = c4 + c5$ from M by firing task B. Therefore, E should not be enabled at M. Now consider whether task E would be enabled at this new marking $M' = c4 + c5$. At M', $c4$ and $c5$ have one token each and there are no other tokens in the net. Hence, it is not possible for $c6$ to be marked in the reachable markings of M'. Task E is enabled at M'.

From the above example, one may conclude that OR-join evaluation depends only on the number of active paths out of an OR-split. If this is true, it is possible to know in advance the number of active paths to wait for before proceeding. However, it is only the case because we are dealing with a structured net whereby all paths from task A lead to task E. Figure 3.2 represents a slight modification to the YAWL net of Fig. 3.1 and shows that this conclusion is not justified. In Fig. 3.2, $c4$ is an input condition of task F and $c5$ and $c6$ are input conditions of task E. Consider a marking $M = c1 + c5$. In this case, there is no reachable marking from M that

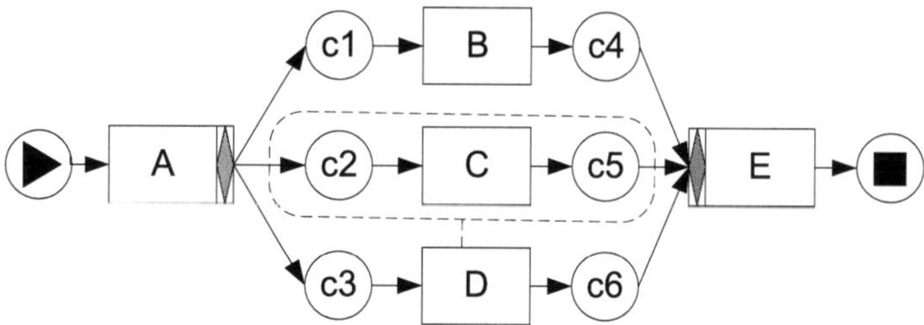

Fig. 3.3 A structured net with an OR-join task and cancelation regions

has any tokens in $c6$, and therefore, E is enabled at M. So, even though two active paths are chosen after OR-split task A, the OR-join evaluation should not wait for tokens from both paths, as it is possible that not all the tokens are on the path to this OR-join task. (Note that Fig. 3.2 is not sound as it can deadlock in $c4$.)

Now, let us consider an OR-join in the context of cancelation. Figure 3.3 is a net with an OR-join task E and task D removing tokens from the conditions $c2$, $c5$ and from the internal condition of task C when firing. Consider a marking $M = c2 + c3 + c4$, which results from the scenario where all outgoing paths are selected after completing task A, and task B has subsequently been performed. At M, there is a token in the input condition $c4$ of OR-join task E. To determine whether task E should be enabled at M, we need to find out whether tokens could be put into $c5$ or $c6$ in the reachable markings of M. As it is possible to reach a new marking $M' = c3 + c4 + c5$ from M by firing task C, E should not be enabled at M. Now consider whether task E would be enabled at marking $M' = c3 + c4 + c5$. At M', $c4$ and $c5$ have one token each and so we need to see if $c6$ can be marked in the reachable markings of M' (i.e., can we reach a marking $c4+c5+c6$ or bigger?). The only reachable marking from M' (without enabling task E itself) is $M'' = c4 + c6$ by firing task D. Note that because of the cancelation region associated with task D, the token from $c5$ has been removed. Hence, task E is enabled at M'. It can be seen from this example that even though the net is a structured net, having cancelation regions can drastically change the behavior of the OR-join and these cancelation regions cannot be ignored while performing an OR-join evaluation. Note also that if task E fires at $M' = c3 + c4 + c5$, we reach a new marking $c3 + o$, which can lead to another marking $c6 + o$ after firing task D where task E is enabled again and then to $2o$ after firing task E a second time. In this case, the marking $2o$ indicates that there are two tokens in the output condition of the net and, as a result, the net is not sound.

3.2.2 The Formal Semantics

Even though YAWL is based on Petri nets, the YAWL language supports complex constructs such as multiple instances, hierarchy, cancelation, and OR-joins that are not easy to model in Petri nets. For OR-join analysis, cancelation plays an important role and it is not possible to abstract from cancelation regions. Hence, we need a formalism that can support the notion of cancelation regions. Reset nets are Petri nets extended with reset arcs that remove tokens from reset places when a transition fires (see Chap. 2) and reset arcs can capture the behavior of the cancelation regions supported by YAWL. Hence, we define the formal semantics of the OR-join using reset nets.

The complexity introduced by a reset arc (when compared with classical Petri nets) is threefold: (1) as the transition removes *all* tokens and not just one when it fires, place invariants do not hold for such nets, (2) the reset action can be *ineffective* if a place does not contain any tokens at the exact time when the transition fires and the reset action is carried out, and (3) a reset arc can affect any place in the entire net (i.e., its effect is global), unlike normal arcs of a transition which can only influence their input and output places (i.e., their effect is local). As a result, the notion of reachability is undecidable for reset nets with more than two reset arcs. Hence, to define the formal semantics of an OR-join, we use the notions of *backwards firing* and *coverability* in reset nets. Coverability is a weaker notion than reachability; a marking is *coverable* from another marking if there is a sequence of tasks that can fire from the first marking to arrive at a marking equal to or bigger than the second marking (for a formal definition see Chap. 2). Normally, a transition fires at a marking to reach a new marking and we can consider the state as moving forward. Backward firing a transition, on the other hand, allows us to start from a desired marking (e.g., a marking where an OR-join is enabled) and transitions can be fired backwards to see if an initial marking is coverable. A new (backwards coverable) marking is created, which puts one token into all input places of the transition which are not reset places, removes one token from all output places of the transition which are not reset places, and puts one token in all reset places of the transition. Backwards firing a transition in a reset net is only possible at a marking if there are no tokens in the reset places of that transition. The backwards firing rule returns a *coverable* marking and not necessarily a *reachable* marking as it is impossible to know how many tokens originally existed in the reset places.

Using this notion of coverability in reset nets, the semantics of an OR-join can be defined as follows. An OR-join in a YAWL net is enabled at a given marking M (where one or more input conditions of the OR-join have tokens) if and only if there is no marking in the future states of the *corresponding reset net* that covers M.

To perform an OR-join evaluation using reset net semantics, it is necessary to first transform a YAWL net into a corresponding reset net for the OR-join of interest. Formal mappings have been proposed to carry out this transformation. Figure 3.4 summarizes how the various YAWL elements are mapped to reset net places and transitions. In general, a YAWL task is mapped to two transitions (one that corresponds to the start of the task (ts) and one that corresponds to the end of the task (te) as well as a place in between (pt)). Various splits and joins in YAWL are also mapped to corresponding places and transitions in reset nets. Conditions and tasks within a cancelation region of a task are mapped to reset places for the corresponding transitions, represented by double-headed arcs in the diagram.

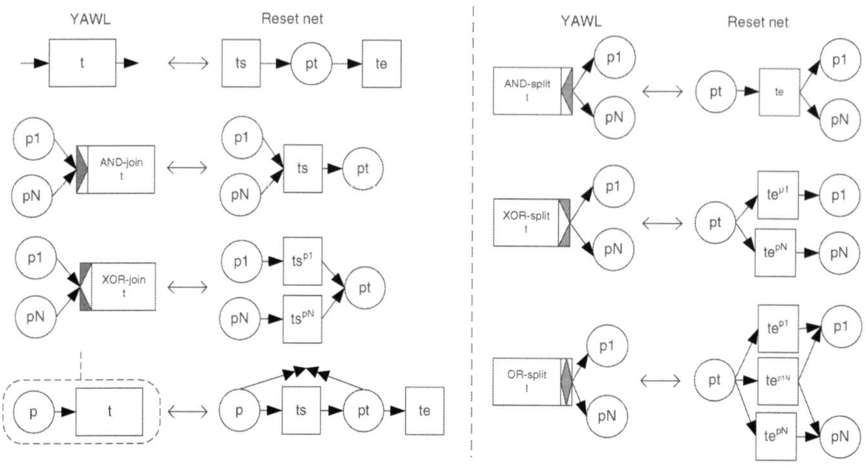

Fig. 3.4 Mapping various YAWL elements to reset net transitions and places

The following steps are taken to convert a YAWL net to a corresponding reset net in the context of a specific OR-join:

1. The YAWL elements that do not affect the OR-join evaluation such as multiple instances and composite tasks are abstracted.
2. All OR-joins in the net (except the one under consideration) are converted into XOR-joins. This is called an *optimistic* approach and it is used to prevent the premature firing of OR-joins. (See the next section for the reasoning behind this conversion.)
3. The resulting YAWL net is transformed into a reset net using the mappings given in Fig. 3.4.

After this transformation, a coverability analysis is carried out on the reset net to determine whether the OR-join should be enabled at a given marking.

In this section, we have described the informal semantics of an OR-join using some example YAWL nets. We have also seen how the existence of cancelation regions in the net can complicate the analysis. Finally, we have briefly explained how the formal semantics is defined in the YAWL language using reset nets. In the next section, we motivate the particular OR-join semantics chosen for YAWL.

3.3 Motivation

In a workflow language like YAWL that supports many complex constructs such as cancelation and loops without restrictions, there are a number of complicating factors when it comes to defining a *general approach* to OR-join semantics. First, for workflows with multiple OR-joins, it is an open issue how a state space analysis for a given OR-join should treat other OR-joins. Second, for workflows with infinite loops (e.g., a continuous monitoring activity), the state space of a sound process could potentially be infinite. Third, cancelation regions complicate the computation of future states. A task that an OR-join is waiting for, that is, in the cancelation region of some other task, may or may not be disabled. Such considerations make state space analysis computationally expensive.

3.3.1 *Different Interpretations and Implementations of an OR-Join*

Even though the OR-join construct is useful in process modeling, its formal semantics are difficult to capture and to implement. Different approaches assign a different (often only intuitive) semantics to this type of join, although they do share the common theme that synchronization is only to be performed for the active paths that are being executed in a given workflow instance.

The Business Process Modeling Notation (BPMN) contains an OR-join like construct called the inclusive OR-join gateway. The semantics of the OR-gateway in BPMN 1.0 is given in the specification as "it will wait for (synchronize) all Tokens that have been produced upstream. If an upstream Inclusive OR produces two out of a possible three Tokens, then a downstream Inclusive OR will synchronize those two Tokens and not wait for another Token, even though there are three incoming Sequence Flow." Note that Wohed et al. have argued that the OR-join gateway does not capture the correct behavior for unstructured BPMN models. It is challenging to select a suitable OR-join semantics and to implement it efficiently.

Several proposals have been put forward to formally define a semantics of OR-joins in Event-driven Process Chains (EPCs) without introducing syntactic restrictions. Van der Aalst et al. highlighted problems associated with the OR-join semantics, especially those of vicious circles. It was suggested that any formal OR-join semantics will impose some restrictions or will deviate from the informal semantics to some extent. This led to a proposal by Kindler to define a nonlocal semantics in EPCs (including the OR-join) in terms of a pair of transition relations using techniques from fixed point theory. Their semantics can then be considered as more general and precise. This approach is forward looking in determining when an OR-join can fire as it generates the entire set of reachable states. Therefore, a deadlock situation can be detected before it happens.

Many workflow systems also struggle with the semantics and implementation of the OR-join. This is because its non-local semantics requires a synchronization depending on an analysis of future execution paths, which requires some non-trivial reasoning. Workflow management systems like InConcert, eProcess, and WebSphere MQ Workflow have solved problems related to the OR-join using syntactical restrictions. IBM WebSphere MQ Workflow (used as a basis for the BPEL standard) appears to offer full support for the OR-join for acyclic workflows. Other systems like Eastman and Domino Workflow also support an OR-join concept with non-local semantics. The use of the non-local semantics may result in poor performance as is stated in the manual of Eastman together with the recommendation to avoid this type of routing. Even the OR-join definition from the Workflow Management Coalition does not support nonlocal semantics. An OR-join is defined as "a point within the workflow where two or more alternative activity(s) workflow branches reconverge to a single common activity as the next step within the workflow. (As no parallel activity execution has occurred at the join point, no synchronization is required.)."

3.3.2 *Factors Affecting the Choice of the OR-Join Semantics*

While defining a suitable semantics for an OR-join, we have chosen an approach such that the resulting semantics are *general*, *formal*, and *decidable*. The criterion for a "general" semantics is that the OR-join semantics should behave correctly without imposing any syntactic restrictions on the net and should also behave as expected even for those nets that are not sound. The semantics should be based on some kind of formal notion so that it can be precisely defined, in this case, on reset nets. The semantics should be decidable, that is, for any net it is always possible to determine whether an OR-join should be enabled or not at a given marking and one should be able to develop an algorithm for this. Note that the notion of vicious circle shows that there is no perfect solution. As a consequence of the paradox associated with the OR-join concept, there are contradicting requirements. However, unlike other workflow systems, in YAWL a best effort is made to address the above requirements.

3.3.2.1 Cancelation Regions

As mentioned previously, the presence of the cancelation feature complicates the formal semantics of the OR-join. The cancelation feature is commonly used to model external events that can change the behavior of a running workflow. It can be used to either disable activities in certain parts of a workflow or to stop currently running activities. Even though it is possible to cancel activities in workflow systems using some sort of abort function, many workflow systems do not provide direct support for this feature in the workflow language. Sometimes, cancelation affects only a selected part of a workflow and other activities can continue after performing a cancelation action. In those cases, an OR-join is the *only* synchronization construct flexible enough to ensure that the process is completed correctly. As cancelation occurs naturally in business scenarios, comprehensive support and a corresponding implementation in workflow systems is required. The concept of cancelation is supported in YAWL through the use of arbitrary cancelation regions that can be associated with tasks. The OR-join semantics take these cancelation regions into account when determining whether or not to enable an OR-join.

3.3.2.2 Loops

The YAWL language supports the existence of both structured and unstructured loops. Together with the presence of cancelation regions, these loops can introduce complex enabling semantics for an OR-join. For instance, a YAWL net can contain an infinite loop together with a cancelation region and can still be sound. Hence, the effect of loops on OR-join operation need to be carefully considered. The OR-join semantics in YAWL have been deliberately designed to work correctly in the presence of arbitrary (possibly infinite) loops.

3.3.2.3 Composite Tasks

The original approach chosen for the OR-join in YAWL took into account the YAWL net at a lower level by unfolding composite tasks (subnets) and subnets are not treated as black boxes. We propose to abstract from constructs that exist in a lower level net (including OR-joins). Hence, for the sake of an OR-join evaluation, a composite task is treated in the same way as an atomic task. So just like an atomic task, a composite task is assumed to complete properly when started. In the current semantics, the OR-join analysis considers only tasks that are at the same net level as the OR-join.

3.3.2.4 Multiple OR-Joins

The informal semantics of an OR-join can be supported reasonably well when there is only one OR-join in a given net. However, when dealing with multiple OR-joins where one precedes the other, the semantics are not well-defined. The question arises as to "how to treat other OR-joins in the workflow when we try to decide whether a certain OR-join should be enabled?."

In the current semantics, we solve this issue by considering one OR-join at a time during the analysis. Two alternative treatments have been proposed for OR-joins: to treat them either as XOR-joins (*optimistic*) or as AND-joins (*pessimistic*). The treatment of an OR-join as an XOR-join is considered *optimistic* as the analysis waits for synchronization if the resulting XOR-join can be enabled. The term "optimistic" refers to the expectation that the preceding OR-join can be enabled when treated as an XOR-join. Consider a marking $M = c1 + c3$ in Fig. 3.5, where an OR-join analysis for task F would be performed. Instead of ignoring the OR-join task E during the analysis, it will be treated as an XOR-join task. It means that the occurrence sequence $c1 + c3 \xrightarrow{C} c3 + c4 \xrightarrow{E} c3 + c7$ would be considered. As a result, F is not enabled at M. This interpretation of OR-join task E as an XOR-join prevents F from being enabled prematurely and it matches closely with the informal semantics of an OR-join.

The treatment of an OR-join as an AND-join is a *pessimistic* approach, as this approach now requires tokens in all input conditions of the AND-join, and if it is

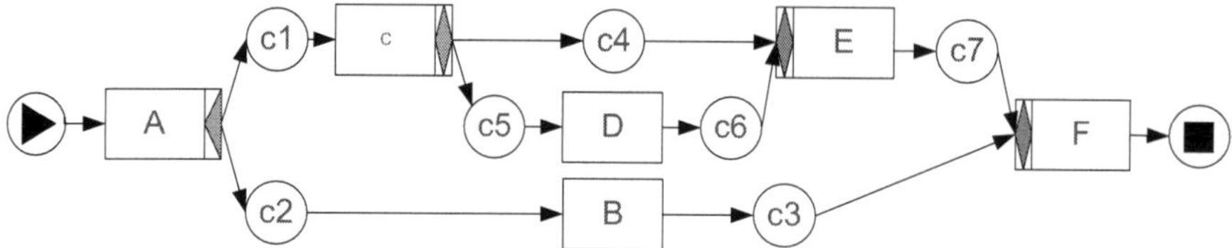

Fig. 3.5 A YAWL net with an OR-join task *E* preceding another OR-join task *F*

not possible, the OR-join will be enabled. Consider again $M = c1 + c3$ in Fig. 3.5, where an OR-join analysis for task *F* would be performed. This time, instead of ignoring task *E*, it will be treated as an AND-join task. Because of the OR-split behavior of task *C*, tokens can be present in $c4$ or $c5$ or both after firing *C*. The occurrence sequence $c1 + c3 \xrightarrow{C} c3 + c4 + c5 \xrightarrow{D} c3 + c4 + c6 \xrightarrow{E} c3 + c7$ is possible. As a token can be put in $c7$ while $c3$ remains marked, *F* is not enabled at M. This preserves the same informal semantics as an optimistic approach, and both approaches result in delaying the enablement of the OR-join task *F*.

Both optimistic and pessimistic approaches support the informal semantics by delaying enablement when there is the possibility of more tokens arriving at the unmarked input conditions of an OR-join. Hence, they are more closely related to the informal semantics of OR-joins and still allow for sound semantics. This strategy is better than ignoring these OR-joins completely during the analysis.

Clearly, any formal semantics impose some restrictions or deviate from the informal semantics to some extent. The motivation behind our semantics and the use of the optimistic approach is to postpone execution of an OR-join for as long as possible. For this reason, the optimistic approach (XOR-join treatment) is chosen during the analysis. The resulting OR-join semantics are well-defined in all circumstances. However, the interpretation of this semantics can sometime lead to a deadlock in the presence of vicious circles, as both OR-joins will wait for each other to fire first.

3.4 Operationalizing the OR-Join

The proposed OR-join semantics as defined in this chapter has been implemented in the YAWL framework. To ensure an efficient implementation of the algorithm, we propose two optimization techniques to speed up the OR-join analysis, which are described in this section.

3.4.1 Algorithm

We now describe the implemented algorithm in a bit more detail. Given a YAWL net N with one or more OR-joins, the following steps are taken to determine whether an OR-join oj is enabled at a marking M of N. Let P_i be a set of input conditions to the OR-join. We first check if M contains tokens for one or more conditions in P_i. If it is false, oj is *not enabled* at M. We also check if all conditions in P_i are marked. If so, oj is enabled at M and an unnecessary call to the expensive algorithm is avoided. Otherwise, the following algorithm is invoked to determine whether oj should be enabled at M.

- The YAWL net is first converted into a reset net using the transformation rules given in Sect. 3.2
- A set of larger markings X that mark more input conditions in P_i is generated for the reset net. This set of markings represents possible enabling markings for the OR-join
- For each marking x in X, we generate a set of backwards coverable markings X' using the backwards firing rule for reset nets. If M (or a smaller marking) is found in X', this indicates that it is possible to cover the marking x in future reachable markings of M and hence, oj should wait to synchronize. In that case, the algorithm returns FALSE
- If M is not coverable by any of the markings generated by markings in X, then the algorithm returns TRUE

As one can imagine, this algorithm could potentially be called multiple times for a particular OR-join as the state of the workflow changes. Hence, it can become very time consuming and may be intractable for YAWL nets with large state spaces. Similarly, its efficiency becomes a major issue when running a workflow with potentially millions of active cases. Next, we present two optimization techniques that can dramatically speed up this analysis.

3.4.2 Optimization

When analyzing an OR-join to determine if it can be enabled, it is possible to consider only a portion of the net that is relevant to the analysis and refrain from exploring those paths that do not affect the OR-join enabling behavior. To improve the performance of the OR-join evaluation algorithm, two forms of restriction are proposed: *structural restriction* and *active projection*. *Structural restriction* involves removing from a net tasks and conditions that are not on the path to the OR-join task under consideration. *Active projection* involves removing tasks and associated conditions that could not be enabled from a given marking. Hence, active projection enables us to stop exploring those parts of the net that can never be reached from a given marking. As a YAWL net with OR-join tasks is translated into a reset net for OR-join analysis, the restriction operations are also performed on the reset net.

3.4.3 Worked Example

Throughout this chapter, several examples have been presented, which indicate that it is a non-trivial task to decide if an OR-join is enabled or not. Clearly, the algorithm can be applied successfully to these situations. To illustrate its inner working in some detail, we use one last example taken from the Order Fulfillment example described in Appendix A. Figure 3.6 represents the Carrier Appointment process with an OR-join task, *Create Bill of Lading* together with its input conditions $c1$, $c3$, and $c4$ (the task and its input conditions have been shaded grey in the figure). Please note that there is also another OR-join task in the figure, *Complete PickUp/Delivery*, but as it is part of a structured OR-split and OR-join combination, we focus our attention on a more interesting scenario involving the *Create Bill of Lading* task.

We know that to invoke the OR-join enabling algorithm, at least one of the input conditions for *Create Bill of Lading* task must have one token. Now, we will observe the results of the algorithm for different markings of the YAWL net.

- First, let us assume that we are at a state in the workflow, where the quote for transportation has been prepared and one token each has been placed in conditions $c1$ and $c2$. At marking $c1+c2$, the algorithm evaluates to FALSE, indicating the OR-join task should not be enabled. This is because it is possible to reach a marking $c1 + c4$, which marks two input conditions of the OR-join by executing the *Arrange Pickup Appointment* task
- Similarly, at marking $c2+c3$, it is possible to reach marking $c3+c4$ in the future and the algorithm returns FALSE
- At marking $c1 + c4$, the algorithm detects that the only reachable marking from $c1 + c4$ is $c3 + c4$. Therefore, it is not possible to cover the larger marking $c1 + c3 + c4$ where all initially marked input conditions $(c1, c4)$ remain marked and a previously unmarked input condition $(c3)$ becomes marked. Hence, the algorithm returns TRUE
- Similarly, at marking $c3 + c4$, the algorithm detects that it is not possible to have any more tokens arriving at $c1$, so it returns TRUE and the OR-join task is enabled at this marking

Let us now more closely examine the effect of the two optimization techniques on the example in Fig. 3.6. Figure 3.7 shows a rough indication of what the application of structural restriction and active projection would result in for the Carrier Appointment process when evaluating the *Create Bill of Lading* task. Please note that the actual restrictions are applied to the underlying reset net and not to the YAWL net. For this example, the structural restriction technique removes approximately half of the tasks under consideration (the bottom half and the left side of Fig. 3.6) as these tasks are not on the path to the OR-join task in question. This is fairly significant as the backwards coverability algorithm needs to consider firing all the transitions in the reset net if there are no restrictions. The active projection technique is then applied depending on the current marking. The figure shows an active projection region for the marking $c1 + c2$. Because of the nature of the AND-split before conditions $c1$ and $c2$, $c1 + c2$ is the first reachable marking where the OR-join evaluation will be performed. So, all active projection regions for other markings will even be smaller than this one. This example clearly demonstrates the efficiency gained from applying these two restriction techniques. Of course, for other nets where most tasks are on the path to the OR-join, the structural restriction technique might not be as effective. Similarly, for nets where there are many reachable states

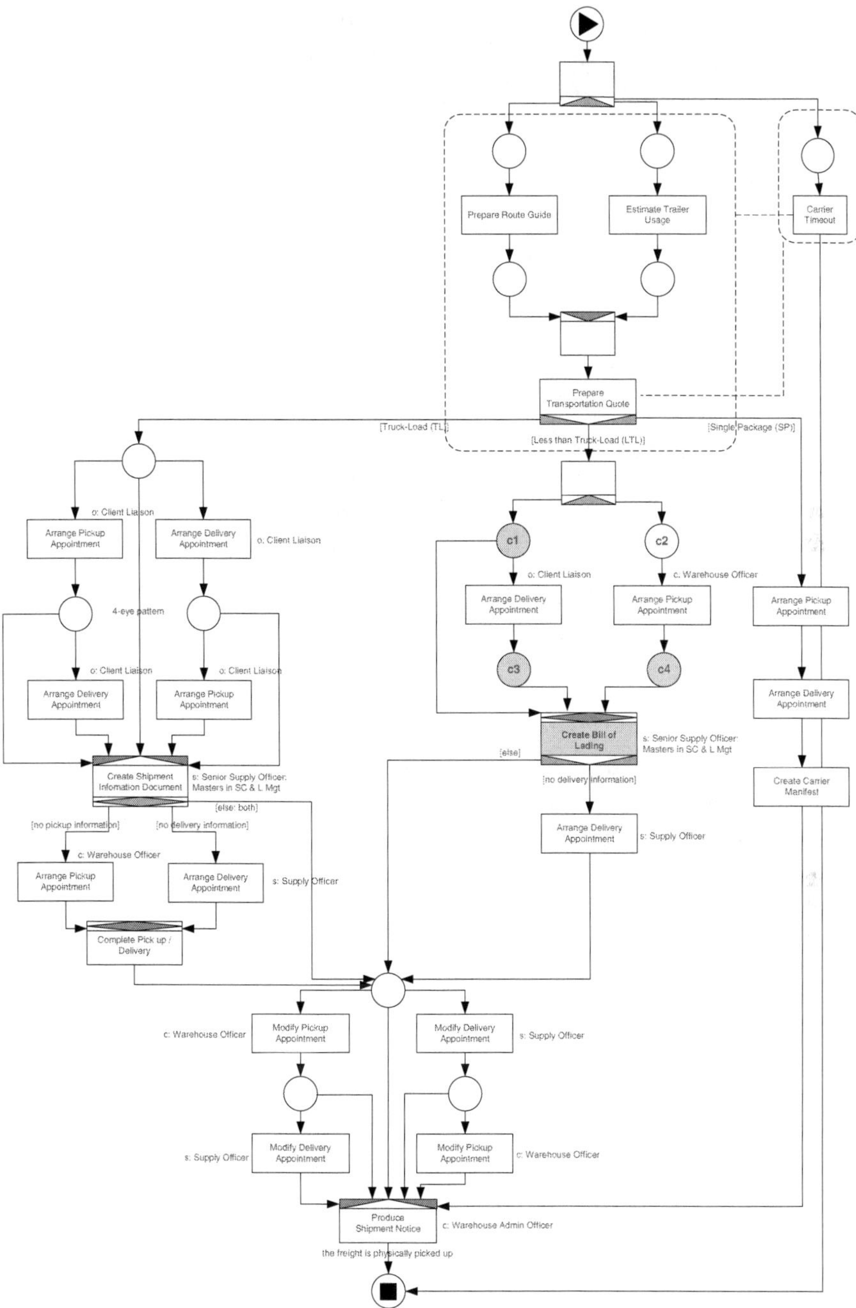

Fig. 3.6 The carrier appointment process with the create bill of lading task modeled as an OR-join

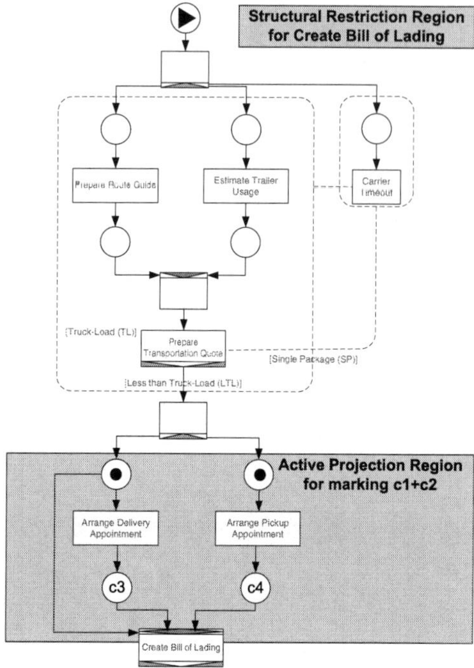

Fig. 3.7 Applying restriction techniques to the carrier appointment process

between the current marking and the enabling markings for an OR-join, the active projection technique might not be as significant. Hence, both techniques need to be applied to achieve a more efficient OR-join evaluation.

3.5 Conclusion

Many workflow management systems and other process-aware information systems, for example, Enterprise Resource Planning (ERP) systems, Customer Relationship Management (CRM) systems, and Product Data Management (PDM) systems, have problems supporting OR-join evaluation without imposing restrictions. Moreover, diagramming notations such as BPMN, EPCs, etc. are not clear about the precise semantics of the OR-join construct. It is surprising to see that standards such as BPMN seem to ignore subtle issues such as the vicious circle and leave the interpretation to users and tool developers. In this chapter, we have presented the semantics for OR-join evaluation in the presence of cancelation regions, other OR-joins, and (infinite) loops without adding structural restrictions. In the YAWL approach, reset nets are used as a formal basis for OR-join analysis to support workflows with cancelation. A transformation function to map a YAWL net with OR-joins into a reset net is provided. An OR-join evaluation algorithm which is based on the backward search techniques for reset nets is then proposed. The proposed semantics uphold the notion that an OR-join waits to synchronize when necessary and continues when appropriate. In addition, we have operationalized this formal semantics and presented an efficient implementation in the workflow language YAWL using two optimization techniques. To the best of our knowledge, no other semantics or workflow system implementation come close to supporting such a general OR-join, especially in the presence of potentially unbounded behavior and arbitrary cancelation regions.

Exercises

Exercise 1. Consider the set of reachable markings for the YAWL net in Fig. 3.1, whereby the split behavior of task *A* is now an XOR-split. Describe at which markings task *E* will be enabled.

Exercise 2. Construct a reachability graph for the YAWL net in Fig. 3.8 and indicate at which markings the OR-joins will be enabled according to the proposed OR-join semantics in this chapter.

Exercise 3. For the YAWL net in Fig. 3.9, please indicate at which of the following markings the OR-joins will be enabled and why.

- Markings $c4$, $c6$, $c4 + c5$, $c4 + c6$ for the OR-join *E*
- Markings $c3$, $c7$, $c1 + c3$, $c2 + c5$, $c3 + c5$, $c3 + c6$, $c3 + c7$ for the OR-join *F*

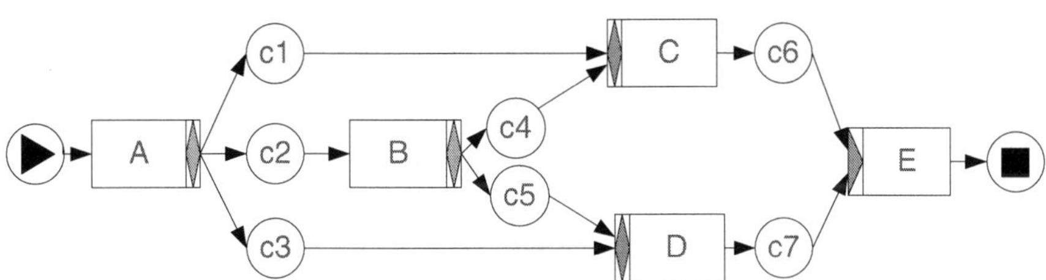

Fig. 3.8 A YAWL net with two OR-joins, *C* and *E*

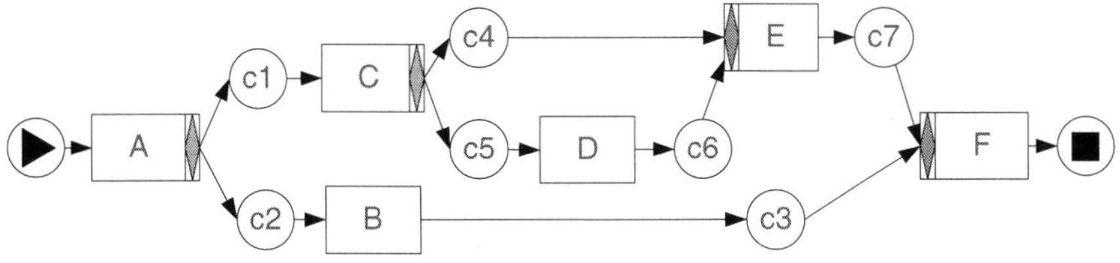

Fig. 3.9 A YAWL net with an OR-join task *E* preceding another OR-join task *F*

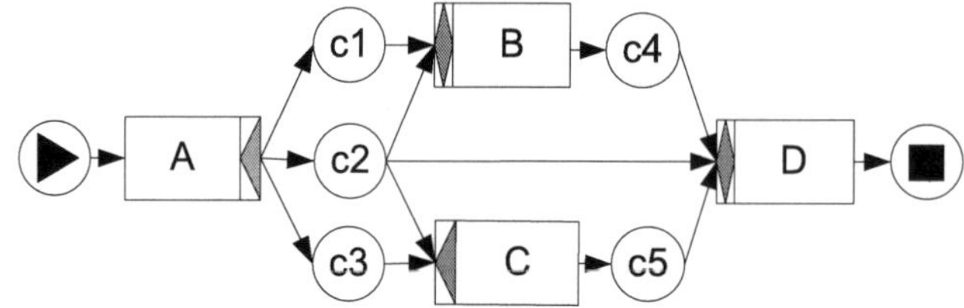

Fig. 3.10 A YAWL net with two OR-joins, *B* and *D*

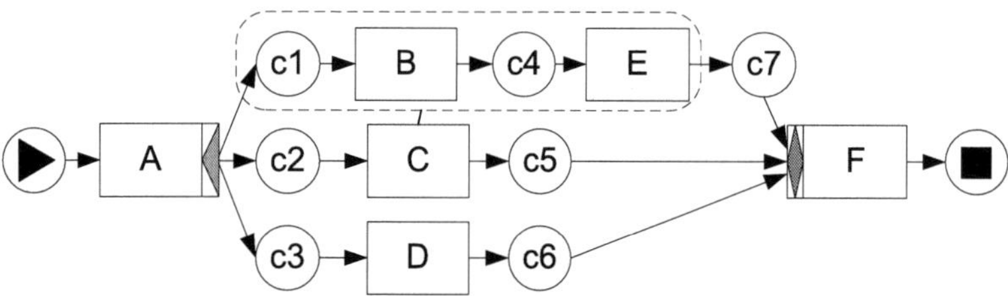

Fig. 3.11 A YAWL net with an OR-join task *F* and cancelation regions

Exercise 4. For the YAWL net in Fig. 3.10, please indicate at which of the following markings the OR-joins will be enabled and why. Also construct a reachability graph and explain why the net may not be sound.

- Markings $c1 + c2 + c3$, $c1 + c3 + c5$ for the OR-join *B*
- Markings $c1 + c2 + c3$, $c1 + c3 + c5$, $c3 + c4 + c5$, $c1 + 2c5$ for the OR-join *D*

Exercise 5. Construct a reachability graph for the YAWL net in Fig. 3.11 and please answer the following questions.

- Is marking $c5+c6$ a reachable marking of the net? If so, is the OR-join *F* enabled at that marking?
- If task *C* fires at marking $c1 + c2 + c6$, what is the next reachable marking? Explain the effect of cancelation region upon firing task *C*.
- If task *C* fires at marking $c2 + c6 + c7$, what is the next reachable marking? Explain the effect of cancelation region upon firing task *C*.
- Is the OR-join *F* enabled at the following markings: $c2 + c6$, $c5 + c6$, $c6 + c7$?

Exercise 6. Explain the main differences between structural restriction and active projection techniques. What are the main characteristics of the YAWL nets that will benefit from each of these two techniques?

Exercise 7. Given the example in Fig. 3.6, determine the different markings at which the OR-join task *Complete PickUp/Delivery* will be enabled. Also indicate the structural restriction region and the active projection region for this OR-join task.

Chapter Notes

The formal semantics defined in this chapter represent a second attempt at defining the OR-join semantics in YAWL. The original semantics of the OR-join in YAWL can be found in this paper [15]. The original semantics have been found to be problematic when there are multiple OR-joins in the net at the same level and in composite tasks. The reasoning behind the current OR-join semantics together with the formal foundations of the approach can be found in the PhD thesis of the first author [272]. The work on the current OR-join semantics was first published in 2005 [273] and the extended version with technical details of the restriction techniques and empirical evidence of their improved performance using several YAWL nets can be found in [275].

An OR-join formalization is proposed for syntactically correct EPCs based on the notion of state (i.e., marking of tokens) and the context (i.e., wait or dead status of an arc) in [162]. The context of an input arc to an OR-join is then used to determine whether an OR-join should be enabled at a given state. The semantics proposed by Mendling, however, will mean that if there is a deadlock preceding an OR-join (e.g., a situation where an AND-join connector upstream cannot propagate tokens due to a deadlock), the OR-join connector downstream will continue to wait for a token. Additional details can be found in Mendling's PhD thesis [160].

The issue of "vicious circles" in EPCs is pointed out in [9]. The proposal by Kindler for a nonlocal semantics of the OR-join in EPCs can be found in [132, 133].

More information on IBM Websphere MQ Workflow can be found in [150]. The Eastman manual that was referred to in this chapter is [85].

The definition of the OR-join by the Workflow Management Coalition is taken from [270].

The reader is referred to [261] for further information about BPMN and to [263] for a discussion on why BPMN's inclusive OR-join gateway does not work correctly in the context of unstructured models. The specification of BPMN 1.0 by the OMG is described in [185]. More on the relationship between BPMN and YAWL can be found in Chap. 13.

Part III
Flexibility and Change

Chapter 4
Dynamic Workflow

Michael Adams

4.1 Introduction

Change is an accepted part of every modern workplace. To remain effective and competitive, organizations must continually adapt their business processes to manage the rapid changes demanded by the dynamic nature of the marketplace or service environment.

However, workflow management systems are generally designed to support the modeling of rigidly structured business processes, which in turn derive well-defined workflow instances. The proprietary process definition frameworks often imposed make it difficult to support (1) dynamic evolution and adaptation (i.e., modifying process definitions during execution) following unexpected or developmental change in the business processes being modeled; and (2) deviations from the prescribed process model at runtime.

The term *flexibility* is used to denote the degree to which a workflow system is able to support or handle expected or unexpected deviations in the execution of process instances, both from within the context of the instance or from the external environment, without negatively impacting on the essence of the process or its expected completion.

Historically, there is generally little or no flexibility provided by systems to accommodate the natural evolution of the work process or organizational goals. Manual interventions into workflow processes become increasingly frequent as staff attempt to manipulate workflow inputs and outputs to conform with changes in workplace practices. These manual intrusions necessitate reduced productivity and increased processing time. Since it is undertaken in an ad-hoc manner, manual handling incurs an added penalty: the corrective actions undertaken are not added to "organizational memory," and so natural process evolution is not incorporated into future iterations of the process. In fact, after initial deployment, the inevitable system changes are often handled so haphazardly that they can lead to major work disruptions and increasing dissatisfaction to the point where the entire system implementation is considered a failure.

M. Adams
Queensland University of Technology, Brisbane, Australia
e-mail: mj.adams@qut.edu.au

A.H.M. ter Hofstede et al. (eds.), *Modern Business Process Automation*,
DOI 10.1007/978-3-642-03121-2_4,

In other work environments, such as those where activities are more creatively focussed, formal representations may provide merely a contingency around which ad-hoc tasks can be formulated. Barthelmess et al. state that "In real life, both in office and scientific lab environments, the enactment of any workcase may deviate significantly from what was planned/modeled" [41]. Thus, adherence to formal representations of task sequences ignores, and may even damage, the informal work practices that also occur in any set of activities.

In summary, a large group of business processes do not easily map to the rigid modeling structures provided due to the lack of flexibility inherent in a framework that, by definition, imposes rigidity. Rather, process models are "system-centric," meaning that work processes are forced into the paradigm supplied, rather than the paradigm reflecting the way work is actually performed. As a result, users are forced to work outside the system, and/or constantly revise the process model, in order to successfully complete their activities, thereby negating the perceived efficiency gains sought by implementing a workflow solution in the first place. Therefore, for a workflow system to be most effective, it must support dynamic change (i.e., during execution).

4.2 YAWL and Dynamic Workflow

Human work is complex and is governed by rules often to a much lesser extent than computerized processing. While some workplaces have strict operating procedures because of the work they do (e.g., air traffic control), many workplaces have few prespecified routines, but successfully complete activities by developing a set of informal tasks that can be flexibly and dynamically combined to solve a large range of problems. Generally, approaches to workflow flexibility usually rely on a high-level of runtime user interactivity, which directly impedes the basic aim of workflow systems (to bring greater efficiencies to work practices) and distracts users from their primary work procedures into process support activities. Another common theme is the complex update, modification, and migration issues required to evolve process models.

The YAWL language supports flexibility through a number of constructs at design time. Like many other languages, YAWL supports parallel branching, choice, and iteration natively, which allow for certain paths to be chosen, executed, and repeated based on conditionals and data values of the instance. In addition (and *unlike* most other languages), YAWL also supports advanced constructs such as multiple-atomic and multiple-composite tasks, where several instances of a task or subnet can be executed concurrently (and dynamically created), and cancelation sets, which allow for arbitrary tasks (or sets of tasks), to canceled or removed from a process instance. Chapter 2 deals with these forms of flexibility in more detail.

The YAWL environment also supports flexibility through its service-oriented architecture (cf. Chap. 7). This means that dedicated services can be built that leverage the power of the YAWL enactment engine to provide flexibility for processes in various ways.

An integral service distributed as part of the YAWL environment that provides for *dynamic* flexibility and exception-handling support for YAWL processes is the *Worklet Service*. The remainder of this chapter will discuss the unique conceptual design of the Worklet Service and how it supports dynamic workflow. The way the Worklet Service handles runtime exceptions is described in the next chapter. The service implementation (with examples) is detailed in Chap. 11. In addition, Chap. 6 describes *Declare*, an approach to workflow flexibility using constraints.

4.3 Worklets: Theoretical Basis

Whenever a series of actions is undertaken with a view of achieving a preconceived result, some plan or set of principles is implemented that guide and shape those actions towards that goal. To be effective, a plan must be described using constructs and language that are relevant to both the actions being performed and the desired result, and be comprehensible by its participants and stakeholders. In workflow terms, analysts seek to model some aspect of the real world by using a metaphor that bears some resemblance to the real world, but also represents an understanding of computational processes. Such metaphors are abstract constructions that form a common reference model, which assist us in representing the external world through computers.

The fundamental and widely understood *computational* metaphor [242] takes a set of inputs, performs a series of functional steps in a strict sequence, and, on completion, produces some output that represents the goal of the process. Thus the computational metaphor describes a single, centralized thread of control, which very much reflects its mathematical ancestry, and reveals the influence of pioneers such as von Neumann and his team, and especially Turing, whose abstract machine proposed "step-at-a-time" processing, and which in turn reflects the influence on thinking of the contemporaneous development of assembly-line manufacturing.

As the prevailing technological advances influenced the structure of early computers, so too has the computational metaphor become a significant model system for the conceptualization and interpretation of complex phenomena, from cognition to economics to ecology. Of particular interest is the way the metaphor has been applied to the definition of organizational behavior issues and the representation of organizational work processes. The computational metaphor remains applicable to well-defined problem domains where goal-directed, sequential, endpoint-driven planning is required. Such domains were the early beneficiaries of workflow management systems. Consequently, workflow systems typically provide support for standardized, repetitive activities that do not vary between execution instances.

Adherence to the metaphor by workflow systems has been an important factor in their acceptance by organizations with structured work practices. Descriptions can be found throughout the workflow literature to the "processing," "manufacturing," and "assembly-line" modeling metaphors that are employed by commercial workflow systems. However, while the Workflow Management Coalition claims that "even office procedures can be processed in an assembly line" [271], there are many aspects where administrative and service processes differ from manufacturing processes.

It may be that while the computational metaphor has been a major enabler of workflow solutions, it may also have played the part of an inhibiting factor in the development of workflow systems able to effectively and dynamically support flexible work practices. Many technologically adept systems fail because they ignore human and social factors. A workflow management system that better supports flexible work environments requires a sound theoretical foundation that describes how work is conceived, carried out, and reflected upon. One such theoretical base can be found in Activity Theory.

4.3.1 Activity Theory – An Overview

This section gives a brief summary of Activity Theory, which forms the theoretical framework for the Worklet Service. Activity Theory is a powerful and clarifying descriptive tool rather than a strongly predictive theory that originated in the former Soviet Union in the 1930s and 1940s as part of the cultural-historical school of psychology.

Before an activity is performed in the real world, it is typically planned using a model. Generally, the better the model, the more successful the activity. However, models and plans are not rigid and accurate descriptions of the execution steps but always incomplete and tentative. Plans can be used as historical data by investigating not adherence to the plan, but the deviations from it. It is the deviations that represent a learning situation, and therefore only the deviations need to be recorded. The experience of using a plan to guide an activity is gained during the instantiation of the activity. In order for plans to become resources for future instantiations of an activity, it is important that the planning tool allows for the ongoing creation and dynamic modification of a plan based on experience gained while operating the plan.

One of the traditional limitations of workflow implementations in less than strictly defined work processes is that it is very difficult if not impossible to incorporate every deviation into the workflow template and therefore future instantiations of the plan. Some deviations may apply to every future instance. Some may only apply once in a while, but these cases should not be left out of the plan. In the normal course of events, these "deviations" are performed externally to the system. But incorporating dynamic change is a fundamental feature of the Worklet Service's design.

To summarize, Activity Theory states that human activity has four basic characteristics:

1. Every activity is directed towards a material or ideal object satisfying a need, which forms the overall motive of the activity. For example, an order fulfillment process is directed toward completing the order and delivering the goods to the customer

2. Every activity is mediated by artifacts, either external (order forms, delivery invoices) or internal (cognitive – compiling freight routes, knowledge, and experience of order staff)
3. Each individual activity is almost always part of collective activities, structured according to the work practice in which they take place. For example, a order cannot be fulfilled without reference to a diversity of other information, such as carrier availability, financial arrangements, stockists, and so on. Thus collective activities are organized according to a division of labor
4. Finally, human activity can be described as a hierarchy with three levels: *activities* realized through chains of *actions*, which may be performed through *operations*:

    - An activity consists of one or more actions, and describes the overall objective or goal
    - An action equates to a single task carried out to achieve some preconceived result. For example, an order fulfillment activity consists of a number of actions: (1) creating an order request, (2) filling the order, (3) appointing a carrier, (4) receiving payment, and (5) final delivery. Each action is achieved through operations determined by the actual conditions in the context of the activity
    - Operations describe the actual performance of the action, and are dependent on the context or conditions that exist for each action. Exactly how the carrier appointment is performed, for example, depends clearly on the concrete conditions, for example, the distance, the type of carrier requested, the size and weight of the freight, availability of appropriate carriers, and so on

4.3.2 Principles Derived from Activity Theory

Ten fundamental principles, representing an interpretation of the central themes of Activity Theory applicable to an understanding of organizational work practices, have been derived and are summarized below.

- **Principle 1 – Activities are hierarchical:** An activity consists of one or more actions. Each action consists of one or more operations.
- **Principle 2 – Activities are communal:** An activity almost always involves a community of participants working towards a common objective.
- **Principle 3 – Activities are contextual:** Contextual conditions and circumstances deeply affect the way the objective is achieved in any activity.
- **Principle 4 – Activities are dynamic:** Activities are never static but evolve asynchronously, and historical analysis is often needed to understand the current context of the activity.
- **Principle 5 – Activities are mediated:** An activity is mediated by tools, rules, and divisions of labor.

- **Principle 6 – Actions are chosen contextually:** A repertoire of actions and operations is created, maintained, and made available to any activity, which may be performed by making contextual choices from the repertoire.
- **Principle 7 – Actions are understood contextually:** The immediate goal of an action may not be identical to the objective of the activity of which the action is a component. It is enough to have an understanding of the overall objective of the activity to motivate successful execution of an action.
- **Principle 8 – Plans guide work:** A plan is not a blueprint or prescription of work to be performed, but merely a guide, which is modified depending on context during the execution of the work.
- **Principle 9 – Exceptions have value:** Exceptions are merely deviations from a preconceived plan. Deviations will occur with almost every execution of the plan, and give rise to a learning experience, which can then be incorporated into future executions.
- **Principle 10 – Granularity based on perspective:** A particular piece of work might be an activity or an action depending on the perspective of the viewer.

Activity Theory offers a number of interesting insights into workflow research domains, particularly the related issues of workflow adaptability, flexibility, evolution, and exception handling. The derived principles above have formed the theoretical foundations for the implementation and deployment of the Worklet Service. Activity Theory was chosen as the theoretical framework because it provides, as demonstrated in this section, a tight fit between actual work practices and the requirements of PAIS designed to support them. This section does not claim Activity Theory to be the only applicable theoretical framework, but merely one from which sound principles of work practice for adaptive business processes could be derived.

4.4 Conceptualization of Worklets

The consideration of the derived principles of Activity Theory formed the conceptual foundations of the Worklet Service, a discrete service that transforms otherwise static workflow processes into fully flexible and dynamically extensible process instances that are also supported by dynamic exception handling capabilities (cf. Chap. 5). This chapter represents a conceptual view of the Worklet Service; the implementation and use of the service is described in detail in Chap. 11.

Fundamentally, a workflow management system that is based on the principles derived from Activity Theory would satisfy the following criteria:

- *A flexible modeling framework*: a process model is to be regarded as a guide to an activity's objective, rather than a prescription for it
- *A repertoire of actions*: extensible at any time, the repertoire would be made available for each task during each execution instance of a process model

- *Dynamic, contextual choice*: to be made dynamically from the repertoire at runtime by considering the specific context of the executing instance
- *Dynamic process evolution*: allow the repertoire to be dynamically extended at runtime, thus providing support for unexpected process deviations, not only for the current instance, but also for other current and future instantiations of the process model, leading to natural process evolution

Thus, to accommodate flexibility, such a system would provide each task of a process instance with the ability to be linked to an extensible repertoire of actions, one of which to be contextually and dynamically chosen at runtime to carry out the task. To accommodate exception handling, such a system would provide an extensible repertoire of exception-handling processes to each process instance, members of which to be contextually and dynamically chosen to handle exceptions as they occur.

To support dynamic workflow, the Worklet Service presents the repertoire-member selection actions as *worklets*. In effect, a worklet is a small, self-contained, complete workflow process, which handles one specific task (action) in a larger, composite process (activity).[1] A top-level or parent process model is developed that describes the workflow at a macro level. From that manager process, worklets may be contextually selected and invoked from the repertoire of each enabled task, using an associated extensible set of selection rules, when the task instance becomes enabled during execution. New worklets for handling a task may be added to the repertoire at any time (even during process execution) as different approaches to completing a task are developed and derived from the context of each process instance. Importantly, the new worklet becomes an implicit part of the process model for all current and future instantiations, avoiding issues of migration and version control. In this way, the process model undergoes a dynamic natural evolution.

The worklet approach provides support for the modeling, analysis, and enactment of business processes, and directly provides for dynamic exception handling, ad-hoc change, and process evolution, without having to resort to off-system intervention and/or system downtime.

4.5 Context, Rules, and Worklet Selection

For any situation, there are multiple situational and personal factors that combine to influence a choice of action. The set of factors that are deemed to be *relevant* to the current situation we call its *context*.

The consideration of context plays a crucial role in many diverse domains, including philosophy, pragmatics, semantics, cognitive psychology, and artificial intelligence. Capturing situated context involves quantifying and recording the

[1] In Activity Theory terms, a worklet may represent one action within an activity, or may represent an entire activity.

relevant influencing factors and relationships between the inner state and the external environment.

A taxonomy of contextual data that may be recorded and applied to a workflow instance may be categorized as follows (examples are drawn from the Order Fulfillment process):

- **Generic (case independent):** Data attributes that can be considered likely to occur within any process (of course, the data values change from case to case). Such data would include descriptors such as when created, created by, times invoked, last invoked, current status; and role or agent descriptors such as experience, skills, rank, history with this process and/or task, and so on. Process execution states and process log data also belong to this category.
- **Case dependent with a priori knowledge:** The set of data that are known to be pertinent to a particular case when it is instantiated. Generally, this data set reflects the data variables of a particular process instance. Examples are customer name, address, and delivery location; freight costs, size, and weight; ordered item names, descriptions, costs, etc.; and deadlines both approaching and expired.
- **Case dependent with no a priori knowledge:** The set of data that only becomes known when the case is active and deviations from the known process occur. Examples in this category may include complications that arise for incorrect payments; unavailable stock, routes, and/or couriers; natural disasters preventing delivery; and so on.

Methods for capturing contextual data typically focus on collecting a complete set of knowledge from an "expert" and representing it in a computationally suitable way. Such approaches depend heavily on the expert's ability to interpret their own expertise and express it in nonabstract forms. However, experts often have difficulty providing information on how they reach a specific judgment, and will offer a justification instead of an explanation. Furthermore, the justification given varies with the context in which the judgement was made.

Theories of context generally fall into two distinct groups: *divide-and-conquer*, a top-down approach that views context as a way of partitioning a global model into simpler pieces (e.g., the "expert" approach described above), and *compose-and-conquer*, a bottom-up approach that holds that there is no tangible global model to begin with, but only local perspectives, and so views context in terms of locality in a (possible or potential) network of relations with other local perspectives.

A top-down, complete representation of knowledge within a given domain is considered by many researchers to be impossible to achieve in practice, and is perhaps not even desirable. In terms of using context as a factor in computational decision making, it is considered more judicious to capture only that subset of the complete contextual state of a particular domain relevant to making a correct and informed decision.

One bottom-up approach to the capture of contextual data that offers an alternative method to global knowledge construction is *Ripple Down Rules* (RDR), which comprise a hierarchical set of rules with associated exceptions.

The fundamental feature of RDR is that it avoids the difficulties inherent in attempting to precompile a systematic understanding, organization, and assembly of all knowledge in a particular domain. The RDR method is well established and fully formalized and has been implemented as the basis for a variety of commercial applications, including systems for reporting DNA test results, environmental testing, intelligent document retrieval, fraud detection based on patterns of behavior, personal information management, and data mining of large and complex data sets. The Worklet Service uses RDR to define rules that allow the correct worklet to be chosen from a repertoire of available worklets for a given task in a process instance, using the particular context of the instance.

An RDR Knowledge Base is a collection of simple rules of the form "if *condition* then *conclusion*" (together with other associated descriptors), conceptually arranged in a binary tree structure (e.g., Fig. 4.1). Each rule node may have a false ("or") branch and/or a true ("exception") branch to another rule node, except for the root node, which contains a default rule and can have a true branch only. If a rule is satisfied, the true branch is taken and the associated rule is evaluated; if it is not

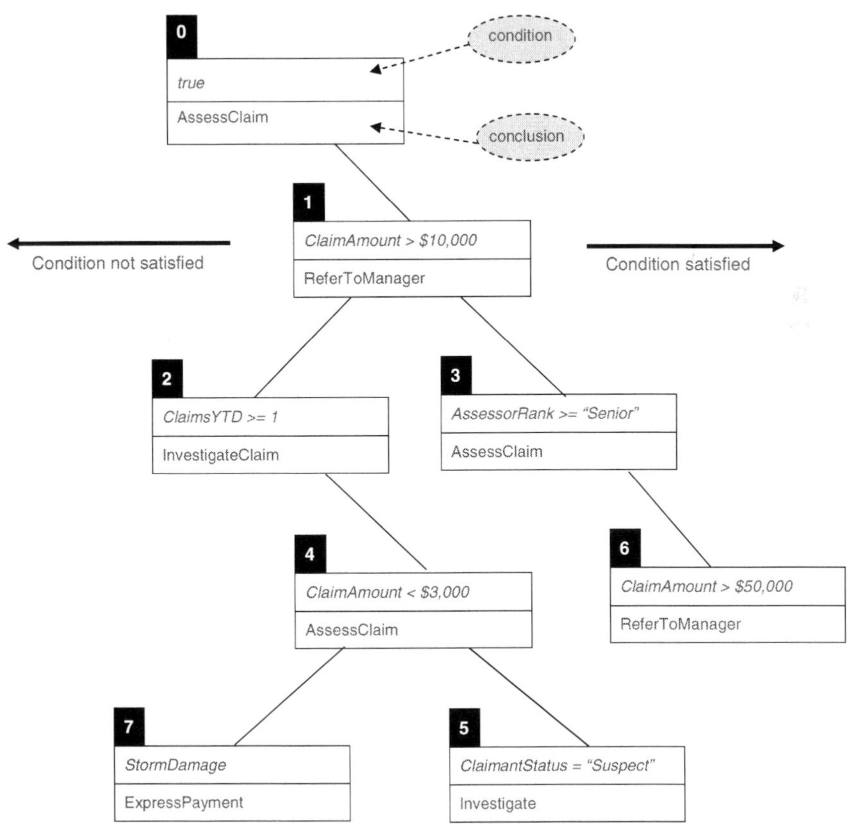

Fig. 4.1 Conceptual structure of a Ripple-Down rule (*assess claim* example)

satisfied, the false branch is taken and its rule evaluated. When a terminal node is reached, if its rule is satisfied, then its conclusion is taken; if its rule is not satisfied, then the conclusion of the last rule satisfied on the path to that node is taken. For terminal nodes on a true branch, if its rule is not satisfied then the last rule satisfied will always be that of its parent (since it must have evaluated to true for the terminal node to be evaluated).

This tree traversal gives RDR implied *locality* – a rule on an exception branch is tested for applicability only if its parent (next-general) rule is also applicable. This feature provides the fundamental benefit of RDR: general rules are defined first, and refinements to those rules added later as the need arises or as knowledge about the domain grows. Thus, there is always a working rule set that extends over time.

For example, the rule tree in Fig. 4.1 represents the following illustrative set of (somewhat artificial) business rules:

1. **Node 0:** By default, assess a claim
2. **Node 1:** An exception to the node 0 rule is that for those cases where the claim amount is greater than $10,000, the claim must be referred to a manager (since the condition of node 0 is always satisfied, the condition of node 1 will always be evaluated)
3. **Node 3:** An exception to the node 1 rule is that cases where the claim amount is greater than $10,000 may be assessed by an assessor of at least *Senior* role (satisfied node 1)
4. **Node 6:** An exception to the node 3 rule is that those cases where the claim amount is greater than $50,000 must always be referred to a manager (regardless of the rank of available assessors (satisfied node 3)
5. **Node 2:** If the claim amount is less than $10,000 (and so the condition of node 1 was not satisfied), then if the claimant has already made a claim in the current year it must be investigated
6. **Node 4:** An exception to the node 2 rule is that if the claim amount is also less than $3,000, simply assess the claim (i.e., satisfied node 2, but the amount is too trivial to warrant an investigation)
7. **Node 5:** An exception to the node 4 rule is that for those cases where the claimant status is set to "suspect," the claim should always be investigated (satisfied node 4)
8. **Node 7:** If the claim amount is less than or equal to $10,000 (unsatisfied node 1) and there has been a claim from this claimant in the current year (satisfied node 2) and the claim amount is also greater than or equal to $3,000 (unsatisfied node 4) and the claim is for storm damage, then the claim should be escalated for express payment

If the conclusion returned is found to be unsuitable for a particular case instance – that is, while the conclusion was correct based on the current rule set, the circumstances of the case instance make the conclusion an inappropriate choice – a new rule is formulated that defines the contextual circumstances of the instance and is added as a new leaf node using the following algorithm:

- If the conclusion returned was that of a satisfied terminal rule, then the new rule is added as a local exception to the exception "chain" via a new true branch from the terminal node
- If the conclusion returned was that of a nonterminal, ancestor node (i.e., the condition of the terminal rule was not satisfied), then the new rule is added via a new false branch from the unsatisfied terminal node

In essence, each added exception rule is a refinement of its parent rule. This method of defining new rules allows the construction and maintenance of the rule set by "subdomain" experts (i.e., those who understand and carry out the work they are responsible for) without regard to any engineering or programming assistance or skill.

Importantly, each rule node also incorporates a set of case descriptors, called the "cornerstone case," which describe the actual case context that was the catalyst for the creation of the rule. When a new rule is added to the rule set, its conditional predicate is determined by comparing the descriptors of the current case to those of the cornerstone case and identifying a subset of differences. Not all differences will be relevant – *it is only necessary to determine the factor or factors that make it necessary to handle the current case in a different fashion to the cornerstone case to define a new rule*. The identified differences are expressed as attribute-value pairs, using the usual conditional operators. The current case descriptors become the cornerstone case for the newly formulated rule; its condition is formed by the identified attribute-value pairs and represents the context of the case instance that caused the addition of the rule.

Rather than impose the need for a closed knowledge base that must be completely constructed a priori, this method allows for the identification of that part of the universe of discourse that differentiates a particular case *as the need arises*. Indeed, the only context of interest is that needed for differentiation, so that rule sets evolve dynamically, from general to specific, through experience gained as they are applied.

Ripple-Down Rules are well suited to the worklet selection processes, since they:

- Provide a method for capturing relevant, localized contextual data
- Provide a hierarchical structuring of contextual rules
- Do not require the top-down construction of a global knowledge base of the particular domain prior to implementation
- Explicitly provide for the definition of exceptions at a local level
- Do not require expert knowledge engineers for its maintenance
- Allow a rule set to evolve and grow, thus providing support for a dynamic learning system

Each worklet is a representation of a particular situated action that relies on the relevant context of each case instance, derived from case data and other (archival) sources, to determine whether it is invoked to fulfill a task in preference to another worklet within the repertoire. When a new rule is added, a worker describes the contextual conditions as a natural part of the work they perform[2]. This level

[2] In practice, the worker's contextual description would be passed to an administrator, who would add the new rule.

of human involvement – at the "coal-face," as it occurs – greatly simplifies the capturing of contextual data. Thus RDR allows the construction of an evolving, highly tailored local knowledge base about a business process.

4.6 The Selection Process

The worklet approach allows for two related but discrete areas of dynamic and flexible workflow to be addressed: dynamic selection of tasks and exception handling with corrective and compensatory action. A conceptual synopsis of the selection process is dealt with in this section; exception handled in Chap. 5.

When a YAWL specification is created in the Editor (cf. Chap. 8), one or more of its tasks may each be associated with a corresponding repertoire of worklets from which one will be selected as a substitute for the task at runtime. Each task associated with the Worklet Service has its own particular repertoire, and its members may be found in a number of other repertoires. Along with the specification, a corresponding RDR rule set is created, which defines the conditions to be evaluated against the contextual data of the case instance. That is, each task may correspond to a particular "tree" of RDR nodes within which are referenced a repertoire of worklets, one of which may be selected and assigned as a substitute for the task dynamically. Not all tasks need be linked to a repertoire – only those for which worklet substitution at runtime is desired.

Each task that is associated with a worklet repertoire is said to be "worklet-enabled." This means that a process may contain both worklet-enabled tasks and non-worklet-enabled (or ordinary) tasks. Any process instance that contains a worklet-enabled task will become the parent process instance for any worklets invoked from it.

Importantly, a worklet-enabled task remains a valid (ordinary) task definition, rather than being considered as merely a vacant "placeholder" for some other activity (i.e., a worklet). The distinction is crucial because, if an appropriate worklet for a worklet-enabled task cannot be found at runtime (based on the context of the case and the rule set associated with the task), the task is allowed to run as an "ordinary" task, as it normally would in a process instance. So, instead of the parent process being conceived as a template schema or as a container for a set of placeholders, it is to be considered as a complete process containing one or more worklet-enabled tasks, each of which *may* be contextually and dynamically substituted at runtime.

It is possible to build for a task an initial RDR rule tree containing many nodes, each containing a reference to a worklet that will be used if the conditional expression for that node and its parent nodes are satisfied; alternately, an initial rule tree can be created that contains a root node only, so that the worklet-enabled task runs as an ordinary task until such time that the rule tree is extended to capture new contextual scenarios (which may not have been known when the process was first defined). Thus, the worklet approach supports the full spectrum of business processes, from the highly structured to the highly unstructured.

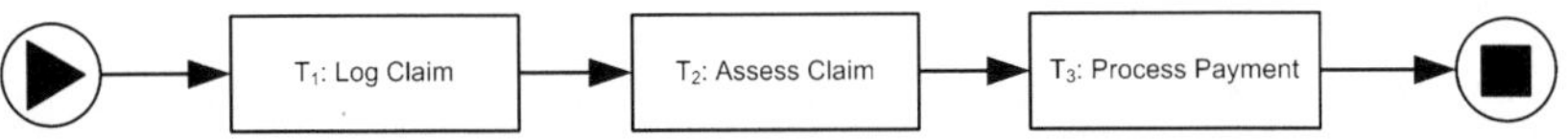

Fig. 4.2 Simple insurance claim model

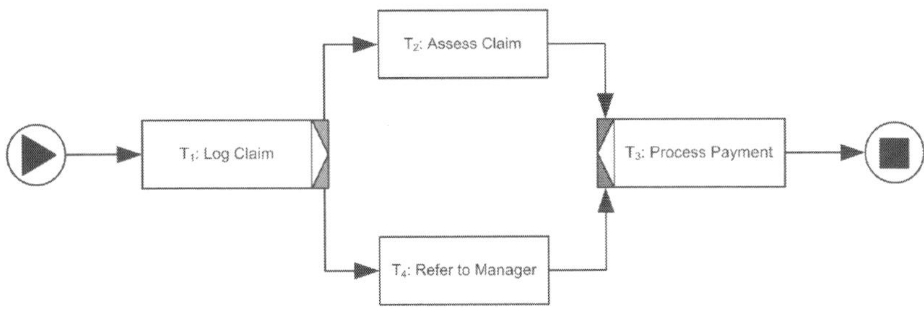

Fig. 4.3 A "view" of Fig. 4.2 extrapolated from the modified RDR for task T_2

Consider the simple insurance claim example in Fig. 4.2. Suppose that, after a while, a new business rule is formulated, which states that when a claim comes to be assessed, if the claim amount is more than $10,000 then it must be referred to a manager. In conventional workflow systems, this would require a redefinition of the model. Using the worklet approach, it simply requires a new worklet to be added to the repertoire for the *Assess Claim* task and a new rule added as a refinement to the appropriate RDR by the administrator. That is, the new business rule is added as a localized refinement of a more general rule (see Fig. 4.1).

The modified RDR tree can be used to extrapolate a view or schematic representation of the model, with the modified rule for the *Assess Claim* represented as XOR choice (Fig. 4.3). That is, a tool can be used to translate the RDR set back into a *view* of a set of tasks and conditional branches within a standard monolithic workflow schema; of course, a translated rule set of a more than trivial size would demonstrate the complexities of describing the entire set of possible branches monolithically. This approach enables the model to be displayed as the derived view in Fig. 4.3, or as the original representation with separate associated worklets, thereby offering layers of granularity depending on factors such as the perspective of the particular stakeholder and the frequency of the occurrence of a condition-set being satisfied. From this it can be seen that an RDR tree may be represented in the modeling notation as a composite set of XOR splits and joins. The advantage of using RDRs is that the correct choice is made dynamically and the available choices grow and refine over time, negating the need to explicitly model the choices and repeatedly update the model (with each iteration increasingly camouflaging the original business logic).

It may also be the case that changes in the way activities are performed are identified, not by an administrator or manager via new business rules, but by a worker who has been allocated a task. Following the example above, after *Log Claim* completes,

Assess Claim is selected and assigned to a worker's worklist for action. The worker may decide that the generic *Assess Claim* task is not appropriate for this particular case, because this claimant resides in an identified storm-damaged location. Thus, the worker *rejects* the *Assess Claim* worklet via a button on their inbox. On doing so, the system refers the rejection to an administrator who is presented with the set of case data for *Assess Claim* (i.e., its cornerstone case), and the set of current case data.

The administrator then compares the two sets of case data to establish which relevant aspects of the current case differ from *Assess Claim's* cornerstone. Note that while many of the values of the two cases may differ, only those that relate directly to the need to handle this case differently are selected (e.g., the location of the claimant and the date the damage occurred). After identifying the differences, the administrator is presented with a list of possible worklet choices, if available, that may suit this particular context. The administrator may choose an appropriate worklet to invoke in this instance, or, if none suit, define a new worklet for the current instance. In either case, the identified differences form the conditional part of a new rule, which is added to the RDR tree for this task using the rule addition algorithm described earlier.

The principles derived from Activity Theory state that all work activities are mediated by rules, tools, and division of labor. Translating that to an organizational work environment, rules refer to business rules, policies and practices; tools to resources and their limitations (physical, financial, staff training, experience and capabilities, and so on); and division of labor to organizational hierarchies, structures, roles, lines of supervision, etc. Of course, these constraints apply to the creation of a new worklet, just as they would in any workflow management system. This means that the authority to create new worklets and add rules to rule sets would rest with the appropriate people in an organization, and the authority to reject inappropriate worklets would reside within the duties of a worker charged with performing the task – the "subdomain expert." Of course, spurious rejection of worklets would be managed in the same way as any other instances of misconduct in the workplace.

In all future instantiations of a specification, the new worklet defined following a rejection would be chosen for that task if the same contextual conditions occur in a new instance's case data. Over time, the RDR rule tree for the task grows towards specificity as refinements are added (cf. Fig. 4.1).

4.7 Service Interface

To enable the Worklet Service to serve the YAWL enactment engine, a number of events and methods must be provided by an interface between them. Being a web-service (cf. Chap. 7), the Worklet Service has been designed to enable remote deployment (i.e., deployed on a web server in a location remote to the YAWL Engine) and to allow a single instance of the service to concurrently manage the flexibility and exception handling management needs for a number of disparate enactment engines that conform to the interface.

The interface requires a number of events and methods, some originating from the service side and others from the engine side. Some require a response, while others do not require any acknowledgment (such as event notifications that do not necessarily require action).

This section describes the selection interface requirements. The event notifications that must be provided to the service by the engine are the following:

- A *work item is enabled* event, where the engine notifies the service that a work item is ready to be executed. The Worklet Service will use this event as a trigger to query the rule set to discover if the enabled work item has a worklet repertoire and, if so, if one is appropriate to act as a substitute for the work item. If an appropriate worklet is found, this event becomes the catalyst for a service selection procedure. If an appropriate worklet is not found, the event is simply ignored, allowing the work item to be executed in the default manner for the engine. Since it is only necessary for an engine to notify the service if there *may* be available worklets for a work item, the service-aware YAWL Engine would allow tasks to be flagged as "service-enabled" and thus would operate more efficiently than an engine that sent events for *every* enabled work item, although this is not a necessary requirement of the enactment engine in general.
- A *work item is canceled* event, which is necessary to accommodate the situation where the service has substituted a worklet for a work item and the work item is canceled (e.g., if it is a member of a cancelation region of the parent process or if the parent process is canceled). In such cases, the service will need to take appropriate action. Again, a service-aware engine would only generate this event if the service has already substituted the work item with a worklet, but it is not a necessary requirement of an enactment engine.
- A *process instance has completed* event, which is required by the service so that it is aware when a worklet instance completes to enable the finalization of the substitution procedure and allow the parent process to continue to the next task.

Each of the events above do not require acknowledgments from the service back to the originating engine – they are simply notifications from the engine that may or may not be acted on.

In addition to the three events, a small number of interface methods are required to be made available to the interface to enable the service to communicate with the engine and take the appropriate action. The required methods are generic in their nature and so would typically be available in most workflow enactment engines. They are the following:

- *Connect to engine*: A connection would typically be required through the interface to enable messaging to pass between the engine and the Worklet Service.
- *Load specification*: The service requires the ability to load a worklet specification into the enactment engine. Since a worklet may be represented as a

"normal" specification of the host enactment engine, this method would already be available within the engine.
- *Launch case*: The service must have the ability to launch an instance of a loaded worklet specification.
- *Cancel case*: The service requires the means to cancel a launched worklet instance (e.g., if the work item it has substituted for is canceled, then the worklet instance would typically need to also be canceled).
- *Check-out work item*: When the service has been notified of an enabled work item, and there is a worklet to act as a substitute for it, the service needs a way to take control of the work item's execution. Thus, by "checking out" the work item from the engine, the engine would pass responsibility for the execution of the work item to the service, and wait for the service to notify it that the execution has completed (i.e., when the worklet case completes).
- *Check-in work item*: When a worklet instance has completed, the service will notify the engine that execution of the original work item it acted as a substitute for has completed. The data gathered by the worklet would be mapped to the work item before it was checked in.

Figure 4.4 summarizes the interface required by the service's selection process.

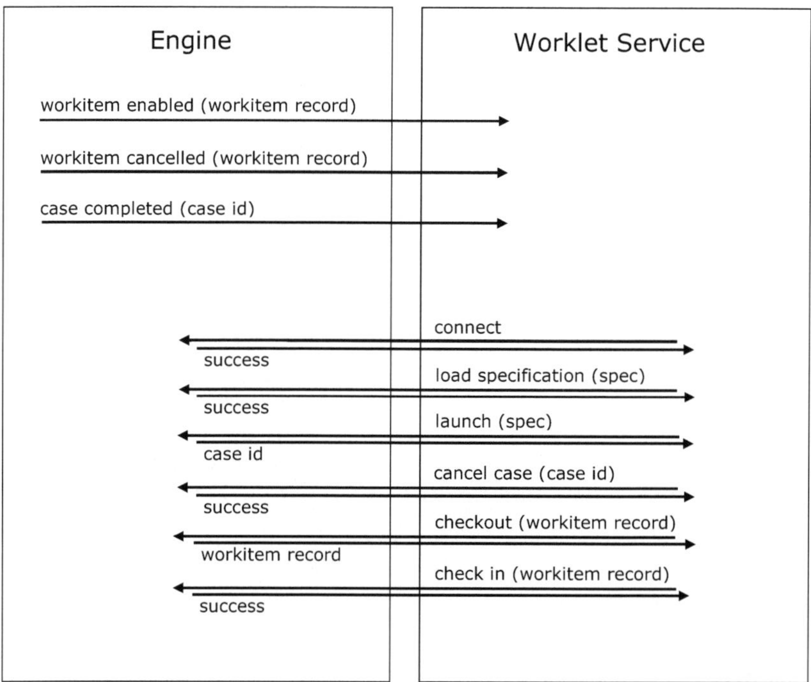

Fig. 4.4 Required selection interface

4.8 Secondary Data Sources

When making a contextual choice of an appropriate worklet, it may be desirable or even necessary to seek data outside the parameters of the task/or case. For example, the current state of the process, or the states of individual tasks, may have some bearing on the choices made; available resources may be an important factor and so on. Thus, choosing the most appropriate worklet for a situation will be achieved by defining rules that use a combination of currently available data attribute values, both case-dependent and independent.

One method of accessing the current set of states for an instantiated process, resource data, and archival histories of previous instantiations of a specification may be deduced by mining the process log file. A series of predicates can be constructed to enable the extraction of the current state set and any relations between active worklets and tasks, as well as archival trends and relations. These predicates may then be used to augment the conditionals in RDR nodes to enable selection of the most appropriate worklet.

The kinds of information that may be extracted from the process log file using these predicates include the current status of a worklet, whether a worklet is a parent or child of another worklet, when a certain state was entered or exited for a particular worklet or task, the resource that triggered a state change, and so on. Chapter 11 provides an explanation of how such predicates may be constructed and used in practice.

Figure 4.5 shows an ORM diagram for the log kept by the Worklet Service. The entity "Parent Case" in the worklet log corresponds to the case identifier of the parent process that a worklet was launched for, which can be drawn from the engine's process logs. Hence, the entity would map to a record of the parent case instance in

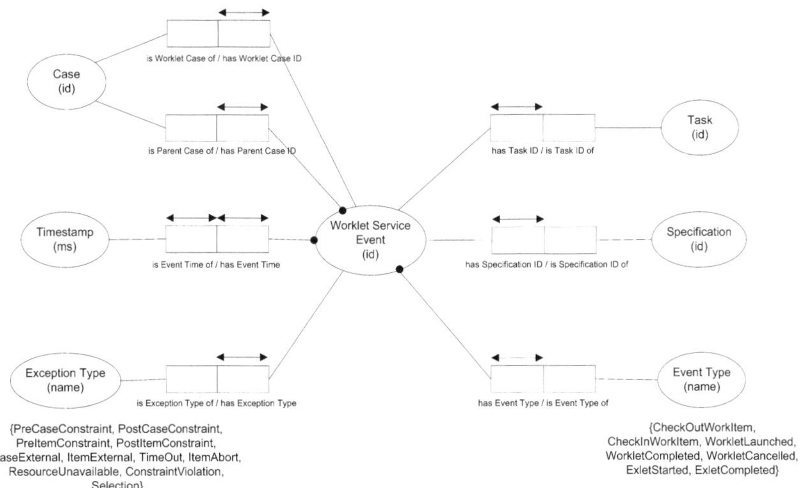

Fig. 4.5 Object role model: process log for the worklet service

the engine process logs. The worklet log entities "Task" and "Case" would also map to corresponding records in the engine process logs. Thus, by mapping those entity values, a complete view of the process, inclusive of any actions taken on behalf of it by the Worklet Service, can be constructed.

It should be noted that the event types for the engine log differ from those of the worklet log. Since worklets are launched as separate cases in the workflow engine, the engine's log records the progress of the process instance, so that it is not necessary to record those types of events in the Worklet Service logs – indeed, to do so would be a duplication. Thus, the worklet log concentrates on the event details that are not recorded on the engine side, such as CheckOutWorkItem, WorkletLaunched, and so on.

4.9 Conclusion

The worklet approach to dynamic workflow presents several key benefits, including the following:

- A process modeler can describe the standard activities and actions for a workflow process and the worklets for particular tasks using the same modeling methodology
- It allows reuse of existing process and worklet components. Removing the differentiation between dynamically inserted worklets and the "normal" workflow aids in the development of fault tolerant workflows out of preexisting building blocks
- Its modularity simplifies the logic and verification of the standard model, since individual worklets are less complex to build and therefore to verify than monolithic models
- It provides for workflow views of differing granularity, which offers ease of comprehensibility for all stakeholders
- It allows for gradual and ongoing evolution of the model, so that global modification to the model each time a business practice changes or an exception occurs is unnecessary
- In the occurrence of an unexpected event, an administrator needs simply to choose an existing worklet or build a new one for the particular context, which can be automatically added to the repertoire for future use as necessary, thus avoiding complexities including downtime, model restructuring, versioning problems, and so on.

Most importantly, the worklet approach is built on the solid theoretical foundations of Activity Theory, and so fully supports the set of derived principles of organizational work practices, and the criteria for a workflow support system based on those principles.

Exercises

Exercise 1. How does a worklet process model differ from a normal YAWL process model?

Exercise 2. Name three principal insights into human work activity offered by Activity Theory that have application in the development of support for dynamic flexibility in Process-aware Information Systems.

Exercise 3. Construct a Ripple-Down Rule tree (on paper) for the following items. In each case, the condition will be the type of item, and the conclusion will be "can fly" if the item can fly, or "can't fly" if it cannot, generally speaking. The rule tree is to be constructed in the sequence in which the items are listed, which proceed from the more general to the more specific.

(a) A bird
(b) A cat
(c) A baby bird
(d) An airplane
(e) A penguin
(f) A penguin inside an airplane

Exercise 4. Construct a Ripple-Down Rule tree (on paper) for the following rules. The rule tree is to be constructed in the sequence in which the items are listed.

(a) If it is warm, then play tennis
(b) If it is raining, then stay indoors
(c) If it is dry, then water the garden
(d) If it is snowing, then go skiing
(e) If it is fine, then go on a picnic
(f) If it is cool, then go to a movie
(g) If it is snowing heavily, then stay indoors
(h) If it is fine and cool, then go shopping
(i) If it is warm and raining, then stay indoors

Exercise 5. Construct a Ripple-Down Rule tree (on paper) for the following (fictional) requirements. "A patient may present to the Casualty Department with one of five general conditions: fever, wound, rash, abdominal/chest pain, or fracture. Each of those conditions should be referred to the corresponding department for treatment. If, however, a patient with abdominal/chest pain has a pulse over 150 bpm, they should be treated for a heart condition. If a patient has a fever *and* a rash, they should be quarantined in the infectious diseases department. If a patient has a wound *and* a fracture, treatment of the wound takes precedence. If a patient with abdominal/chest pain is pregnant, they are to be referred to the maternity ward. If a patient has a wound with high blood loss, they should be immediately referred to the ER. If a patient's fever is mild, they should be given a pill and sent home."

Chapter Notes

Worklets

The worklet approach arose from an investigation into why workflow systems had difficulties supporting flexible processes. A strong disconnect was found between the modeling and execution frameworks supplied, and the way human work was actually performed. Simply put, existing systems had grown from programming or linear, sequential execution bases, and most work activities are far from linear. These findings were first reported in [24], which also detailed a set of criteria against which the systems could be measured for their ability to support flexible processes. The idea of *worklets* was first published in [26], and the worklet approach to dynamic flexibility was further detailed in [27]. A full exploration of the worklet approach, including a complete formalization and exemplary studies, can be found in [23].

For the interested reader, a discussion of the use of worklets in very creative working environments may be found in [237].

Ripple-Down Rules

Ripple-Down Rules were first devised by Compton and Jansen [62]. While on the surface it may seem that the quality of RDR sets would be dependent on the insertion sequence of new rules, and may be open to the introduction of redundant and/or repeated rules, this has been shown to be not the case. In terms of the correct selection of the appropriate rule based on case context, it is *always* the case that the correct rule is chosen, regardless of the insertion sequence of rules into the tree. In terms of the potential for redundancy and repetition of rules throughout the tree, studies have shown that the issue is far less serious than first perceived [63, 156] and that the size of an RDR set which includes a normal distribution of redundant rules compares favorably in size with other various inductively built Knowledge Based Systems.

In terms of the number of computational steps required to reach the finally chosen rule, it has been empirically shown that Ripple-Down Rules are able to describe complex knowledge systems using less rules than conventional "flat" rule lists [62, 99, 234]. So, when comparing insertion sequences between RDR sets and traditional decision lists, RDR sets will have the higher quality in this regard. In terms of the potential for degradation of computational time taken to traverse a rule set due to the growth of rule sets over time, a number of algorithms exist for the reordering and optimization of rule trees (cf. [99, 211, 234]). Thus, trees may occasionally be optimized and thus the highest quality can be maintained over time.

Activity Theory

For more information on Activity Theory, the interested reader is directed to [24, 39, 139].

Other Approaches

Most workflow systems use their own unique conceptual framework, which is usually based on programming constructs rather than founded on theories of work practice. Consequently, since the mid-nineties, much research has been carried out on issues related to dynamic flexibility and exception handling in workflow management systems. Such research was initiated because, generally, commercial workflow management systems required the model to be fully defined before it could be instantiated, and that any changes must be incorporated by modifying the model statically. These typically flat, monolithic, single-schema architectures make it difficult to fully capture flexible business processes [46, 117].

While there have been many proposed and/or implemented approaches to flexibility, this section discusses a sample of the more interesting ones.

An optional component of the Tibco iProcess Suite is the *Process Orchestrator* [102], which allows for the dynamic allocation of subprocesses at runtime. It requires a construct called a "dynamic event" to be explicitly modeled that will execute a number of subprocesses listed in an "array" when execution reaches that event. Which subprocesses execute depend on predefined data conditionals matching the current case. The listed subprocesses are statically defined as are the conditionals. There is no scope for dynamically refining conditionals, nor adding subprocesses at runtime.

COSA (version 5.4) [66] provides for the definition of external "triggers" or events that may be used to start a subprocess. All events and subprocesses must be defined at design time, although models can be modified at runtime (but only for future instantiations). COSA also allows *manual* ad-hoc runtime adaptations such as reordering, skipping, repeating, postponing, or terminating steps. *SAP Workflow* (version 6.20) [227] supports conditional branching, where a list of conditions (each linked to a process branch) is parsed and the first evaluating to true is taken; all branches are predefined. *FLOWer* (version 2.1) [45, 191] is described as a "case-handling" system; the process model (or "plan") describes only the preferred way of doing things and a variety of mechanisms are offered to allow users to deviate in a controlled manner [22].

There have been a number of academic prototypes developed in the last decade (although activity was greater during the first half); very few have had any impact on the offerings of commercial systems [174]. Several of the more widely acknowledged are discussed here.

The *eFlow* system [56] supports flexibility in e-Services by defining compensation rules for regions, although they are static and cannot be defined separately to the standard model. The system allows changes to be made to process models, but such changes introduce the common difficulties of migration, verification, consistency and state modifications.

ADEPT [118,206] supports modification of a process during execution (i.e., add, delete, and change the sequence of tasks) both at the model (dynamic evolution) and at the instance levels (ad-hoc changes). Such changes are made to a traditional monolithic model and must be achieved via manual intervention, abstracted to a high level interaction. The system also supports forward and backward "jumps" through a process instance, but only by authorized staff who instigate the skips manually [204].

The *AdaptFlow* prototype [111] provides a hybrid approach to flexibility. It supports the dynamic adaptation of process instances, although each adaptation must be confirmed manually by an authorized user before it is applied (alternate manual handling to override the dynamic adaptation offered is also supported). Also, the rule classifications and available exception handling actions are limited to medical treatment scenarios. The prototype has been designed as an overlay to the *ADEPT* system, providing dynamic extensions.

The *ADOME* system [57] provides templates that can be used to build a workflow model, and provides some support for (manual) dynamic change; it uses a centralized control and coordination execution model to initiate problem solving agents to carry out assigned tasks. A catalog of "skeleton" patterns that can be instantiated or specialized at design time is supported by the *WERDE* system [54]. Again, there is no scope for specialization changes to be made at runtime.

AgentWork [176] provides the ability to modify process instances by dropping and adding individual tasks based on events and ECA rules. However, the rules do not offer the flexibility or extensibility of the YAWL approach, and changes are limited to individual tasks, rather than the task-process-specification hierarchy. Also, the possibility exists for conflicting rules to generate incompatible actions, which requires manual intervention and resolution.

The *ActivityFlow* specification language described in [151] divides workflows into different types at design time (including ad-hoc, administrative, or production), and provides an open architecture that supports interaction and collaboration of different workflow systems. The system, like *ADEPT*, advocates the use of a dedicated (human) workflow coordinator/administrator to monitor workflows with an eye on deadlines, handle exceptions, prioritize, stop, resume and abort processes, dynamically restructure running processes, or change a specification.

An approach that uses a "society of intelligent agents" that work together to execute flexible processes is found in [257], and another that uses *BPBots (Business Process Robots)* to perform the roles of service requesters, providers, and brokers in the formation of a hierarchical community for the execution of a process instance is introduced in [279]. A further approach using incompletely specified process definitions is found in the *SwinDeW (Swinburne Decentralized Workflow)* project [278]. SwinDew is a peer-to-peer based decentralized model, where a process definition is split into a set of task partitions and distributed to peers, and on-the-fly process elaboration is performed at runtime. Thus, a multi-tiered process modeling and execution framework is provided.

CBRFlow [258] uses a case-based reasoning approach to support adaptation of predefined workflow models to changing circumstances by allowing (manual) annotation of business rules during runtime via incremental evaluation by the user. Users must be actively involved in the inference process during each case. An approach, which integrates *CBRFlow* into the *ADEPT* framework, is described in [213]. In doing so, semantic information about the reasons for change, and traceability data, are presented to the *ADEPT* user/administrator to support decision making processes. The information can also be used to facilitate reuse of ad-hoc changes from similar scenarios. When deviations from a process schema are required, the case-based reasoning component assists the user to find similar previous cases through a series of questions and answers, one of which may then be applied to the current instance [213]. While the process is quite user-intensive, the approach does provide a good example of the combination of contextual information with flexibility techniques.

Chapter 5
Exception Handling

Michael Adams and Nick Russell

5.1 Overview

Translating abstract concepts and descriptions of business practices and rules into process models is a far from trivial exercise. Even for highly structured processes (such as medical and banking environments), it is difficult (if not impossible) to successfully capture all work activities, and in particular all of the task sequences possible, in a workflow model, at least not without producing some very complex models.

In fact, it is because of the discrepancies between real-world activities and formal representations of them that workflow process instances typically experience *exceptions* during their execution.[1] Traditionally, exceptions are understood to be events that by definition occur rarely. But in a workflow process, an exception can be described as an event that is deemed to be outside "normal" behavior for that process. Rather than being an error, it is simply an event that is considered to be a deviation from the expected control-flow or was unaccounted for in the original process model. Such events happen frequently in real working environments.

Exceptions are a fundamental part of most organizational processes; in fact, a substantial proportion of the everyday tasks carried out in a business can be categorized as exception handling work. Consequently, every executing instance of a work process will be likely to incorporate some deviation from the plan. In this sense, a work plan can be seen to be just another resource or tool that mediates the activities of workers towards their objective, rather than a prescriptive blueprint that must be strictly adhered to. Such deviations from the plan should not be considered as errors, but as a natural and valuable part of the work activity, which provides the opportunity for learning and thus evolving the plan for future instantiations.

Historically, exception handling within PAIS has fallen well short, particularly after execution has commenced. It is assumed that if an exception is expected,

[1] The focus of this chapter is exception handling at the conceptual level – that is, within the control-flow, data, and resourcing perspectives.

M. Adams (✉)
Queensland University of Technology, Brisbane, Australia

A.H.M. ter Hofstede et al. (eds.), *Modern Business Process Automation*,
DOI 10.1007/978-3-642-03121-2_5,

or even could conceivably have been anticipated, then the modeler has *a priori* knowledge of such an event, and it should therefore have been built into the model. However, if a modeler builds all possible *a priori* exception scenarios into a model, it can lead to very complex models, much of which will never be executed in most cases. Also, mixing business logic with exception handling routines complicates the verification and modification of both, in addition to rendering the model almost unintelligible to most stakeholders.

Conversely, if the exception is unexpected, or could not possibly have been anticipated, given the scope of knowledge of the work process available at the time the model was designed, then the model is deemed to be simply deficient, and thus needs to be amended to include this previously unimagined event. This view, however, tends to gloss over the frequency of such events, or the costs involved with their correction. Such exceptions occur more frequently in processes that are very complex and/or with a high variability. When they occur, they are normally captured and managed manually, typically by halting process execution, which naturally has a negative impact on business efficiency. Since most "real-world" workflow processes are long and complex, neither human intervention nor terminating the process are satisfactory solutions.

5.2 A General Framework for Exception Handling

This section describes a conceptual framework for exceptions that allows their format and operation to be described in a precise manner. This framework focuses on the notion of a workflow exception in a general sense and the various ways in which they can be triggered and dealt with. It considers an exception to be a distinct, identifiable event that occurs at a specific point in time during the execution of a process. Generally, the occurrence of a specific exception will be detected in the context of a work item that is currently executing. Such an occurrence is assumed to be immediately detectable and to have a specific type. The action of dealing with an exception that has been identified is known as *exception handling* and specific strategies can be defined describing the actions that should be taken to mitigate their effects. At the lowest level, these strategies may be bound to an individual task. Exception handling strategies can also be recognized in relation to groups of tasks (often described as *scopes*), blocks (i.e., a set of tasks at the same overall decomposition level in a process), or entire process models. In all of these situations, the same handling considerations apply to all the tasks encompassed within the given process component in which the exception is experienced. The strategy for handling an exception depends on four main factors:

1. The type of exception that has been detected
2. How the work item that detected the exception will be handled
3. How the other work items in the case will be handled
4. What recovery action will be taken to resolve the effects of the exception

By describing these factors in detail for a given exception, it is possible to precisely describe a strategy to handle it with, both in terms of dealing with the affected work item and also more generally in terms of mitigating its overall effects. Each of the possible configurations for each of these factors follow.

5.2.1 Exception Types

The first factor that is relevant to how an exception will be handled is the type of the exception. A comprehensive review of the workflow literature and current commercial offerings indicate that there are five distinct types of exceptional events that can occur during the execution of a business process that are able to be effectively remedied. These are as follows:

Work Item Failure (WIF)

Work item failure is characterized by the inability of a currently executing work item to either continue executing or to progress any further in its current execution state. This may be a consequence of a variety of distinct causes such as user-initiated termination of the program that implements the work item; the failure of a hardware, software, or network resource associated with the work item; or a user indication that the associated work item should considered as having failed rather than having completed successfully. Often this form of failure cannot easily be dealt with within the context of a process model or the effects of such failure are not localized to a specific part of the process and an exception-based handling strategy offers an effective means of managing the situation such that both later work items and the process as a whole continue to behave correctly.

Deadline Expiry (DEX)

The use of deadlines within business processes are a common means of enforcing organizational performance requirements within a business process. Generally, they are associated with a specific work item and indicate when it should be completed, although commencement deadlines are also possible. Often when specifying a deadline, it is also useful to define the action that will be pursued if the deadline is reached and the work item has not been started or completed.

Resource Unavailability (RUN)

It is often the case that a work item requires access to one or more resources during its execution. These may be data resources that contain information required for the work item to execute or physical resources (human or non-human) that are necessary in order to actually perform the work item. If these resources are not available to the work item at initiation, then it is usually not possible for it to proceed. In most workflow systems, it is necessary for a work item to be allocated to a specific human resource who is responsible for managing its associated work activities and signaling when it is complete. Problems with work item allocation can arise if (1) at distribution time, no suitable human resource can be found who meets the specified allocation criteria for the work item or (2) at some time after allocation, the resource is no longer able to undertake or complete the work item. Although the occurrence of these issues can be automatically detected, they often cannot be resolved within the context of the executing process and may involve some form of escalation or manual intervention. For this reason, they are ideally suited to resolution via exception handling.

External Trigger (EXT)

Triggers from sources in the environment in which a process executes are a common means of signaling a work item that an event has occurred that affects the work item and requires some form of handling. These triggers can originate from internal as well as external sources, and it is possible that nonlinked work items (i.e., work items that are not directly linked to the work item in question by a control edge) elsewhere within the process model or even in other process models may also be the source of such notifications. Although a work item can anticipate events such as triggers and make provision for dealing with them, it is not predictable if or when such events will occur. Consequently, the handling of triggers is not suited to normal processing within the work item implementation and is often better dealt with via exception handling mechanisms. Generally, triggers indicate that an *out-of-bound* condition has arisen that needs to be dealt with. Managing such a situation may require that the current work item be halted, possibly undone, and some alternative form of (compensatory) action be taken.

Constraint Violation (CVI)

The use of constraints within process model are a common means of specifying operational invariants over control-flow, data or resource elements in a process that need to be maintained to ensure that its integrity and operational consistency is preserved. They are generally monitored on an ongoing basis to ensure that they are enforced. The implementation of mechanisms that can detect and deal with constraint violations within the context of an operational process is similar to the issue of dealing with external triggers. Generally, it is a work item that will be needed to detect and deal with a violation, although it is conceivable that constraints could be specified and enforced at block or process level. Constraints may be specified over data, resources, or other work items within a process model, hence the approach chosen for handling them needs to be as generic as possible to ensure that it has broadest applicability.

5.2.2 *Exception Handling at Work Item Level*

The second factor that is associated with any exception handling strategy is the way in which the work item that detected the exception will be handled. There are a multitude of ways in which an exception may be dealt with; however, the actual approach pursued will depend on the current execution state of the detecting work item and its state after the exception has been dealt with.

To quantify the range of possible options, we first need to consider a generalized set of states that are associated with the execution life-cycle for a work item. Figure 5.1 illustrates as solid arrows the states through which a work item progresses during normal execution from a resourcing perspective (cf. Chap. 2). It is initially *offered* to one or more resources for execution. A resource issues an *allocate* command to indicate that it wishes to execute the work item at some future time; the work item is then *allocated* to that resource. Typically this involves adding the work item to the resource's work queue and removing any references to the work item that other resources may have received, either on their work queues or via other means. When the resource wishes to commence the work item, it issues a *start* command and the state of the work item changes to *started*. Finally, once the work item is finished, the resource issues a *complete* command and the state of the work item is changed to *completed*. Note that there are two possible variations to this course of events, shown as dotted arcs in Fig. 5.1, (1) where a work item offered to a resource is *selected* by another resource, it is *withdrawn* from the first resource's worklist and

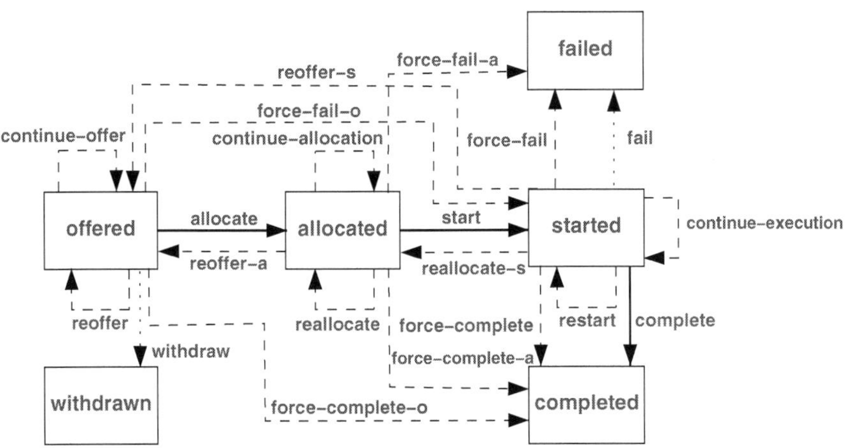

Fig. 5.1 Options for handling work items

(2) where an executing work item is detected as having *failed*, its state is changed accordingly.

Figure 5.1 also illustrates the 15 possible state transitions that may occur for a work item when it is subject to exception handling, which it depicts as dashed arcs. The details of each of these are summarized in Table 5.1.

5.2.3 Exception Handling at Case Level

As exceptions can always occur in the context of a system in which more than one case is currently executing, it is possible that their occurrence may also affect other current work items in the same and possibly other cases. Therefore, the third consideration that arises is how to handle the exception at the level of the case. There are three possible alternative options for this:

1. **Continue workflow case (CWC)** – The workflow case can be continued, and no intervention occurs in the execution of any other work items
2. **Remove current case (RCC)** – Selected or all remaining work items in the current case are removed
3. **Remove all cases (RAC)** – Selected or all remaining work items in both the current all other executing cases which correspond to the same process model are removed

In the latter two scenarios, a selection of work items to be removed can be specified using both static design time information relating to the corresponding task definition (e.g., original role allocation) as well as relevant runtime information (e.g., actual resource allocated to, start time).

5.2.4 Recovery Action

The final consideration in regard to exception handling is what action is undertaken to mitigate the effects of the exception that has been detected. There are three alternate courses of action that can be taken to deal with the exception:

1. **No action (NIL)** – Do nothing
2. **Rollback (RBK)** – Rollback the effects of the exception by undoing the preceding work item(s) based on the state changes recorded in an execution log. To specify a rollback action, the point in the process (i.e., the task) to which the process should be undone needs be stated. By default this is just the current work item
3. **Compensate (COM)** – Compensate for the effects of the exception by executing an alternative task or subprocess. To specify a compensation action in a process definition, the corresponding compensation task(s) or subprocess needs to be identified

Table 5.1 Possible state transitions for work items during exception handling

Abbrev	State change	Description
OCO	Continue-offer	The work item has been offered to one or more resources and there is no change in its state as a consequence of the exception
ORO	Reoffer	The work item has been offered to one or more resources and as a consequence of the exception, these offers are withdrawn and the work item is once again offered to one or more resources (these resources may not necessarily be the same as those to which it was offered previously)
OFF	Force-fail-o	The work item has been offered to one or more resources, these offers are withdrawn, and the state of the work item is changed to failed. No subsequent work items on this path are triggered
OFC	Force-complete-o	The work item has been offered to one or more resources, these offers are withdrawn, and the state of the work item is changed to completed. All subsequent work items are triggered
ACA	Continue-allocation	The work item has been allocated to a specific resource that will execute it at some future time and there is no change in its state as a consequence of the exception
ARA	Reallocate	The work item has been allocated to a resource, this allocation is withdrawn and the work item is allocated to a different resource
ARO	Reoffer-a	The work item has been allocated to a resource, this allocation is withdrawn, and the work item is offered to one or more resources (this group may not necessarily include the resource to which it was previously allocated)
AFF	Force-fail-a	The work item has been allocated to a resource, this allocation is withdrawn, and the state of the work item is changed to failed. No subsequent work items are triggered
AFC	Force-complete-a	The work item has been allocated to a resource, this allocation is withdrawn, and the state of the work item is changed to completed. All subsequent work items are triggered
SCE	Continue-execution	The work item has been started and there is no change in its state as a consequence of the exception
SRS	Restart	The work item has been started, progress on the current execution instance is halted, and the work item is restarted from the beginning by the same resource that was executing it previously
SRA	Reallocate-s	The work item has been started, progress on the current execution instance is halted, and the work item is reallocated to a different resource for later execution
SRO	Reoffer-s	The work item has been started, progress on the current execution instance is halted, and it is offered to one or more resources (this group may not necessarily include the resource that was executing it)
SFF	Force-fail	The work item is being executed, any further progress on it is halted, and its state is changed to failed. No subsequent work items are triggered
SFC	Force-complete	The work item is being executed, further progress on it is halted, and its state is changed to completed. All subsequent work items are triggered

5.2.5 Characterizing Exception Handling Strategies

Having described the various factors that are relevant when handling a specific exception, it is now possible to provide a precise means of categorizing an exception handling strategy in the form of a pattern that succinctly describes the form of recovery that will be attempted. Exception patterns take the form of four-tuples comprising the following elements:

- The type of exception detected
- How the task on which the exception is based should be handled
- How the case and other related cases in the process model in which the exception is raised should be handled
- What recovery action (if any) is to be undertaken

For example, the pattern WIF-SFF-CWC-COM specified for a work item failure exception for the *Investigate Customer Complaint* task in the *Respond to Billing Dispute* process indicates that if failure of a work item corresponding to the *Investigate Customer Complaint* task is detected when it has been started, then the work item should be terminated, have its state changed to failed and the nominated compensation task should be invoked. No action should be taken with other work items in the same case. It is important to note that this pattern only applies to instances of the work item that fail once started; it does not apply to instances in the offered or allocated states (which if required should have distinct patterns nominated for them). From the various alternatives identified for each of these elements in Sect. 5.2.2–5.2.4, there are *135 possible patterns* that can be conceived. Not all patterns apply to a given exception type, however, and Table 5.2 identifies those that apply to each of the exception types identified in Sect. 5.2.1.

5.3 YAWLeX: A Graphical Exception Handling Language

YAWLeX is an exception handling language that allows resolution strategies defined in terms of the exception patterns to be described using a graphical notation. YAWLeX is intended to be both compact and generic in nature. It is comprised of a set of primitives that support various aspects of exception handling. The various primitives that make up the language are illustrated in Fig. 5.2. These primitives can be assembled into sequences of actions that define exception handling strategies. These sequences may theoretically also contain standard YAWL constructs, although that capability is not illustrated here.

The interlinkage of exception handling strategies based on these primitives and the overall process model is illustrated in Fig. 5.3. A clear distinction is drawn between the process model and the exception handling strategies. This is based on the premise that the process model should depict the normal sequence of activities

Table 5.2 Exceptions patterns support by exception type

Work item failure	Work item deadline	Resource unavailable	External trigger	Constraint violation
WIF-OFF-CWC-NIL	DEX-OCO-CWC-NIL	RUN-ORO-CWC-NIL	EXT-OCO-CWC-NIL	CVI-SCE-CWC-NIL
WIF-OFF-CWC-COM	DEX-ORO-CWC-NIL	RUN-OFF-CWC-NIL	EXT-OFF-CWC-NIL	CVI-SRS-CWC-NIL
WIF-OFC-CWC-NIL	DEX-OFF-CWC-NIL	RUN-OFF-RCC-NIL	EXT-OFF-RCC-NIL	CVI-SRS-CWC-COM
WIF-OFC-CWC-COM	DEX-OFF-RCC-NIL	RUN-OFC-CWC-NIL	EXT-OFC-CWC-NIL	CVI-SRS-CWC-RBK
WIF-AFF-CWC-NIL	DEX-OFC-CWC-NIL	RUN-ARO-CWC-NIL	EXT-ACA-CWC-NIL	CVI-SFF-CWC-NIL
WIF-AFF-CWC-COM	DEX-ACA-CWC-NIL	RUN-ARA-CWC-NIL	EXT-AFF-CWC-NIL	CVI-SFF-CWC-COM
WIF-AFC-CWC-NIL	DEX-ARA-CWC-NIL	RUN-AFF-CWC-NIL	EXT-AFF-RCC-NIL	CVI-SFF-CWC-RBK
WIF-AFC-CWC-COM	DEX-ARO-CWC-NIL	RUN-AFF-RCC-NIL	EXT-AFC-CWC-NIL	CVI-SFF-RCC-NIL
WIF-SRS-CWC-NIL	DEX-AFF-CWC-NIL	RUN-AFC-CWC-NIL	EXT-SCE-CWC-NIL	CVI-SFF-RCC-COM
WIF-SRS-CWC-COM	DEX-AFF-RCC-NIL	RUN-SRA-CWC-NIL	EXT-SRS-CWC-NIL	CVI-SFF-RCC-RBK
WIF-SRS-CWC-RBK	DEX-AFC-CWC-NIL	RUN-SRA-CWC-COM	EXT-SRS-CWC-COM	CVI-SFF-RAC-NIL
WIF-SFF-CWC-NIL	DEX-SCE-CWC-NIL	RUN-SRA-CWC-RBK	EXT-SRS-CWC-RBK	CVI-SFC-CWC-NIL
WIF-SFF-CWC-COM	DEX-SCE-CWC-COM	RUN-SRO-CWC-NIL	EXT-SFF-CWC-NIL	CVI-SFC-CWC-COM
WIF-SFF-CWC-RBK	DEX-SRS-CWC-NIL	RUN-SRO-CWC-COM	EXT-SFF-CWC-COM	
WIF-SFF-RCC-NIL	DEX-SRS-CWC-COM	RUN-SRO-CWC-RBK	EXT-SFF-CWC-RBK	
WIF-SFF-RCC-COM	DEX-SRS-CWC-RBK	RUN-SFF-CWC-NIL	EXT-SFF-RCC-NIL	
WIF-SFF-RCC-RBK	DEX-SRA-CWC-NIL	RUN-SFF-CWC-COM	EXT-SFF-RCC-COM	
WIF-SFC-CWC-NIL	DEX-SRA-CWC-COM	RUN-SFF-CWC-RBK	EXT-SFF-RCC-RBK	
WIF-SFC-CWC-COM	DEX-SRA-CWC-RBK	RUN-SFF-RCC-NIL	EXT-SFF-RAC-NIL	
WIF-SFC-CWC-RBK	DEX-SRO-CWC-NIL	RUN-SFF-RCC-COM	EXT-SFC-CWC-NIL	
	DEX-SRO-CWC-COM	RUN-SFF-RCC-RBK	EXT-SFC-CWC-COM	
	DEX-SRO-CWC-RBK	RUN-SFF-RAC-NIL		
	DEX-SFF-CWC-NIL	RUN-SFC-CWC-NIL		
	DEX-SFF-CWC-COM	RUN-SFC-CWC-COM		
	DEX-SFF-CWC-RBK			
	DEX-SFF-RCC-NIL			
	DEX-SFF-RCC-COM			
	DEX-SFF-RCC-RBK			
	DEX-SFC-CWC-NIL			
	DEX-SFC-CWC-COM			

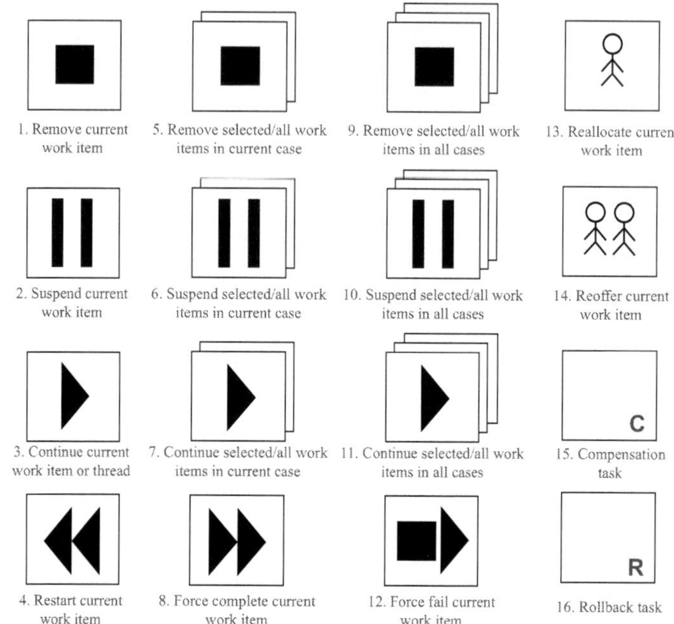

Fig. 5.2 Exception handling primitives

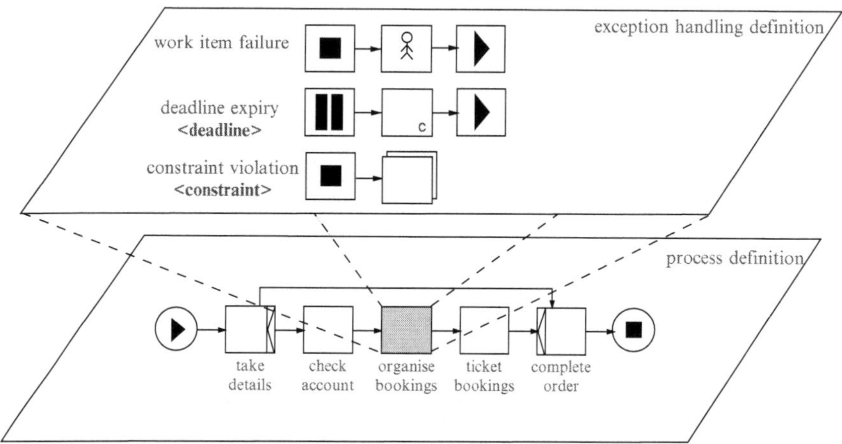

Fig. 5.3 Exception handling in relation to workflow processes

associated with a business process and should aim to present these activities precisely without becoming overburdened by excessive consideration of unexpected events that might arise during execution.

Exception handling strategies are able to be bound to one of five distinct workflow constructs: individual tasks, a scope (i.e., a group of tasks), a block, a process

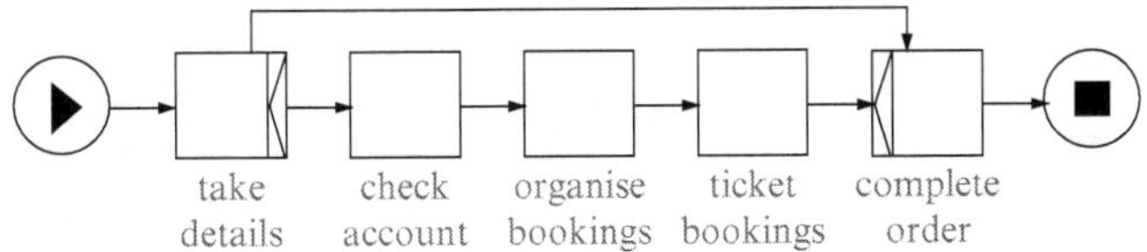

Fig. 5.4 Travel bookings process

(i.e., all of the tasks in a process model), and a workflow (i.e., all of the process models in a given workflow environment). The binding is specific to one particular type of exception, for example, work item failure or constraint violation. It may also be further specialized using conditions based on elements from the data perspective, for example, there may be two exception handling strategies for a task, one for work items concerned with financial limits below $5,000, and the other with limits above that figure. Exception handling strategies defined for more specific constructs take precedence to those defined at a higher level, for example, where a task has a work item failure exception strategy defined and there is also a strategy defined at the process-level for the same exception type, then the task-level definition is utilized should it experience such an exception.

To illustrate the application of these concepts, we present an example based on a travel booking process illustrated in Fig. 5.4 using the YAWL process modeling notation. In this process, booking requests are taken from customers for travel requirements. Each customer has an account that is checked for available credit, and assuming that this exists, their travel requirements are booked and ticketed before their order is completed. In the event that they have insufficient credit, their order is immediately finalized. In either case, the customer is advised of the outcome to their travel request.

Figure 5.5a illustrates two alternate exception handling strategies for the *check account* work item. The first of these is used when the account is in terms. It involves suspending the work item, advancing to the next work item, and starting it. In summary, the check account work item is skipped and control is returned to the process at the commencement of the next work item. For situations where the account is overdue, the current work item is suspended, the execution point is rewound to the beginning of the work item, and it is recommenced.

Figure 5.5b shows the exception handling strategy for the *organize bookings* work item where the deadline for its completion is exceeded. In general, the aim is to handle travel requests within 48 hours of them being received. Where this deadline is not met, recovery involves suspending the current work item, reassigning it to another resource, running a compensation task that determines if the booking can be resolved within 24 hours (and if not applies a small credit to the account), then the *organize bookings* work item is restarted with the new resource.

Figure 5.5c illustrates the resource unavailable handling strategy. Where the required resource is a data resource, this involves stopping the current work item, going back to its beginning, and restarting it. This strategy is bound to the process model, that is, by default, it applies to all work items. In the event where the

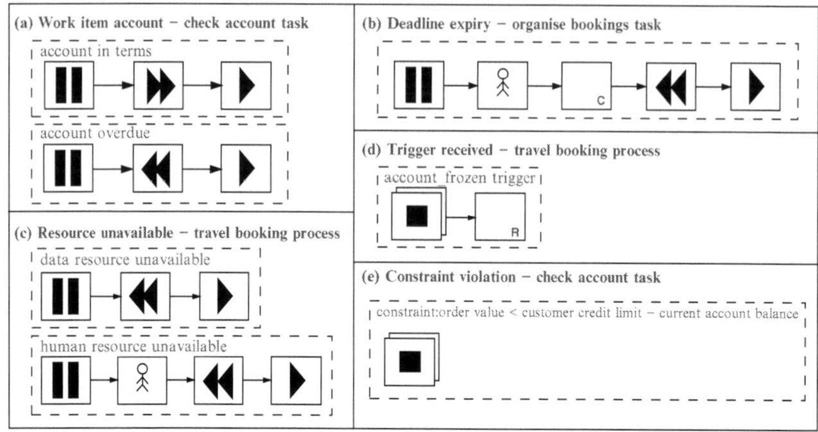

Fig. 5.5 Exception handling strategies – travel booking process

unavailable resource is a human resource (i.e., the person undertaking the work item), the recovery action is also shown in Fig. 5.5c and involves suspending the work item, reassigning it to another person and then restarting it from the beginning.

Figure 5.5d indicates the exception handling strategy when an *account frozen* trigger is received by any of the tasks in the current process. In this situation, the recovery action is to stop the current work item and all other work items in the process and to undertake a rollback action, which involves undoing all changes that have occurred right the way back to the beginning of the case. In other words, any work that has been undertaken on dispatching goods to the customer is completely undone.

Finally, Fig. 5.5e illustrates the recovery action that is taken when the `travel order value` constraint is exceeded for the *organize bookings* task. In this case, the strategy is simply to stop the work item and all other work items (if any) in the process.

5.4 Exception Handling in YAWL

From the discussion above, it can be seen that exception handling requirements form a specific domain drawn from the more general topic areas of flexibility and adaptation. It makes sense then that a solution for exception handling would extend from ways of making workflows more flexible. Chapter 4 described the YAWL environment's solution to the issues of flexibility and adaptation using the Worklet Service. An extension of the Worklet Service, called the *Worklet Exception Service*, extends the capabilities of the Worklet Service to provide dynamic exception handling with corrective and compensatory actions. This section provides a conceptual overview and a description of the general framework of the Exception Service; a detailed description of how it is implemented and deployed (including examples) can be found in Chap. 11.

5.4.1 The Worklet Exception Service

The Worklet Exception Service uses the same repertoire and Ripple-Down Rule (RDR) set framework as the Worklet Selection Service (see Chap. 4 for a detailed description of the framework), and can handle both expected and unexpected exceptions when they occur at runtime. For each anticipated exception (an event that is not expected to occur in most instances and so is not defined as part of the "main" business logic of the specification), a set of repertoire-member exception handling processes, known as *exlets* (which are defined in the YAWLeX language using a graphical editor and may include worklets as compensation processes), may be defined for handling the event, to be dynamically incorporated into a running workflow instance on an as-needed basis. That is, for any exception that may occur at the task, case, or specification level, a repertoire of exlets may be defined and maintained, with the most appropriate one system-selected at runtime based on the context of the case and the type of exception that has occurred. Further, worklets that are invoked as compensation processes as part of an exception handling process are constructed *in exactly the same way* as those created to support flexibility, which in turn are constructed in the same way as ordinary, static YAWL process models.

In the occurrence of an unanticipated exception (i.e., an event for which a handling exlet has not yet been defined), either an existing exlet can be manually selected (reused) from the repertoire, one may be adapted on the fly to handle the immediate situation, or a new exlet constructed and immediately deployed while the parent workflow instance is still active, so in each case allowing execution of the process that raised the exception to take the necessary action and either continue unhindered, or, if specified in the exception handler, to terminate, as required. Crucially, the method used to handle the new exception and a record of its context are captured by the system and immediately become an implicit part of the parent process model, and so a history of the event and the method used to handle it is recorded for future instantiations, which provides for continuous evolution of the process while avoiding any need to modify the original process definition.

The Exception Service has been designed so that the enactment engine, besides providing notifications at certain points in the life-cycle of a process instance, needs no knowledge of an exception occurring, or of any invocation of handling processes – all exception checking and handling is provided by the service.

Although the Exception Service uses the same repertoire and dynamic rules approach as the Selection Service, and extends from the Selection Service and so is built on the same framework, there are, however, two fundamental differences between the two subservices. First, where the Selection Service selects a *worklet* as the result of satisfying a rule in a rule set, the result of an Exception Service rule being satisfied is an *exlet*. Second, while the Selection Service is invoked for certain

Table 5.3 Summary of service actions

Cause	Interface	Selection	Action returned
Work item enabled	B	Case and item context data	Worklet
Internal exception	X	Exception type and case and item context data	Exlet
External exception	–	Exception type and case and item context data	Exlet

nominated tasks in a process, the Exception Service, when enabled, is invoked for *every* case and task executed by the enactment engine, and will detect and handle up to ten different kinds of process exceptions (those exception types are described in Sect. 5.4.3). Table 5.3 summarizes the differences between the two subservices (the interfaces are described in the next section).

To construct an exlet, a process designer may choose from various actions (such as canceling, suspending, completing, failing, and restarting) and apply them at a work item, case, and/or specification level. And, as the exlets can include compensatory worklets, the original parent process model only needs to reveal the actual business logic for the process, while the repertoire of exlets grows as new exceptions arise or different ways of handling exceptions are formulated.

An extensible repertoire of exlets is maintained by the service for each type of potential exception within each workflow specification. Each time the service is notified of an exception event, either actual or potential (i.e., a constraint check), the service first determines whether an exception has in fact occurred, and if so, where a rule tree for that exception type has been defined, makes a choice from the repertoire based on the type of exception and the data attributes and values associated with the work item/case, using a set of rules to select the most appropriate exlet to execute (cf. Chaps. 4 and 11).

If an exlet that is executed by the Exception Service contains a compensation action (i.e., a worklet to be executed as a compensatory process), then as for a worklet launched by the Selection Service, it is run as a separate case in the enactment engine, so that from an engine perspective, the worklet and its "parent" (i.e., the process that invoked the exception) are two distinct, unrelated cases. Figure 5.6 shows the relationship between a "parent" process, an exlet repertoire, and a compensatory worklet, using an *Organize Concert* process as an example. As a compensatory worklet is launched as a separate case, it may have, in turn, its own worklet/exlet repertoire, so that a hierarchy of executing worklets may sometimes exist.

Any number of exlets can form the repertoire of an individual task or case. An exlet may be a member of one or more repertoires – that is, it may be reused for several distinct tasks or cases within and across process specifications. Like the selection service, the exception handling repertoire for a task or case can be added

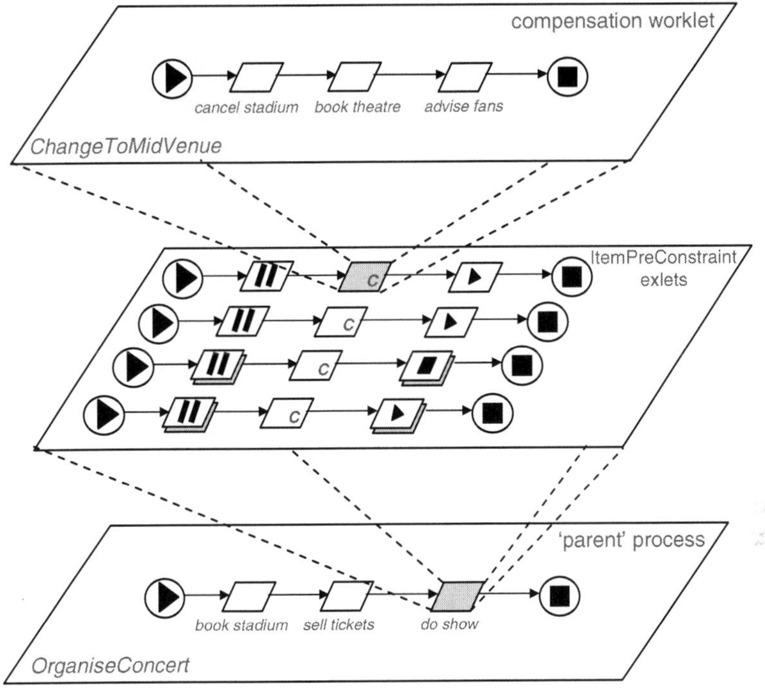

Fig. 5.6 Process – exlet – worklet hierarchy

to at any time, as can the rules base used, including while the parent process is executing.

The Selection and Exception subservices can be used in combination within particular case instances to achieve dynamic flexibility *and* exception handling simultaneously. The Worklet Service is extremely adaptable and multifaceted, and allows a designer to provide tailor-made solutions to runtime process exceptions and requirements for flexibility.

5.4.2 Architectural Overview

This section provides a brief overview of the technical attributes and structure of the Worklet Service (for a much more detailed treatment, see Chap. 11). The service is constructed as a web service and so consists of a number of J2EE classes and servlet pages, organized in a series of packages, for deployment via a servlet container (such as *Apache Tomcat*).

The external architecture of the Worklet Service is shown in Fig. 5.7. The entities "Worklet Specs," "Rules," and "Logs" comprise the *Worklet Repository*. The service uses the repository to store rule sets, compensatory worklet specifications

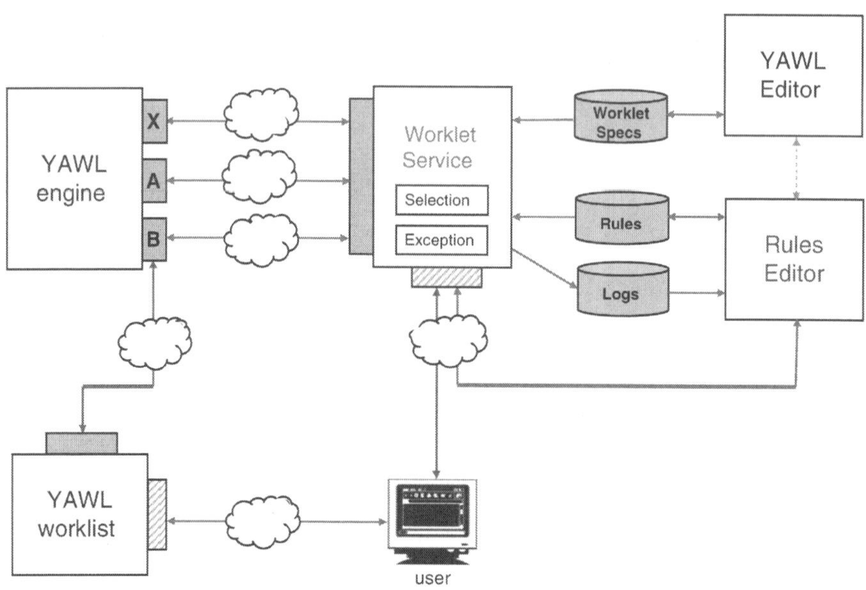

Fig. 5.7 External architecture of the worklet service

for uploading to the engine, and generated process and audit logs. The YAWL Editor is used to create new worklet specifications, and may be invoked from the Rules Editor, which is used to create new or augment existing rule sets, making use of certain selection logs to do so, and to graphically define exlets, and may communicate with the Worklet Service via a JSP/Servlet interface to override worklet/exlet selections following rule set additions. The service also provides servlet pages that allow users to directly communicate with the service to raise external exceptions and to create and carry out administration tasks.

Any YAWL specification may have an associated rule set. The rule set for each specification is stored as XML data in a disk file that has the same name as the specification's identifier, but with an ".xrs" extension (signifying an "XML Rule Set"). Each YAWL specification has a single rule-set file for all selection and exception rules defined for it. All rule set files are stored in the *rules* folder of the Worklet Repository.

To enable the Worklet Service to serve the YAWL workflow enactment engine, a number of events and methods must be provided by an interface between them, some originating from the service side and others from the engine side. Some require a response, while others do not require any acknowledgment (such as event notifications that do not necessarily require action).

The first four methods described for the selection interface in Chap. 4 are also used by the exception handling procedure. In addition, the exception interface requires a further set of events and methods. The enactment engine must notify the service for the following events:

- *Check constraints events*, which notify the service at the launch and completion of each process instance, and at the enabling and completion of each work item – thus, four unique constraint type events are required. The service uses these events to check whether constraint rules have been violated, using available contextual data, and if so, launch the corresponding exlet for that constraint type. Therefore, the constraint event types do not notify of exception *per se*, but are checkpoints that allow the service to determine if a constraint violation has occurred.
- A *time-out event*, which would notify the service when a deadline has been reached, allowing the service to take the appropriate action, if necessary.
- *Unavailable resource*, *item abort*, and *constraint violation during execution* events. For these event types, the enactment engine must determine if and when these exceptions occur, as opposed to those above where the determination rests with the Worklet Service.
- *Case canceled event*, required so that the service can take the appropriate action if a worklet instance, or a parent instance for which a worklet is currently executing as a compensation, is canceled. Note that a case canceled event is different from a case completed event described in Chap. 4.

The following methods are specifically required to be available to the interface for the exception handling service (existing methods of other interfaces are also used):

- *Suspend work item*: To pause or prevent the execution of a work item, until such time as it is continued (unsuspended). Through this method, entire process instances may also be suspended, by suspending each "active" work item of the process instance as a list.
- *Continue work item*: To unsuspend a previously suspended work item, or unsuspend each suspended work item for a previously suspended process instance.
- *Update work item data*: For those situations where a worklet has been run as a compensation process for a work item that has generated an exception, this method would enable any data gathered during the compensation execution to be passed back to the original work item.
- *Update case data*: As above, but for case-level exceptions as opposed to work item-level exceptions.
- *Restart, cancel, force-complete, fail work item*: To perform an action as specified in an exlet definition.

Figure 5.8 shows the interface requirements for the exception handling procedures of the Worklet Service (see also Fig. 4.4).

5.4.3 Exception Handling Types and Primitives

To recap, there are several different types of exception that may occur during the life of a YAWL process instance. At various points, constraints may be checked

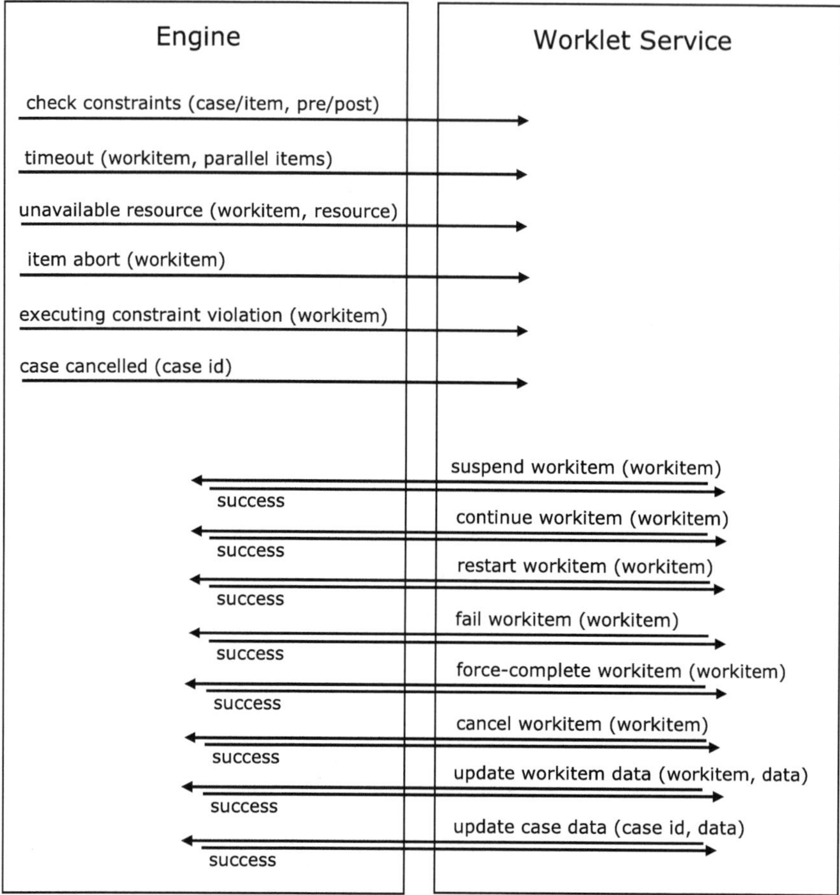

Fig. 5.8 Required exception interface

to determine if an exception has occurred. Also, certain tasks may have associated deadlines that have passed, a task being performed by an application or process may fail, or an exception may be raised externally. With all of these events, the context of the current instance is important in determining the appropriate action to take. That action may include suspending, canceling, restarting, or completing a task and/or case instance, as well as performing some compensatory activities. Often a series of such actions will be needed to successfully handle an exception. Also, action may need to be taken at the task level, the case instance level, or may even affect all of the current case instances for a specification.

In the Worklet Service, all of the necessary actions are captured in an exception-handling process (exlet). As mentioned previously, an exlet may contain a series of steps incorporating the actions described earlier, and may include one or more worklets to perform compensatory activities as part of an exception-handling process.

Each exception type may have an associated RDR rule tree in the rule set for a specification – that is, an RDR set for a specification may contain up to 11 rule trees, one for each exception type and one for selections. So, depending on the exception type that has occurred and the context of the case instance, an appropriate exlet can be chosen and executed to handle the exception.

If there are no rules defined for a certain exception type in the rule set for a specification, the exception event is simply ignored by the service. Thus rules are needed only for those exception events that are desired to be handled for a particular task and/or specification.

An invoked exlet may suspend its parent case instance when it is activated, but there may be some occasions when the exlet can (or needs to) operate in parallel to its parent. For example, it is undesirable to suspend an entire *Travel Bookings* process when a request to upgrade a single room booking has been received – it is far easier to handle the request as a (local) externally raised exception while allowing the parent process to continue. Exlets may be defined to accommodate whatever action is desired to be taken.

5.4.3.1 Exception Types

There are ten types of exception that are defined by the service for handling, as detailed below.

Constraint Types

Constraints are rules that are applied to a work item or case immediately before and after execution of that work item or case. Thus, there are four types of constraint exception handled by the service:

- *CasePreConstraint* – Case-level preconstraint rules are checked when each case instance begins execution
- *ItemPreConstraint* – Item-level preconstraint rules are checked when each work-item in a case becomes enabled (i.e., ready to be checked out or executed)
- *ItemPostConstraint* – Item-level postconstraint rules are checked when each work item reaches a finalized status (e.g., completed, canceled, failed)
- *CasePostConstraint* – Case-level postconstraint rules are checked when a case completes

The service receives notification from the workflow engine when each of these life-cycle events are reached within each case and work item instance, then checks the rule set associated with the specification to determine, first, if there are any rules of that exception type defined for the specification, and if so, if any of the rules evaluate to true using the contextual data of the case or work item. If the rule set finds a rule that evaluates to true for the exception type and data, an associated exlet is selected and invoked.

TimeOut

A timeout event occurs when a work item has an enabled timer and the deadline set for that timer is reached. In this case, the workflow engine notifies the service of the timeout event, and passes to the service a reference to the work item and each of the other work items that were running in parallel with it. Therefore, separate timeout rules may be defined (or indeed not defined) for each of the work items affected by the timeout, including the actual timed out work item itself. Thus, separate actions may be taken for each affected work item individually.

Externally Triggered Types

Externally triggered exceptions occur, not through the case's data parameters or via an engine initiated event, but rather because of the occurrence of an event in the external environment, outside of the process instance, that has an effect on the continuing execution of the process. Thus, these events are triggered by a user or administrator. Depending on the actual event and the context of the case or work item, a particular exlet will be invoked. There are two types of external exceptions: *CaseExternalTrigger* (for case-level events) and *ItemExternalTrigger* (for item-level events).

ItemAbort

An ItemAbort event occurs when a work item being handled by an external program (as opposed to a human user) reports that the program has aborted before completion.

ResourceUnavailable

This event is triggered by the Resource Service (cf. Chap. 10) when an attempt has been made to allocate a work item to a resource and the resource reports that it is unable to accept the allocation or the allocation cannot otherwise proceed.

ConstraintViolation

This event occurs when a data constraint has been violated for a work item *during* its execution (as opposed to pre- or post-execution).

Note that the ItemAbort and ConstraintViolation exception types are supported by the framework, but are not yet operational, although it is envisaged that they will be incorporated into YAWL 2.1.

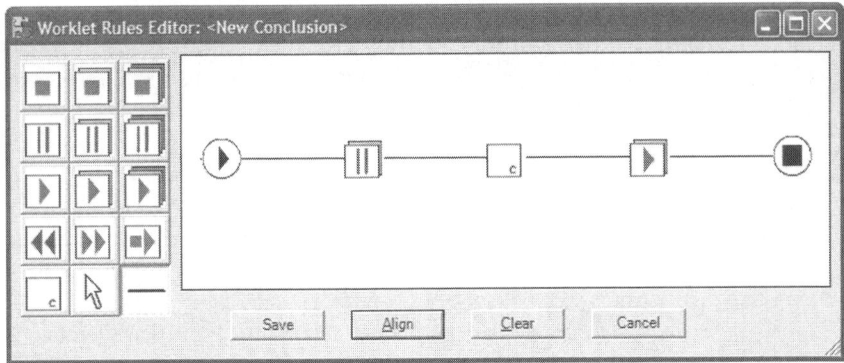

Fig. 5.9 Example handler process in the rules editor

5.4.3.2 Exception Handling Primitives

When any of the above exception event notifications occur, an appropriate exlet for that event, if defined, will be invoked. Each exlet may contain any number of steps, or *primitives*, and is defined graphically using a Rules Editor (described in detail in Chap. 11).

An example of an exlet definition in the Rules Editor can be seen in Fig. 5.9. On the left of the graphical editor is the set of primitives that may be used to construct an exlet. The available primitives (reading left-to-right, top-to-bottom) are the following:

- *Remove Work Item*: Removes (or cancels) the work item, execution ends, and the work item is marked with a status of canceled. No further execution occurs on the process path that contains the work item.
- *Remove Case*: Removes the case. Case execution ends.
- *Remove All Cases*: Removes all case instances for the specification in which the task of which the work item is an instance is defined, or of which the case is an instance.
- *Suspend Work Item*: Suspends (or pauses) execution of a work item, until it is continued, restarted, canceled, failed, or completed, or the case that contains the work item is canceled or completed.
- *Suspend Case*: Suspends all "live" workitems in the current case instance (a live work item has a status of fired, enabled, or executing), effectively suspending execution of the entire case.
- *Suspend All Cases*: Suspends all "live" workitems in all of the currently executing instances of the specification in which the task of which the work item is an instance is defined, effectively suspending all running cases of the specification.
- *Continue Work Item*: Unsuspends (or continues) execution of the previously suspended work item.
- *Continue Case*: Unsuspends execution of all previously suspended work items for the case, effectively continuing case execution.

- *Continue All Cases*: Unsuspends execution of all work items previously suspended for all cases of the specification in which the task of which the work item is an instance is defined or of which the case is an instance, effectively continuing all previously suspended cases of the specification.
- *Restart Work Item*: Rewinds work item execution back to its start. Resets the work item's data values to those it had when it began execution.
- *Force Complete Work Item*: Completes a "live" work item. Execution of the work-item ends, and the work item is marked with a status of *ForcedComplete*, which is regarded as a successful completion, rather than a cancelation or failure. Execution proceeds to the next work item on the process path.
- *Force Fail Work Item*: Fails a "live" work item. Execution of the work item ends, and the work item is marked with a status of *Failed*, which is regarded as an unsuccessful completion, but not as a cancelation – execution proceeds to the next work item on the process path.
- *Compensate*: Run one or more compensatory processes (i.e., worklets). Depending on previous primitives, the worklets may execute simultaneously to the parent case, or execute while the parent is suspended.

A number of compensatory worklets may be executed consecutively by adding a sequence of compensation primitives to an exlet. Optionally, a particular compensation primitive may contain an *array* of worklets – when multiple worklets are defined for a compensation primitive via the Rules Editor, they are launched concurrently as a composite compensatory action when the exlet is executed.

As mentioned in Chap. 4 with regards to the Selection Service, worklets can in turn invoke child worklets to any depth – this also applies for worklets that are executed as compensatory processes within exlets. The primitives "Suspend All Cases," "Continue All Cases," and "Remove All Cases" may be flagged when being added to an exlet definition via an option in the Rules Editor so that their action is restricted to ancestor cases only. Ancestor cases are those in a hierarchy of worklets back to the original parent case – that is, where a process invokes an exlet that invokes a compensatory worklet, which in turn invokes another worklet and/or an exlet, and so on (see the "TimeOut" example in Chap. 11). Also, the "Continue" primitives are applied only to those corresponding work items and cases that were previously suspended by the same exlet.

Execution moves to the next primitive in the exlet when all worklets launched from a compensation primitive have completed.

In the same manner as the Selection Service, the Exception Service also supports data mapping from a case to a compensatory worklet and back again. For example, if a certain variable has a value that prevents a case instance from continuing, a worklet can be run as a compensation, during which a new value can be assigned to the variable and that new value mapped back to the parent case, so that it may continue execution.

Referring back to Fig. 5.6, the center tier shows the exlets defined for ItemPreConstraint violations. As mentioned earlier, there may actually be up to eight different members of this tier. Also, each exlet may refer to a different set of compensatory processes, or worklets, and so at any point there may be several worklets operating on the upper tier for a particular case instance.

5.4.4 Local and Global Perspectives on Exception Handling

The Worklet Exception Service operates from both local and global perspectives. It is important at this point to clearly distinguish between them.

The Worklet Service deals with both flexibility and exception handling locally, in that additions to a repertoire are instigated by the person charged with performing the task (or an administrator for non-human resourced tasks). With each addition, a rule node is added, which becomes, in effect, a localized exception to the more general parent rule – the use of the term "exception" here refers to an "exception to the rule," rather than that which refers to a control-flow deviation during process instance execution. This locality of change is an essential benefit of the approach. So, in this regard, deviations referring to both flexibility and exception handling are created locally and their effects are propagated upwards.

In addition, the Worklet Service also provides the ability to handle exceptions at a global level – that is, at case or specification level. As discussed in Sect. 5.4.3.1, case level exceptions based on pre- and post-constraints can occur when a case starts and completes. As well, externally triggered exceptions can be raised at the case level at any time during an instance's life cycle. Thus, these kinds of exceptions are raised at a global level and their effects are propagated downwards.

Further, any exception, no matter whether it was raised locally or globally, can be applied to affect the task, case, and/or specification levels, as the exception and the context of the instance demand.

Thus, the Worklet Service provides support for deviations that occur at both local and global levels; they can be triggered automatically (via RDRs) and/or manually (via a selection from a list when externally triggered); and they are able to apply their effects at any execution level.

Consequently, the scope of the worklet solution provides improved efficiencies over manual handling systems in a vast number of cases. But, importantly, it also provides for manual handling in those cases where it is appropriate.

A critical position of the worklet approach can be briefly summarized thus:

> *If, for a particular case having a certain context, a method of handling an exception at a particular point in the life-cycle of the case has been previously defined and deemed the most suitable method, then it is appropriate to assume that, given another instance of the same case which has* exactly *the same context, that same exception handling method can be correctly applied to that instance.*

There can be only two situations where such a position does not apply: (1) the context of a new case instance is not exactly the same as the one for which the handler was previously defined; or (2) an improved way of handling the exception is conceived. Of course, if an appropriate handler has not been previously defined, and the exception should be handled, then a handler must be defined and added to the repertoire.

There is no scenario in which it would make sense for two case instances with *identical* contexts to have a particular exception handled in two different ways. Here, the benefit of automated selection of the handler over manual selection is twofold: (1) it is vastly more efficient; and (2) manual selection introduces the capacity for human error – that is, if the case contexts are identical, then it is simply at the whim of the user as to which handler is selected. Or, to put it another way, for a decision to be correctly made about the invocation of a certain handler, the decision of whether to choose one handler over another depends directly on the context of the case instance. It cannot be asserted that two case instances would require two different handlers if their contexts were identical. It is precisely the differences in context that define and require differences in exception handling methods.

In the chapter notes, examples are given of approaches to exception handling; generally those approaches depend on manual intervention and selection of handlers. It may be said that those kinds of approaches are *not context-aware*, and thus the decision making is left to a (human) administrator.

The worklet approach *is* context-aware, and so the majority of decisions are system-based, with reference to the context of the case instance. However, within the Worklet Service manual intervention is required when there is no handler defined that matches the context of a particular case instance. This is manifested in two ways: either a handler cannot be matched to the current case context (i.e., a handler has not yet been defined) or a handler has been returned that is correct for the general context but deemed inappropriate because of differences in the context detail. In both cases, a manual interaction involving a user and an administrator is undertaken to define a new handler for that particular case instance.

Manual intervention is also required when externally triggered exceptions occur. In such scenarios, a user is presented with a list of possible exception events, from which they make a choice.

Therefore, while in the majority of instances manual intervention is not required, the Worklet Service also provides support for manual intervention in certain scenarios as required.

5.5 Epilogue

While there have been many approaches to exception handling for workflows, most are built as extensions to an underlying framework that imposes rigidity in process expression and enactment. Such approaches usually involve a reliance on manual intervention and runtime user interactivity. Thus, they are in opposition to the perceived gains offered by workflow systems. Users are required to incorporate "unnatural" actions into their work routines so that their work processes match that of the workflow specification, or take their work off-system until such time as the process model can be made to more closely reflect their actual work practices, which involves complex update, modification, and migration phases.

The overarching benefit of the worklet approach is that it allows a primary (or top-level) process model to be defined *once* – change can thereafter be introduced to the process *without changing the original parent model in any way*. This immediately exhibits many advantages over monolithic approaches. In particular, it solves the problem of updating a (monolithic) process model whenever a change is required, and thus overcomes the issues of downtime, versioning, maintenance, verification, and correctness that plague more traditional approaches.

While many rule-based systems provide a single "fixed" control-flow of alternative steps associated with a particular exception type, the Ripple-Down Rules approach used in the Worklet Service provides the capability to offer many (in fact, unbounded) different control-flow-step sets for any exception instance, and because the repertoire is dynamically extensible, it allows nuance within rule sets for even the smallest of contextual differences. That is, it offers far greater benefits than the traditional rule-based approaches and clear advantages over monolithic approaches.

Exercises

Exercise 1. What are the four main factors that need to be considered when deciding on an appropriate handling strategy for an exception?

Exercise 2. Explain why one YAWL Custom Service (the Worklet Service) is designed to handle both dynamic flexibility and exception handling, rather than separating the functionality into two distinct services.

Exercise 3. Name at least three reasons why is it desirable to separate exception handling processes from those processes where exceptions may occur.

Exercise 4. How are exception handling processes defined in the YAWL environment? How is the appropriate exception handling process chosen and invoked for a particular exception instance?

Exercise 5. What would happen if a worklet, currently executing within a compensation primitive of an exlet, experiences an exception of its own?

Exercise 6. What is the exlet defined in Fig. 5.9 designed to do?

Exercise 7. Give two benefits offered by the Worklet Service's approach to exception handling over more manual approaches.

Exercise 8. Can you think of any runtime exceptions that may occur that cannot be captured by the approach discussed in this chapter?

Chapter Notes

Exception Patterns

The application of a pattern-based approach to defining exception handling strategies and their use for assessing the exception handling capabilities of PAIS was first described in [222]. This paper also proposed the notion of a graphical exception handling language for defining exception handling strategies, a notion that ultimately led to the development of YAWLeX, the language on which the exlet strategies in YAWL were realized. A comprehensive assessment of the exception handling capabilities of a number of leading workflow and case handling systems, business process modeling notations, and business process execution languages is contained in [220].

Worklets/Exlets

The use of the *worklets* paradigm for exception handling was first discussed in [26], and a full description of the conceptual framework of worklets, exlets, and dynamic exception handling was further detailed in [25]. A full exploration of the worklet approach, including a complete formalization and exemplary studies, can be found in [23].

Other Approaches

The need for reliable, resilient, and consistent workflow operation has long been recognized [101]. Early work in the area [90,268] was essentially a logical continuation of database transaction theory and focussed on developing extensions to the classic ACID transaction model that would be applicable in application areas requiring the use of long duration and more flexible transactions. As the field of workflow technology matured, the applicability of exceptions to this problem was also recognized [226, 244]. In [87], Eder and Liebhart presented the first significant discussion on workflow recovery that incorporated exceptions, and gave the now classic categorization of workflow exceptions into four groups: basic failures, application failures, expected exceptions, and unexpected exceptions. Subsequent research efforts into workflow exceptions have mainly concentrated on the last two of these classes, and on this basis, the field has essentially bifurcated into two research areas. Investigations into expected exceptions have focussed previous work on transactional workflow into mechanisms for introducing exception handling frameworks into workflow systems. Research into unexpected exceptions has established the areas of adaptive workflow and workflow evolution [212].

Although it is not possible to comprehensively survey these research areas in the confines of this chapter, it is worthwhile identifying some of the major contributions in these areas that have influenced subsequent research efforts and have a bearing on this research initiative. Significant attempts to include advanced transactional concepts and exception handling capabilities in workflow systems include WAMO [86], which provided the ability to specify transactional properties for tasks that identified how failures should be dealt with; ConTracts [210], which proposed a coordinated, nested transaction model for workflow execution allowing for forward, backward, and partial recovery in the event of failure; and Exotica [34], which provided a mechanism for incorporating Sagas and Flexible transactions in the commercial FlowMark workflow product. OPERA [113, 114] was one of the first initiatives to incorporate language primitives for exception handling into a workflow system, and it also allowed exception handling strategies to be modeled in the same notation as that used for representing workflow processes. TREX [243] proposed a transaction model that involves treating all types of workflow failures as exceptions. A series of exception types were delineated and the exception handler utilized in a given situation was determined by a combination of the task and the exception experienced. WIDE [55] developed a comprehensive language – Chimera-Exc – for specifying exception handling strategies in the form of Event-Condition-Action (ECA) rules.

Other important contributions include the concepts of compensation spheres and atomicity spheres and their applicability to workflow systems [149], a proposal for independent models for the specification of workflow and transaction properties using generic rules to integrate the workflow specification and the atomicity specification into a single model based on Petri Nets [76], modeling workflow systems as a set of reified objects with associated constraints and conceptualizing exceptions as violations of those constraints that are capable of being detected and managed [47], a high-level framework for handling exceptions in web-based workflow applications together with an implementation based on extensions to WebML [48] and the identification of the pivot, retriable, and compensation transaction concepts widely used in subsequent research [159].

Identifying potential exceptions and suitable handling strategies is a significant problem for large, complex workflows and recent attempts [112, 123] to address this have centered on mining previous execution history to gain an understanding of past exception occurrences and using this knowledge (either at design or runtime) in order to determine a suitable handling strategy. Klein et al. [134] propose a knowledge-based solution to the issue based on the establishment of a shared, generic, and reusable taxonomy of exceptions. Luo et al. [153] use a case-based reasoning approach to match exception occurrences with suitable handling strategies.

Until recently, the area of unexpected exceptions has mainly been investigated in the context of adaptive or evolutionary workflow [212], which center on dynamic change of the process model. A detailed review of this area is beyond the scope of this chapter; however, an initiative that offers the potential to address both expected and unexpected exceptions simultaneously is ADOME-WFMS [58], which provides an adaptive workflow execution model in which exception handlers are specified generically using ECA rules, providing the opportunity for reuse in multiple scenarios and user-guided adaptation where they are not found to be suitable.

Chapter 6
Declarative Workflow

Maja Pesic, Helen Schonenberg, and Wil van der Aalst

6.1 Introduction

During the design of any information system, it is important to balance between *flexibility* and *support*. This is of particular importance when designing process-aware information systems. On the one hand, users want to have support from the system to conduct their daily activities in a more efficient and effective manner. On the other hand, the same users want to have flexibility, that is, the freedom to do whatever they want and without being "bothered" by the system. Sometimes it is impossible to provide both flexibility and support because of conflicting requirements. The continuous struggle between flexibility and support is illustrated by Fig. 6.1.

The right-hand-side of Fig. 6.1 shows the part of the spectrum covered by classical *workflow management systems*. These systems focus on processes that are repeatedly executed in some predefined manner and are driven by *procedural languages*. Note that in procedural workflow models there may be alternative paths controlled by (X)OR-splits/joins. However, the basic idea is that the completion of one task triggers other tasks. The YAWL nets described in earlier chapters provide such a procedural language. Although the YAWL language is highly expressive, its token-based semantics is most suitable for repetitive processes with tight control.

The left-hand-side of Fig. 6.1 shows the other end of the spectrum. Here processes are less repetitive and the emphasis is on flexibility and user empowerment. Here it is difficult to envision all possible paths and the process is driven by user decisions rather than system decisions. *Groupware* systems (e.g., "enhanced" electronic mail, group conferencing systems, etc.) support such processes and focus on supporting human collaboration and co-decision. Groupware systems do not offer support when it comes to ordering and coordination of tasks. Instead, the high degree of flexibility of these systems allows users to control the ordering and coordination of tasks while executing the process (i.e., "on the fly").

The trade-off between flexibility and support is an important issue in workflow technology. Despite the many interesting theoretical results and the availability of

---

M. Pesic
Eindhoven University of Technology, Eindhoven, the Netherlands
e-mail: m.pesic@tue.nl

A.H.M. ter Hofstede et al. (eds.), *Modern Business Process Automation*,
DOI 10.1007/978-3-642-03121-2_6,

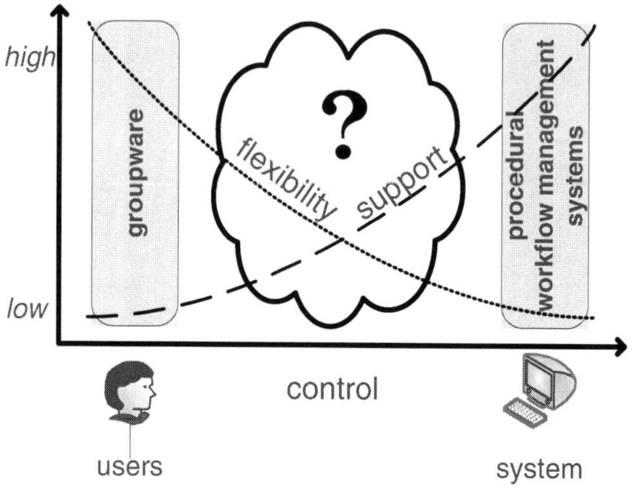

Fig. 6.1 Flexibility vs. support, adapted from [79]

innovative research prototypes, few of the research ideas have been adopted in commercial systems. This is the reason why there is a question mark in the middle of Fig. 6.1.

This chapter presents a completely new way of looking at workflow support. Instead of a procedural language, we describe a *declarative language based on constraints*. The basic idea is that anything is allowed and possible unless explicitly forbidden. To implement this idea we use a temporal logic: *Linear Temporal Logic* (LTL). Although LTL has been around for decades, it has never been made accessible for workflow design. Therefore, we provide a graphical notation and a set of supporting tools. This nicely complements the core engine and design environment of YAWL.

Before elaborating on our constraint-based approach, it is important to stress that this is not a silver bullet. In reality, both flexibility and support are needed when it comes to the computer-facilitated execution of processes and the various paradigms offer different advantages. On the one hand, flexibility is needed for unpredictable processes, where users can quickly react to exceptional situations and execute the process in the most appropriate manner. On the other hand, support is needed when it comes to processes that are repeatedly executed in the same manner, in situations that are too complex for humans to handle and where human mistakes must be minimized. For example, the processing of insurance claims, customer orders, tax declarations, etc. can benefit from a high degree of support because cases are repeatedly executed in a similar manner. Thus, an optimal balance between flexibility and support is needed in order to be able to facilitate processes of various kinds. Moreover, in a large process there may be parts that require more flexibility and parts that require more support. Therefore, procedural and flexible workflows should not exclude each other. Instead, arbitrary decompositions of process models developed

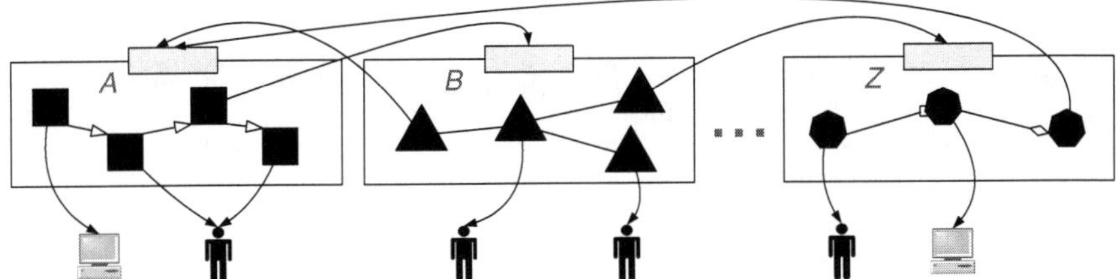

Fig. 6.2 Decomposing processes using various modeling languages

in procedural and declarative languages should be possible. These decompositions can be achieved by the seamless integration of different process engines based on different paradigms. Figure 6.2 illustrates how various processes can be combined in arbitrary decompositions. Instead of being executed as a simple (manual or automatic) unit of work, a task can be decomposed into a process modeled in an arbitrary language. The declarative language based on constraints presented in this chapter can be seen as one of the modeling languages in Fig. 6.2, while the YAWL nets, and worklets (cf. Chap. 11) can be seen as another alternative.

The basic language used by YAWL provides excellent support for procedural workflow modeling as it supports most of the workflow patterns (cf. Chap. 2). However, in this chapter we focus on adding flexibility to workflow models. A workflow model specifies the order in which tasks can be executed, that is, possible execution alternatives. An execution alternative is a unique ordering of tasks. More execution alternatives mean more flexibility. There are several ways to increase the flexibility of workflow management systems:

1. *Flexibility by design* is the ability to specify alternative execution alternatives (task orderings) in the workflow model, such that users can select the most appropriate alternative at runtime for each workflow instance.
2. *Flexibility by underspecification* is the ability to leave parts of a workflow model unspecified. These parts are later specified during the execution of workflow instances. In this way, parts of the execution alternatives are left unspecified in the workflow model, and are specified later during the execution.
3. *Flexibility by change* is the ability to modify a workflow model at runtime, such that one or several of the currently running workflow instances are migrated to the new model. Change enables adding one or more execution alternatives during execution of workflow instances.
4. *Flexibility by deviation* is the ability to deviate at runtime from the execution alternatives specified in the workflow model, without changing the workflow model. Deviation enables users to "ignore" execution alternatives prescribed by the workflow model by executing an alternative not prescribed in the model.

Flexibility of workflow management systems can be increased by using *declarative languages for workflow specification*. This chapter describes how an optimal balance between flexibility and support can be achieved with declarative workflow management systems that use *constraint* workflow models. This chapter shows how

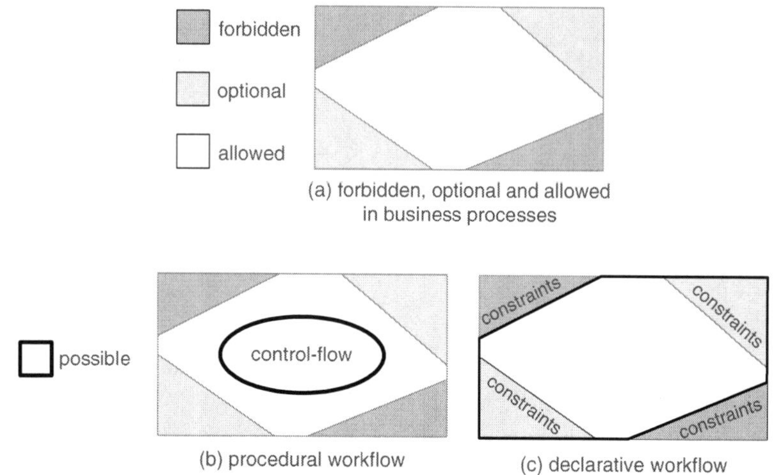

Fig. 6.3 Declarative vs. procedural workflows

the constraint-based approach can provide for all types of flexibility listed in the previous paragraph. A concrete implementation that enables creating decompositions of YAWL (e.g., procedural) and declarative models is presented in Chap. 12 of this book. Figure 6.3 illustrates the difference between procedural and declarative process models.

Starting point for the declarative constraint-based approach is the observation that only three types of "execution alternatives" can exist in a process: (1) *forbidden* alternatives should never occur in practice, (2) *optional* alternatives are allowed, but should be avoided in most of the cases, and (3) *allowed* alternatives can be executed without any concerns. This is illustrated in Fig. 6.3a. Procedural workflow models (e.g., YAWL nets) *explicitly* specify the ordering of tasks, that is, the *control-flow* of a workflow. In other words, during the execution of the model, it will be *possible* to execute a process only as explicitly specified in the control-flow, as shown in Fig. 6.3b. Because of the high level of unpredictability of processes, many allowed and optional executions often cannot be anticipated and explicitly included in the control-flow. Therefore, in traditional systems it is *not possible* to execute a substantial part of all potentially *allowed* alternatives, that is, users are unnecessarily limited in their work and, hence, these systems lack flexibility by design.

Our declarative *constraint-based approach* to workflow models makes it *possible* to execute both *allowed* and *optional* alternatives in processes. Instead of explicitly specifying the procedure, constraint workflow models are declarative: they specify constraints, that is, rules that should be followed during the execution, as shown in Fig. 6.3c. Moreover, there are two types of constraints: (1) *mandatory* constraints focus on the forbidden alternatives, and (2) *optional* constraints specify the optional ones. Constraint-based models are declarative: anything that does not violate mandatory constraints is *possible* during execution. The declarative nature of constraint models allows for a wide range of possible execution alternatives, which enhances flexibility.

The remainder of this chapter is organized as follows. First, our constraint-based language for workflow specification is presented in Sect. 6.2. This section also discusses the possibility to verify constraint-based models and the ability to define alternative languages and templates. Section 6.3 shows how declarative workflows can be enacted, that is, how to create a runtime environment for execution. The possibility to change models while they are being executed (i.e., the so-called dynamic change) is described in Sect. 6.4. Section 6.5 concludes the chapter, followed by exercises and chapter notes.

6.2 Constraint-based Workflow Specification

A constraint-based workflow is defined by the specification of tasks and constraints. In this section, we first show the formal foundation of a constraint specification language. Then we introduce the concept of *constraint templates*, that is, reusable constructs linking graphical notations to temporal logic. These templates also allow for branching, that is, a construct can be associated to a variable number of tasks. Then we introduce *ConDec* as an example language. This language is used in the remainder of this chapter. We conclude this section by showing that such models can be verified.

6.2.1 LTL-based Constraints

Various temporal logics have been proposed in the literature: Computational Tree Logic (CTL), Linear Temporal Logic (LTL), etc. All these logics could be used to specify constraints. In this chapter, we focus on *Linear Temporal logic* LTL.

LTL can be used to define properties of traces, for example, "after every occurrence of X eventually Y should occur." In the context of workflow models, traces correspond to sequences of tasks. Every trace represents an executed alternative where the tasks in the trace occur exactly in the order in which they appear.

Definition 1 (Trace). Let $\mathcal{T}$ denote the universe of task identifiers, that is, the set of all *tasks*. Trace $\sigma \in \mathcal{T}^*$ is a sequence of tasks, where $\mathcal{T}^*$ is the set of all traces composed of zero or more elements of $\mathcal{T}$. We use $\sigma = \langle t_1, t_2, \ldots, t_n \rangle$ to denote a trace consisting of tasks $t_1, t_2, \ldots, t_n$. σ_i denotes the ith element of the trace, that is, $\sigma_i = t_i$. $e \in \sigma$ denotes $\exists_{1 \leq i < |\sigma|}[\sigma_i = e]$. We use $+$ to concatenate two traces into a new trace: $\langle e_1, e_2, \ldots, e_n \rangle + \langle f_1, f_2, \ldots, f_m \rangle = \langle e_1, e_2, \ldots, e_n, f_1, f_2, \ldots, f_m \rangle$.

The concept of traces will be used to explain the semantics of LTL. However, we first give the syntax of LTL. LTL provides the classical logic operators ($\neg$, $\wedge$, $\vee$, $\Rightarrow$, and $\Leftrightarrow$), and uses several temporal operators ($\bigcirc$, $\Diamond$, $\Box$, U, and W) that can be used to specify constraints over the sequencing of workflow tasks. Definition 2 defines the syntax of LTL.

Definition 2 (LTL syntax). Each atomic proposition p is an LTL formula. Using various operators, more complex expressions can be constructed. If Φ and Ψ are LTL formulas, then $\neg\Phi$, $\Phi \vee \Psi$, $\bigcirc\Phi$, and $\Phi U \Psi$ are also LTL formulas. The semantics of these constructs are given in Definition 3. Moreover, the following operators are so-called derived LTL operators. These operators can be expressed in terms of the basic operators.

$$\begin{aligned}
\Phi \wedge \Psi &\equiv \neg(\neg\Phi \vee \neg\Psi)\\
\Phi \Rightarrow \Psi &\equiv \neg\Phi \vee \Psi\\
\Phi \Leftrightarrow \Psi &\equiv (\Phi \Rightarrow \Psi) \wedge (\Psi \Rightarrow \Phi)\\
\texttt{true} &\equiv \Phi \vee \neg\Phi\\
\texttt{false} &\equiv \neg\texttt{true}\\
\Diamond\Phi &\equiv \texttt{true}U\Phi\\
\Box\Phi &\equiv \neg\Diamond\neg\Phi\\
\Phi W \Psi &\equiv (\Phi U \Psi) \vee \Box\Phi
\end{aligned}$$

LTL formulas can be used to express that (1) Φ should hold at the next element of the trace ($\bigcirc\Phi$), (2) Φ should hold eventually ($\Diamond\Phi$), (3) Φ should always hold ($\Box\Phi$), (4) Φ should hold until eventually Ψ holds ($\Phi U \Psi$), and (5) Φ should hold until Ψ holds, but Ψ is not required to hold eventually ($\Psi W \Phi$). The formal semantics of the basic temporal operators are given in Definition 3.

Definition 3 (LTL semantics). Let p be an atomic proposition, $\sigma = \langle\sigma_1, \sigma_2, \sigma_3, \ldots\rangle$ be a trace. Then

$$\begin{aligned}
\sigma \models p &\Leftrightarrow \sigma_1 = \mathrm{p}\\
\sigma \models \neg\Phi &\Leftrightarrow \text{not } \sigma \models \Phi\\
\sigma \models \Phi \vee \Psi &\Leftrightarrow \sigma \models \Phi \text{ or } \sigma \models \Psi\\
\sigma \models \bigcirc\Phi &\Leftrightarrow \langle\sigma_2, \sigma_3, \ldots\rangle \models \Phi\\
\sigma \models \Phi U \Psi &\Leftrightarrow \exists_{1 \le i}[\langle\sigma_i, \sigma_{i+1}, \sigma_{i+2}, \ldots\rangle \models \Psi \wedge \forall_{1 \le j < i}[\langle\sigma_j, \sigma_{j+1}, \ldots\rangle \models \Phi]]
\end{aligned}$$

Note that the semantics of the other operators are provided in an indirect manner, for example, $\Diamond\Phi$ is semantically equivalent to $\texttt{true}U\Phi$, etc. Figure 6.4 shows some example traces to clarify some of the constructs. Note that LTL is typically defined on infinite traces, but all results for infinite sequences can be mapped onto finite sequences.

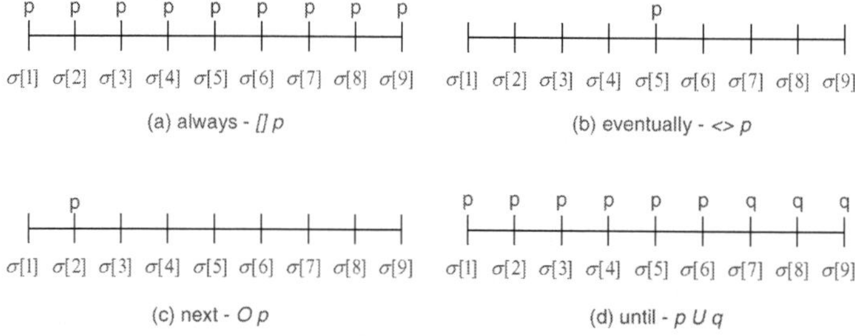

Fig. 6.4 Semantics of some LTL operators

Table 6.1 Formal specification of constraints with LTL

LTL formula for constraint c	Automaton representing $\mathcal{T}^*_{\models c}$	Semantics
$\Diamond$ *bill*	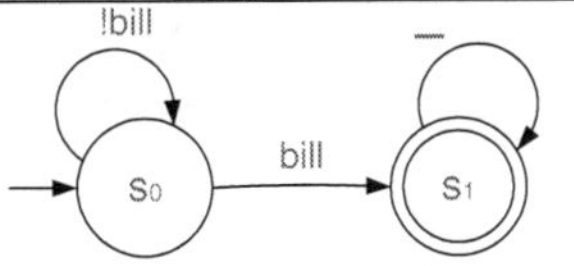 !bill, S0, bill, S1, _	Task *bill* must be executed at least once
!(*bill*) *U* *pickup*	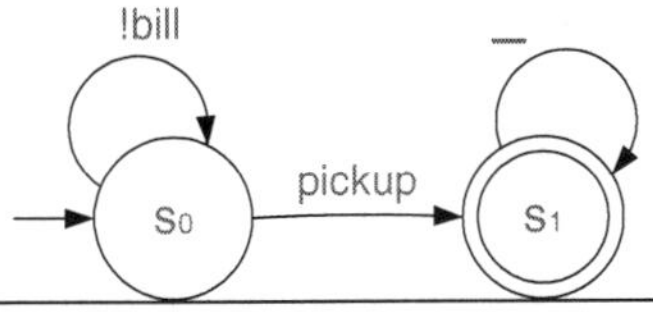 !bill, S0, pickup, S1, _	Task *bill* cannot be executed until task *pickup* is executed

The set of possible execution traces satisfying an LTL constraint is defined as follows.

Definition 4 (Constraint satisfying traces $\mathcal{T}^*_{\models c}$). Let constraint c be an LTL formula. Then the set of traces satisfying c is defined as $\mathcal{T}^*_{\models c} = \{\sigma \in \mathcal{T}^* \mid \sigma \models c\}$.

In the field of model checking, LTL is extensively used to check whether a system satisfies properties specified by LTL formulae. Using the results from this field, it is possible to generate a nondeterministic finite state automaton that represents *exactly all traces that satisfy the LTL formula*. For a constraint c, specified by an LTL formula, the set of all traces satisfying c ($\mathcal{T}^*_{\models c}$) is exactly the language produced by the automaton generated for this formula using the algorithm of D. Giannakopoulou and K. Havelund. Table 6.1 shows two examples illustrating the general idea. It is not important to understand how the automata are generated; however, to understand our approach it is important to understand that for any LTL constraint a corresponding automaton can be generated. This automaton will be used to enforce constraints, detect violations, and to check the consistency of a specification. Section 6.3 describes how the LTL-based automata can be used for declarative workflow execution.

The next example gives the LTL constraints for a process involving tasks such as *shipment* (i.e., *Create Shipment Information Document*), *bill* (i.e., *Create Bill of Lading*), and *pickup* (i.e., *Arrange Pickup Appointment*). In this example, the number of constraints is small and it is not difficult to determine whether a trace satisfies the given constraints, or not. For larger models this is less intuitive. Therefore, the next subsection proposes *constraint templates* as a graphical way to represent LTL formulas.

Example 1 (Constraint satisfying traces). Consider the two constraints described in Table 6.1 for the execution of tasks *bill* and *pickup*; (c_1) $\Diamond$*bill* and (c_2) !(*bill*) *U* *pickup*. Examples of traces satisfying constraints c_1 and c_2 are $\langle pickup, bill\rangle \models c_1 \wedge c_2$ and $\langle pickup, pickup, bill\rangle \models c_1 \wedge c_2$. Traces not satisfying constraints c_1 and c_2 are $\varepsilon \not\models c_1 \wedge c_2$, $\langle pickup\rangle \not\models c_1 \wedge c_2$, and $\langle bill, pickup\rangle \not\models c_1 \wedge c_2$. Note that $\varepsilon \models c_2$.

6.2.2 Constraint Templates

Procedural workflow languages such as YAWL provide constructs such as AND-split, AND-join, cancelation region, etc. These constructs aim at supporting frequently needed workflow patterns. For our declarative approach, we also aim at supporting frequently needed patterns to model relationships (constraints) between tasks. However, we aim at different types of patterns, that is, declarative constructs aiming at more flexibility. In principle, LTL offers everything needed to support such patterns. However, because LTL formulas can be difficult to understand by nonexperts, we provide a graphical representation of constraints that hides the associated LTL formulas from users of declarative workflow, in the form of *constraint templates*. Each template has (1) a name, (2) an LTL formula, and (3) a graphical representation. A constraint inherits the name, graphical representation, and the LTL formula from its template. Figure 6.5 depicts some example constraint templates. The template parameters are depicted by boxes. Templates can be used to create actual constraints for a specific process. In actual constraints, tasks replace template parameters, both in the graphical representation and the associated LTL formula.

The *existence* template is graphically represented with the annotation "1..∗" above the task. This indicates that A is executed at least once in each instance. The template $init(A)$ can be used to specify that task A must be the first executed task in an instance. Templates *response*, *precedence*, and *succession* consider the ordering of tasks. Template *response* requires that every time task A executes, task B has to be executed afterwards. Note that this is a very relaxed interpretation of the notion of *response*, because B does not have to execute straight after A, and another A can be

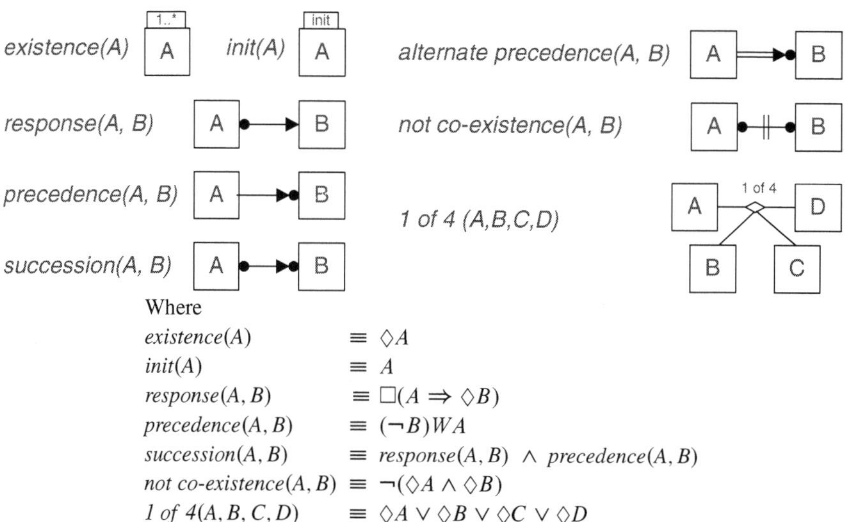

Fig. 6.5 Constraint templates

executed between the first A and the subsequent B. The template *precedence* specifies that task B can be executed only after task A is executed. Just like in the *response* template, other tasks can be executed between tasks A and B. The combination of the *response* and *precedence* templates defines a bi-directional execution order of two tasks and is called *succession*. In this template, both *response* and *precedence* relations have to hold between the tasks A and B. Template *alternate precedence* strengthens the *precedence* template: tasks A and B have to alternate. The *not coexistence* template specifies that tasks A to B cannot both be executed in the same instance. The *1 of 4* template specifies that at least one of the three tasks A, B, C, and D has to be executed, but all four tasks can be executed an arbitrary number of times as long as one of them occurs at least once.

6.2.3 Branching of Templates

The behavior of models containing multiple constraints is given by the conjunction of all constraints. In some cases it might be necessary to specify the disjunction of constraints. This can be done by assigning more than two tasks to one parameter in a template. When a template parameter is replaced by more than one task in a constraint, then we say that this parameter *branches*. An example of a branched *precedence* constraint is shown in Fig. 6.6, where parameter A has been replaced by the task *bill* and parameter B is branched on tasks *pickup* and *delivery*. In case of branching, the parameter is replaced (1) by multiple arcs to all branched tasks in the graphical representation and (2) by a disjunction of branched tasks in the LTL formula. The semantics of branching can vary from template to template, depending on the LTL formula of the template. For example, the branched constraint in Fig. 6.6 specifies that each occurrence of *bill* must be preceded by at least one occurrence of the task *pickup* or the task *delivery*. Note that it is possible to branch all parameters, one parameter, or none of the parameters. The number of possible branches in constraints is unlimited. For example, it is possible to branch the parameter B in the *response* template to N alternatives, as shown in Fig. 6.6.

Another way of branching templates are *choice* templates, which can be used to specify that one must choose between tasks. An example of such a template is the *1 of 4* template (see Fig. 6.5), which is used to specify that at least one of the four given tasks has to be executed.

6.2.4 An Example Language: ConDec

As explained before, we use LTL as a basis to define constraint templates. The template is defined by specifying the corresponding LTL formula and graphical representation. This makes it easy to create various languages, as any collection of constraint templates forms a language. In this chapter, the focus is on the

ConDec language. ConDec provides more than 20 constraint templates, among which are the ones depicted in Fig. 6.5. ConDec templates are structured into three groups: (1) *Existence* templates specify how many times or when a task can be executed, for example, *init* and *existence* templates; (2) *relation* templates define some relation between two (or more) tasks, for example, *response*, *precedence*, and *succession*; (3) *negation* templates define a negative relation between tasks, for example, *not coexistence*; and (4) choice templates define a choice between tasks, for example, *1 of 4*.

Figure 6.7 depicts the relation between constraint templates, constraints, and models for ConDec. At the template abstraction level, templates are defined on a predefined number of abstract parameters (e.g., parameter A and B). When a constraint is specified, the abstract parameter is replaced by a concrete one, for example, in Fig. 6.7 the left *precedence* constraint concretizes abstract parameter A from the *precedence* template with concrete parameter *bill*. A "plain" constraint is a concretization of the predefined parameters of a constraint template (cf. Fig. 6.6). Each constraint can easily be extended to deal with more parameters than defined by its template by the means of branching.

Every ConDec template involves a specific number of tasks. For example, templates *existence(A)*, *precedence(A,B)*, and *1 of 4(A,B,C,D)* involve one, two, and four tasks, respectively. When a constraint is created based on a template, the constraint will involve as many real tasks as predefined in the template. Templates can be reused to specify different constraints, as shown for the *precedence* template in Fig. 6.7. Finally, constraints from the ConDec constraint templates can be used to specify ConDec models.

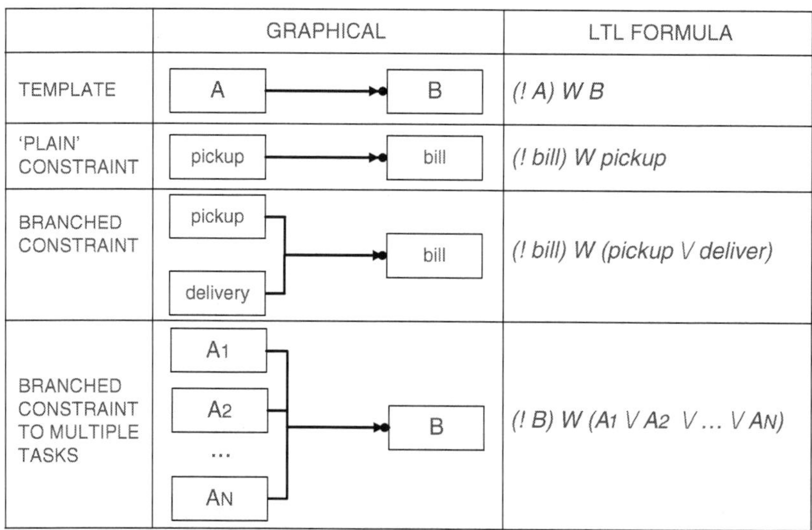

Fig. 6.6 Branching the *response* template

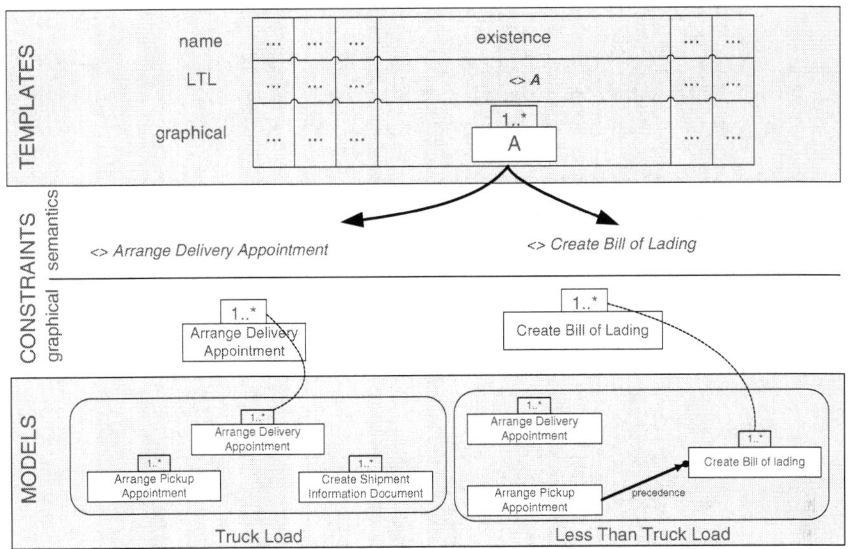

Fig. 6.7 ConDec templates, constraints and models

Example 2 (Carrier Appointment examples in ConDec). Consider, for example, the two highlighted constraints in different models shown in Fig. 6.7. First, model *Truck Load* contains the constraint specifying that "Task *Arrange Delivery Appointment* must be executed at least once." Second, the *Less Than Truck Load* model contains a constraint specifying that "Task *Create Bill of Lading* must be executed at least once." The first constraint can be specified with formula $\Diamond$ *Arrange Delivery Appointment* and the second one with a similar formula $\Diamond$ *Create Bill of Lading*. Both constraints use the same (*existence*) template, their LTL specifications are similar: $\Diamond\ A$, but the constraints are specified on different tasks. Instead of having to individually specify formulas for every constraint, constraint templates provide a way to reuse formulas.

6.2.5 Constraint Workflow Models

Constraint workflow models consist of tasks and constraints. Moreover, there are two types of constraints: *mandatory constraints* represent rules that must be followed during execution, while *optional constraints* can be violated.

Definition 5 (Constraint model cm**).** Let $\mathcal{T}$ be the universe of all tasks and $\mathcal{C}$ be the set of all constraints. A constraint model cm is defined as a triple $cm = (T, C_M, C_O)$, where:

- $T \subseteq \mathcal{T}$ is a set of *tasks* in the model
- $C_M \subseteq \mathcal{C}$ is a set of *mandatory constraints*, where every element $c \in C_M$ is a well-formed LTL formula over T

- $C_O \subseteq C$ is a set of *optional constraints*, where every element $c \in C_O$ is a well-formed LTL formula over T

Note that each mandatory and optional constraint is a well-formed LTL formula over tasks from cm, that is, a constraint cannot involve tasks that are not in the model.

During execution, mandatory constraints must be followed. Therefore, the set of traces that satisfy constraint model cm contains all traces that satisfy *all* mandatory constraints in cm. The conjunction of all mandatory formulas is defined as the *mandatory formula of the model.*

Definition 6 (Mandatory formula f_{cm}). Let $cm \in \mathcal{U}_{cm}$ be a constraint model where $cm = (T, C_M, C_O)$, then the mandatory formula for model cm is defined as

$$f_{cm} = \begin{cases} \texttt{true} & \text{if } C_M = \varnothing; \\ \bigwedge_{f \in C_M} f & \text{otherwise.} \end{cases}$$

The automaton generated for the mandatory formula (f_{cm}) of a model cm generates the language $\mathcal{T}^*_{\models cm}$, that is, the set of traces that satisfy cm. In the absence of mandatory constraints, all traces over tasks in the model are allowed.

Definition 7 (Constraint model satisfying traces). Let f_{cm} be the mandatory formula for constraint model cm.

$$\mathcal{T}^*_{\models cm} = \begin{cases} \mathcal{T}^* & \text{if } C_M = \varnothing; \\ \mathcal{T}^*_{\models f_{cm}} & \text{otherwise.} \end{cases}$$

Furthermore, we say a trace $\sigma \in \mathcal{T}^*$ *satisfies* model cm, if and only if, $\sigma \in \mathcal{T}^*_{\models cm}$, and σ *violates* cm, if and only if, $\sigma \notin \mathcal{T}^*_{\models cm}$.

Consider, for example, the *Carrier Appointment* subprocess of the Order Fulfillment process model described in Appendix A. In particular, consider the parts (i.e., branches) of the *Carrier Appointment* net for handling the shipments that require *the truck load* (TL) and *less than truck load* (LTTL).[1] Figure 6.8a shows the *the truck load* part, and Fig. 6.8b shows the *less than the truck load* part of the *Carrier Appointment* net. For the purpose of simplicity, in the remainder of this chapter, we will use shorter names for the relevant tasks: *pickup* for *Arrange Pickup Appointment*, *delivery* for *Arrange Delivery Appointment*, *bill* for *Create Bill of Lading* and *shipment* for *Create Shipment Information Document*. Examples 3 and 4 show ConDec models for these two parts of the *Carrier Appointment* net.

Example 3 (The ConDec model for the LTTL process (cm_{LTTL})). In the LTTL branch of the process, the goal is to execute tasks *pickup*, *delivery*, and *bill* in such

[1] Note that in the original *Carrier Appointment* net, this branch of the net is called LTL. In this chapter we will use the LTTL abbreviation to avoid confusion with Linear Temporal Logic (LTL).

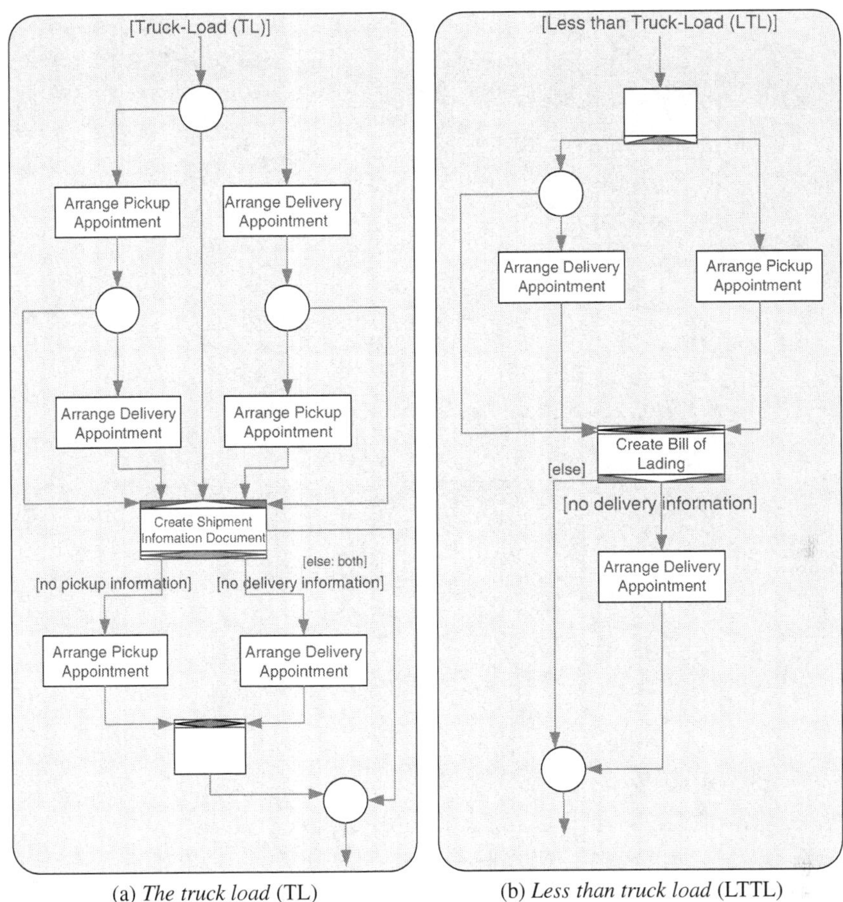

(a) *The truck load* (TL)

(b) *Less than truck load* (LTTL)

Fig. 6.8 Two parts of the *carrier appointment* YAWL subnet of the order fulfillment process

a way that task *pickup* must be executed before the task *bill*, and task *delivery* can be executed either before or after the task *bill*. While the YAWL net for the *Carrier Appointment* subprocess explicitly specifies all possible orderings of the related tasks, these orderings can be implicitly specified by using ConDec constraints, as shown in Fig. 6.9a. Constraints *1..** on tasks *delivery* and *bill* specify that each of these tasks must be executed at least once. The *precedence* constraint specifies that task *bill* can be executed only after task *pickup* is executed. The automaton presented in Fig. 6.9b is generated for the mandatory formula of this model, and it represents exactly all traces that satisfy this model. From the initial state (S_0), the accepting state (S_5) can only be reached if *bill* and *delivery* are executed and *bill* is not executed before *pickup*.

Example 4 (The ConDec model for the TL process (cm_{TL})). In the TL branch of the process, the goal is to execute tasks *pickup*, *delivery*, and shipment in any order.

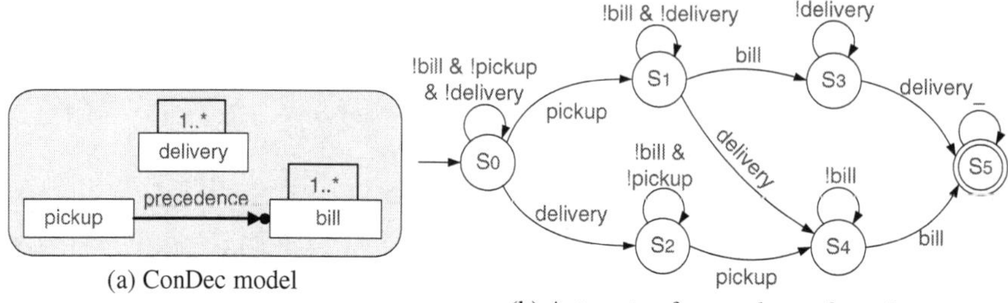

(a) ConDec model

(b) Automaton for mandatory formula $f_{cm_{LTTL}} = (\Diamond\ delivery) \wedge (\Diamond\ bill) \wedge ((!\ bill)\ W\ pickup)$ represents the set of satisfying traces $\mathcal{T}^*_{\models cm_{LTTL}}$

Fig. 6.9 The *LTTL* model $cm_{LTTL} = (\{pickup,delivery,bill\},\{\Diamond\ delivery, \Diamond\ bill,(!\ bill)\ W pickup\},\varnothing)$

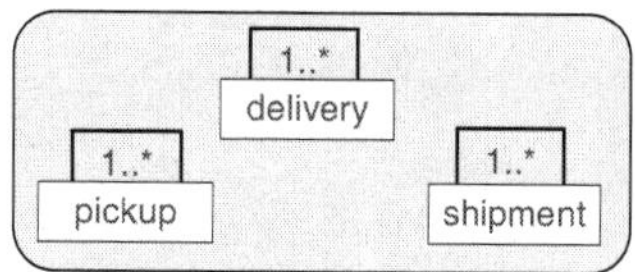

Fig. 6.10 The *TL* model $cm_{TL} = (\{pickup,delivery,\text{shipment}\},\{\Diamond\ pickup, \Diamond\ delivery, \Diamond shipment\},\varnothing)$

While the YAWL net for the *Carrier Appointment* subprocess explicitly specifies all possible ordering of the related tasks, these orderings can be implicitly specified by using ConDec constraints, as shown in Fig. 6.9a. Constraints *1..** on tasks *delivery* and shipment specify that each of these tasks must be executed at least once. Note that Fig. 6.10 does not show the automaton generated for the mandatory formula of this model. This is because the generated automata is too complex (i.e., it has eight states and 20 transitions) to be presented as an illustrative example.

Note that both the *TL* and the *LTTL* subprocess just contain a small number of constraints, nonetheless there are infinitely many traces that satisfy the constraint model. In fact, both models have infinitely many possible execution alternatives, and so are very flexible by design (cf. Sect. 6.1). *The main difference between procedural and declarative languages is that in procedural languages it takes effort (specifying all paths) to extend the behavior of a process, whereas in declarative languages it takes effort (specifying constraints) to limit the behavior of a process.* In this sense, declarative languages offer a more convenient way to express flexibility than procedural ones. Consider, for example, the ConDec models for the *TL* and *LTTL* parts of the *Carrier Appointment* net. Not only that ConDec models define the possible execution alternatives in a more concise way, but they also allow that tasks are executed multiple times when necessary.

6.2.6 Verification of Constraint Models

Verification techniques can be used for detection of possible errors in models. The verification capabilities of YAWL will be described in Chap. 20. Just like for procedural languages having formal semantics, constraint-based model expressed in LTL allow for various kinds of analysis.

Certain combinations of constraints can cause errors in constraint models that lead to undesirable effects at execution-time. We distinguish two types of constraint model errors: *dead tasks* and *conflicts*.

A task in a constraint model is dead if none of the traces satisfying the model contains this task, that is, the task cannot be executed in any instance of the model without violating the specified (mandatory) constraints.

Definition 8 (Dead task). Let $cm \in \mathcal{U}_{cm}$ be a constraint model. Task $t \in \mathcal{T}$ is *dead* in model cm, if and only if $\nexists \sigma \in \mathcal{T}^*_{\models cm} : t \in \sigma$.

Example 5 (Dead task). Figure 6.11a shows a ConDec model where tasks *pickup* and *bill* are dead. This error is caused by the combination of the two *precedence* constraints. While one *precedence* constraint specifies that task *bill* cannot be executed before task *pickup*, the other *precedence* constraint specifies that task *pickup* cannot be executed before task *bill*. Therefore, it will never be possible to execute tasks *pickup* and *bill* in any instance of the model shown in Fig. 6.11a.

Dead tasks can easily be detected by analyzing the automaton generated for the mandatory formula of a ConDec model. A task is dead if none of the transitions in the automaton allow the execution of this task. Figure 6.11b shows the automaton generated for the model in Fig. 6.11a. Indeed, this automaton does not contain transitions that allow for the execution of tasks *pickup* and *bill*.

A constraint model contains a conflict if there are no traces that can satisfy the model, that is, instances of the model can never become *satisfied*.

Definition 9 (Conflict). Model $cm \in \mathcal{U}_{cm}$ has a *conflict* if and only if $\mathcal{T}^*_{\models cm} = \emptyset$.

Example 6 (Conflict). Figure 6.12a shows a ConDec model containing a conflict. In addition to the constraints described in Example 5, there now is an additional

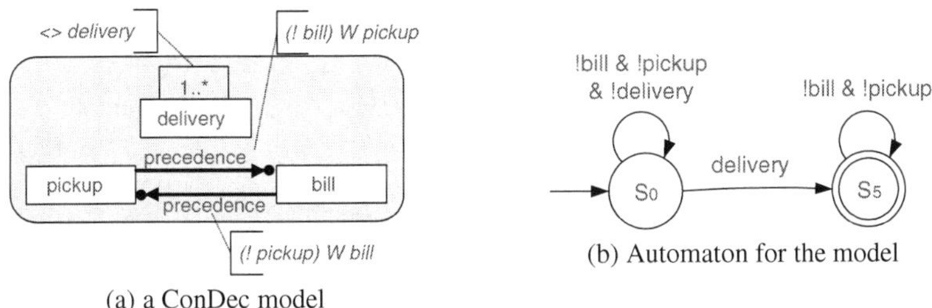

(a) a ConDec model

(b) Automaton for the model

Fig. 6.11 Tasks *pickup* and *bill* are dead

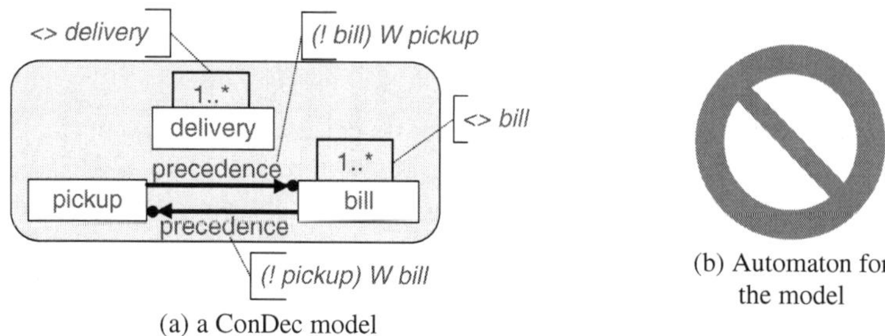

(a) a ConDec model

(b) Automaton for the model

Fig. 6.12 The ConDec model has a conflict as shown by the empty automaton

constraint *1..** requiring that task *bill* should be executed at least once. However, task *bill* is a dead task (see Figure 6.11), that is, it cannot be executed without violating constraints. So, there are no traces that can satisfy all constraints in this model.

Detection of conflicts can also be done using the automaton generated for the mandatory formula of the model. If the automaton is empty (i.e., it has no states), then this model has a conflict, otherwise it is conflict free. Figure 6.12b shows the automaton generated for the model from Fig. 6.12a. The automaton is empty, thus indicating that Fig. 6.12a has a conflict.

In the previous examples, dead tasks and conflicts were not caused by the combination of all constraints in the model. Instead, they were caused by a specific group of constraints. Dead tasks *pickup* and *bill* in Example 5 are caused by the two *precedence* constraints and the conflict in Example 6 by the two *precedence* constraints and the *1..** constraint on task *bill*. The smallest subset of mandatory constraints that causes the error is called *the cause of the error*. In ConDec models, the cause of error can be found by searching through the powerset of mandatory constraints. For each element (i.e., subset of mandatory constraints) in the powerset, the automaton for the mandatory formula can be analyzed for the presence of dead tasks and conflicts as described in previous paragraphs. Detecting the smallest group of mandatory constraints that causes an error supports the user in developing error-free constraint models.

6.3 Enactment of Constraint Model Instances

Thus far, we focused on the modeling and analysis of constraint models. In this section, the focus is on the *enactment* of these models, that is, the actual execution environment. Section 6.3.1 introduces the notion of constraint model instances, Sects. 6.3.2 and 6.3.3 describe how states of constraints and instances can be monitored during the execution, respectively. Enforcing a correct instance execution and completion is discussed in Sects. 6.3.4 and 6.3.5, respectively.

6.3.1 Constraint Model Instances

The execution of constraint models is driven by the constraints. These constraints should ensure the correct execution of instances of the model, that is, it should not be possible to violate mandatory constraints during execution and all mandatory constraints should be satisfied after execution. To ensure the correct execution of an instance, it is necessary to keep track of the execution trace of that instance. Formally, a *constraint workflow model instance* is defined as follows.

Definition 10 (Constraint model instance ci**).** A constraint model instance ci is defined as a pair $ci = (\sigma, cm)$, where:

- $\sigma \in \mathcal{T}^*$ is the instance's trace
- $cm \in \mathcal{U}_{cm}$ is a constraint model

We use $\mathcal{U}_{ci}$ to denote the set of all constraint instances.

6.3.2 Monitoring Constraint States

The sequence of executed tasks for an instance is kept in its instance trace. Depending on the instance trace, constraints from the model can be *satisfied*, *violated*,[2] and *temporarily violated*, which means that the constraint is currently violated, but can still be satisfied. Consider, for example, the *LTTL* ConDec model shown in Fig. 6.9 on page 188. Although trace $\langle pickup \rangle$ violates both *1..** constraints, once *pickup* and *bill* are executed, all constraints will become satisfied. Hence, both *1..** constraints are *temporarily violated.* As long as the execution is in progress and more tasks can be executed, we distinguish these three states.

Definition 11 (Constraint state ν**).** Let $c \in \mathcal{C}$ be a constraint and $\sigma \in \mathcal{T}^*$ an execution trace. The function $\nu : (\mathcal{C} \times \mathcal{T}^*) \rightarrow \{satisfied, temporarily\ violated, violated\}$ is defined as:

$$\nu(\sigma, c) = \begin{cases} satisfied & \text{if } \sigma \in \mathcal{T}^*_{\models c}; \\ temporarily\ violated & \text{if } (\sigma \notin \mathcal{T}^*_{\models c}) \wedge (\exists \gamma \in \mathcal{T}^* : \sigma + \gamma \in \mathcal{T}^*_{\models c}); \\ violated & \text{otherwise.} \end{cases}$$

The state of a constraint can continuously be monitored via the automaton generated for its LTL formula. Every time a user executes a task, a transition in the automaton is triggered, and the automaton changes state. When the automaton is in an accepting state, then the constraint is *satisfied*. If the automaton is in a nonaccepting state from which an accepting state is reachable, then the constraint is *temporarily violated.* Finally, if the trace cannot be "replayed" on the automaton at all, then the constraint is and will continue to be *violated* for this instance.

[2] In Sect. 6.3.4, we explain how this state is avoided during execution.

6.3.3 Monitoring Instance States

During execution, the state of an instance is also monitored. The state of an instance depends on the satisfaction of the mandatory constraints in the model.

Definition 12 (Instance state ω). Let $ci \in \mathcal{U}_{ci}$ be an instance where $ci = (\sigma, cm)$. The function $\omega : \mathcal{U}_{ci} \rightarrow \{satisfied, temporarily\ violated, violated\}$ of instance ci is defined as:

$$\omega(ci) = \begin{cases} satisfied & \text{if } \sigma \in \mathcal{T}^*_{\models cm}; \\ temporarily\ violated & \text{if } (\sigma \notin \mathcal{T}^*_{\models cm}) \wedge (\exists \gamma \in \mathcal{T}^* : \sigma + \gamma \in \mathcal{T}^*_{\models cm}); \\ violated & \text{otherwise.} \end{cases}$$

Recall that $\mathcal{T}^*_{\models cm}$ is the set of all traces that satisfy the conjunction of all mandatory formulas (cf. Definition 6 on page 186). Similar to the monitoring of constraint states, the state of an instance can be monitored using the automaton generated for the mandatory formula of the model.

Example 7 (Instance states). Table 6.2 shows how three instances (ci_1, ci_2, and ci_3) of the model shown in Fig. 6.9 on page 188 change state during execution. The instance state is always deduced from the automaton's current state (Fig. 6.9b). As long as the automaton stays in nonaccepting states (i.e., $S_0, S_1, \ldots, S_4$), the instance is *temporarily violated* (tv). The instance becomes *satisfied* (s) only when the automaton reaches its accepting state S_5. If the instance trace cannot be replayed in the automaton, the instance is *violated* (v).

The instance state cannot always be directly induced from the states of its individual constraints. When all mandatory constraints are *satisfied*, the instance is also *satisfied*, and if at least one mandatory constraint is *violated*, the instance is also *violated*. However, in the presence of *temporarily violated* constraints and the absence of *violated* constraints, a deeper analysis of the constraints is needed to induce the instance state. Consider, for example, the instance of the ConDec model shown in Fig. 6.13. The state of the individual constraints is depicted for execution trace $\langle C, A, B \rangle$. Observe that none of the mandatory constraints is *violated*. The *precedence* constraint is *satisfied*, because task A is executed before task B.

Table 6.2 State change for instances of the ConDec model in Fig. 6.9

Instance $ci_1 = (\sigma, cm)$			Instance $ci_2 = (\gamma, cm)$			Instance $ci_3 = (\delta, cm)$		
σ_i	State	$\omega(ci_1)$	γ_i	State	$\omega(ci_2)$	δ_i	State	$\omega(ci_3)$
`initial`	S_0	tv	`initial`	S_0	tv	`initial`	S_0	tv
delivery	S_2	tv	*pickup*	S_1	tv	*delivery*	S_2	tv
pickup	S_4	tv	*delivery*	S_4	tv	*bill*	⊠	v
bill	S_5	s	*bill*	S_5	tv			
pickup	S_5	s						

"s"=*satisfied*, "tv"=*temporarily violated*, "v"=*violated*

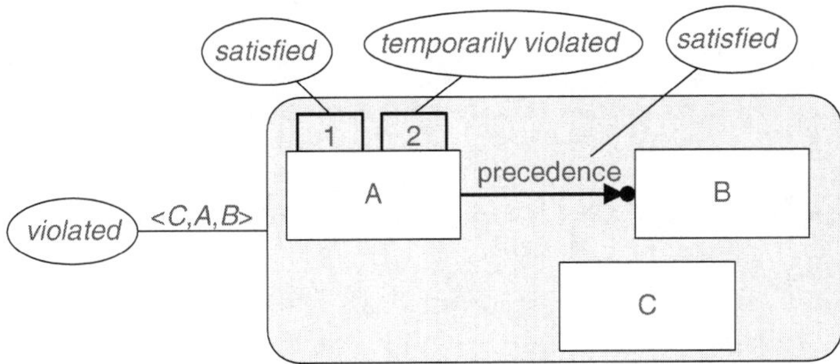

Fig. 6.13 An instance of ConDec model with trace $\langle C, A, B \rangle$

Constraint *1*, which specifies that task *A* must be executed exactly once, is also *satisfied*. Constraint *2*, which specifies that task *A* must be executed exactly twice, is *temporarily violated*, and would become *satisfied* if task *A* would be executed once again. However, if task *A* would be executed once again, then the *1* constraint would become *violated*. As it is not possible to satisfy the mandatory constraints, the instance state is *violated*.

Note that the instance presented in Fig. 6.13 cannot be *satisfied* because the instance model contains a conflict. Recall from Sect. 6.2.6 that when the model contains a conflict, then there is no execution that will satisfy the constraints from the model. The cause of the error is the combination of constraints *1* and *2*. Verification can be used to detect and correct such errors. In the next section, we explain how instances of correct constraint models can be executed without becoming violated.

6.3.4 Enforcing Correct Instance Execution

The goal of instance execution is to finish the execution in such a way that the instance is *satisfied*. In order for an instance to be executed in a correct manner, the execution of tasks that would eventually lead to the *violated* state of the instance should be prohibited. This can be done using the automaton generated for the mandatory formula of the instance. During execution, the automaton changes state when tasks are executed. Given the current state in the automaton, tasks that are not implied by labels on transitions from this state are prohibited, and other tasks are allowed.

Example 8 (Instance execution). Again, consider the ConDec model shown in Fig. 6.9 on page 188. The automaton represents exactly all traces that satisfy this model. Initially, no tasks have been executed and the automaton is in state S_0. Because of the *precedence* constraint, only task *bill* should be prohibited, and other tasks should be allowed to execute. This can be detected in the automaton by the absence of a *bill* transition from S_0 and the presence of *pickup* and *delivery* transitions. Naturally, the same approach is applied again to all states, as the automaton changes states due to execution of tasks.

6.3.5 Enforcing Correct Instance Completion

The goal of the execution of a constraint workflow model instance is to satisfy all mandatory constraints, and so the state of the instance should be *satisfied* at completion. The automaton generated for the mandatory formula of the model is used to monitor the state of the instance. Recall that the state of the instance is *satisfied* if the automaton is in an accepting state. Now the completion of an instance is allowed, if and only if, the generated automaton is in an accepting state.

Enforcing the correct execution and completion of instances is done in the following way: (1) The states of all constraints are constantly monitored; (2) The state of the instance is constantly monitored; (3) The correct execution of instance is enforced by prohibiting the execution of tasks that *would eventually* bring the instance into the *violated* state; and (4) The completion of the instance is allowed if and only if the instance state is *satisfied*. All this can be done using the automaton generated for the mandatory formula of the instance's constraint model.

6.4 Dynamic Instance Change

In some cases, it is necessary that the model of the instance changes (i.e., by adding and removing tasks or constraints), although the instance is already executing and the instance trace might not be empty. We refer to such a change as a *dynamic instance change*. Workflow management systems that support dynamic change are called adaptive systems. For example, ADEPT is a workflow management system that uses powerful mechanisms to support dynamic change of procedural process models by allowing to add, remove, and move tasks at runtime. The constraint-based approach offers a simple method for dynamic change that is based on a single requirement: the instance should not become *violated* after the change.

Because of the fact that in dynamic instance change the trace of an instance remains the same whereas the model changes, it might happen that the instance state changes according to the new model (cf. Definition 12). For example, it is possible that after adding a mandatory constraint, the instance state changes from *satisfied* to *violated*, which is an undesired state. Therefore, instance change can only be successfully applied if the change does not bring the instance into the *violated* state. After a successful change, the instance continues execution with an updated model and the original trace. Automata generated for ConDec models enable easy implementation of dynamic change of ConDec instances: dynamic change of a ConDec instance is successful if the instance trace can be "replayed" on the mandatory automaton of the new model.

Consider, for example, two instances $ci_1 = (\langle pickup, delivery, bill\rangle, cm_{LTTL})$ and $ci_2 = (\langle pickup, bill, delivery\rangle, cm_{LTTL})$ of the model cm_{LTTL} shown in Fig. 6.9 on page 188. Figure 6.14 shows that ci_1 is *satisfied* by replaying their traces on the model automaton.

Assume now that we attempt to dynamically change the cm_{LTTL} model in both instances by removing the *pickup* task and the *1..** constraint from task *delivery*, removing the *precedence* constraint, and adding a new *precedence* constraint between tasks *delivery* and *bill*. The new model and its automaton are shown in Fig. 6.15.

Figure 6.16 shows that this dynamic change involving is *successful* for ci_1 and *not successful* for ci_2. This figure shows how the instance trace is "replayed" on the automaton. On the one hand, the trace of instance ci_1 can be replayed on the new automaton and, although the instance is *temporarily violated*, this is a valid dynamic change. On the other hand, the trace of instance ci_2 cannot be "replayed" on the automaton because it is not possible to execute task *bill* from the state S_0, that is, this change would *violate* the instance. Therefore, the dynamic change for instance ci_2 is not possible.

Note that, even though the task *pickup* is removed from the model after it was executed (the instance trace contains task *pickup*), the dynamic change for instance ci_1 is successful. This is due to the property that a trace that satisfies a model can contain tasks that are *not* in the model. The only consequence of removing task *pickup* from the model is the fact that it will not be possible to execute this task again in the future.

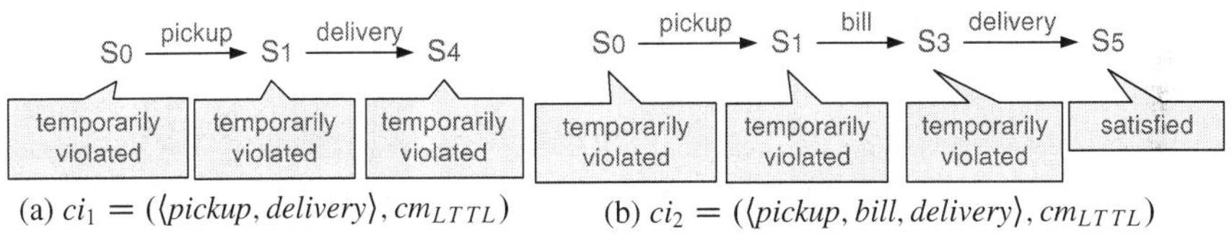

(a) $ci_1 = (\langle pickup, delivery\rangle, cm_{LTTL})$ (b) $ci_2 = (\langle pickup, bill, delivery\rangle, cm_{LTTL})$

Fig. 6.14 Replaying traces of two instances of model cm_{LTTL} shown in Fig. 6.9 on page 188 on the model automaton

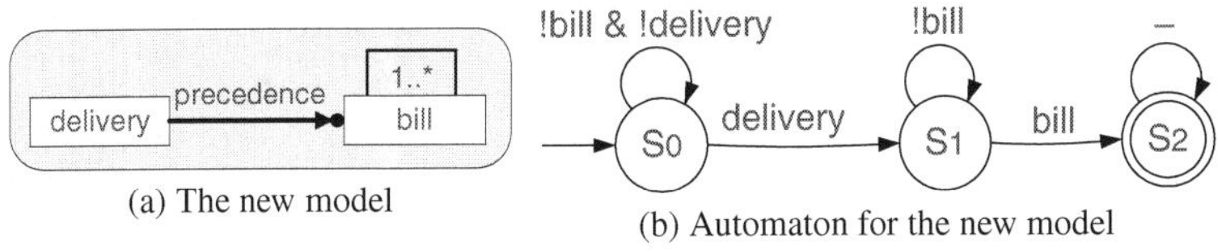

(a) The new model (b) Automaton for the new model

Fig. 6.15 Applying a dynamic change to the model cm_{LTTL} shown in Fig. 6.9 on page 188

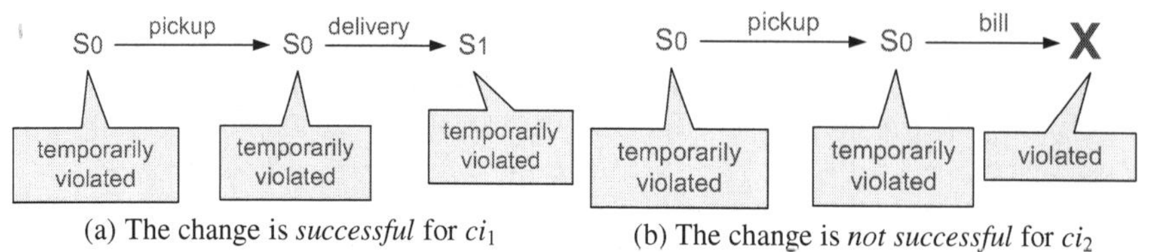

(a) The change is *successful* for ci_1 (b) The change is *not successful* for ci_2

Fig. 6.16 The dynamic change involving the model shown in Fig. 6.15

6.5 Conclusions

To conclude this chapter, we first reflect on the differences between procedural and declarative languages. Then, we summarize the main capabilities of the approach presented in this chapter by using the flexibility taxonomy given in Sect. 6.1.

Neither procedural nor declarative workflows represent a better solution when it comes to automation of business processes. On the one hand, business processes that are repeatedly executed in the same manner can clearly benefit from a procedural workflow specification. For example, for the handling of insurance claims, tax declarations, customer orders, etc., a process where always the same procedure is followed is desirable. On the other hand, some processes must be executed in a way that fits best with frequently changing and personalized circumstances. These processes are better specified in a declarative manner because they need a high degree of flexibility, that is, it is desirable that they can be executed in various manners. For example, many processes from the health care domain need flexibility because they must be adjusted to specific circumstances and patients.

Moreover, for processes that are partly highly structured and partly loosely structured, a combination of both approaches can be chosen to specify the overall business process. For example, the declarative ConDec workflows *Truck Load* and *Less than truck Load* can be subprocesses of the *Carrier Appointment* net, as shown in Fig. 6.17. Also, when necessary, a YAWL workflow can be a part of a declarative ConDec model. Using the service orientation it is possible to mix various modeling styles. A concrete implementation that enables creating decompositions of YAWL (e.g., procedural) and declarative models is presented in Chap. 12 of this book.

This chapter introduced a new way of modeling and enacting workflows. The described approach aims to be more flexible than the existing procedural approaches. Table 6.3 shows that declarative ConDec workflows can be used to support multiple types of flexibility.

First, because they can easily include a wide range of execution alternatives, ConDec workflows have a high degree of *flexibility by design*. For example, the ConDec models shown in Figs. 6.9–6.11 and 6.15 all have infinitely many execution alternatives (i.e., the sets of satisfying traces of these models are infinite). Second, decomposing ConDec tasks to sub-workflows (e.g., YAWL workflows) allows for flexibility by *underspecification*. Moreover, this type of flexibility can be achieved either by explicitly specifying the decomposed sub-workflow in the

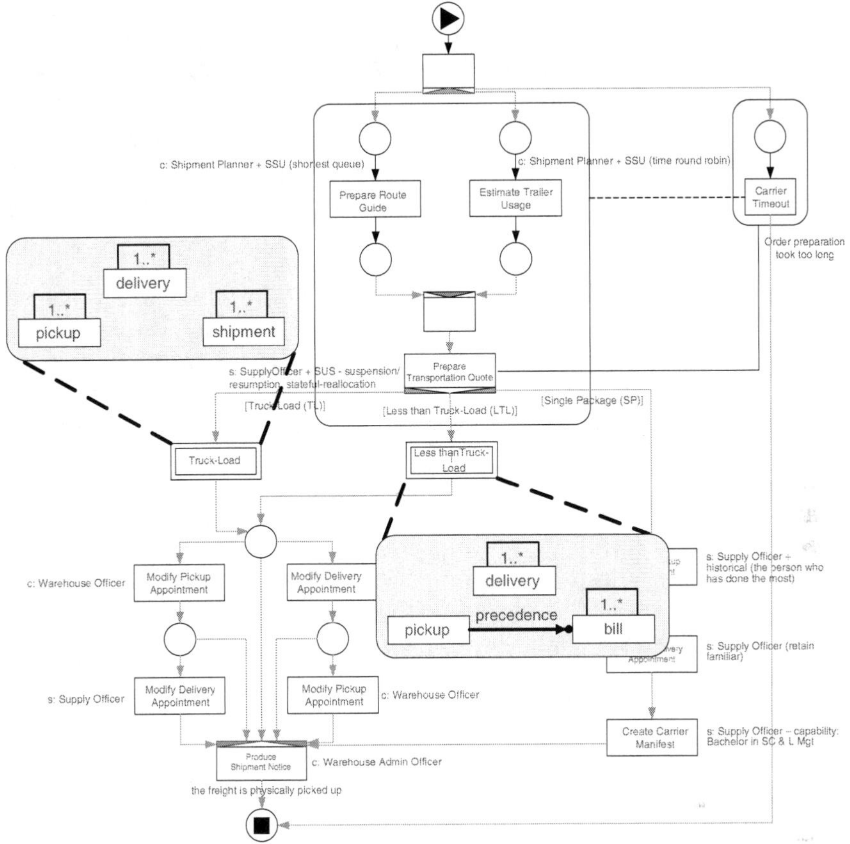

Fig. 6.17 Combining YAWL and declarative workflows

Table 6.3 Various types of flexibility and support of constraint-based workflows

Flexibility	Support
Flexibility by design	Model verification
Flexibility by underspecification	Monitoring instance states
Flexibility by change	Ensuring correct instance execution
Flexibility by deviation	Ensuring correct instance completion
	Monitoring states of constraints

parent-workflow, or by allowing for late specification (at runtime). Third, dynamic change of instances of ConDec models, which is described in Sect. 6.4, allows for *flexibility by change*. Fourth, the existence of optional constraints in ConDec models allows for flexibility by *deviation*.

Apart from being flexible, declarative workflows should and can support users in several ways. Constraint models can contain an arbitrary number of constraints that interfere in subtle ways, which can cause errors in models. Verification of constraint models provides an automated mechanism for detecting dead tasks and conflicts, as described in Sect. 6.2.6. In addition to the automated verification of constraint models, it is also possible to detect the exact combination of constraints that causes (each of) the error(s), which helps developing error-free ConDec models.

Execution of instances of constraint models is driven by constraints. To execute an instance in a correct way, it is necessary that, at the end of the execution, all mandatory constraints are *satisfied*, that is, that the instance is *satisfied*. Executing activities in an instance may change the state of one or more constraints and the instance itself. It is important that the instance state and states of all its constraints are constantly monitored and presented to users throughout the execution of the instance (cf. Sects. 6.3.2 and 6.3.3). This helps users who execute instances of ConDec models to understand what is going on and which actions are necessary in order to execute the instance in a correct way. Also, to ensure a correct instance completion, it is necessary to allow completion only if the instance is *satisfied* (cf. Sect. 6.3.5).

Some actions of users might cause an instance (and its constraints) to leave the *satisfied* state. In some of these cases it is possible to take some actions that will eventually lead to a correct, that is, *satisfied*, instance. We refer to this type of violations as *temporary violations*. In other cases, the instance becomes *permanently violated*, that is, it becomes impossible to *satisfy* the instance in the future. Especially, for instances with multiple constraints, it is very difficult for users to be aware of actions that will permanently violate the instance. Therefore, it is important to *prevent* users from taking actions that lead to permanent violation of instances of ConDec models (cf. Sect. 6.3.4).

Exercises

Exercise 1. Tasks *bless*, *pray*, *curse*, and *become holy* are four tasks in the Religion process. Users of this process must obey two important rules. First, one should *pray* at least once. Second, every time one *curses*, one must eventually *pray* for forgiveness afterwards.

(a) Develop a ConDec model for this process using the constraint templates shown in Fig. 6.5 on page 182.
(b) Write down the mandatory LTL formula for the Religion model.

Exercise 2. Figure 6.18 shows the automaton generated for the mandatory formula of the Religion model described in the previous exercise. Using this automaton, mark which of the following execution alternatives are possible in instances of the Religion model:

(a) $\langle become\ holy, curse, bless\rangle$
(b) $\langle pray, bless, pray\rangle$
(c) $\langle curse, bless, curse, bless, pray\rangle$

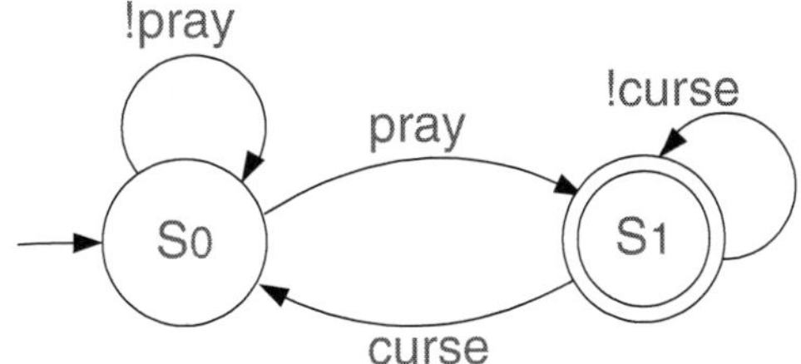

Fig. 6.18 The automaton generated for the mandatory formula of the Religion model

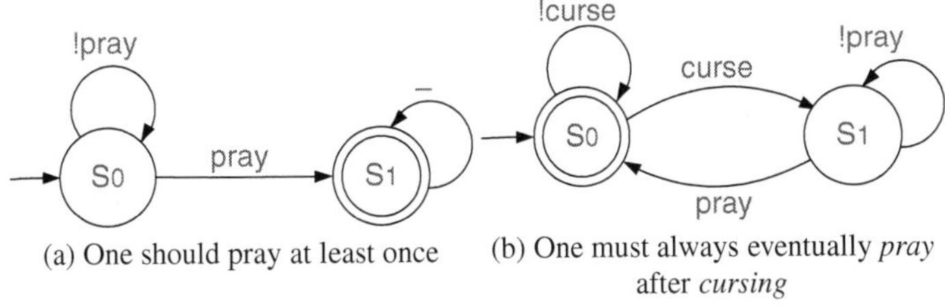

(a) One should pray at least once

(b) One must always eventually *pray* after *cursing*

Fig. 6.19 Automata generated for the two constraints from the religion model

(d) $\langle bless, become\ holy\rangle$

Exercise 3. Figure 6.19 shows automata generated for the two constraints from the Religion model from the first exercise. Using these two automata and the automaton generated for the mandatory formula of this model (cf. Fig. 6.18), write down how states of the instance and both constraints change while executing trace $\langle bless, curse, bless, bless, curse, become\ holy, pray, curse, pray\rangle$.

Exercise 4. Consider an instance of the Religion model described in the first exercise, where the instance trace is $\langle bless, curse, bless, bless, curse, become\ holy, pray, curse, pray\rangle$. Assume that we attempt to change the model of this instance in a way that:

- The constraint specifying that *one should pray at least once* is removed
- Task *become holy* is removed.

Can this dynamic change be applied to the instance at this moment? Explain why. Hint: use the automata generated for the mandatory formula of the new (i.e., changed) model shown in Fig. 6.20.

Exercise 5. Consider an instance of the Religion model described in the first exercise, were the instance trace is $\langle bless, curse, bless, bless, curse, become\ holy, pray, curse, pray\rangle$. Assume that we attempt to change the model of this instance in a way that a constraint specifying that *task become holy cannot be executed before task pray* is added.

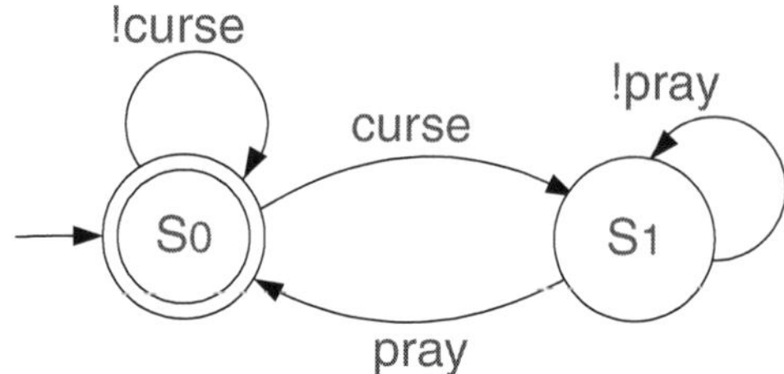

Fig. 6.20 The automaton generated for the mandatory formula of the religion model after removing task *become holy* and the constraint specifying that *one should pray at least once*

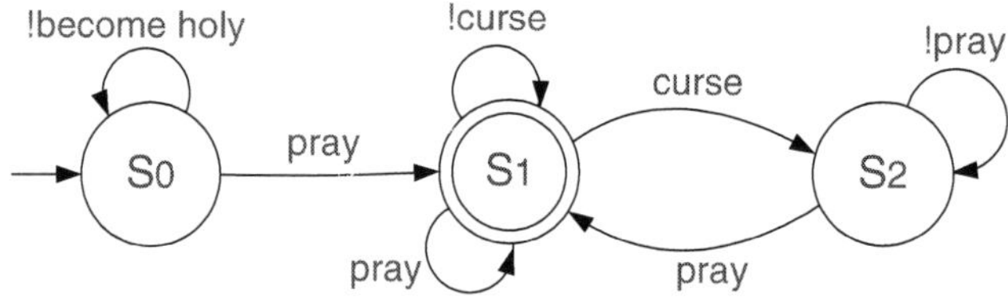

Fig. 6.21 The automaton generated for the mandatory formula of the religion model after adding a constraint specifying that *task become holy cannot be executed before task pray*

(a) Develop a ConDec mode for the new model using the appropriate constraint templates from Fig. 6.5 on page 182.
(b) Can this dynamic change be applied to the instance at this moment? Explain why. Hint: use the automata generated for the mandatory formula of the new (i.e., changed) model shown in Fig. 6.21.

Exercise 6. Develop a ConDec model for the process containing tasks *bless*, *pray*, *curse*, and *become holy*, where three rules must be followed:

(a) One should *become holy* at least once.
(b) Every time one *curses*, one must eventually *pray* for forgiveness afterwards.
(c) Tasks *become holy* and *curse* cannot both be executed in the same instance.

Does this model have errors? Explain why. Hint: use the automata generated for the mandatory formula of this model shown in Fig. 6.22.

Exercise 7. Develop a ConDec model for the process containing tasks *bless*, *pray*, *curse*, and *become holy*, where four rules must be followed:

(a) One should *become holy* at least once.
(b) Every time one *curses*, one must eventually *pray* for forgiveness afterwards.
(c) Tasks *become holy* and *curse* cannot be both executed in the same instance.
(d) One should *curse* at least once.

Does this model have errors? Explain why. Hint: the automata generated for the mandatory formula of this model is empty, that is, it does not have any states or transitions.

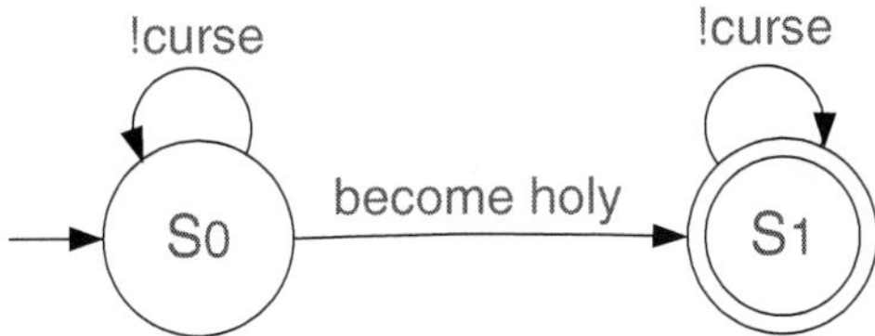

Fig. 6.22 The automaton generated for the mandatory formula of the religion model described in this exercise

Exercise 8. Which types of flexibility and support are available in declarative workflows? Explain which properties of declarative workflows enable different types of flexibility and support.

Chapter Notes

The trade off between flexibility and support and the taxonomy of flexibility used in this chapter has been addressed in [117, 236]. The positive influence of *declarative languages for workflow specification* of workflow flexibility has been discussed in [73, 117, 152, 200, 256].

More information on temporal logic, automata generation, and model checking can be found in [60, 103]. A more elaborate discussion on the choice for LTL and usage of automata is given in [192]. For details on how LTL can be used for finite sequences, see [104, 105, 192].

In this chapter, the focus was on the ConDec language, and several ConDec templates were explained and used. The complete list of all ConDec templates can be found in [192]. Other similar constraint languages can be defined if needed. For example, more information about a constraint-based language for modeling web services DecSerFlow can be found in [19, 20].

More information about the procedural workflow management system ADEPT and its mechanisms to support dynamic change can be found in [205].

Part IV
The Core System

Chapter 7
The Architecture

Michael Adams, Marlon Dumas, and Marcello La Rosa

7.1 Architectural and Implementation Considerations

The YAWL System is structured as a service-oriented architecture. It is composed of an extensible set of *YAWL Services* (see, for example Chaps. 10–12), each of which is deployed at a certain endpoint and offers one or multiple interfaces. Some of these services are user-facing, meaning that they offer interfaces to end users, while others offer interfaces to applications or other services.

Among the different styles of service-oriented architectures, the YAWL System adopts the Representational State Transfer (REST) style. In line with the principles of REST (or RESTful) architectures, YAWL services provide access to *resources*, each with a Unique Resource Identifier (URI). For example, each workflow specification is exposed by the Engine as a resource, and when an instance of a workflow specification (i.e., a *case*) is created, this case is also represented as a resource. Also, in line with the principles of RESTful architectures, YAWL services interact with one another using the basic operations defined by the HyperText Transfer Protocol (HTTP). Specifically, YAWL services interact by means of HTTP's GET and POST operations: GET being used to obtain information about a given resource, and POST being used to create resources and to update existing resources. Messages exchanged through these operations are encoded using the eXtensible Markup Language (XML) and (with some exceptions) conform to predefined XML schemas. Messages may contain references to resources, which are concretely represented as Uniform Resource Locators (URLs).

RESTful architectures are conceptually simple and universal. Indeed, HTTP is a mature and pervasive web protocol and XML is supported by libraries and frameworks available for almost every contemporary programming environment.

---

[1] Here, the term *resource* is used to designate an information resource accessible over the Web. Later in the chapter, the term is used in a different setting to denote human actors or software applications that perform tasks during the execution of a workflow.

M. Adams (✉)
Queensland University of Technology, Brisbane, Australia

A.H.M. ter Hofstede et al. (eds.), *Modern Business Process Automation*,
DOI 10.1007/978-3-642-03121-2_7,

These characteristics contribute to keeping the YAWL System lightweight and platform-neutral, which have been two of the key underlying design principles.

The YAWL System has been developed using Java technology and uses several third-party open-source libraries: JDOM[2] for XML processing and for evaluation of XPath expressions – found, for example, in conditional branches within the workflow specifications; Saxon[3] for evaluation of XQuery expressions, which are used to transform data transferred into and out of tasks and to define mappings amongst instances of multiple instance tasks ; and Xerces[4] for XML Schema validation. The core services of the YAWL System are packaged as web archives (or *war*) files that can be deployed in a Java Servlet container. The current implementation of the YAWL System is intended to be deployed into the Apache Tomcat[5] Servlet container, but it can in principle be deployed into other Servlet containers. The process Editor is packaged separately in the form of a Java desktop application.

The rest of this chapter gives an overview of the services composing the YAWL System and of the interfaces they provide.

7.2 A Three-Tier View of the YAWL System

Figure 7.1 presents a three-tier view of the YAWL System. It shows how YAWL services at the *business logic* layer encapsulate resources in the *data* layer, and provide functionality to user-facing services (or applications) at the *presentation* layers.

The inner layer in this figure is the data layer, which among other things, stores workflow specifications. A workflow specification essentially covers three aspects: (1) the control-flow logic; (2) the data definitions (XML schemas, input and output mappings for each task, and boolean conditions for conditional flows); and (3) The resources required to execute the various tasks. Resources are linked to an *organizational model*, which specifies information about the participants, such as roles, capabilities, and privileges. The data layer also contains the *execution data*, which includes *case data* and the *execution logs*. The term "case data" refers to data associated with individual cases, specifically values of variables defined at the level of the root net of a workflow specification and values of variables defined at the level of tasks or subnets thereof. The execution logs are composed of entries, each representing an event such as the creation or completion of a case or the start or end of a task. Log entries include values such as start/end times, input/output data, and end-status. Case data and execution logs may be persisted into a database for subsequent analysis or to allow the recovery of workflow cases in the event of a system failure.

The core service of the YAWL System is the *workflow engine* (or *Engine* for short). This service is responsible for creating, routing, and synchronizing the

[2] www.jdom.org

[3] saxon.sourceforge.net

[4] xerces.apache.org/

[5] tomcat.apache.org/

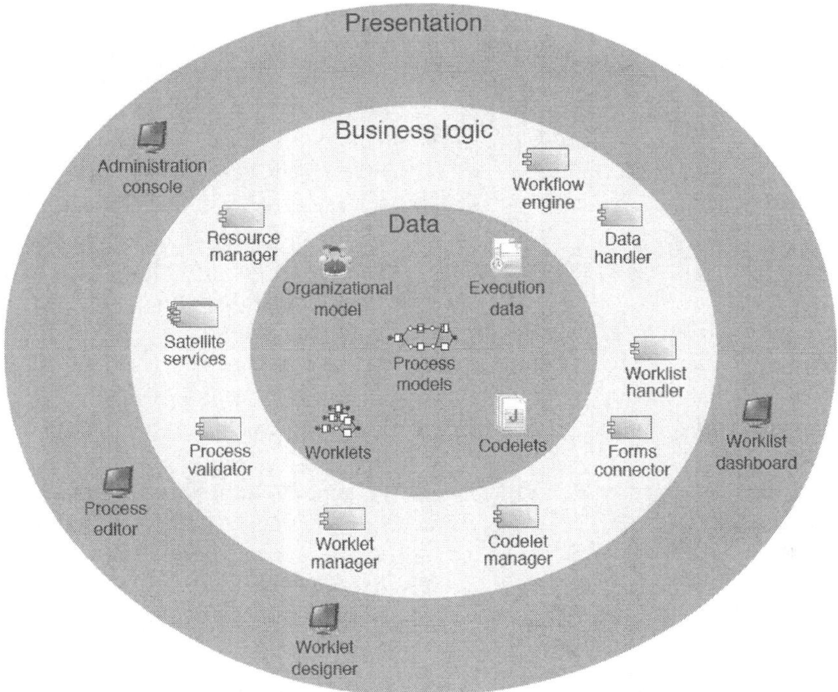

Fig. 7.1 The YAWL system architecture: the YAWL services in the business logic layer access the data layer and interact with users via the presentation layer

execution of work items according to a workflow specification. Every other service must first establish a valid session with the Engine before further communication can proceed. The Engine, on receiving a connection request from a service will, if the service is deemed to be authorized, pass back a session handle token. That session handle is then passed as a parameter to the Engine for all subsequent requests. A request that is accompanied by a missing or invalid session handle will be rejected by the Engine. Thus, only authorized services may communicate with the Engine. A session handle expires after a certain period of inactivity. By default, this allowed period of inactivity is set to 60 minutes but it can be configured differently.

A number of services communicate with the Engine to extend the environment with functionality related to the management of individual tasks. To understand these *task-related services*, it is useful to recall the concept of a *work item*. A work item is a runtime instantiation of a task defined in a workflow specification together with its associated data, and is instantiated from its task definition when control-flow reaches the task during execution of the process instance. When a work item is instantiated, it is said to be *enabled*. Each enabled work item is handled (or processed) in some way either by a task-related service that was associated with the task at design time, or, if no such service association is defined, its handling defaults to the resourcing process described below.

For resourcing purposes, work items can be defined as *manual* or *automated*. Manual work items are performed directly by a human worker, while automated work items are routed to a software application for automated processing.[6] Manual work items are routed to the *resource manager*, while automated work items are routed to the *codelet manager*. The distribution of manual work items to organizational participants is governed by a resource-based access control mechanism handled via the *resource manager*. This service resolves the task–resource associations by identifying a set of participants to whom a work item should be offered for execution. This information is transferred to the *worklist* service, which handles interactions with the end user. All operations performed by the human resources participating in a workflow case are stored in the execution logs (e.g., allocation and starting time of a work item). Furthermore, the resource manager offers access to system administrators to create and modify the organizational model via the Administration console (e.g., to add or remove a participant or to assign roles to participants).

The *worklist handler* is responsible for offering and allocating manual work items to users and transferring the associated business data through a Web form. This service provides a dashboard through which each user can query the set of work items being offered to them, can allocate a specific work item to themselves (thus locking it), start a work item, or complete it. When a work item is started (or "checked out"), the worklist handler generates a Web form, allowing users to view and enter data related to the work item. When a work item is completed (or "checked in"), the worklist handler submits to the Engine any data gathered during the execution of the work item.

As communication with the worklist handler (and other services of the YAWL System) is via XML over HTTP, it is not difficult to build customized Web applications on top of the worklist handler to expose work lists and work items to the end user. The system is shipped with a default renderer that generates Web forms with a basic layout. Alternatively, the *forms connector* can be combined to the worklist handler to allow the connection of manual work items to custom-made Web forms (e.g., JSP or ASP forms). The path of a manual work item through the involved services is illustrated in Fig. 7.2.

Automated tasks are handled by codelet executions. A codelet is essentially a Java class that complies to a predefined Java interface, and is programmed against other predefined Java interfaces for controlling a work item and accessing work item data. The *codelet service* is a special type of service to which the Engine delegates the execution of codelets. The codelet model offers the expressiveness and versatility of a programming language, and is therefore suitable for the implementation of operations that involve complex data manipulations.

A task, whether manual or automated, may additionally be linked to the *Worklet Service* for its execution. This service allows users to dynamically change a running

[6] Note that the distinction between manual and automated work items is only meaningful for a task handled by the Resource Service – it is ignored for those work items that are associated with other services.

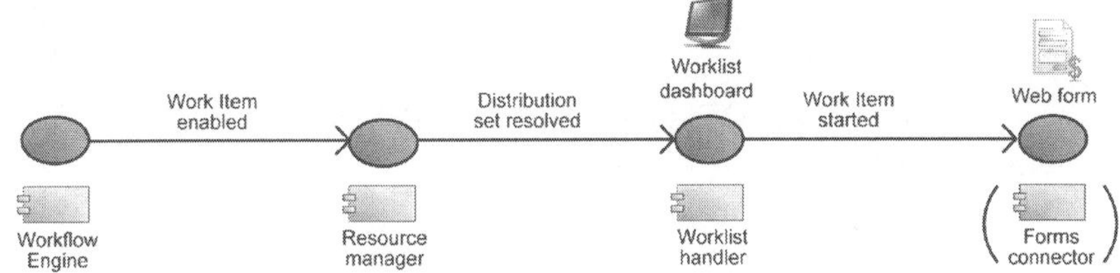

Fig. 7.2 The path of a manual work item through the YAWL services

work item by plugging self-contained subprocesses (worklets) drawn from a Worklet Repository, thus providing dynamic flexibility to otherwise static process instances. The Worklet Service is also called upon to handle expected and unexpected exceptions and to store information, allowing users to better deal with such exceptions in future occasions. The Worklet Service provides such capabilities through a Graphical User Interface (GUI) component that allows users to create rules and exception handling processes (or *exlets*) and link them to tasks through a Ripple-Down Rule strategy.

The YAWL Editor allows users to create and edit workflow specifications. The Editor relies on the *process validator* to handle the validation of the workflow specifications, both syntactically and semantically.

Developers can extend the YAWL System by introducing *custom services*. A custom service is any service that interacts with the Engine or with the task-related YAWL services presented above (e.g., resource manager, worklist handler). By default, the YAWL system is shipped with a number of custom services. For example, the execution of a task can also be delegated to the *declare engine*, the *notification* service, the *WS-invoker*, and the *digital signature* service. These services are grouped under the label *custom services* in Fig. 7.1. The list of custom services included in the YAWL System is intended to be extended. One could think, for example, of a custom services providing RSS feeds to track the creation and execution of work items. Another example of a YAWL Custom Service could be a special type of worklist visualizer, which would display work items based on spatial attributes as described in de Leoni et al.

Having provided an overview of the YAWL System and its underlying services, the following section introduces the interfaces that these services offer and consume.

7.3 YAWL Services and Interfaces

The interfaces of the YAWL Architecture are inspired by those defined in the *Workflow Reference Model*[7] (WRM) of the Workflow Management Coalition (WfMC). The WRM describes a core Workflow Enactment Service (or Engine) interacting with a number of generic components via a defined set of standardized interfaces and

[7] www.wfmc.org/standards/docs/tc003v11.pdf

data interchange formats. In addition to the core Engine, the Workflow Reference Model identifies five major component types and their interfaces:

- Interface 1 is for Process Definition Import and Export, through which process definitions are loaded into, or unloaded from, the engine
- Interface 2 is the Client Application interface, through which work items are passed from the engine to an application (e.g., a worklist) for processing, and passed back to the engine when processing is complete
- Interface 3 is the Invoked Application interface, through which the engine may directly activate an external application to process a work item
- Interface 4 provides interoperability with other workflow enactment engines
- Interface 5 is for Administration and Monitoring, through which an external administrative tool may administrate processes running in the engine, and the engine may pass administrative information to such tools

Unlike the YAWL System, the WRM is not structured as a service-oriented architecture. Instead, the WRM is a more abstract description of a core enactment engine and an interoperating set of components, which could be implemented using a variety of architectures. The YAWL System provides a specific embodiment of the WRM interfaces in the setting of a service-oriented architecture. Also, in many respects, the YAWL interface extends and deviates from the WRM interfaces. Thus, it cannot be said that the YAWL System complies with the WRM. It is merely inspired by the WRM, but significantly differs from it in many respects.

The relations between the core services of the YAWL System and their interfaces are captured in Fig. 7.3 and discussed one-by-one below.

7.3.1 Engine

Process specifications are loaded into the Engine and stored in a repository, from where they may be instantiated to produce cases. The Engine manages the execution of cases by progressing each case according to its current state and control-flow description, and by performing the specified data mappings between the case and its tasks as required. In doing so, at each stage of a process, the Engine determines which work items should be offered for processing, and which events should be announced to the environment. Each task in a process instance is associated at design time with at least one YAWL Service (either explicitly or, if not specified, is implicitly associated with the Resource Service).

An overarching design principle of the YAWL System was that the Enactment Engine should be completely agnostic with regards to the services interacting with it; that is, totally unaware of the operations of external services, so that each could be served in a generic way. From an engine perspective, each service is a "black-box" that avails itself to process data served by the engine through its interfaces. Thus, the Engine needs no knowledge of its external environment. This principle is in contrast to traditional workflow systems where, for example, the worklist handler is treated

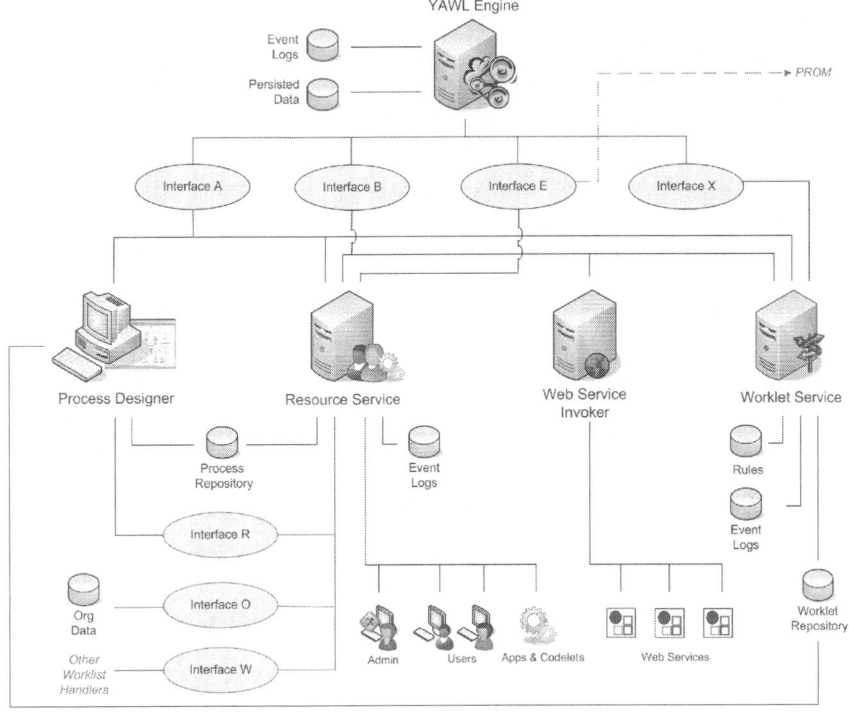

Fig. 7.3 The YAWL System's core services and their interfaces

as being part of the engine, and hence, it is not feasible to replace the worklist handler with a different one. This approach makes the Engine more lightweight, while providing a flexible and extensible framework for plugging in additional (custom) services into the YAWL System.

The Engine interacts with other services in the YAWL System through four interfaces, three of which correspond to the WRM interfaces described earlier. The Engine interfaces are the following:

- *Interface A*, which corresponds to Interface 1 (and partially to Interface 5) of the WRM, and provides endpoints for uploading and unloading process specifications, registering or removing references to external services and basic user connections and disconnections
- *Interface B*, which corresponds to Interfaces 2, 3, and 4 of the WRM, and provides endpoints for services to establish a session with the Engine, launch process instances, check work items in and out of the Engine, and retrieve process data and state information
- *Interface E*, which (partially) corresponds to Interface 5, and provides endpoints for the retrieval and analysis of process logs

- *Interface X* (for eXception), which does not correspond to the WRM, but provides endpoints for the detection and handling of runtime process-level exceptions

The decision to combine the WRM interfaces 2, 3, and 4 into one YAWL Interface (B) was taken to ensure that the Engine remained agnostic to its external services; rather than requiring separate interfaces for invoked applications and interoperability, this decision delegates management of the interaction of those component types to a YAWL Service. The result is that all communications between the Engine and such components are handled through a single, generic interface that provides an additional abstraction layer between the Engine and other services, as opposed to requiring particular interfaces for particular services. For example, the Web Service Invoker Service acts as an abstraction layer between the engine and external web services (such as delegating a booking task to an online travel agency). Without this layer, the engine would have to provide specific interfaces for all the different types of web services and intercommunication protocols available.

Interfaces B and X also provide for the notification of certain events that occur during the life-cycle of a process instance. Interface B notifies all registered services of the following events (Interface X is discussed in Sect. 7.3.4):

- *EnabledWorkItemEvent*: this event is fired when a work item becomes *enabled*, that is, ready to be executed. A service that receives notification of an enabled work item event has the option to check-out the work item (via an interface method), thus taking responsibility for its processing (e.g., displaying its data on a worklist form for editing or passing its data as input to an application). On checkout, the engine marks the work item as *executing*. When the service completes its processing of the work item, it checks it back in to the engine (again via an interface method), the engine maps the updated data (if any) back to its parent net, marks the work item as *completed*, and then enables the next task or tasks in the control-flow.
- *CanceledWorkItemEvent*: this event fires when an enabled or executing work item is canceled (via a cancelation region or exception handler, or withdrawn when a nonselected member of a deferred choice set). A service that receives notification of a canceled work item event may need to undertake remedial action for the work item if it has previously checked it out (e.g., to remove it from a worklist, or cancel an application running on its behalf).
- *CompletedCaseEvent*: this event fires when a process instance completes and is useful to services that started the case so that any finalize actions can be performed.
- *TimerExpiryEvent*: this event occurs when a timed work item reaches its deadline.

Note that a service receiving notifications may choose to ignore them, for example, a service may have no interest in a CompletedCaseEvent if the case was launched by another service.

7.3.2 The Editor

The YAWL Editor (sometimes referred to as the *Process Designer*) is a design environment for the creation and verification of YAWL process specifications. It is packaged as a Java desktop application, as opposed to the other YAWL System services that are packaged as web services. It communicates with a running Engine through Interface A to obtain a list of the YAWL Services currently registered with the Engine, to be used when associating a task with a service. It also communicates with a running Resource Service through Interface R (described in Sect. 7.3.3) to obtain lists of the various organizational resources and codelets currently available, so that selected resources can be associated with particular tasks.

The Editor provides a tool palette from which modeling elements (such as tasks and conditions) may be chosen for placement on the design canvas. Routing constructs may be added to tasks and arcs added to join tasks and conditions to form a complete workflow graph representing a particular business process. At any time, a process may be verified and analyzed using various algorithms to ensure completeness, soundness, and so on.

When complete, the process definition may be saved to a disk file, which will contain an XML representation of the specification conforming to the YAWL specification schema. A specification file may be selected by a YAWL Service for deployment into the Engine, which is achieved by either passing a reference to the file, or its actual contents, via the appropriate method of Interface A.

The Editor is described in more detail in Chap. 8.

7.3.3 Resource Service

The basic role of the Resource Service is to allocate enabled work items to resources so that they can be processed. To fully enable all of the resource management patterns supported by the YAWL System, there are four sub-services falling under the umbrella of the Resource Service:

- a *Resource Manager*, which manages the allocation of resources to work items
- a *Worklist Handler*, a web-form based user interface that provides users with the ability to interact with and process work items
- a *Forms Connector*; while the Worklist Handler includes a dynamic web-form generator for the display and editing of work item data, designers may choose to implement specialized forms for particular work items – these forms are managed by this service
- a *Codelet Service* that maintains and executes codelets selected for automated tasks

Resources may be people or may be an application, service, or codelet of some kind. At design time, each task in a specification may be marked as *manual* (requiring a human resource) or *automated* (requiring a non-human resource). In the

Resource Service, a human resource is referred to as a *Participant*. Each participant may perform one or more *Roles*, hold one or more *Positions* (each of which belongs to an *Org Group*), and possess a number of *Capabilities*. Workflow tasks that require resourcing at runtime have their resourcing requirements specified at design time concomitantly with the design of the process control-flow and data perspectives, using the Editor.

In addition to communicating with the Engine through Interfaces A and B, the Resource Service exposes functionality through three interfaces of its own:

- *Interface O* provides an interface to organizational data sources. In addition to providing a default organizational data model, which may be populated post-installation, this interface allows preexisting organizational data sources to be directly "plugged in" to the service. The interface is generic enough to allow the use of not only organizational data existing in other relational database systems, but also other types of data sources, such as LDAP servers, web services, spreadsheets, or even plain text files, requiring only a thin translation layer to be implemented
- *Interface R* provides access to the organizational data by authorized external clients (such as, but not limited to, the Editor). This interface provides sets of both human resource and codelet descriptors
- *Interface W* exposes the entire worklist routing functionality to allow external, specialized worklists to be developed and implemented. Such worklists may be used in conjunction with, or completely replace, the default worklist handler

To access the Resource Service via Interfaces R and W, a session must first be established with the service in a similar fashion to sessions established with the Engine. The Resource Service maintains its own session tokens with external clients.

At runtime, the Resource Service receives *EnabledWorkItemEvent* notifications from the Engine via Interface B for all enabled work items that are not specifically registered with another YAWL Service. For manual tasks, the service then retrieves the resourcing specification (defined at design time) for the task from which the work item was created. Based on that specification, it will determine the initial *distribution set* of participants, and then apply all the defined filters, constraints, and allocation strategies to that set before distributing the work item to the appropriate participant(s). For automated tasks, the service retrieves a reference to the specified codelet (if any) and then executes it using the work item's data as inputs.

The sets of filters, constraints, allocation strategies, and codelets available to a designer for a task are each "pluggable," that is, they can be easily extended so that developers can add new members to the set of each by implementing the appropriate (Java) interface and adding the new class to the Resource Service's repository. Such additions are immediately available to designers via Interface R.

The Resource Service is described in detail in Chap. 10.

7.3.4 Worklet Service

The Worklet Service is a term used to refer to two separate but complementary services: a Selection Service, which enables dynamic flexibility for process instances, and an Exception Service, which provides facilities to handle both expected and unexpected process exceptions (i.e., events and occurrences that may happen during the life of a parent process instance that are not explicitly modeled) at runtime.

In essence, a *worklet* is a (typically small) YAWL workflow specification that has been designed to execute as a substitute for an enabled work item, so that it contextually handles one specific task in a larger process instance. At design time, any task may be associated with the Worklet Selection Service to enable dynamic flexibility for that task. At runtime, on receiving notification from the Engine of a work item-enabled event, the Worklet Selection Service selects an appropriate worklet from a repertoire of worklets defined for that task, by applying a set of extensible rules to the context of the work item and its parent case. The work item is checked out of the Engine via Interface B, the corresponding data inputs of the work item are mapped to the selected worklet, which is then loaded (via Interface A) and launched (via Interface B) in the Engine as a separate case (the Engine has no knowledge of the relation between the parent case and the worklet instance). When the worklet completes, the service receives a *CaseCompleted* notification, and then maps the worklet's output data to the output data of the original work item, which is then checked back into the engine, allowing the original process instance to continue.

The Worklet Exception Service extends from the Selection Service and uses the same repertoire and dynamic rules approach to provide detection and handling support for process exceptions that may occur during the life-cycle of a case. Instead of worklets, the Exception Service maintains a repertoire of exception handling processes (called *exlets*) for each specification, which are selected and invoked in the event of an exception occurring, to perform any required actions and compensations defined. While the Selection Service is invoked for certain nominated tasks in a process, the Exception Service, when enabled, is invoked for *every* case and task executed by the Engine, and will detect and handle up to ten different kinds of process exception.

As part of an exlet definition, a process modeler may choose from various actions (such as canceling, suspending, completing, failing, and restarting) and apply them at a work item, case, and/or specification level, using a graphical tool. Also, any compensatory actions can be performed by adding a worklet to the exlet definition. By providing an external exception handling service, the original parent process model only needs to reveal the actual business logic for the process, while the repertoire of exlets grows as new exceptions arise or different ways of handling exceptions are formulated. Table 7.1 summarizes the differences between the two subservices in terms of the interface used and the action returned from a service selection.

In addition to Interface B, the Worklet Service also listens for notifications passed by the Engine through Interface X when the Exception subservice is configured to

Table 7.1 Summary of service actions

Cause	Interface	Selection	Action Returned
work item enabled	B	Case & item context data	Worklet
Internal exception	X	Exception type and case & item context data	Exlet
External exception	–	Exception type and case & item context data	Exlet

be on (enabling and disabling an Interface X listening service is achieved via a parameter setting in a configuration file). Interface X provides the following events:

- *CheckCaseConstraintEvent:* This event is fired when a case begins, and again when it ends. If there are constraints (in the rule set) associated with the particular specification, they are checked against the case data. If the constraints are violated, an appropriate exlet is selected based on the context of the case and/or work item, and then executed
- *CheckWorkItemConstraintEvent:* This event is fired when a work item is enabled, and again when it completes, and works in a similar way to the case-level constraint events
- *WorkItemAbortException:* This event is fired when the processing of a work item is being performed by an external application and the application fails before processing is completed
- *TimeoutEvent*: This event occurs when a work item has an invoked timer associated with it and the deadline set for that work item is reached
- *ResourceUnavailableException:* This event occurs when an attempt has been made to allocate a work item to a resource and the resource reports that it is unable to accept the allocation or the allocation cannot otherwise proceed
- *ConstraintViolationException:* This event occurs when a data constraint has been violated for a work item during its execution (as opposed to pre- or post-execution)
- *CaseCancelationEvent:* This event is fired when a case is canceled (as opposed to the *CanceledWorkItemEvent* of Interface B)

It is through these events that the Exception Service detects exceptions and provides handling processes for them. Note again that the Engine has no knowledge of an exception occurring, or of how it is (or is not) handled, but simply announces milestones in a process instance life-cycle or relays exception triggering events from other services to the Worklet Exception Service (and/or any other Interface X listeners). A further exception type, the external exception, is triggered not via Interface X, but by the environment (e.g., via an administrator interacting with a web form). The Worklet Service is described in more detail in Chap. 11.

7.3.5 Web Service Invoker Service

As mentioned earlier, automated work items need to be routed to software applications. It is common nowadays to expose software applications as SOAP web services. The interface and binding information of such web services is normally captured using the Web Service Description Language (WSDL). The Web Service (WS) Invoker Service provides a mediation layer between the Engine and external SOAP web services. In this way, a task can be associated with an operation of a SOAP web service at design-time, and at runtime work items associated to this task are routed to the corresponding SOAP web service through the WS-Invoker. Without this layer, a Custom Service would have to be developed between each SOAP web service and the Engine.

Architecturally, the WS-Invoker Service is quite straightforward. It listens for *EnabledWorkItemEvent* notifications on Interface B, which occur for tasks that have been linked with the WS-Invoker Service at design time. Upon receiving such an event, the Invoker checks out the work item and uses its data to call the specified web service. When a reply is received from the service, the work item's output data is populated and the work item is checked back into the Engine. All of the interactions between the Invoker Service and the Engine are thus through Interface B.

For a task to be registered with the Invoker Service, its data inputs require that values for three attributes be defined at design-time:

- *YawlWSInvokerWSDLLocation*: specifies the URI of the WSDL file describing the web service to invoked
- *YawlWSInvokerPortName*: specifies the port binding that the web service listens on for interaction with external clients and protocols
- *YawlWSInvokerOperationName*: specifies the name of the operation to be executed within the web service

In addition, any parameters that are required for the web service's operation must also be specified within the task definition; these are passed to the web service on invocation. The Invoker Service returns a mapping of resultant data to the work item's output data parameters. Note that when a task is registered with the Invoker Service at design time, the Editor automatically populates the task's input parameters with the required data attributes above.

At present, the WS-Invoker Service supports only SOAP over HTTP and request-response and one-way interactions (out-in and out-only message exchange patterns). It is expected that future versions of the WS-Invoker will support other protocols and possibly other message exchange patterns.

7.3.6 Other Custom Services

As previously mentioned, the YAWL System can be extended by implementing custom services. These services may be designed, for example, to handle specific types of work items (e.g., to display work items with geo-referenced attributes in a map) or to add monitoring functionality on top of the Engine. Each of the component services described earlier is in fact an example of a custom service. The following custom services are also included in the current distribution of the YAWL System:

- *Declare Service*. Declare is an approach to business process modeling that is intended for use for the specification and enactment of loosely structured processes. Rather than explicitly modeling the control-flow, Declare models define constraints over the activities of a process, so that the control-flow is specified implicitly, and any execution path that does not violate the specified constraints is valid. The Declare Service receives notifications of registered enabled work items via Interface B and lists them in an administration tool, from which an administrator may choose an appropriate, loosely structured Declare subprocess to execute in its place.The Declare Service is described in more detail in Chap. 12
- *SMS Service*. The SMS service uses a properly configured SMS Gateway web service to send and receive SMS messages and pass the data into and out of the Engine (again via Interface B). In this way, participants can view, update, and complete work items via mobile phones and other SMS capable devices
- *Digital Signature Service*. The purpose of this service is to ensure the authenticity of the information entered on a submitted custom form. It is composed of two parts: first, it passes the web form through the service to capture and append the digital signature as XML, then it checks the validity of the signature before passing the form back to the worklist handler, where its contents are mapped to the work item's output parameters for checking back into the Engine
- *Email Sender Service*. A task registered with this service will create a work item at runtime that presents the user with a custom form for the composing and sending of an email

7.4 Summary

This chapter introduced the YAWL System and its service-oriented architecture. The key building block of the system, namely the Engine, provides a number of interfaces that are used and enhanced in various ways by task-related services. These interfaces also allow workflow specifications to be deployed into the Engine by the YAWL default Editor or other custom-made designers such as the one provided by the Declare service. Additionally, using these interfaces, the execution of tasks can be monitored, controlled, or channeled by custom services, such as the SMS sender service or the WS-Invoker service. The Custom Service approach provides a powerful point of extensibility, allowing developers to customize the Engine to their requirements.

The choice of a RESTful architectural style, with messages exchanged in Plain Old XML (POX) over HTTP, makes the YAWL System simple to program against. An alternative SOAP-based Web services style would have introduced additional complexity which, while enabling additional infrastructure-level functionality to be exploited, was not found to be strictly necessary in the context of the YAWL System. A potential drawback of the adopted architecture is that it ties the YAWL System to the HTTP protocol. The ubiquity of HTTP, however, compensates for this potential drawback. In addition, the YAWL WS-Invoker enables SOAP-based services to be invoked during the performance of designated tasks.

The interfaces provided by the YAWL System's services are described in its technical documentation. In particular, XML Schemas for a number of the YAWL services are provided with the system. A more complete and machine-processable specification of these interfaces could be achieved using WSDL, especially given that WSDL 2.0 supports the description of RESTful services. It is expected that such specifications will be added in future releases of the system.

The overview of the YAWL System provided in this chapter sets the ground for more in-depth descriptions of specific parts of the system given in Chaps. 8–12.

Exercises

Exercise 1. The choice of a RESTful architectural style for the YAWL environment did not come without certain trade-offs between its features and those of other network-based software architectures. Research REST and the available alternative architectures in terms of the features and capabilities each offers. Discuss why you think the RESTful architectural style was chosen for the YAWL environment, with particular reference to the following:

(a) Does it make it easier to create and manage communication between custom services and the engine?
(b) Does it impose any limitations on the environment?

Exercise 2. YAWL uses a number of interfaces to interact with external services and applications.

(a) Compare the YAWL interfaces with those defined in the Workflow Reference Model. How closely do the functions of the YAWL interfaces map to those proposed by the reference model?
(b) Why does YAWL not have a direct implementation of the reference model's Interface 3? How is the functionality of Interface 3 implemented in YAWL?
(c) Which YAWL interface is used:
   (i) To load a specification?
   (ii) To "check-out" a work item?
   (iii) To query the process logs?
   (iv) To start a case instance?
   (v) To detect a runtime exception?

Exercise 3. What are the similarities and differences between a YAWL Custom Service and a Web Service?

Exercise 4. YAWL Custom Services can be implemented to handle work announced by the engine in a wide variety of ways. Think of an idea for a YAWL Custom Service. What function would it have? What interface(s) would it use?

Chapter Notes

The initial release of the YAWL environment was a prototype version released late in 2003. The first full open-source release was the "beta2" version, released in July 2004. Since then, there have been around two dozen full releases of the environment to date.

The seminal paper on Representational State Transfer (REST) style network-based software architecture, on which the service-oriented architecture of YAWL is based, can be found in [94], while a good resource for more information on SOAP and WSDL is provided by Curbera et al. [68].

The work item visualizer service referred to in Sect. 7.2 is detailed in de Leoni et al. [148].

Chapter 8
The Design Environment

Stephan Clemens, Marcello La Rosa, and Arthur ter Hofstede

8.1 Introduction

The core modeling component of the YAWL System is the *Editor*. This tool enables workflow designers to graphically define complex process models, and to analyze and export these models to the Engine.

Specifically, the Editor is a rich client application offering visual support for the definition of the process control logic, the data associated with the process and its tasks, and the organizational resources participating in the process. An important aspect of the Editor is the provision of sophisticated logic to verify the produced models. Through this capability, a designer can pinpoint syntactical and semantic issues at a mouse click, so as to avoid potentially costly mistakes before deploying a process to a workflow engine.

The main driver behind the development of a visual editor for YAWL was the necessity to speed up the creation of process models and to foster the uptake of the language by nontechnical users. In light of this, the Editor had to fulfill the requirements of *free availability*, *portability*, *ease of use*, and *interoperability*. The tool had to be freely available; therefore, it was decided to release it under the open source LGPL license. Portability was guaranteed by developing the Editor in Java™ and avoiding any code dependency on OS-specific libraries. Ease of use was achieved by providing the Editor with an intuitive user interface on top of a core graphical component based on extensions to the JGraph libraries.[1] Interoperability was supported by the definition of a common XML format and a set of API calls for the exchange of workflow specifications between an editor and the runtime environment (cf. Chap. 9). The interaction between design and runtime environment in the YAWL System is depicted in Fig. 8.1.

The decision to minimize the dependencies between the design environment and the runtime environment via the use of a common interchange format and a set of

[1] www.jgraph.com

S. Clemens (✉)
Queensland University of Technology, Brisbane, Australia
e-mail: stephan.clemens@qut.edu.au

A.H.M. ter Hofstede et al. (eds.), *Modern Business Process Automation*,
DOI 10.1007/978-3-642-03121-2_8,

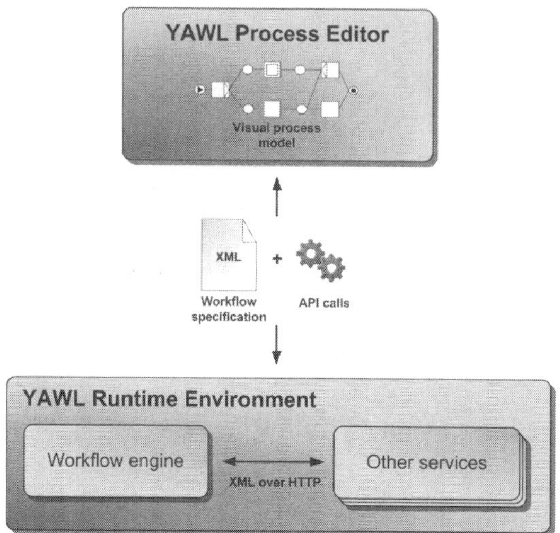

Fig. 8.1 Workflow specifications created with an editor are deployed to the runtime environment

APIs is inspired by the Workflow Reference Model proposed by the Workflow Management Coalition (WfMC).[2] In WfMC terminology, an editor embodies a *Process definition tool* for modeling and analysis of workflows, while the runtime environment represents a *Workflow Enactment Service* for the execution of such workflows. The interaction between these two components is achieved via the *Workflow Definition Interchange* (Interface 1), which is embodied by the XML format and the set of APIs in the YAWL System.

According to the WfMC, the advantage in using a standard interaction format is twofold. First, it defines a clear point of separation between design and runtime environments. Second, it enables the use of different design tools and workflow engines. For example, a workflow specification exported by an editor could be executed by a number of independent workflow engines that cooperate to provide a distributed runtime environment. Similarly, it is also conceivable to think of a development environment in which a number of design tools can export workflow specifications to the same runtime environment. For example, these tools could differ in their look "n"feel or preferences to address different classes of users.

8.2 Setting up the Process Control Logic

The Editor user interface comprises a modeling canvas located in the center, two toolbars – one on top and the other on the left of the canvas, a bottom panel for notes/problems, and a status bar reflecting the status of the Engine, Resource Service, and general messages. Figure 8.2 shows a screenshot of the user interface with

[2] www.wfmc.org

the starting net *Overall* of the Order Fulfillment process model (see Appendix A) displayed on the canvas.

The control-flow of a YAWL specification can be defined by using elements from the palette, located at the top of the left toolbar. The palette *Wf-Elements & Tools* contains seven buttons, which assist with creation, selection, and positioning of workflow elements on the canvas (see Fig. 8.2). The first five buttons are used to place workflow elements on the net, such as an atomic task or a condition. The marquee button allows the selection of individual or multiple elements that can be moved within the net, while the Net Drag button (indicated by a cross) is used to drag or modify single workflow elements on the canvas.

The palette *Task Icons* contains a set of predefined icons to decorate tasks (see Fig. 8.2). For example, icons can be used to distinguish tasks that are executed by a human resource from those executed by an application, as in the Order Fulfillment process model. This set of standard icons can be enriched with custom-made icons, added through a plug-in mechanism. The use of icons can increase the overall understanding of a process, but it will not influence the task's behavior.

A YAWL workflow specification is composed of one starting net and zero or more subnets. The starting net captures the behavior of the overall process and is the first net to be executed when a case is launched. Each subnet captures the behavior of a composite task and is executed once the respective task fires. For example, in the Order Fulfillment process model, the net labeled *Overall* is the starting net, while nets *Ordering*, *Payment*, and *Freight in Transit* are subnets mapped to the homonymous composite tasks (see Fig. 8.2). The division of a process model into smaller parts by means of subnets can facilitate maintenance and readability, especially in the case of complex process models such as the Order Fulfillment example.

Each net features two mandatory elements: the input and the output condition, which cannot be removed from the canvas. Tasks can be arranged between these two nodes to describe the control-flow of the process and are connected to each other by means of arrows, which represent order dependencies.

While a composite task is a placeholder for a subnet, an atomic task captures a standalone piece of work, namely a work item. There are three types of atomic task: Manual, automated, and routing task. A manual task is executed by a human resource, for example, a particular employee or an organization participant with a certain role or position. In the subnet *Ordering* (Fig. A.3) of the Order Fulfillment process model, tasks *Create Purchase Order* and *Confirm Purchase Order* are manual tasks that are performed by the Purchase Order Manager. The Editor supports the workflow designer in setting up resource related aspects, for example, which manual task will be performed by which resource or what strategy to use when allocating a work item. This can be done via the *Manage Resourcing...* dialogue, which is available from the context menu that pops up by right-clicking on the task. More details on resource assignments are provided in Sect. 8.4.

An atomic task associated with the default worklist handler is manual by default. If set as automated, a designer may assign a codelet or an external application to the task via the *Update Task Decomposition* dialogue available from the task's context menu. Codelets are code snippets that are executed internally by the Engine.

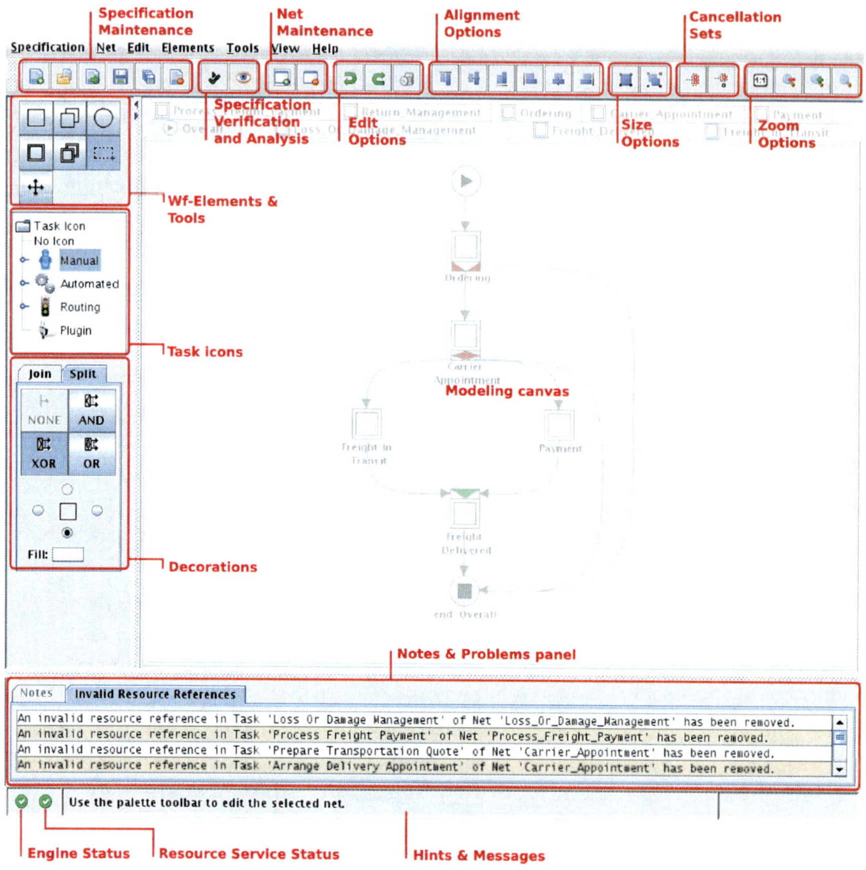

Fig. 8.2 Graphical user interface of the Editor

To select a codelet, the Editor needs to be connected to the Resource Service. A predefined set of codelets is provided with the Editor. These include a codelet for executing shell commands and a codelet for evaluating XQuery expressions. In addition, workflow designers can plug their own codelets into the YAWL environment. Similarly, to assign an external application to an atomic task, the application needs to be exposed as a Web Service and the latter be registered with the YAWL environment. The YAWL System provides a number of custom services already (cf. Chap. 7). To associate a Custom Service with a task, the Editor needs to be connected to the Engine. If the Engine is active but the status icon in the Editor (see bottom left of Fig. 8.2) states it as offline, a connection can be manually established by invoking the dialogue *Engine Connection Settings* via menu entry *Tools*.

If an atomic task is neither manual nor automated, it is an empty task. An empty task may be a routing task, which is essentially a silent task that is used only for routing purposes. Routing tasks are executed internally by the Engine.

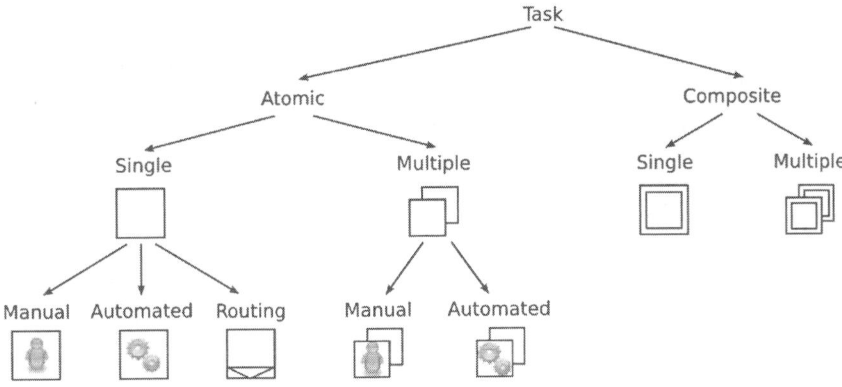

Fig. 8.3 Tasks hierarchy in the Editor

Another type of task is the multiple instance task, which captures an atomic or composite task to be executed multiple times in parallel. The upper and lower bounds for the number of instances, the threshold for completion, and the way of instance creation can be specified via the *Set Instance detail...* dialogue available from a multiple instance task's context menu. Figure 8.3 shows the tasks hierarchy in the Editor.

An atomic single instance task can have a timer set if it needs to be executed within a given timeframe. The parameters of a timer task can be specified via the *Set Task Timer...* dialogue accessible from the task's context-menu. These are the expiry value and the activation type. The expiry value indicates when the timeout should expire. This can be after a period of time (e.g., 3 hours) or at a specific point in time (e.g., 6:00 am, 25 December 2008). The activation type depends on whether the task is manual or automated. In the case of a manual task, the designer can specify whether the timer should be activated when work items of that task are enabled or when they are started. In the case of automated tasks, the timeout behaves as a delay, that is, the execution of the task is delayed by the timeout value. Upon timeout expiry, the task is started. In contrast to manual tasks, automated tasks are not assigned to any human resource. Therefore, the timeout is always activated upon work item enablement.

In the Order Fulfillment process model, the *Ordering* subnet has an automated timer task *Order Timeout* (see Fig. A.3 in Appendix A). This task is set to a duration of 3 days as shown in Fig. 8.4. After the purchase order has been approved, it can be either confirmed (task *Confirm Purchase Order*) or modified (task *Modify Purchase Order*). Should none of these two tasks be executed within 3 days, the *Order Timeout* task will then be triggered.

In the case of multiple incoming flows, the task needs to be decorated with a join construct, while a split construct needs to be applied in the case of multiple outgoing flows. The type of split and join for a task can be chosen from the *Decorations* palette (see Fig. 8.2), and appears as soon as the task is selected on the canvas. As long as a task has no join or split decoration, the Editor does not allow the connection

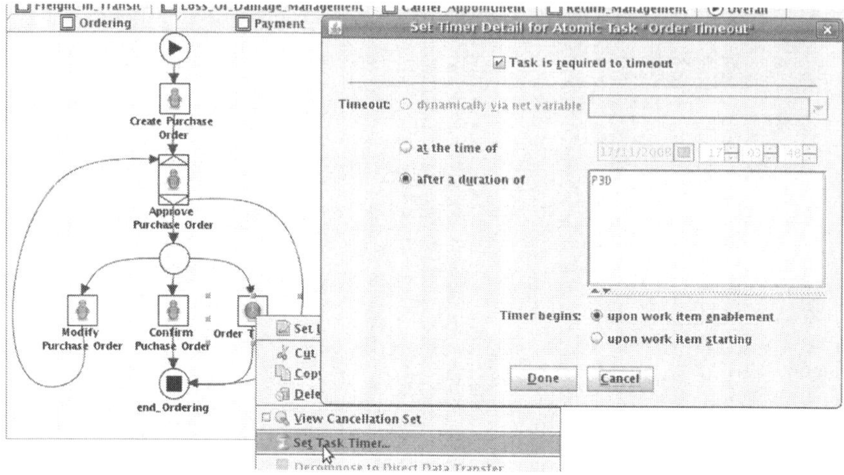

Fig. 8.4 Setting up the parameters of the timer task *Order Timeout*

of more than one incoming or outgoing arc to that task to prevent creating structural issues.

If a task is decorated with an (X)OR split, the workflow designer can specify routing conditions for each outgoing arc of the split. These conditions are predicates expressed in XPath, which can be set via the *Update flow detail...* dialogue available from the task's context-menu (see Fig. 8.5). This dialogue shows the list of outgoing arcs of the split, where each arc is referred to by the label of the subsequent node. In the case of an XOR split, the list of arcs needs to be ordered to determine the evaluation order of the predicates at runtime. Control will be passed along the first arc whose predicate evaluates to true. In this way, only one branch will be executed even if more than one predicate evaluates to true. In the case of an OR split, control will be passed along each arc whose predicate evaluation is true. If all predicates of an (X)OR split evaluate to false, control will be passed along the designated default arc (the last arc of the list) irrespective of the predicate's result.

The atomic task *Approve Purchase Order* in the subnet *Ordering* (Fig. A.3) is decorated with a split and a join. The XOR construct at the bottom of the task splits the control-flow: If the purchase order is not approved, the flow is routed to the end-condition of the subnet. Otherwise, the flow is routed to the subsequent condition leading to the atomic tasks *Modify Purchase Order*, *Confirm Purchase Order*, and *Order Timeout.*

A task can be associated with a cancelation set, which may include a number of tasks, conditions, and/or arcs (in the latter case the implicit condition in the arc is included). The cancelation set of a task can be visualized on the canvas by clicking on *View cancelation set* from the task's context-menu. This enables two buttons in the top toolbar (*Cancelation Sets*, see Fig. 8.2), which can be used to add in elements or remove them from the cancelation set. The task that initiates a cancelation set is grayed out, while the elements in the cancelation set are indicated with a red border.

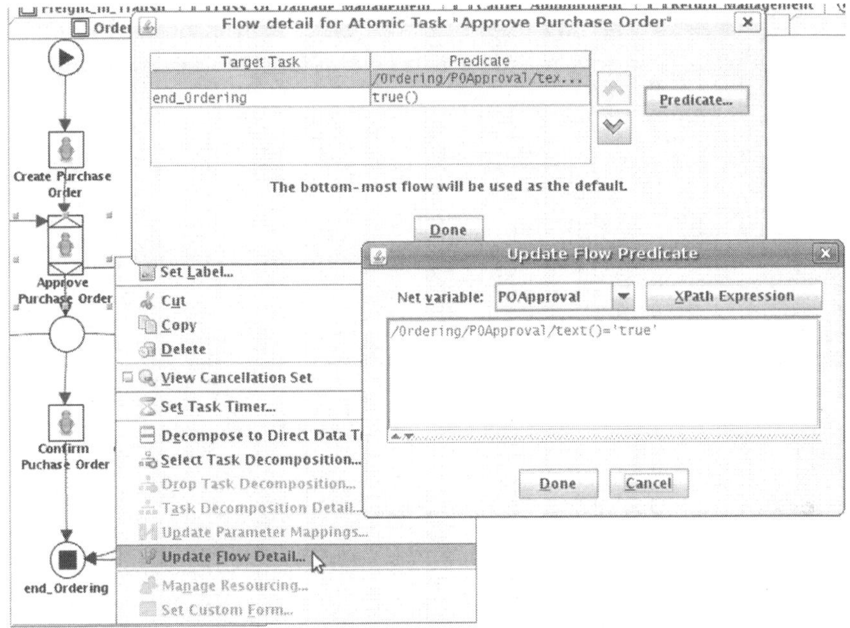

Fig. 8.5 Setting up routing conditions for an XOR split

8.3 Defining Data Aspects

In a YAWL specification, the data flow is captured by means of net variables. The passage of data to/from tasks is achieved by mapping net variables to tasks' variables. At runtime, variables are used to compose the content of a task's work item and to determine the routing behavior of splits.

A variable acts as a container in which a value of a certain XML Schema data type is stored. The Editor supports the full set of simple XML Schema types, for example, *long*, *string*, *boolean*. Complex XML types can be defined as compositions of simple and complex types via the *Update Data Type Definitions* dialogue, which can be found under the *File* menu. For example, *PurchaseOrderType* is a complex type used in the Order Fulfillment process model to describe the content of a purchase order (see Fig. 8.6). It consists of further complex type variables such as *Company* and *Order*, and of simple type variables such as *DeliveryLocation* (*string*) and *PrePaid* (*boolean*).

Before nets and tasks are able to read or modify data, they need to be assigned a decomposition (cf. Chap. 2). A decomposition has one decomposition label (a unique identifier within the whole workflow specification) and may have one or more variables. The Editor facilitates the assignment of the same decomposition to multiple tasks and prevents the workflow designer from creating more than one decomposition with the same label.

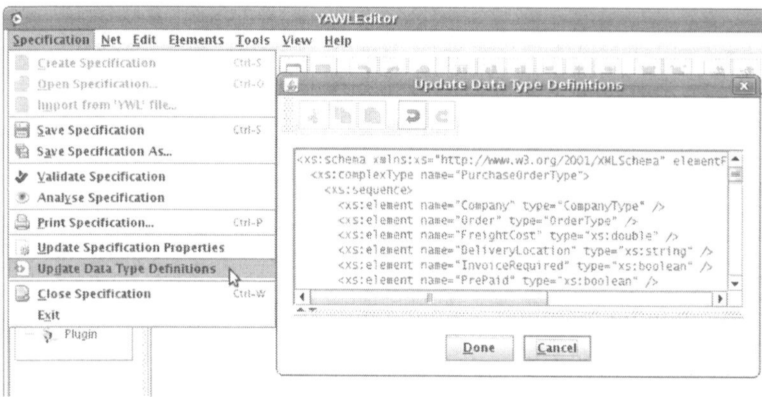

Fig. 8.6 Complex XML Schema type definition of *PurchaseOrderType*

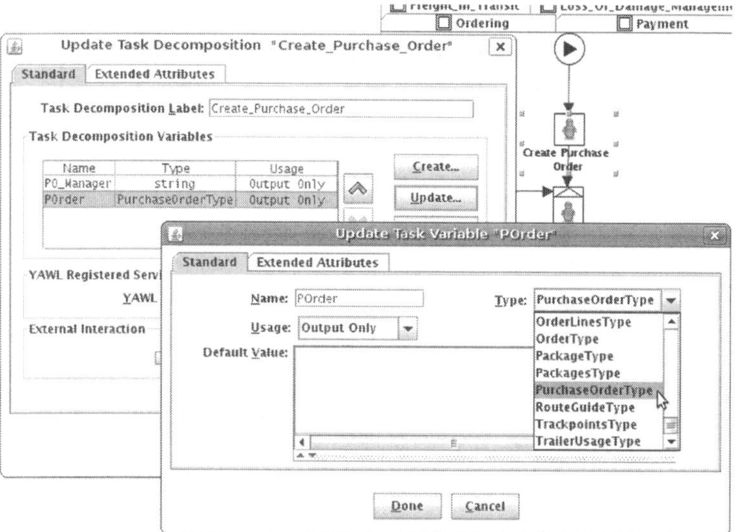

Fig. 8.7 Decomposition of task *Create_Purchase_Order* and set up of variable *POrder*

The details of a net decomposition (i.e., its label and variables) can be set up through the *Update net detail...* dialogue, under the *Net* menu, whereas task decompositions can be created via the *Select Task Decomposition* dialogue and modified via the *Task Decomposition detail...*, both available from the task's context-menu. Figure 8.7 shows the decomposition of task *Create_Purchase_Order*. Here variable *PO_Manager* is of simple data type *string* to contain the username of the purchase order manager, while *POrder* has been assigned the complex type *PurchaseOrderType* as defined in Fig. 8.6.

The Editor supports the workflow designer in setting up inbound and outbound mappings to specify how data is transferred between net and task variables. In

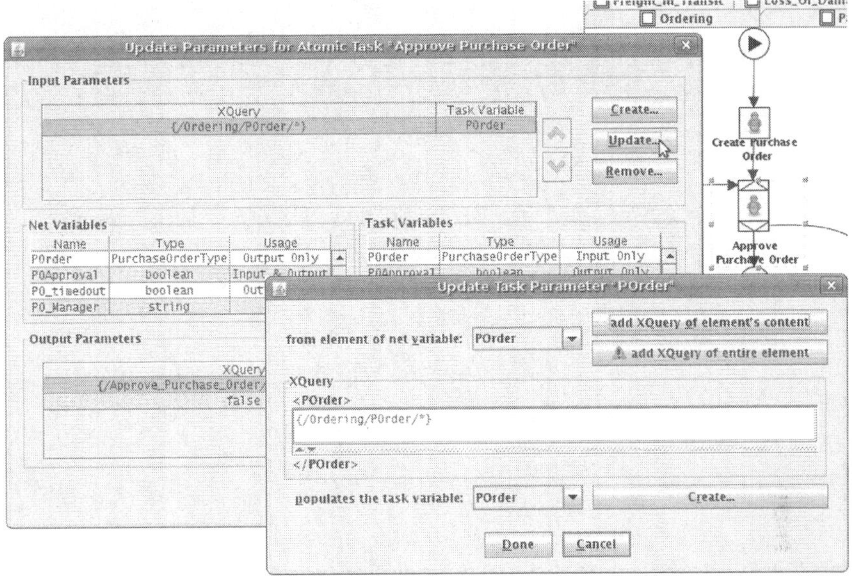

Fig. 8.8 The input parameter to transfer the content of net variable *POrder* to task variable *POrder*

particular, for each task's decomposition, input parameters are used to transfer the content of net variables into task variables (inbound mapping), while output parameters are used to transfer the content of task variables into net variables (outbound mapping). These parameters are defined as XQuery expressions via the *Update Parameters* dialogue from the task's context-menu. Figure 8.8 shows the mapping for task *Approve Purchase Order* of the subnet *Ordering* (Fig. A.3). Here an input parameter is used to copy the content of the variable *POrder* of the net *Ordering* (indicated by the XQuery `/Ordering/POrder/*`) into the task variable *POrder*. Even though both variables are labeled the same, they are assigned to different decompositions and therefore different scopes. This copy operation does not create any issue. On the other hand, the two output parameters of task *Approve Purchase Order* take care of copying the content of the task variables *PO_timeout* and *POApproval* to the net variables bearing the same names.

The creation of decompositions can be labor intensive in the case of large workflow specifications. Complex types may need to be created, variables need to be typed, and mappings need to be defined between net and task variables. The Editor is able to facilitate this process as long as no decomposition has been assigned to an atomic task yet. This can be done via dialogue *Decomposition to direct data transfer*, accessible from a task's context-menu. Here the designer can use net variables as a template to create a task's decomposition. Specifically, the designer can select a set of net variables to be used as input to the task, and similarly a set of net variables to be used as output from the task. Then the Editor creates task variables bearing the same type of the net variables, input parameters to map input net variables onto task

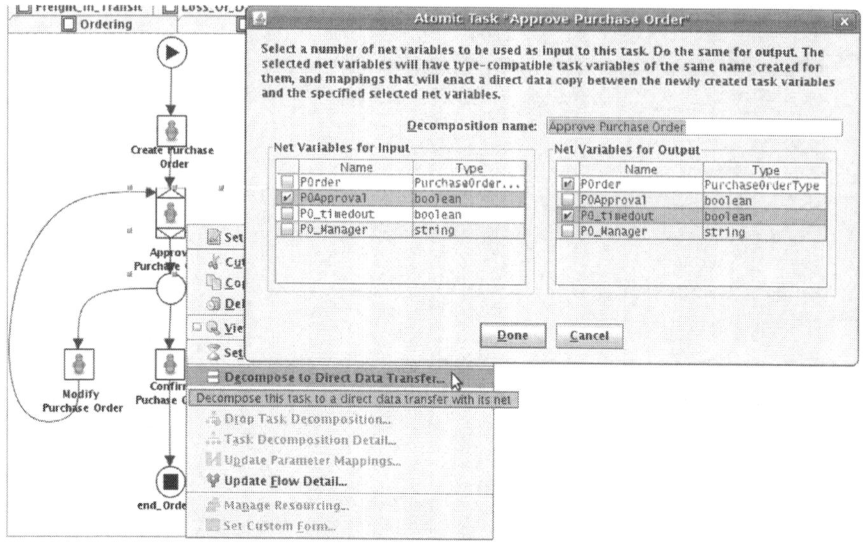

Fig. 8.9 Decomposition to direct data transfer

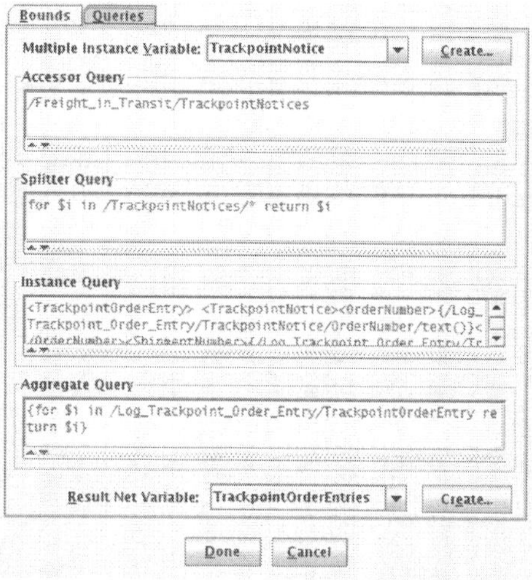

Bounds Queries
Multiple Instance Variable: TrackpointNotice Create...
Accessor Query
/Freight_in_Transit/TrackpointNotices
Splitter Query
for $i in /TrackpointNotices/* return $i
Instance Query
<TrackpointOrderEntry> <TrackpointNotice><OrderNumber>{/Log_Trackpoint_Order_Entry/TrackpointNotice/OrderNumber/text()}</OrderNumber><ShipmentNumber>{/Log_Trackpoint_Order_Entry/Tr
Aggregate Query
{for $i in /Log_Trackpoint_Order_Entry/TrackpointOrderEntry return $i}
Result Net Variable: TrackpointOrderEntries Create...
Done Cancel

Fig. 8.10 Four XQuery expressions for a multiple instance task

variables, and output parameters to map task variables onto output net variables. Figure 8.9 shows the direct data mapping for task *Approve Purchase Order*.

Besides input and output parameters, multiple instance tasks need further information for data manipulation. This information is specified via four XQuery expressions. An example of these expressions is provided in Fig. 8.10, which refers to the multiple instance task *Log Trackpoint Order Entry* in the Order Fulfillment example (cf. Fig. A.6 in Appendix A). The designer needs to define a *Multiple Instance Variable* that will contain the data to be distributed to the various task instances at runtime. The content of this variable can be taken from net variables using input parameters. The data contained in this variable can be separated with the *Splitter query* to pass a unique value to each instance task. The *Accessor query* can be used to manipulate the content of the variable before the unique values are split out (e.g., if some format conversions are needed). Upon completion of all instances, the *Instance query* can be used to transform the returned XML document from each instance to a form suitable for the *Aggregate query* to generate an overall result. A *Result Net Variable* needs also to be specified to contain the overall result. This variable can then be mapped onto net variables via output parameters.

Finally, data aspects also concern the timer task. This task in fact allows designers to late-bind its expiry value to a period of time or fixed date via the use of a variable of type *YTimerType*. In this way, the expiry value will be dynamically determined at runtime.

8.4 Assigning Human Resources to the Process

The Editor provides a *Resource Manager Wizard*, which allows designers to assign participants to manual tasks. This wizard can be invoked by clicking on *Manage Resourcing...* from the task's context-menu after participants have been created in the YAWL workflow environment (see Chap. 10), the task has been assigned a decomposition, and the task has been associated with the default worklist (i.e., the Resource Service). It need to be ensured that the Resource Service is running so that resources can be associated with a task. The status of the Resource Service is reflected by the second right icon in the bottom left corner of Editor (see Fig. 8.2). If the Resource Service is active but the status icon states this service offline, a connection can be manually established by invoking the dialogue *Resource Service Connection Settings* via menu entry *Tools*.

The first step of the wizard allows designers to set up the way a manual task should be offered, allocated, and started. These are called interaction points. The interaction points for task *Create Purchase Order* are shown in Fig. 8.11. At the first interaction point, the work item gets offered to one or more participants. The offering can be set up as system-initiated or user-initiated. In the case of a system-initiated offering, it is required to specify the group of participants at design time. This can be done in the second and third steps of the wizard.

Step two enables a workflow designer to select individual participants by their name, for example, Peter Clemenza, and/or participants of a certain role, for example, PO Manager (see Fig. 8.12).

In step three, it is possible to restrict the group of participants to those who belong to a certain organizational group or positions or have appropriate capabilities (see Fig. 8.13).

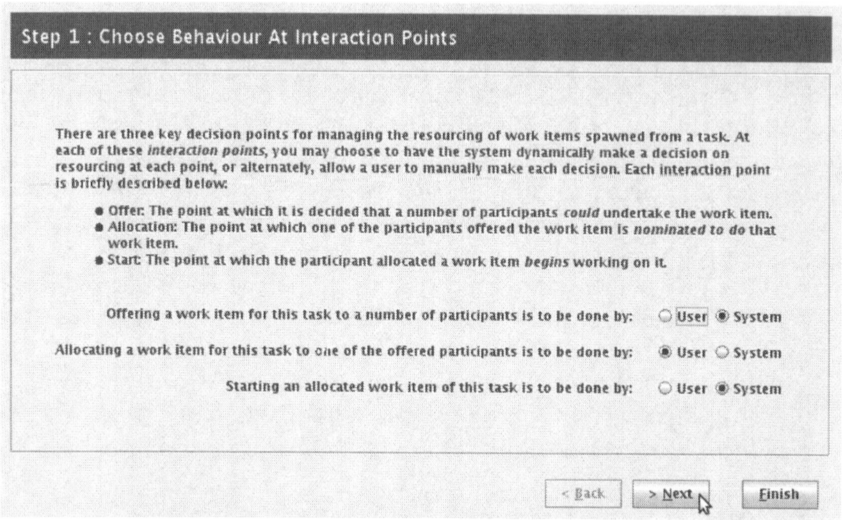

Step 1 : Choose Behaviour At Interaction Points

There are three key decision points for managing the resourcing of work items spawned from a task. At each of these *interaction points*, you may choose to have the system dynamically make a decision on resourcing at each point, or alternately, allow a user to manually make each decision. Each interaction point is briefly described below:

- Offer: The point at which it is decided that a number of participants *could* undertake the work item.
- Allocation: The point at which one of the participants offered the work item is *nominated to do* that work item.
- Start: The point at which the participant allocated a work item *begins* working on it.

Offering a work item for this task to a number of participants is to be done by: ○ User ◉ System

Allocating a work item for this task to one of the offered participants is to be done by: ◉ User ○ System

Starting an allocated work item of this task is to be done by: ○ User ◉ System

< Back | > Next | Finish

Fig. 8.11 Step one: choosing how to offer, allocate, and start a work item

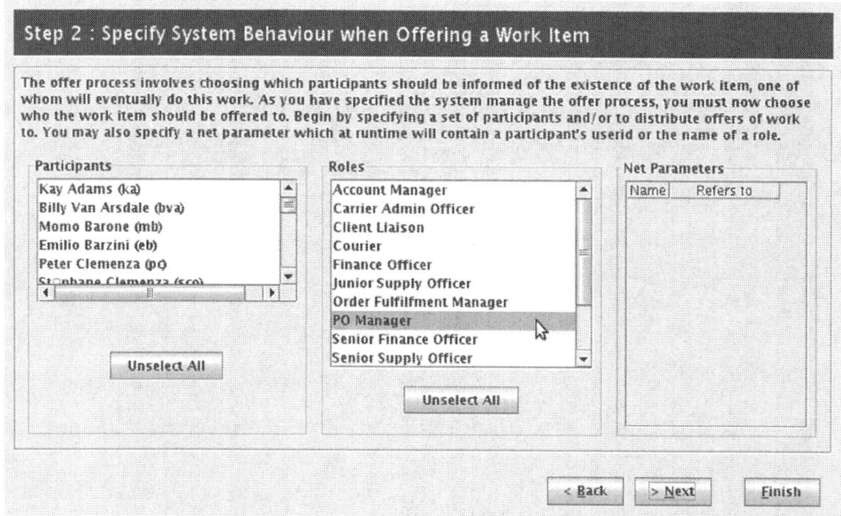

Fig. 8.12 Step two: specifying the distribution set for work item offering

In some cases it may be required to defer the choice of which participant will get offered a particular work item until runtime. This can be done by setting the offering to user-initiated, to allow a user with offering rights (e.g., an administrator) to determine whom the work item is to be offered to at runtime. Alternatively, the Editor enables the use of a variable to store the participant's user identification (late-binding). The variable can be specified within step two of the wizard. In the subnet

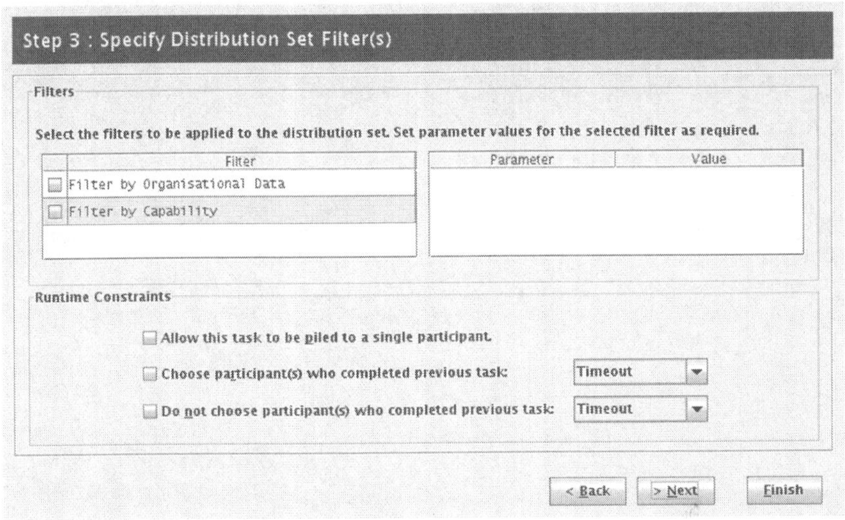

Fig. 8.13 Step three: filtering the distribution set

Ordering, task *Create Purchase Order* allows one to specify a participant who will work on the purchase order if the latter needs to be modified. So, should the task *Modify Purchase Order* be executed, the participant in the net variable *PO_Manager* will get offered the work item.

At the second interaction point, the work item gets allocated to a participant out of the group of participants who have been selected for this work item. Once a work item is allocated to a specific participant, the work item offer is withdrawn from all other participants. In the case of a user-initiated allocation, the participant who is offered the work item can decide whether or not to accept it. On the other hand, in a system-initiated allocation, the workflow system decides which participant the work item should be allocated to. However, in this case the workflow designer needs to define the appropriate allocation strategy during design time. The fourth step of the Resource Manager Wizard provides several strategies to choose from, for example, Round Robin or Shortest Queue (see Fig. 8.14).

At the third interaction point, the work item gets started (i.e., executed) by a participant. This is usually done via a Web form that is linked to the manual task. The Editor allows a designer to specify the URL of this form by clicking on *Set Custom Form...* from the task's context-menu. If no custom form is specified, a default form is dynamically generated for the work item at runtime. If the third interaction point is user-initiated, the participant whom the work item has been allocated to can decide when to start the execution of that work item. In the case of a system-initiated start, it is the workflow environment that places the work item in the participant's started queue.

Finally, step five of the Resource Manager Wizard allows the specification of several participant privileges for runtime, such as suspending or skipping the execution of a work item (see Fig. 8.15).

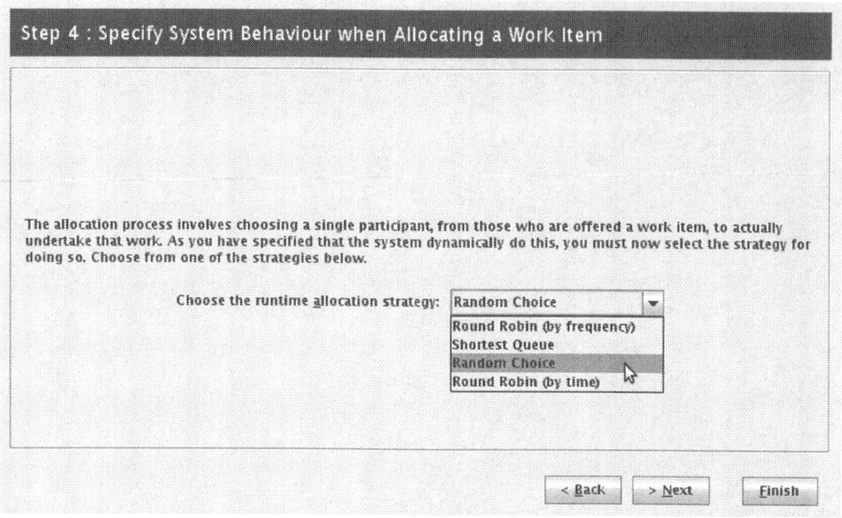

Fig. 8.14 Step four: specifying the allocation strategy

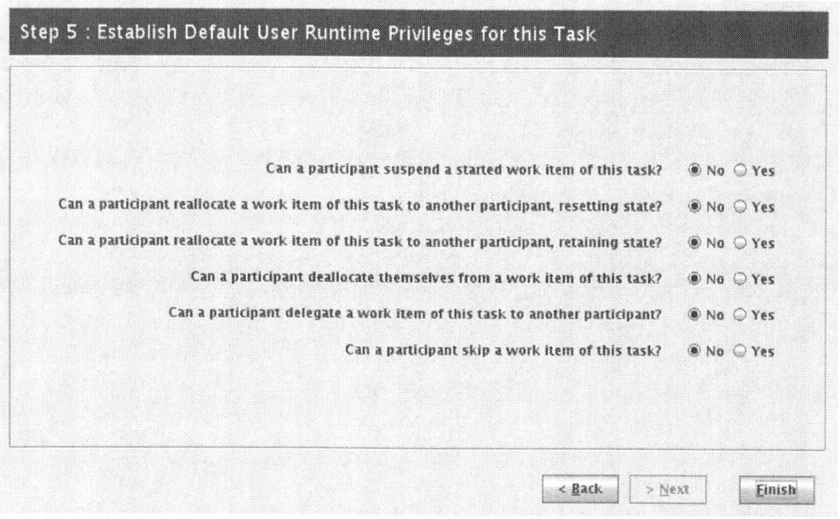

Fig. 8.15 Step five: configuring participant's privileges for execution

8.5 Error Reporting

The underlying philosophy of the Editor is to support the workflow designer in detecting certain undesirable characteristics in the workflow specification at design time. As shown before, XQuery and XPath expressions are strongly used in YAWL for data handling. However, those expressions are not very intuitive to build and can be complex. An incorrect expression can have undesired side effects during workflow execution. Therefore, each XQuery or XPath expression is verified by the Editor when it is entered. Correct expressions are shown in green font color, incorrect expression are shown in red. Furthermore, in the case of an incorrect expression, a suggestion is provided on how the error can be solved. By validating the specification via the button *Validate this Specification* in the top toolbar, a table with listed problems appears in the *Specification Validation Problem* panel, located at the bottom of the Editor interface. The entries show details about inconsistencies that will prevent the specification from running in a workflow engine. For example, by validating the specification in Fig. 8.16, the workflow designer will be informed about the missing inbound mapping for variable *bar* of task *Task A*.

Another error reporting feature of the Editor is the analysis of a specification. This can be trigged via button *Analyze this specification* from the top toolbar (*Specification Verification and Analysis*, see Fig. 8.2). Analysis allows the workflow designer to detect potential behavioral problems of the workflow specification. That is, for example, deadlock situations, unnecessary cancelations set members or unnecessary, OR joins (see Chap. 20 for further information). Analysis results are reported in the *Specification Analysis Problem* panel, displayed at the bottom of the Editor interface (see Fig. 8.2). Figure 8.17 shows a workflow with a potential deadlock and the respective analysis result.

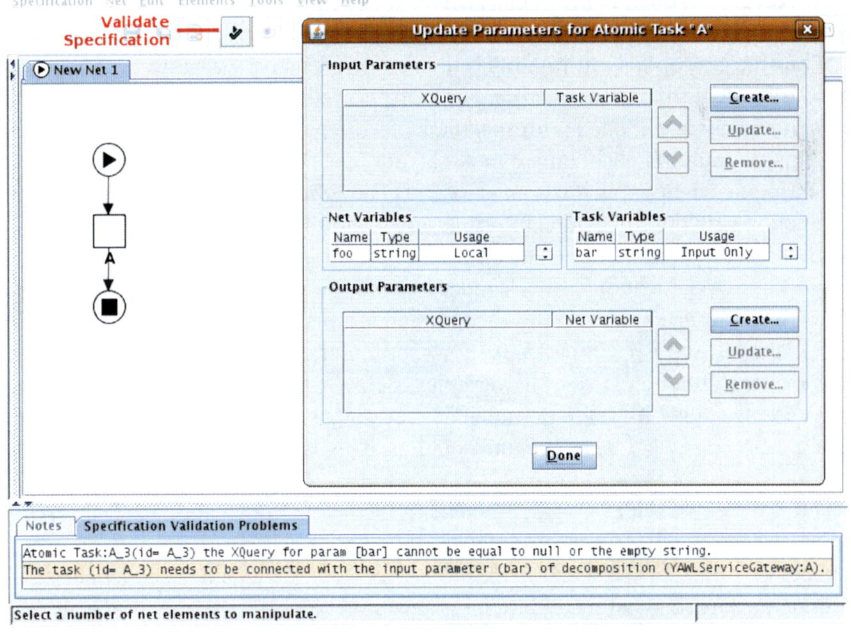

Fig. 8.16 The missing inbound mapping for variable *bar* causes the listed validation problems

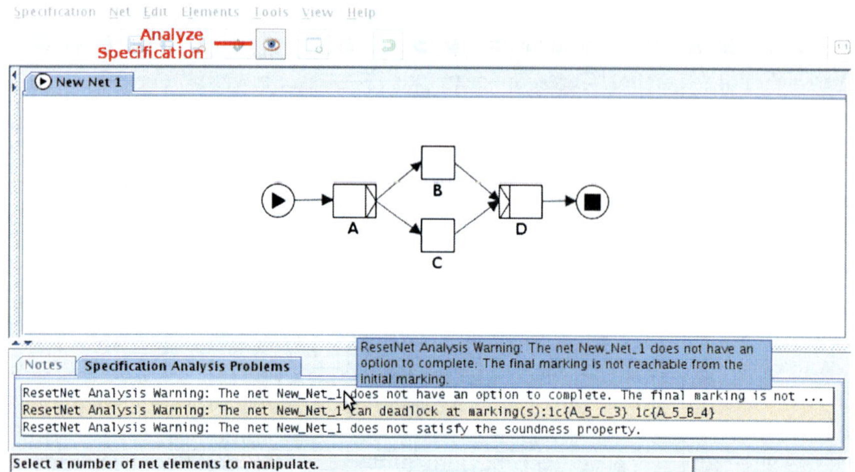

Fig. 8.17 The Editor detects a potential deadlock

8.6 Specification File

The Editor serializes a workflow specification into an XML document with extension `.yawl`. This document contains a mandatory conceptual part with information about decompositions, workflow elements, and their relations.

While the conceptual information is required to execute a workflow, its visualization is important for communication purposes. Workflow elements might be placed in a certain way or enriched with specific icons to improve the process understanding. This information is contained in the layout part, which is optional and ignored by the Engine when a specification is loaded. The Editor uses layout information in conjunction with the conceptual information to represent a workflow. However, if no layout information is enclosed within the XML document, the Editor will visualize the workflow specification with a standard layout.

Listing 8.1 is an extract of the conceptional information for the Order Fulfillment process model (see Appendix A). It shows a decomposition with its identifier *Create_Purchase_Order* and two of its variables. Below, the task *Create Purchase Order* is represented with its reference to its successor task *Approve Purchase Order*, related mappings, and resource information. This task refers to the mentioned decomposition *Create_Purchase_Order*.

The XML extract in Listing 8.2 represents the layout information for task *Create Purchase Order* and for its incoming and outgoing arcs. Layout information for a task includes the size and location of the task, as well as the path of the icon used to decorate the task. Layout information for an arc includes the location of the arc and a code for its line style, which can be orthogonal (encoded as `11`), bezier (`12`), or spline (`13`).

```
<decomposition id="Create_Purchase_Order"...>
 <outputParam>
    <name>PO_Manager</name>
    <type>string</type>...
  </outputParam>
  <outputParam>
    <name>POrder</name>
    <type>PurchaseOrderType</type>...
  </outputParam>
  <externalInteraction>manual</externalInteraction>
</decomposition>
...
<task id="Create_Purchase_Order_104">
  <name>Create Purchase Order</name>
  <flowsInto>
    <nextElementRef id="Approve_Purchase_Order_1901" />
  </flowsInto>
  ...
  <completedMappings>
    <mapping>
      <expression query=
            "&lt;POrder&gt;{/Create_Purchase_Order/POrder/*}&lt;/POrder&gt;"/>
      <mapsTo>POrder</mapsTo>
    </mapping>
    ...
  </completedMappings>
  <resourcing>
    <offer initiator="system">
    ...
  </resourcing>
  <decomposesTo id="Create_Purchase_Order" />
</task>
```

Listing 8.1 Conceptual information for task *Create Purchase Order*

```
<container id="Create_Purchase_Order_104">
  <vertex>
    <iconpath>.../Manual.png</iconpath>
    <attributes>
       <bounds x="177.0" y="80.0" w="32.0" h="32.0" />
    </attributes>
  </vertex>
  <label>
    <attributes>
      <bounds x="144.0" y="112.0" w="96.0" h="28.0" />
    </attributes>
  </label>
</container>

...

<flow source="InputCondition_16" target="Create_Purchase_Order_104">
  ...
  <attributes>
    <lineStyle>11</lineStyle>
    <points>
      <value x="188.5" y="31.0" />
      <value x="192.5" y="80.0" />
    </points>
  </attributes>
</flow>

...
```

```
<flow source="Create_Purchase_Order_104" target="Approve_Purchase_Order_1901">
  ...
  <attributes>
    <lineStyle>11</lineStyle>
    <points>
      <value x="192.5" y="111.0" />
      <value x="191.5" y="153.0" />
    </points>
  </attributes>
</flow>
```

Listing 8.2 Layout information for task *Create Purchase Order*

8.7 Summary

This chapter introduced the main features of the YAWL Editor. It showed how to create and analyze the control logic of a YAWL process, how to specify the involved data, and how to link organizational resources to process tasks.

Moreover, the chapter demonstrated the error reporting feature of the Editor and provided an overview of the YAWL interchange format.

Exercises

Exercise 1. Download the Editor from SourceForge (more details in the Chapter Notes) and start the tool. Open the Order Fulfillment example available with the distribution and browse the various subnets (cf. Appendix A). Specifically, check how data and resourcing requirements have been specified and how different icons were used to identify the various task types.

Exercise 2. Create a workflow specification with two atomic tasks: *Enter command* and *Display command*. Through the first task the user should be able to enter a sequence of characters into a variable *Command* of type *string*. The value of *Command* should then be displayed via the second task. Enrich the specification with icons from the *Task Icons* palette.

Exercise 3. Extend the above example such that after displaying a command, the user should be asked to enter another command. However, if the user enters the command *quit*, the process should terminate straightaway without displaying the command *quit*.

Exercise 4. Introduce a timer task *Timeout*, which cancels task *Enter command* if no command is given within 20 s. In this case, ensure that the message *timeout has been triggered* is visualized in task *Display command*. What happens to task *Timeout* if *Enter command* is completed in time?

Exercise 5. Create a YAWL workflow specification, which represents the following credit card application process:

A typical credit card application process begins when a customer submits an application with a proposed amount of credit. Next, a credit clerk verifies the status of the application. If the application is incomplete (e.g., the customer's credit history is missing), the clerk requests additional information and waits until the customer provides this information. However, if no information is received within a given period of time (e.g., 3 days), a request is sent again. This is done at most 3 times, then the process terminates assuming the customer is no longer interested in the application. If a complete application is received on time, the clerk verifies the customer's income and credit history. Different checks need to be performed depending on the amount of credit (e.g., more stringent requirements may apply to amounts greater than $1,000). After this, the application is passed on to a manager who decides whether to accept or reject it. In either case, the customer is notified of the decision and the process finishes.

Exercise 6. Extend this workflow by allowing the possibility to cancel the application process at any time after a complete application is received but before the manager decides on the application.

Exercise 7. Apply the following change to the credit card application process: *When an approved application is notified, the customer is asked to choose any extra features they may want to add. These include customized card, reward program, and secondary cardholder(s).*

Chapter Notes

In this chapter, we used the YAWL Editor version 2.0. The tool, along with the other components of the YAWL System, can be downloaded from the YAWL project Website hosted by SourceForge.[3] Here the reader can find documentation on various aspects of the Editor and consult the mailing lists and forums to learn the latest news about the system.

The *YAWL user manual* is the complete reference documentation for the YAWL user. It contains information on how to install YAWL and dedicated guides for the Editor and the YAWL runtime environment. In addition, it includes the tutorial *Getting started with YAWL*, which provides a brief overview for the reader interested in experimenting with the Editor. On the other hand, the reader interested in advanced data aspects can consult the section *How to manipulate data in YAWL* of this manual.

The development aspects behind the YAWL System are illustrated in the *Technical manual*. Finally, software bug reports and feature requests can be made through the Google code project.[4]

[3] sourceforge.net/projects/yawl

[4] code.google.com/p/yawl

Chapter 9
The Runtime Environment

Michael Adams

9.1 Introduction

Runtime, with regards to the YAWL environment, refers to the period when the YAWL System is active and executing – the host server is operational and the Engine is running in its servlet container (Apache Tomcat by default), accepting requests from Custom Services and applications to load specifications, start process instances (or *cases*), check out work items, and so on, generating events and progressing cases as per their control-flow towards conclusion. While much of what occurs during runtime is described in other chapters, most of the internal operations have been hidden. This chapter emphasizes the internal execution mechanisms of the YAWL Engine.

9.2 Basic Operations

The primary responsibility of the Engine is to execute process instances. However, there is a distinct demarcation between the work performed by the Engine and that handled by custom services. Throughout the life-cycle of a process instance, the Engine will prepare work items (i.e., task instances) for execution, at the appropriate times according to the specified control-flow of the process. But, it is important to realize that:

- *The Engine is not responsible for how the work of a task instance is performed.* Each and every task is associated at design time with a chosen Custom Service that will be responsible for performing the work of the task instance. If the association of a task with a Custom Service is not explicitly specified at design time, the Resource Service is associated implicitly as a default. The Engine announces the enablement of each task to the specified Custom Service at the designated

M. Adams
Queensland University of Technology, Brisbane, Australia
e-mail: mj.adams@qut.edu.au

A.H.M. ter Hofstede et al. (eds.), *Modern Business Process Automation*,
DOI 10.1007/978-3-642-03121-2_9,

time with regards to the control-flow of the process, then awaits further requests from the service as to obtaining ownership of the task instance (via a "check-out" call) and later passing ownership back to the Engine (via a "check-in" call); what happens in relation to the task in the meantime is of no concern to the Engine. At check-in, the Engine will validate the data values assigned to the task instance by the Custom Service against the data schema of the task to ensure that the data is valid before the check-in is accepted, but it is not concerned with how that data was actually assigned to the task instance by the Custom Service (e.g., through user input, database lookup, web service response values, and so on).

- *The Engine knows nothing about users*, but only of its registered custom services. It may be that a Custom Service manages a set of users (e.g., the Resource Service) but the Engine allows only custom services to connect to it. Thus, it is said to be completely *agnostic* to "physical" users.
- *The Engine is unconcerned about how a task is resourced.* This is a corollary to the second point above: the Engine will pass responsibility for a task instance to a Custom Service, but has no concern for which "physical" user actually performs the work of the task instance, if any. In other words, the entire resource perspective is handled externally to the Engine and wholly within the Resource Service.

The two perspectives that the Engine *is* aware of and manages for each process are the control-flow perspective (determining which task(s) are enabled at certain times during the life-cycle of a process, based on arcs, conditions, splits, joins, and so on), and the data perspective (mapping data values to and from tasks and their parent nets, performing transformations, and evaluating expressions using the specified XPath and XQuery predicates). How it manages these two perspectives at runtime is detailed later in this chapter.

The Engine can execute a number of process instances concurrently, each one an instance of a *process specification*. A specification is expressed as an XML document, typically stored in a disk file, that describes the structure, format, data, and layout of a process that has been expressed graphically in the Editor. An XML specification file is produced when the process is saved in the Editor (the layout information stored is used by the Editor when a specification file is reopened, but is ignored by the Engine). Each specification is loaded into the Engine via Interface A – the Resource Service provides a web form that allows for the loading of specifications (see Fig. 9.1). Once loaded, the Engine stores the specification until such time it is manually unloaded (which will succeed only if there are no process instances based on it currently running).

Several versions of the same specification can coexist in the Engine at the same time (i.e., having the same specification identifier but different version numbers). Figure 9.1 shows two versions of the Order Fulfillment specification loaded and running, for example. This allows for currently running process instances to continue to completion, even though a newer version of the specification is loaded into the Engine; however, new case instances may only be started from the most recent version loaded.

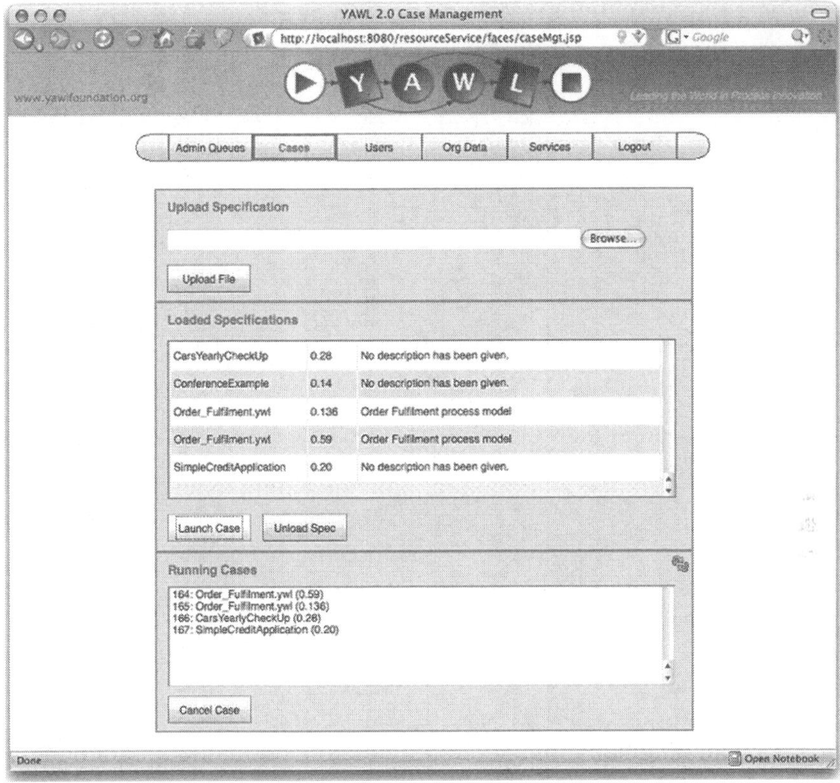

Fig. 9.1 The Case Management Administration web form

9.3 Internal Architecture

Figure 9.2 presents a view of the internal architecture of the Engine, its major components, and the relationships between them. At the core is the **Engine Kernel**, which is responsible for the coordinated operation of the various components, and manages interface communications with external services and applications. Through the kernel, specifications are loaded, case instances are started and canceled, work items are managed, services are registered, and so on. The kernel also takes care of the data-perspective, mapping values to and from tasks and their parent nets, evaluating XQuery predicates, and storing those results.

The kernel communicates with custom services via the Engine's four **interfaces**: A (specification and session management); B (case and work item instance management); E (process logging); and X (exception handling management) – see Chap. 7 for more details.

Each loaded specification is inserted into the **Specification Store**, which serves as a repository for specification "templates" from which process instances may be created. The store holds the specification until such time as it is manually removed from the Engine. The store is persisted across server sessions (cf. Sect. 9.6).

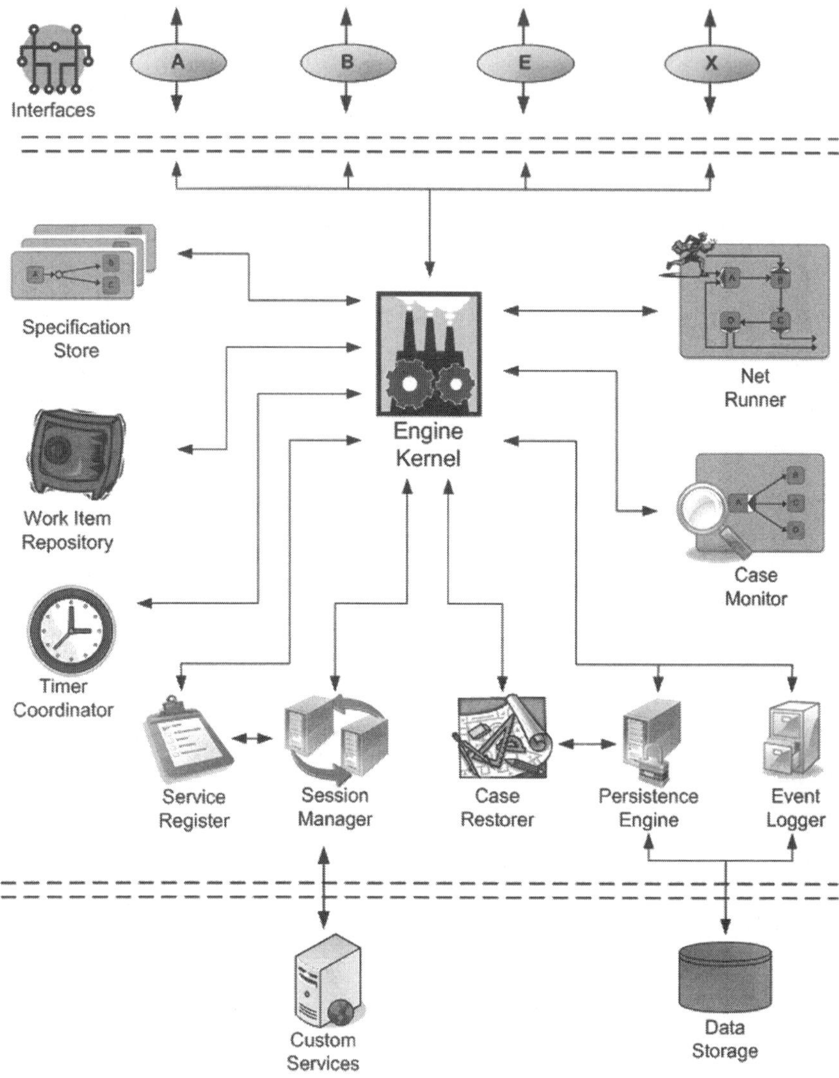

Fig. 9.2 YAWL Engine internal architecture

The **Net Runner** is responsible for progressing a process instance in terms of its control-flow, and is arguably the most critical Engine component. A net runner is created for the root net and each subnet (if any) of a process instance as required, and so a case may have a number of net runners associated with it at any one time, each responsible for exactly one net. Conceptually, a net runner traverses the net from start condition to end condition, enabling tasks along the path determined by the net's control-flow. At each progress iteration (i.e., when the net begins and after each task reaches a completion status), a net runner examines the current state of its net, and creates enabled task instances (for atomic tasks) or new "child" net runners (for composite tasks) for all those tasks enabled for the current net marking. The net runner notifies the Engine of these enablings, which in turn announces the tasks via Interface B to the appropriate custom services. Net runners are also persisted.

All "live" atomic task instances (i.e., work items) are cached in the **Work Item Repository**, which orders work items into "families" for each task (a task may have a number of work items created for it). The repository also groups items by their status (enabled, fired, executing, etc.), their parent net runner, and their case instance for access by a number of Engine components. Work items are removed from the repository when a case completes, and all work items in the repository are persisted across server sessions.

The **Case Monitor** tracks the status of all current process instances, caching their details for easy access by custom services and external applications via Interface B. The monitor maintains details of each case, including time started and execution times, details of the complete set of all work items created for each process, and details of all data modifications for each data parameter of each work item, which it delivers on demand as XML data, so that the current state of all processes in the entire Engine may be examined.

Any timers that have been defined for tasks are managed by the **Timer Coordinator**. A timer can be defined for a task at design time, may be initialized either when a work item of the task is enabled or when it begins execution (cf. Sect. 9.5), and will expire either after a period of time has elapsed or at a precise moment, as specified. There is one Timer Coordinator within the Engine that manages all timers for all tasks – conceptually it is a timer with a dynamic table of work items and their expiry times; when an expiry time is reached, the timer takes the appropriate action to skip the work item (if enabled) or complete the work item (if executing), then removes the work item from its dynamic table. Timer entries are persisted.

All custom services that require communication with the Engine must first be registered with the Engine via Interface A – the default worklist handler of the Resource Service contains a web form to facilitate this (Fig. 9.3). The **Service Register** maintains the list of currently registered services. An error will occur if a Custom Service has been nominated at design time to handle a task and that Custom Service is not registered with the Engine when the task is enabled at runtime.

Additionally, all custom services that require communication with the Engine must create a session with (i.e., logon to) the Engine at runtime. These sessions are maintained by the **Session Manager**, which on receiving a logon request from a service will first check that the service is correctly registered (via the Service Register), then validate the logon credentials for the service via those stored, and if valid, will record and return a "session handle" (or key) for the session. The service will then pass this session handle with all interface requests; the Engine will respond to the request only after first verifying that the session handle is both valid and active. A session handle will by default expire after 60 minutes of inactivity, although the period can be changed via a configuration file.

The **Persistence Engine** manages runtime object persistence (see Sect. 9.6 for more details). All runtime requests to persist objects, and to retrieve objects from

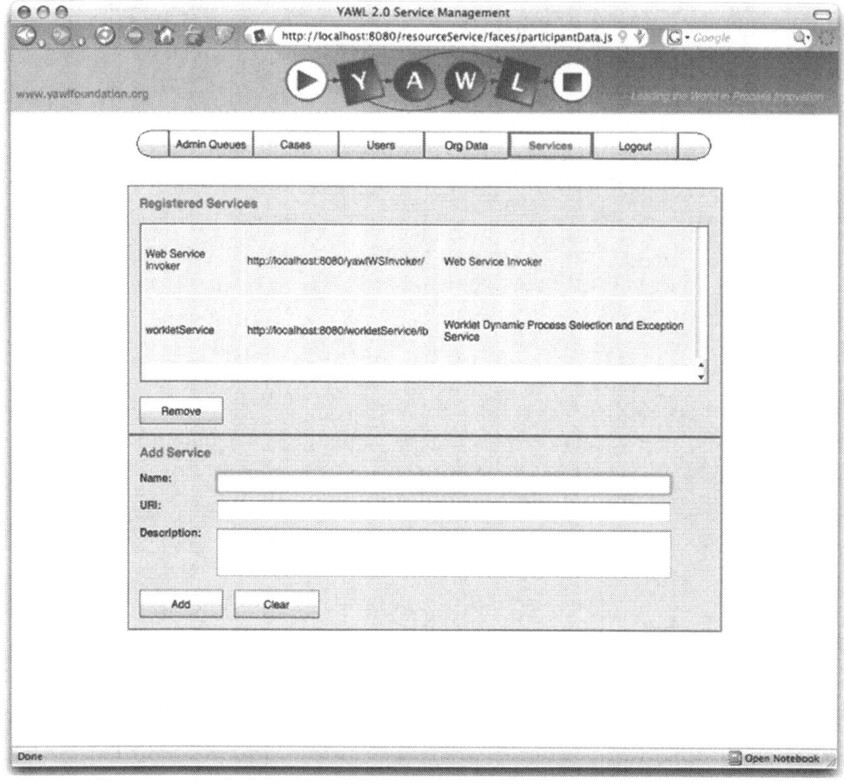

Fig. 9.3 The Service Management Administration web form

persistence, pass through this component. The **Case Restorer** is activated when the Engine is (re)started; its job is to restore all the case instances that were running when the Engine was shutdown. It retrieves all the data for each case via the Persistence Engine, then reassembles the data into case instances, which are then set running again.

Finally, the **Event Logger** handles the logging of all process events for later examination. See Sect. 9.7 for more information about process logging.

9.4 The Life-Cycle of a Case

A typical case (or *instance*) will iterate through a sequence of steps, recursing within that sequence until such time as the case reaches a completed state. While there are a number of exceptions to the general rule (such as certain control-flow constructs, runtime exceptions, timers, and environmental intrusions), the typical sequence is described in Fig. 9.4 and below.

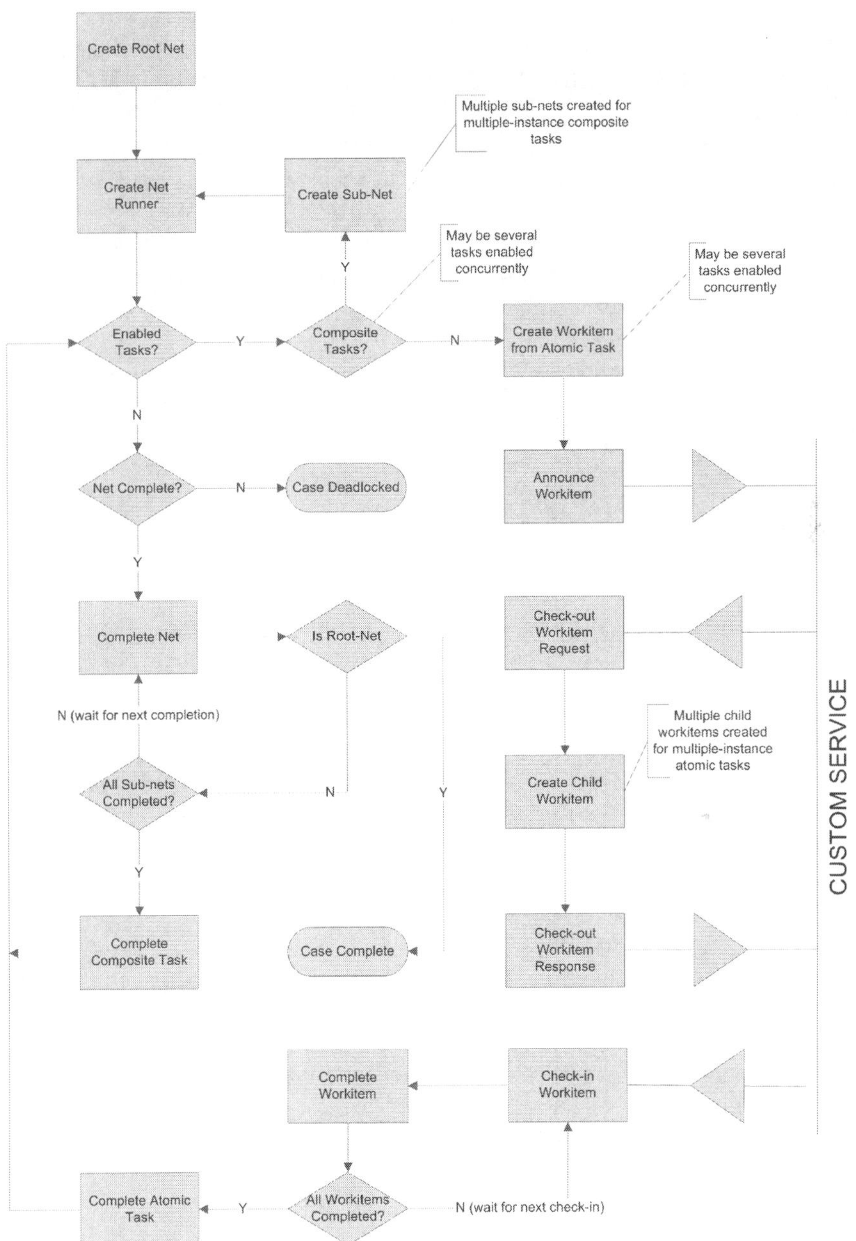

Fig. 9.4 Case life-cycle

1. The Engine starts a case. The internal objects of the case instance are created and initialized using the process definition (i.e., the specification) as a template.
2. A root net is created and a token is placed in its initial input condition.
3. A net runner is created for the net.
4. The net runner determines which task(s) are enabled by the token in the input condition(s). For each enabled composite task, a new subnet is created (or, in the case of multiple instance composite tasks, a specified number of subnets are created), a token is placed in each net's initial input condition, and the cycle returns to Step 3 for each subnet. For each enabled atomic task, a work item is created, and the Engine announces an enabled work item event to the appropriate Custom Service.
5. The Custom Service responds to the event by requesting to check-out the work item from the Engine (via Interface B).
6. For a single instance atomic task, the Engine spawns a copy (or *child*) of the work item. For a multiple instance atomic task, the specified number of child work items are created.
7. The Engine maps data values from the net-level data to each data variable of each work item, according to the XQuery defined for each variable's input predicate.
8. The Engine moves one child work item to *Executing* status (see Sect. 9.5) and passes that work item back to the Custom Service, as a response to the check-out request.
9. For multiple instance atomic tasks, the remaining children are created with *Fired* status. The Custom Service sends a check-out request for each of the remaining child work items; the Engine responds by moving each to *Executing* status and passing it to the Custom Service.
10. The Engine waits (with regards to that particular process) until such time as the Custom Service checks-in each of the checked-out work items back to the Engine. The Engine is unconcerned with exactly how the work item's work has been performed by the service.
11. For each checked-in work item, the Engine maps data values from each of its data variables back to the net-level data parameters, according to the XQuery for each variable's output predicate.
12. The Engine completes the work item. If there are still outstanding child work items for the task, the cycle returns to Step 10.
13. When all child work items have been checked in to the Engine, the Engine completes the task and produces a token in each of its output conditions.
14. The cycle returns to Step 4 and continues until either a token in the net has reached the final output condition (normal completion); there are tokens left in the net but no tasks are enabled (deadlock); or the case is canceled.
15. If a token has reached the net's final output condition, the Engine completes the net.
16. If the completing net is a subnet (i.e., created from a composite task), the final net-level data parameter values are mapped back to the parent net via the task's output parameters, according to the XQuery for each parameter's output

predicate. If there are still outstanding subnets for the task, the cycle returns to Step 14. If there are no more outstanding subnets, the Engine completes the parent task, produces a token in each of its output conditions, and the cycle returns to Step 4.

17. If the completing net is the root net, the case completes.

9.5 The Life-Cycle of a Work item

A work item is a runtime instance of a task definition, and is created by the Engine at the time determined by the control-flow of the process specification from which a process instance is derived. If the task definition is associated with a decomposition, its work items are created to allow interaction with the external environment (i.e., custom services and applications). If there is no decomposition associated with the task (a so-called placeholder or empty task), it is automatically created and immediately completed internally within the Engine.

When a case starts, and when each task within it completes, the Engine analyses the current state of the case to determine which task or tasks it should next make available for execution – or in workflow net terms, it examines the current marking of the net to determine which transitions (tasks) have tokens in each of their input places (conditions), taking into account what type of join decorator (if any) the task has. If the Engine finds there are no tasks available for execution, it means one of two things: either the case has completed normally (if there is/are token(s) in the final output condition) or the case is deadlocked (if not).

Between a work item's creation and its completion, it may move through a number of states; for each state it enters, it is assigned a corresponding status that describes it. There are ten distinct statuses that a work item may be assigned: four *live* statuses (Enabled, Fired, Executing, and Suspended); five *completed* status (Complete, ForcedComplete, Canceled, Failed, and Deadlocked); and one "special" status (IsParent). Figure 9.5 shows the work item life-cycle, and how it may progress through the various statuses from creation to completion. Each status is described below.

When a work item is created by the Engine, it will be initialized with one of three statuses:

- **Enabled:** In most scenarios, a newly created work item will be given *Enabled* status. In fact, the first-created work item of every task will initially be *Enabled*. Technically, an enabled work item represents a task that has a token in each of its input places, or if it has a join decorator, a token in each of those input places sufficient to satisfy the join. In practical terms, it means the work item is available to be started or "checked-out" of the Engine
- **Fired:** For a previously enabled work item, this status denotes that the work item has been requested to start, but has not yet started, and so represents a brief, transient stage between *Enabled* and *Executing*. For a multi-instance task, the

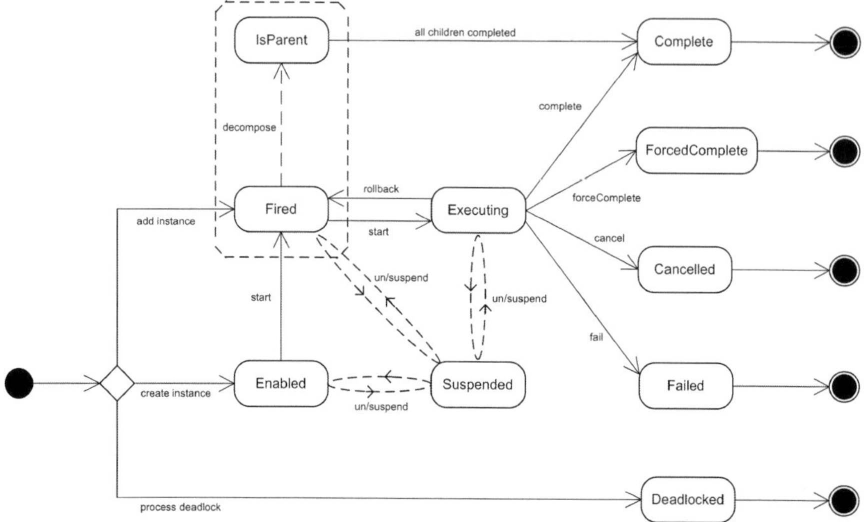

Fig. 9.5 Work item status transition chart

initial status of all created instances, besides the first, is *Fired* (as only one work item instance of each task may have *Enabled* status)

- **Deadlocked:** When there are no enabled tasks in a net, and tokens still exist in the net in places other than the final output condition, the net is said to be deadlocked. When this occurs, the Engine will report the deadlock and remove the case. Part of the reporting process is to determine where there are remaining tokens in the net. For each condition where a token resides, the name of the task on its outgoing arc is recorded in the process log. To facilitate this logging, each of those tasks has a work item created with the status of *Deadlocked*, which is actually a completed status type. To clarify, a work item is created and given this status only to facilitate a logging event, immediately before the entire deadlocked process is removed from the Engine

One point of possible confusion is the use of the term *enabled*, which has a slightly different meaning depending on whether it is used in workflow net terms or work item terms. In workflow net terms, a YAWL task is enabled if there is a token in each of its input conditions (after taking its join decorator, if any, into account). In the Engine, a work item has the status *Enabled* when it is available to be started. Each time a task completes, a new work item is created and initialized with *Enabled* status for each task that is newly enabled in the workflow net.

Every task definition contains a reference to the Custom Service that will be responsible for processing it (or the Resource Service by default if none are specified). Each time a work item is created for a newly enabled task (with a decomposition) and assigned a status of *Enabled*, that enabled work item is "announced" to the environment. That is, the Engine generates a "handleEnabledWorkItemEvent" and propagates it to the Custom Service that is specified at design time as the listener (or handler) of the event. The Custom Service will typically deal with the announcement event by requesting that the work item be *checked-out* of the Engine. Both the event and the checkout request are performed via Interface B. When the Engine receives a check-out request from the service, it will attempt to start the enabled work item.

When an enabled work item is started, it *decomposes* or splits into two or more discrete work items. First, a *parent* work item is created and maintained by the Engine; its purpose is to manage the progress of all the work item instances created for the task (referred to as *child* instances). For single instance tasks, exactly one child is created, while for multiple instance tasks, as many child instances are created as required for the parameters set for the task at design time. The parent work item is given the status *IsParent*. As each child instance completes, the parent instance takes note of its completion – when all of the child work items have reached a completion status and, for multiple instance tasks, when the *completion threshold* has been reached (a parameter that describes the minimum number of child instances that must complete before the task can be considered complete), the parent work item instance also completes, at which point the task is marked as completed and a token is produced in its output place, allowing the process control-flow to continue.

Further, when an enabled work item of a multiple instance task is started and decomposed into a parent and a number of children, the children are each given the status *Fired*. Then, exactly one child is moved from *Fired* to *Executing* status (where there are multiple children, one is chosen at random to become *Executing*). This means that, regardless of whether the task is single or multiple instance, only one child is ever started for a check-out request, while the remaining instances (i.e., in the case of a multiple instance task) are left at the intermediate *Fired* status, and must each be explicitly started before they are moved to *Executing* status.

To summarize, the "execution" statuses are:

- **IsParent:** Denotes that instance of a work item that oversees the execution of the other (child) instances. When all child instances have completed, and in the case of multiple instance tasks, when the completion threshold has been reached, the parent also completes, allowing the process to continue as per its control-flow.
- **Executing:** Each work item that is successfully started is given *Executing* status. For multiple instance tasks, only one (indiscriminate) child is initially started via a check-out request; the other child instances are initialized at that time and given *Fired* status, and must be individually started to move each to *Executing*. Thus, *Executing* status denotes that the work item has started and is being processed by a designated Custom Service.
- **Suspended:** A work item with a status of *Enabled*, *Fired*, or *Executing* may be suspended. Basically, a status of *Suspended* blocks a work item from being progressed to a status subsequent to the status it had at the time it was suspended. So, an enabled work item, when suspended, cannot be fired; a fired work item cannot be executed; an executing work item cannot be completed. Suspension is typically used to "hold" a process in its current state, while, for example, some compensatory action is taken following an exception. Note that

when a suspended work item is unsuspended, it always reverts to the status it held immediately before it was suspended.

Every work item created for a task will eventually reach a completion status. As stated earlier, once all (or the specified threshold) of the child instances have reached a completion status, the parent instance also completes, and the task itself is said to have completed. The five completion statuses are the following:

- **Complete:** The work item has completed normally, typically via a check-in request from a Custom Service
- **ForcedComplete:** The work item was executing and has been forced to complete, typically by an exception handler. This status implies that the work item was not completed in the normal manner, but the process should move to the next task(s) in the control-flow (i.e., when all work items of the task are completed, the task should produce a token in each of its output places, depending on any split decorator the task may have)
- **Canceled:** The work item has been canceled, perhaps because the case has been canceled, or the work item is a member of another task's cancelation set, or via an exception handler
- **Failed:** The work item has been forced to fail. This status is similar to canceled, except that the task from which the work item was created should not produce a token in any of its output places (i.e., control-flow on that arc of the process that contains the task is blocked)
- **Deadlocked:** This is a "special" status that is given to work items that may otherwise have eventually been enabled (i.e., the task has tokens in one or more of its input places, but not all of the places sufficient to enable it), except for the fact that the overall process has reached a deadlock state (i.e., it has not completed, but cannot continue). Such work items are created with a status of *Deadlocked* solely for the purposes of logging the fact that a deadlock has occurred. Once this status change has been logged, they are immediately removed from the Engine

9.6 Persistence

A single instance of a workflow process will take a certain amount of time to complete. Some may be short-lived, but others may run for days or weeks (e.g., an order fulfillment process) or even months (conference organization) or, for some processes, years (a course of study). However, like any computerized process, workflow instances are susceptible to interruptions or failures on the hardware hosting them. For all processes, it is imperative that if such an outage occurs, the work performed thus far for a process is not lost. To guard against the potential loss of process information as a result of server downtime, all YAWL process instances may be *persisted*.

Data persistence refers to the mechanism of saving to permanent data storage (i.e., a disk) all data relevant to a process, so that in the event of an outage, on restart the process can be "reconstituted" from the stored data, and work can continue on the process as if the outage had not occurred. In YAWL, persistence is configured on by default and recommended, although it can be configured off if desired.

For a successful restart, it is important that all the necessary data and descriptors of a process are stored, and can be reassembled from that stored data. This requires that data be inserted, updated, or removed at every event and milestone in the process life-cycle.

The YAWL environment uses the Hibernate[1] framework to support persistence. Amongst a number of benefits, Hibernate offers two main advantages. First, a Java class can be configured to store exactly those data members that are of value to the persisting application; by instructing Hibernate to store an object, only those data members are stored, and when instructed to retrieve the data, Hibernate reconstructs the object with the stored data. This approach simplifies object persistence, as the application does not have to concern itself with deconstructing and reconstructing objects. Second, the process Hibernate uses to store and retrieve data is abstracted from the data source used. This means the application can easily configure (through a properties file) the data source to be used, selected from a large number of data sources and types, without changing anything within the application itself. For example, the YAWL Enterprise release uses Postgres[2] as its default data source, while the YAWL4Study release uses Derby.[3] In each case, only a small change to the configuration file was required. YAWL has also been successfully run using MySQL and Oracle. With Hibernate, processing overheads for data storage can be kept to a minimum, through its use of optimized data structures that are written out to database only at times of low load.

All relevant aspects of a process instance are persisted and updated if those values change. When an instance (case, task, or token) completes, all references to it are removed from the persistence tables. Those tables are the following:

- **CaseDataDocument:** Stores the current net-level data parameters and values for each net. Each net of a process, whether the root net or a subnet, has its net-level data stored in this table
- **CaseNbrStore:** Stores the next available case number that may be assigned to a case, which ensures all cases have a unique identifier and that duplications cannot occur
- **Locations:** This table stores the locations of all tokens within each net, that is, each net's current marking
- **RunnerBusyTasks:** The set of all currently executing tasks
- **RunnerEnabledTasks:** The set of all currently enabled tasks
- **RunnerStates:** Stores the current state of each *YNetRunner* object responsible for the execution of each net
- **Services:** The set of custom YAWL services currently registered with the Engine

[1] www.hibernate.org

[2] www.postgres.org

[3] db.apache.org/derby/

- **Specs:** The set of specifications currently loaded in the Engine
- **WorkItems:** Stores a full set of internal descriptors for each active work item
- **WorkItemTimer:** This table stores a record of each timer that has been started for a work item, but has not yet expired
- **YIdentifiers:** A YIdentifier object is essentially a token in a YAWL net, and stores information identifying itself with its parent case instance

When a server restarts, case instances are reconstructed from the data held in these tables by the Case Restorer component. Each case is rebuilt in a predefined sequence to ensure that all necessary objects and data structures are available, so that it can successfully start executing from the point of last persistence.

9.7 Logging

An important consideration for any workflow environment is the generation of *process logs*, that is, the recording of data and events that describe the execution of each process. Process logging plays two fundamental roles: (1) it provides an audit trail of interactions between the workflow Engine and its environment; and (2) it represents a historical archive of process executions, which may later be analyzed for insights into the operation of specifications (cf. Chap. 17).

The Engine generates process logs that detail each executing case from commencement to completion (or cancelation), the enabling of each work item and each of its status changes through to completion (or cancelation), and all changes made to work item and net data values. It also provides for *configurable* logging, that is, the logging of user-defined data attributes, values, and descriptors associated with certain parts of the process data that may prove of interest at a particular site or for a particular process specification.

The process of writing data to the process logs is achieved via a Hibernate interface, in the same manner as that used by the persistence tables. This means that the process logs can be written to any data source that can be configured for the Hibernate framework. Each log entry is formulated by creating a log object, populating it with the appropriate data, and then passing the object to the framework for insertion into the data source.

The physical structure of the process logs, and the methods made available for its interrogation via Interface E, are built with a view towards both the requirements of log analysis by external tools such as ProM[4], and for ease of data retrieval by custom services and applications.

9.7.1 Process Log Relational Schema

Figure 9.6 depicts a relational schema for the Engine's process log, and shows primary key fields as bolded text. The unique identifiers for the NetInstance and

[4] www.processmining.org

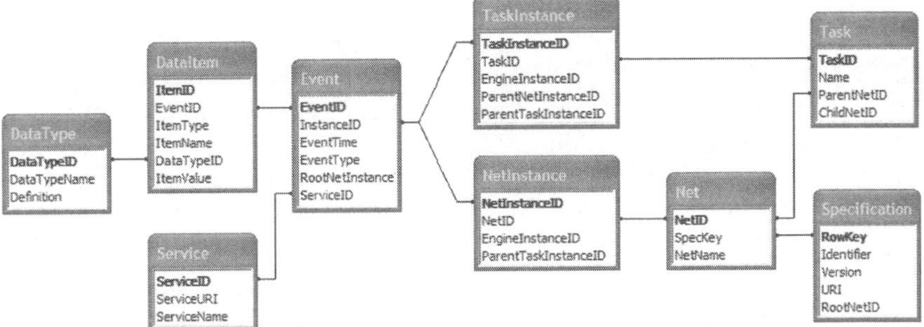

Fig. 9.6 Relational schema for Engine log (main relations shown)

TaskInstance tables are supplied by the Engine, which produces an identifier (referred to generally as a "caseID") that is unique across all instantiations of all nets and tasks. It is formulated by first assigning the next available case number to a new case instance, then using a dotted number formation to refer to its child instances. So, for example, the root net of a new case instance might receive a caseID of "123"; its first child task instance (i.e., a work item) will receive a caseID of "123.1," its second "123.2," and so on. For a composite task, the subnet instance might be assigned "123.3," and its first work item "123.3.1". This convention ensures that all instance identifiers are unique.

The process log consists of nine tables:

- **Specification:** A YAWL specification is uniquely identifiable through its internal specification identifier (Identifier) and version number (Version) – all versions of a specification share the same identifier. It also has a non-unique name assigned to it at design time, which is perhaps more meaningful to a human reader than the identifier. In addition, this table stores, for convenience, a reference to its root net via a foreign key to the *Net* table
- **Net:** Each specification contains one or more nets; this table stores data of each net of each specification, from which net instances may be created
- **NetInstance:** Each execution of a net is recorded in this table. This may represent the start of a new case instance (i.e., if it is a root net) or the decomposition of a composite task (i.e., a subnet). This table has fields to record the foreign keys for the net from which the net instance has been created and, if the net has been created from a composite task, the key of that task instance. It also stores a unique identifier (NetInstanceID), and the "case ID" assigned to the net by the Engine (EngineInstanceID)
- **Task:** This table stores a reference to each task started by a net instance, and contains a unique identifier, the name of the task and a foreign key reference to its parent net. For composite tasks, the key of the net (decomposition) of the task is also stored. The entries in this table refer to task definitions from which task instances are created (and stored in the TaskInstance table), in the same way as the *Net* table refers to process definitions from which process instances are created (and stored in the *NetInstance* table)

- **TaskInstance:** Each task instance is assigned a unique identifier (TaskInstanceID) and the "caseID" of the task instance (EngineInstanceID). There is also a foreign key reference to the *Task* table record from which this instance was created. Note that the task instance stored in this table may represent a work item (for atomic tasks) or to a subprocess (for composite tasks), in which case it will also be recorded in the NetInstance table. There are also foreign keys to the net of which this task is a member (ParentNetInstanceID) and, for child work items, to its parent (ParentTaskInstanceID)
- **Service:** A loggable event may be instigated by a Custom Service. A record of each service that has communicated with the Engine is kept in this table; each record of the *Event* table has a foreign key relation to this table
- **Event:** The *Event* table is arguably the key table in the process logging database, as it records all events for all runtime instances, whether task or net instances. Each record contains a unique identifier, a reference to the task or net instance, a timestamp of when the event occurred, and the type of event. For a net instance, the event types will include starting, completing and canceling, and updating of values of any net-level data parameters; for a task instance, the event types will include any of the work item statuses described in Sect. 9.5 and any event that updates the value of a work item-level data parameter. Each event may also have a number of configurable data entries (see Sect. 9.7.2). A foreign key reference to the instigating Custom Service is included, as is the key of the root net instance, for querying convenience
- **DataItem:** For all events that modify a data parameter's value, or include configured data entries, a record is written to the *DataItem* table. Each record contains a unique identifier, and the item's name, data type, and value. There are also two foreign key references, one to the identifier of the event that contained the data item, and one to the *DataType* table
- **DataType:** All entries in the *DataItem* table must refer to a certain data type. While some are simple types (e.g., string, integer, boolean) others will refer to complex data type definitions, which may change over time. To accommodate this, each data type definition is stored in the *DataType* table, and consists of a unique identifier, a name, and a definition

9.7.2 Configurable Logging

Configurable logging refers to the facility for custom services and applications to append user-defined data attributes and values to events for logging purposes. Such data items may contain descriptive information pertinent to all events (such as the resource that triggered the event, e.g., who started a case) or may refer to particular needs for particular processes (e.g., temperature readings, associated costs and so on). In any case, the facility is provided to allow process designers and custom services to store data associated with processes as desired, without the Engine being aware of what the data actually represents. That is, it is a facility to store data items and values in the Engine's process logs for later retrieval and analysis, so that all descriptive data of a process can be stored in one place, without the need for special handling by the Engine.

A special object called *LogDataItem* is available to custom services and applications to store a data item, its type and value. These objects may be passed with any interface method that will eventuate in a log entry being written.

9.8 Summary

This chapter described the primary functions performed by the YAWL Engine at runtime, with particular reference to its internal machinations and those details of the runtime environment that have not been covered in other chapters.

The Engine is responsible for managing the control-flow and data perspectives of processes. The resourcing perspective, and the actual performance of task instances, is delegated to custom services, as discussed in the following chapters.

Exercises

Exercise 1. Why do you think it is important that the Engine remain "agnostic" to both the resource perspective and the actual performance of task instances?

Exercise 2. What does it mean when it is said that a task instance is enabled by the Engine? How does that meaning differ in relation to workflow net terminology?

Exercise 3. How many child instances of a task can have *Enabled* status? Why?

Exercise 4. For what purpose is a work item assigned a status of *Deadlocked*?

Exercise 5. What is the difference between a work item being canceled and it being "failed"?

Exercise 6. What two steps are required for a Custom Service to successfully connect to the Engine?

Exercise 7. Why is it important to persist currently executing cases?

Part V
Services

Chapter 10
The Resource Service

Michael Adams

10.1 Introduction

A workflow comprises three main perspectives: control-flow, data, and resources. The resource perspective is particularly important because, for the most part, workflow tasks are designed to be performed by people, and so a workflow environment should support efficient and flexible ways to associate work with the people who have the required skills and authorizations to carry it out. Thus, the resource perspective is primarily responsible for modeling an organizational structure, and the people who populate it, in a computational form, so that a person may be coupled with tasks and data emanating from the control-flow and data perspectives.

While control-flow and data-flow are necessarily tightly coupled within a workflow enactment engine, in the YAWL environment the resource perspective is supported by a discrete Custom Service called the *Resource Service*, in line with YAWL's Service Oriented Architecture. Consequently, the Engine is oblivious to the assignment of resources to tasks (i.e., it is said to be *agnostic* with regards to resourcing). The YAWL Resource Service provides full support for 37 of the 43 identified resource patterns (the remaining six being particular to the case-handling paradigm) and so may be considered the preeminent implementer of workflow resource pattern support.

10.2 Functional Overview

The YAWL Resource Service is the largest and most complex Custom Service implemented for the YAWL environment and consists of several distinct components. This section gives a broad functional overview of the service, followed by more detailed discussion of the various components in later sections.

M. Adams
Queensland University of Technology, Brisbane, Australia
e-mail: mj.adams@qut.edu.au

A.H.M. ter Hofstede et al. (eds.), *Modern Business Process Automation*,
DOI 10.1007/978-3-642-03121-2_10,

Using the YAWL Editor, a process designer, concomitantly with the design of the process control-flow and data perspectives, is able to designate how each task will be resourced at runtime. First, a decision will be made for each task whether to have it handled at runtime by a specified Custom Service, in which case that service is responsible for its successful performance, or whether it will by default be handled by the Resource Service. Each task coupled with the Resource Service may be designated as a *manual* (the default) or an *automated* task (and whether a task is manual or automated is only of relevance to the Resource Service). A manual task is a task that will be performed by a human resource; an automated task by definition does not require human resourcing, but instead may execute a specified *codelet* when it enables at runtime. The Resource Service receives notification from the engine of all task enablings (via Interface B, as per other Interface B services) for all manual tasks not associated with other custom services and for all automated tasks (i.e., the Resource Service handles all automated tasks).

In YAWL, a human resource is referred to as a *Participant*. Each participant may perform zero or more *Roles*, hold zero or more *Positions* (each of which may belong to an *Org Group*), and possess zero or more *Capabilities*. For a manual task, a designer may provide details of a *distribution set* of resources to which the task should be offered at runtime. A distribution set may consist of the union of zero or more individual participants, zero or more roles, and zero or more dynamic variables (which at runtime will be supplied with details of participants and/or roles to which the task should be offered, thereby supporting the *late binding* of resources to tasks). The resultant distribution set may be further filtered by specifying that only those participants with certain capabilities, occupying certain positions and/or being members of certain org groups, be included.

A designer may also specify certain constraints to apply, for example, that a certain work item must not be performed by the same participant who completed an earlier specified work item in a process (called the *Four Eyes Principle* or *Separation of Duties*), or that if a participant who is a member of the distribution set of a work item is the same participant who completed a particular previous work item in the process, then they must also be allocated the new work item (called *Retain Familiar*).

Tasks vs. Work Items

A YAWL process specification will contain a number of *task* definitions, which may be considered to be descriptions of a piece of work that forms part of the overall process. Thus, control-flow, data, and resourcing specifications are all defined with reference to tasks at design time.

At runtime, each task acts as a template or contract for the instantiation of one or more *work items*. That is, a work item is a runtime instance derived from a task definition. An analogy might be: a task is to a work item as a blueprint is to a house constructed from that blueprint.

Technically, when a task is enabled at runtime, at least two work items are instantiated for it. One is referred to as the *parent* work item, the others as its child or children. The parent is held internally within the engine, and oversees the execution of its children. For a single instance atomic task, one child work item is instantiated, and for a multiple instance atomic task, the specified number of child work items are instantiated. It is these work items that are actually executed by resources (e.g., a participant). When all of its children have completed, the parent itself then completes, allowing the process to progress to the next task, and so on.

It is important to understand the distinction between a task and a work item derived from it. When we talk about tasks, we refer to design time definitions; at runtime, we refer to work items as instantiations of tasks.

At the base of the resourcing framework are three interaction points, which are places in the life-cycle of a work item where resourcing decisions are to be made, up to and including the moment the work item is placed in a *work queue* (cf. Chap. 2). At each interaction point, the decision may be specified to be *system-initiated* (automatically performed by the system, using parameters set at design time) or *user-initiated* (manually performed by a participant or administrator at runtime). The three interaction points for each work item are:

- **Offer:** The work item is offered to one or more participants for execution. There is no implied obligation (from a system perspective) for the participant to accept the offer.
- **Allocate:** The work item is allocated to a single participant, so that the participant is committed (willingly or not) to performing that work item at a later time. If the work item was previously offered to several other participants, the offer is withdrawn from them at this time.
- **Start:** The work item is started by the allocated participant (i.e., enters executing state).

If a work item's offer interaction is user-initiated, it is passed to the administrator's *Unoffered* queue so that it can be manually offered to one or more participants. If an offer interaction is system-initiated, the service will offer the work item to one or more participants based on the resourcing parameters defined for it at design time, by placing it on the participants' *Offered* queue.

If a work item's allocation interaction is user-initiated, one of the participants who is offered the work item may manually choose to allocate the task to him/herself, at which point the work item is placed on the participant's *Allocated* queue and removed from the *Offered* work queues of all other offered participants. If the allocation interaction is system-initiated, an allocation strategy (e.g., random choice, round robin, shortest queue), as defined at design time, is invoked that selects a single participant from those offered; the work item is placed on that participant's *Allocated* work queue and removed from the offered work queues of all other *Offered* participants.

Table 10.1 Summary of interaction point actions

Offer	Allocate	Start	Action
U	U	U	An administrator offers the work item to one or more participants. One participant self-allocates the work item; it is withdrawn from the other offered participants. The participant manually starts the work item.
U	U	S	An administrator offers the work item to one or more participants. One participant self-allocates the work item; it is withdrawn from the other offered participants. The work item immediately starts.
U	S	U	An administrator chooses one or more participants to offer the work item to. One participant from those offered is immediately allocated the work item. The participant manually starts the work item.
U	S	S	An administrator chooses one or more participants to offer the work item to. One participant from those offered is immediately allocated the work item. The work item immediately starts.
S	U	U	The system offers the work item to one or more participants. One participant self-allocates the work item; it is withdrawn from the other offered participants. The participant manually starts the work item.
S	U	S	The system offers the work item to one or more participants. One participant self-allocates the work item; it is withdrawn from the other offered participants. The work item immediately starts.
S	S	U	The system offers then immediately allocates the work item to a participant. The participant manually starts the work item.
S	S	S	The system immediately offers, allocates, and starts the work item for one participant.

Finally, if a work item's start interaction is user-initiated, a participant must manually select it from their *Allocated* queue to start its execution. If a start interaction is system-initiated, the work item is automatically started and placed on the participant's *Started* queue for action.

Thus, there are eight different interaction point combinations handled by the Resource Service, summarized in Table 10.1. Corresponding to these three interaction points, each participant may have, at any particular time, work items in any of the three personal work queues, one for each of the interaction points (a fourth queue, *Suspended*, is a derivative of the *Started* queue).

After the initial distribution of a work item has completed, a participant, having the required authorizations (or *privileges*), can further affect the allocation and execution of work items. If a work item is allocated to them, he/she may *deallocate* it (re-initiate the distribution, excluding themselves from the distribution set); *delegate* it (to a member of their "team" – those who occupy positions that ultimately report to the participant's position); or *skip* the work item (complete it immediately without first starting it). If the work item has been started, a participant may *reallocate* it (to a member of their team), and in doing so may preserve the work done within the work item thus far (*stateful* reallocation), or to reset the work item data to its original values (*stateless* reallocation).

Further, at runtime, a participant with the necessary privileges may choose to *pile* a task, so that all future instances of work items of the task across all current and future cases of the process are directly allocated to the participant, overriding any design time resourcing specifications; and/or *chain* a case, which means that for all future work items in the same process instance where the distribution set specified includes the participant as a member, each of those work items is to be automatically allocated to the participant and started.

Finally, an administrator has access to a *Worklisted* queue, which includes all of the currently active work items of all participants, whether offered, allocated, started, or suspended, from which a task can be manually reoffered, reallocated, or restarted to another participant.

10.3 Organizational Model

Before work items can be assigned to (human) resources, a data source that describes those resources must first be established. That is, at a minimum, the Resource Service needs access to data that details the attributes of a set of human resources and the relationships (if any) between them. Then, a list of such resource descriptors can be provided to the Editor, from which particular resources may be chosen at design time, for each task in a process specification, to potentially perform the execution of work items derived from those tasks at runtime.

In the Resource Service, a human resource is referred to as a participant, that is, someone who willingly participates in the performance of tasks within a workflow instance, progressing it towards completion. Although a set of discrete participants is all that is required to allocate work items to resources, more typically a participant is a member of some kind of organization, and therefore is defined in various ways within an *organizational model*. An organizational model describes the relationships between the participants of an organization, and their jobs, roles, duties, managerial hierarchies (lines-of-reporting), and so on. The Resource Service provides a default organizational model database and tools to administrate it. On the other hand, organizations with existing organizational data sources (such as other RMDSs, LDAP, text files, XML, and so on) may use that data directly instead, by importing it through an interface provided by the service and mapping the data into the model (see Sect. 10.4 for details).

10.3.1 The YAWL Organizational Model

The Resource Service defines a typical, generic organizational model into which participants and their various relationships may be placed. The service's organizational model is an implementation of the ORM diagram shown in Chap. 2, Fig. 2.33.

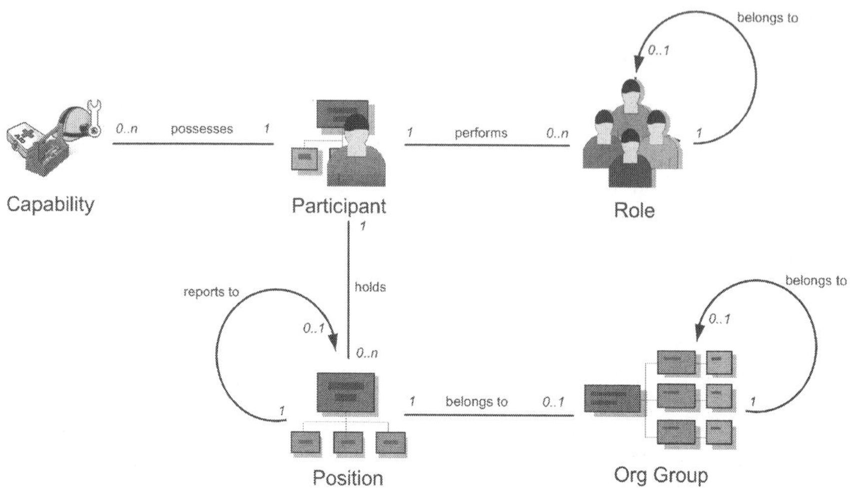

Fig. 10.1 The basic YAWL organizational model structure (from the perspective of a unique participant)

In contrast, Fig. 10.1 shows a representation of the model from the perspective of one (unique) participant, so that the relationship between a particular participant and the various organizational entities can be seen (these relationships are further explained below). To maintain flexibility in the model, a participant's relationship to the other entities is not enforced; thus, a simple set of participants is sufficient to allow resourcing of work items in YAWL. However, several of the resource patterns cannot be achieved without the use of the other entities in the model. Those entities are:

- **Role:** Generally, a role is a duty or set of duties that are performed by one or more participants. For example, bank teller, police constable, credit officer, auditor, properties manager, and junior programmer are all examples of roles that may be carried out within an organization. There may be several participants performing the same role (e.g., a bank may have a number of tellers), so a typical role in an organizational model may contain a number of participants. Conversely, a certain participant may perform multiple roles. Therefore, a role may be considered as a grouping of participants who share the same (or similar) duties within an organization. Further, a role may belong to a larger, more general role (e.g., the roles junior teller and senior teller may both belong to a more general role called 'teller'). In the model, a participant may perform zero or more roles, and a role may belong to zero or one other roles. A role may be included in the distribution set for a task at design time, meaning that all of the participants performing that role (or any of its sub-roles) are to be considered as potential recipients of a work item created from the task at runtime.

- **Capability:** A capability is some desired skill or ability that a participant may possess. For example, first aid skills, health and safety training, a forklift license, or a second language may all be considered as capabilities that a participant may possess that may be useful to an organization. There may be several participants within an organization possessing the same capability, and a certain participant may possess a number of capabilities. In the YAWL model, a participant may possess zero or more capabilities. A capability (or capabilities) may be included in a filter defined at design time that is run over the distribution set for a task at runtime, meaning that only those participants within the distribution set that possess the specified capability or capabilities are to be considered as potential recipients of a work item created from the task.
- **Position:** A position typically refers to a unique job within an organization for the purposes of defining lines-of-reporting within the organizational model. Examples might include CEO or Bank Manager, or may be internal job codes (such as "TEL0123"). Although generally a participant will hold exactly one position, and each position in the model will contain exactly one participant, to maximize flexibility these restrictions are not enforced in the YAWL model; within YAWL, a participant may hold zero or more positions. Importantly, a position may report to zero or one other positions (e.g., bank teller "TEL0123" may report to the Bank Manager), and may belong to zero or one Org Groups (see below). Like capabilities, a position (or positions) may be included in a filter defined at design time that is run over the distribution set for a task at runtime. Positions are also used at runtime to enable resource patterns such as delegation, reallocation, and viewing of team work queues (see Sect. 10.7 for more details).
- **Org Group:** An organizational group (org group) is a functional grouping of positions. Common examples might include Marketing, Sales, Human Resources, and so on, but may be any grouping relevant to an organization. In the YAWL model, each position may belong to zero or one org groups. Further, like roles, an org group may belong to a larger, more general org group (e.g., the groups Marketing and Sales may each belong to the more general Production group). Org groups are often also based on location. Like positions, org groups may be included in a filter defined at design time that is run over the distribution set for a task at runtime.

While the descriptions of the various entities in the YAWL model above discuss the typical uses of each, it should be clear that they represent, at the most basic level, merely various ways to group participants. Therefore, they can equally be used to assemble participants into any kinds of groupings that are meaningful to the implementing organization. The main point of distinction between them is that only roles can be used to populate a distribution set, the other three are used to perform restrictions over the set.

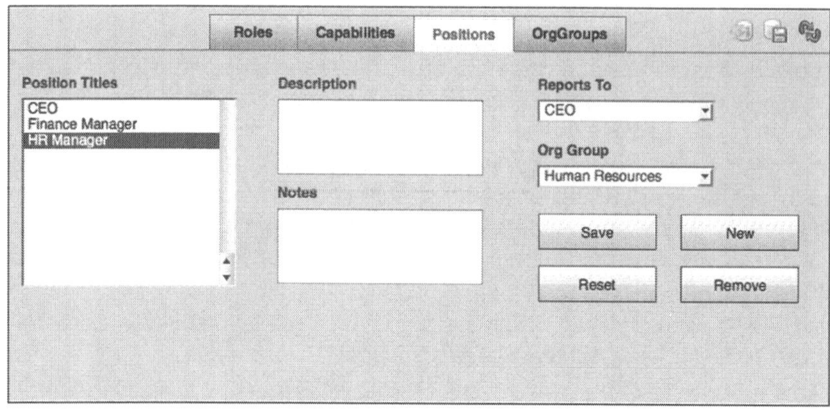

Fig. 10.2 The Organizational Data Management form (Position tab selected)

10.3.2 Org Model Maintenance

As previously mentioned, the Resource Service maintains a set of database tables to internally store the various entities and relationships of the default organizational model. These tables may be populated and maintained via the *User Management* and *Organizational Data Management* forms, part of the toolset available to Resource Service administrators. Hybrid models are also supported: if an external data source provides some entities of the model but not others, the service's data structures may be incorporated to augment the external source with the missing entities. For example, if an external data source provides participants and their position numbers, they may be incorporated into the service's Role, Capability, and Org Group structures to enhance the resource pattern support for their original data. Note that only organization data created through the service's administrative forms may be edited through those forms – data supplied and maintained externally is considered to be "read-only."

Figure 10.2 shows an example of the *Organizational Data Management* form, with the *Positions* tab selected. Entities can be added, removed, and modified, and relationships between entities can be established or removed using this form. In this example, it can be seen that the *Position Titles* refer to the names of various positions that are unique in the organization (e.g., there can only be one *CEO*), but might just as easily refer to actual position numbers, as mentioned earlier, depending on the particular needs and structure of the organization. The example shows that the *HR Manager* position reports directly to the *CEO*, and is a member of the *Human Resources* org group. In this way, positions are added to org groups, and organizational hierarchies can be created.

An example of the *User Management* form is shown in Fig. 10.3. Participants may be created, modified, or removed using this form, and may be assigned or removed from roles, positions, and capabilities. Note that an Org Group tab

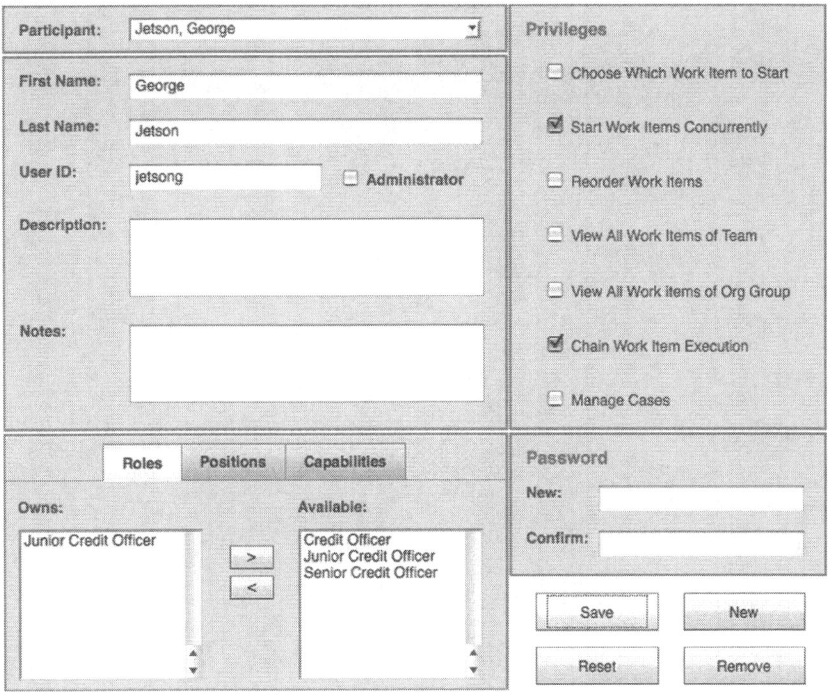

Fig. 10.3 The User Management form

does not feature on this form because individual participants cannot be assigned directly to org groups, but must instead be assigned to a position, which in turn is assigned to an org group via the *Organizational Data Management* form. User Privileges may also be set for participants using this form (see Sect. 10.6.1 for more details).

10.4 Architecture

The Resource Service is the largest and the most complex of the YAWL Custom Services, and consists of a number of components. The core service manages the allocation of resources to work items as designated at design time, in addition to runtime manipulations. To support the various patterns, a dedicated worklist is incorporated. Forms for editing work items are dynamically generated and displayed via the dynamic forms component, while "custom" forms are also supported. An administrative toolset is made available to maintain the organizational model, to manage case instances, to manage custom services, and so on. Much of the service has *pluggable* interfaces, where end-user organizations can add their own customizations.

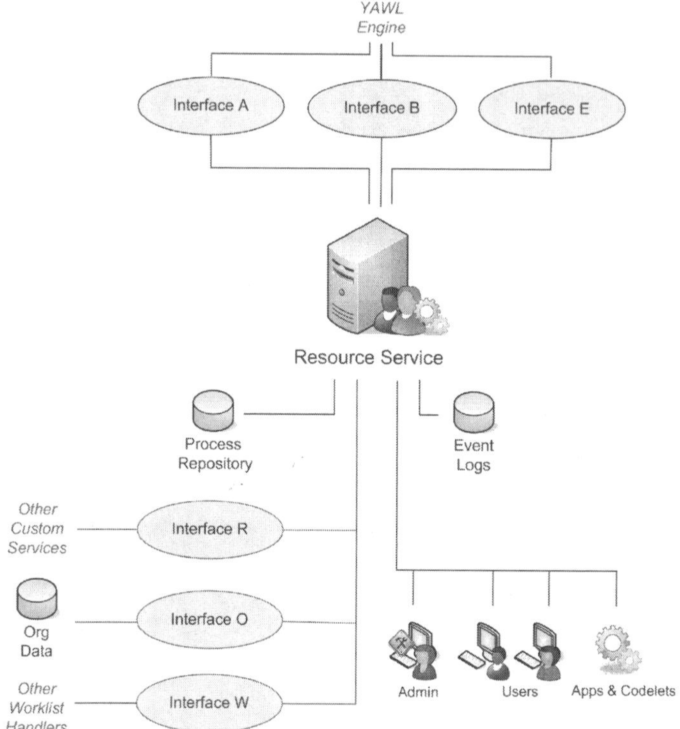

Fig. 10.4 Resource Service external architecture

10.4.1 External Architecture

Figure 10.4 shows the service's external architecture, and reveals no less than six separate interfaces that it uses and/or provides to the external environment. The service communicates with the engine via three interfaces:

- **Interface A** is used to check the service's connection to the Engine, to upload and unload specifications (via the *Case Management* form), and to add and remove custom services (via the *Service Management* form).
- **Interface B** is where the bulk of the communication between the Engine and Resource Service takes place. It is primarily used to handle work item and case events (enablings, cancelations, timer expiries, and so on); to launch and cancel cases, and maintain a running cases list; to checkout and check-in work items; to suspend, unsuspend, and skip work items; to create new work item instances (for dynamic multiple instance tasks); and to retrieve data schemas and other metadata for tasks and specifications.
- **Interface E** is used to query the Engine's process logs when formulating an allocation strategy reliant on historical data (e.g., the *Round–Robin* strategies).

Even though there is necessarily a certain volume of data passed from the engine to the service and back again, one of the fundamental design goals of the service was to minimize data passing to the greatest possible extent. One way it has been minimized is through the utilization of various data caches on the service side – see Sect. 10.4.2 for more details.

In addition to the engine interfaces, the Resource Service provides three interfaces of its own:

- **Interface O** is the Organizational Model Interface, providing the means to use organizational data held in external data sources, rather than the default internal YAWL Org Model. Any type of external data format may be used; an abstract *DataSource* class is provided, which must be extended by each client to map the external org data into the format required by the service, called a *ResourceDataSet* – essentially a set of five maps, each a mapping of unique identifier – populated YAWL resource object pairs (i.e., Participant, Role, Position, OrgGroup, and Capability).
- **Interface R** is the Resourcing Interface, which externalizes the organizational data managed by the service via a set of API methods, thus providing custom services (and particularly the YAWL Process Editor) with a means of accessing the YAWL default org model contents and codelet list (all data are exposed as read-only through this interface).
- **Interface W** is the Workqueue Interface, which externalizes the mechanisms of the internal worklist handler, allowing external applications and custom services the means to provide alternate worklist visualizations, rather than the default worklist forms provided with the service. This interface supports and exposes the full set of resource patterns handled by the service.

The *Process Repository* is a data store (i.e., local disk or URL) holding the set of process specifications created via the Editor, from where they may be loaded through the service's administration tools into the Engine. The *Event Logs* is a data store (again, local disk or URL) where a process log is written containing all resourcing decisions made for each work item handled by the service. The service supports administrators and users via a number of web forms.

10.4.2 Internal Architecture

Figure 10.5 is a representation of the internal structure of the Resource Service and shows its primary components and the major relationships between them. At the core of the service is the *Resource Coordinator*, which is the hub for all traffic into and out of the service via all six interfaces, and is responsible for communication between, and coordination of, all of the composite parts of the internal service.

To minimize data traffic across the interfaces (described in Sect. 10.4.1), the service makes extensive use of data caching. First, when the service starts, the entire organizational data model is loaded into memory-cached data structures, which

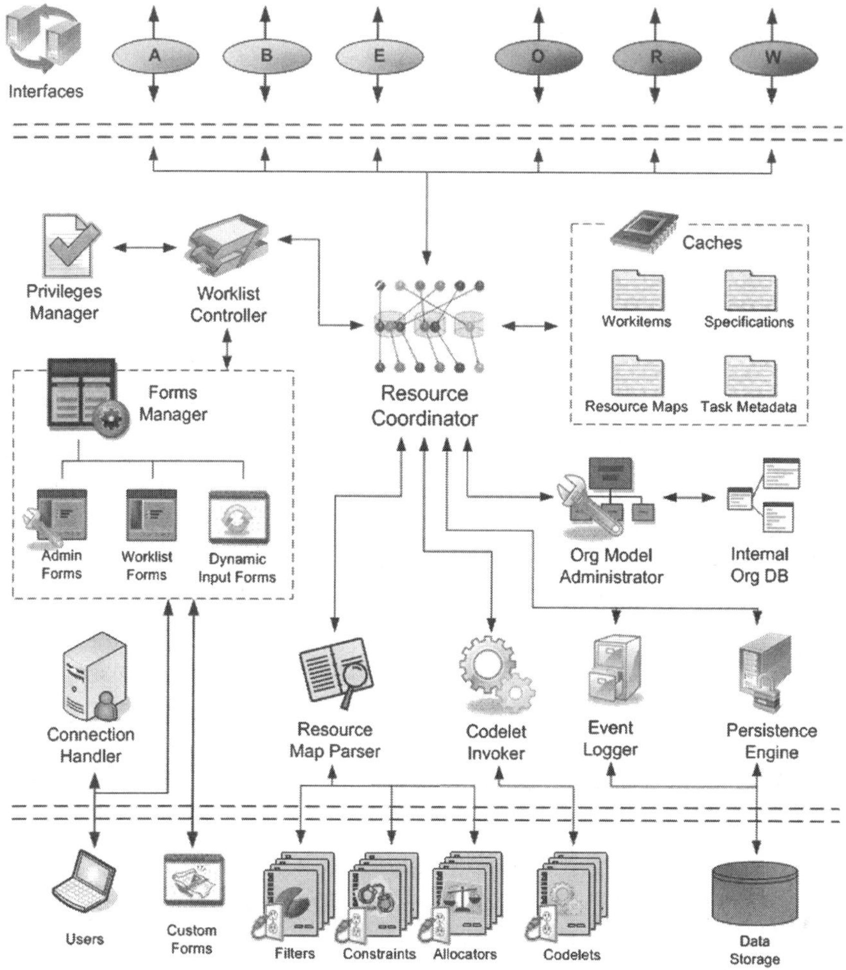

Fig. 10.5 Resource Service internal architecture

maximizes response times while minimizing load on the interface and persisted data sources. Modifications made to the internal YAWL org model are written to both cache and the persistent source (at times of low load). If the org model is loaded from an external source, then as stated in Sect. 10.3, it is treated as read-only, and so modifications to the data are assumed to be also carried out externally. While it is only necessary to load internal organizational data into the cache once on start-up, a service configuration value may be set for external sources so that the caches are updated at a regular, specified interval to enable synchronization with external modifications.

Other objects are passed from the engine to the service at runtime, and immediately stored in caches, so that the internal caches of the engine are not relied upon for regular updates and synchronizations. The cached objects include:

- **Work items**: Whenever the service receives notification from the engine of an enabled (manual) work item via Interface B, the work item is immediately added to the *Work item Cache*. In this way, the service is always aware of the set of "live" work items at any particular time, rather than having to repeatedly poll the engine for this information (which for large sets of work items would place unnecessary load on the engine). When a work item is started, the (parent) work item in the cache is replaced with the child (or children) work items resulting from the start process. When a work item completes, it is removed from the cache. This cache is persisted so that it is reloaded when the service restarts; when the engine restarts, the service's work item cache is synchronized against the engine's own caches – this synchronization process may also be manually triggered via a button on the *Administration Queues* form, if desired. All work items in all work queues are actually references to work items in the cache (i.e., only one copy of each work item ever exists), which also minimizes the size of objects and data structures created by the service.
- **Specifications**: For each specification that is loaded via the *Case Management* form, a copy is placed in the *Specifications Cache*. Some specification data, particularly schemas of user-defined data types, are required for dynamic form generation. If a work item is received from the engine that belongs to a specification that is not in the service's cache (i.e., it has been loaded via another Custom Service or the engine has been restarted), a copy is retrieved from the engine via Interface B. The specification cache is not persisted, as a noncached specification can be retrieved by a once-only request from the engine.
- **Task Metadata**: Information describing a task (e.g., input and output parameters and their various descriptors, and data schemas describing the entire data structure of the work item) are also required for dynamic forms generation, and thus are cached to avoid requests across the interface each time a work item's dynamic form is displayed. Task metadata is cached for all tasks not associated with another Custom Service, for all versions of all specifications for which a case instance is currently executing.
- **Resource Maps**: A resource map is an object that is created from a task's resourcing specification, and is used to perform the distribution of a task's work items to resources at runtime. The design time resourcing parameters for each task are stored in the specification file, as XML (see Listing 10.1). When a work item is enabled at runtime, the service retrieves the resource map for it from the cache, and uses it to distribute the work item. If the resource map is not currently contained in the cache, the XML parameters for the task are retrieved and parsed by the *Resource Map Parser* into a resource map object and placed in the cache (see below). Thus, there is a resource map for each task in each version of each specification for which a work item has been enabled.

```
<resourcing>
  <offer initiator="system">
    <distributionSet>
      <initialSet>
        <role>RO-7a1f36e8-8b47-46f4-b8b5-af07dac6174a</role>
      </initialSet>
    </distributionSet>
  </offer>
  <allocate initiator="system">
    <allocator>
      <name>RandomChoice</name>
    </allocator>
  </allocate>
  <start initiator="user"/>
  <privileges>
    <privilege>
      <name>canDeallocate</name>
      <allowall>true</allowall>
    </privilege>
  </privileges>
</resourcing>
```

Listing 10.1 Example of a resourcing specification (extract from *OrderFulFilment* process *Confirm Purchase Order* task)

Other Internal Components

The ***Forms Manager*** component provides a user interface to the service via a number of web-based forms, which can be categorized as worklist forms (those providing worklist views and functionality to manipulate both the resourcing of the listed work items and to progress them to completion – see Sect. 10.7 for more details); administration forms; and dynamic forms (dynamically generated input forms to capture values for work item parameters). The Forms Manager is responsible for displaying the appropriate form, to fill a form's content (e.g., to display a certain work queue for the user requesting it), and dynamic form generation. All forms are created and controlled using Java Server Faces (JSF) technology.

A complete forms rendering engine is encapsulated by the service, and is responsible for taking a work item's data parameter set, expressed as an *xsd:schema*, and rendering it as a web-based form with an appropriate set of input fields. Listing 10.2 shows an example of the data schema and Fig. 10.6 the resultant dynamic form for the *Estimate Trailer Usage* task of the Order Fulfillment process. The data schema is compiled from the data type definitions declared in the specification for each data parameter of the task, into a data schema encapsulating the entire task. From there, a Forms Manager "factory" class creates a number of *FormField* objects, which are hierarchical (i.e., may have *FormField* children), and take into account any restrictions, enumerations, and unions declared for each data type. Each field object may represent one field (e.g., *OrderNumber*), or, if they have children, a *SubPanel* of child fields (e.g., *Line*). These field objects are assembled into a set, then sent to the *ComponentBuilder*, which creates a JSF input field for each of them. That set of JSF field components are then rendered in a browser window.

```
<?xml version="1.0" encoding="UTF-8"?>
<xsd:schema xmlns:xsd="http://www.w3.org/2001/XMLSchema" elementFormDefault="
   qualified">
   <xsd:element name="Estimate_Trailer_Usage">
      <xsd:complexType>
         <xsd:sequence>
            <xsd:element maxOccurs="1" minOccurs="0" name="TrailerUsage">
               <xsd:complexType>
                  <xsd:sequence>
                     <xsd:element name="OrderNumber" type="xsd:string"/>
                     <xsd:element name="OrderLines">
                        <xsd:complexType>
                           <xsd:sequence>
                              <xsd:element maxOccurs="unbounded" name="Line">
                                 <xsd:complexType>
                                    <xsd:sequence>
                                       <xsd:element name="LineNumber" type="
                                          xsd:integer"/>
                                       <xsd:element name="UnitCode" type="
                                          xsd:string"/>
                                       <xsd:element name="UnitDescription"
                                          type="xsd:string"/>
                                       <xsd:element name="UnitQuantity" type="
                                          xsd:integer"/>
                                       <xsd:element name="Action" type="
                                          xsd:string"/>
                                    </xsd:sequence>
                                 </xsd:complexType>
                              </xsd:element>
                           </xsd:sequence>
                        </xsd:complexType>
                     </xsd:element>
                     <xsd:element name="Packages">
                        <xsd:complexType>
                           <xsd:sequence>
                              <xsd:element maxOccurs="unbounded" name="Package
                                 ">
                                 <xsd:complexType>
                                    <xsd:sequence>
                                       <xsd:element name="PackageID" type="
                                          xsd:string"/>
                                       <xsd:element name="Volume" type="
                                          xsd:integer"/>
                                    </xsd:sequence>
                                 </xsd:complexType>
                              </xsd:element>
                           </xsd:sequence>
                        </xsd:complexType>
                     </xsd:element>
                  </xsd:sequence>
               </xsd:complexType>
            </xsd:element>
         </xsd:sequence>
      </xsd:complexType>
   </xsd:element>
</xsd:schema>
```

Listing 10.2 Compiled data schema for task *Estimate Trailer Usage*

Different types of JSF components are built depending on the base data type of the field (as referenced from the data schema). Fields that expect boolean values are rendered as check-boxes, while enumerations become drop-down combo boxes. Dates are rendered as calendar inputs – a drop-down calendar widget allows users to easily select appropriate date values. All other types are rendered as text

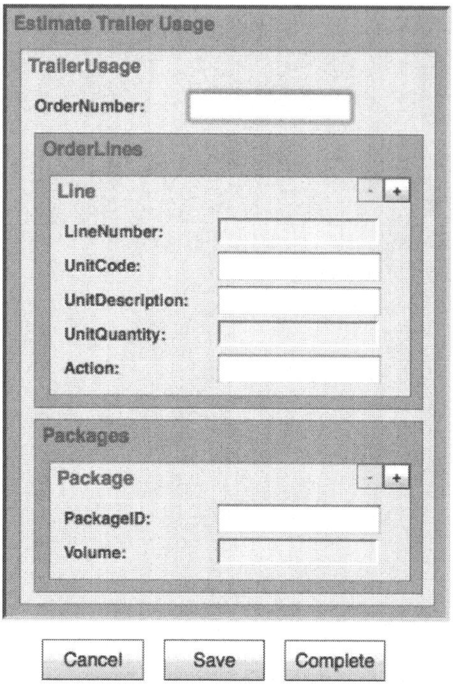

Fig. 10.6 The resultant dynamic form (derived from Listing 10.2)

inputs. In addition, mandatory fields (i.e., those that require a value to be entered) are shown with a yellow background, while optional fields have a white background, and every field has a mouse-over tool-tip that displays a prompt about the type of data expected.

When a user attempts to save a dynamic form, each user-provided value is validated against its data type, along with any further restrictions specified in its schema. Fields with invalid values change to a red background, and an appropriate error message is also displayed; the form cannot be saved until all fields validate.

Referring again to Listing 10.2 and Fig. 10.6, the definition of a complex type in the schema results in the creation of a form panel, within which its child fields are contained. Such definitions that have a *maxOccurs* value greater than its *minOccurs* value (if omitted, the default values for these attributes is 1), or is "unbounded," will also be rendered with small *increase* and *decrease* buttons in the top right of their sub-panel. These buttons allow multiple sets of those fields to be dynamically created or removed by the user. In the example, an order may have any number of lines (one for each item ordered) and any number of packages; a user can easily increase or decrease the number of order lines or packages as appropriate for each order.

Dynamic forms are designed for maximum flexibility and can display work item parameters of any data type. However, their generic look and feel may not be appropriate in all cases, for example where an organization has a standardized set of forms for their business processes, and would like their web-based forms to match the standard. In such cases, a *Custom Form* may be user-defined and associated with a task at design time by specifying a URL to the form. At runtime for such a task, the Resource Service will package up the work item and its data and send it to the custom form to populate its fields, and then will display it. On submission of the form by the user, the work item data is extracted from the form by the service and passed back to the work item in the same manner as dynamic forms. Custom forms may be built using any web-based technology, such as JSF, Javascript, .NET, PHP, or any other environment that can receive data, use it to populate form fields, update the data with user inputs, and pass control back to the calling service.

A ***Worklist Controller*** component maintains the references to work items within each work queue for each participant, and responds with the appropriate work item list when requested by the Forms Manager. The work queues managed by the Controller are updated after each occurrence of a user or system action affecting them by the Resource Coordinator. The Worklist Controller in turn queries the ***Privileges Manager*** to ensure only those actions for which a participant is authorized are enabled on worklist forms. The Privileges Manager handles both *User* privileges (sourced from the Organizational Data for each participant) and *Task* privileges (sourced from the Resource Map for each task) – discussed in detail in Sect. 10.6.

The ***Org Model Administrator*** is responsible for loading and maintaining the organizational model and its incorporated data, whether from the internal YAWL Organizational Model database tables or from an external data source via Interface O. It also manages the updating of data via the *Org Data Management* and *User Management* forms (cf. Sect. 10.3.2), and provides internal table access to external sources for objects that are not loaded via Interface O.

The ***Resource Map Parser*** is invoked when the service is notified of an enabled work item, and is used to parse the resourcing specification XML (Fig. 10.1) into a *ResourceMap* object (if it is not already in cache), and then carries out the initial resource distribution (cf. Sect. 10.5). ResourceMap objects also store secondary resourcing information and participate in later user-triggered resourcing changes. Each task definition has its own discrete Resource Map. As part of the initial distribution, the parser may make use of various filters, constraints, and allocation strategies as defined at design-time – each of these are also "pluggable," so that user-organizations can add their own particular methodologies to the way work items are resourced.

The ***Codelet Invoker*** is responsible for the execution of codelets. Codelets may be thought of as non-human resources for executing tasks, and each is literally a Java class, implementing a generic interface, that performs a certain action. A task is associated with a codelet at design time using the Editor; the task must first be marked as automated (automated tasks are those that do not require human resourcing). It is not mandatory for an automated task to be associated with a codelet – data transformations may also be performed via the XQuery predicates of a task's output data parameters.

Codelets are "pluggable" – new codelets may be easily added to the codelet repertoire, and are immediately available to be associated with tasks in the Editor (the Editor retrieves the list of available codelets though Interface R). A number of codelets are distributed as part of the default environment. Perhaps the most interesting of these is *ShellExecution*, which provides for any program to be executed in the external operating system environment (i.e., outside the immediate YAWL environment), using application names and parameters passed as data values within the automated task associated with the codelet.

The ***Connection Handler*** manages user sessions. This component checks logon credentials, and creates and maintains sessions with end users. A timer is also maintained for each session, and times out after a period of inactivity (by default 60 min, but this value may be modified via a configuration setting). Each time a user interacts with the service, their session is checked for currency, and if valid, the inactivity timer is reset.

Every runtime action taken by a user or the service on a work item is logged by the ***Event Logger*** component, so that a process history is kept for all cases. All changes in resource status (e.g., "Allocated" to "Started") are logged, as well as user actions such as delegation and reallocation. Data values describing the participant, the work item, the action, and the timestamp are kept. Log records can be accessed via Interface W, and may be incorporated with the engine's process logs to reveal a detailed process execution history for each case.

Finally, the ***Persistence Engine*** saves all runtime objects, updating them as necessary, to ensure a copy of "live" cases and their associated data are permanently backed up so that the entire runtime environment can be recreated in the event of a server shutdown and restart.

10.5 Initial Distribution

When the engine notifies the service of an enabled work item, the service undertakes to distribute the work item to resource(s) using the resourcing specifications for the task from which the work item was created, as specified at design time. The process followed by the service is shown in the flowchart in Fig. 10.7.

The first step is to retrieve the resource map applicable to the work item. A *Resource Map* is an object created by parsing the task's resourcing specification, and describes how the work item's resourcing decisions are to be made, as well as performing the actual distribution process. For all specifications created in YAWL versions prior to 2.0, there will be no resource map available, as specifications based on older schemas had no knowledge of the Resource Service. In such cases, the work item is offered to all participants (i.e., each participant in the organizational model will see the work item in their *Offered* queue) and the distribution process completes. If the specification is based on the YAWL 2.0 schema, then each task will have a designated resourcing specification (a default specification of *User–User–User* initiated interaction points is applied to any task that has not been explicitly designated a resourcing specification at design time).

Once the resource map is available, it is first checked to determine if a participant has *piled* the work item. If so, the work item is immediately started and placed on

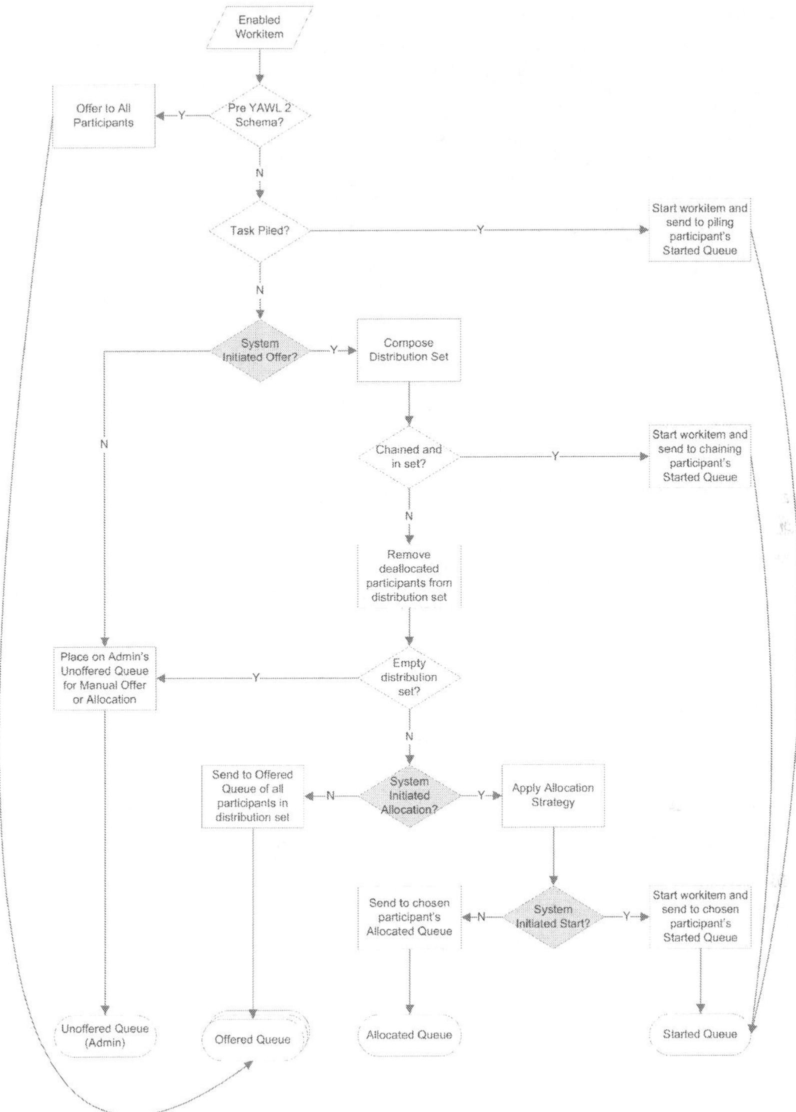

Fig. 10.7 Runtime initial distribution flowchart

the piling participant's *Started* queue, and the distribution process completes. Piling overrides all design time resourcing specifications for the work item. If the work item is not piled, the process moves on to the *Offer Interaction* stage.

If the offer interaction is *user* initiated, the work item is placed on the administrator's *Unoffered* queue to await manual offering (to a number of participants) or allocation or starting (to a single participant) and the distribution process completes. If it is *system* initiated, the designated distribution set is assembled as follows:

1. First, the resource map is checked to determine whether a *familiar task* constraint is set. If so, any participants who have previously completed work items of the designated familiar task for the current case are added to the distribution set, and the set composition completes (all other offer interaction settings are ignored – a familiar task constraint takes precedence over specified participants, roles, and net-level variables). If there is no familiar task constraint set, then the process moves to step 2.
2. Each specified participant is added to the distribution set. Note that it is a set in the strict sense – all duplicate participants are discarded.
3. Each specified role is deconstructed into its corresponding set of participants, as are each of its descendant roles (i.e., any roles that *belong to* the role specified). Each of the resultant participant-sets are added to the distribution set. Again, all duplicates are discarded, which may be a quite common occurrence as participants may be members of more than one role or may have been individually specified in step 2.
4. Each specified net-level variable is inspected for its corresponding participant or role value. Net-level variables are nominated to contain values that are populated at runtime with either a participant or role reference. Participant values are determined and added to the distribution set as per step 2; similarly, role values as per step 3.
5. Each specified filter is applied to the distribution set. Filters are set at design time to ensure that the distribution set includes only those members who have some specified capability and/or hold a certain position and/or belong to a certain org group. Filters are also "pluggable," so that user-defined filters may also be specified.
6. Each specified constraint is applied to the distribution set. One constraint is provided by default, the so-called *four-eyes* or *separation of duties* constraint, which, conversely to *familiar task*, will remove from the distribution set any participants who have previously completed work items of the specified task for the current case. Like filters, constraints are "pluggable" and so may include user-defined constraint methods.

The final distribution set is then checked for a *chaining* participant. If a particular participant has chosen to chain the work items of the current case, *and* the participant is a member of the distribution set for a work item, the work item is immediately started and placed on the participant's started queue, and the distribution process completes; chaining overrides all subsequent design time resourcing specifications for the work item in the current distribution process. If the case is not being chained, or the chaining participant is not a member of the distribution set, the process continues by removing from the set any participants who have previously been allocated the work item and have chosen to deallocate it.

At this stage, the distribution set is checked to ensure it is not empty, whether as a result of the initial set composition process or after the removal of deallocated participants. If it is empty, then the work item is placed on the administrator's *Unoffered* queue to await manual offering (as per a user-initiated start interaction) and the distribution process completes. If the set has members, the process moves to the *Allocate Interaction* stage.

If the allocate interaction is specified as user-initiated, then the work item is offered to all participants in the distribution set (i.e., placed on each of their *Offered* queues) and the distribution process completes. If it is system-initiated, then the specified allocation strategy is applied. All allocation strategies take the distribution set and select exactly one participant from it. Allocators also are "pluggable," and so user-defined strategies for choosing a participant from the set may be made available to this stage.

The chosen participant is then passed to the *Start Interaction* stage, which is relatively straightforward: if the interaction is user-initiated, the work item is placed on the participant's *Allocated* queue, to be manually started by the participant at some future time; if it is system-initiated, the work item is immediately started and placed on the participant's *Started* queue. The distribution process for the work item is now complete.

This initial distribution process describes that followed by the service, commencing with the notification of a new enabled work item, and completing with the placement of the work item in one of the four queue types: on the *Allocated* or *Started* queue of a particular participant; on the *Unoffered* queue of the administrator (all administrators access the same queue); or on the *Offered* queue of one or more participants (multiple work item references on multiple queues). Of course, following the completion of this design-time specified process, participants and administrators may manually move a work item off, on, and between queues, using the various methods described in Sect. 10.7.

10.6 Privileges

Some of the actions that a participant may take for a work item are allowed only when the participant has the necessary privileges. Thus, privileges are a way of controlling the access of participants to certain functionalities of the Resource Service. There are three different categories of privileges.

The first type is the level of logon access that a participant may hold: *user* or *administrator*. Simply, a participant with default user access privileges may access their own work queues and may view their own profile. They may also be granted certain privileges of the other two types (see below), which extend to some degree their access to particular system actions. Conversely, if a participant is granted administrator privileges, unrestricted access to the system is permitted, overriding any individually set privileges of the two types below, in addition to permitting the various administration activities such as user management, organizational data management, loading and unloading specifications, starting and canceling cases, loading and unloading custom services, and access to the administration work queues. Note that the generic "admin" userid gains administration access to the system, but is not considered a participant, and so does not have any work queues or profile. All users with administrator privileges share a single *Unoffered* queue and a single *Worklisted* queue.

10.6.1 User Privileges

User privileges are granted by an administrator to a participant, on an individual basis, via the User Management form (Fig. 10.3). Such privileges apply to the participant at all times, for all case instances – that is, they are privileges that are permanently owned by each individual participant. User privileges extend the default access levels of a participant. The User privileges are the following:

- **Choose Which Work Item to Start:** When granted, this privilege allows a participant to choose any work item listed on their allocated queue to start. When denied (the default) only the first listed work item may be chosen. Work items are listed in order of age, with the oldest work item at the top of the list. While normally participants are obliged to start the oldest work item in their queue, granting this privilege overrides that obligation.
- **Start Work Items Concurrently:** When granted, this privilege allows a participant to have a number of work items executing concurrently on their started queue (or, more accurately, may choose to start additional work items from their allocated queue while other previously started work items have not yet been completed). When denied (the default), a work item on the participant's allocated queue may not be started while there is a previously started work item on their started queue (i.e., one that has not yet completed).
- **Reorder Work Items:** When granted, the participant may choose a work item to start from anywhere in the list of allocated work items. When denied, only the first listed work item may be chosen. In the YAWL environment, there is essentially no difference between this privilege and *Choose Which Work Item to Start*.
- **View All Work Items of Team:** When granted, this privilege gives a participant access to the *Team Queues* form, and displays on that form a consolidated list of all work items on all work queues of all participants subordinate to the participant who has been granted the privilege (i.e., participants holding positions that report to a position held by the granted participant, either directly or through a hierarchy of positions). When denied (the default), the *Team Queues* form is not available to the participant.
- **View All Work Items of Org Group:** When granted, this privilege gives a participant access to the *Team Queues* form, and displays on that form a consolidated list of all work items on all work queues of all participants in the same Org Group as the granted participant. When denied (the default), the *Team Queues* form is not available to the participant.
- **Chain Work Item Execution:** When granted, this privilege allows a participant to chain work items for a case. When denied (the default), the participant may not chain cases.

- **Manage Cases:** When granted, this privilege gives a participant access to the *Case Mgt* form, providing the ability to load process specifications and start and cancel case instances. When denied (the default), the *Case Mgt* form is not available to the participant.

10.6.2 Task Privileges

Task privileges (or, more precisely, *User-Task* privileges), unlike the *User* privileges described in the previous Section, are set at design time via the Editor (cf. Chap. 8, Fig. 8.15) on an individual task basis. That is, for each task in a process specification, task privileges may be set for individual participants, roles, or may be "blanket" granted to all participants.

Broadly speaking, task privileges grant or deny the ability to affect in various ways how work items are resourced after initial distribution has completed. There are seven task privileges (see Sect. 10.7 for more details of their application):

- **Can Suspend:** When granted, allows a participant to suspend the execution of a work item after it has been started.
- **Can Reallocate Stateless:** When granted, allows a participant to transfer responsibility for the execution of a work item from themselves to another participant, with the data parameters of the work item reset to the values held when the work item was first started.
- **Can Reallocate Stateful:** When granted, allows a participant to transfer responsibility for the execution of a work item from themselves to another participant, with the data parameters of the work item having their current values maintained.
- **Can Deallocate:** When granted, allows a participant to reject or rollback the allocation of a work item to their allocated queue. The work item is redistributed using the original resourcing specification, but with the participant removed from the distribution set.
- **Can Delegate:** When granted, allows a participant to delegate the responsibility for the execution of a work item to a subordinate member of their work team, as defined by the organizational model.
- **Can Skip:** When granted, allows a participant to have the execution of a work item skipped – that is, immediately completed without performing its work.
- **Can Pile:** When granted, allows a participant to demand that all future instances of work items derived from this task, in all future instances of the specification of which the task is a member, are immediately directly routed to the participant and started.

All task privileges are denied by default, and so must be set explicitly for each task as required.

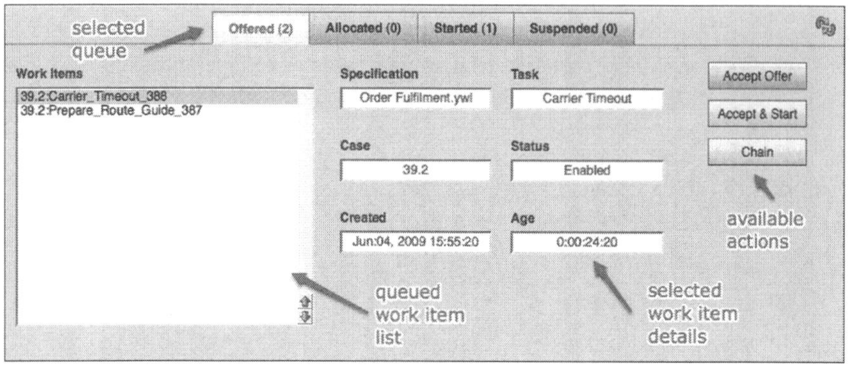

Fig. 10.8 The Offered work queue

10.7 The Worklist

Each participant has access to their own worklist, which is a graphical representation of their work queues via a series of web forms. It is important to recognize that each participant's queues are held internally by the service, and the default worklist is simply one representation of them, allowing a participant to interact with the YAWL environment; other worklist representations and visualizations may be constructed by developers accessing work queue data via Interface W.

As mentioned earlier, each worklist consists of four work queues: *Offered*, *Allocated*, *Started*, and *Suspended*. Depending on a participant's privileges, there are a number of actions that can be performed on a work item in each queue. Some are concerned with processing the work item, while others provide for changes to the work item's resourcing.

The layout of each work queue is similar; Fig. 10.8 shows an example of the Offered queue. On the left is a list of work items currently held in that queue. In the center are some fields that describe the currently selected work item. On the right are a set of buttons representing the actions that may or may not be taken on that queue for the currently selected work item. Note that whether or not some buttons are enabled depends on whether the logged on participant has the appropriate privileges to perform that action on the work item.

10.7.1 The Offered Queue

The Offered queue lists the work items that have been offered to a participant. Each work item in an offered queue may have potentially been offered to a number of participants. A work item may be offered to a participant via one of three paths: the task from which the work item was created had a system-initiated offer interaction *and* a user-initiated allocate interaction specified at design time, and the participant is a member of the specified distribution set; the task from which the work item was created had a user-initiated offer interaction specified at design time or had an empty or invalid resourcing specification, and an administrator has manually offered the work item to the participant; or an administrator has manually reoffered the work item to the participant, the work item having previously been offered to, allocated to, or started by another participant.

As a work item on an offered queue may have been offered to a number of participants, there is no implied obligation to accept the offer, rather it is understood that the participant is one of a group, any one who *may* choose to perform the work item.

A participant may take the following actions on a work item in an offered queue:

- **Accept Offer**: By accepting an offer, a participant takes responsibility for the execution of the work item. The work item is moved from the offered queue and, if the start interaction is user-initiated, placed on the participant's allocated queue, or if the start interaction is system-initiated, the work item is immediately started and placed on the participant's started queue. This action removes the work item from the offered queues of all other participants who had been previously offered the work item.
- **Accept & Start**: This action works similarly to *Accept Offer*, except that if the work item's start interaction is user-initiated, the work item will instead be immediately started and placed on the participant's started queue. Effectively, this concatenates two user actions into one, simply as a convenience for the user.
- **Chain**: This action will chain all the eligible work items of the case of which the work item is a member to this participant. Chaining means that, when a participant chooses to enact it, each remaining work item for the case is routed to the participant and immediately started, but *only* if the participant is a member of the distribution set for the work item. Chaining is effectively a short-circuiting of a resource specification for a task, where the participant chooses to automatically and immediately allocate and start any work item offered to him/her within the chosen case. Chaining of work items for a case continues until the case completes, or the participant turns off chaining via the *View Profile* form. A participant must have the "Chain Work Item Execution" user privilege to enable chaining.

10.7.2 The Allocated Queue

The Allocated queue lists the work items that have been allocated to a participant. Unlike an offer, a work item on an allocated queue means that it has been allocated to that participant alone, and comes with the understanding that the participant will at some time start the work item and perform its work. A work item may be allocated to a participant via one of the five paths: the task from which the work item was created had a system-initiated offer interaction, a system-initiated allocate interaction, *and* a user-initiated start interaction specified at design time, and the participant is a member of the specified distribution set and the specified allocation strategy employed has resulted in the participant being chosen to be allocated the work item; the task from which the work item was created had a user-initiated start interaction specified at design time, and the participant has chosen the *Accept Offer* action for the work item on the offered queue; the task from which the work item was created had a user-initiated offer interaction specified at design time or had an empty or invalid resourcing specification, and an administrator has manually allocated the work item to the participant; the work item has been delegated by another participant (see below); or an administrator has manually reallocated the work item to the participant, the work item having previously been allocated to or started by another participant.

A participant may take the following actions on a work item in an allocated queue:

- **Start**: The work item is started (i.e., begins executing), and moved to the participant's started queue.
- **Deallocate**: This action provides an authorized participant with a means of rejecting the allocation of a work item. The work item is removed from the participant's allocated queue, the participant is removed from the original distribution set (if any), and the work item is redistributed as per the resourcing specification. If there is no distribution set or there is an empty or invalid resourcing specification for the task from which the work item was created, the work item is placed on the administrator's unoffered queue for manual distribution. A participant must have the task privilege "Can Deallocate" to enable deallocation.
- **Delegate**: This action allows a participant to delegate responsibility for a work-item to another participant. The receiving participant must be subordinate to the delegating participant in the organizational model, that is, the receiving participant must hold a position that reports to a position held by the delegating participant, either directly or through a hierarchy of positions. The work item is moved from the allocated queue of the delegator to the allocated queue of the receiver. A participant must have the task privilege "Can Delegate" and have subordinate staff to successfully deallocate a work item.
- **Skip**: This action skips the execution of the work item, that is, the work item is immediately started and then completed, allowing the process to continue according to its subsequent control-flow. A participant must have the task privilege "Can Skip" to enable the skipping of a work item.
- **Pile**: When a work item is piled, the work item is immediately started and placed in the participant's started queue. Furthermore, each and every future instance of the work item across all cases (of the same specification) is automatically allocated to the participant and started, completely ignoring any resourcing specification for the task from which the work item is created. To put it another way, by piling a work item, a participant is entering into a contract with the Resource Service, asking that this work item, and all future occurrences of such work items created from the same task description as the original work item was created from, be immediately allocated and started to him/her. Piling of such work items

continues until the participant turns off piling for the task via the *View Profile* form, or the participant logs out (although a configuration value may be set to persist the piling across user sessions). A participant must have the "Can Pile" task privilege to enable piling.

10.7.3 The Started Queue

The Started queue lists the work items that have been started by or for a participant. Each work item on a started queue has begun execution in a system sense, but may or may not have had any actual work begun for it by the participant – such work is performed by the participant viewing, editing, and finally completing the work item. A work item may arrive on a participant's started queue via one of the following paths: the task from which the work item was created had a system-initiated offer interaction, a system-initiated allocate interaction, *and* a system-initiated start interaction specified at design time, and the participant is a member of the specified distribution set, and the specified allocation strategy employed has resulted in the participant being chosen to be allocated the work item; the task from which the work item was created had a user-initiated start interaction specified at design time, and the participant has chosen the *Start* action for the work item on the allocated queue; the task from which the work item was created had a system-initiated start interaction specified at design time, and the participant has chosen the *Accept Offer* or *Accept & Start* action for the work item on the offered queue; the task from which the work item was created had a user-initiated start interaction specified at design time, and the participant has chosen the *Accept & Start* action for the work item on the offered queue; the task from which the work item was created had a user-initiated offer interaction specified at design time or had an empty or invalid resourcing specification, and an administrator has manually started the work item for the participant; the work item was on the participant's suspended queue and has been subsequently unsuspended; the work item was on the started queue of another participant and reallocated to this participant (see below); or an administrator has manually restarted the work item for the participant, the work item having previously been started by another participant.

A participant may take the following actions on a work item in a started queue:

- **View/Edit**: This action will display the data parameters and their current values for the selected work item, either on a dynamically generated form or, if specified, a custom form, allowing the participant to view and/or edit the form's values. Any modified values are stored so that this action can be repeated for a particular work item a number of times before completion, allowing the work item to be processed by the participant in a progressive manner, if required. This action is disabled if the work item has no data parameters to display or gather values for.

- **Suspend**: This action suspends the selected work item. The work item is removed from the started queue and placed on the participant's suspended queue. A participant must have the task privilege "Can Suspend" to successfully suspend a work item.
- **Reallocate Stateless**: This action allows a participant to reallocate a work item to another participant. The receiving participant must be subordinate to the reallocating participant in the organizational model, that is, the receiving participant must hold a position that reports to a position held by the reallocating participant, either directly or through a hierarchy of positions. The work item's data values are reset to the values that existed when the work item was first started (i.e., stateless reallocation), and it is moved from the started queue of the reallocator to the *started* queue of the receiver. A participant must have the task privilege "Can Reallocate Stateless" and have subordinate staff to successfully reallocate a work item.
- **Reallocate Stateful**: Similar to "Reallocate Stateless," except that any modified data values are maintained when the work item is reallocated. A participant must have the task privilege "Can Reallocate Stateful" and have subordinate staff to successfully reallocate a work item.
- **New Instance**: This action creates a new instance of the selected work item; it is enabled only for a work item of a multiple instance atomic task that allows dynamic creation of additional work item instances.
- **Complete**: Completes the selected work item. The work item is posted back to the engine, which then progresses the case according to its control-flow. This action is initially disabled if the work item contains mandatory editable data parameters, and becomes enabled after the first view/edit of the work item.

10.7.4 The Suspended Queue

The Suspended queue lists executing work items that have been suspended. Note that suspended work items must have already been started and not yet completed, and so this queue may be seen as an extension of the started queue. There is only one path for a work item to appear on the suspended queue, and that is via the *Suspend* action on the started queue. This queue contains one action, **Unsuspend**, which resumes the work item, removing it from the suspended queue and returning it to the started queue.

10.8 Conclusion

This chapter gave a broad overview of the YAWL Resource Service. The service is completely autonomous to the engine, fully realizing the Service Oriented Architecture of the YAWL environment, and supports *all* the identified resource patterns applicable to non-case-handling paradigms. Although it is the default resource enabler for YAWL, its architecture means that it may be substituted with other work handlers, if desired. Additionally, its pattern support is fully externalized, so that other worklist representations and visualizations can be constructed that make full use of, and extend from, the service's capabilities. Also, the service's default dynamic forms generation can easily be superseded by custom designed forms that meet the implementing organizations own requirements. The service supports the importing of organizational data from external sources, in addition to supplying a default data structure for that purpose. The sets of filters, constraints, allocators, and codelets may be added to at any time, to extend the capabilities of the service to individual requirements. While providing extensive support for the resourcing of work, the service is also flexible and extensible, and so meets a wide variety of organizational needs.

Exercises

Exercise 1. What is the difference between a *manual* and an *automated* task? If an automated task is handled by a custom service other than the Resource Service, what will be the result?

Exercise 2. What is meant by the *late binding* of tasks to resources? How is it achieved by the Resource Service?

Exercise 3. Log on as an administrator to YAWL, and go to the "Org Data Management" form. On the "Roles" tab, add two new roles called "Client Service" and "Public Relations." Make the "Client Service" role report to "Public Relations." Now, go to the "User Mgt" form and add five new participants: Alan Black, Bobby Brown, Carly Blue, Doris Orange, and Elvis Green. Give Alan and Bobby the "Client Service" role, add the rest to "Public Relations." In the YAWL Process Editor, Create the simple process shown in Fig. 10.9; assign the "Client Service" role to A and "Public Relations" to B. Save the specification. Back at the YAWL web forms, go to the "Case Mgt" form, load the specification you just created, and start a case. Now, go to the "Admin Queues" form. A and B should appear in the 'Work-

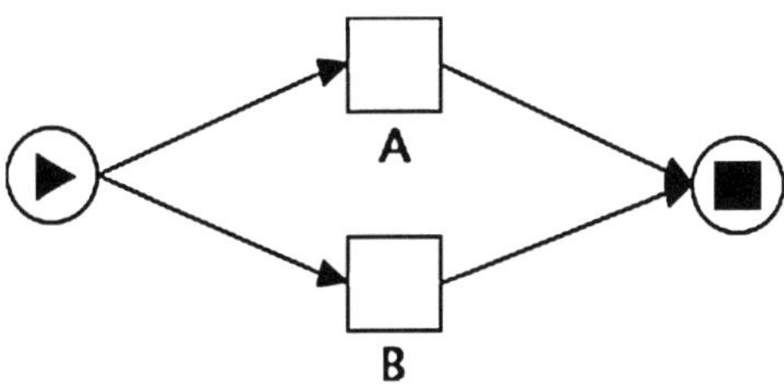

Fig. 10.9 A simple process model (see Exercise 3)

listed' queue. Which participants has A been assigned to? Which participants has B been assigned to? Why?

Exercise 4. In terms of their effect on a distribution set, what is the distinction between Roles and the other three grouping entities Positions, Capabilities, and Org Groups?

Exercise 5. Open the Order Fulfillment process in the Editor. Pick a subnet (say, "Payment") and inspect the resourcing settings for each task via the "Manage Resourcing" menu item.

(a) Why is the "Manage Resourcing" menu item disabled for some tasks?
(b) Start an instance of the process in the engine, and follow the assignment of tasks to resources, reviewing each task in the editor as it appears in the admin queues. Check that each of the runtime assignments match those specified at design time.

Exercise 6. What is a 'codelet'? How is a codelet invoked? How does it differ from an external application or web service?

Exercise 7. What is the main difference between User Privileges and Task Privileges? Is there any overlap between them?

Chapter Notes

For a survey of the various levels of support for the resource perspective amongst Process-aware Information Systems and modeling languages, the interested reader is directed to [222] or Sect. 4.10 of [219].

An example of an application that uses Interface W to display worklists in a different visual style as an alternative to the default worklist handler can be found in [148].

Chapter 11
The Worklet Service

Michael Adams

11.1 Introduction

This chapter discusses the implementation and use of the Worklet Service (a conceptual overview of the service can be found in Chaps. 4 and 5). The Worklet Service has been implemented as a YAWL Custom Service (cf. Chap. 7). In the discussion that follows, a reference to the "Worklet Service" applies to the entire Custom Service, while a reference to the "Selection Service" or the "Exception Service" applies to that particular subcomponent of the service.

All of the worklet process models in this chapter are expressed using YAWL notation – the language is used to model static YAWL processes, and the worklets used both for selection and exception compensations – thus, all worklet specifications are examples of standard, complete YAWL process models.

11.2 Service Overview

The *Worklet Service* comprises two discrete but complementary subservices: a *Selection Service*, which enables dynamic flexibility for process instances, and an *Exception Service*, which provides facilities to handle both expected and unexpected process exceptions (i.e., events and occurrences that may happen during the life of a parent process instance that are not explicitly modeled) at runtime.

11.2.1 The Selection Service

The Selection Service enables flexibility by allowing a process designer to designate certain work items to each be substituted at runtime with a dynamically selected

M. Adams (✉)
Queensland University of Technology, Brisbane, Australia
e-mail: mj.adams@qut.edu.au

A.H.M. ter Hofstede et al. (eds.), *Modern Business Process Automation*,
DOI 10.1007/978-3-642-03121-2_11,

worklet, which contextually handles one specific task in a larger, composite process activity. Each worklet is a complete extended workflow net (EWF-net) compliant with Definition 1 of the YAWL semantics. Each worklet instance is dynamically selected and invoked at runtime and may be designed and provided to the Selection Service at any time, *even while a parent process instance is executing*, as opposed to a static subprocess that must be defined at the same time as, and remains a static part of, the main process model.

An extensible repertoire of worklets is maintained by the service for each task in a specification. Each time the service is invoked for a work item, a choice is made from the repertoire based on the contextual data values within the work item, using an extensible set of Ripple-Down Rules (cf. Sect. 11.10) to determine the most appropriate substitution.

The work item is checked out of the Engine, the corresponding data inputs of the original work item are mapped to the net-level inputs of the worklet, and the selected worklet is launched in the Engine as a separate case. When the worklet has completed, its output data is mapped back to the original work item, which is then checked back into the Engine, allowing the original process to continue.

The worklet executed for a task is run as a separate case in the Engine, so that, from an Engine perspective, the worklet and its parent are two distinct, unrelated cases. The Worklet Service tracks the relationships, data mappings, and synchronizations between cases, and creates a process log that may be combined with the Engine's process logs via case identifiers to provide a complete operational history of each process instance.

Worklets may be associated with either a single instance or a multiple instance atomic task. Any number of worklets can form the repertoire of an individual task, and any number of tasks in a particular specification can be associated with the Worklet Service. A worklet may be a member of one or more repertoires, that is, it may be reused for several distinct tasks within and across process specifications. In the case of multiple instance tasks, a separate worklet is launched for each child work item. Because each child work item may contain different data, the worklets that substitute for them are individually selected, and so may all be instances of different worklet specifications.

11.2.2 The Exception Service

Virtually every process instance (even if it follows a highly structured process definition) will experience some kind of exception (or deviation) during its execution. It may be that these events are known to occur in a small number of cases, but not often enough to warrant their inclusion in the process model; or they may be things that were never expected to occur (or perhaps never even imagined could occur). In any case, when they do happen, since they are not included in the process model, they must be handled "off-line" before processing can continue (and the way they are handled is rarely recorded). In some cases, the process model will be later modified to capture this unforeseen event, which involves an often large organizational cost (downtime, remodeling, testing, and so on); in certain circumstances the entire process must be aborted.

Alternately, an attempt might be made to include every possible twist and turn into the process model so that when such events occur, there is a branch in the process to take care of it. But this approach often leads to very complex models where much of the original business logic is obscured, and does not avoid the same problems arising when the next unexpected exception occurs.

The Exception Service addresses these problems by allowing designers to define exception handling processes (called *exlets*) for parent workflow instances, to be invoked when certain events occur and thereby allowing execution of the parent process to continue unhindered. It has been designed so that the enactment engine, besides providing notifications at certain points in the life-cycle of a process instance, needs no knowledge of an exception occurring or of any invocation of handling processes – all exception checking and handling is provided by the service. Additionally, exlets for unexpected exceptions may be added at runtime, and such handling methods automatically become an implicit part of the process specification for all current and future instances of the process, which provides for continuous evolution of the process while avoiding any need to modify the original process definition.

The Exception Service uses the same repertoire and dynamic rules approach as the Selection Service. An extensible repertoire of exlets is maintained by the service for each type of potential exception within each workflow specification. Each time the service is notified of an exception event, either actual or potential (i.e., a constraint check), the service first determines whether an exception has in fact occurred, and if so, where a rule tree for that exception type has been defined makes a choice from the repertoire based on the type of exception and the data attributes and values associated with the work item/case, using a set of rules to select the most appropriate exlet to execute.

Any number of exlets can form the repertoire of an individual task or case. An exlet may be a member of one or more repertoires, that is, it may be reused for several distinct tasks or cases within and across process specifications. Like the selection service, the exception handling repertoire for a task or case can be added to at any time, as can the rules base used, including while the parent process is executing.

The Selection and Exception subservices can be used in combination within particular case instances to achieve dynamic flexibility *and* exception handling simultaneously. The Worklet Service is extremely adaptable and multifaceted, and allows a designer to provide tailor-made solutions to runtime process exceptions and requirements for flexibility.

11.3 Service Oriented Architecture

The Worklet Service has been implemented as a YAWL Custom Service. However, the deployment of the Worklet Service is in no way limited to the YAWL environment, but may be ported to other environments (e.g., BPEL engines) by making the necessary associations in the service interface. As such, the implementation may also be seen as a case study in service-oriented computing whereby dynamic flexibility and exception handling in workflows, orthogonal to the underlying workflow language, is provided.

As detailed in Chap. 7, Custom YAWL Services interact with the Engine through XML/HTTP messages via certain interface endpoints, some located on the Engine side and others on the service side. Specifically, custom services may elect to be notified by the engine when certain events occur in the life-cycle of nominated process instantiations (i.e., when a work item becomes enabled, when a work item is canceled, when a case completes or is canceled), to signal the creation and completion of process instances and work items, or to notify certain events or changes in the status of existing work items and cases.

For example, on receiving notification from the engine of a work item-enabled event, a Custom Service may elect to "check out" the work item from the engine. On doing so, the engine marks the work item as *executing* and effectively passes operational control for the work item to the Custom Service. When the custom service has finished processing the work item, it will check it back in to the engine, at which point the engine will mark the work item as *completed*, and proceed with the process execution. It is this interaction that is the fundamental enabler of the Worklet Selection Service.

Three interfaces of the Engine are used by the Worklet Service (cf. Fig. 11.2):

- *Interface A* provides endpoints for process definition, administration, and monitoring – the Worklet Service uses Interface A to upload worklet specifications to the engine
- *Interface B* provides endpoints for client and invoked applications and workflow interoperability – used by the Worklet Service for connecting to the engine, to start and cancel case instances, and to check work items in and out of the engine after interrogating their associated data
- *Interface X* ("X" for "e**X**ception"), which has been designed to allow the engine to notify custom services of certain events and checkpoints during the execution of each process instance where process exceptions either may have occurred or should be tested for. Thus, Interface X provides the Worklet Service with the necessary triggers to dynamically capture and handle process exceptions.

The logical layout of Interface X can be seen schematically in Fig. 11.1, which shows that a Custom Service (in this case, the Exception Service) implements the *InterfaceXService* (Java) interface, which defines seven methods to be instantiated that enable the handling of notifications that are sent from the *EngineSideClient* object to the *ServiceSideServer* object. The Custom Service may use the methods of the *ServiceSideClient* object to call complementary methods in the *EngineSideServer* object, thus enabling inter-service communication, using XML over HTTP. The *ExceptionGateway* (Java) interface describes the methods that the engine side client must implement to provide the necessary notifications to the Custom Service.

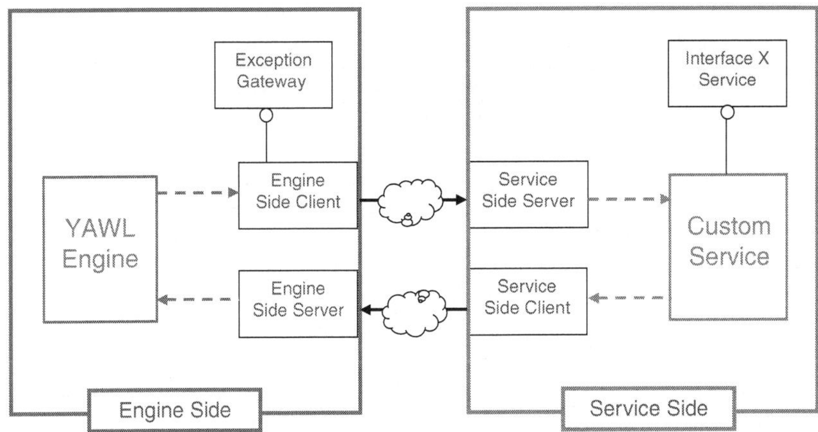

Fig. 11.1 Interface X architecture

While the implementation of the Exception Service was the catalyst for the creation of Interface X, one of the overriding design objectives was that the interface should be structured for generic application, that is, it can be applied by a variety of services that wish to make use of checkpoint and/or event notifications during process executions. For example, in addition to exception handling, the interface's methods provide the tools to custom services to enable ad-hoc or permanent adaptations to process schemas, such as redoing, skipping, replacing, and looping of tasks.

As it only makes sense to have one Custom Service acting as an exception handling service at any one time, services that implement Interface X have two distinct states: *enabled* and *disabled*. When enabled, the engine generates notifications for *every* process instance it executes, that is, the engine makes no decisions about whether a particular process should generate the notifications or not. Thus it is the responsibility of the designer of the Custom Service to determine how best to deal with (or indeed ignore) the notifications. When the service is disabled, the engine generates no notifications across the interface. Enabling and disabling an Interface X Custom Service is achieved via parameter setting in a configuration file (cf. Sect. 11.5).

11.4 Worklet Service Architecture

This section describes the technical attributes and structure of the Worklet Service. The service is constructed as a web service and so consists of a number of J2EE classes and servlet pages, organized in a series of Java packages.

A schematic of the external architecture of YAWL, showing the relation of the Worklet Service within it, is shown in Fig. 11.2. The entities "Worklet Specs,"

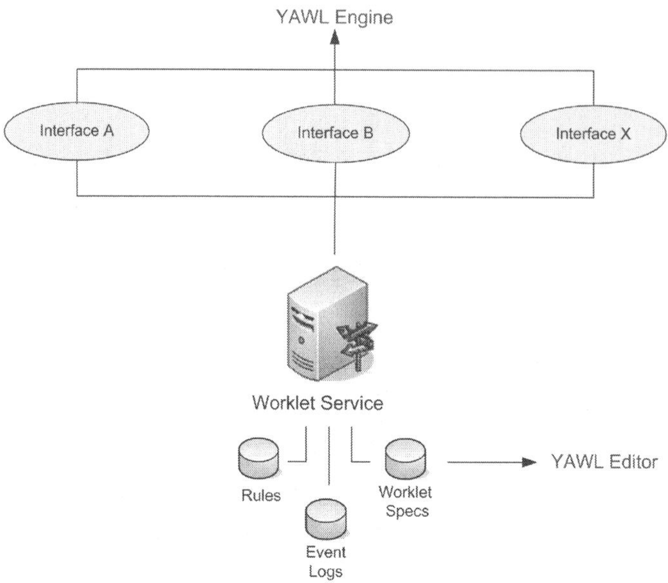

Fig. 11.2 Worklet Service external architecture

"Rules," and "Event Logs" comprise the *Worklet Repository*. The service uses the repository to store rule sets, worklet specifications for uploading to the engine, and generated process and audit logs. The YAWL Editor is used to create new worklet specifications, and may be invoked from the Rules Editor (cf. Sect. 11.10). The Rules Editor is used to create new or augment existing rule sets, making use of certain selection logs to do so, and may communicate with the Worklet Service via a JSP/Servlet interface to override worklet selections following rule set additions (see Sect. 11.10.2). The service also provides servlet pages that allow users to directly communicate with the service to raise external exceptions and to create and carry out administration tasks.

Any YAWL specification may have an associated rule set. The rule set for each specification is stored as XML data in a disk file that has the same name as the specification, but with an ".xrs" extension (signifying an "XML Rule Set"). All rule set files are stored in the *rules* folder of the Worklet Repository. Listing 11.1 shows an excerpt from a rules file for a *Casualty Treatment* process.

Figure 11.3 shows a representation of the internal architecture of the Worklet Service. The obvious hub of the service is the *WorkletService* class, which administrates the execution of the service and handles interactions with the Engine via Interfaces A and B.

For each work item that is checked out of the engine, the service creates a *CheckedOutItem* object. In the Engine, each work item is a "parent" of one or more child items – one if it is an atomic task, or a number of child items in the case of a multiple instance atomic task. Thus, the role of each *CheckedOutItem* object is

```
        ...
      </ruleNode>
    </task>
  </selection>
  <constraints>
    <case>
      <pre>
         <ruleNode>
            <id>0</id>
            <parent>-1</parent>
            <trueChild>1</trueChild>
            <falseChild>-1</falseChild>
            <condition>True</condition>
            <conclusion>null</conclusion>
            <cornerstone></cornerstone>
            <description>root level default node</description>
         </ruleNode>
         <ruleNode>
            <id>1</id>
            <parent>0</parent>
            <trueChild>-1</trueChild>
            <falseChild>-1</falseChild>
            <condition>paid = true</condition>
            <conclusion>
               <_1>
                  <action>remove</action>
                  <target>case</target>
               </_1>
            </conclusion>
        ...
    </case>
  </constraints>
</spec>
```

Listing 11.1 Excerpt of a rules file

to create and manage one or more *CheckedOutChildItems*, which hold information about worklet selection, data associated with the work item, and the results of rules searches.

The *WorkletService* object, for each work item that is checked out from the engine, also loads from file the set of rules pertaining to the specification of which the work item is a member into an *RdrSet* object. At any time, there may be a number of RdrSets loaded into the service, one for each specification for which a work item has been checked out. Each RdrSet manages one or more *RdrTree* objects, each tree representing the rule tree for a particular task within the specification, of which this work item is an instance. In turn, each RdrTree owns a number of *RdrNode* objects, which contain the actual rules, conclusions, and other data for each node of the rule tree.

When a rule tree is evaluated against the data set of a work item, each of the associated nodes of that tree has its condition evaluated by the *ConditionEvaluator* object, which returns the boolean result to the node, allowing it to traverse to its true or false branch as necessary. Finally, the *WorkletGateway* object provides communications via a JSP/Servlet interface between the service, the various servlet pages, and the Rules Editor (see Sect. 11.10.2 for more details).

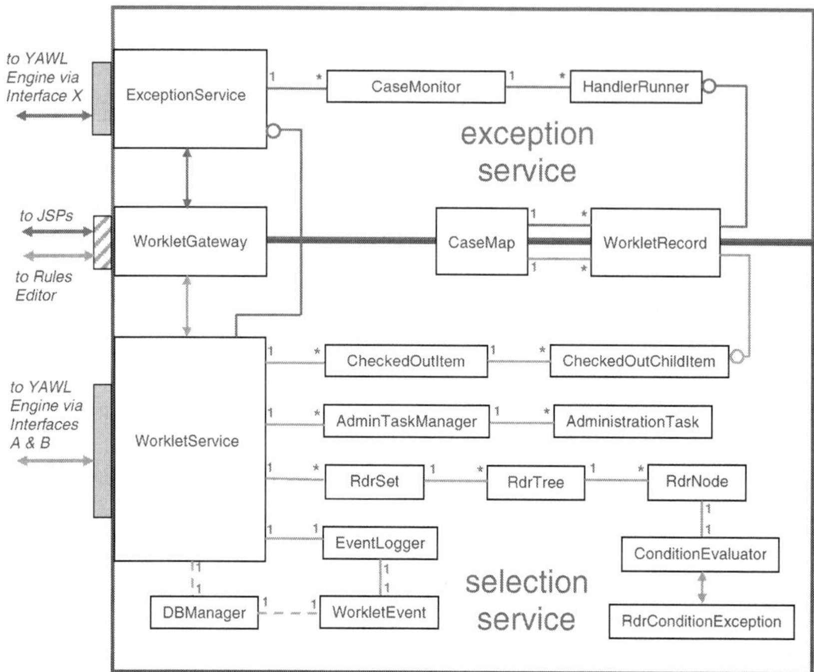

Fig. 11.3 Internal architecture of the Worklet Service

11.5 Service Installation and Configuration

Like other YAWL Custom Services, the Worklet Service is a Java-based web service and so must be hosted within a *servlet container*. By default, an installation of the open-source servlet container *Apache Tomcat* is used to host the YAWL environment, and as a consequence the Worklet Service is also hosted on Tomcat. Further, as the Worklet Service is distributed as part of the YAWL installation, it resides within the same host (again by default).

However, as it is a totally discrete service, the Worklet Service may be installed on and hosted by any local or remote server running a servlet container. In fact, it is possible for one instance of the Worklet Service to manage the selection and exception handling requirements of a number of YAWL engines across domains simultaneously (or potentially other types of enactment engines that conform to the interface).

The Worklet Service is distributed as a Java web archive (or *.war* file), which includes compiled versions of all of the classes of the service, ancillary classes, the Worklet Repository (with samples), and the Rules Editor. The service is installed by placing the war file in the appropriate applications folder of the servlet container.

Each web service, including the Engine, makes use of a configuration file (in Tomcat terminology, a "deployment descriptor"), which sets the parameters for the instantiation and operation of the service instance. Individual YAWL services use these parameters to specify settings such as the location of services, enabling of the Exception Service, enabling of persistence, location of the Worklet Repository, and so on. Setting values for parameters can be done using any text editor.

This section discusses the various parameters that can or may be set to successfully deploy the Worklet Service.

11.5.1 Configuration: Service Side

The configuration file for the Worklet Service has parameters for specifying the URL of the Engine's Interface B, the location of the WorkletService and ExceptionService classes (i.e., the relative location of those classes within the web archive), the locations of the interface classes that send notifications from the engine to the service, and mappings for resolving the URLs of the service's jsps to the internal location of the page being requested. All these parameters are set to their defaults in the deployed web archive, which assumes that the Worklet Service is installed in the same servlet container as the Engine, but of course these can be changed to their appropriate values where the service is installed remotely to the engine.

The configuration file also has two parameters that must be set for each individual installation (see Listing 11.2):

- The *Repository* parameter, which maps the disk location of the Worklet Repository for that installation (the default location is within the servlet container's deployed applications folder, but it can be any valid location on the local server), allowing the repository to be installed in an appropriate place of the administrator's choosing depending on local needs
- The *EnablePersistence* parameter, which when set to *true* enables persistence of instantiated service objects to a database so that items and cases currently being handled by the service may be persisted across sessions, and to allow log records to be written to a database table. If the parameter is set to false, service states are not persisted, and log entries are instead written to a comma-delimited file in the Worklet Repository.

11.5.2 Configuration: Engine Side

When the Worklet Service has been installed, it is automatically configured to receive notifications from the engine for worklet-enabled tasks as part of the selection process, that is, there are no explicit configuration tasks necessary to enable the Selection Service. For the Exception Service, there are two relevant parameters in the engine's configuration file; the first sets the URL of the Exception Service (set by default to a location relative to the same server as the engine is installed on).

```
<context-param>
   <param-name>Repository</param-name>
   <param-value>
     ${catalina.base}\webapps\workletService\repository\
   </param-value>
   <description>
      The path where the worklet repository is installed.
   </description>
</context-param>

<context-param>
   <param-name>EnablePersistence</param-name>
   <param-value>true</param-value>
   <description>
     'true' to enable persistence and logging
     'false' to disable
   </description>
</context-param>
```

Listing 11.2 The Worklet Service configuration file (detail)

```
<!-- PARAMS FOR EXCEPTION SERVICE -->

<context-param>
   <param-name>EnableExceptionService</param-name>
   <param-value>false</param-value>
   <description>
      Set this value to 'true' to enable monitoring by an Exception Service
      (specified by the URI param below).
      Set it to 'false' to disable the Exception Service.
   </description>
</context-param>

<context-param>
   <param-name>ExceptionObserverURI</param-name>
   <param-value>http://localhost:8080/workletService/ix</param-value>
   <description>
      This value provides the URI of an Exception Service that monitors
      for process exceptions through Interface X.
   </description>
</context-param>
```

Listing 11.3 The Engine configuration file (detail)

The second parameter enables or disables the service (Listing 11.3); when the parameter is set to *true*, the engine notifies the service at various points when exceptions have (or may have) occurred throughout the life-cycle of every case launched in the engine.

In addition to the parameter settings in the engine's configuration file, the Exception Service makes use of extensions (or "hooks") built into the Resource Service's worklist handler (cf. Chap. 10) to provide methods for users to interact with the service (e.g., allowing the raising of an external exception); so, if the Exception Service is enabled via the engine's configuration file, then the extensions to the worklist must also be enabled via its configuration file. The Resource Service has been deployed as a discrete service also, and so has its own configuration file.

Within that file, a parameter is supplied to specify the URL of the Exception Service; by default it is set to a URL of the service relative to the same servlet container

```
<!-- This param, when available, enables the worklet exception
     service add-ins to the worklist. If the exception service
     is enabled in the engine, then this param should also be
     made available. If it is disabled in the engine, the
     entire param should be commented out. -->
<!--
<context-param>
   <param-name>InterfaceX_BackEnd</param-name>
   <param-value>http://localhost:8080/workletService</param-value>
   <description>
      The URL location of the worklet exception service.
   </description>
</context-param>
-->\vspace*{-6pt}
```

Listing 11.4 The Resource Service configuration file (detail)

as the engine, but again it can be modified if the Exception Service is installed remotely. Also by default, the parameter is commented out, as the Exception Service is disabled by default in the engine's configuration when first deployed. When the comment tags are removed, the worklist becomes aware that the Exception Service is enabled and so enables the appropriate methods and items available via its user interface. See Listing 11.4 for the relevant part of the worklist's configuration file.

11.6 Worklet Process Definition

Fundamentally, a worklet is nothing more than a YAWL process specification that has been designed to perform one part of a larger, parent specification. However, it differs from a decomposition or subnet, in that it is dynamically assigned to perform a particular task at runtime, while subnets are statically assigned at design time. So, rather than being forced to define all possible branches in a specification, the Worklet Service allows the definition of a much simpler specification that will evolve dynamically as more worklets are added to the repertoire for particular tasks within it.

To illustrate the worklet approach by means of a simple example, the *Payment* subnet of the Order Fulfillment process is used in the following discussion. Figure 11.4 shows the original *Payment* subnet, Fig. 11.5 shows the simplified, worklet enhanced version, where the entire left-hand side of the process has been replaced with a single, worklet-enabled task. Together, the figures show a sort-of "before" and "after" view of how *worklets can be used to simplify a parent process model*, so that the primary business logic is clearly revealed, while the detail is handled at a lower level of granularity.

Even though the process model in Fig. 11.5 has a task that is worklet-enabled, it remains an ordinary YAWL process, which, without interaction from the Worklet Service, would execute as a standard, static process. However, because the *Issue*

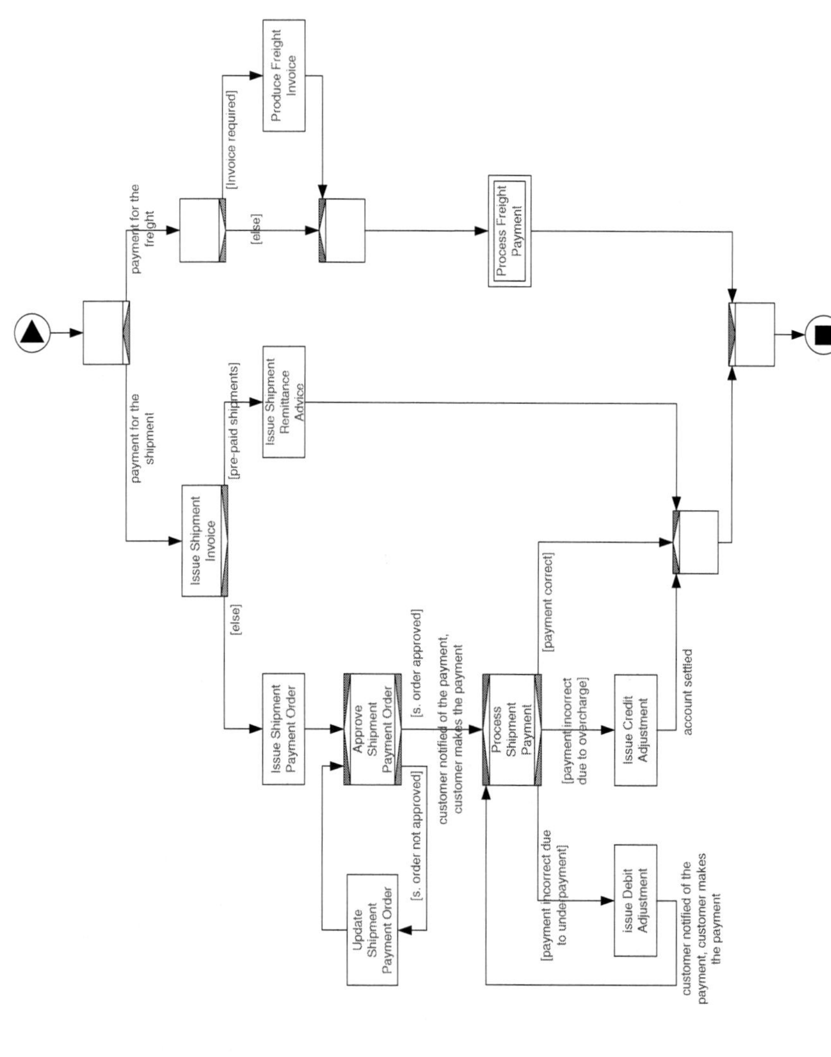

Fig. 11.4 The Order Fulfillment *Payment* subnet

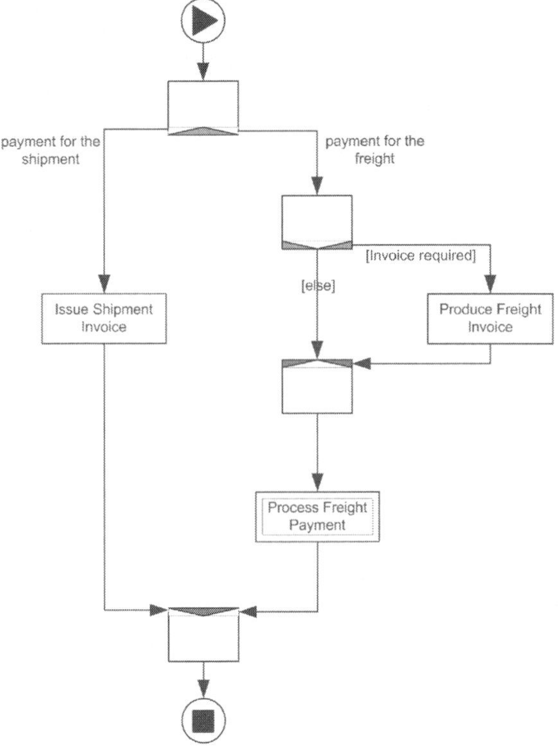

Fig. 11.5 The worklet-enabled Order Fulfillment *Payment* subnet

Shipment Invoice task is worklet-enabled, it will be substituted at runtime with the appropriate worklet based on the order data collected earlier in the process.

In this example, only the *Issue Shipment Invoice* task is worklet-enabled; the other tasks will be handled directly by the YAWL environment. So, when an instance of the Order Fulfillment process is executed, the Engine will notify the Worklet Service when the *Issue Shipment Invoice* task becomes enabled. The Worklet Service will then examine the data of the task and use it to determine which worklet to execute as a substitute for the task.

As a worklet specification is a standard YAWL process specification, it is created in the Editor in the usual manner. Figure 11.6 shows the trivial worklet to be substituted for the *Issue Shipment Invoice* top-level task when an order has been prepaid, while Fig. 11.7 shows the only slightly more elaborate worklet that will be chosen when the order has payment outstanding.

In themselves, there is nothing special about these worklets. Even though each is considered by the Worklet Service as a member of the worklet repertoire for the *Issue Shipment Invoice* task and may thus be considered a "worklet," each also remains a standard YAWL specification and as such may be executed directly by the Engine without any reference to the Worklet Service, if desired.

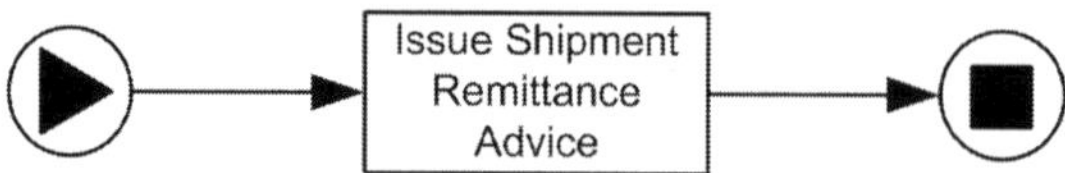

Fig. 11.6 The Issue Shipment Remittance worklet

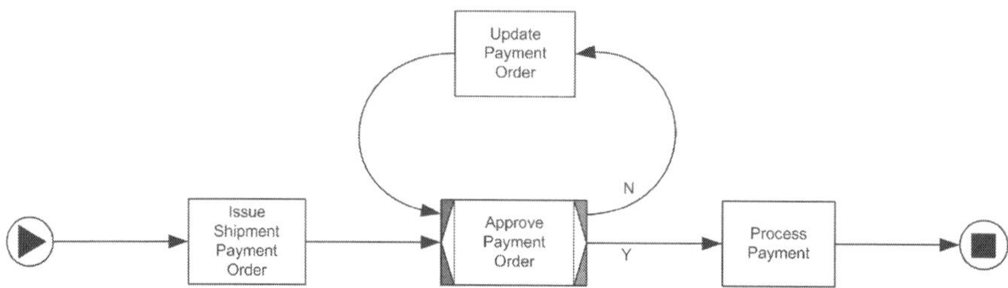

Fig. 11.7 The Prepare Payment Order worklet

At this stage, two worklets have been defined as possible substitutes for a task (either payment has already been received, or it hasn't). But now consider the situation where a *partial* payment has already been received. In the non-worklet version of the *Payment* subnet, this new scenario would require a reworking of the entire net, with a new branch introduced to accommodate it. In the worklet-enabled version, all that is required is for a new worklet to be defined and added to the repertoire for the task, which demonstrates the advantages the worklet approach offers over monolithically designed processes, or statically defined subnets. Any number of worklets can be created for a particular task.

The *Issue Shipment Invoice* task is associated with the Worklet Service (i.e., the task is "worklet-enabled") via the Editor's *Update Task Decomposition* dialog (Fig. 11.8). This dialog shows the variables defined for the task – each one of these maps to a net-level variable, so that the data collected for an order earlier in the process may be made available to this task. The result is that all of the *relevant* current case data for this process instance can be used by the Worklet Service to enable a contextual decision to be made. Note that it is not necessary to map all available case data to a worklet-enabled task, only those data required by the service to make an appropriate decision via the conditional expressions in the rule nodes defined for the task. The list of task variables in Fig. 11.8 shows that the *POrder* variable is defined as "Input Only," indicating that the value for that variable will not be modified by any of the worklets that may be executed for this task, but it may be used for the selection process. The other variables are defined as "Input & Output" or "Output Only," so that the worklet can modify and return (i.e., map back) to those variable data values that are captured during the worklet's execution. The dialog includes a section at the bottom called *YAWL Registered Service Detail*. It is here that the task is associated with the Worklet Service (i.e., made worklet-enabled) by choosing the Worklet Service from the list of available services.

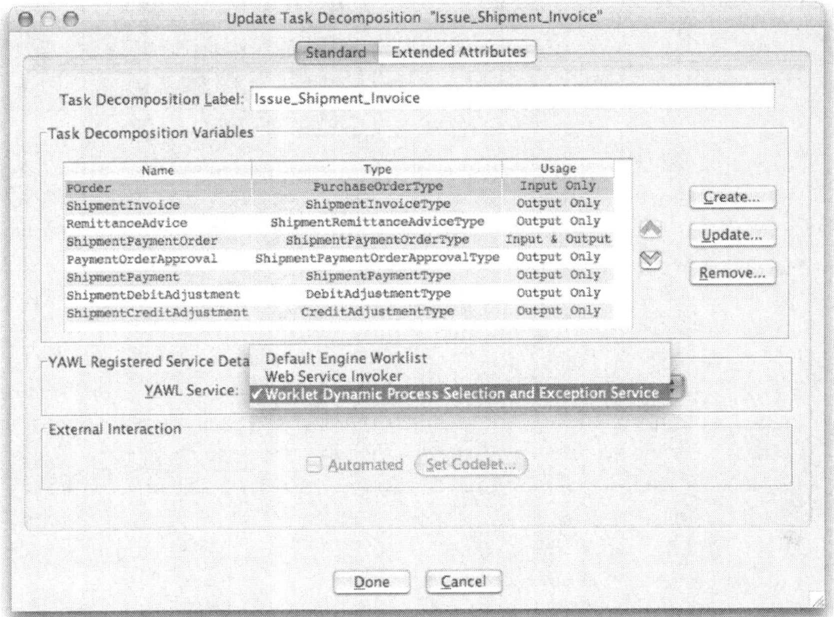

Fig. 11.8 The Update Task Decomposition dialog for the *Issue Shipment Invoice* task

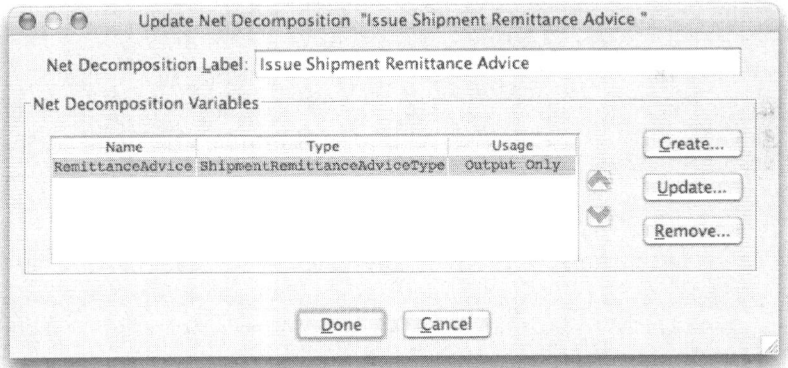

Fig. 11.9 The net-level variable for the Issue Shipment Remittance Advice worklet

Those data values that are required to be mapped from the parent task to the worklet need to be defined as net-level variables in each of the worklet specifications made available to this task. Figure 11.9 shows the net-level variables for the *Issue Shipment Remittance Advice* worklet. Note the following:

- Only a subset of the variables defined in the parent *Issue Shipment Invoice* task (see Fig. 11.8) are defined here. It is only necessary to map from the parent task to the worklet those variables that contain values to be displayed to the user, and/or

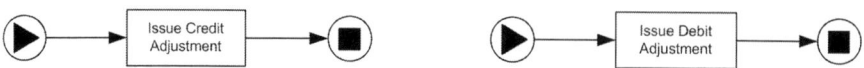

Fig. 11.10 The worklet repertoire members for the *Process Payment* task

those variables that the user will supply values for to be passed back to the parent task when the worklet completes.

- The definition of variables is not restricted to those defined in the parent task. Any additional variables required for the operation of the worklet may also be defined here; their values will not be passed back to the parent task when the worklet completes.
- Only those variables that have been defined with an identical name and data type to variables in the parent task and with a *Usage* of "Input Only" or "Input & Output" will have data passed into them from the corresponding variables of the parent task by the Worklet Service when the worklet is launched.
- Only those variables that have been defined with an identical name and data type to variables in the parent task and with a *Usage* of "Output Only" or "Input & Output" will have their data values mapped back to the corresponding variables of the parent task by the Worklet Service when the worklet completes.

For this example (cf. Fig. 11.9), only the "RemittanceAdvice" variable is required to be passed from the worklet back to the parent net.

The association of tasks with the Worklet Service is not restricted to top-level specifications. Worklet specifications also may contain tasks that are associated with the Worklet Service and so may have worklets substituted for them, so that a hierarchy of executing worklets may sometimes exist. It is also possible to recursively define worklet substitutions, that is, a worklet may contain a task that, while certain conditions hold true, is substituted by another instance of the same worklet specification that contains the task.

Referring back to the worklet definition shown in Fig. 11.7, the *Process Payment* task is also worklet-enabled, that is, it has its own worklet repertoire, which effectively are "grand-children" or the original, parent *Payment* process. At runtime, the appropriate worklet will be chosen for execution depending on the current value held in the *ShipmentPayment* variable. Figure 11.10 shows the two trivial worklets defined for the *Process Payment* task.

At runtime, if the order has been underpaid, the *Issue Debit Adjustment* worklet is executed; if there has been an overpayment, the *Issue Credit Adjustment* worklet is executed. Importantly, if the payment is correct (i.e., neither underpaid or overpaid), an "empty" worklet will be returned by the rule set, which means no further action is required in that instance.

11.7 Exlet Process Definition

This section discusses the definition of exlets, and describes each of the handling primitives that may be used to form an exlet.

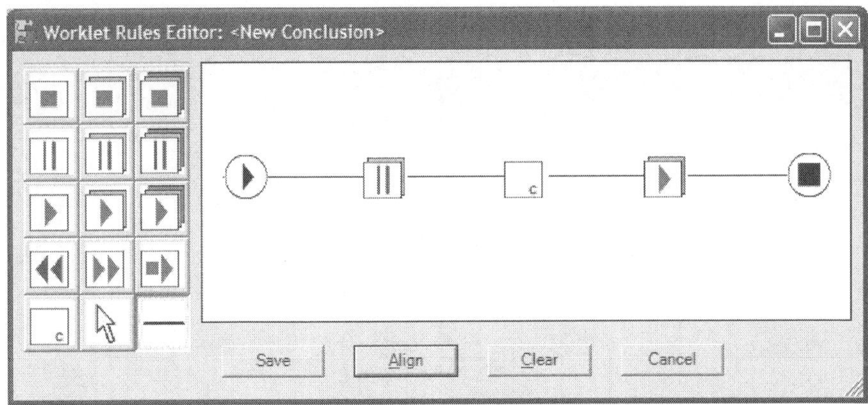

Fig. 11.11 Example handler process in the Rules Editor

The Exception Service maintains a set of Ripple-Down Rules that is used to determine which exlet, if any, to invoke. If there are no rules defined for a certain exception type in the rule set for a specification, the exception event is simply ignored by the service. Thus rules are needed only for those exception events that are desired to be handled for a particular task and/or specification.

An example of a definition of an exlet in the Rules Editor can be seen in Fig. 11.11 (see Sect. 11.10 for more details regarding the Rules Editor and its use). On the left of the graphical editor is the set of primitives that may be used to construct an exlet (see Chap. 5 for a detailed description of each primitive and its effects). The available primitives (reading left-to-right, top-to-bottom) are:

- *Remove* Work Item, Case, All Cases
- *Suspend* Work Item, Case, All Cases
- *Continue* Work Item, Case, All Cases
- *Restart Work Item*
- *Force Complete Work Item*
- *Force Fail Work Item*
- *Compensate*

To demonstrate a few of the major features of the Exception Service, this section will use a small portion of the *Carrier Appointment* subnet of the Order Fulfillment process, specifically the branch that deals with the delivery of a single package (Fig. 11.12).

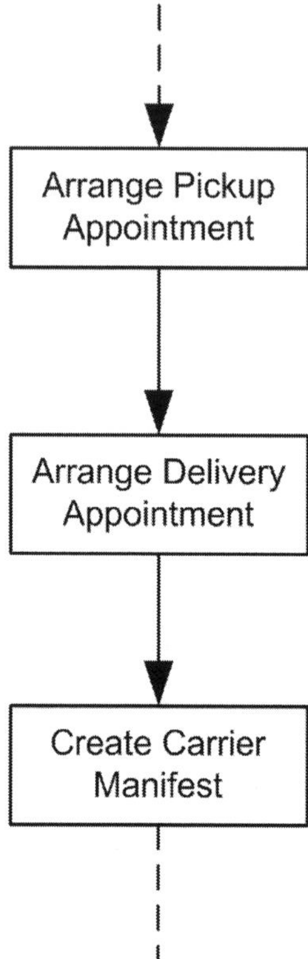

Fig. 11.12 The Carrier Appointment subnet, Order Fulfillment process (portion)

11.7.1 Constraints Example

As soon as the Engine launches a case, it notifies the Exception Service of a new case instance via a PreCaseConstraint event. If the rule set for Order Fulfillment process contained a rule tree for pre-case constraints, that tree will be queried using the initial case data to determine whether there are any pre-constraints not met by the case. In this example, there are no precase constraint rules defined, and so the notification is simply ignored.

Pre-case constraints can be used, amongst other things, to ensure that case data is valid or is within certain ranges before the case proceeds; can be used to run compensatory worklets to correct any invalid data; or may even be used to cancel the case as soon as it starts (in certain circumstances, if required).

As each work item becomes enabled during the execution of a process, the engine notifies the service of a PreItemConstraint for it. Like pre-case constraint rules, preitem rules can be used to ensure that work items have valid data before they are executed. With each constraint notification, the entire set of case data is made available to the Exception Service – thus the values of any case variables may be queried in the Ripple-Down Rule tree sets for any constraint type rule. When the work item is completed, a PostItemConstraint event is generated by the engine.

In this example, the *Arrange Pickup Appointment* work item does have a pre-item constraint rule tree, and so when it becomes enabled, the rule tree is queried. One particular rule in the pre-item constraint rule tree for the task checks that valid values have been supplied for the *Pickup Instructions* and *Pickup Spot* fields of the *PickupInstruction* variable. If the values are empty or invalid, the *Get Pickup Instructions* exlet is invoked, which will first suspend the process instance, then run a compensatory worklet to get the pickup instructions from a user, then continue the case. Note that since exlet compensations are always worklet executions, the pickup instruction data gathered by the worklet is automatically mapped back to the corresponding variables of the task on which the exception was raised, so that the case can continue with the required data inserted in the appropriate place.

The actions taken by the exlet are also reflected in the worklist. The *Arrange Pickup Appointment* work item is marked with a status of "Suspended" and thus is unable to be selected for starting, while the compensatory worklet has been launched and its first work item would be found in the *Offered* queue of the appropriate participants. The compensatory worklet is treated by the exception service as just another case, and so the service receives notifications from the engine for pre-case and pre-item constraint events of the worklet also.

When the last work item of the worklet has completed, the engine completes the worklet case and notifies the Exception Service of the completion, at which time the service completes the third and final part of the exlet, that is, continuing (unsuspending) the *Arrange Pickup Appointment* work item so that the parent case can continue. On the worklist, that work item will now be able to be checked out, with the relevant data values entered in the worklet's work items mapped back to its variables.

11.7.2 External Trigger Example

It has been stated previously that almost every case instance involves some deviation from the standard process model. Sometimes, events occur completely removed from the actual process model itself, but affect the way the process instance proceeds. Typically, these kinds of events are handled "off-system," and so there is no record of them, or the way they were handled, kept for future executions of the process specification.

The Exception Service allows for such events to be handled on-system by providing a means for exceptions to be raised by users externally to the process itself. The Order Fulfillment specification will again be used to illustrate how an external exception can be triggered by a user or an administrator.

A case-level external exception can be raised via extensions added to YAWL's Resource Service worklist handler. Figure 11.13 shows detail of the worklist's *Case Management* screen, and indicates that an instance of the Order Fulfillment spec-

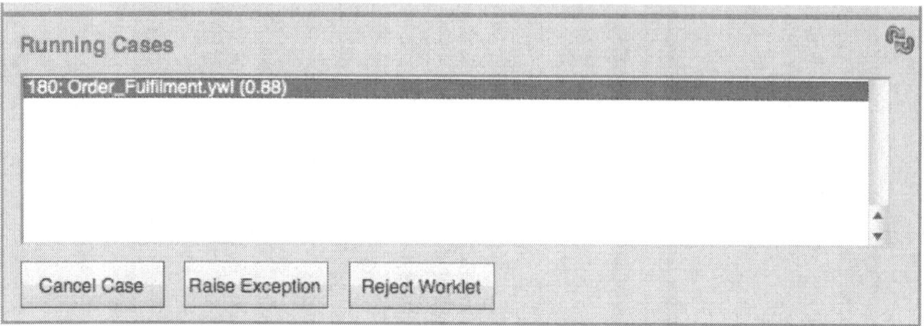

Fig. 11.13 Workflow Specifications screen, Order Fulfillment instance running (detail)

ification is currently executing. The two buttons "Raise Exception" and "Reject Worklet" are worklist extensions; the first appears only when the Exception Service is configured as "enabled" and the second when the Worklet Service is loaded.

When the "Raise Exception" button is clicked, a *Raise Case Level Exception* screen is displayed. This screen is a member of the Worklet Service's Java Servlet Pages (jsp). Before this screen is displayed, it directly calls a method of the Exception Service over HTTP and retrieves from the rule set for the selected case the list of existing external exception triggers (if any) for the case's specification. This list contains all of the external exception "triggers" that were either conceived when the specification was first designed or added later as new kinds of exceptional events occurred and were added to the rule set. Notice that at the bottom of the list, the option to add a New External Exception is provided – that option is explained in detail in Sect. 11.10.2. For this example, we will assume that a client has requested a change in the delivery date and time. Selecting that exception trigger passes it as the trigger value to the specification's CaseExternalException rule tree and the conclusion for that trigger's rule (i.e., the exlet) is invoked by the service as an exception handling process for the current case.

After the selection is made, the user's worklist will show that the parent case's active work items have been suspended and the first work item of the compensatory worklet, *Update Delivery Appointment*, has been enabled. Once the worklet is completed, the parent case is continued (via the final primitive of the exlet).

Item-level external exceptions work in much the same way, and can be raised from the worklist by selecting the relevant work item, and then clicking the *Raise Exception* button (another extension to the worklist provided by the Exception Service when "enabled"), which invokes the *Raise Item Level Exception* screen. From there, the procedure is identical to that described for case-level exceptions, except that the item-level external exception triggers, if any, will be displayed.

External exceptions can be raised at any time during the execution of a case – the way they are handled may depend on how far the process has progressed via the defining of the conditional expressions of the nodes in the appropriate rule tree or trees that query changes in data values and/or state changes of work items within the case instance.

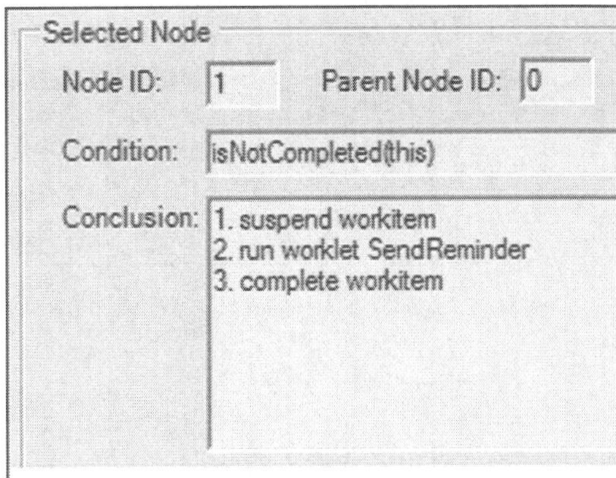

Fig. 11.14 Rule detail for Arrange Delivery Appointment timeout

11.7.3 Timeout Example

YAWL allows for each work item to be associated with a timer. When a work item timer expires, the engine notifies the Exception Service. Then, the Timeout rule tree (if any) for the task is parsed for the appropriate response to the timer expiry event. For example, if the *Arrange Delivery Appointment* task were to expire before a suitable delivery time could be agreed upon, the timeout tree set is queried for the task, and the appropriate exlet is invoked. There may be a number of exlets defined for a timeout on this task, which may take various actions such as extending the timer, setting a default delivery time, make other arrangements with the client, or even cancel the order, depending on the context of the case.

Figure 11.14 shows the rule tree defined for the TimeOut exception type for the *Arrange Delivery Appointment* task. Notice the rule's condition *isNotCompleted(this)*:

- **isNotCompleted** is an example of a defined function that may be used as (or as part of) a conditional expression in a rule node.
- **this** is a special variable created by the Worklet Service that refers to the work item that the rule is defined for and contains, amongst other things, all of the work-item's data attributes and values.

In this example, the condition tests if the work item has not yet been completed (i.e., if it has a status of *Fired*, *Enabled*, *Executing*, or *Suspended*). If it has not completed (thus arrangements for delivery have not yet been completed), then the conclusion (exlet) will be executed, which includes the launching of the compensatory worklet *SendReminder*. The *SendReminder* worklet consists of three tasks: *Send Request*, and the parallel tasks *wait* and *Receive Reply* – the *wait* task in this specification is again associated with the timer. When its *wait* task times out, the Exception Service is again notified. The rule set for the SendReminder specification also contains a single timeout rule for the *Receive Reply* task – its condition is again *isNotCompleted(this)*, but this time the rule's conclusion is that shown in Fig. 11.15.

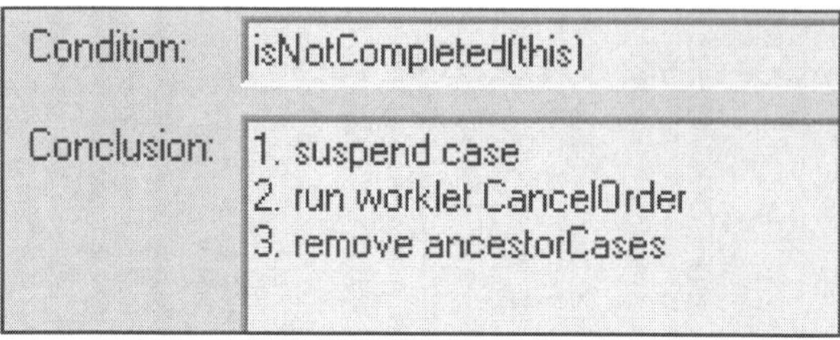

Fig. 11.15 Rule detail for Receive Reply

When this conclusion's exlet is executed, the worklist now lists work items from all three cases, reflecting a hierarchy of worklets: an Order Fulfillment instance is suspended, pending completion of worklet instance SendReminder, which itself is suspended, pending completion of worklet instance CancelOrder. Thus, this example shows that worklets can invoke child worklets to any depth. Notice the third part of the handling process: "remove ancestorCases." Ancestor Cases are all cases from the current worklet case back up the hierarchy to the original parent case that began the exception chain (as opposed to "allCases," which refers to all currently executing cases of the same specification as the case that generated the exception). So, when the CancelOrder worklet completes, the SendReminder instance and the original parent Order Fulfillment instance are both canceled by the Exception Service as a result of the "remove ancestorCases" primitive.

11.8 Ripple-Down Rule Sets

A process specification may contain a number of tasks, one or more of which may be associated with the Worklet Selection Service. For each specification that contains a worklet-enabled task, the Worklet Service maintains a corresponding set of Ripple-Down Rules that determine which worklet will be selected as a substitute for the task at runtime, based on the current case data of that particular instance. Each worklet-enabled task in a specification has its own discrete rule tree.

Further, one or more exlets may be defined for a specification and associated with the Worklet Exception Service. A repertoire of exlets may be formed for each exception type. Each specification has a unique rule set (if any), which may contain between one and eight tree sets (or sets of rule trees), one for selection rules (used by the Selection subservice) and one for each of the seven implemented exception types (see Chap. 5, Sect. 5.4.3.1 for details of the different exception types). Three of those eight relate to case-level exceptions (i.e., CasePreConstraint, CasePostConstraint and CaseExternalTrigger) and so each of these will have at most one rule tree in the tree set. The other five tree sets relate to work item-level events (four exception types plus selection), and so may have one rule tree for each task in the specification, that is, the tree sets for these eight rule types may consist of a number of rule trees.

The rule set for each specification is stored as XML data in a discrete disk file. All rule set files are stored in the Worklet Repository.

Thus, the hierarchy of a Worklet Rule Set is (from the bottom up) are:

- *Rule Node*: contains the details (condition, conclusion, id, parent, and so on) of one discrete rule
- *Rule Tree*: consists of a number of rule nodes conceptually linked in a binary tree structure
- *Tree Set*: a set of one or more rule trees. Each tree set is specific to a particular rule type (Timeout, ExternalTrigger, etc.). The tree set of a case-level exception rule type will contain exactly one tree. The tree set of an item-level rule type will contain one rule tree for each task of the specification that has rules defined for it (not all tasks in the specification need to have a rule tree defined)
- *Rule Set*: a set of one or more tree sets representing the entire set of rules defined for one specification. Each rule set is specific to a particular specification. A rule set will contain one tree set for each rule type for which rules have been defined.

It is not necessary to define rules for all eight exception/selection types for each specification, only for those types that are required to be handled; the occurrence of any exception types that are not defined in the rule set file are simply ignored. So, for example, if an analyst is interested only in capturing pre- and post-constraints at the work item level, then only the ItemPreConstraint and ItemPostConstraint tree

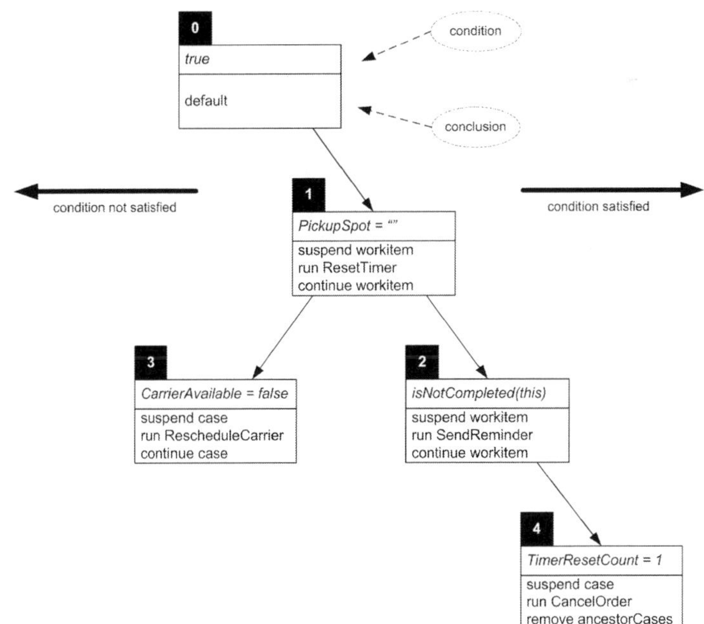

Fig. 11.16 Example rule tree (TimeOut tree for Arrange Delivery Appointment task of the *Payment subnet*)

sets need to be defined (i.e., rules defined within those tree sets). Of course, rules for other event types can be added later if required. Figure 11.16 shows the Timeout rule tree for the or *Arrange Delivery Appointment* task of the *Payment* subnet (cf. Sect. 11.7.1).

11.9 Extending the Available Conditionals

The Worklet Service provides a discrete class that enables developers to extend the set of functions available to be used as part of the conditional expressions of rule sets. The class is called *RdrConditionFunction*, the source code for which currently contains a small number of examples to give developers an indication of the kinds of things that can be achieved with the class and how to create additional functions. The class is freely extensible into any domain space, and thus data can be drawn from diverse sources, such as process mining, resource logs, and external databases, and queried through user-defined functions as part of the evaluation of the conditional expressions of rule nodes. The ConditionEvaluator class has been designed to call the RdrConditionFunction object to resolve function references in conditional expressions to their evaluated values. As the interface between this class and the rest of the service is very small, it can easily be extended and separately recompiled, avoiding the overhead of rebuilding the entire service.

The class code is split into four sections:

- *Header*: containing a listing of available functions and a method to test for the existence of a function
- *Execute Method*: a method called by the ConditionEvaluator object when evaluating a condition containing a user-defined function
- *Function Definitions*: the functions themselves
- *Implementation*: where additional support methods may be created.

Once a function is added, it can be used in any rule's conditional expression. One example method of the class is *isNotCompleted*, which is designed to return a boolean true value if a work item does not have a completed status.

The objective of the *RdrConditionFunction* class is to allow developers to easily extend the capabilities of the Worklet Service by providing the means to test for things in the conditional expressions of rule nodes other than the process instance's data attributes and values. This allows data from any source to be tested during rule evaluations and also provides the service with ease-of-use capabilities when formulating new conditions.

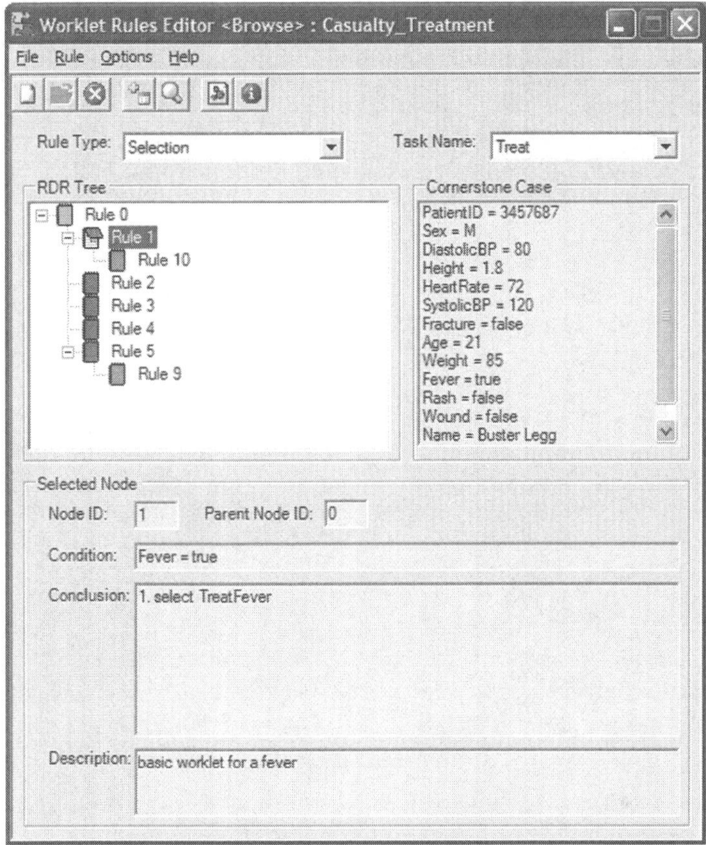

Fig. 11.17 Rules Editor main screen

11.10 The Rules Editor

The *Worklet Rules Editor* is a purpose-built tool, deployed as part of the Worklet Repository, that enables the addition of new rules to existing rule sets of specifications, and the creation of new rule sets. It has been developed as a Microsoft .NET 2005 based application; the primary reason for choosing a .NET platform was to provide a demonstration of the interoperability of the web- and Java-based Worklet Service with a Windows-based administration tool using HTTP messaging.

The main screen of the editor allows users to graphically view the various parts of each rule node of each rule tree defined for a particular specification. From this screen users can also add new rules to the current rule set, create new tree sets for an existing rule set, and create entirely new rule sets for new specifications. Figure 11.17 shows the main screen with a rule set loaded.

The main features of the screen are explained below.

- *Current Rule Type*: This drop-down list displays each rule type that has a tree set defined for it in the opened rules file. Selecting a rule type from the list displays in the Tree View panel the associated rules tree from the tree set. Works in conjunction with the *Task Name* drop-down list.
- *Current Task Name*: This drop-down list displays the name of each task that has a rules tree associated with it for the rule type selected in the *Rule Type* list. Selecting a task name will display the rules tree for that task in the Tree View. This drop-down list is disabled for case-level rules types.
- *Tree View*: This area displays the currently selected rules tree in a graphical tree structure. Selecting a node in the tree will display the details of that node in the Selected Node and Cornerstone Case panels. Nodes are color coded for easier identification:
  - A blue node represents the root node of the tree
  - Green nodes are *true* or *exception* nodes (i.e., they are on a true branch from their parent node)
  - Red nodes are *false* or *else* nodes (i.e., they are on a false branch from their parent node).
- *Selected Node*: Displays the details of the node currently selected in the Tree View
- *Cornerstone Case*: displays the complete set of case data that, in effect, caused the creation of the currently selected rule (see Sect. 11.10.1 for more details).

11.10.1 Adding a New Rule

There are occasions when the worklet launched for a particular case, while the correct choice based on the current rule set, is an inappropriate choice for the case. For example, if a valued client in the Process Payment worklet example has underpaid by a trivial amount, while the current rule set correctly returns the *Raise Debit Adjustment* worklet, it may be desirable to waive the underpayment in such a context to avoid upsetting the client. In such a case, as the Worklet Service begins execution of an instance of the Raise Debit Adjustment process, it becomes obvious to the user (in this case, a sales manager) that a new rule needs to be added to the rule set so that cases of that context (both now and in the future) will be handled correctly.

Every time the Worklet Service selects a worklet or exlet to execute for a specification instance, a log file is created that contains certain descriptive data about the worklet selection process. These files are stored in the Worklet Repository. The data stored in these files are in XML format, and the files are named according to the following format:

CaseID_SpecificationID_RuleType_WorkItemID.xws

For example: *12_OrderFulfillment_Payment_Selection_ProcessPayment.xws* (xws for XML Worklet Selection). The identifiers in the filename refer to the parent specification instance, not the worklet case instance. Also, the WorkItemID identifier part will not appear for case-level rule types. So, to add a new rule after an inappropriate Worklet choice, the particular selected log file for the case that was the catalyst for the rule addition must be located and loaded into the rules editor's *Add Rule* form. Before a rule can be added, the appropriate rule set must first be loaded into the editor. Note that the selected file chosen must be for an instance of the specification that matches the specification rule set loaded in the editor (in other words, you cannot attempt to add a new rule to a rule set that bears no relation to the xws file opened here). If the specifications do not match, an error message will display.

The Rules Editor takes care of where the new rule node is to be attached in the rule tree, and whether it is to be added as a true child or false child node, using the algorithm described in Chap. 4. The parent node of the new rule node is the node that was returned as the "last searched node" included in the result of the original rule tree evaluation.

The *Add Rule* form has two panels: the *Cornerstone Case* panel contains the case data that existed for the creation of the original rule that resulted in the selection. The *Current Case* panel contains the case data for the current case, that is, the case that is the catalyst for the addition of the new rule. The *New Rule Node* panel is where the details of the new rule may be added.

As the case data for the original rule and the case data for the new rule are both displayed, to define a condition for the new rule it is only necessary to determine what it is about the current case that makes it necessary for the new rule to be added. That is, it is only where the values for the corresponding case data attributes differ that distinguishes one case from the other, and further, only a *subset* of that differing data is relevant to the reason why the original selection was inappropriate. For example, while there may be many data items that differ between the two case data sets, there are only two differing data items of relevance here: underpaid amount and client rating – those are the only data items that, in this case, make the selection of the Issue Debit Adjustment worklet inappropriate.

Clicking on those lines in the *Current Case* panel copies them to the *Condition* field in the *New Rule Node* panel. Thus, a condition for the new rule has been easily created, based on the differing data attributes and values that has caused the original worklet selection to be invalid for this case.

The *Condition* input of the *Add Rule* form allows direct editing of the inserted condition. Conditions may be expressed as strings of operands and operators of any complexity, and subexpressions may be parenthesized, or alternately XQuery expressions may be used. Table 11.1 shows the supported operators and their order of precedence.

All conditions must finally evaluate to a Boolean value.

After defining a condition for the new rule, the name of the appropriate worklet or exlet to be executed when this condition evaluates to true must be entered in the *Conclusion* field of the *New Rule Node* panel. If the new rule is to be added

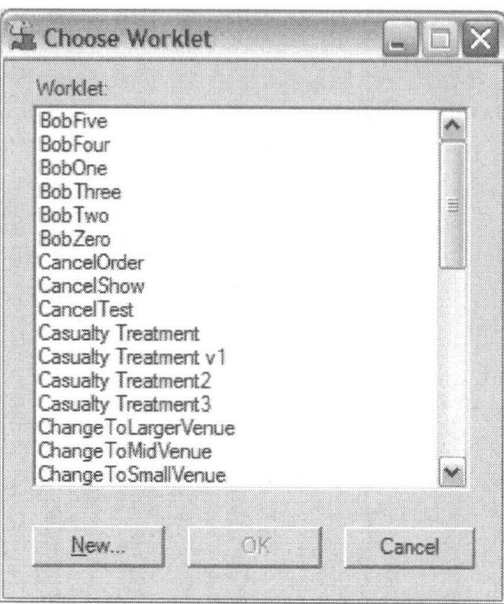

Fig. 11.18 The Choose Worklet dialog

to a selection tree, the process to add the rule is that explained below. Refer to Sect. 11.10.3 for details on adding a conclusion for the exception rule types.

For a selection rule tree, when the *New* button is clicked, a dialog is displayed that comprises a drop-down list containing the names of all the worklets in the Worklet Repository (refer Fig. 11.18). An appropriate worklet for this rule may be chosen from the list, or, if none of the existing worklets are suitable, a new worklet specification may be created.

Clicking the *New* button on this dialog will open the YAWL Editor so that a new worklet specification can be created. When the new worklet is saved and the Editor is closed, the name of the newly created worklet will be displayed and selected in the worklet drop-down list. Once all the fields for the new rule are complete and valid, clicking the *Save* button adds the new rule to the rule tree in the appropriate location.

Table 11.1 Operator order of precedence

Precedence	Operators	Type
1	* /	Arithmetic
2	+ −	
3	= < <=	Comparison
	> >= !=	
4	&	Logical AND
	|	Logical OR
	!	Logical NOT

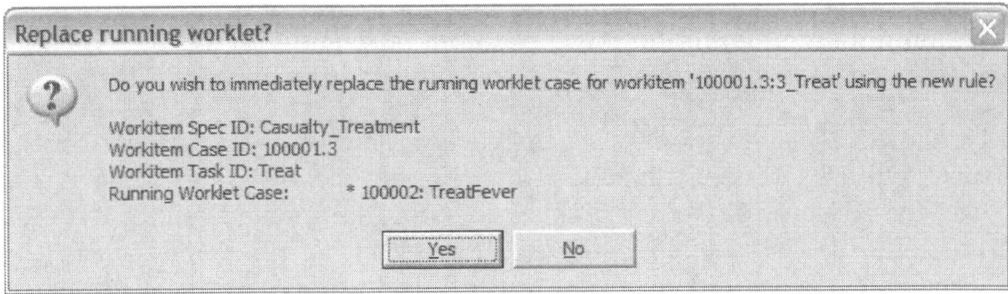

Fig. 11.19 Dialog offering to replace the running worklet

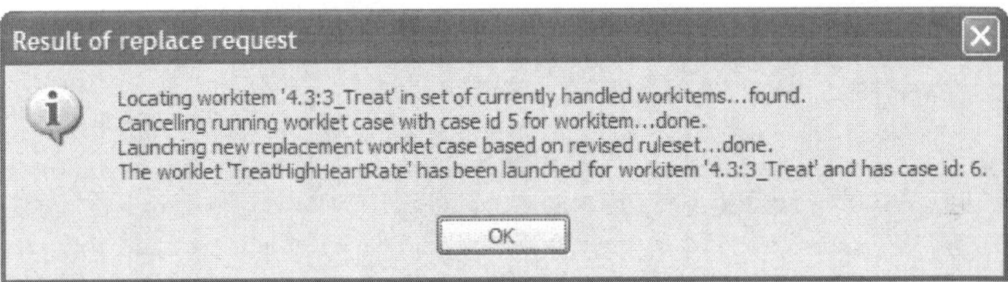

Fig. 11.20 Example dialog showing a successful dynamic replacement

11.10.2 Dynamic Replacement of an Executing Worklet

Creation of the new rule in the running example above was triggered by the selection and execution of a worklet that was deemed an inappropriate choice for the current case. So, when a new rule is added, administrators are given the choice of replacing the executing (inappropriate) worklet instance with an instance of the worklet defined in the new rule.[1] After saving the new rule, a dialog similar to that in Fig. 11.19 is shown, providing the option to replace the executing worklet using the new rule. The message also lists the specification and case identifiers of the original work item, and the name and case id of the launched worklet instance.

If *Yes* is clicked, then in addition to adding the new rule to the rule set, the rules editor will contact the Worklet Service using HTTP messaging and request the change. A further message dialog will be shown soon after, with the results of the replacement process sent from the service back to the editor, similar to that in Fig. 11.20. If the *No* button is clicked, then the new rule is simply added to the rule set, and the original process unsuspended. An administrator would choose to replace the rejected worklet in almost every instance, but the option remains to effectively overrule the worker's worklet rejection depending on organizational rules, authorities, and constraints.

[1] When a worklet instance is rejected, it is automatically suspended pending replacement.

11.10.3 Creating a New Rule Set and/or Tree Set

As mentioned previously, it is not necessary to create tree sets for all the rule types, nor a rule tree for an item-level rule type for each and every task in a specification. So, most typically, working rule sets will have rule trees defined for a few rule types, while other types will not have rules trees defined in the set (any events that do not have associated rules for that type of event are simply ignored). It follows that there will be occasions where it becomes necessary to add a new tree set to a rule set for a previously undefined rule type, or add a new tree, for a task that has no rule tree for a particular rule type previously defined, to an existing tree set. Also, when a new specification has been created, a corresponding base rule set will also need to be created (if selections and exceptions for the new specification are to be handled by the service).

For each of these situations, the rules editor provides a *Create New Rule Set* form (see Fig. 11.21), which allows for the definition of new rule trees (with any number of rule nodes) for existing tree sets (where there is a task of the specification that has not yet had a tree defined for it within the tree set); the definition of new tree sets for specifications that have not yet had a tree set defined for a particular rule type; and entirely new rule sets for specifications that have not yet had a rule set created for them. The form allows administrators to create a rule set, one rule tree at a time (for the selected specification, rule type and, if applicable, task name); its use varies slightly depending on whether it is working with a new rule set or an existing rule set.

This section describes the features of the *Create New Rule Set* form for adding a new rule set, and points out how the process differs for existing rule sets where required. The creation of a new rule set begins by clicking the *New Rule* toolbar button. On the form:

- The *Process Identifiers* panel is where the names of the specification, rule type and, if applicable, task name for the new tree are defined. The *Specification Name* input is read-only – for new rule sets, it is the specification chosen via a *Specification Location* dialog displayed when the form is first opened; for existing rule sets, it is the specification for the rule set currently loaded into the editor. The *Rule Type* drop-down list contains all of the available rule types (i.e., all the rule types for which no or incomplete tree sets exist). For new rule sets, all rule types are available. The *Task Name* drop-down list contains all the available tasks for the selected rule type (i.e., tasks for which no tree exists in the tree set for this rule type). The names of all tasks defined for a specification are automatically gathered by the editor's parsing of the particular specification file. The *Task Name* list is disabled for case-level rule types.
- The *New Rule Node* panel is identical to the panel on the *Add New Rule* form. Here a condition and optional description can be entered, and the conclusion for the new rule created or selected from the list (depending on the rule type – see below).

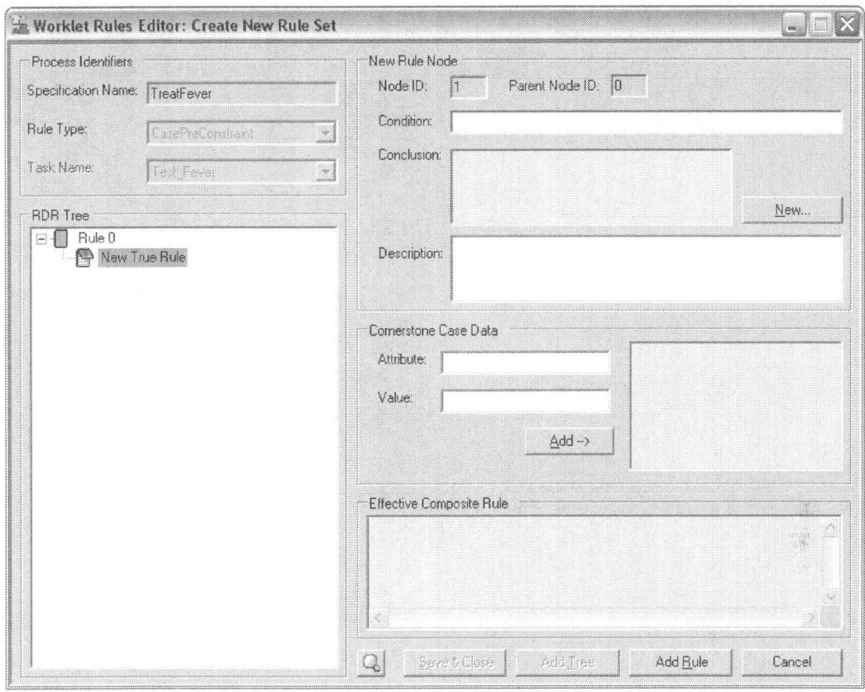

Fig. 11.21 Create New Rule Set Form

- The *Cornerstone Case Data* panel allows a set of cornerstone data for the new rule to be defined, via the Attribute and Value inputs, and the Add button. The standard XML naming rules apply to the data attributes: the attribute name must begin with an alpha character or underscore and contain no spaces or special characters.
- The *Effective Composite Rule* panel displays a properly indented textual interpretation of the composite condition comprising the conditions of the selected node and all ancestor nodes back to the root node – in other words, the entire composite condition that must evaluate to true for the conclusion of the selected node to be returned.
- The *RDR Tree* panel dynamically and graphically displays the new rule tree as it is being created.

New rule nodes can be added wherever there is a node on the tree that does not have a child node on both its true and false branches (except of course the root node, which can have a true branch only). To identify available locations where a new rule node may be added, special "potential" nodes are displayed in the *RDR Tree* panel, called "New True Node" or "New False Node." These potential nodes are colored yellow for easy identification. Selecting a yellow new rule node enables the various inputs for the new rule on the form (they are disabled or read-only by

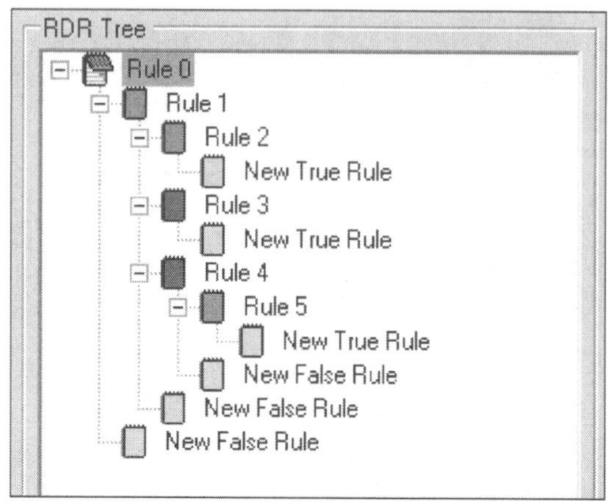

Fig. 11.22 Creating a new Rule Tree

default). Clicking the *New* button adds a conclusion for the new rule. If the currently selected rule type is "Selection," a worklet can be added as a conclusion in the manner described in Sect. 11.10.1.

If it is one of the exception rule types, the *New* button will display the *Draw Conclusion* dialog, allowing for the graphical creation of an exlet process model comprising a sequence of tasks (or primitives), as explained in detail in Sect. 11.10.4 below. When the conclusion sequence has been defined and the dialog closed, a textual representation of it will be displayed in the *Conclusion* panel.

Once the new rule node has been given a condition and conclusion, and optionally some cornerstone data and a description, clicking the *Add Rule* button adds the rule to the tree. The new node will be displayed at the selected position on the tree with the relevant colored node icon indicating whether it is a true or false node of its parent. New potential node add-points will also be displayed. Figure 11.22 is an example of a newly created tree that has had several nodes added.

The add rule procedure may be repeated for however many rule nodes are to be added by clicking on the appropriate yellow (potential) nodes. When node addition is complete, clicking the *Add Tree* button will add the tree just created to the tree set selected (via the selected Rule Type and, if applicable, Task Name lists).

Once the newly created tree has been added to the selected tree set, administrators will no longer be able to add nodes to the tree via the *New Rule Set* form. This is to protect the integrity of the rule set. As each subsequent rule will be added because of an exceptional case or where the selected worklet does not fit the context of a case instance, the preferred method is to create the base rule set and then add rules as they become necessary via the *Add Rule* form as described earlier. In this way, added rules are based on real case data and so are guaranteed to be valid. In a similar vein, there is no option to modify or delete a rule node within a tree once the tree has been added to the rule set, since to allow it would destroy the integrity of the rule set, because the validity of a child rule node depends on the conditional expressions of its ancestors.[2]

When a tree is added to the tree set:

- If it is a case-level tree, the rule type that the tree represents will be removed from the *Rule Type* list. That is, the rule type now has a tree defined for it and so is no longer available for selection on the *New Rule* form.
- If it is an item-level tree, the task name that the tree represents will be removed from the *Task Name* list. That is, the task now has a rule tree defined for it (for the selected rule type) and so is no longer available.
- If it is an item-level tree, and all tasks of the specification now have trees defined for them for the selected rule type (i.e., this was the final task of the specification for which a tree was defined), the rule type that the tree represents will be removed from the *Rule Type* list.

This approach ensures that rule trees can only be added where there are currently no trees defined for the selected specification. Once the tree is added, the form resets to allow the addition of another new tree as required, by repeating the process above for a new rule type (or rule type/task name for item-level trees).

11.10.4 Drawing a New Exlet

As described in Sect. 11.10.1, adding a conclusion to a selection rule is a matter of choosing a worklet from the list or creating a new worklet in the Editor. However, when adding a conclusion for a rule type other than "Selection" (i.e., an exception rule type), an exlet needs to be defined that will be invoked when the rule is returned as the last satisfied. Chapter 5 detailed the various actions that make up the available set of exception handling primitives (or tasks) that may be sequenced to form an entire handling process. The *Draw Conclusion* dialog makes the process of defining an exception handling sequence easier by allowing administrators to create an exlet graphically by selecting the appropriate primitive from the toolbox on the left, and then clicking on the drawing canvas to place the selected primitive. Figure 11.11 shows an example of the *Draw Conclusion* dialog.

In addition to the various primitive tools, the *Arc Tool* is used to define the sequence order. For a conclusion to be valid (and thus permitted to be saved), there must be a direct, unbroken path from the start node to the end node (the start and end nodes are always displayed on the drawing canvas). Also, the conclusion will be considered invalid if there are any nodes on the canvas that are not attached to the sequence when a save is attempted.

The *Select Tool* is used to move placed primitives around the canvas. The *Align* button will immediately align the nodes horizontally and equidistantly between the

[2] There are algorithms defined in the literature for the reduction and transformation of Ripple-Down Rule sets that may be applied from time to time (e.g., see [234]).

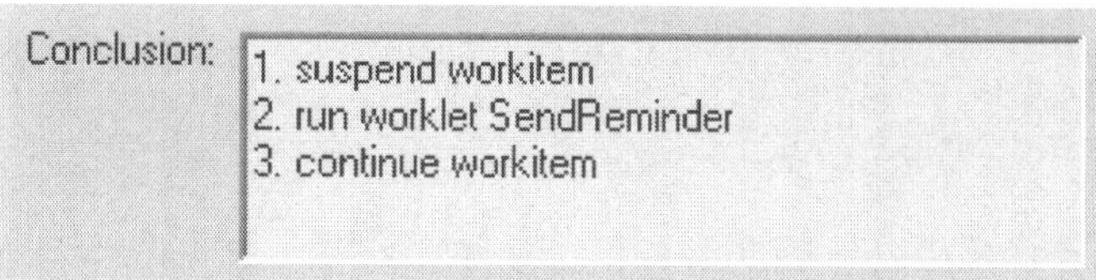

Fig. 11.23 A Conclusion Sequence shown as text (detail)

start and end nodes (as in Fig. 11.11). The *Clear* button will remove all added nodes to allow a restart of the drawing process. The *Cancel* button discards all work and returns to the previous form. The *Save* button will save the conclusion and return to the previous form (as long as the exlet is deemed valid).

The *Compensate* primitive will, when invoked at runtime, execute a worklet as a compensation process as part of the handling exlet process. To specify which worklet to run for a selected compensate primitive, a popup menu is provided, which invokes the *Choose Worklet* dialog (identical to the dialog shown for a "Selection" conclusion process), allowing the selection of an existing worklet or the definition of a new worklet to run as a compensatory process. Selecting the appropriate worklet adds it to the compensation primitive. An exlet will be considered invalid (and thus unable to be saved) if it contains a compensate primitive for which a worklet has not yet been defined.

The primitives *SuspendAllCases*, *RemoveAllCases*, and *ContinueAllCases* may be optionally limited to ancestor cases only via the popup menu associated with those kinds of primitives. Ancestor hierarchies occur where a worklet is invoked for a case, which in turn invokes a worklet, and so on. When a primitive is limited to ancestor cases, it applies the primitive's action to all cases in the hierarchy from the current case back to the original parent case, rather than all running cases of the specification.

When a valid exlet is saved, the editor returns to the previous form (i.e., either the *Add Rule* or *New Rule* form depending on from where it was invoked). The conclusion will be displayed textually as a sequential list of action-target pairs (an example can be seen in Fig. 11.23).

Summary

Workflow management systems impose a certain rigidity on process definition and enactment because they generally use frameworks based on assembly line metaphors rather than on ways the work is actually planned and carried out. An analysis of Activity Theory in Chap. 4 provided principles of work practices that were used as a template on which a workflow service has been built that provides innovative techniques that directly provide for process evolution, flexibility and dynamic exception handling, and mirror accepted work practices.

This implementation uses the open-source, service-oriented architecture of YAWL to develop a service for flexibility and dynamic exception handling completely independent to the core engine. Thus, the implementation may be viewed as a successful case study in service-oriented computing. As such, the approach and resultant software can also be used in the context of other process engines (e.g., BPEL based systems, classical workflow systems, and the Windows Workflow Foundation).

For a demonstration of the use of worklets in a real-world scenario, see Chap. 21.

Exercises

Exercise 1. Using the Editor, create a process specification for a treatment plan at a generic hospital casualty department. The process should contain tasks to examine the patient and collect medical data, to diagnose the problem, to refer the patient to the appropriate department, and then to discharge the patient. Then, start the Rules Editor and create a new, properly constructed rule set for the specification, using the rule set described in Chap. 4, Exercise 5. Worklets for each new rule can be created by triggering the Editor from within the Rules Editor. Go back to the original specification and enable the appropriate task(s) so that they are associated with the Worklet Service. Execute an instance of the process in the Engine, taking care to monitor the results. Devise at least three scenarios where an addition to the rule set is required, and add the new rule and worklet.

Exercise 2. Using the process specification created in Exercise 1, imagine ways in which exceptions may be raised during an instance of the process. Devise at least two scenarios for three different exception types. Add rules to capture those exceptions to the rule set, and for each define an exlet to handle the exception. Extend the rule tree for one particular exception type by adding two additional rules, so that an exception may be handled in different ways depending on its context. Run as many instances of the process in the Engine as required, taking care to watch for the invocation of exlets at the appropriate time(s) and their effect on the parent process (based, of course, on how you defined them).

Chapter Notes

The Worklet Service was originally implemented in two phases: the Selection subservice was implemented first, in part to ensure proof of concept and also to construct the framework for the overall approach, and was first deployed as a member component of the YAWL Beta 7 release. The Exception subservice was then added as an extension to the Selection Service and was first deployed as a component member of the YAWL Beta 8 release. See [23, 25–27] for more details.

An interesting approach, which leverages the Service-Oriented Architecture of the YAWL environment to provide support for dynamic flexibility using the Worklet Service and the Declare Service (cf. Chap. 12) in tandem, can be found in [5].

The Worklet Service (including the rules editor), its source code and accompanying documentation, can be freely downloaded from the YAWL project pages at www.sourceforge.net/projects/yawl.

Chapter 12
The Declare Service

Maja Pesic, Helen Schonenberg, and Wil van der Aalst

12.1 Introduction

The Declare Service is a YAWL Custom Service that enables decomposing YAWL tasks into DECLARE workflows, that is, workflows supported by the workflow management system (WfMS) called DECLARE. The goal of this service is to enable a particular kind of flexibility. Chapter 6 describes a constraint-based approach to workflow models and the ConDec language. This approach, supported by the DECLARE WfMS, allows for more flexibility, that is, execution of tasks is allowed if it is not explicitly forbidden by some constraint. This chapter describes DECLARE and the Declare Service for YAWL.

Sometimes it is easier to express a process in a procedural language (e.g., the native workflow language of YAWL) and sometimes a declarative approach is more suitable. Moreover, in a larger process it may be useful to express parts of the process in a procedural language and specify other parts in terms of constraints. Using the service-oriented architecture of YAWL, this can easily be realized. A YAWL task may decompose into a DECLARE process and a task in DECLARE can be decomposed into a YAWL process. Arbitrary decompositions of DECLARE and YAWL models allow for the integration of declarative and YAWL workflows on different abstraction levels.[1] This way the designer is not forced to make a binary choice between a declarative and a procedural way of modeling. Hence, a seamless integration can be achieved, where parts of the workflow that need a high degree of flexibility are supported by declarative DECLARE models, and parts of the processes that need centralized control of the system are supported by YAWL models.

Consider, for example, the decomposition of the Carrier Appointment process shown in Fig. 12.1. At the highest level of decomposition, the main process is modeled using a procedural YAWL *Carrier Appointment* net, which is described

[1] Note that the service oriented architecture also allows for decompositions involving worklets, which are described in Chaps. 4 and 11.

M. Pesic (✉)
Eindhoven University of Technology, Eindhoven, the Netherlands
e-mail: m.pesic@tue.nl

A.H.M. ter Hofstede et al. (eds.), *Modern Business Process Automation*,
DOI 10.1007/978-3-642-03121-2_12, © Springer-Verlag Berlin Heidelberg 2010

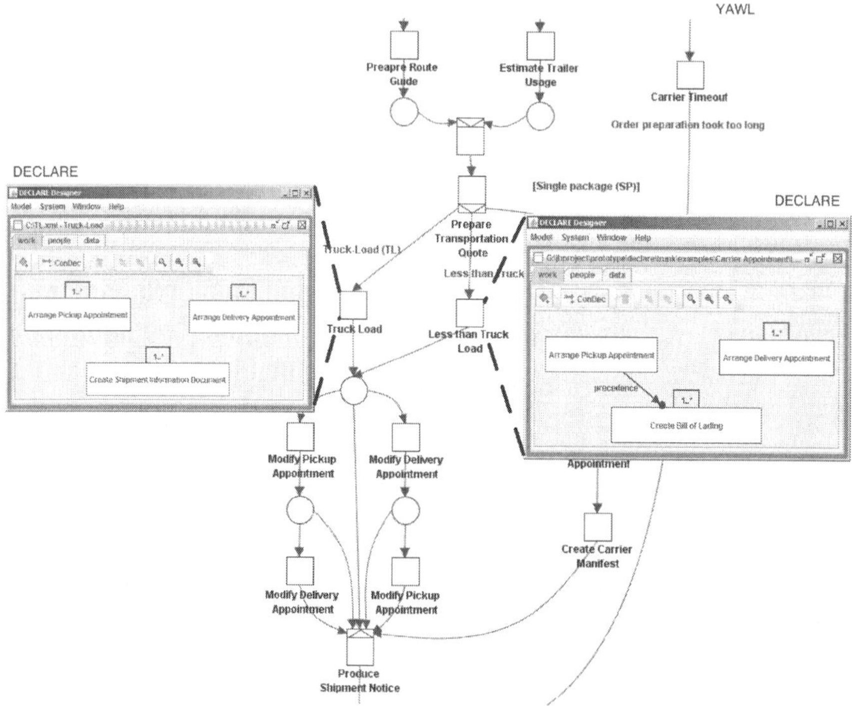

Fig. 12.1 An example illustrating the service-oriented architecture: tasks *Truck Load* and *Less than Truck-Load* in the top-level YAWL net are decomposed into DECLARE processes

in Appendix A. However, parts of the net that refer to the Truck Load and Less than Truck Load scenarios are now decomposed and specified as declarative workflows, as described in Chap. 6. Although this example presents only decomposition of YAWL nets into DECLARE submodels, it is also possible to decompose DECLARE models into YAWL subnets. This way, procedural and declarative workflows can be combined.

12.2 Service Architecture

As Fig. 12.2 shows, the Declare Service uses Interface B in two ways. First, YAWL can delegate a task to the service, instead of to the Resource Service. For example, task *S* in Fig. 12.2a can be delegated to a Custom Service. Second, a Custom Service can request the launch of a new YAWL instance. In both cases, relevant instance and task data elements are exchanged between YAWL and the service. The Resource Service is a Custom Service, which is described in Chap. 10. By default, all tasks in YAWL models are delegated to the Resource Service, that is, all tasks are by default executed by users through their YAWL worklists. If a task should be delegated to

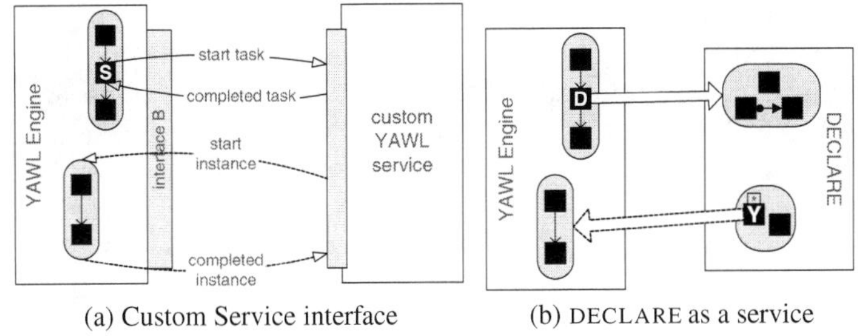

(a) Custom Service interface (b) DECLARE as a service

Fig. 12.2 DECLARE as a YAWL Custom Service

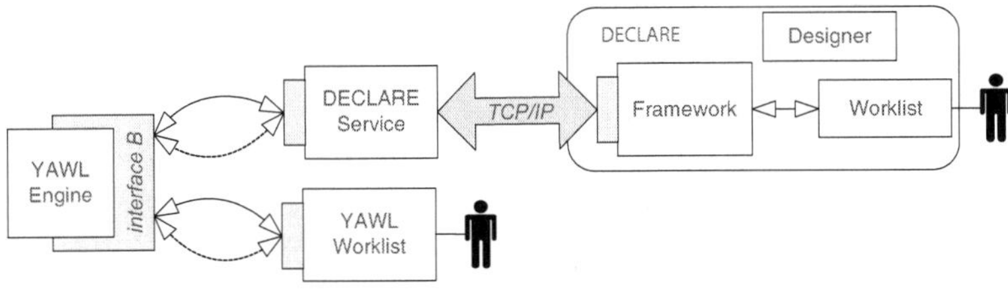

Fig. 12.3 Interface between YAWL, Declare Service, and DECLARE

another Custom Service, then this must be explicitly specified in the model. The Declare Service enables arbitrary decompositions of DECLARE and models in two ways, as shown in Fig. 12.2b. First, it is possible that a YAWL task triggers the execution of a DECLARE instance: when the YAWL task *D* becomes enabled, DECLARE will initiate the required constraint model. YAWL will consider the completion of the launched DECLARE instance as a completion of its task. Second, DECLARE task *Y* can trigger the execution of a YAWL instance.

As Fig. 12.3 shows, DECLARE consists of three components; the *Designer*, the *Framework*, and the *Worklist*. The *Designer* component is used for creating constraint templates, constraint models, and by verification of constraint models. For example, Fig. 12.1 shows the *Designer* component containing the Carrier Appointment model. The *Framework* component facilitates instance creation and change of constraint models. Finally, the *Worklist* component offers users access to active instances of constraint models. Also, users can execute tasks of active instances in their *Worklist*. The Declare Service acts as a bridge between the YAWL Engine and the DECLARE *Framework*, as shown in Fig. 12.3. Note that users execute "normal" tasks (i.e., tasks that are not decomposed or subcontracted to another system or process) of DECLARE and YAWL instances in a default manner by using the respective worklists.

After presenting the architecture of DECLARE and the way it is embedded into the YAWL environment, we now focus on the functionality provided by DECLARE. In the remainder, we show the following capabilities:

- The possibility to define *constraint templates*. DECLARE is multilingual and can support multiple constraint languages at the same time. Moreover, it is possible to define new languages using constraint templates.

- The possibility to define *constraint models*. Using the constraint templates it is possible to specify specific constraints for a particular process.
- The *verification* capabilities of DECLARE. While modeling or redesigning workflows errors may be introduced. Hence, DECLARE offers a wide range of verification capabilities.
- The *enactment* of constraint model instances. Based on such constraint languages, DECLARE can automatically generate the support needed, that is, mandatory constraints can not be violated while optional constraints are used to generate warnings.
- The possibility to *change* constraint models and to *migrate* instances. Although DECLARE allows for a declarative way of modeling, other flexibility issues such as migrating cases are supported.
- The *integration* of YAWL and DECLARE. As explained using Fig. 12.1, it is rather easy to support an arbitrary nesting of workflow languages.

12.3 Constraint Templates

An arbitrary number of constraint-based languages can be defined in DECLARE. For example, Fig. 12.4 shows how the ConDec language (cf. Chap. 6) is defined in the *Designer* tool. A tree with the language templates is shown under the selected language. the selected template is presented graphically on the panel on the right side of the screen.

An arbitrary number of templates can be created for each language. Figure 12.4 shows a screenshot of the *Designer* while defining the *precedence* template. First, the template name and additional label are entered. Next, it is possible to define an arbitrary number of parameters in the template. The *precedence* template has two parameters: *A* and *B*. For each parameter it is specified whether it can be *branched* or not. When creating a constraint from a template in a model, a task replaces each of the template's parameters. If a parameter is branchable, then it is possible to replace the parameter with more tasks. In this case, the parameter will be replaced by a *disjunction* of selected tasks in the formula (cf. Chap. 6). The graphical representation of the template is defined by selecting the kind of symbol that will be drawn next to each parameter and the style of the line. Figure 12.4 shows that the *response* template is graphically represented by a single line with a filled arrow-and-circle symbol next to the second task (*B*). Furthermore, a textual description and the LTL formula are given.

12.4 Constraint Workflow Models

Constraint models can be developed in the *Designer* tool after selecting one of the languages defined in the system. In this chapter, we use only the ConDec templates described in Chap. 6. However, as explained before, DECLARE is extensible and multilingual. As an example, we consider the Less than Truck Load (LTTL)

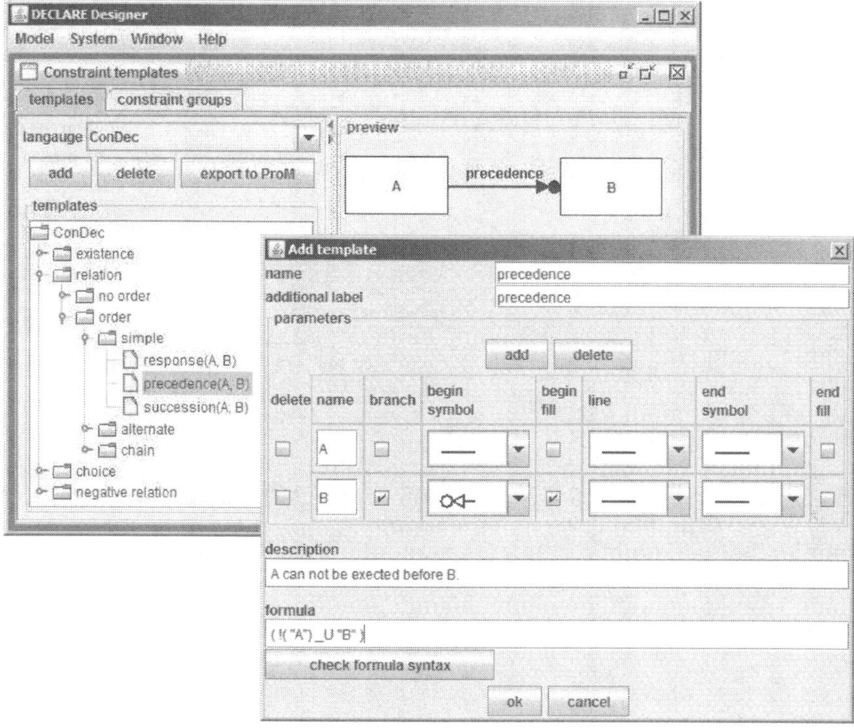

Fig. 12.4 Creating a language by defining constraint templates using the DECLARE *Designer* tool

process described in Chap. 6. Figure 12.5 shows the ConDec LTTL model in DECLARE.

Each constraint in the model in Fig. 12.5 is created using a ConDec template. For example, the constraint between tasks *pickup* and *bill* is created by applying the *precedence* template.[2] The creation of this template was already shown in Fig. 12.4. However, when making a constraint model, there is no need to define new constraints from scratch. Instead, constraints are defined by applying templates. Figure 12.6 shows the application of the *precedence* template. The template is selected in the top left corner of the screen. Underneath the template, all its parameters are shown. Tasks are assigned to parameters by selecting a task (or multiple tasks in case of branching) from the model. Some additional information can be given on the right side of the screen. Initially, constraints have the template name, but constraints can have arbitrary names and can be renamed. Constraints can have a condition involving some data elements from the model. For example, condition "totalPrice

[2] For the purposes of simplicity, in this chapter we will use shorter names of the relevant tasks: *pickup* for *Arrange Pickup Appointment*, *delivery* for *Arrange Delivery Appointment*, *bill* for *Create Bill of Lading*, and *shipment* for *Create Shipment Information Document*.

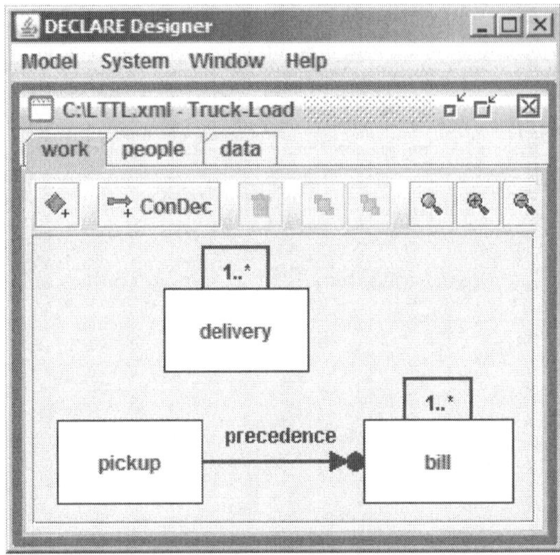

Fig. 12.5 The LTTL model in DECLARE

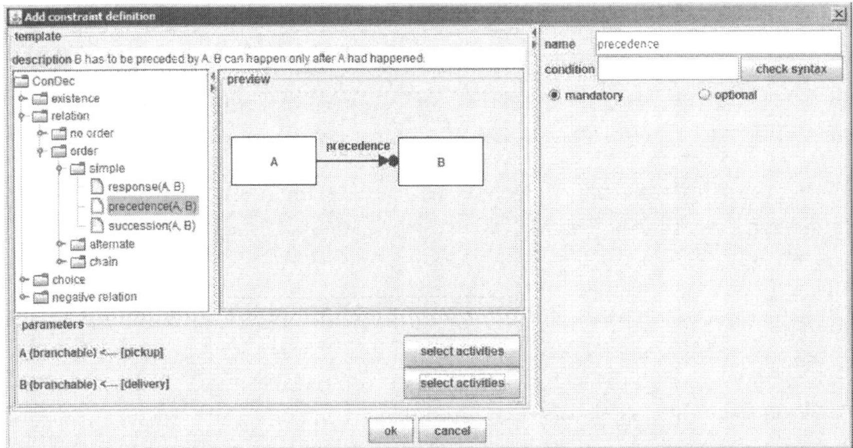

Fig. 12.6 Defining a constraint in DECLARE based on the *precedence* template constructed in Fig. 12.4

< 1,000" on a constraint would mean that the constraint should hold only if the data element *totalPrice* has a value less than 1,000. Finally, the type (mandatory or optional) of the constraint must be specified. The constraint shown in Fig. 12.6 is mandatory.

12.5 Verification of Constraint Models

DECLARE uses the methods presented in Chap. 6 to detect dead tasks, conflicts, and the smallest set of constraints that cause them. To illustrate this, we use the example for a ConDec model with a conflict from Chap. 6. Figure 12.7a shows this model in DECLARE: the two *1..** constraints specify that each of the tasks *delivery* and *bill* must be executed at least once. The two *precedence* constraints specify that (1) task *bill* cannot be executed before task *pickup* and (2) task *pickup* cannot be executed before task *bill*. Besides, for detecting error(s) in models, for each error DECLARE searches through (the powerset) mandatory constraints to find the smallest group of constraints that causes the error. DECLARE applies the verification methods described in Chap. 6 for each group of constraints to detect the smallest group of constraints that causes the error.

Figure 12.7b–d shows that the DECLARE verification procedure reports three errors in this model. First, the two *precedence* constraints cause the *bill* task to be dead (cf. Fig. 12.7b). Second, the two *precedence* constraints cause the *pickup* task to be dead (cf. Fig. 12.7c). Finally, the two *precedence* constraints and the *1..** constraint on task *bill* cause a conflict (cf. Fig. 12.7d). Because of this conflict, it is not possible to satisfy all constraints at the same time. At least one of the three related constraints mentioned needs to be dropped.

Detailed verification reports in DECLARE aim at helping model developers to understand error(s) in the model. The goal is to assist the resolution of such problems. As discussed in Chap. 6, errors can be eliminated from a model by removing at least one constraint from the group of constraints that together cause the error. For example, if one of the two *precedence* constraints would be removed from the DECLARE model in Fig. 12.7a, there would no longer be errors in this model. Also, removing the *1..** constraint on task *bill* would remove the conflict, but tasks *bill* and *pickup* would still be dead.

12.6 Execution of Constraint Model Instances

DECLARE determines which tasks can be executed next, that is, determining the enabled tasks, the state of an instance, and states of constraints using the approach presented in Chap. 6. Each instance is launched (i.e., created) in the *Framework* tool and users can work on instances via their *Worklists* (cf. Fig. 12.3 on page 329). A user can work on instances from his/her *Worklist*. Figure 12.8 shows a screenshot of a *Worklist*. All available instances are shown in the list on the left side of the screen, underneath the "instances" header. In Fig. 12.8, there are two instances of the *Less than Truck Load* model presented in Fig. 12.5 on page 332: "(1) Less than Truck Load" and "(2) Less than Truck Load." The model of the selected instance is shown on the right side of the screen. After the user starts a task by double-clicking it, the task will be opened "in the task panel" under the model.

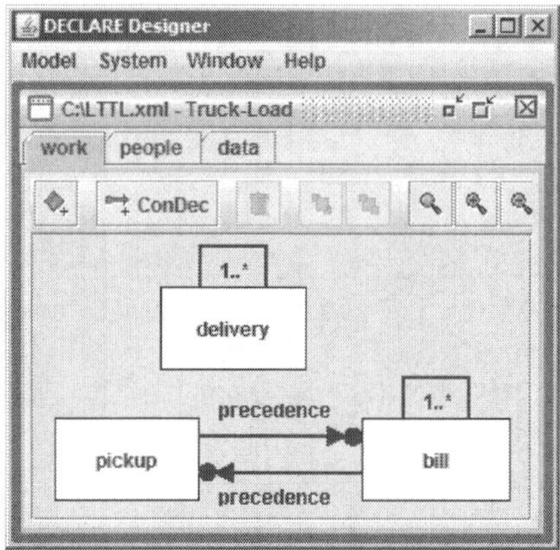

(a) DECLARE model

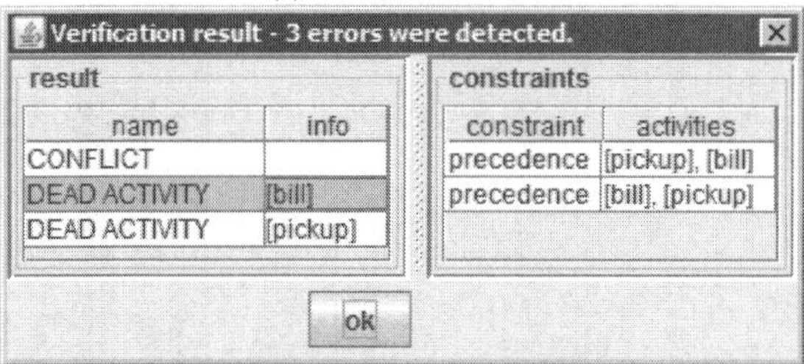

(b) task *bill* is dead

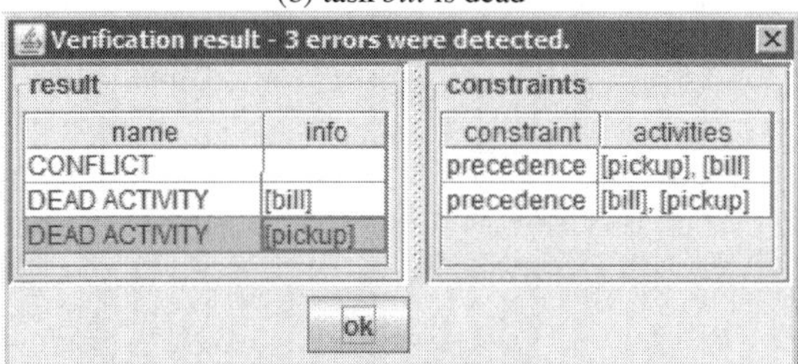

(c) task *pickup* is dead

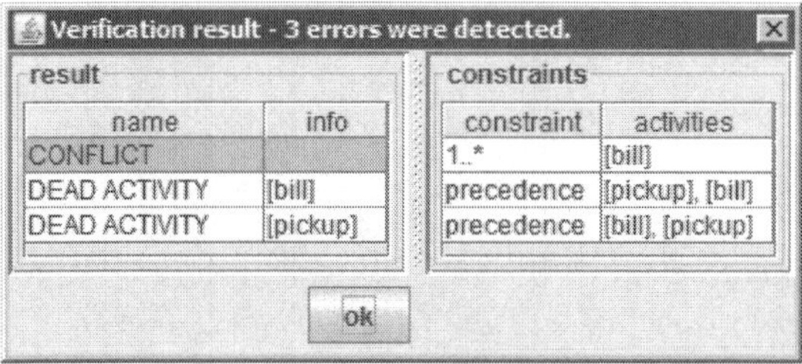

(d) a conflict

Fig. 12.7 A DECLARE model with a conflict and two dead tasks

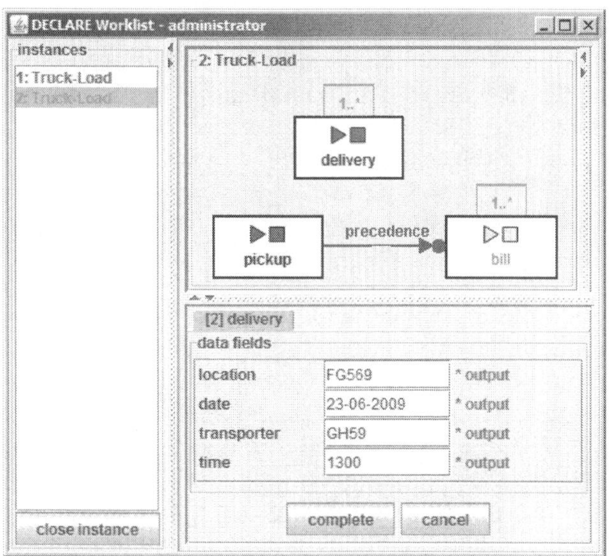

Fig. 12.8 The DECLARE *Worklist* showing details of a selected instance

Although the structure of the process model is the same as in the *Designer* (cf. Fig. 12.5), the *Worklist* presents some additional symbols and colors to users to indicate which tasks are enabled, the current state of the instance and each constraint. Each task contains "start" (play) and "complete" (stop) icons, which indicate whether users can start/complete the task at the moment. State *satisfied*, *violated*, and *temporarily violated* are indicated by the colors green, red, and orange, respectively. This color scheme is used for both constraint states and instance states.

The initial state of the process instance in Fig. 12.8 shows that it is only possible to start tasks *delivery* and *pickup*, because the corresponding start symbols are enabled. Starting and completing task *bill* is not possible, as indicated by the disabled start icon. In addition, all currently disabled tasks are colored grey (cf. task *bill*). This initial state of the process instance is influenced by the *precedence* constraint, which specifies that task *bill* cannot be executed until task *pickup* is executed. Also note that each constraint is colored to indicate its state. The two *1..** constraints are *temporarily violated* (i.e., *orange*), while the *precedence* constraint is *satisfied* (i.e., *green*).

Figure 12.8 shows the instance after starting task *delivery*. This task is now "open" in the "task panel" on the bottom of the screen. Data elements that are used in this task are presented in the "task panel." In this case, four data elements are available – *location*, *date*, *transporter*, and *time*. In this way, users can manipulate data elements while executing tasks. The task can be completed or canceled by clicking on the appropriate buttons on the "task panel."

12.7 Optional Constraints

So far in this chapter, we have used models where all constraints are mandatory. However, as discussed in Chap. 6, constraints can also be optional, that is, they can be violated. If the user is about to violate an optional constraint, DECLARE issues a warning. This warning contains additional information about the constraint, and it should help the user when deciding whether to proceed and violate the constraint, or to abort and not violate it. As Fig. 12.9 shows, additional information for optional constraints contains (1) a group; (2) the importance level on a scale from 1 (for low importance) to 10 (for high importance), and (3) some local message that should be displayed. This additional information for an optional constraint is presented in the warning given when a user is about to violate this constraint.

Groups for optional constraints are defined in the *Designer* component and can be used to define corporate policies, etc. Each group has a name and a short description, as illustrated in Fig. 12.10, and end-users should be able to interpret them. For example, it should be easier to decide to violate a constraint belonging to the *Ordering Policy*, than a constraint that belongs to the *Billing Policy*.

In some cases, triggering an event or closing the instance can violate optional constraint(s). For example, consider an instance of the DECLARE model shown in Fig. 12.11, where the user attempts to start executing task *bill*. Note that, despite the *precedence* constraint, it is possible to start task *bill* in this instance. This is because this constraint is optional. If the user would now attempt to start executing task *bill*, DECLARE would first issue a warning associated with the optional constraint, as shown in Fig. 12.11. The user can now decide based on the information presented in the warning whether to start the execution of task *bill* and violate the constraint or not. Note that, in case of mandatory constraints, this is not possible, that is, if this would be a mandatory constraint, it would not be possible to start executing task *bill* at this moment.

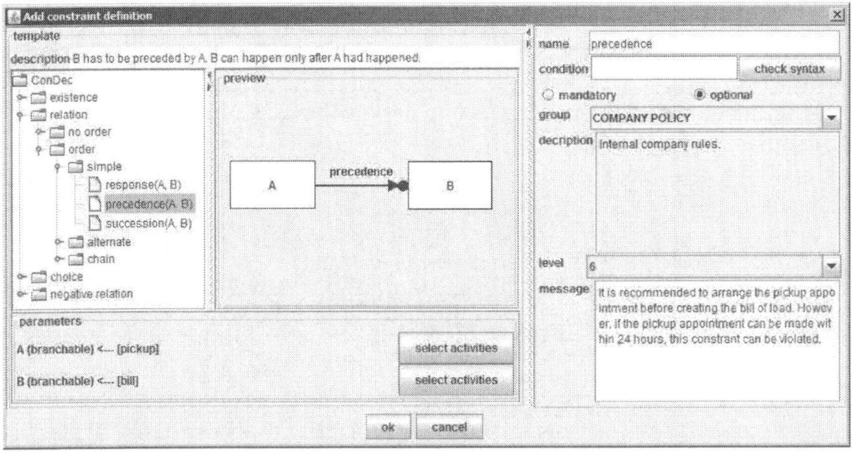

Fig. 12.9 Defining the optional constraint *precedence*

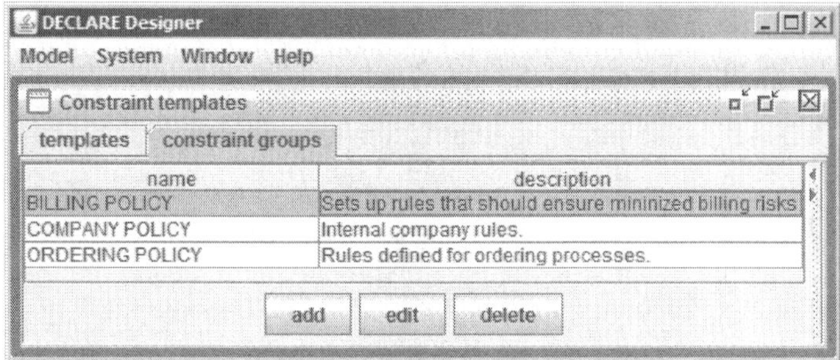

Fig. 12.10 Warnings related to possible violations are linked to so-called constraint groups

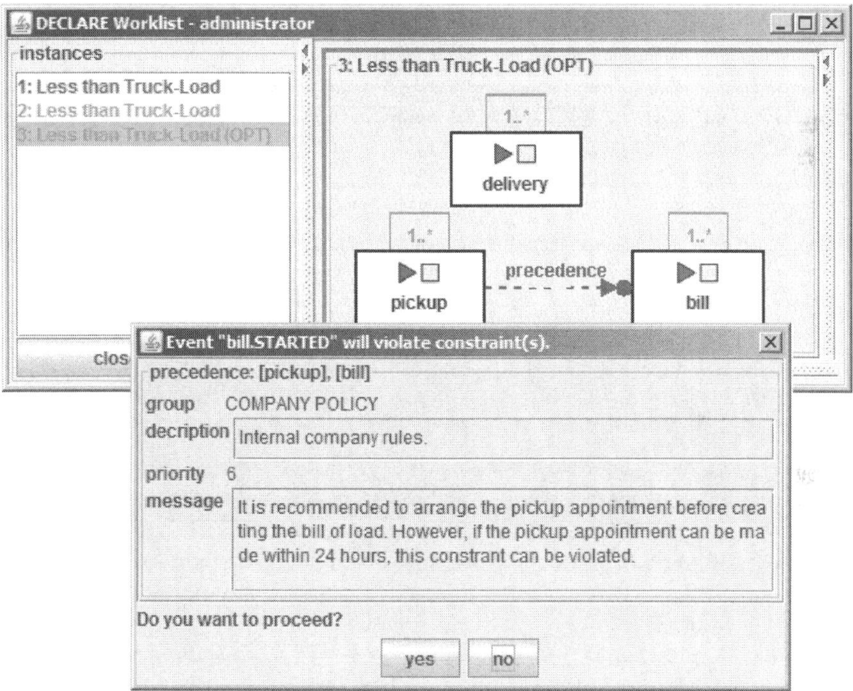

Fig. 12.11 Warning: starting task *bill* violates the optional constraint *precedence*

12.8 Dynamic Instance Change

Instances in DECLARE can be changed dynamically by adding and removing tasks and constraints. After the change, DECLARE creates an automaton for the whole mandatory formula (i.e., the conjunction of all mandatory constraints expressed in LTL) of the new model. If the instance trace can be "replayed" on this automaton, the dynamic change is accepted. If not, the error is reported and the instance continues

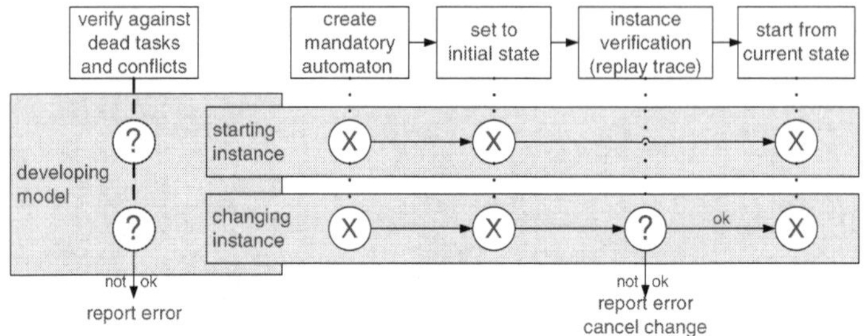

Fig. 12.12 Procedure for starting and changing instances in DECLARE

its execution based on the old model. Naturally, when applying a dynamic change, it is also possible to verify the new model against dead tasks and conflicts but these errors will not disable the change. Actually, the procedure for dynamic change is very similar to the procedure for starting instances in DECLARE, as Fig. 12.12 shows. It is possible to perform basic model verification in both cases. The only difference is in the execution of the automaton. When an instance is started, the execution of the automaton begins from the initial state. In case of a dynamic change, DECLARE first makes an attempt to "replay" the current trace of the instance on the new automaton, that is, the new model is verified against the current trace. If this is possible, the dynamic change is successful and the execution continues from the current set of possible states in the new automaton, that is, the instance state, enabled tasks, and states of constraints are determined using the new automaton and the current trace. If this is not possible, the dynamic change fails and the instance must continue using the old model.

Besides changing a single instance, DECLARE offers two additional options: *migration* of all instances and *changing* the original constraint model. It is possible to request a migration of all instances, that is, that the dynamic change is applied to all running instances of the same constraint model. DECLARE performs migration by applying the same procedure for dynamic change to all instances of the same constraint model, that is, only instances with traces that can be replayed on the new automaton are migrated. It is also possible to change the original constraint model. In this case, all instances created in the future will be based on the new model.

Consider, for example, the dynamic change scenario described in Chap. 6 with two instances of the ConDec model for the *Less than Truck Load* process shown in Fig. 12.13a. For the first instance, tasks $\langle$*pickup*, *delivery*$\rangle$ have been executed, and tasks $\langle$*pickup*, *bill*, *delivery*$\rangle$ have been executed for the second instance. Figure 12.13b shows a DECLARE screenshot of a changed model where task *pickup*, constraint *1..** on the *delivery* task, and the *precedence* constraint are removed, and a new *precedence* constraint is added between *delivery* and *bill*. As a consequence of adding the *precedence* constraint, task *bill* can now be executed only after task *delivery*. The migration of all active instances of this model are requested. Figure 12.13c shows the DECLARE report for the requested dynamic change. The migration is applied to the two currently running instances. The change failed

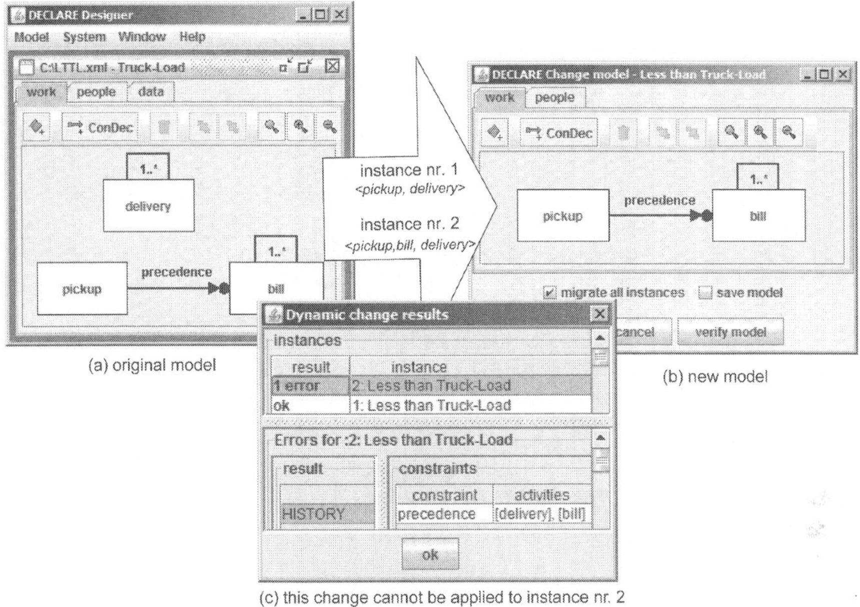

Fig. 12.13 Dynamic instance change in DECLARE

for the first instance due to the violation of the *precedence* constraint as task *bill* already occurs before task *delivery* in the trace ⟨*pickup*, *bill*, *delivery*⟩. However, the dynamic change is successfully applied to the second instance, which has trace ⟨*pickup*, *delivery*⟩. Hence only the second instance is migrated and the first instance remains in the original process.

12.9 Decompositions of YAWL and Declarative Workflows

In both YAWL and DECLARE models, tasks are offered by default to users to manually execute them. If a task should be delegated to an external application, then this must be explicitly defined in the process model. Figure 12.14 shows how the *delivery* task from the DECLARE model shown in Fig. 12.5 could be decomposed into a YAWL model, for example, a model named Arrange Delivery Appointment. In this case, task *delivery* would be graphically presented with a special "YAWL" symbol in the DECLARE model. Note that, although task *delivery* will not be executed manually by a user, the user will decide *when* this task (i.e., the referring YAWL model) will be executed. However, when the user starts this task, it will not be opened in the "task panel" in the *Worklist*. Instead, it will be automatically delegated to YAWL, which will launch a new instance of its Arrange Delivery Appointment model.

Similarly, in a YAWL process model it can be specified that a task should be delegated to a YAWL Custom Service, for example, DECLARE. Figure 12.15 illustrates a

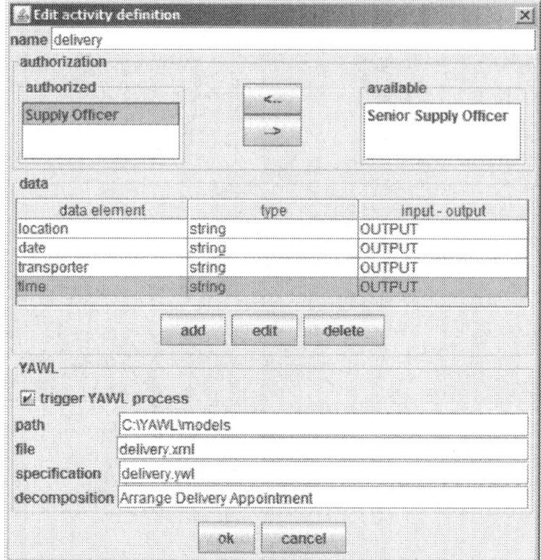

Fig. 12.14 DECLARE task *delivery* launches a YAWL instance

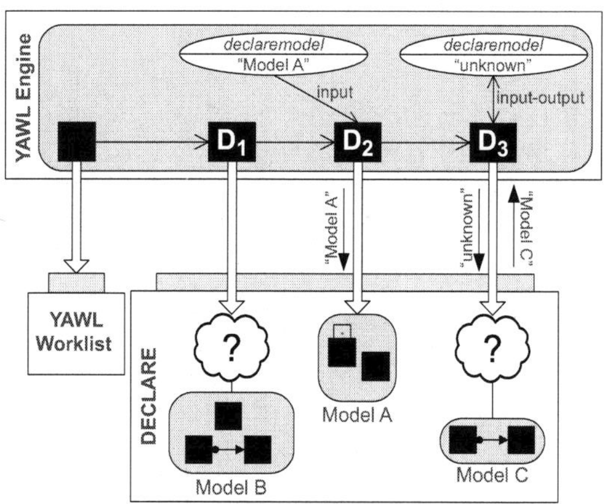

Fig. 12.15 YAWL tasks D_1, D_2, and D_3 launch DECLARE instances

YAWL instance where tasks D_1, D_2, and D_3 are delegated to DECLARE. In the general scenario, DECLARE users must manually select which DECLARE model should be executed for each YAWL request. For example, DECLARE users can choose to execute Model B for task D_1. If the decomposed YAWL task contains an input data element named "declaremodel," then DECLARE automatically launches a new instance of the referring model. For example, in this particular YAWL process, task D_2 launches a new instance of Model A in DECLARE. If the specified model cannot be found, DECLARE users must manually select a DECLARE process models to be executed. For example, users can select to execute an instance of model Model C for task D_3. Because the "declaremodel" data element is also an output data element in task D_3, DECLARE will return the name of the executed model to YAWL. In this manner, YAWL users are informed about the subprocess that was executed for the decomposed YAWL task.

12.10 Conclusions

Depending on the nature of a business process, a higher or lower degree of flexibility is needed when it comes to workflow support. For example, while procedural YAWL processes are appropriate for many of the more structured processes encountered in practice, they may be less suitable for less structured processes (such as our TL and LTTL subprocesses) that require the flexibility of declarative workflow models. Moreover, it might be the case that one business process contains some subprocesses that require the procedural approach and some subprocesses that require the declarative approach.

The Declare Service enables combining the two approaches by allowing for arbitrary decompositions of procedural YAWL and declarative DECLARE workflows. This is achieved by allowing the decomposition of both YAWL tasks into DECLARE workflows, and DECLARE tasks into YAWL nets. This way the workflow designer is not forced to make a binary choice between declarative and procedural workflows. Instead, an integration can be achieved, where parts of the workflow that need a high degree of flexibility are supported by declarative DECLARE models and parts of the processes that need centralized control of the system are supported by more procedural YAWL models. Note that YAWL also provides flexibility via the so-called Worklet Service. The Worklet Service (cf. Chap. 11) offers a different kind of flexibility that is complementary to DECLARE. As the Declare Service and the Worklet Service are embedded in a similar manner, it is possible to arbitrarily mix procedural YAWL models, declarative models, and worklets.

Exercises

Exercise 1. Tasks *bless*, *pray*, *curse*, and *become holy* are four tasks in the Religion process. Users of this process must obey two important rules. First, one should *pray* at least once. Second, every time one *curses*, one must eventually *pray* for forgiveness afterwards. Develop a model in DECLARE for this process using ConDec constraint templates.

Exercise 2. Consider the following four scenarios:

(a) $\langle become\ holy, curse, bless\rangle$
(b) $\langle pray, bless, pray\rangle$

(c) $\langle curse, bless, curse, bless, pray\rangle$
(d) $\langle bless, become\ holy\rangle$

For each scenario, create an instance of the Religion model developed in the previous exercise. Try to execute these four scenarios using DECLARE. Which traces were possible to execute and which were not? Describe why it was not possible to execute some of the given scenarios.

Exercise 3. Launch one instance of the Religion model described in the first exercise and monitor how states of the instance and constraints change while executing tasks $\langle bless, curse, bless, bless, curse, become\ holy, pray, curse, pray\rangle$.

Exercise 4. Launch one instance of the Religion model described in the first exercise and execute tasks ⟨*bless, curse, bless, bless, curse, holy, pray, curse, pray*⟩. Now try to change the model of this instance in a way that:

- The constraint specifying that *one should pray at least once* is removed, and
- Task *become holy* is removed.

Did DECLARE accept this dynamic change? Explain why or why not.

Exercise 5. Launch one instance of the Religion model described in the first exercise and execute tasks ⟨*bless, curse, bless, bless, curse, holy, pray, curse, pray*⟩. Now try to change the model of this instance in a way that a constraint specifying that *task become holy cannot be executed before task pray* is added. Did DECLARE accept this dynamic change? Explain why or why not.

Exercise 6. Develop a ConDec model in DECLARE for the process containing tasks *bless*, *pray*, *curse*, and *become holy*, where three rules must be followed:

(a) One should *become holy* at least once.
(b) Every time one *curses*, one must eventually *pray* for forgiveness afterwards.
(c) Tasks *become holy* and *curse* cannot be both executed in the same instance.

Verify this model in DECLARE. Does this model have errors? Explain why or why not.

Exercise 7. Develop a ConDec model in DECLARE for the process containing tasks *bless*, *pray*, *curse*, and *become holy*, where four rules must be followed:

(a) One should *become holy* at least once.
(b) Every time one *curses*, one must eventually *pray* for forgiveness afterwards.
(c) Tasks *become holy* and *curse* cannot be both executed in the same instance.
(d) One should *curse* at least once.

Verify this model in DECLARE. Does this model have errors? Explain why or why not.

Chapter Notes

This chapter described the Declare Service in YAWL. More details about the service and the DECLARE tool can be found in [19, 20, 192, 193, 195]. The combined use of YAWL, the Declare Service, and the Worklet Service, possible due to the service-oriented architecture of YAWL, is described in [5]. More details about the DECLARE tool in relation to the YAWL architecture can be found in [194].

Part VI
Positioning

Chapter 13
The Business Process Modeling Notation

Gero Decker, Remco Dijkman, Marlon Dumas, and Luciano García-Bañuelos

13.1 Introduction

Business processes may be analyzed and designed at different levels of abstraction. In this respect, it is common to distinguish between business process models intended for business analysis and improvement, and those intended for automation by means, for example, of a workflow engine such as YAWL. At the business analysis level, stakeholders focus on strategic and tactical issues such as cost, risks, resource utilization, and other nonfunctional aspects of process models. At the automation level, stakeholders are interested in making their models executable, which entails the need to provide detailed specifications of data types, data extraction and conversion steps, application bindings, resource allocation, and distribution policies, among others.

The requirements for process modeling notations at these two levels of abstraction are significantly different. This in turn has resulted in different languages being advocated at the business analysis level and at the execution level. Common languages used at the business analysis level include flowcharts, UML activity diagrams, the Business Process Modeling Notation (BPMN), and Event-driven Process Chains (EPCs). In this chapter, we consider BPMN, and specifically, version 1.0 of the BPMN standard specification.

In general, the main purpose of BPMN models is to facilitate communication between domain analysts and to support decision-making based on techniques such as cost analysis, scenario analysis, and simulation. However, BPMN models are also used as a basis for specifying software system requirements, and in such cases, they are handed over to software developers. This handover raises the following question: How can developers fully exploit BPMN process models produced by domain analysts?

One way to achieve a seamless handover is by transforming, either manually or automatically, business process models obtained from business analysis (e.g.,

G. Decker (✉)
Signavio, Berlin, Germany
e-mail: gero.decker@signavio.com

A.H.M. ter Hofstede et al. (eds.), *Modern Business Process Automation*,
DOI 10.1007/978-3-642-03121-2_13,

BPMN models) into "equivalent" process models defined in an executable language, such as YAWL. Accordingly, this chapter will discuss relationships between BPMN and YAWL, and will show how BPMN models can be transformed to YAWL nets.

The remainder of this chapter is organized as follows. Section 13.2 introduces the Business Process Modeling Notation (BPMN). Section 13.3 explains the transformation from BPMN models to YAWL nets, and Sect. 13.4 presents the tool support that was developed to automate the transformation.

13.2 BPMN

This section explains BPMN by example, distinguishing control-flow, data manipulation, resource management, and exception handling aspects. The presentation of BPMN is meant to be didactical rather than exhaustive.

13.2.1 Control-Flow

In BPMN, a process model is represented as a Business Process Diagram (BPD). With reference to the running example of the book, Fig. 13.1 depicts a BPD corresponding to a Freight In Transit process. This BPD consists of nodes of three types: *events* (represented as circles), *activities* (represented as rectangles), and *gateways* (represented as diamonds). Events denote things that happen at a particular point in time, activities denote work that needs to be performed, and gateways serve to route the flow of control along the branches of the BPD. Nodes are connected by means of directed edges called *sequence flows*. A sequence flow basically says that the flow of control can pass from the source node to the target node.

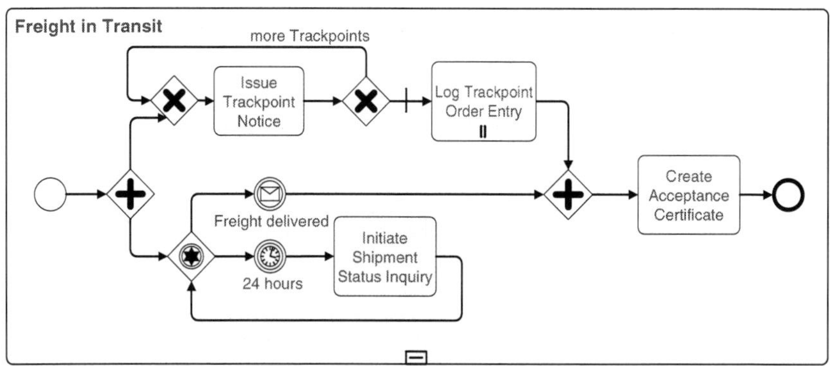

Fig. 13.1 Example of a business process model in BPMN

There are three types of gateways in BPMN: XOR gateways (represented by an "X"), AND gateways (represented by a "+"), and OR gateways (represented by an "O"). Moreover, a gateway is said to be a *split* gateway if it has multiple outgoing flows or a *join* gateway if it has multiple incoming flows. It may happen that a gateway is both a split and a join gateway, but this essentially means that there are two gateways in a single one. If we put together the distinction between the three types of gateways (XOR, AND, OR) and the distinction between split and join gateways, we obtain six different types of gateways: XOR-split, XOR-join, AND-split, AND-join, OR-split, and OR-join.[1] These six types of gateways are similar to the decorators in YAWL.

Like in YAWL, an XOR-split is a decision point. Based on the evaluation conditions, the flow of control will go to exactly one of the outgoing flows of the XOR-split. For example, Fig. 13.1 has one XOR-split gateway. One of the flows going out of this gateway is labeled with the condition *more trackpoints are needed*, while the other outgoing flow is the default flow, denoted by a stripe through the flow. The default flow is taken if the condition(s) attached to the other conditional flow(s) is(are) not fulfilled – in this example the default flow is taken if the condition *more trackpoints are needed* is not fulfilled. An XOR-join, on the other hand, merges two incoming branches into a single one. Figure 13.1 features an XOR-join just before the task *Issue Trackpoint Notice*, which merges two incoming paths.

An AND-split splits one thread of execution into two or more parallel threads. For example, the AND-split on the left-hand-side of Fig. 13.1 starts two threads in parallel. These threads then join in an AND-join just before task *Create Acceptance Certificate*. The AND-join is a synchronization point – it waits for both threads to complete.

The OR-split (which we do not exemplify) is a hybrid between the XOR-split and the AND-split, allowing a split into *any* number of outgoing flows instead of *one* (XOR-split) or *all* (AND-split). Similarly, the OR-join is a hybrid between an XOR-join and an AND-join, synchronizing all flows that have the control. These constructs correspond to those of OR-split and OR-join decorators in YAWL. Unlike YAWL, however, the exact semantics of the OR-join in BPMN is underspecified in the standard specification. There is also another type of gateway in BPMN called "complex gateway," which is not discussed in this chapter.

Finally, BPMN introduces a type of split gateway that corresponds to the "Deferred Choice" workflow pattern. This is called the "event-based exclusive gateway." The event-based exclusive gateway is similar to the XOR-split (which is also called data-based exclusive gateway in BPMN). However, instead of the choice being driven by conditions (which are evaluated based on data), the choice is driven by two or more events, such that the choice for an outgoing flow is made when the event on that outgoing flow occurs. Concretely, an event-based exclusive gateway is

[1] The BPMN specification uses alternative terms: XOR-splits are *exclusive decision gateways*, XOR-joins are *exclusive merge gateways*, OR-splits are *inclusive decision gateways*, OR-joins are *inclusive merge gateways*, AND-splits are *parallel forking gateways*, and AND-joins are *parallel joining gateways*. Here, we adopt a simpler and more uniform terminology.

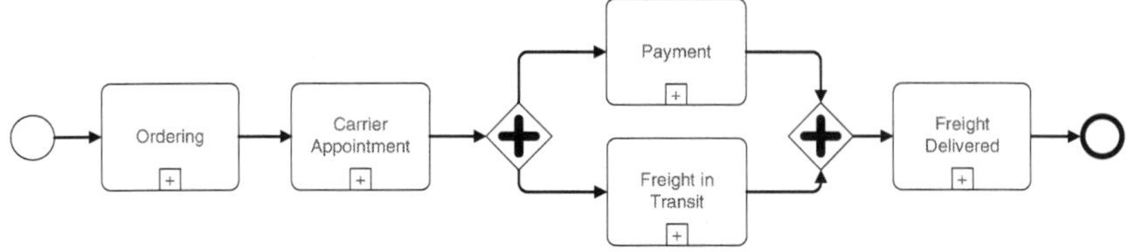

Fig. 13.2 Example of a business process model in BPMN

represented using a star symbol. Figure 13.1 features an event-based exclusive gateway. When the execution of the process arrives at this point (in other words, when a token arrives at this gateway), the execution of that thread stops until either the message event or the timer event occur. Whichever occurs first will determine which way the execution will proceed. If the timer event occurs first, a shipment status inquiry is initiated and the execution flow comes back to the event-based exclusive gateway. If the message signaling the freight delivery is received first, the execution flow proceeds along the sequence flow that leads to the AND-join. Note that in Fig. 13.1, the event-based exclusive gateway is also a join, as it has two incoming flows. A token may arrive to this gateway from either of these incoming flows.

Events are classified along two criteria. First, an event can be a start event, an end event, or an intermediate event. The sample BPD contains a single start (leftmost element) and a single end event (rightmost element), and two intermediate events in the middle of the BPD. Second, events are classified according to their trigger. In this chapter, we consider four types of triggers: messages, timers, conditions, and errors. A message event is represented using an envelope icon. Such an event is triggered by the receipt of a message, and can also be used to represent the event of sending a message. Figure 13.1 shows a message event that is triggered by the receipt of a message (presumably from a system located in the warehouse) representing the arrival of the freight. A timer event is represented using a clock icon. Figure 13.1 shows a timer event that is triggered when 24 h have elapsed. A rule event is represented by an icon corresponding to a sheet of lined paper. It has a condition attached to it and the event is triggered when this condition becomes true. An error event is represented by a lightning icon. It is triggered when an error occurs, but it can also represent the event of throwing an error.[2]

The process model shown in Fig. 13.1 is "flat," meaning that all the activities in the model correspond to atomic units of work – also called *atomic tasks* in BPMN. BPMN also supports activities that correspond to the execution of entire *subprocesses* – like composite tasks in YAWL. Figure 13.2 shows a fragment of a BPD consisting of a number of subprocesses, one of which corresponds to the Freight In Transit process shown in Fig. 13.1. Subprocesses are distinguished from atomic tasks using the + marker.

[2] Specifically, an *end error event* corresponds to throwing an error, while an *intermediate error event* corresponds to catching an error.

There are several types of subprocesses in BPMN, including *embedded subprocess* and *independent subprocess*. An embedded subprocess is part of the parent process, while an independent subprocess exists independently of the parent process. This chapter does not go into the details of these nuances and instead treats both embedded and independent subprocesses equally.

13.2.2 Data

There are two ways in which the manipulation of data by processes and tasks can be specified in BPMN: by means of *data objects* and by means of *properties*.

Data objects represent documents, data, or other objects that are used in the process. They have a name and, optionally, a state. Data objects can be associated with activities, showing that they are input for (by an arrow pointing to that task) or output of that activity (by an arrow originating in that task). They can also be associated with flows, showing that they are transferred from one activity to another via that flow. Figure 13.3 illustrates these cases, annotating the business process diagram from Fig. 13.1 with data objects. In the BPMN standard, data objects are used only informally for documentation purposes. According to the BPMN specification, they provide additional information for the reader of the BPMN Diagram, but do not directly affect the execution of the process. Therefore, data objects do not play any role when mapping BPMN to an executable workflow language. Indeed, in the mapping from BPMN to the Business Process Execution Language (BPEL) presented in an appendix of the BPMN specification, data objects are simply not mapped.

Properties in BPMN can roughly be equated to variables in YAWL. They have a name and a type and can be associated with (sub-)processes and tasks (and data

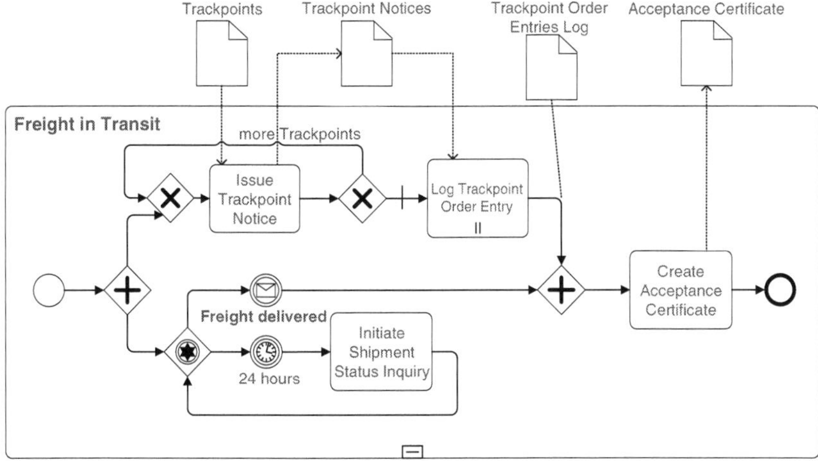

Fig. 13.3 Example of a business process model with data in BPMN

objects). In BPMN, standard properties are defined more precisely than data objects and they are proposed to be used for the purpose of defining transformations from BPMN to executable languages. Indeed, in the mapping from BPMN to BPEL presented in the BPMN specification, they are mapped to the data variables in BPEL. However, the drawback of properties is that they are not visible in the graphical notation of BPMN. To counter these drawbacks, properties and data objects can be used in combination: data objects to document the data aspect of the process and properties to precisely define the data variables that are used in the process. This is the approach we follow below.

BPMN leaves the modeler with much freedom to model data and data manipulation. In particular, it leaves the modeler with the freedom to choose the (textual) language in which property types and data manipulation are modeled. A benefit of this is that BPMN does not favor a particular workflow language for implementing BPMN models. Another benefit is that the modeler can now use BPMN to draw a model, without having to worry about the complexities of a data modeling language such as XQuery. However, a drawback of having freedom to choose a data modeling language is that it precludes a general mapping from BPMN to a workflow language, because such a general mapping would require the definition of a mapping from *any* data modeling language (including natural language) to the data modeling language used by the workflow system. Therefore, we will impose constraints in this chapter on what BPMN models we consider to be *well-formed* for mapping to YAWL.

The YAWL-constrained version of BPMN that we consider uses XQuery to model data and data manipulation (meaning that both the *QueryLanguage* and *ExpressionLanguage* attributes of the BPMN diagram must be set to *XQuery*). Properties of activities and processes must be used to represent data. Data types of properties must be specified in XML Schema. Assignments must be attached to activities in order to represent data manipulation, and conditions on sequence flows must be used to represent conditional passing of control. Assignments and conditions must also be expressed in XQuery. Conditions with *"ConditionType = None"* are not accepted. Figure 13.4 shows an example, in which the details of the data objects from Fig. 13.3 are added. Note that, while this figure uses XML to specify those details, this is done here only for the example, because the concrete syntax for specifying properties, conditions, and assignments is not defined in BPMN. Figure 13.4 defines that the process Freight in Transit has the property *trackpoints* of type *Trackpoints* as well as some other properties. The *trackpoints* property corresponds to the *Trackpoints* data object that is shown in Fig. 13.3. However, it is defined in more detail (assuming that the structure of the *Trackpoints* type is defined in an XML Schema elsewhere). Similarly, the task *Issue Trackpoint Notice* has some properties. This task also has some assignments. One assignment assigns the value of the *trackpoints* property of the Freight in Transit process to the *trackpoints* property of the task, when the task is started. Another assignment adds the *trackpointNotice* that is generated by the task (if one is generated at all) to the *trackpointNotices* of the process. Finally, Fig. 13.4 refines the condition *more Trackpoints* on one of the sequence flows from Fig. 13.3.

```
<Process Id='p1' Name='Freight in Transit'>
  <Properties>
    <Property Name='trackpoints' Type='Trackpoints'/>
    <Property Name='currentTrackpoint' Type='Trackpoint'/>
    <Property Name='trackpointNotices' Type='TrackpointNotices'/>
...

<Task Id='a1' Name='Issue Trackpoint Notice'>
  <Properties>
    <Property Name='trackpoints' Type='Trackpoints'/>
    <Property Name='trackpointNotice' Type='TrackpointNotice'/>
    <Property Name='skipTrackpoint' Type='xsd:boolean'/>
    ...
  </Properties>
  <Assignments>
    <Assignment AssignTime='Start' To='trackpoints' From='{p1.trackpoints}'/>
    <Assignment AssignTime='End' To='p1.trackpointNotices'
      From="{p1.trackpointNotices}
        {if (skipTrackpoint/@value='false') then (trackpointNotice)}"
...

<SequenceFlow Source='g1' Target='g2' Condition='Expression'
  ConditionExpression='{p1.currentTrackpoint!=""}'/>
```

Fig. 13.4 Example of detailed data specification

In the YAWL-constrained version of BPMN, properties of a subprocess invocation *must* be identical to the properties of the process that is invoked. The *InputPropertyMaps* and *OutputPropertyMaps* of the subprocess invocation are ignored. Instead, we map properties of the invoking activity to identically named properties of the invoked process and vice-versa. For example, suppose a process loan application invokes a process objection, which has the property *application number*. The activity that invokes the process must then also have the property *application number*. When performing the subprocess invocation, this activity assigns the value of its *application number* property to the *application number* property of the objection subprocess. When the subprocess returns, the value of the *application number* property of the subprocess is then assigned to the *application number* property of the activity.

13.2.3 Resources

BPMN allows modelers to indicate the resources that may perform a given task. Typically, "lanes" are used for that purpose, although BPMN does not prescribe that lanes *must* necessarily be used for this purpose. A lane is a box in which graphical objects can be placed. The placement of a task inside a lane can represent that the resource represented by the lane performs the task. Lanes can represent individual actors, roles, groups, or any type of resource that the designer requires. Alternatively, the *performers* property of "user" or "manual" tasks can be used to indicate which resources may perform a task. In the remainder, we assume that lanes have been used to represent the resources that perform a task.

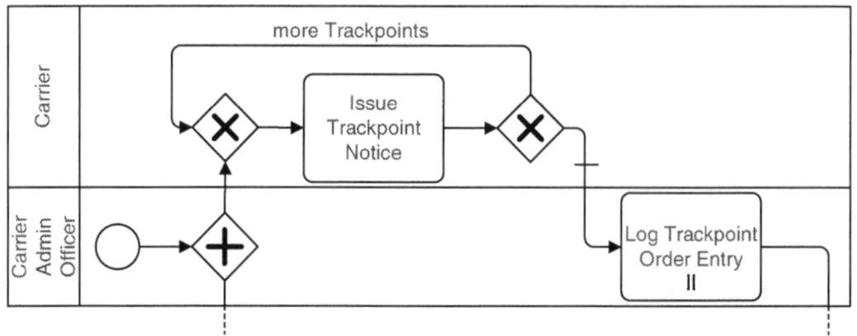

Fig. 13.5 Example of a business process model with lanes

Clearly, BPMN is rather limited when it comes to representing resources. Many of the constructs that are possible in YAWL are not possible in BPMN. Examples of constructs for which YAWL has specific support, but BPMN does not, include the following: representing that a resource can delegate a task to another resource; representing that a check must be performed by another individual than the individual who performed the previous task (the "separation of duties" resource pattern); and representing that a task should be performed by the individual who has done the most of these tasks in the past.

Figure 13.5 shows an extension of the example from Fig. 13.1, in which lanes have been added to represent the roles that are authorized to perform tasks. It shows that the *Carrier* is authorized to perform *Issue Trackpoint Notice* and that the *Carrier Admin Officer* is authorized to perform *Log Trackpoint Order Entry*.

13.2.4 Exceptions

BPMN supports exceptions by allowing intermediate events to be attached to tasks and subprocesses. An intermediate event is attached to a task or a subprocess represents that the task or the subprocess (and everything in it) can be interrupted by the occurrence of the intermediate event. Upon occurrence of the intermediate event, the task or subprocess stops executing and the flow continues along the sequence flow that leaves the intermediate event. The "type" of the intermediate event specifies what kind of exception it reacts to. The timer, message, error, and rule intermediate events discussed previously can be attached to tasks and subprocesses to represent an exception. An error event on a subprocess can have a counterpart (an error event with the same name) in that subprocess. When this is the case, the error event "catches" the error when it is "thrown" in the subprocess.

In addition to these types, "cancel" and "compensate" events can be attached to tasks and subprocesses. These events correspond to cancelation and compensation of transactions. However, in BPMN 1.0, their semantics is underspecified. Therefore, we do not discuss them here.

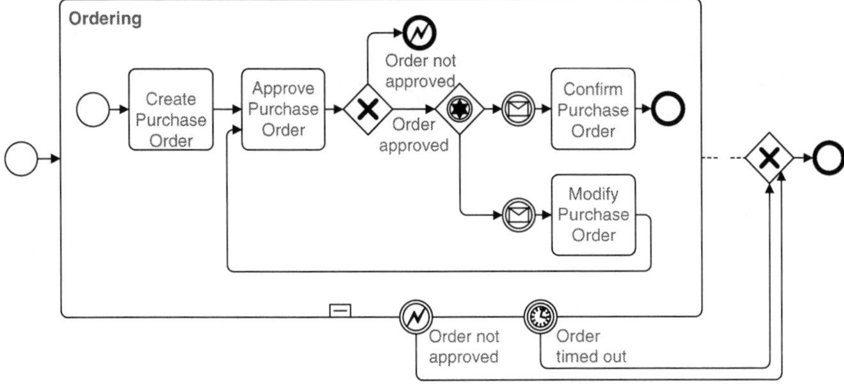

Fig. 13.6 Example of a business process model with exceptions

Figure 13.6 shows a BPMN model that contains exceptions. After the *approve purchase order*, there is a choice not to approve the order. When this choice is made, an error event *order not approved* is generated, upon which the subprocess continues along the flow that leaves the corresponding error exception event. The error event triggers the *order not approved* error intermediate event that is attached to the subprocess activity, which leads to the completion of the main process (via the XOR-join gateway). Similarly, the main process completes upon the occurrence of the *order timed out* timer exception event.

13.3 Mapping BPMN to YAWL

This section shows how the BPMN concepts and constructs introduced in the previous section can be mapped into YAWL. We first discuss the mapping of control-flow aspects before moving into the data and the resource perspective.

13.3.1 Control-Flow

At their core, BPMN and YAWL share many common concepts. For example, the concept of task in BPMN matches the concept of task in YAWL, the concept of gateway in BPMN matches the concept of decorator in YAWL, and the concept of flow in BPMN matches the same concept in YAWL.

For illustration, Fig. 13.7 shows side-by-side the BPMN diagram previously shown in Fig. 13.1 and the corresponding YAWL net. Note that this YAWL net differs slightly from the Freight In Transit YAWL net shown in other chapters. Specifically, the decision on whether to issue more trackpoints or not is captured using an XOR-split decorator rather than using an explicit condition. This slight

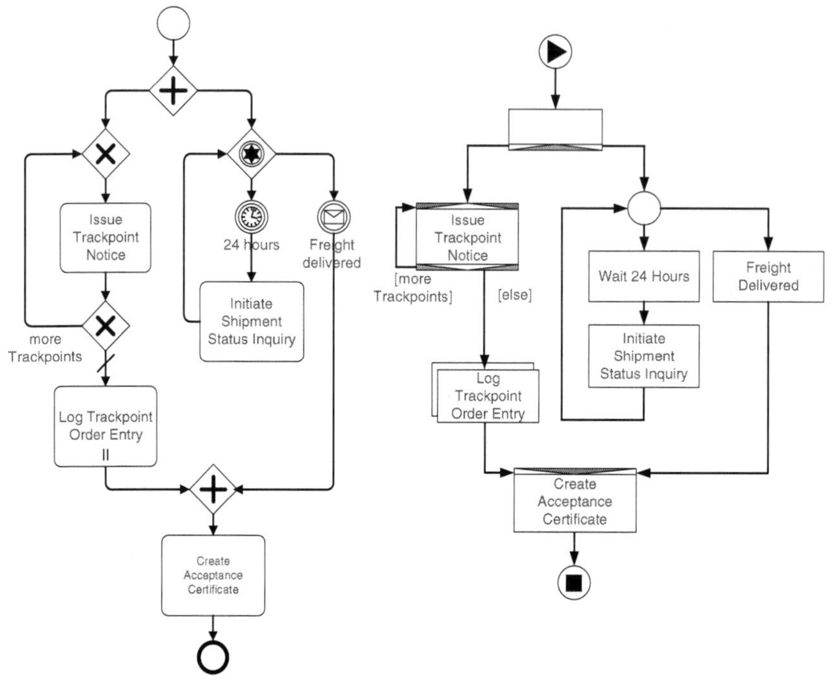

Fig. 13.7 Side-by-side BPMN diagram and corresponding YAWL net

change has been introduced so that the same example can be used to illustrate the mapping of data-driven and of event-based decision gateways.

This example highlights the one-to-one correspondences between BPMN and YAWL. Each task and each event in the BPMN diagram has a corresponding task in the YAWL net. Similarly, each gateway is mapped to a decorator. For example, the XOR-split and XOR-join gateways surrounding the *Issue Trackpoint Notice* task become XOR-split and XOR-merge decorators in the corresponding YAWL net, while the AND-join just before task *Create Acceptance Certificate* becomes an AND-join decorator attached to the corresponding task in the YAWL net.

In some cases, empty tasks need to be inserted in the YAWL net for the purpose of holding decorators that cannot be attached to other nonempty tasks. For example, the initial AND-split in the BPMN diagram is mapped to an AND-split decorator attached to an empty YAWL task. Indeed, as this AND-split decorator is needed right at the start of the net, it is not possible to attach it to any of the nonempty tasks in the net. Another scenario where empty tasks need to be introduced in a YAWL net in order to hold a decorator is when multiple gateways in the original BPMN diagram are chained together. This scenario is illustrated in Fig. 13.8, which shows a fragment of a BPMN diagram and the corresponding YAWL fragment. The YAWL fragment contains an empty task with an AND-split decorator. This AND-split decorator cannot be attached to the previous task (*Prepare Transportation Quote*) because this previous task already has an XOR-split decorator.

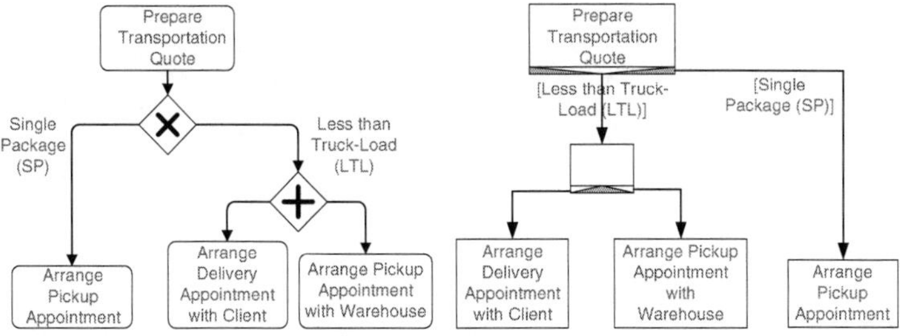

Fig. 13.8 AND-split decorator attached to an empty task

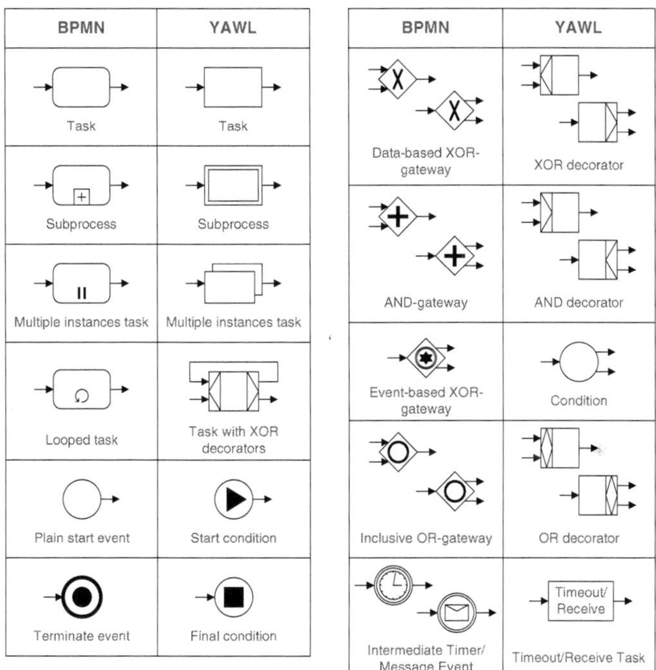

Fig. 13.9 Mapping of tasks, events, and gateways

Figure 13.9 provides an overview of the mapping between tasks, events, and gateways in BPMN and their corresponding constructs in YAWL. Note that although every BPMN control-flow construct shown in this figure can be mapped to a corresponding construct in YAWL, the reverse is not always true. Specifically, an event-based exclusive gateway in BPMN can always be mapped into an explicit condition in YAWL. But in the reverse direction, there are situations where an explicit condition in YAWL cannot be mapped to an event-based exclusive gateway in BPMN. The reason for this is that an event-based exclusive gateway in BPMN must be followed by either an event or a task of type "receive." It cannot be followed immediately by a gateway nor by a task of another type. This restriction does not

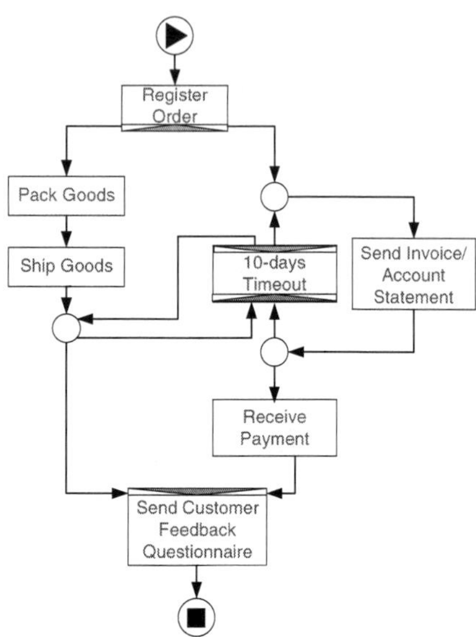

Fig. 13.10 YAWL net with explicit conditions that cannot be mapped to event-based gateways

exist in YAWL. To illustrate the impact of this restriction in the BPMN notation, consider the diagram in Fig. 13.10. Here, task *10-days timeout* should only fire if there is a token in each of its input conditions – meaning that the 10-days timer is only armed once the goods have been shipped and the invoice/account statement has been sent. Ten days after the latest of these two tasks has completed, a new invoice/account statement is sent to the customer as a reminder that the payment is due. Hence, there is an AND-join decorator attached to this timeout task. Because of the above restriction, the two explicit conditions preceding the timer (marked in bold in the figure) cannot be directly mapped to event-based gateways in BPMN.

Indeed, if one tried to map these explicit conditions into event-based gateways, a problem would arise, because the resulting event-based gateway would be followed immediately by an AND-join gateway. This configuration is forbidden in BPMN, because event-based gateways can only be immediately followed by intermediate events. Fundamentally, the problem stems from the fact that BPMN does not offer a concept of "state", which would allow one to capture "the state after the goods have been shipped" or "the state after the invoice has been sent." This state can of course be captured in the form of variables, but then it is not directly visible in the diagram.

Events

In Fig. 13.9, BPMN message and timer events are mapped to YAWL tasks. BPMN message events are mapped to YAWL tasks linked to a YAWL Custom Service that sends and/or receives messages as required. Meanwhile, a BPMN timer event is mapped to a YAWL empty task with an attached timer.[3]

[3] Specifically, a YAWL "on-enablement" timer needs to be used here as the timer is armed as soon as the task becomes enabled.

This way of mapping events works well for intermediate message/timer events and for final message/timer events. But it does not work for start events. Start message/timer events in BPMN have the effect of creating new process instances. For example, a start message event in BPMN means that a new instance of a process is created every time a message (of the designated type) is received. This feature is called implicit instantiation – because a process instance is created when a given type of event happens. Meanwhile, YAWL supports *explicit instantiation*, meaning that to create an instance of a YAWL net, an external application needs to send a "create instance" message to the process engine with the corresponding net identifier and the required input data.

From an application development perspective, the above difference is rather minor. It mainly implies that in YAWL, one has to follow a strict separation between the instantiation of a net and the net itself. For example, if we wanted to create an instance of a YAWL net every time an invoice is received, or every Saturday at 2 am, we would need to develop an application that receives the invoice message or that is triggered every Saturday at 2 am. This application would then send a "create instance" message to the Engine to start the execution of the corresponding YAWL net.

A BPMN diagram may have multiple start events, in which case at least one of these events must have a trigger (e.g., message/timer) event. This situation cannot be captured directly in YAWL because YAWL supports only plain start events (i.e., explicit instantiation). As there is only one way of creating an instance of a YAWL net, it is natural that YAWL nets only have one start (input) condition.

Similarly, a BPMN diagram may have multiple end events, while a YAWL net always has a single final (output) condition. Also, the termination policy of end events in YAWL is such that when a token reaches the output condition, the process instance immediately completes, even if there are still active or enabled tasks. Meanwhile, if the execution of a branch in a BPMN diagram reaches an end event, but there are other enabled or active tasks, the process instance does not terminate immediately (implicit termination policy). For BPMN diagrams that are 1-safe, meaning that no two tokens can accumulate in the same flow, this is not a limitation. Such models can be translated into YAWL by mapping all the final flows to YAWL arcs that lead to a single empty task with an OR-join. This empty task with an OR-join is then connected to the single output condition in YAWL. The OR-join waits for at least one token in each branch (except for branches that need not wait for tokens) before producing one token in the output condition. This transformation is illustrated in Fig. 13.11.

Multiple Instance Tasks

As shown in Fig. 13.9, a multiple instance task in BPMN is mapped into a multiple instance task in YAWL – and similarly, a multiple instance subprocess invocation in BPMN is mapped to a multiple instance composite task in YAWL. In BPMN, any number of instances of a multiple instance activity may be started (without

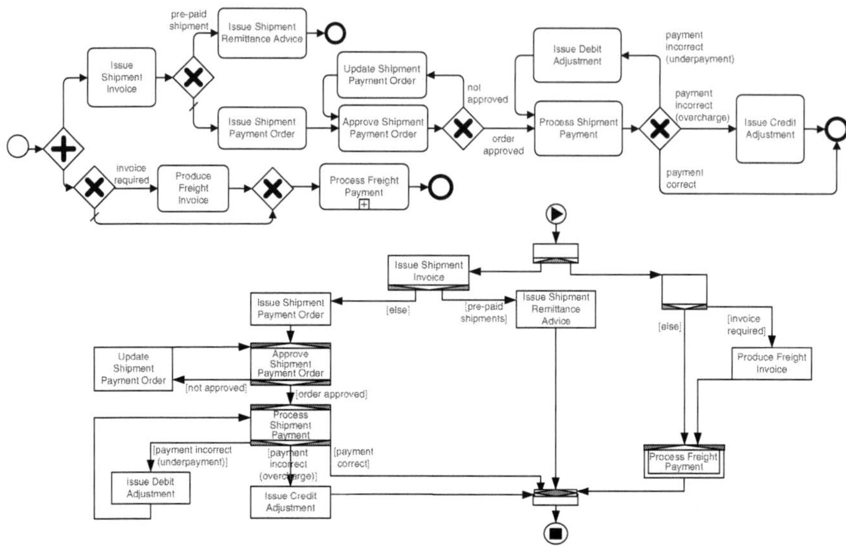

Fig. 13.11 BPMN diagram with multiple end events and corresponding YAWL net

any fixed upper bound), and this number is determined immediately when the multiple instance activity is started. In other words, when mapping a BPMN multiple instance activity to YAWL, the *Minimum Instances* attribute is set to *1*, the *Maximum Instances* attribute bound is set to *unbounded*, and the *Instance Creation Type* is set to *static*. The number of instances of a BPMN multiple instance activity is determined by an attribute called *MI_Condition*. This attribute corresponds to the splitter query attached to YAWL multiple instance tasks. However, the splitter query not only determines how many instances need to be started, but also what data should be given as input to each of these instances.

Furthermore, BPMN multiple instance activities have an attribute *MI_Flow-Condition* that determines how many instances should have completed before the flow of control is passed on to the following task(s) or event(s) in the process model. This attribute corresponds to the *Continuation Threshold* in YAWL.

13.3.2 Data

Table 13.1 shows the details of mapping of data concepts from BPMN to YAWL. The main assumption of this mapping is that BPMN properties are used to represent data, while BPMN assignments are used to capture data manipulation. Accordingly, properties are mapped to YAWL variables and parameters while assignments are mapped to inbound and outbound parameter mappings in YAWL.

The mapping from BPMN properties to YAWL variables is straightforward. A property's name maps to a variable's name and a property's type maps to a

Table 13.1 Mapping data from BPMN to YAWL

BPMN	YAWL
Property	Variable
Process property reference	Process variable reference
`<process name>.<property name>`	`<process decomposition name>/<variable name>`
Task property reference	Task variable reference
`<process name>.<task name>.<property name>`	`<task decomposition name>/<variable name>`
Assignment (Assigntime = Start)	Input parameter
Assignment (Assigntime = End)	Output parameter
Condition (ConditionType = Expression)	Predicate
Condition (ConditionType = Default)	Last predicate is "true()"
Property visible in embedded subprocess	Variable of composite task
Property visible in independent subprocess	Variable of composite task

variable's type. This principle applies whether the property is defined at the level of a process model or at the level of an individual task. For example, properties *trackpoints*, *currentTrackpoint*, and *trackpointNoticesdefined* at the top of Fig. 13.4 are defined at the level of the process model. They will therefore be mapped into variables attached to the corresponding YAWL net. Meanwhile, in the same figure, three properties (*trackpoints*, *trackpointNotice*, and *skipTrackpoint*) are defined under the task *Issue Trackpoint Notice*. These properties are mapped to variables defined at the level of the corresponding YAWL task. A subtle difference here is that, in YAWL, variables defined at the level of a task are grouped inside a decomposition, while in BPMN, properties are assigned directly to a task. Therefore, a decomposition needs to be defined under task *Issue Trackpoint Notice* and the above three variables must be attached to this decomposition.

BPMN assignments that are executed when a task starts (i.e. that have *"Assigntime = Start"*) assign values from a process property to a task property. Such assignments map to YAWL input parameters. The *from* part of the assignment is an XQuery expression that refers to properties of the process, and the *to* part is a reference to a property of the task. In the *from* and the *to* parts, references to BPMN properties must be mapped to references to YAWL variables. Assignments that are executed when a task completes (i.e., that have *"Assigntime = End"*) assign values from a task property to a process property. Such assignments map to YAWL output parameters in a similar manner. For example, Fig. 13.4 contains two assignments attached to task *Issue Trackpoint Notice*. The *AssignTime* of the first assignment is *Start*. Accordingly, this assignment is mapped into an input parameter in the corresponding YAWL task decomposition. This input parameter copies the value of the net variable *trackpoints* into the task variable with the same name. Meanwhile, the *AssignTime* of the second assignment is *End*. Hence, this assignment is mapped into an output parameter in YAWL.

A BPMN condition on a sequence flow is mapped to a YAWL predicate on a flow. There are two possible cases: either the condition is represented as an (XQuery) expression, or the condition is set to *Default*, meaning that the flow is the default flow. Expression conditions are directly mapped to predicates subject to differences in the naming conventions used. Meanwhile, a default condition is transformed into the predicate *"true()"* and listed as the last predicate of a "split," meaning that it is checked (and fulfilled, because it is *"true()"*) when no other predicate is fulfilled. Therefore, it has the same behavior as the BPMN default condition.

Tasks in a BPMN embedded process can access the properties of the process that contains the embedded process. However, in YAWL, subnets cannot access data from a parent process. Therefore, a transformation must be performed to make this possible. An embedded BPMN subprocess is mapped to a YAWL composite task. To give the tasks within the composite task access to the properties of the parent processes, the composite task is given all the variables that the parent process has. These variables are assigned the values of the corresponding process variables by input parameters when the composite task is started. The values of the task variables are assigned to the corresponding net variables when the composite task is completed. In this way, the tasks within the composite task have access to the net variables.

Properties of a BPMN independent subprocess map to variables of the composite task to which the subprocess is mapped. As per the assumption introduced in Sect. 13.2, the properties of a task that invokes a subprocess and those of the subprocess itself are identical. As this is also the case for a YAWL composite task that invokes a subnet and the invoked subnet, no special mapping rules are necessary to map properties of independent subprocesses.

13.3.3 Resources

As we used the assignment of tasks to lanes to represent the assignment of resources, we map the assignment of a task to a lane to the assignment of a task to the corresponding resource. This means that BPMN is not used to model the resources themselves, but only to model assignment of tasks to resources. Hence, BPMN is used in the same way as the YAWL Editor currently is. The YAWL Editor is also only used to model resource assignment, the resources themselves are defined in the YAWL administration console.

13.3.4 Exceptions

Exceptions in BPMN can be classified into two categories:

- Exceptions generated because the execution of a subprocess reaches a specific element that throws the event (specifically a node of type "end error event")
- Exceptions that may occur at any point during the execution of an activity due to an external factor (a timeout, the receipt of a message, or a rule).

The mapping of an error event generated by a subprocess is shown in Fig. 13.12(i). The end error event node (which raises the exception) is mapped to a YAWL task

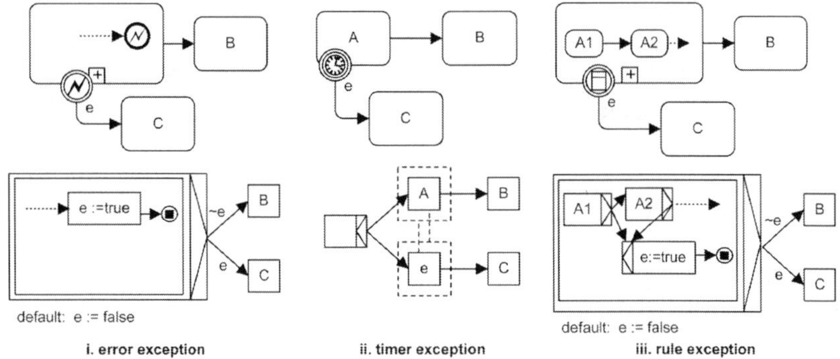

Fig. 13.12 Mapping of exceptions

that sets a boolean variable (say *e*) to *True*, followed by an output condition. In YAWL, when a token reaches the output condition of the (sub-)net, the entire execution of the subnet is stopped. This matches the behavior of the end error event in BPMN, which effectively stops the execution of all activities in the encompassing (sub-)process instance. The error variable *e* is initially set to *False*. Setting it to *True* signifies that an exception has occurred. The value of the error variable is passed to the parent net. Recall that a subprocess in BPMN is mapped to a composite task in YAWL. Accordingly, a subprocess in BPMN with an attached error event is mapped to a composite task in YAWL with an XOR-split decorator. This XOR-split tests whether the error variable is set to *True* or *False*. If it is set to *True*, it means that the error event has occurred. Thus, the exception flow should be taken (in the example at hand, task *C* should be executed). Meanwhile, if the error variable is set to *False*, it means that the error event has not occurred. Thus, the normal flow should be taken (in the example, task *B* should be executed).

The mapping of the second type of error events is illustrated in Fig. 13.12(ii) and 13.12(iii). Figure 13.12(ii) illustrates the mapping of a timer event attached to an activity, regardless of whether this activity is an atomic activity or a subprocess invocation. A timer event in BPMN maps to a task with an attached timer in YAWL, cf. Task *e* in Fig. 13.12(ii). Meanwhile, the BPMN activity to which the timer event is attached (task *A* in the figure) is also mapped to a YAWL task (which may be simple or composite). These two YAWL tasks are enabled in parallel, but they cancel one another, that is, the firing of task *e* cancels task *A* and vice-versa. Thus, the expiration of the timer causes task *A* to be canceled and the corresponding "exception path" to be taken. Reciprocally, completion of task *A* causes the timer task to become irrelevant and the "normal path" to be pursued. A message event attached to a BPMN activity is mapped to YAWL in a similar way, except that the main task (i.e., task *A* in the example) is mapped to a task that listens for the corresponding message type.

Figure 13.12(iii) shows a mapping of a rule event attached to a subprocess. A rule is evaluated immediately after completion of each activity in the subprocess. Specifically, every task in the generated YAWL subnet has an XOR-split attached to it with two branches: one corresponds to the rule evaluating to true, the other corresponds to the rule evaluating to false. If the rule evaluates to true, a variable corresponding to the rule event is set to true (variable *e* in the figure), and the subprocess instance is terminated. On the other hand, if the rule evaluates to false, the normal flow of control is followed. The rule is also evaluated when the subprocess starts.

Assuming that the rule event depends only on data within the scope of the subprocess, this mapping ensures that the satisfaction of the rule will be detected as early as possible. On the other hand, this mapping can lead to spaghetti-like models as an XOR-split and an additional arc need to be inserted for every activity in the subprocess. Also, this mapping corresponds to one interpretation of the BPMN specification. The BPMN specification does not state exactly when a rule event should be raised. One possible interpretation of BPMN is that an attached rule event is raised when changing from a state of nonsatisfaction of the condition to a state of satisfaction of the condition during the execution of the subprocess. Another interpretation is that the event is raised whenever the activity to which it is attached is enabled or running, and the rule evaluates to true (in particular, the event is raised if the rule evaluates to true when the activity is enabled or started). Thus, what we present is one possible interpretation of the BPMN specification rather than a definite mapping of rule events to YAWL.

An alternative mapping of attached rule events could be obtained by introducing a task within the subprocess that evaluates the rule periodically (e.g., every 5 min). However, this introduces a potentially heavy execution overhead and does not ensure that the rule event is raised as soon as the condition becomes true.

Yet another approach to mapping attached events from BPMN to YAWL is by making use of the concept of exlets, as described in Chaps. 5 and 11.

13.4 Tool Support

The BPMN2YAWL tool supports the transformation described earlier. It is implemented as an Eclipse plug-in. The tool itself and installation instructions are available for download through the YAWL web site, together with the other components of the YAWL System.

The tool takes as input BPMN diagrams produced by the STP BPMN Modeler.[4] The models produced by this tool are split into two files: one contains a representation of the model in XMI (XML Interchange Format), while the other contains layout information (e.g., coordinates of graphical elements). Once installed, the BPMN2YAWL plug-in provides a menu item that transforms the XMI file (.bpmn file). It then produces a YAWL Engine file that does not contain layout information. This file can be imported into the YAWL Editor, which applies an automated layout algorithm.

[4] The STP BPMN Modeler is available at www.eclipse.org/stp/bpmn/

The STP BPMN modeler does not support certain features of BPMN. Specifically, it does not support the markers and properties for multiinstance activities and ad-hoc activities. To overcome this limitation, the BPMN2YAWL tool is able to detect special types of text annotations: one for multiinstance activities and one for ad hoc activities. The text annotations for multiinstance activities include parameters for specifying minimum and maximum amount of instances to be started, and number of instances that need to complete before proceeding (also called the *threshold*).

In addition to these limitations imposed by the STP BPMN modeler, the transformation tool is not able to map data aspects (properties) nor resource allocation aspect (lanes and pools) that are described in this chapter. It also does not support features of BPMN that are not relevant for execution purposes such as link events, groups, associations, and text annotations (other than the special text annotations for multiinstance and ad hoc activities mentioned above). Also, as the purpose of the tool is to transform one BPMN diagram at a time (as opposed to a collection of interacting BPMN diagrams comprising multiple pools), it does not take into account message flows. Finally, the tool does not deal with advanced control-features that are not covered in this chapter, such as complex gateways, compensation subprocesses, and "compensate" events.

It is planned that the tool will in future be complemented by a reverse transformation or a model co-evolution mechanism, in order to help developers who modify a YAWL net produced by the BPMN2YAWL tool to propagate these changes back to the BPMN diagram. This feature would allow business analysts working with BPMN models and developers working with YAWL models to maintain their models synchronized.

In addition to assisting in the transformation from high-level BPMN diagrams to executable workflows in YAWL, the BPMN2YAWL tool, used in conjunction with the WofYAWL component of the YAWL System, enables the analysis of BPMN diagrams in order to detect deadlocks and other forms of unsafe behavior.

13.5 Summary

This chapter presents the Business Process Modeling Notation (BPMN). The BPMN is a notation for modeling business processes at a more conceptual level as opposed to a level at which the business process models can be executed directly by a workflow engine. At the conceptual level, business process models are used, among other things, to facilitate communication between stakeholders, to analyze business problems by means of analysis or simulation of the models, and to serve as specifications towards software development.

This chapter also presents the relation of BPMN process models to YAWL workflow models. The relation is described in such a way that automated transformation of BPMN process models into YAWL models is possible. This is especially useful when using BPMN models as a specification towards software development, because it allows partial automation of the software development process to the extent to which the transformation can transform BPMN models into YAWL models. The transformation outlined in this paper has been implemented in a freely available tool, namely BPMN2YAWL.

The chapter relies on version 1.0 of the BPMN standard specification. At the time of writing this chapter, version 2 of the BPMN specification is being finalized. This new version will bring some significant changes with respect to the previous one. In particular, a new type of event will be introduced, namely escalation. An escalation event is raised by a user in order to signal that an exception has occurred and that appropriate actions need to be taken, for example, canceling the task or initiating another task in parallel. This concept is similar to the concept of exlet in YAWL. Also, a stricter approach to define data attached to tasks will be introduced in BPMN 2.0, inspired in great part by the data definition approach used in BPEL. Such changes will have some impact on the mapping between BPMN and YAWL. Still, most of the mapping principles outlined in this chapter will remain applicable.

Exercises

Exercise 1. Transform the model below into YAWL.

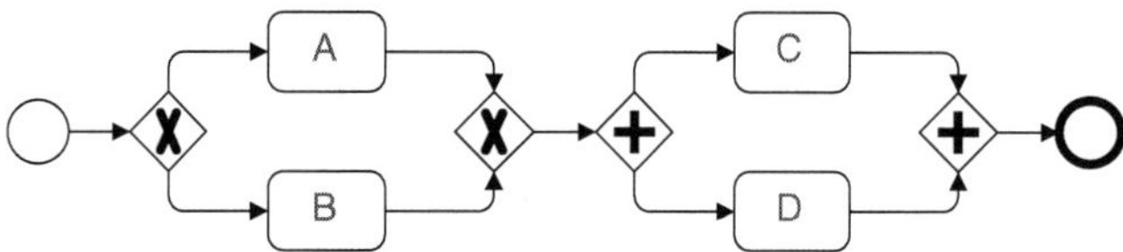

Exercise 2. Transform the model below into YAWL.

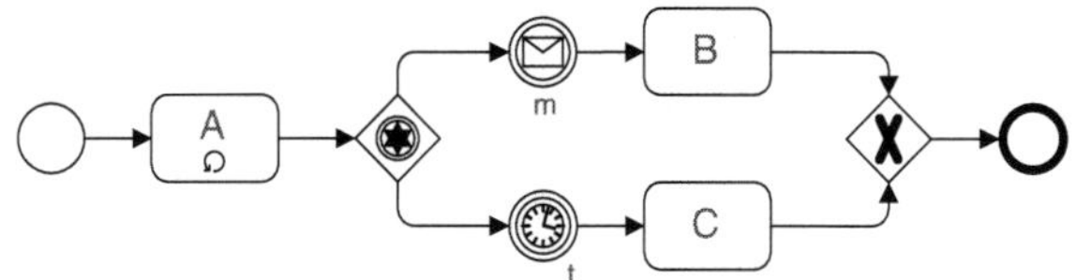

Exercise 3. Transform the model below into YAWL.

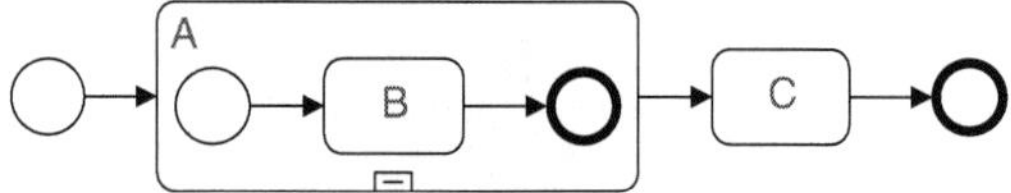

Exercise 4. Transform the model below into YAWL.

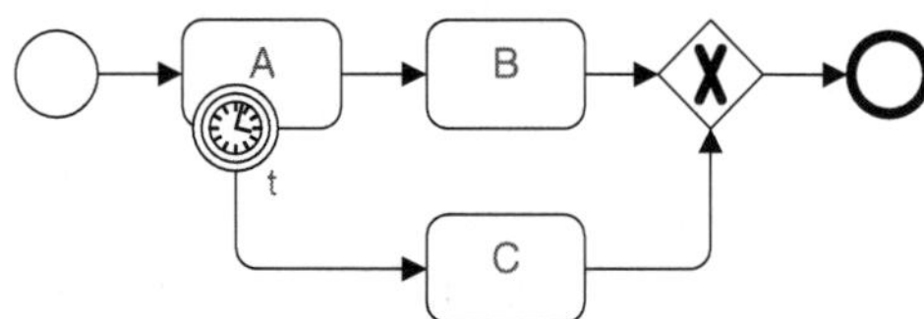

Exercise 5. Transform the model below into YAWL.

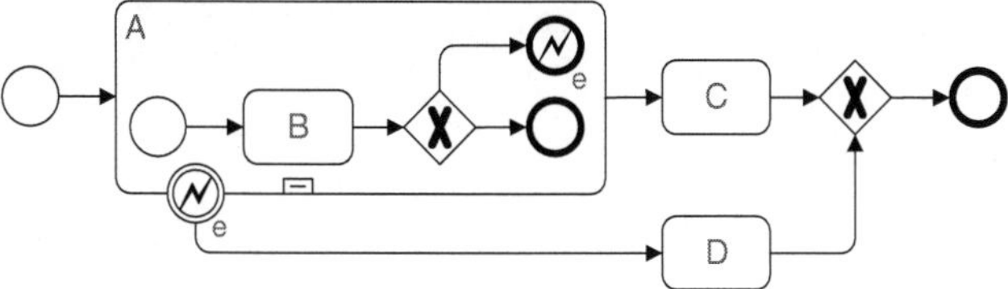

Exercise 6. Transform the model below into YAWL.

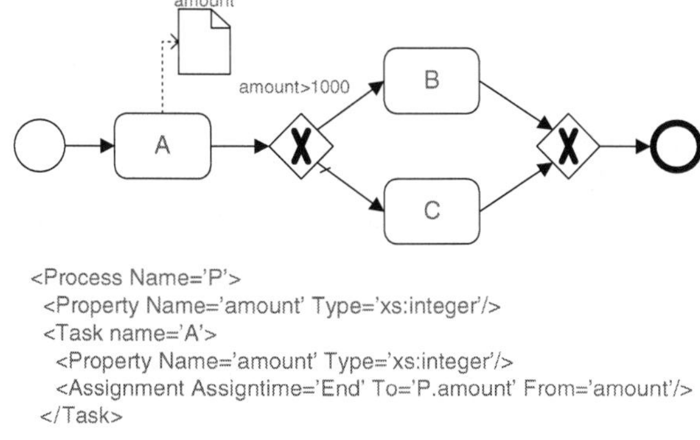

Exercise 7. Transform the model below into YAWL.

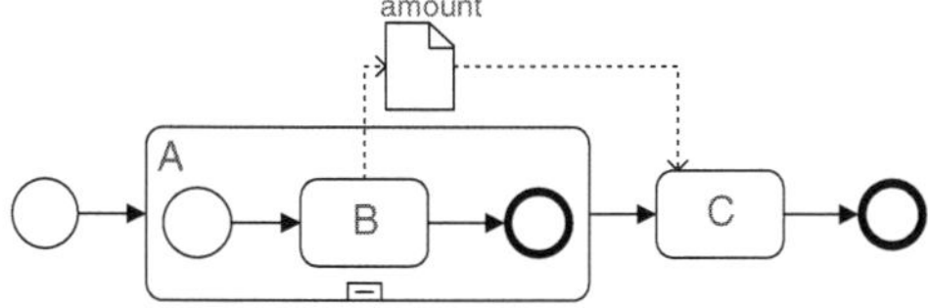

Chapter Notes

The Business Process Modeling Notation is standardized by the Object Management Group [185]. In general, the main purpose of BPMN models is to facilitate communication between domain analysts and to support decision-making based on techniques such as cost analysis, scenario analysis, and simulation [43, 202]. However, BPMN models are also used as a basis for specifying software system requirements.

Alternative notations for modeling business processes include UML Activity Diagrams [183], Event-driven Process Chains [129], and Business Process Execution Language [36]. The latter two will be explained in the next two chapters along with their relation to YAWL.

The chapter discusses an open issue with the underspecification of the OR-join in BPMN, which has given rise to many proposals outside the standardization process [83]. More open issues with BPMN are described in a number of papers [78, 202, 264]

Finally, it is worth mentioning that, while this chapter discusses the relation between BPMN and YAWL, relations between BPMN and other process modeling languages and formalisms have been discussed in the literature. For instance, a transformation from BPMN diagrams to plain Petri nets is presented in [78], while a transformation from BPMN to the process algebra (CSP) is presented in [267]. The purpose of these transformations is to provide an execution semantics of BPMN and to enable the analysis of BPMN diagrams using existing static analysis techniques. Transformations from BPMN to BPEL have also been studied in prior work [187] and they have been implemented in various tools, including commercial tools. The purpose of the BPMN to BPEL transformation is to enable the execution of BPMN diagrams using existing BPEL orchestration engines. However, in some cases, this transformation may lead to complex and unreadable BPEL process definitions that may subsequently be hard to debug and to maintain. In contrast, the transformation from BPMN to YAWL discussed in this chapter is very direct. One could conceive collaborative modeling environments in which some stakeholders will see a BPMN view of a process model, while others would see an equivalent YAWL net. Given the direct relations between BPMN and YAWL, it would be possible to maintain these views "in sync." The only issue would be to restrict changes in the YAWL view so that any change can be propagated into the BPMN view.

Chapter 14
EPCs

Jan Mendling

14.1 Introduction

This chapter discusses the relationship between YAWL and Event-Driven Process Chains (EPCs). EPCs were introduced in the early 1990s when major software vendors identified the need for a business-oriented representation of system functionality in order to speed up the rollout of business software in companies. Back then, the Institute of Information Systems (IWi) in Saarbrücken, Germany, collaborated with SAP on a project to define a suitable business process modeling language for documenting processes of the SAP R/3 enterprise resource planning system. There were two results from this joint effort: the definition of EPCs as a modeling language and the documentation of the SAP system in the SAP Reference Model as a collection of EPC process models. This SAP Reference Model had a huge impact on industry and research, with several publications referring to it. It also motivated the creation of further EPC reference models, for example, in computer integrated manufacturing, logistics, and retail. EPCs are frequently used in real-world projects due to a high user acceptance and extensive tool support. Some examples of tools that support EPCs are *ARIS Toolset* by IDS Scheer AG, *ADONIS* by BOC GmbH, and *Visio* by Microsoft Corp. There is also a tool-neutral interchange format called EPC Markup Language (EPML).

While EPCs and YAWL share most of their simple and advanced routing elements, there are some subtle differences that are important to know before mapping them. The aim of this chapter is to clarify commonalities and differences and to discuss how transformations can be specified. Against this background, it is organized as follows. Section 14.2 revisits the Carrier Appointment process as an example to introduce EPCs. Section 14.3 then uses the 20 original workflow patterns to highlight commonalities and some important differences between both languages. Section 14.4 presents how a mapping from EPCs to YAWL can be specified, while Sect. 14.5 discusses the transformation in the reverse direction. An alternative to

J. Mendling (✉)
Humboldt-University zu Berlin, Berlin, Germany
e-mail: jan.mendling@wiwi.hu-berlin.de

A.H.M. ter Hofstede et al. (eds.), *Modern Business Process Automation*,
DOI 10.1007/978-3-642-03121-2_14,

these mappings using concepts from Petri net synthesis is presented in Sect. 14.6. Section 14.7 summarizes the chapter.

14.2 Event-Driven Process Chains

The EPC is a business process modeling language for the representation of temporal and logical dependencies of activities in a business process. It is utilized by Scheer in the Architecture of Integrated Information Systems (ARIS) as the central method for the conceptual integration of the functional, organizational, data, and output perspective in information systems design. EPCs offer *function type* elements to capture activities of a process and *event type* elements describing preconditions and postconditions of functions. There are three kinds of *connector types* including AND (symbol $\wedge$), OR (symbol $\vee$), and XOR (symbol $\times$) for the definition of complex routing rules. Connectors have either multiple incoming and one outgoing arc (join connectors) or one incoming and multiple outgoing arcs (split connectors). As a syntax rule, functions and events have to alternate either directly or indirectly when they are linked via one or more connectors. OR- and XOR-splits after events are not allowed as events cannot make decisions. Control-flow arcs are used to link these elements.

The informal (or intended) semantics of an EPC can be described as follows. The AND-split activates all subsequent branches concurrently; the XOR-split represents a choice between one of several alternative branches; the OR-split triggers one, two, or up to all of multiple branches. In both cases of the XOR- and OR-split, the activation conditions are given in events subsequent to the connector. Accordingly, splits from an event to multiple functions should be avoided with XOR and OR, as the activation conditions do not become clear in the model. The AND-join waits for all incoming branches to complete and then propagates control to the subsequent EPC element. The XOR-join merges alternative branches. The OR-join synchronizes all active incoming branches. This feature is called nonlocality as the state of all transitive predecessor nodes is considered.

Figure 14.1 shows the *Carrier Appointment* process from Fig. A.4 as an EPC business process model. It is started when a carrier appointment is needed (start event at the top). It activates the *Prepare Route Guide* and *Estimate Trailer Usage* functions via an AND-split. The AND-join synchronizes the two branches and triggers the *Prepare Transportation Quote*. In contrast to YAWL, EPC split-connectors and join-connectors are elements of their own. A separate XOR-split describes the following choice and the subsequent events define the respective conditions. In case of *Single Packaging* (right branch), the *Pickup Appointment* is arranged first, then the *Delivery Appointment*, and finally the *Carrier Manifest* is created. In case of *Less than Truck-Load* (*LTL*), the *Pickup Arrangement* is done in parallel to a decision whether the *Delivery Appointment* is done now or later. The OR-join synchronizes the two active of the three incoming branches before the *Bill of Lading* is created. The *Delivery Appointment* can then be done afterwards if it has not

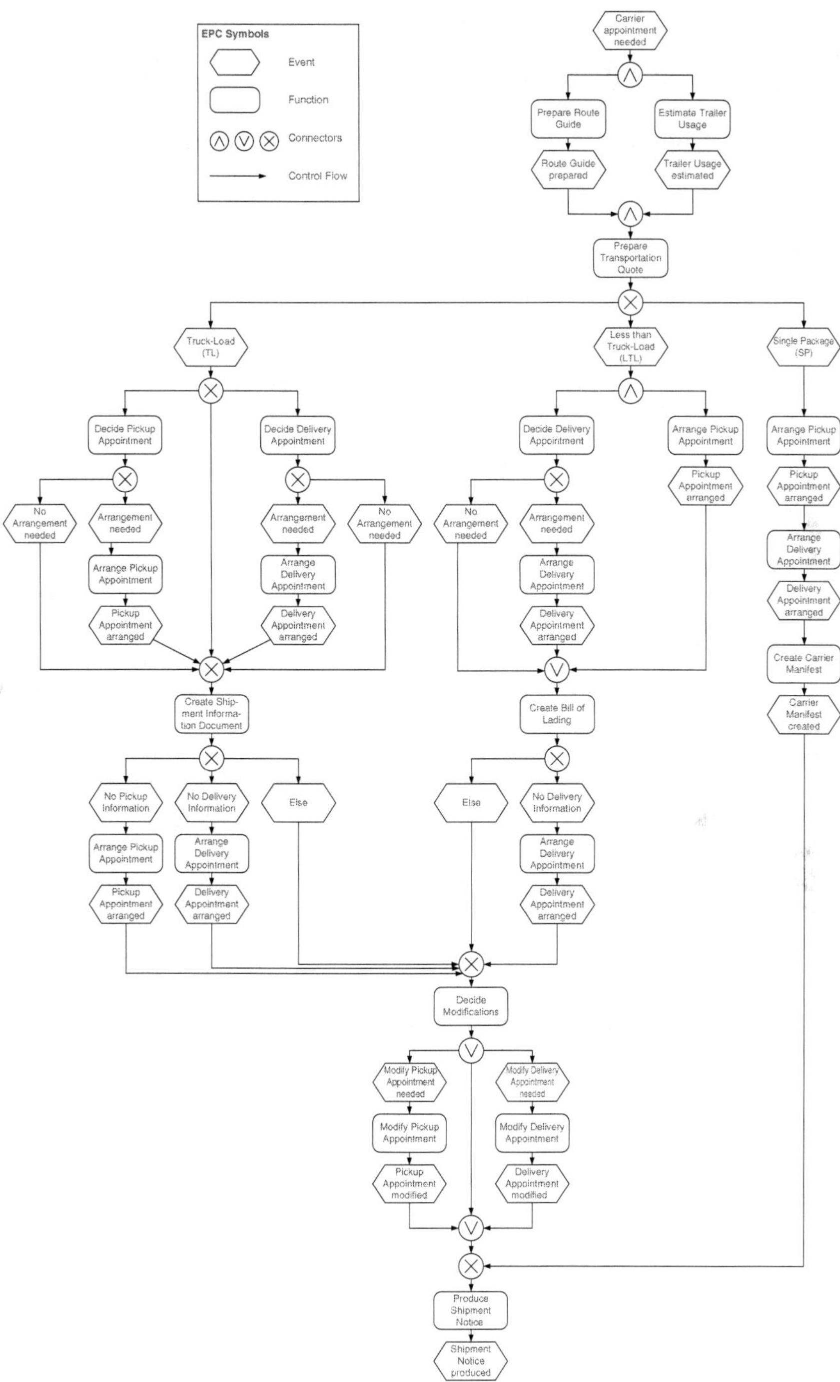

Fig. 14.1 The *Carrier Appointment* process as an EPC

been done before. The *Truck-Load* case (*TL*) is even more flexible. The *Pickup* and *Delivery Appointments* can be done first or after the creation of the *Shipment Information Document*. All paths of *TL* and *LTL* lead to a *Decide Modifications* function. If required, *Pickup Appointment* and *Delivery Appointment* can still be changed. Finally, all three cases (*TL*, *LTL*, and *Single Package*) lead to the production of the *Shipment Notice*, which triggers the end event of the process.

This EPC model introduces some adaptations for deferred choices and cancelation regions of the original YAWL model. Both concepts are not directly supported by EPCs as the following section discusses.

14.3 Pattern Comparison of YAWL and EPCs

This section compares YAWL and EPCs using the 20 original workflow patterns (see Appendix C).

Table 14.1 summarizes the Workflow Patterns support of EPCs and YAWL. While YAWL was particularly tailored to support most of the patterns,[1] EPCs supports less patterns. They offer AND-connectors and XOR-connectors to represent all basic control-flow patterns. The OR-connector can implement Multiple Choice and Synchronizing Merge, but Multi Merge (depending on the formalization) and Discriminator are not covered. The two Structural Patterns can be modeled. In contrast, there is no direct support for the Patterns involving Multiple Instantiation although workarounds can be defined for (13) MI with a priori Design Time Knowledge. EPCs also lack an explicit notion of state. Accordingly, they cannot directly capture the State-based Patterns. There is also no support for the Cancelation Patterns.

The missing support for the Multi Merge requires some comments. Some formalization approaches for EPCs define the XOR-join semantics as synchronizing, for

Table 14.1 Workflow Pattern support of EPCs and YAWL

	EPCs	YAWL		EPCs	YAWL
(1) Sequence	+	+	(11) Implicit Termination	+	–
(2) Parallel Split	+	+	(12) MI without Synchronization	–	+
(3) Synchronization	+	+	(13) MI with Design Time Knwl.	+	+
(4) Exclusive Choice	+	+	(14) MI with Runtime Knwl.	–	+
(5) Simple Merge	+	+	(15) MI without Runtime Knwl.	–	+
(6) Multiple Choice	+	+	(16) Deferred Choice	–	+
(7) Synchr. Merge	+	+	(17) Interleaved Parallel Routing	–	+
(8) Multi Merge	–	+	(18) Milestone	–	+
(9) Discriminator	–	+	(19) Cancel Activity	–	+
(10) Arbitrary Cycles	+	+	(20) Cancel Case	–	+

[1] Intentionally, the Implicit Termination pattern is not supported in order to force the designer to carefully consider the termination condition.

instance the one by Nüttgens and Rump. Other formalizations define the XOR-join similar to a Petri net place with multiple incoming transitions, for example, the one by van der Aalst. Another comment is required on the Synchronizing Merge pattern. There are competing OR-join formalizations with different characteristics (see Chap. 3). None of them fulfills all requirements perfectly. For this reason, the following discussion is abstracted from different implementations of the OR-join and assumes that the EPC OR-join and the YAWL OR-join describe the same behavior. This feature is used in the following sections on mapping.

14.4 Mapping EPCs to YAWL

This section discusses a transformation from EPCs to YAWL. Most of the elements can be mapped directly as EPCs and YAWL support similar routing concepts. Even though being similar, there are three major differences that have to be considered in the transformation: (1) state representation, (2) connector chains, and (3) multiple start and end events.

EPC functions can be mapped to YAWL tasks following mapping rule (a) of Fig. 14.2. The first difference between EPCs and YAWL is related to *state representation*. EPC events can often be associated with states but they mainly define preconditions for the start of functions and postconditions after their completion. Therefore, a direct mapping of events to YAWL conditions is not an option. Consider, for example, a syntactically correct EPC that models *one event* followed by an AND-connector that splits control-flow to two functions. In YAWL there are *two conditions* required as preconditions for the two functions. As a consequence, most EPC formalizations define the state as a mapping of the arcs. Accordingly, rule (b) in Fig. 14.2 defines that events are not mapped to YAWL, taking advantage of the fact that arcs in YAWL represent implicit conditions if they connect two tasks. In EPCs, connectors are independent elements. Therefore, it is allowed to build the so-called *connector chains*, that is, paths of two or more consecutive connectors as at the bottom of Fig. 14.1. In YAWL, there are no connector chains as splits and joins are part of tasks. The mapping rules (c) to (h) map every connector to a dummy task with the matching join or split condition (see Fig. 14.2). The third difference stems from *multiple start and end events*. An EPC is allowed to have more than one start event. Multiple end events represent implicit termination: the triggering of an end event does not terminate the process as long as there is another path still active. In YAWL, there must be exactly one start condition and one end condition. Therefore, the mapping rules (i) and (j) generate an OR-split for multiple starts and an OR-join for multiple ends. This implies that any combination of start and end events is considered to be correct even if only a restricted set of combinations is meaningful. By using such an interpretation, this mapping yields a YAWL model that includes all execution paths that can be taken in the EPC. Some additional paths may be covered that were not intended by the original EPC though. Figure 14.3 shows the example of a simple EPC model with multiple start events. Following the described mapping, the resulting YAWL model has a unique start condition and a subsequent

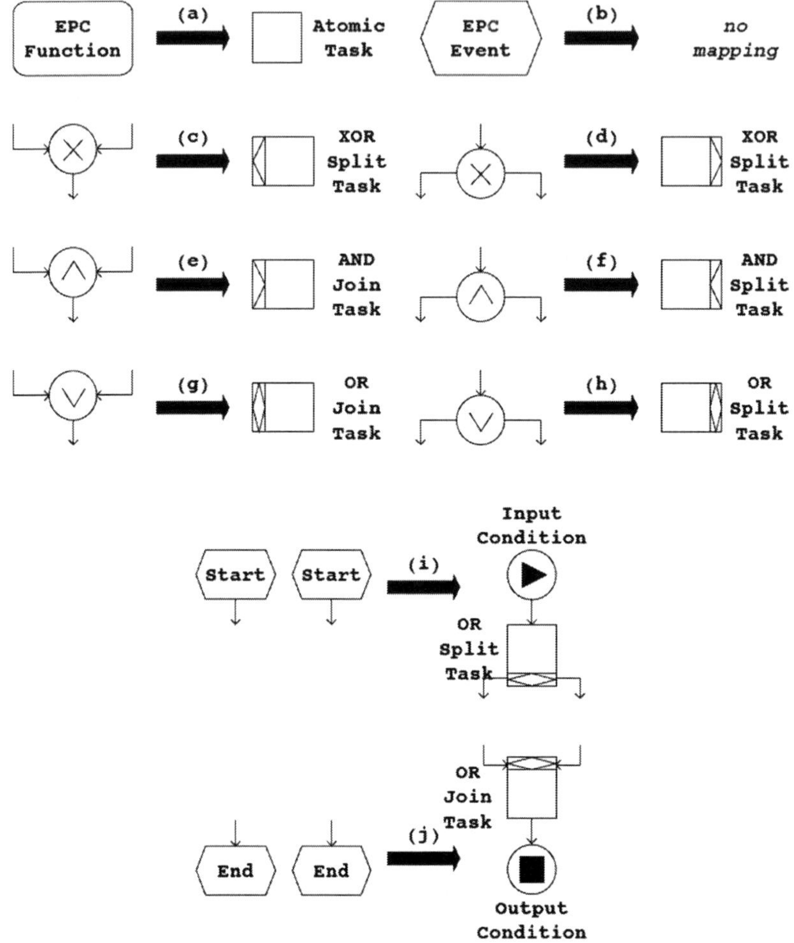

Fig. 14.2 Overview of the EPC to YAWL mapping

OR-split. Each outgoing branch of this OR-split corresponds to a start event in the EPC. As can be seen, the OR-split allows to execute both branches although the following XOR-join is not able to synchronize them appropriately. Such problems can be identified using verification techniques, and they need to be fixed before using the YAWL model in a workflow engine.

14.5 Mapping YAWL to EPCs

The previous section has already shown that some elements can be easily mapped from EPCs and YAWL. This also holds in the opposite direction. The problems of mapping stem from the gaps in the workflow pattern support of EPCs and from

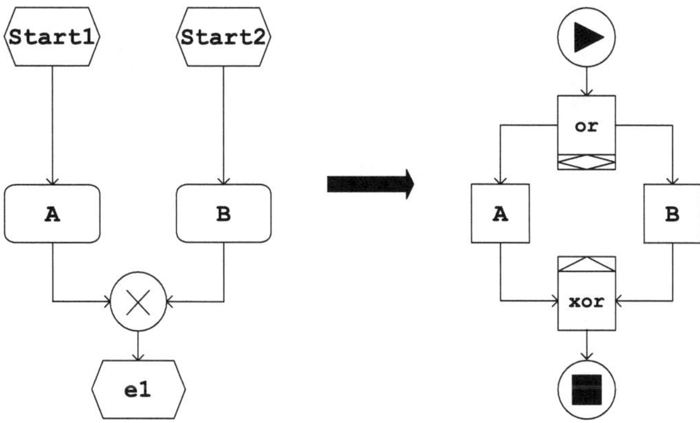

Fig. 14.3 Mapping of an EPC model to YAWL

the fact that EPCs are free-choice. This property has been extensively discussed for Petri nets. In essence, a *free-choice net* is a Petri net in which a place that is in the preset of multiple transitions is the only place in all these presets. Therefore, the choice is "free" in a sense that it can be made without considering other places.

Figure 14.4 shows an example of a non-free-choice net. In Fig. 14.4a, an initial choice is made to proceed either with *t1* or *t2*. It is assumed that *t2* is fired, followed by *t4* and *t5*. This produces the marking that is shown in Fig. 14.4b. Although the token on the place after *t4* could potentially be consumed by either *t6* or *t7*, it is only possible to fire *t7*. As one can see from the structure of the net, there is an overlap in the preset of *t6* and *t7*, but it is not covering all preset places. This causes the net to be non-free-choice, with the effect that places with multiple outgoing arcs are not necessarily choices. The corresponding EPC is depicted in Fig. 14.4c. While the non-free-choice semantics of YAWL guarantee that the firing of *t4* and *t5* leads to *t7*, there is a free-choice XOR-split in the EPC that also allows to progress towards the AND-join before *t6* (which is a deadlock). In the following, the mapping from YAWL to EPCs is discussed, in particular, the problems of (1) non-free-choice behavior, (2) function-event alternation, (3) cancelation regions, and (4) multiple instantiation.

YAWL tasks can be mapped to EPC functions using mapping (a) of Fig. 14.5. The first challenge is the mapping of YAWL conditions. Mapping (b) shows that an XOR-join and an XOR-split have to be introduced for multiple conditions. The problem of this mapping is that it preserves behavior only for those YAWL nets that are free-choice. Think of Fig. 14.4 as a corresponding YAWL net: while the net forbids to fire *t6* in marking (b), such behavior cannot be enforced by any concept of an EPC. Mapping (b) can therefore lead to unsound EPC models if the YAWL net is non-free-choice. This problem relates to mapping all state-based workflow patterns and the deferred choice. Figure 14.1 introduces additional functions to make the deferral explicit, yet no corresponding mapping is proposed here. Mappings (c) to (h) describe the obvious mapping of split and join conditions. Please note

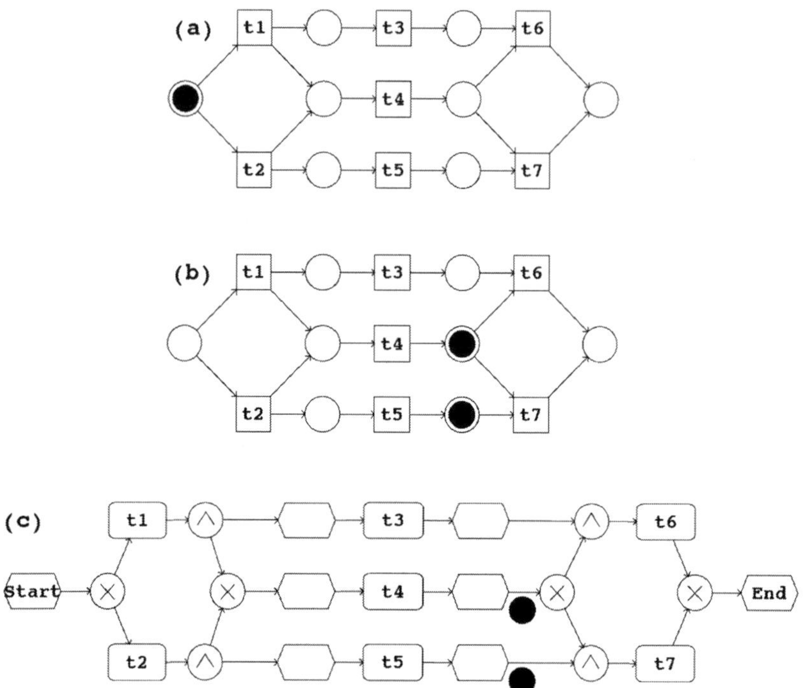

Fig. 14.4 Example of a non-free-choice Petri net and corresponding EPC

that a task produces a separate EPC function. These mappings obviously do not guarantee the required alternation of functions and events in an EPC. Therefore, an additional postprocessing must introduce events at the appropriate positions. Cancelation regions cannot be mapped to EPCs; there is no corresponding concept. The same holds for multiple instantiation of tasks. EPCs can only express that a task is executed but not its multiple instantiation attributes. As EPCs support pattern 13 (MI with a priori Design Time Knowledge), it is possible to duplicate the tasks in an AND-block for each required instance.

Figure 14.6 depicts three examples to illustrate the mapping of YAWL conditions to EPCs as described in Mapping (b). Quite different YAWL models can result in the same EPC in this way. Figure 14.6a shows that a condition with two outgoing arcs is mapped to an XOR-split. Figure 14.6b includes an end condition with two incoming arcs. This is mapped to an XOR-join that merges these incoming arcs. As a consequence, Fig. 14.6a and Fig. 14.6b result in the same EPC. Figure 14.6c shows a part of a YAWL model having a condition with multiple incoming and outgoing arcs. This model yields an EPC with an XOR-join merging the incoming branches and an XOR-split for each outgoing arc.

The limitations in workflow pattern support has motivated research on extending EPCs such that all YAWL behavior can be directly expressed. The respective extension is called Yet Another EPC (yEPC). yEPCs include three additional concepts

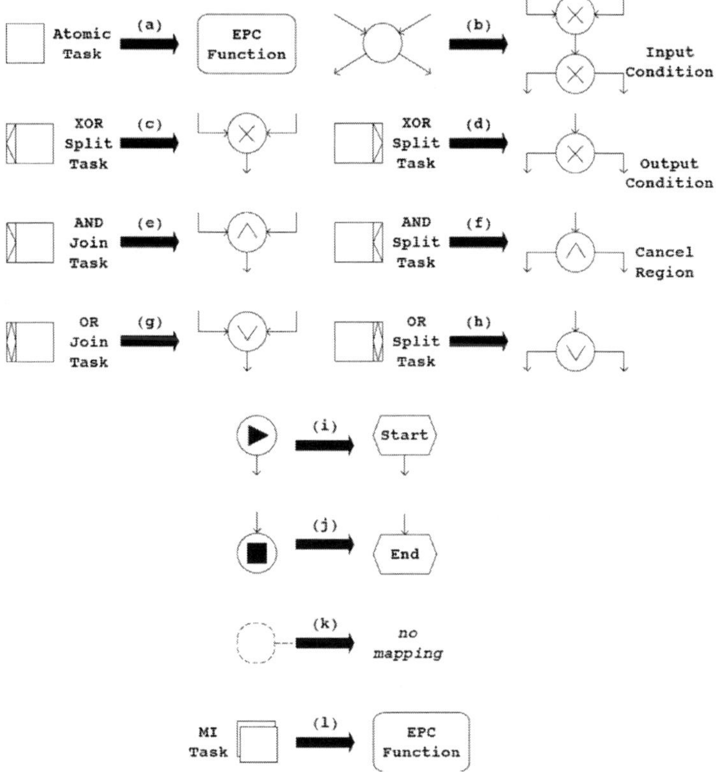

Fig. 14.5 Overview of the YAWL to EPC mapping

compared to EPCs: the empty connector, multiple instantiation parameters, and cancelation regions. The missing support for state-based workflow patterns is resolved by introducing a new connector type that is called *empty connector*. This connector is represented by a cycle, just like the other connectors, but without any symbol inside. Semantically, the empty connector represents a join or a split without imposing a rule. Consider an event that is followed by an empty split that links to multiple functions. The empty split allows all subsequent functions to pick up the event. As a consequence, there is a run between the functions: the first function to consume the event causes the other functions to be active no more. These split semantics match the deferred choice pattern. Consider the other case of an empty join with multiple input events. The subsequent function is activated when one of these events has been reached. This behavior matches the multiple merge pattern. In yEPCs, *multiple instantiation* is supported similarly as in YAWL using a quadruple of parameters. The parameters `min` and `max` define the minimum and maximum cardinality of instances that may be created. The `required` parameter specifies an integer number of instances that need to have finished in order to complete multiple instantiation. The `creation` parameter may take the values

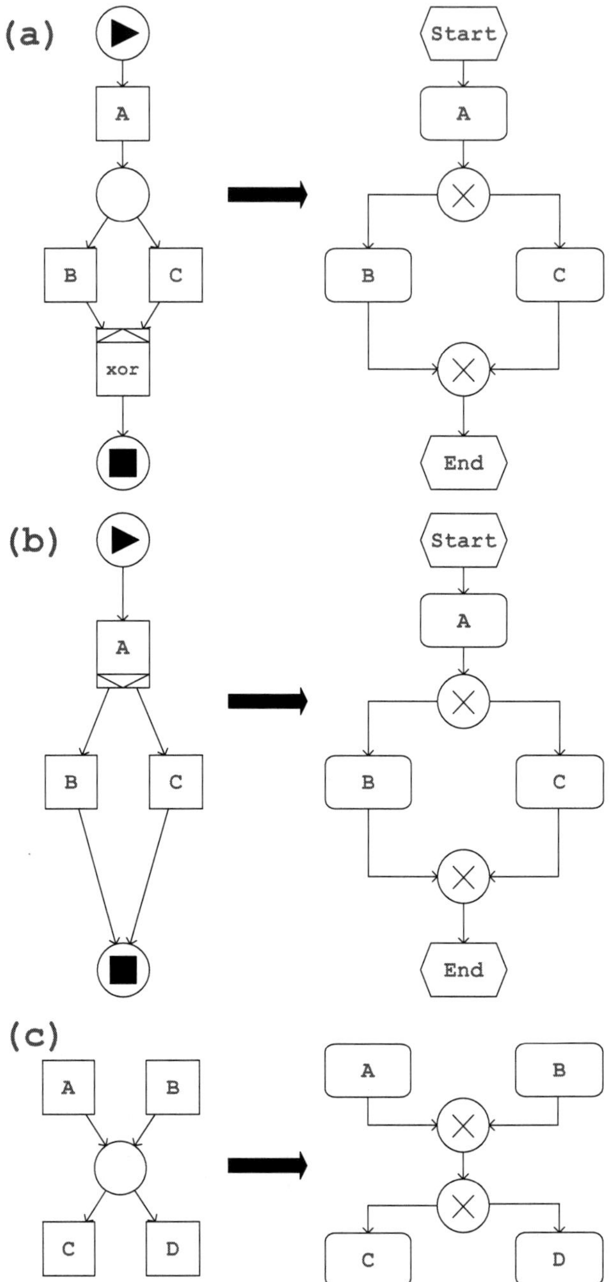

Fig. 14.6 Three examples of mapping YAWL to EPCs

`static` or `dynamic`, which specifies whether further instances may be created at runtime (`dynamic`) or not (`static`). *Cancelation* patterns are supported the same way as in YAWL. Cancelation areas (symbolized by a lariat) can include several functions and events. The end of the lariat has to be connected with a function. When this function completes, all functions and events in the lariat are canceled.

Up until now, there has been little uptake of yEPCs. Therefore, it is a good idea to consider the rules of Fig. 14.5 when mapping YAWL to EPCs. An alternative to an element-wise transformation builds on techniques from Petri net synthesis. This approach is explained in the next section.

14.6 Transformation by Synthesis

An alternative to an element-wise transformation between EPCs and YAWL is to use the formal semantics. An implication of formal semantics is that the behavior of an EPC or a YAWL net can be described in terms of a reachability graph. At this stage, it is abstracted from some formal problems of reachability in a YAWL formalization based on reset nets. Wynn et al. use reset nets to formalize the behavior of cancelation regions. It has been shown that coverability (generalization of reachability) is then only decidable using backward search techniques. A corresponding verification approach for YAWL nets has been defined, too. See also Chaps. 3 and 20. As a second step, synthesis techniques can be used to derive a process model back from the reachability graph. This section describes this approach in an informal way.

Formal semantics of a process modeling language expresses which states are defined by the model and which transitions are allowed between which states. For EPCs, the concept of state is typically defined as a mapping from the arcs to 0 and 1. A particular state assigns a 1 to all arcs that are currently active and a 0 to all other arcs. State changes are formalized by a *transition relation*. A node is called *enabled* if there are enough tokens on its incoming arcs that it can fire. That is, a state change defined by a transition can be applied. The enabling rule differs for the different types of nodes in an EPC and a YAWL net. For example, the AND-join requires all input arcs to have a token while the XOR-join needs only one. Changing the state at a node is also called *firing*. Firing of a node n consumes tokens from its input arcs n_{in} and produces tokens for its output arcs n_{out}. The formalization of whether an OR-join is enabled is a nontrivial issue, as not only the incoming arcs must be considered (cf. Chap. 3). The sequence $\tau = n_1 n_2 \ldots n_m$ is called a firing sequence if it is possible to execute a sequence of steps; that is, after firing n_1, it is possible to fire n_2, etc. Through a sequence of firings, the EPC or the YAWL net move from one *reachable* state to the next. The *reachability graph* represents how states can be reached from other states. Figure 14.7 shows the reachability graph for the Carrier Appointment process as depicted in the beginning of this chapter. The picture illustrates the complexity of the state space. The part directly after the start event is magnified at the top on the righthand side of the figure, showing the concurrency of the first AND-block as a diamond shape.

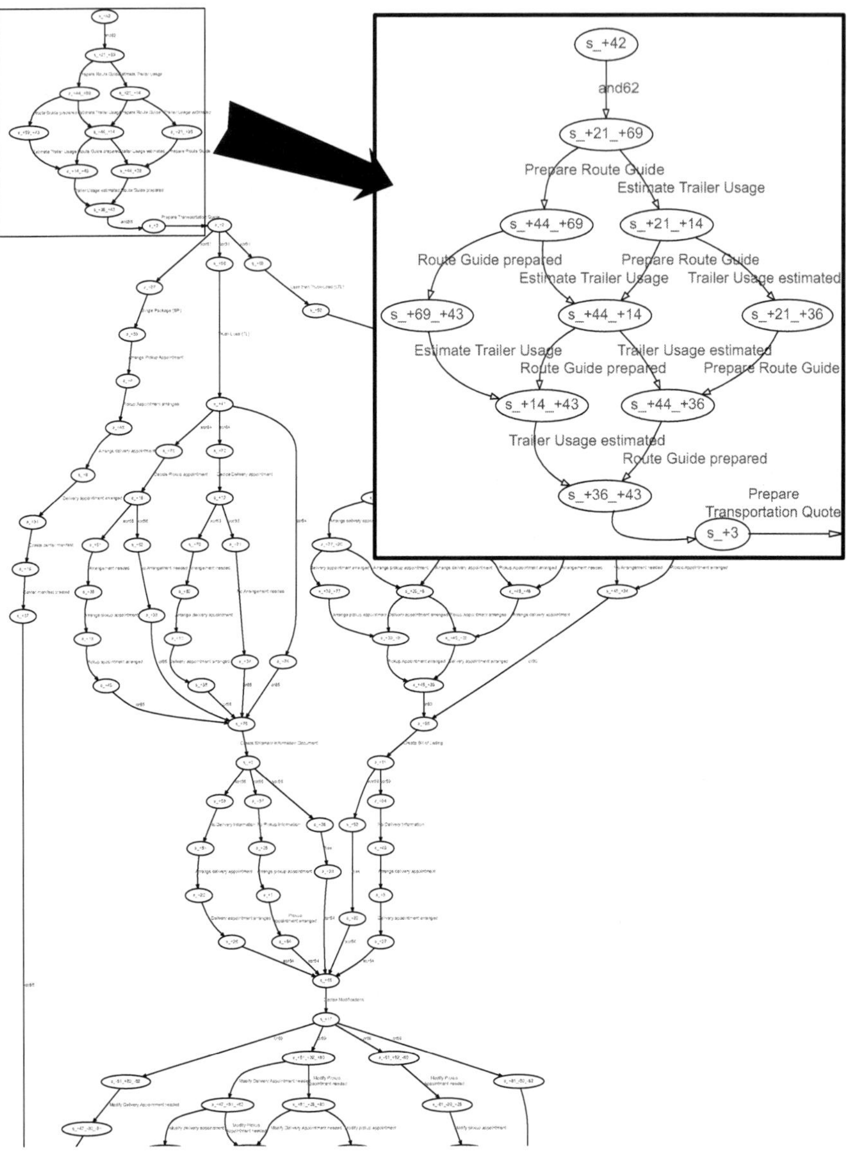

Fig. 14.7 Reachability Graph of the *Carrier Appointment* process, partially magnified

It is well-known in Petri net theory that a reachability graph can be translated back into a Petri net. The concept of this Petri net synthesis builds on the theory of regions by Ehrenfeucht and Rozenberg. There is also a software tool called *Petrify* that implements this theory. The theory of regions has been used to transform between different process modeling languages. The different set of routing elements in EPCs and Petri nets has been a serious problem for any element-wise transformation approaches. Mendling et al. use the formal semantics of EPCs and the Petrify tool to derive a behavior-equivalent free-choice Petri net. Using, for example, the ProM tool, such a Petri net can easily be transformed into a YAWL net without loss of information. The same approach can also be applied from YAWL to EPCs: generate the reachability graph, synthesize a free-choice Petri net, and generate an EPC without loss of information. The required functionality is available in the ProM tool.[2]

14.7 Conclusion

This chapter discussed the relationship between EPCs and YAWL. It identified that a great portion of YAWL and EPC elements can easily be mapped, yet there are problems with element-wise mappings. From EPCs to YAWL, there are three challenges: EPCs can have multiple start and end events, their state representation is not explicit, and connector chains have to be mapped to tasks with corresponding routing semantics. From YAWL to EPCs, there are four impediments: EPCs can only express free-choice behavior, functions and events must alternate, cancelation regions are not supported, and multiple instantiation is missing. To overcome these mapping problems, an approach based on reachability graph calculation and synthesis was described. In this way, the formal gap between EPCs and YAWL can be closed, though the resulting models are a lot less compact.

It has to be stressed that the pragmatic gap between EPCs and YAWL is much broader than the formal one. Even if behavior is correctly transformed between both languages, it has to be taken into consideration that EPCs are typically used to describe processes at a much more abstract and business-oriented level than YAWL nets. This implies that several aspects that are important for workflow implementation are not supported. A corresponding YAWL net implementation model, therefore, has to be extended with interface declarations, data-flow, and resource assignment. While a transformation of control-flow is desirable, it is also clear that technical aspects have to be added by a workflow designer who is familiar with software engineering techniques.

Exercises

Exercise 1. Please have a look at the EPC of Fig. 14.8. Which of the following traces is valid for this EPC, and if not why:

1. *t1*, *t4*
2. *t2*, *t5*
3. *t1*, *t3*, *t4*
4. *t1*, *t2*, *t3*
5. *t2*, *t1*, *t3*, *t5*

---

[2] www.processmining.org

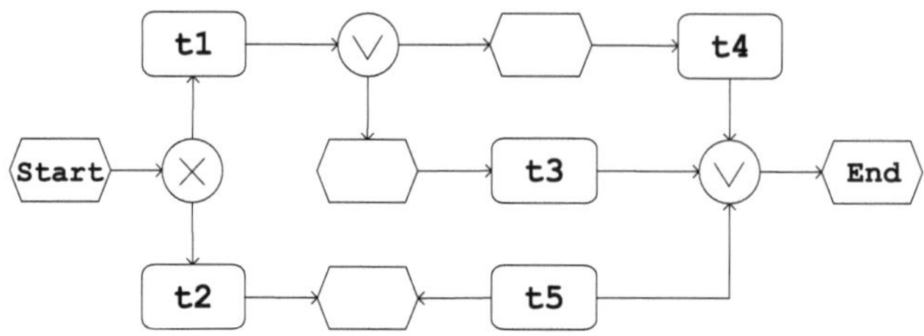

Fig. 14.8 An EPC model

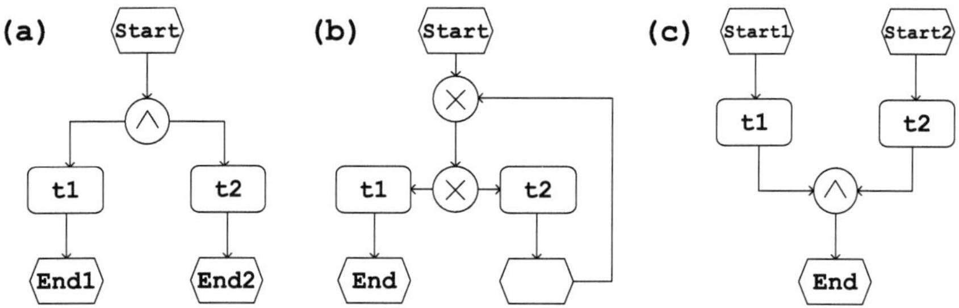

Fig. 14.9 Three EPC models

Exercise 2. Please map the EPCs of Fig. 14.9 to YAWL. Describe the mapping challenges for each of the examples.

Exercise 3. Please map the YAWL models of Fig. 14.10 to EPCs. Describe the mapping challenges for each of the examples.

Chapter Notes

EPCs are introduced as a modeling language in [129] and described in different books [231–233]. The documentation of the SAP system in the SAP Reference Model is described in [70, 130]. Research referring to the SAP Reference Model includes [92, 145, 167, 214, 240, 247, 280]. Further, EPC reference models were constructed for computer integrated manufacturing [229, 230], logistics [138], and retail [44]. The tool-neutral interchange format called EPC Markup Language (EPML) is defined in [169].

The nonlocality of OR-joins is formalized among others in [133, 162]. OR-joins in YAWL are discussed in [273, 274]. Some formalization approaches for EPCs define the XOR-join semantics as synchronizing, for instance [133, 182]. This is reflected in the evaluation reported in [166]. Other formalizations define the XOR-join similar to a Petri net place with multiple incoming transitions, for example,

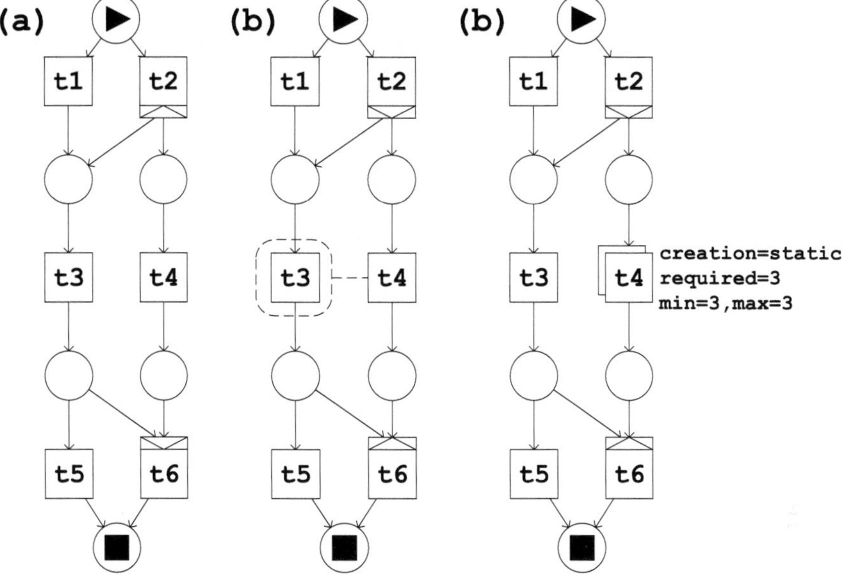

Fig. 14.10 Three YAWL models

[3, 162]. Chapter 2 of [161] provides a detailed discussion of OR-join semantics. Instantiation semantics for EPCs, YAWL, and other languages are discussed in [74].

Mappings between EPCs and YAWL are discussed in [165,167,168]. These mappings have been used for the analysis of the SAP Reference Model in [170]. The theory of regions is introduced by Ehrenfeucht and Rozenberg [37,88]. There is also a software tool called *Petrify* that implements this theory [64,65]. Mendling et al. use the formal semantics of EPCs and the Petrify tool to derive a behavior-equivalent free-choice Petri net [163, 164].

Chapter 15
The Business Process Execution Language

Chun Ouyang, Marlon Dumas, and Petia Wohed

15.1 Introduction

The number of commercial tools for workflow and business process management that have emerged over the past two decades is considerable. In the past, lack of standardization led to a situation where each of these tools implemented a different language. As shown by the tool evaluations undertaken using the workflow patterns, many of these languages suffered from clear limitations in terms of their expressiveness and suitability to capture nontrivial business processes. By now, industry consolidation has rendered most of these languages obsolete. In parallel, standardization efforts initiated by tool vendors have led to the definition of two standard languages for capturing business processes, namely the Business Process Modeling Notation (BPMN) and the Web Services Business Process Execution Language (WS-BPEL or BPEL for short).

BPMN is a business process modeling notation intended to facilitate communication between domain analysts and to support decision-making based on techniques such as cost analysis, scenario analysis, and simulation. BPMN models are not meant to be directly executable, although they may serve as input to software development projects. A comparison between BPMN and YAWL and a method for mapping BPMN diagrams into YAWL nets are given in Chap. 13.

Meanwhile, BPEL is a language for defining executable business processes. In this respect, it can be seen as "competing" in the same space as YAWL. Yet, the original motivation and design of YAWL are very different from those of BPEL, and this fact transpires in profound differences between these two languages.

BPEL is intended to support the definition of a class of business processes known as *Web service orchestrations*. In a Web service orchestration, one service (the orchestrator) interacts with a number of other services. The focus of BPEL is therefore on capturing interactions between a given process and a set of partners (represented as Web services). The logic of the interactions is described as

---

C. Ouyang (✉)
Queensland University of Technology, Brisbane, Australia,
e-mail: c.ouyang@qut.edu.au

A.H.M. ter Hofstede et al. (eds.), *Modern Business Process Automation*,
DOI 10.1007/978-3-642-03121-2_15,

a composition of communication actions (send, receive, send/receive, etc.). These communication actions are interrelated by control-flow dependencies expressed through constructs corresponding to parallel, sequential, and conditional execution, event, and exception handling.

YAWL, on the other hand, is intended to support the definition of workflows composed of interrelated tasks that may be performed by humans or delegated to software applications according to resource allocation policies. The software applications with which YAWL workflows interact may be implemented as RESTful services or as SOAP Web services as outlined in Chap. 7, but in any case, the basic abstraction in YAWL is that of *task* as opposed to that of *message exchange* as in BPEL.

Second, both BPEL and YAWL possess an XML syntax. However, YAWL additionally possesses a graphical notation, whereas BPEL lacks a standardized graphical notation.[1] Unlike YAWL, BPEL lacks formal semantics in its standard specification. Several researchers have defined formal semantics of BPEL using various formalisms, including Petri nets, process algebra, abstract state machines, and finite state machines. This has paved the way for the development of static analysis tools for BPEL that can compete with the static analysis capabilities provided by the YAWL System.

As a result of BPEL being driven by a standardization committee supported by several IT industry players, a number of different implementations exist (both proprietary and open source),[2] contrary to YAWL, which at present has a single implementation. However, the standard BPEL specification does not provide support for human tasks, that is, tasks that are allocated to human actors and that require these actors to complete actions, possibly involving a physical performance of tasks. A number of BPEL engines already provide extensions to BPEL for managing human tasks (i.e., BPEL4People, WS-HumanTask), but these extensions are yet to be standardized. In contrast, the YAWL System provides external interfaces for worklists and resources based on Web standards. This interface supports a majority of the workflow resource patterns, the management of the organizational data, and worklist handling.

Finally, another major difference between BPEL and YAWL lies in the nature of their modeling constructs. In BPEL, most language constructs are block-structured, which means that if a segment of a BPEL model starts with a branching construct (e.g., AND-split), it ends with the corresponding synchronization construct (e.g., AND-join). As a result, BPEL supports only a restricted set of the control-flow structures. In contrast, YAWL is graph-oriented, and therefore supports arbitrary control-flow structures.

This chapter introduces BPEL and compares it to YAWL. The purpose is not to claim one outperforms the other, but to investigate, as in the above, the differences between these two languages. The chapter focuses on the differences between BPEL

---

[1] Most tools that support BPEL provide a graphical editor, but the standard specification of BPEL does not provide a graphical notation.

[2] See en.wikipedia.org/wiki/Comparison_of_BPEL_engines for a compendium.

and YAWL in terms of the modeling constructs. An overview of BPEL and its comparison to YAWL is presented in two ways: first by means of examples, and second in terms of how these two languages support (or fail to directly support) the original 20 control-flow workflow patterns (see Appendix C). A brief discussion about the related work on translation from BPMN to BPEL and translation from BPEL to YAWL is also included. Note that there are three versions of BPEL: 1.0, 1.1, and 2.0. This chapter is based on BPEL 2.0, which is the standard specification of BPEL.

15.2 Overview of BPEL through the YAWL Prism

BPEL is a language with an XML syntax, which combines features found in classical imperative programming languages with constructs for capturing concurrent execution and constructs specific to Web service implementation. It is used to describe business collaborations (both intra- and inter-organizational) and to implement them as composite Web services. Composite Web services resemble executable processes, in that they specify the execution order between *activities*, the *partners* involved in them, the *data* needed for and resulting from the execution of the different activities, the *messages* exchanged between the partners, and the *fault handling* to be carried out in cases of errors and exceptions.

A simplified structure of a BPEL process is shown in Fig. 15.1a, where *partnerLinks* are used for the definition of different partners,[3] *variables* are used for the specification of process data, and *activity* for denoting the flow of activities specified in the process.

A process can be decomposed into a number of smaller coherent parts. This is done with the help of the *scope* construct, which one could think of as a *kind* of a subprocess defined inside the main process. However, it should be noted that, in contrast to subprocesses that are defined in one place of the process and potentially invoked from several other places, the definition of scopes is intermixed in the code of the process and scope invocation from other parts of the process is not supported.

A simplified structure of a scope is shown in Fig. 15.1b. Similar to programming languages, variables (partnerLinks, etc.) defined inside a scope are local to that scope, while variables (partnerLinks, etc.) defined for the process are global and hence usable to all scopes and activities defined for the process.

Given that BPEL was created to orchestrate different partner Web services, it offers constructs for capturing interactions between Web services. Three such constructs are *invoke*, *receive*, and *reply*. Invoke is used for synchronous or asynchronous communication when the process calls a partner's Web service. The specification of an *invoke* action includes references to the partner link (capturing the service provider), as well as the operation and possible input/output variables

[3] More precisely, a partnerLink represents a communication channel to a partner. The communication with a partner can be realized through more than one channel.

```
<process name="..."
  targetNamespace="http://www..." >
  ...
  <partnerLinks> ...
  <messageExchanges> ...
  <variables> ...
  ...
  <faultHandlers> ...
  <eventHandlers> ...
  ...
  activity
</process>
```
(a)

```
<scope name="..." >
  <variables> ...
  <partnerLinks> ...
  <messageExchanges>...
  ....
  <eventHanlders>...
  <faultHandlers>...
  <compensationHanlders>...
  ...
  activity
</scope>
```
(b)

```
<invoke name = "..."
  partnerLink = "..."
  operation = " ..."
  inputVariable = "..."
  outputVariable = "..." />
```
(c)

```
<receive name = "..."
  partnerLink = "..."
  operation = " ..."
  variable = "..."
  createInstance = "yes/no"
  messageExchange = "..." />
```
(d)

Fig. 15.1 (**a**) A simplified structure of a BPEL Process; (**b**) A simplified scope structure; (**c**) The Invoke construct; (**d**) The Receive construct

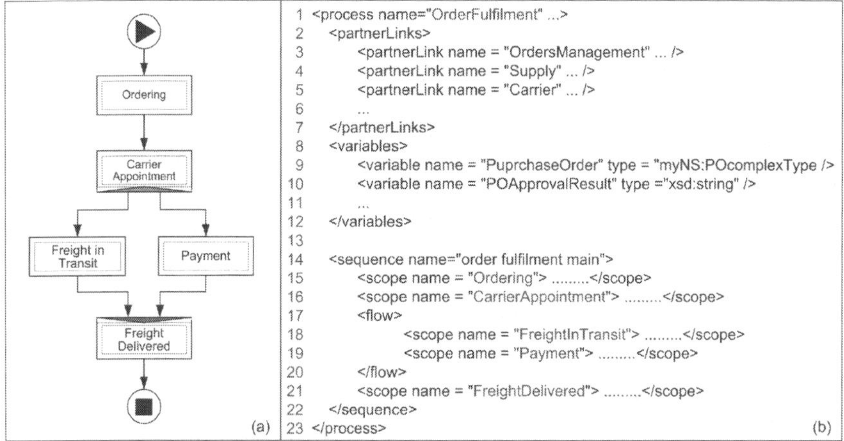


```
<process name="OrderFulfilment" ...>
  <partnerLinks>
    <partnerLink name = "OrdersManagement" ... />
    <partnerLink name = "Supply" ... />
    <partnerLink name = "Carrier" ... />
    ...
  </partnerLinks>
  <variables>
    <variable name = "PuprchaseOrder" type = "myNS:POcomplexType />
    <variable name = "POApprovalResult" type ="xsd:string" />
    ...
  </variables>

  <sequence name="order fulfilment main">
    <scope name = "Ordering"> .........</scope>
    <scope name = "CarrierAppointment"> .........</scope>
    <flow>
      <scope name = "FreightInTransit"> .........</scope>
      <scope name = "Payment"> .........</scope>
    </flow>
    <scope name = "FreightDelivered"> .........</scope>
  </sequence>
</process>
```

Fig. 15.2 A simplified version of the Order Fulfillment process in (**a**) YAWL; (**b**) BPEL

(see Fig. 15.1c). If no output variables are specified, the invoking process will not wait for a reply from the invoked process. Receive and reply are used for capturing asynchronous communication initiated by partner Web services. The elements denoting *invoke*, *receive*, and *reply* actions have similar structure (see Fig. 15.1d). The *receive* action has an additional attribute *createInstance* that specifies whether a new instance of the process is created or the communication will be related to an already existing instance (created for instance through an earlier call).

Now, let us go to the Order Fulfillment process shown in Appendix A, simplify it as shown in Fig. 15.2a, and model it in BPEL. Then, a structure can be obtained as the one shown in Fig. 15.2b. In this BPEL process definition, the variables *PurchaseOrder* and *POApprovalResult* are defined as global variables, while the partnerLinks *OrdersManagement*, *Supply*, and *Carrier* are defined to capture the different partners in the process. The scopes *Ordering*, *CarrierAppointment*, *FreightInTransit*, *Payment*, and *FreightDelivered* capture the corresponding subprocesses that appear in the YAWL net. The reader will note that in Fig. 15.2b there are two more constructs: *sequence* and *flow*. These are called *structured activities* and are used to define control-flow dependencies in a process model. The following type of structured activities are available in BPEL:

- *sequence*, as the name indicates, defines that the activities specified inside it will be executed sequentially in the order in which they appear in the definition
- *flow*, on the other hand, does not imply any order of execution, which means that the activities inside a flow construct are executed in parallel
- *if*, for capturing conditional routing (i.e., XOR split and merge) between activities
- *pick*, capturing race conditions based on timing or external triggers such as a message receipt event
- *while*, for structured looping where the loop condition is evaluated at the *beginning* of each iteration
- *repeatUntil*, for structured looping where the loop condition is evaluated at the *end* of each iteration
- *forEach*, for executions of multiple instances of the contained activity with synchronization
- *scope*, as discussed earlier, for grouping activities into blocks. Within a scope, one can define local variables and partners links, as well as *fault handlers*, *event handlers*, *compensation handlers*, and *termination handlers*. These different types of handlers will be explained later

The structured activities can be nested and combined in arbitrary ways. This enables the presentation of complex structures in a BPEL process. In addition to the structured activities, BPEL offers a set of *basic activities*. Basic activities correspond to atomic actions. These include the following:

- *invoke*, as already mentioned, for invoking operations offered by the (external) partner Web services
- *receive*, waiting for messages from partners Web services
- *reply*, sending responses to requests previously accepted via a receive activity
- *wait*, delaying the process execution either for a period of time or until a certain deadline is reached
- *assign*, updating variables
- *throw*, signaling faults (i.e., exceptions) during the process execution
- *rethrow*, propagating the faults that are not solved
- *compensate*, triggering a compensation handler
- *empty*, doing nothing
- *exit*, immediately ending a process

To demonstrate some of these basic and structured activities let us continue with the Order Fulfillment example. In Fig. 15.3, the process is extended with an XOR split related to the *Ordering* subprocess (the extended code is written in bold style). In BPEL, this is captured through an *if* structured activity. An if activity contains a *condition*, an *activity* to be performed if the condition is fulfilled, as well as alternative path(s) of execution to be taken when the condition is not fulfilled. The alternative path(s) are specified through *elseif* and/or *else* constructs. In the example, if a purchase order is approved (as specified by *POApprovalResult* = *"approved"*), the

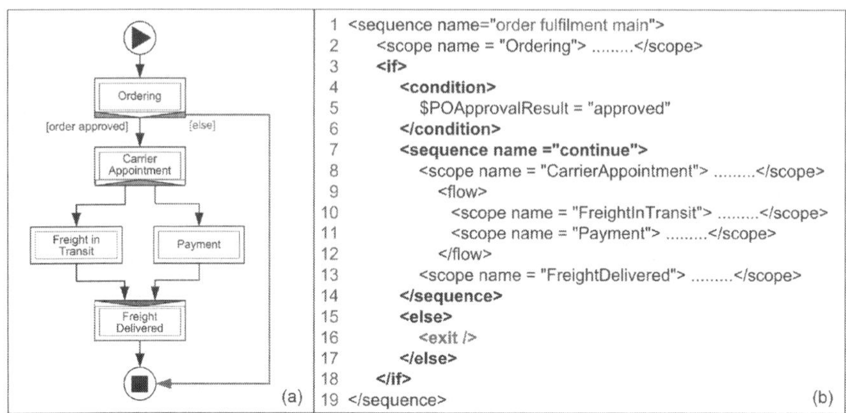


```
1 <sequence name="order fulfilment main">
2     <scope name = "Ordering"> .........</scope>
3     <if>
4        <condition>
5           $POApprovalResult = "approved"
6        </condition>
7        <sequence name ="continue">
8           <scope name = "CarrierAppointment"> .........</scope>
9              <flow>
10                <scope name = "FreightInTransit"> .........</scope>
11                <scope name = "Payment"> .........</scope>
12             </flow>
13          <scope name = "FreightDelivered"> .........</scope>
14       </sequence>
15       <else>
16          <exit />
17       </else>
18    </if>
19 </sequence>
```

Fig. 15.3 An extended version of the Order Fulfillment process in (**a**) YAWL; (**b**) BPEL

process will continue, otherwise it will be interrupted. The interruption is captured with the *exit* basic activity. Note, the *<sequence name="continue">* activity (lines 7–14) is introduced to frame the activities inside the *if* construct.

An essential difference between BPEL and YAWL is that BPEL was initially created for orchestrating Web services, while YAWL was created for automating business processes. A BPEL process definition essentially defines how a given service sends and receives messages to/from other services. The core BPEL language is not concerned with human activities. Thus, it can be assumed that each interaction with a user is provided through a designated Web service. When a user interaction is needed, the corresponding Web service has to be invoked. For our working example this means that every task in the YAWL process (except for routing tasks and automatic tasks) will be delegated to a Web service. The partnerLinks defined in Fig. 15.2a serve as the communication channels between these Web services. For clarity, these partnerLinks are named according to the groups/roles in YAWL, that is, the *Supply* department in the YAWL model is captured through a *Supply* partnerLink in BPEL.

Consider, for instance, the *Ordering* subprocess depicted in Fig. A.3 (which for convenience is reprinted in Fig. 15.4a). In BPEL, this subprocess is captured through the scope *Ordering* (see Fig. 15.4b). The task *Create Purchase Order* in YAWL, which also is the first task in the whole Order Fulfillment process, is performed by *OrdersManagement*. It is captured in BPEL as a *receive* activity during which the message *PurchaseOrder* from *OrdersManagement* is received. As this is the first activity, it will result in the creation of a new instance of the Order Fulfillment process. This is captured with the setting *createInstance = "yes"* (on line 11). Later on, the *Approve PO* operation provided through the *Supply* partnerLink's Web service is invoked. The communication is synchronous, which means that the process will not continue until the operation has finished and a result from it has been received. This is implied by the specification of an output variable from the operation. The result is then stored in the *POApprovalResult* variable (which as mentioned before

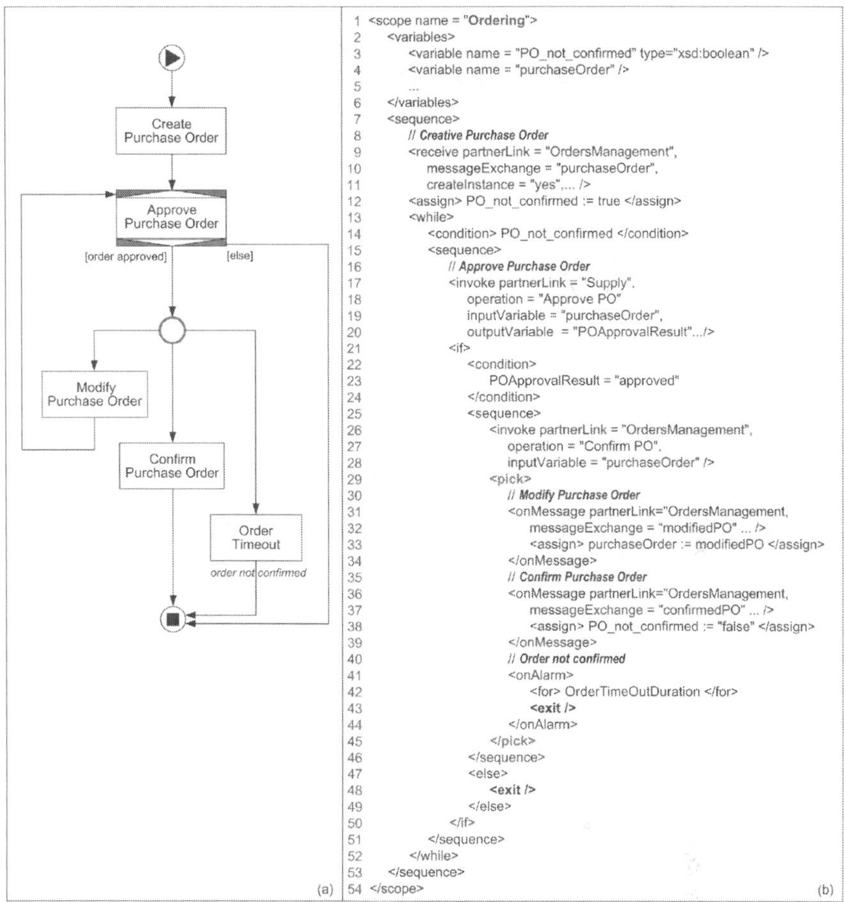

Fig. 15.4 The *Ordering* subprocess in (**a**) YAWL; (**b**) BPEL

is a global process variable). The next service invocation goes to the *Confirm PO* operation provided by *OrdersManagement*. This time the communication is asynchronous (as no output variables are specified in the invoke activity) and the process continues without the receipt of any feedback.

Let us now have a look at the control-flow of the process and how it is captured in BPEL. The *Ordering* subprocess in YAWL contains the following constructs: *sequence* between the tasks *Create PO* (purchase order), *Approve PO*, and *Confirm PO*; *exclusive choice* at the *Approve PO* task; *deadline* through the *Order Timeout* task; *loop* allowing for *Approve PO*, *Modify PO*, *Approve PO* iteration; and *deferred choice* between the tasks *Modify PO*, *Confirm PO*, and *Order Timeout*. These are captured in BPEL as follows.

- *Sequence*. Through the *sequence* activity (see lines 7–51) surrounding the definitions of the remaining activities in the subprocess.

- *Exclusive Choice.* Through an *if* activity testing the value of a variable called *POApprovalResult*. The variable *POApprovalResult* is a global variable, and its value is obtained during the execution of the operation *Approve PO* in the service provided by *OrdersManager*. If the order is not approved, the process instance will terminate, which is specified through the *exit* activity in the *else* path (on lines 47–49).
- *Loop.* The loop is captured through a *while* activity surrounding the invocations of the *Approve PO* and *Modify PO* operations (lines 13–52). The loop will be exited when the purchase order gets confirmed, that is, the variable *PO_not_confirmed* gets the value "false" (this loop condition is specified on line 14).
- *Deferred Choice.* A deferred choice is a choice made by the user or environment when a work item is allocated or started. In BPEL, a deferred choice can be captured through the *pick* activity. Pick specifies the behavior of a process based on timing or external triggers such as a message arrival. In the example, either a message containing a modified purchase order is received from the *OrdersManagement* (lines 30–34), or a message containing a confirmed purchase order is received (lines 35–39) or a set deadline expires.
- *Deadline.* The deadline is captured through an *onAlarm* element of the pick activity (see lines 40–44). An *onAlarm* specifies a deadline and the action to be taken in case the deadline expires. In the example, when the deadline expires the process will be canceled. The *onAlarm* part of the pick activity in this BPEL example corresponds to the *Order Timeout* task in YAWL.

As mentioned earlier, different types of handlers, for example, event handlers, fault handlers, and compensation handlers, can be attached to scope activities. An *event handler* is responsible for handling a message receipt event or a time-out event that occurs in parallel to the main activity within an active scope. It represents an *event–action rule* associated with the scope. An event handler is enabled while the scope is under execution. When an occurrence of the event associated with an enabled event handler is registered, the body of the handler is executed concurrently with the execution of the scope's main activity. The completion of the scope as a whole is delayed until all active event handlers have completed.

Fault handlers and *compensation handlers* are designed for exception handling in BPEL. Fault handlers define reactions to internal or external faults that occur during the execution of a scope. Some of these faults may be raised explicitly using a *throw* activity. Unlike event handlers, fault handlers do not execute concurrently with the scope's main activity. Instead, this main activity is interrupted before the body of the fault handler is executed. Compensation handlers, in conjunction with a compensate activity, enable a process to undo the effect of a successfully completed scope. In contrast to event and fault handlers, which can be defined for a whole process, compensation handlers can only be defined inside the boundary of a scope.

The event handler construct is exemplified in the Carrier Appointment subprocess (see Fig. 15.5). The subprocess starts with an order preparation phase. The order

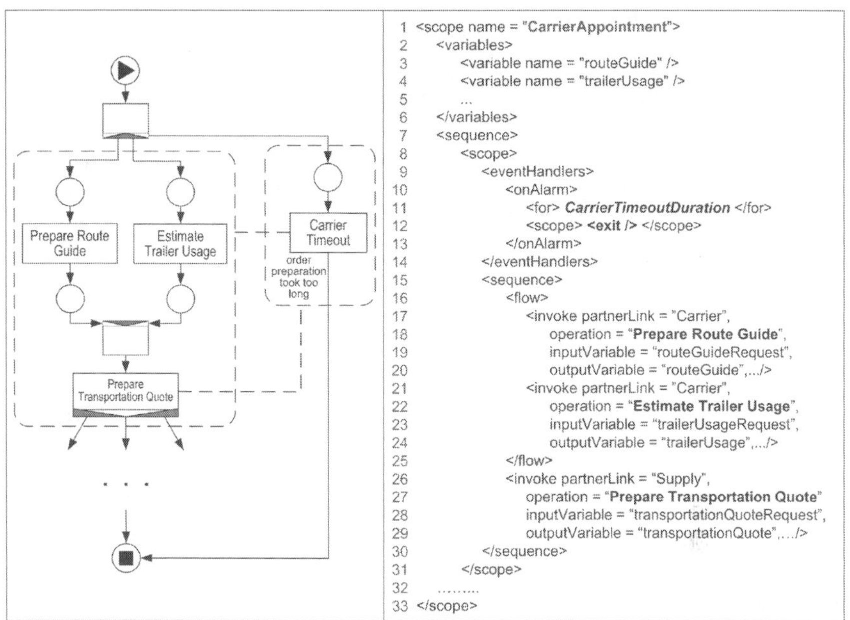

Fig. 15.5 The order preparation phase of the *Carrier Appointment* subprocess in (**a**) YAWL; (**b**) BPEL

preparation needs to be completed within a certain period of time; otherwise, upon a time-out the process instance will terminate. Such a timer is enabled once the order preparation starts and then runs *in parallel* with the preparation. In the BPEL process definition in Fig. 15.5, this is captured by a *scope* activity (lines 8–31) associated with an *event handler* (lines 9–14). The main activity of the scope is a *sequence* of activities for order preparation (lines 15–30), and the event handler captures the deadline upon the occurrence of an *onAlarm* event. The event handler is enabled once the scope is started. In case that the order preparation takes too long and the timer expires before the preparation is completed, the alarm event will be triggered and the *exit* activity will occur, terminating the current process instance.

Finally, in addition to block-structured constructs, BPEL defines *control links* which, together with the associated notions of *join condition* and *transition condition*, allow the definition of directed *acyclic* graphs of activities. A control link between activities X and Y indicates that Y cannot start before X has either completed or been skipped. Moreover, Y can only be executed if its associated join condition evaluates to true, otherwise Y is skipped. Although control links may be used to connect various constructs, their usage is restricted in a number of ways, for example, they cannot form a loop nor cross the boundary of a loop. The use of control links will be discussed with an example in the next section.

15.3 Workflow Patterns Support

This section focuses on the comparison between BPEL and YAWL in terms of their support for control-flow patterns. Table 15.1 lists the 20 original control-flow patterns (see Appendix C) and illustrates their implementations in BPEL and YAWL. Given that the reader is already knowledgeable in YAWL, the solutions provided by YAWL in the table can be used as explanations of the patterns.

YAWL supports all the 20 original control-flow patterns, except for the Implicit Termination pattern (Pattern 11). Implicit Termination captures a scenario of process termination that does not require an explicit end activity. In such a case, a process instance is terminated when there is nothing left to do. However, this pattern can hide design errors as it is not possible to detect deadlocks. To prevent the designer from making such errors, YAWL deliberately does not provide support for Implicit Termination.

For BPEL, version 1.1 offers support for 13 patterns and the standard version 2.0, by introducing the *forEach* construct, extends the support to three more patterns. Hence BPEL currently supports 16 patterns in total. Because of its block-structured nature, BPEL does not offer direct support for Multiple Merge (Pattern 8), Arbitrary Cycles (Pattern 10), and Milestone (Pattern 18). Furthermore, it does not provide support for Multiple Instances without a priori Runtime Knowledge (Pattern 15). In the following, let us look in detail at the patterns support by BPEL as shown in Table 15.1.

The structured activities *sequence*, *flow*, and *if* (as introduced in Sect. 15.2) capture Patterns 1–5, the set of basic control-flow patterns. The block-structure of BPEL can be observed in the table by the fact that two patterns are captured in BPEL through one construct and that every structured activity is defined through a pair of tags, a start tag and an end tag, delineating a "block." This implies that some unstructured scenarios (e.g., when there is no one-to-one correspondence between a split point and a synchronization point) cannot be captured in BPEL. As mentioned earlier, links can be used in addition to structured activities to capture control-flow dependencies between activities. However, the usage of links is governed by a number of restrictions, for example, they are not allowed to form loops nor to cross the boundary of a loop, a scope and any handler attached to the scope, etc. As a results, BPEL provides limited support for unstructured modeling.

The *control links* are used to capture the Multiple Choice and Synchronizing Merge patterns. As the BPEL code in Table 15.1 shows, the activities *A1* and *A2* are defined to run in parallel. They are enabled if the conditions *C1* or *C2* specified on the links going into *A1* and *A2* evaluate to true. Several of these conditions can simultaneously evaluate to true, which means that one or both *A1* and *A2* will be executed, capturing the Multiple Choice behavior. To capture the Synchronizing Merge pattern, an empty activity is introduced as the target of the two control links: one leading from activity *A1*, the other from activity *A2*. Furthermore, a join condition specifying an OR expression is specified. The join condition is evaluated when the values of all links included in it can be determined, that is, in the example after both *A1* and *A2* have been completed. If one of *C1* or *C2* evaluates to false and as a

Table 15.1 A Comparison between BPEL and YAWL using the original 20 control-flow patterns

No.	*Pattern*	*YAWL*	*BPEL*
1	**Sequence**	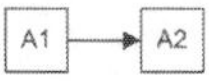 A1 A2	<sequence> activity A1 activity A2 </sequence>
2	**Parallel Split**	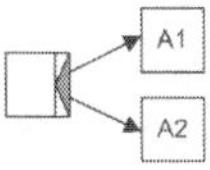 A1 A2	<flow> activity A1 activity A2 </flow>
3	**Synchronization**	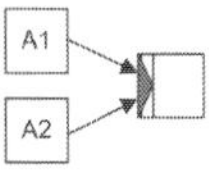 A1 A2	
4	**Exclusive Choice**	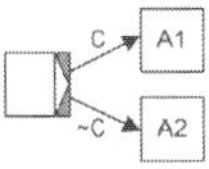 C A1 ~C A2	<if> <condition>C</condition> activity A1 <else> activity A2 </else> </if>
5	**Simple Merge**	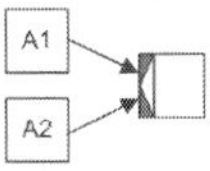 A1 A2	
6	**Multi Choice**	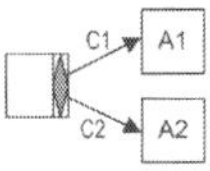 C1 A1 C2 A2	<flow> <links> <link name="SplitToA1"/> <link name="SplitToA2"/> <link name="A1ToMerge"/> <link name="A2ToMerge"/> </links> <empty name="Split"> <sources> <source linkName="SplitToA1"> <transitionCondition>C1</transitionCondition> </source> <source linkName="SplitToA2"> <transitionCondition>C2</transitionCondition> </source> </sources> </empty> activity A1 <targets><target linkName="SplitToA1"/></targets> <sources><source linkName="A1ToMerge"/></sources> activity A2 <targets><target linkName="SplitToA2"/></targets> <sources><source linkName="A2ToMerge"/></sources> <empty name="Merge"> <targets> <joinCondition> $A1ToMerge OR $A2ToMerge </joinCondition> <target linkName="A1ToMerge"/> <target linkName="A2ToMerge"/> </targets> </empty> </flow>
7	**Synchronizing Merge**	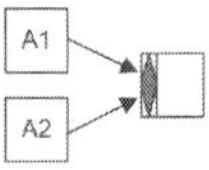 A1 A2	
8	**Multi Merge**	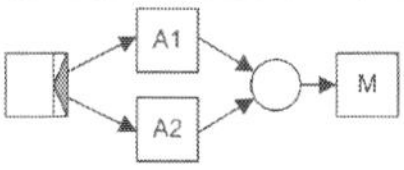 A1 A2 M	*(not supported)*
9	**Discriminator**	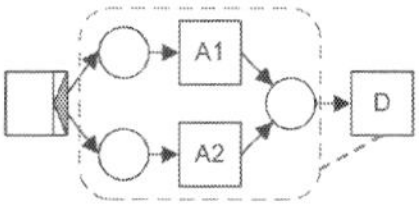 A1 A2 D	*(not supported)*
		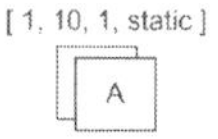 [1, 10, 1, static] A	<forEach counterName="i" parallel="yes"> <startCounterValue>1</startCounterValue> <finalCounterValue>10</finalCounterValue> <completionCondition>1</completionCondition> <scope> activity A </scope> </forEach>
10	**Arbitrary Cycle**	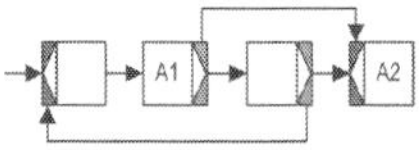 A1 A2	*(not supported)*

(continued)

Table 15.1 (continued)

11	**Implicit Termination**	*(not supported)*	<flow> <links> <link name="A1toA3"/> <link name="A2toA4"/> </links> activity A1 <sources> <source linkName="A1toA3"/> </sources> activity A2 <sources> <source linkName="A2toA4"/> </sources> activity A3 <targets> <target linkName="A1toA3"/> </targets> activity A4 <targets> <target linkName="A2toA4"/> </targets> </flow>
12	**MI without Synchronization**	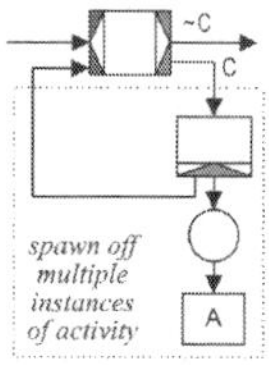 ~C C *spawn off multiple instances of activity* A	<process name="Main"> ... <while> <condition>C</condition> <invoke process-MI ... /> </while> ... </process> <process name="MI"> <receive process-Main ... creativeInstance="yes"/> activity A </process>
13	**MI with a *Priori* Design-Time Knowledge**	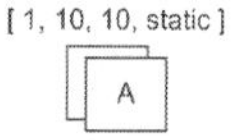 [1, 10, 10, static] A	<forEach counterName="i" parallel="yes"> <startCounterValue>1</startCounterValue> <finalCounterValue>10</finalCounterValue> <scope> activity A </scope> </forEach>
14	**MI with a *Priori* Run-Time Knowledge**	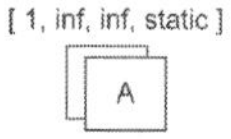 [1, inf, inf, static] A	<forEach counterName="i" parallel="yes"> <startCounterValue>1</startCounterValue> <finalCounterValue>x</finalCounterValue> <scope> activity A </scope> </forEach>
15	**MI without a *Priori* Run-Time Knowledge**	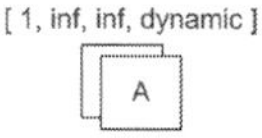 [1, inf, inf, dynamic] A	*(not supported)*
16	**Deferred Choice**	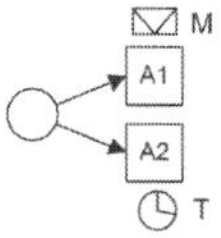 M A1 A2 T	<pick> <onMessage M> activity A1 </onMessage> <onAlarm> <until>T</until> activity A2 </onMessage> </pick>
17	**Interleaved Parallel Routing**	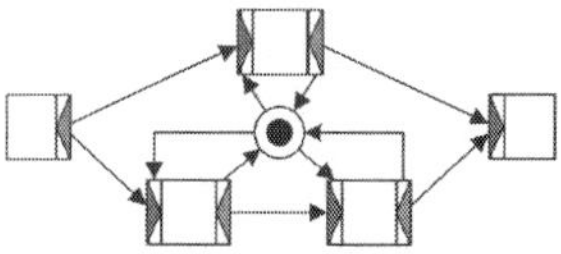	*(indirectly supported by serializable scopes but limited to activities within the same scope)*
18	**Milestone**	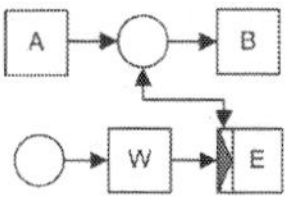 A B W E	*(not supported)*
19	**Cancel Activity**	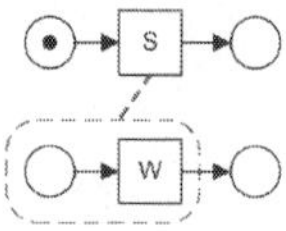 S W	<scope> <faultHandlers> <catch faultName="Wrong"> activity S </catch> </faultHandlers> <sequence> <throw faultName="Wrong"/> activity W </sequence> </process>
20	**Cancel Case**	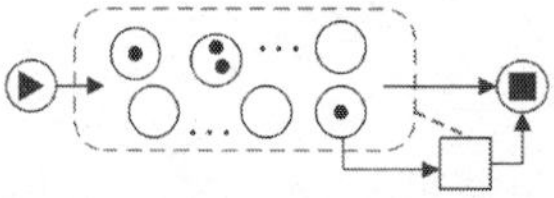	<process> ... <exit/> ... </process>

consequence its corresponding activity *A1* or *A2* is not executed, the false value will be propagated further so that the join condition can be evaluated.

In BPEL, none of the structured constructs can support the Multiple Merge pattern. This pattern resembles a point in a process where two or more branches reconverge without synchronization. If more than one branch gets activated, the activity following the merge will be performed for every action of every incoming branch. In YAWL, this can easily be captured by an XOR-join task, which has multiple incoming branches that are activated, for example, as a result of a preceding AND-split task. However, as BPEL does not allow for two threads of execution to run through the same path in a single process instance, the control links cannot implement this pattern either.

The *forEach* activity in BPEL captures the Discriminator pattern. A forEach activity has a multiple instance counter and will execute its child activity $N + 1$ times, where N equals the final counter value minus the start counter value. Both the start and the final counter values can be specified at design time or they can be evaluated (from expressions) when the forEach activity starts. To ensure parallel execution of the children, the setting *parallel* = *"yes"* has to be used. Also, to capture the discriminator, the value *"1"* has to be specified as the *completionCondition*, which implies that the first instance to complete will continue the flow. This attribute is similar to the *threshold* parameter of a multiple instance task in YAWL. Note, however, that through the forEach activity only a special case of the Discriminator pattern is captured, that is, the case in which all the activities running in parallel are the same (and captured as children of forEach).

In contrast to YAWL, BPEL provides support for Implicit Termination but does not support Arbitrary Cycles. An Arbitrary Cycle is also called an unstructured loop, that is, a loop that has more than one entry point and/or more than one exit point. In YAWL, this can be directly captured by a combination of join and split tasks. In BPEL, the *while*, *repeatUntil*, and *forEach* (with *parallel*=*"no"* setting) activities allow for structured loops only, that is, loops with exactly one entry and one exit point. Furthermore, control links offer no support in this case due to the restriction that they cannot form a loop nor cross the boundary of a loop. On the other hand, BPEL supports Implicit Termination through the flow construct. In a flow, multiple activities can be defined, which are not source of any link, and the process termination depends on the termination of these multiple activities instead of one unique termination activity.

To capture MI without Synchronization requires a facility for spawning off new threads of control that are independent of each other. BPEL offers a way of embedding an invoke activity in a while loop for initiating a separate process (i.e., process-MI). The invoked process, by setting the attribute *createInstance* to *"yes"* in its first receive activity, will be instantiated upon each execution of the invoke activity in the calling process (e.g., process-Main). Also, BPEL uses forEach constructs to implement MI with a priori Design Time/Runtime Knowledge. However, as both the start and the final counter values in a forEach activity have to remain constant till the activity is completed, it is not possible to capture MI without a priori Runtime Knowledge.

In BPEL, the *pick* constructs implement the Deferred Choice pattern, as shown in the example in Sect. 15.2. Using the concept of *isolated scopes*,[4] BPEL provides limited support for the Interleaved Parallel Routing pattern. Isolated scopes are scopes with the attribute *isolated* set to *"yes"* to provide control of concurrent access to shared resources such as variables and partner links. Suppose two concurrent isolated scopes, *S1* and *S2*, access a common set of external variables and partner links for read/write operations. When there is a conflict between accessing the shared variables and partner links, the read/write operations will be reordered so that either all operations within *S1* are completed before any in *S2* or vice versa, and the results will remain the same irrespective of the reordering. However, the semantics of isolated scopes are complicated and not clearly defined in BPEL, and thereby no further discussion is given. Next, BPEL does not provide direct support for Milestone. This pattern is defined as a point in the process where a given activity (A) has finished and an activity (B) following it has not yet started. Another given activity (E) can only be enabled if a milestone has been reached, which has not yet expired (see the YAWL net of the Milestone pattern in Table 15.1).

Finally, BPEL supports Cancel Activity through the use of *fault handlers* and *compensation handlers*, specifying the course of action in cases of faults and cancelation. The Cancel Case pattern is directly realized by the *exit* activity, which can be seen from the example of the Order Fulfillment process in Sect. 15.2.

15.4 Epilogue

In summary, three main differences separate YAWL and BPEL. The first one is the integrated support offered in YAWL for routing of both automated and human tasks. In BPEL, support for human tasks is the realm of a separate specification, namely BPEL4People, while in YAWL, the control-flow and the resource allocation perspectives are tightly integrated. This difference essentially comes from the basic abstractions upon which YAWL and BPEL rely. YAWL relies on the concept of task as the elementary building block, while BPEL's building block is that of action, with an emphasis on communication actions (send/receive) and data manipulation actions (variable assignment). This is natural as BPEL was designed from the start as a language for implementing Web service orchestrations.

Second, BPEL is primarily a block-structured language, while YAWL is a graph-oriented language. As a result, BPEL has some limitations when it comes to capturing certain control-flow structures. Specifically, with respect to the Synchronizing Merge (OR-join) pattern, BPEL provides support for OR-joins only in process model fragments that do not contain cycles, while YAWL supports the OR-join pattern in any topology. Similarly, exception handling in BPEL is supported in the context of single-entry single-exit structures (scopes), whereas in

[4] These were known as *serializable scopes* in BPEL 1.1.

YAWL, cancelation regions can take arbitrary forms.[5] In general, translating from graph-oriented languages to block-structured languages is a challenging task due to structural mismatches between these two classes of languages. Consider an example of translation from BPMN to BPEL (which enables the execution of a BPMN process model in a BPEL engine). A model captured in BPMN can have an arbitrary topology, while in BPEL, if a segment of a BPEL model starts with a branching construct, it ends with the corresponding synchronization construct. A mapping from BPMN to BPEL needs to handle structural mismatches properly and may still result in BPEL code that is hard to understand.

Third, right from the start, YAWL has been designed with the aim of having a formal grounding that enables static analysis of YAWL nets. While some formalizations of BPEL have been developed in the literature, these efforts fall outside the BPEL standardization effort. In particular, a translation from BPEL to YAWL has been developed to perform behavioral analysis of BPEL process definitions using the analysis techniques available for YAWL. However, the translation basically provides execution semantics of each BPEL construct in YAWL. In other words, every BPEL activity (even an elementary activity) is mapped to a YAWL net capturing both the normal execution and the possible exception handling behavior of the activity. This approach has a negative impact on the readability of the resulting YAWL nets. Thus, while the translation enables the static analysis of BPEL process definitions, it would not be usable if the purpose was to migrate an application built in BPEL to YAWL.

On the other hand, BPEL is the result of a substantial standardization effort and it is supported by sophisticated tools that are integrated into commercial application development environments. Also, BPEL skills are becoming a commodity and there is a significant community of BPEL developers.

Exercises

Exercise 1. List three major differences between BPEL and YAWL.

Exercise 2. Complete the BPEL specification of the Order Fulfillment process (see Sect. 15.2) following its definition in YAWL (see Appendix A).

Exercise 3. Is there a general solution in BPEL for capturing Multiple Merge, Discriminator, Arbitrary Cycles, Multiple Instances without a Priori Runtime Knowledge, and Deferred Choice? Provide the reasons and examples to support your answer.

Exercise 4. YAWL provides a solution of cancelation region, which can be used to support both Cancel Activity and Cancel Case patterns. How does BPEL support

[5] YAWLeX discussed in Chap. 5 is another way of dealing with exceptions in YAWL.

these cancelation patterns? Give an example of a cancelation scenario that can be captured in YAWL, but not in BPEL. Provide the solution for YAWL and explain why BPEL cannot express it.

Exercise 5. Translate the seven BPMN models in the Exercises in Chap. 13 to BPEL processes.

Chapter Notes

BPEL emerged as a compromise between IBM's WSFL (Web Services Flow Language) and Microsoft's XLANG (Web Services for Business Process Design), superseding the corresponding specifications. The first version of BPEL (1.0) was published in August 2002, the second version (1.1) was released in May 2003 as input for the standardization within OASIS, and the third version, that is, BPEL 2.0 [35], became an OASIS standard in April 2007.

The BPEL 2.0 specification defines only the kernel of BPEL, which mainly involves the control-flow logic, limited definitions of data handling, and even fewer definitions of communication aspects. A number of BPEL extensions have been developed to support human tasks and subprocesses in a BPEL process. To enable the modeling of human participants in a Web service composition setting, the BPEL4People specification [29] was developed as an add-on to BPEL. WS-HumanTask [28] is another BPEL extension that provides support to human tasks. A joint proposal "BPEL Extensions for Subprocesses" (BPEL-SPE) [136] formulates BPEL extensions for defining subprocesses, which are fragments of BPEL code that can be reused within the same or across multiple BPEL processes. However, this extension has not been standardized.

In [50], van Breugel and Koshkina wrote a survey about several research efforts on defining formal semantics of BPEL using various formalisms. Wohed et al. [262] reported a detailed evaluation of BPEL 1.1 using the original 20 control-flow patterns and also discussed possible workaround solutions for patterns that are not (directly) supported in BPEL 1.1.

Ouyang et al. [187] developed a systematic translation technique for mapping BPMN models to BPEL processes without imposing restrictions on the source BPMN models. The technique also addressed issues that arise when translating from graph-oriented to block-structured flow definition languages. Brogi and Popescu [51] proposed an approach for translating BPEL processes into YAWL workflows in order to perform behavioral analysis of BPEL using the analysis techniques available for YAWL.

Chapter 16
Open Source Workflow Systems

Petia Wohed, Birger Andersson, and Paul Johannesson

16.1 Introduction

The goal of this chapter is to broaden the reader's knowledge in the area of open source WfMS. To achieve this we introduce three other open source WfMSs. These are OpenWFE, jBPM and Enhydra Shark, which according to download statistics (July 2008) are the open source systems with the largest number of downloads (closely followed by YAWL). The purpose of the presentation is not to provide detailed insight into each of these systems, but rather to expose the reader to different approaches and to discuss the similarities and differences of these approaches with regard to YAWL.

The chapter is divided into three parts, each describing one system. The descriptions follow the same format as much as possible. First, some background information is given. Subsequently, the architecture is described. Then an introduction to the underlying process modeling language is given from control-flow, data, and resource perspectives. After that, a part of the Order Fulfillment case is modeled and the solution briefly discussed. Each description concludes with a brief comparison of the system and YAWL. All files containing the discussed examples are distributed for test-runs with the electronic supplement of the book.

16.2 OpenWFEru: Ruote

OpenWFEru, also called Ruote, is a workflow management system written in Ruby. Of the open source projects presented in this book, this was the first project to be registered at an open source repository (May 2002 on SourceForge). The initial development was done in Java and the tool/project was called OpenWFE (or sometimes OpenWFEja). In November 2006, the development migrated to Ruby and, as

[1] openwferu.rubyforge.org

P. Wohed (✉)
Stockholm University and The Royal Institute of Technology, Stockholm, Sweden,
e-mail: petia@dsv.su.se

A.H.M. ter Hofstede et al. (eds.), *Modern Business Process Automation*,
DOI 10.1007/978-3-642-03121-2_16, © Springer-Verlag Berlin Heidelberg 2010

a result, the distribution moved to RubyForge. The tool is intended for developers and distributed under the BSD License. The current (July 2008) development status is 4-Beta.

16.2.1 Architecture

The architecture of the system, following the Workflow Management Coalition's reference model,[2] is shown in Fig. 16.1. It consists of a number of components: a workflow engine, workflow client(s), administration and monitoring tool(s), and process definition tool(s). We sometimes use OpenWFEru or Ruote to denote the entire system, but for the sake of precision, it should be noted that *OpenWFEru (Ruote)* solely refers to the workflow engine component. The engine interfaces with end users and administrators through a workflow client. For this chapter, we have used the *ruote-web* (also called *Densha*) client, which is a web-based application providing end-user and administrator functionality to the engine.

Two central parts of Densha are the *worklist window* and the *stores window*, shown in Figs. 16.2 and 16.3. The *worklist window* provides administration functionality and is available only to administrators of the system. Note that the name "worklist" is used in a rather untraditional way, as no actual worklists are displayed there. The second figure displays the *stores window*. A *store* is a storage space for work items and in a way resembles a worklist in YAWL. In contrast with YAWL, however, where worklists are personal, several users in Ruote may have access to the same store and one user may access several stores. For instance, in Fig. 16.2, Tom, who is a *Senior Supply Officer*, is granted read–write-delegate privileges to

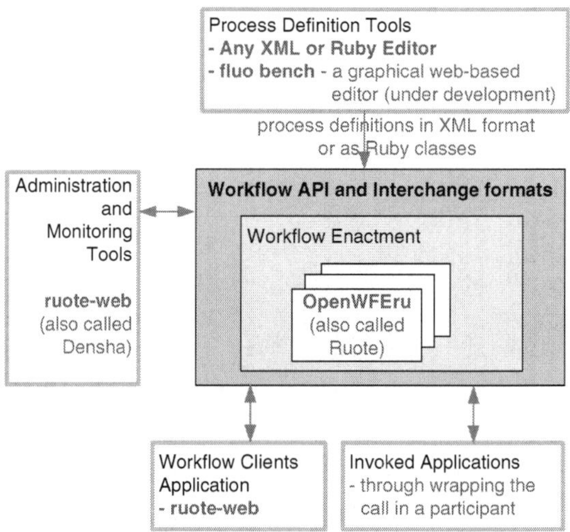

Fig. 16.1 OpenWFEru – architecture

[2] www.wfmc.org/standards/referencemodel.htm

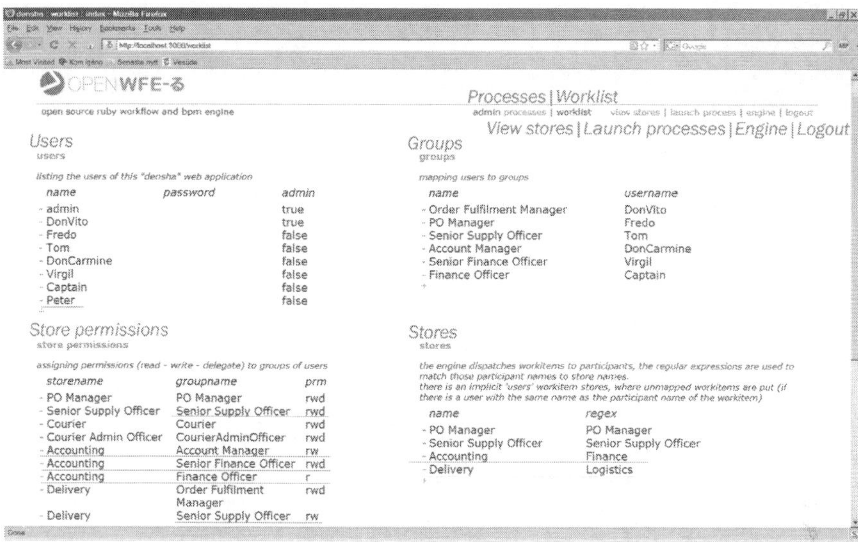

Fig. 16.2 Ruote-web (Densha) – the *worklist* menu available to administrators (The italicized annotations and the underlinings are added to the screenshot to increase its readability or to stress issues discussed in the text)

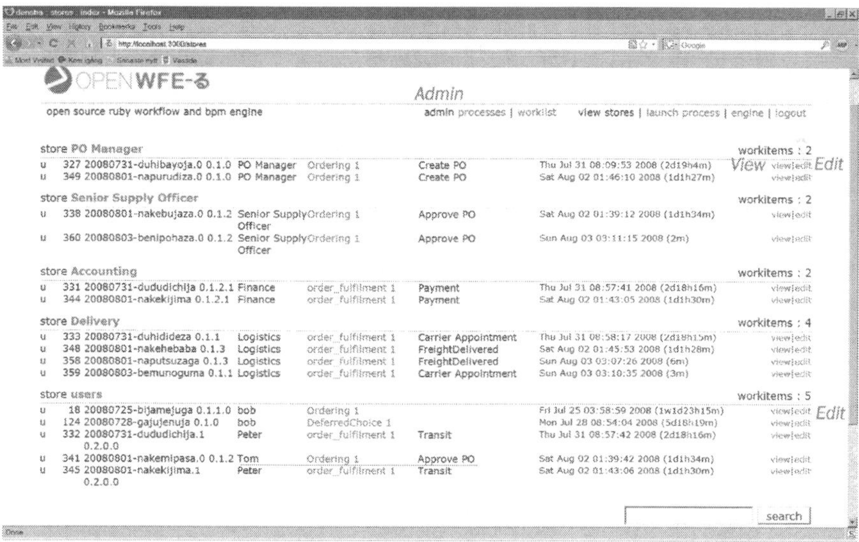

Fig. 16.3 Ruote-web (Densha) – annotated *view stores* window (administrator's view)

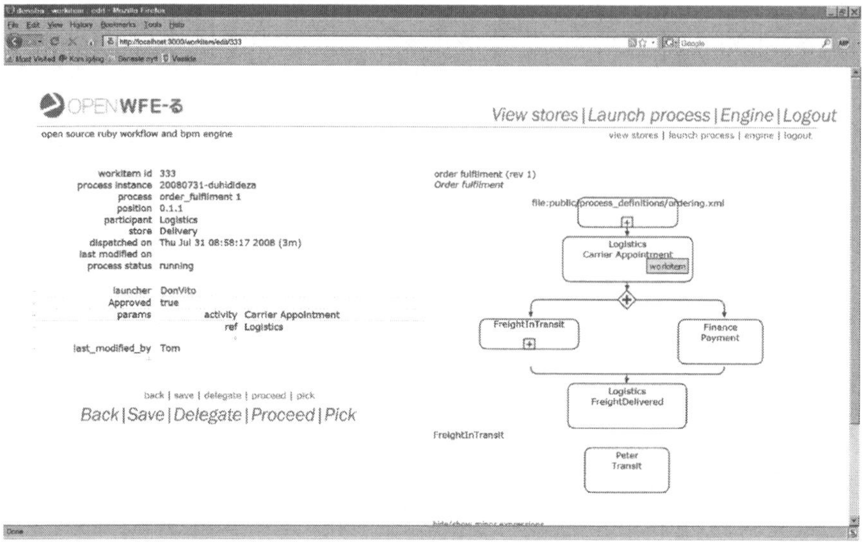

Fig. 16.4 Ruote-web (Densha) – execution of the Carrier Appointment task (annotated window)

the Senior Supply Officer store and read–write privileges to the Delivery store. Furthermore, the store Accounting has different privilege levels accessible to several groups.

A WfMS is configured based on process models, which are typically produced by process analysts and designers and defined in some language through an editor. Ruote uses its own process definition language (which is introduced in the following sections) and accepts processes defined in XML format or as Ruby classes. In contrast with YAWL, which is a graphical language, Ruote's process language is a structured textual notation resembling a programming language. Therefore, any (XML or Ruby) editor can be used as a process definition tool. During execution, Ruote transforms the textual definitions into graphical models and displays them to users as navigation maps. Figure 16.4 shows a screenshot of the execution of the task *Carrier Appointment*, where a corresponding navigation map is displayed along with other data relevant to the execution. Note the *work item* label placed in the model, which denotes the executing task.

16.2.2 The Participant Concept

A fundamental concept in Ruote is that of participant. A *participant* is an abstract entity to which work can be distributed (see Fig. 16.5a, lines 6, 9, 11, and 16). Participants can be related to users, groups, or applications. The purpose of using participants is to keep the definition of process control-flow models separate from the organizational models of enterprises. This increases flexibility, as one and the

```
1 <process-definition name="Order fulfilment" revision="1">
2   <description> Order fulfilment </description>
3
4   <sequence>
5     <subprocess ref="file:public/processes/ordering.xml"/>
6     <participant ref="Logistics" activity="Carrier Appointment"/>
7     <concurrence>
8       <subprocess ref="FreightInTransit" />
9       <participant ref="Finance" activity="Payment" />
10    </concurrence>
11    <participant ref="Logistics" activity="FreightDelivered" />
12  </sequence>
13
14  <!-- Subprocess defintion -->
15  <process-definition name="FreightInTransit">
16    <participant ref="Peter" activity="Transit" />
17  </process-definition>
18
19 </process-definition>
```

```
1 <process-definition name="Ordering" revision="3">
2   <description>Ordering sub-process</description>
3
4   <sequence>
5     <participant ref="PO Manager"
6                  activity="Create PO" />
7     <set field="Approved" value="true"/>
8     <participant ref="Senior Supply Officer"
9                  activity="Approve PO" />
10  </sequence>
11
12 </process-definition>
```

Fig. 16.5 Simplified Order Fulfillment process in Ruote's process language: (**a**) main process; (**b**) Ordering subprocess defined in file ordering.xml. The text written in *gray* is optional and can be omitted from the process definition

same process definition can be deployed in different organizational settings and the participants defined in the process model can be mapped in different ways to corresponding organizational units. Note the difference here from YAWL, where the specification of a task distribution in a process model presupposes an existing organizational model.

A number of predefined participant types are available and additional types can be defined on demand. Out of the box, the most important participant types are *ActiveStoreParticipant* and *ProcessParticipant*. *ActiveStoreParticipants* are the participants whose work items are collected in stores (think of these stores as worklists for the moment) and managed by the engine. Typically they are mapped to (human) users or groups. This mapping is done implicitly or explicitly. An implicit mapping is done when the name of a participant in the process model is identical to the name of a user defined in Densha. An example of this is the user and participant Peter shown in Fig. 16.2 and line 16 in Fig. 16.5a. An explicit mapping is necessary when participants are mapped to groups or when distinct names for participants and users/-groups have been used. It is done through the *stores* menu in Densha. See the bottom right corner in Fig. 16.2, where, for instance, the participant Finance (present in the process definition in Fig. 16.5) is mapped to the store Accounting, which is accessible to the *Accounting Manager*, *Senior Finance Officer*, and *Finance Officer* groups.

The second important type of participant is *ProcessParticipant*. A *ProcessParticipant* is a participant who invokes a subprocess (see Fig. 16.5a lines 5 and 8). It is the construct used for decomposing process models. Subprocesses can either be specified within the definition of the main process (Fig. 16.5a, lines 15–17) or they can be defined as independent process models residing separately from the main process files, which also enables invocation from several main processes (see Fig. 16.5b and Fig. 16.5a, line 5).

Examples of other predefined participant types are *FileParticipant* for exporting work items to files; *MailParticipant* for sending email notifications; *SoapParticipant* for web-services calls; and *BlockParticipant* for specifying executable Ruby code in a process model, hence allowing for the specification of automated participants. Naturally, automated behavior can also be specified through external applications. As the invocation of an application is considered as work distribution to that application, a designated participant has to be included to realize the call. Such a participant is defined on demand.

16.2.3 Control-Flow

Ruote's process definition language is block-structured. Some of the most frequently used expressions are *sequence*, *concurrence*, *if*, and *loop*, with their usual semantics. Furthermore, the following expressions are common: *cursor*, which is like a sequence, with the additional characteristics that the sequence of activities can be aborted (which is done through a *break* expression) or the execution reversed to an earlier activity in that sequence (which is defined through a *rewind* expression); *case* for specifying branching in a process with more than two alternative threads of execution, where the first condition that evaluates to true will continue the flow; *redo* for capturing repetitive execution of a segment of a process model; and *timeout* expressions, which allow for restricting the execution of process segments to predefined time frames. Every expression may be named through an optional *tag* attribute and through this name referred to from other places or expressions in the process, for instance through a *redo* expression.

The occurrence of sequence and concurrence are highlighted in the XML process definition shown in Fig. 16.5. The structure mimics the Order Fulfillment example in YAWL shown in Fig. A.1 on page 600 and contains the tasks/subprocesses *Ordering*, *Carrier Appointment*, *FreightInTransit*, *Payment*, and *FreightDelivered*.

16.2.4 Data

In OpenWFE, data handling is realized through variables and fields. *Fields* hold data values relevant for a work item, which are populated and/or accessed by end users, while *variables* hold data values that are used for internal processing and are not visible to or accessible by end users. Figure 16.5b, line 7, shows how a field, called *Approved*, is defined.

After being defined, a field is accessible to all subsequent tasks, which means that it is visible to end users when executing these tasks. For instance, the definition of the field *Approved* in Fig. 16.5b preceding the activity *Approve PO* causes the field to be displayed during the execution of this activity and every activity following it; see, for example, the field *Approved* in Fig. 16.4 displayed during the execution of the activity *Carrier Appointment*. Furthermore, there are a number of predefined fields. We will exemplify one of them later. Variables are defined in the same way as fields but the system word *variable* is used instead. A distinction is made between local (subprocess) variables, process variables, and engine variables.

Data can be transferred between fields and variables at any time during process execution except during the execution of tasks. Operations on data elements can be done either through *BlockParticipants* (which, as mentioned earlier, enable the definition of Ruby code), or when data elements are passed between tasks.

16.2.5 Resources

The life-cycle of a work item in Ruote is depicted in Fig. 16.6. When a work item is *created*, it is put into the store associated with its participant. As many users may have access to a store, this transition is considered to lead to the work item being *offered to multiple resources*. Alternatively, a participant may implicitly be associated with a user through the use of identical names. The work item is then put into the worklist of that user, which is shown under a predefined store called *users*. Hence the work item enters the *offered to a single resource* state. For an example of this, see the store in Fig. 16.3 and the work items for the task *Transit* offered to Peter. Note that as this screenshot shows an administrator's view, the work items of all users are listed under the user's store. If Peter was logged in, only the work items allocated to him would have been displayed there.

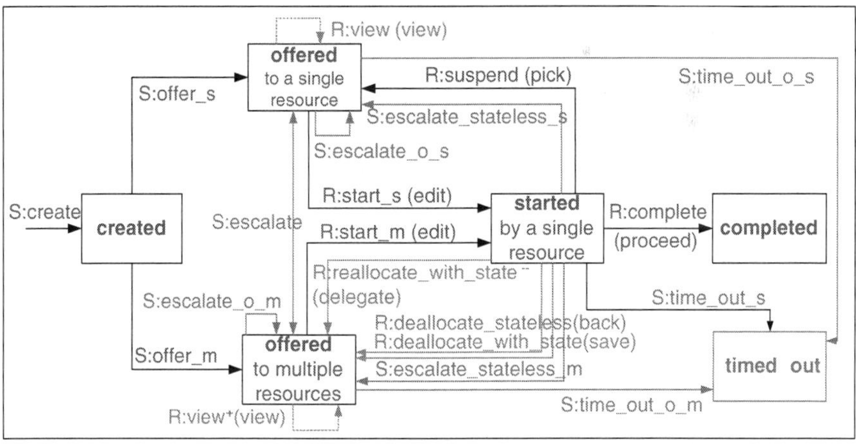

Fig. 16.6 The life-cycle of a work item in Ruote. *Black* is used for states and transitions, as in the YAWL work item life cycle diagram Fig. 2.34 on page 73. Recall that the prefixes S (for system) and R (for resource) indicate the initiators of the different transitions. A *plus symbol* indicates that certain privileges are required to initiate the corresponding transition. The *dotted arrow* for R:view indicates that viewing is not a real state transition as the state of a work item does not actually change. The *minus symbol* for R:reallocated_with_state (delegate) indicates that only users with delegate authority can trigger this transition

Users can view work items from stores they have read access to, which is captured by two corresponding view transitions in Fig. 16.6. A user starts the execution of a work item by selecting *Edit* for that work item, either from their private worklist or from a store for which they have write privileges. In contrast to YAWL, Ruote does not have a notion for the state in which a work item has been accepted by a user but not yet started. This means that the state *allocated* from the work item life-cycle in YAWL (shown in Fig. 2.34, page 73) is not present in Fig. 16.6.

A user who has started a work item (see Fig. 16.4) may perform one of the following activities: modify the data fields offered and complete the work by selecting the *proceed* tab; suspend the execution to a later point in time by selecting the *pick* tab (which will result in the work item being moved to the user's personal worklist); save the data changes made and return the work item to its store by selecting the *save* tab; reject the updates made and return the work item to its store by selecting *back*; if authorized, reallocate the work item to a different store by selecting the *delegate* option. All these transitions are depicted in the life-cycle diagram. They are named with YAWL terminology. The corresponding Ruote commands are given in brackets.

Through its notion of timers, Ruote can time out offered as well as started work items (which in Fig. 16.6 is depicted by three gray S:time_out transitions). A timed out activity can either be skipped or reoffered to a different store or user. When reoffered, the behavior corresponds to that of escalation in YAWL (see the gray S:escalate transitions in Fig. 16.6).

16.2.6 Example

Let us extend the process definitions in Fig. 16.5 and model the processes as defined in YAWL (see Fig. A.1 on page 600 and Fig. A.3 on page 603). In doing so, we will demonstrate some of the more advanced modeling constructs in Ruote. The solution is shown in Fig. 16.7. To follow the definition in YAWL closely, we start with the introduction of five subprocesses: Ordering, CarrierAppointment, FreightInTransit, Payment, and FreightDelivered (see Fig. 16.7a). The first of these subprocesses is defined in Fig. 16.7b and will be explained shortly; the other four are left as exercises for the reader.

To capture the deadlines for the completion of the Ordering and CarrierAppointment subprocesses, two timeout expressions surrounding their invocations are used (see Fig. 16.7a, lines 5–7 and 10–12). The abbreviations 1d and 1w for 1 day and 1 week show the time intervals within which the corresponding subprocesses have to be completed. If a timeout expires, Ruote sets a field *__timed_out_* to true. To capture the fact that the expiry of a deadline will lead to immediate completion of the process, the *break* expression is used. As break can only be used within a cursor construct, the sequence from the simplified solution is replaced with a cursor expression (see Fig. 16.5a lines 4–20).

```
<process-definition name="order fulfilment" revision="2">
  <description> Order fulfilment </description>

  <cursor>
    <timeout after="1d">
      <subprocess ref="file:public/.../ordering2.xml"/>
    </timeout>
    <break if="${field:__timed_out__}" />

    <timeout after="1w">
      <subprocess ref="CarrierAppointment" />
    </timeout>
    <break if="${field:__timed_out__} == true" />

    <concurrence>
      <subprocess ref="FreightInTransit" />
      <subprocess ref="Payment" />
    </concurrence>
    <subprocess ref="FreightDelivered" />
  </cursor>

  <!-- Subprocesses -->
  <process-definition name="FreightInTransit"> ...
  <process-definition name="CarrierAppointment"> ...
  <process-definition name="Payment"> ..
  <process-definition name="FreightDelivered"> ...
</process-definition>
```

```
1 <process-definition name="Ordering2" revision="1">
2   <description>Ordering sub-process </description>
3 <sequence>
4   <participant ref="PO Manager" activity="Create PO"/>
5   <sequence tag="Approve">
6     <set field="Approved" value="true"/>
7     <participant ref="Senior Supply Officer" activity="ApprovePO"/>
8     <if> <equals field-value="Approved" other-value = "true"/>
9       <timeout after="120s">
10        <sequence>
11          <set field="POmodified" value="Modify or Confirm"/>
12          <participant ref="PO Manager" activity="?ModifyPO"/>
13          <case>
14            <equals field-value="POmodified" other-value="Modify"/>
15              <sequence>
16                <participant ref="PO Manager" activity="ModifyPO"/>
17                <redo ref="Approve"/>
18              </sequence>
19            <equals field-value="POmodified" other-value="Confirm"/>
20              <participant ref="PO Manager" activity="ConfirmPO"/>
21          </case>
22        </sequence>
23      </timeout>
24    </if>
25  </sequence>
26 </sequence>
27 </process-definition>
```

Fig. 16.7 Order Fulfillment process in Ruote's process language: (**a**) main process; (**b**) Ordering subprocess

The Ordering subprocess in YAWL (recall Fig. A.3) contains the following constructs: *sequence* between the tasks *Create PO* allocated to a PO Manager, *Approve PO* carried out by Senior Supply Officer, and *Confirm PO* performed by a PO Manager; *exclusive choice* at the *Approve PO* task; *deadline* through the *Order Timeout* task; *loop* allowing for iteration of *Approve PO*, *Modify PO*, *Approve PO* tasks; and *deferred choice* among the tasks *Modify PO*, *Confirm PO*, and *Order Timeout*. These are captured in Ruote as follows.

- *Sequence.* Through a *sequence* expression (see Fig. 16.7b, lines 3–26) surrounding the definitions of the corresponding tasks (see lines 4, 7, and 20).
- *Exclusive choice.* Through an *if* expression, testing the value of a field called *Approved*. The field *Approved* is defined prior to the task *Approve PO* (see Fig. 16.7b, line 6) and its value entered during the execution of the task. In the example, this field has been assigned a default value *true*, which could also have been omitted.
- *Deadline.* Similar to the example in Fig. 16.7a, a timeout expression is used (see lines 9–23). Instead of using a break construct, this time we define the timeout to apply to a whole sequence segment. This implies that if this sequence is not completed within the set up time, the process will continue according to the specification following the deadline expression, which in this case means that the process will end.
- *Loop.* The loop is implemented through a *redo* expression following the activity *ModifyPO* (on line 17) and reversing the flow back to the sequence tagged *Approved* (on line 5). A peculiarity with the redo expression is that when it reverses the execution, any data gathered between the reverse and the rewind points is lost. Therefore, the semantics for the data treatment of the solution

proposed in Fig. 16.7b deviate from the solution in YAWL. Furthermore, for simplicity, the deferred allocation for the *Modify PO* task is not captured here. This behavior can be captured through the use of variables.

- *Deferred choice.* A point in a process where one of the several branches is chosen based on interaction with the operating environment (recall the Deferred choice pattern from Sect. 2.2). Many systems, including Ruote, do not support this construct. The behavior of the deferred choice is therefore replaced by the behavior of exclusive choice, where the choice is made by the system. It is based on data entered into the system before the moment of choice. In the example, a special activity *?ModifyPO* is introduced (see line 12) for gathering this data from the user (see the field *POmodified*, defined on line 11). To capture the two alternative branches of execution, that is, the activity *Modify PO* or *Confirm PO*, a *case* expression is used (see lines 13–21). Note that the replacement of the deferred choice construct with an exclusive choice leads to another difference between the solutions in Ruote and YAWL.

16.2.7 Discussion

After this brief introduction to Ruote, a few differences from YAWL can be outlined. The most obvious difference is that while YAWL provides a graphical, Petri net inspired notation for the control-flow perspective, Ruote is basically a textual block-structured language. (Here we use the term Ruote to refer to Ruote's process definition language.)

Another fundamental difference lies in the key concepts used by both languages. Inspired by Petri nets, YAWL builds on the concepts of *task* and *condition*, while Ruote is centered around the concept of *participant*. Hitherto participants were only associated with users or group of users, but the concept can be given a broader interpretation. The construct <participant ref="X"> was originally only meant to refer to a piece of work to be carried out. The attribute *activity* (e.g., <participant ref="X" activity="Y">) was originally not present in the language and is now only an optional attribute. In our examples, participant expressions of the format <participant ref="User/GroupName" activity="ActivityName"> were consistently used. A different approach, however, would be to use the format <participant ref="ActivityName"> and to create a store per participant (i.e., introduce a store per task). Then, the stores will no longer act as personal or group worklists, but rather as containers collecting logically similar work. For an example of this approach see Fig. 16.8, where the Order Fulfillment example is remodeled according to this strategy. Note that this approach rather resembles the "work station" approach, according to which work is performed at a number of workstations. Instead of moving work items between different users (which is the traditional approach adopted for WfMSs), the users "move" between different stations, pick up work items, and perform them.

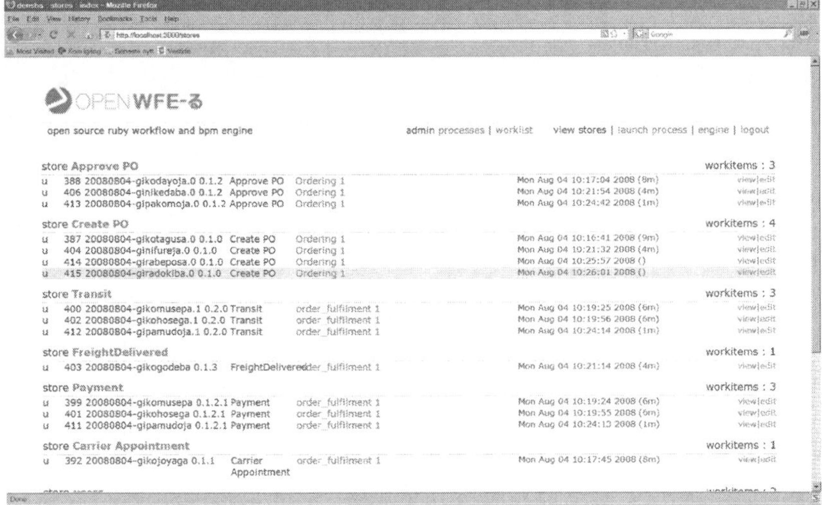

Fig. 16.8 Ruote-web (Densha) – the Order Fulfillment process – a different approach

A consequence of this wider interpretation of participant is that only one work distribution strategy is supported in Ruote. This strategy is that work items are offered to stores (i.e., groups of users) whose members themselves select the work items they decide to perform. No mechanisms for designating a user from a group such as Random Distribution, Shortest Queue, and Round-Robin present in YAWL (cf. Sect. 2.6.4, page 76) are supported in Ruote.

Another difference is the distinction YAWL makes between the *offered*, *allocated*, and *started* states and the lack of a corresponding notion to the allocated state in Ruote. While the states offered and started may be sufficient for many workflow scenarios, the state allocated is essential for some processes. Processes whose activities require thorough planning and early resource allocation fall into this group. For instance, a course delivery process at an education institute is often characterized by course planning and resource allocation activities performed way ahead of the actual course. Hence the allocation of a course to a professor is clearly distinguished from the start of the course. If a system does not distinguish between these two states, some workarounds are needed to capture this difference.

16.3 jBPM

JBoss jBPM[3] is an abbreviation of Java for Business Process Management. It is JBoss's (RedHat's) workflow management system and it is written in Java as the name suggests. According to the download statistics, it is by far the most popular,

[3] www.jboss.com/products/jbpm

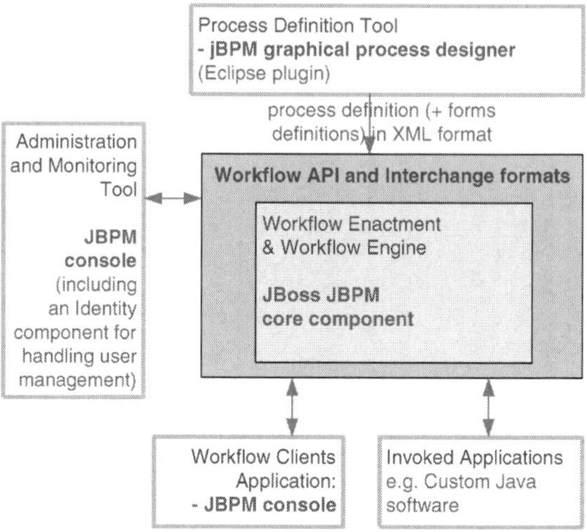

Fig. 16.9 jBPM – architecture

that is, the most downloaded open source WfMS from a code repository. The tool is distributed through SourceForge under the LGPL license. It is aimed at developers and its current development status is 5-Production/Stable. For this book, we used JBoss jBPM version 3.2.3, which at the time of writing was the latest released version. In the remainder of this chapter, we will refer to it as jBPM.

16.3.1 Architecture

The architecture of the system is shown in Fig. 16.9. It contains the following main components:

- A workflow engine called JBoss JBPM core component (also referred to as core process engine), which takes care of the execution of process instances.
- A process definition tool called JBoss jBPM Graphical Process Designer (GPD). This is an Eclipse plug-in (see Fig. 16.13), which provides support for defining processes in jPDL both in a graphical format and in XML. jPDL (jBPM Process Definition Language) is the process language used by the system.
- A workflow client and administration tool, called JBoss jBPM console. It is a web-based client through which users can initiate processes and execute activities (see Fig. 16.11 and 16.12). It is also an administration and monitoring tool through which administrators can observe and intervene in executing process instances. Furthermore, it contains an Identity component, which takes care of the definition of organizational information, such as users, groups, and roles (see Fig. 16.10).

jBPM provides two kinds of access mechanisms: security-roles and organizational groups. The *security-roles* are predefined while the organizational groups

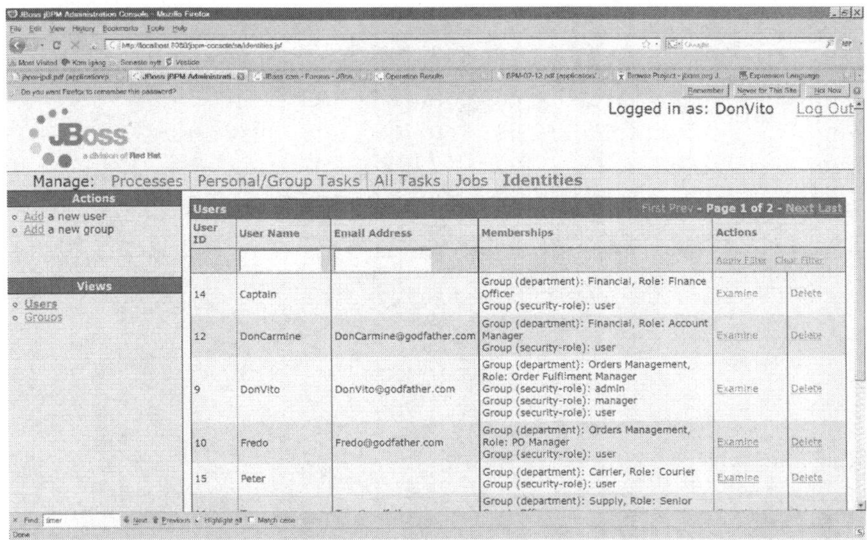

Fig. 16.10 jBPM console – the identities menu

are user-defined. The security-roles are user, manager, and administrator. The role *user* gives the users the privileges to start process instances and execute work items assigned to them. The role *manager* gives the authority to examine and execute work items assigned to other users. The role *administrator* allows a user to define groups and/or users. Note that the word *user* denotes both a user of the system and a security-role. Therefore, in Fig. 16.10 it appears both in the table and column headings, as well as in table cells. In the example in Fig. 16.10 everyone has been assigned the *user* role. In addition, Don Vito has also been assigned the *manager* and *administrator* roles. Furthermore, different *organizational groups* can be defined to capture the organizational structure of an enterprise. For every user, the groups he belongs to and optionally the roles he plays within these groups can be defined. In the example in Fig. 16.10 Don Vito has been assigned to the group *Orders Management* where he plays the role *Order Fulfillment Manager*.

The jBPM console not only includes the Identity component but also provides a user interface for the system. The user interface shows the worklist for a user as illustrated in Fig. 16.11. In the top right-hand corner, the user logged into the system is displayed; in this case, Don Vito has access to the *All Tasks* menu as he is an administrator. When a nonadministrator is logged in, only their personal worklist is displayed.

To execute a work item, a user can bring up its execution window as shown in Fig. 16.12, where the name of the task and other relevant information are displayed in the left panel. During execution, a work item can be suspended or reallocated to a different user (see the menus under *Actions* heading in the window). A number of different *Views* for a work item are available, the most important of which is the *Task Form* view. This view displays the form defined for a task and is the only means of

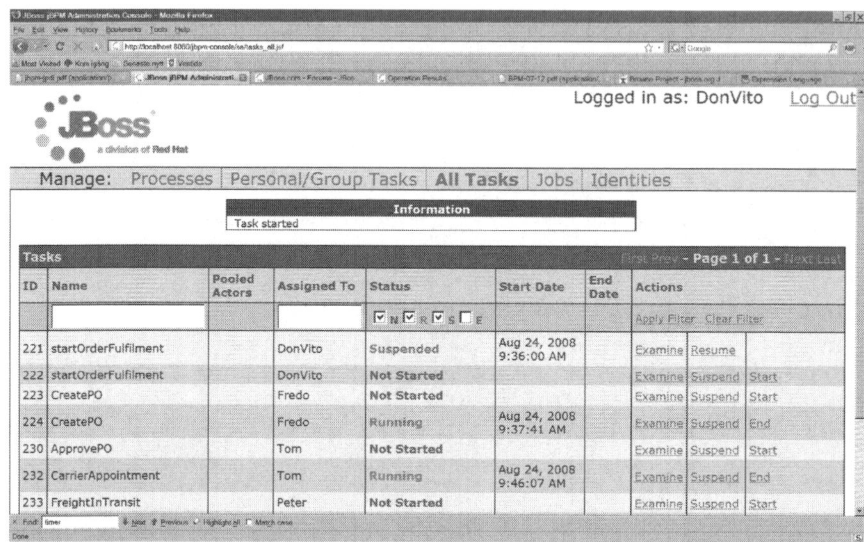

Fig. 16.11 jBPM console – administrator view of all work items (called tasks in jBPM) available in the system

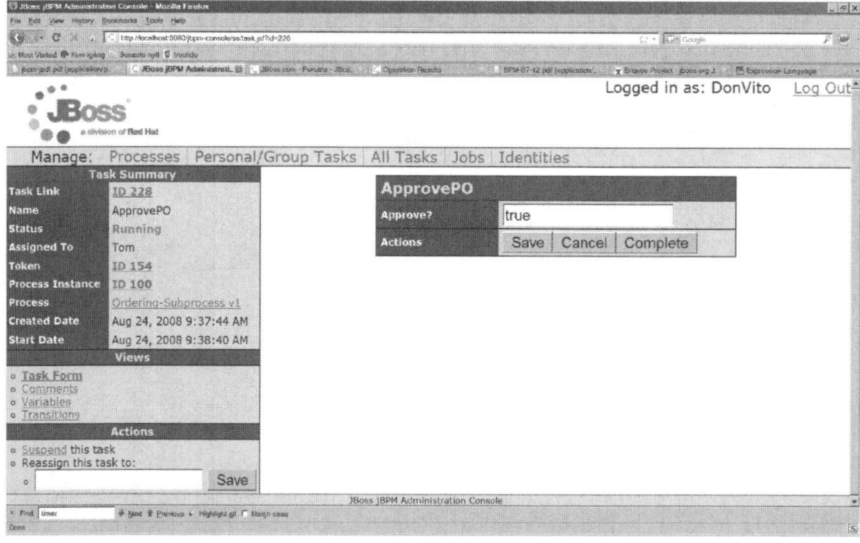

Fig. 16.12 jBPM console – work item execution

entering process instance data. In contrast with YAWL (and Ruote), automatic form generation is not supported by jBPM. Instead, every form used in a process has to be manually defined. The Graphical Process Designer provides some support for this.

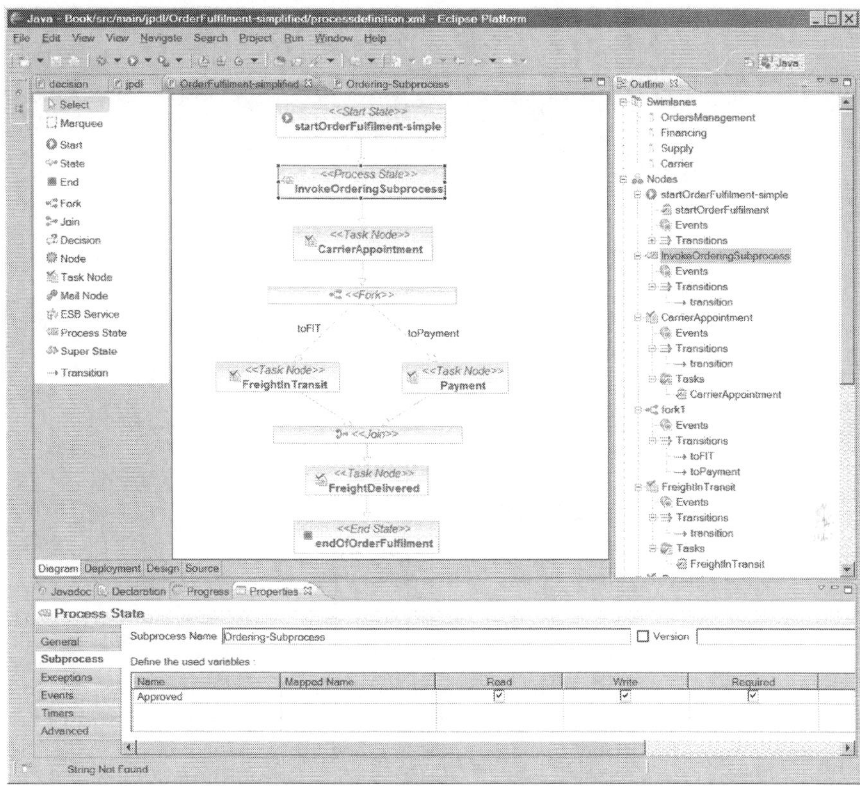

Fig. 16.13 jBPM's graphical process designer (a plug-in to Eclipse)

16.3.2 Control-Flow

The control-flow of a process definition in jPDL consists of a number of *nodes* and *transitions* between these nodes (the graphical symbols in the language are displayed in the left panel in Fig. 16.13). The most important nodes are the *Start* and *End State* nodes representing the beginning and the completion of a process; the *State* node for modeling wait states; the *Task Node* for capturing work packages; and the routing nodes such as *Fork*, *Join*, and *Decision* for performing parallel execution, synchronization, and choice. A task node consists of one or more *Task*s, where a task is an atomic piece of work distributed to a user. Graphically, only the task nodes are shown but not the tasks they consist of. While in YAWL, a distinction is made between the terms *task* and *work item*, that is, an instance of a task in a case, in jBPM, the term task is used to denote both.

To illustrate these constructs, we have used the Order Fulfillment simplified example used in the previous subsection. The graphical model of the process is displayed in Fig. 16.13 and the XML definition is shown in Fig. 16.14. The process

```
1 <process-definition xmlns="urn:jbpm.org:jpdl-3.2"
2         name="OrderFulfilment-simplified">
3   <swimlane name="OrdersManagement">
4       <assignment actor-id="DonVito"></assignment>
5   </swimlane>
6   <swimlane name="Financing">
7       <assignment actor-id="Virgil"></assignment>
8   </swimlane>
9   <swimlane name="Supply">
10      <assignment actor-id="Tom"></assignment>
11  </swimlane>
12  <swimlane name="Carrier">
13      <assignment actor-id="Peter"></assignment>
14  </swimlane>
15  <start-state name="startOrderFulfilment-simple">
16      <task name="startOrderFulfilment">
17          <assignment actor-id="DonVito"></assignment>
18      </task>
19      <transition to="InvokeOrderingSubprocess"></transition>
20  </start-state>
21  <process-state name="InvokeOrderingSubprocess">
22      <sub-process name="Ordering-Subprocess"/>
23      <variable access="read,write,required"
24            name="Approved"></variable>
25      <transition to="CarrierAppointment"></transition>
26  </process-state>
27  <task-node name="CarrierAppointment">
28      <task name="CarrierAppointment" swimlane="Supply" />
29      <transition to="fork1"></transition>
30  </task-node>
31  <fork name="fork1">
32      <transition to="FreightInTransit" name="toFIT"></transition>
33      <transition to="Payment" name="toPayment"></transition>
34  </fork>
35  <task-node name="FreightInTransit">
36      <task name="FreightInTransit" swimlane="Carrier"></task>
37      <transition to="join1"></transition>
38  </task-node>
39  <task-node name="Payment">
40      <task name="Payment">
41          <assignment actor-id="DonCarmine"></assignment>
42      </task>
43      <transition to="join1"></transition>
44  </task-node>
45  <join name="join1">
46      <transition to="FreightDelivered"></transition>
47  </join>
48  <task-node name="FreightDelivered">
49      <task name="FreightDelivered" swimlane="Supply"></task>
50      <transition to="endOfOrderFulfilment"></transition>
51  </task-node>
52  <end-state name="endOfOrderFulfilment"></end-state>
53 </process-definition>
```

Fig. 16.14 jPDL – XML format of the simplified Order Fulfillment process

includes the Start and End States: *startOrderFulfillment-simple* and *endOrderFulfillment*; the Task Nodes: *CarrierAppointment*, *FreightInTransit*, *Payment*, and *FreightDelivered*; and one Join and one Fork node to enable parallel execution of the *FreightInTransit* and *Payment* tasks. To preserve visibility, every task node consists of only one task, carrying the name of the task node (see, for instance, lines 27–30 in Fig. 16.14, or the *CarrierAppointment* task in the *Outline* panel in Fig. 16.13).

A process model can also contain states. A *state* denotes a waiting point for the occurrence of an external event which triggers continued execution. A Start State is a kind of state. To allow a process to be started through a task and not only through external events, the start state, in contrast with other states, can contain a task. In our example, *StartOrderFulfillment* is such a task (see lines 15–18 in Fig. 16.14). Note that there is no equivalent notion to this task in either the YAWL or the Ruote solution.

The main construct used for managing hierarchical process decomposition is a *Process State*. Process States are used for capturing process invocations of subprocesses that are defined outside the process from which they are invoked. In our example, the process state *InvokeOrderingSubprocess* invoking the Ordering subprocess demonstrates this (see Fig. 16.13 and lines 21 and 22 in Fig. 16.14). The subprocess is shown in Fig. 16.15. Another way of structuring a process and grouping logically related nodes is through the construct of *Super State*, but we will not elaborate on it here.

There is also a special node type called *Node* through which custom code for new desired behaviors can be defined. This construct can be used both for invoking external applications and for adding new control-flow behavior to a model. An alternative way to introduce customized behavior is through *Actions*. Actions are

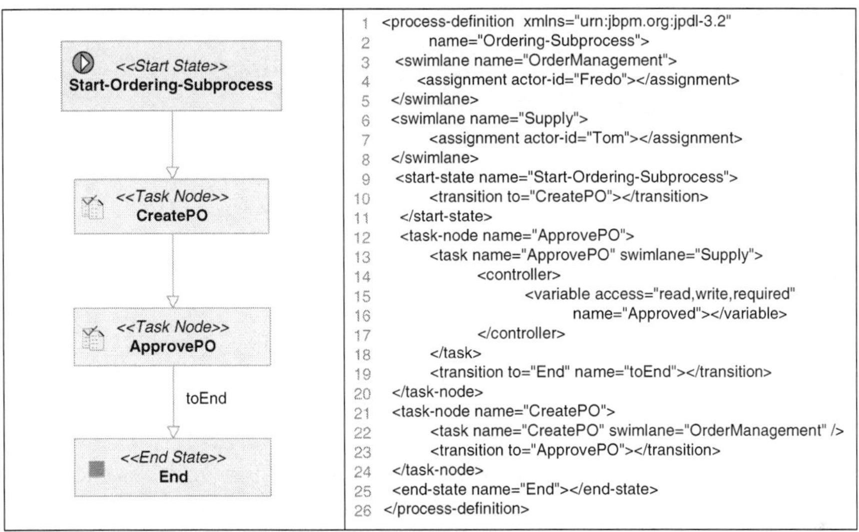


```
1  <process-definition xmlns="urn:jbpm.org:jpdl-3.2"
2          name="Ordering-Subprocess">
3    <swimlane name="OrderManagement">
4        <assignment actor-id="Fredo"></assignment>
5    </swimlane>
6    <swimlane name="Supply">
7          <assignment actor-id="Tom"></assignment>
8    </swimlane>
9    <start-state name="Start-Ordering-Subprocess">
10         <transition to="CreatePO"></transition>
11    </start-state>
12    <task-node name="ApprovePO">
13         <task name="ApprovePO" swimlane="Supply">
14                <controller>
15                       <variable access="read,write,required"
16                              name="Approved"></variable>
17                </controller>
18         </task>
19         <transition to="End" name="toEnd"></transition>
20   </task-node>
21   <task-node name="CreatePO">
22         <task name="CreatePO" swimlane="OrderManagement" />
23         <transition to="ApprovePO"></transition>
24   </task-node>
25   <end-state name="End"></end-state>
26 </process-definition>
```

Fig. 16.15 Definition of the Ordering subprocess in jPDL

defined through Java classes and specify customized behavior, which is hidden in transitions or nodes and is meant to implement behavior important from a technical, but not from a business perspective. In contrast to Nodes, Actions cannot influence the specified control-flow. They are executed when the transition they are attached to is taken, or when the node they are defined for reaches a state where a trigger is defined (which is one of the following: node-enter, node-leave, before-signal, and after-signal).

16.3.3 Data

In jBPM, data is handled through variables. There are two types of variables: process and task variables. Task variables are always mapped to process variables, which is done through the jPDL construct *controller*, that is, controllers are the places where variables are defined. By default, each task variable is mapped to exactly one process variable and carries its name. Figure 16.16 shows the graphical interface of a controller and the definition of a variable *Approved* for the *ApprovePO* task. Lines 14–17 in Fig. 16.15 show the corresponding XML representation. *Mapped Name* defines the task variable name if different from the process variable name defined under *Name*. The settings *Read*, *Write*, and *Required* define whether the value in the task variable will be read from the corresponding process variable (if the process variable has not yet been created through a previous task, it will be created now), written back to it (if so, any previous value is overwritten), and whether the input of a value for the variable is compulsory.

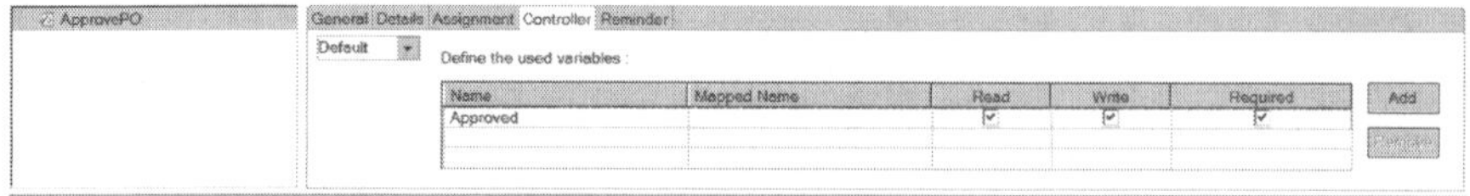

Fig. 16.16 Definition of variables in jBPM graphical process designer

Data binding between the variables of a process and its subprocesses is explicitly specified. The logic and the interface are similar to those for definitions of task variables and their mapping to process variables. In our example, the *Approved* process variable of the Ordering subprocess is mapped to a variable with the same name in the Order Fulfillment process (see the bottom frame in Fig. 16.13, as well as lines 23–24 in Fig. 16.14).

Variables are made visible to and their values editable by users through task forms (recall the form for the *ApprovePO* task displayed in Fig. 16.12). As mentioned earlier, automated form generation is not supported, which means that a form needs to be manually generated for every task. Support for this is provided in the Graphical Process Designer.

16.3.4 Resources

jBPM aims at supporting two work distribution strategies. A task is either directly assigned to (a group of) users (see lines 40–42 in Fig. 16.14), or it is associated with a swimlane (see line 36 in Fig. 16.14). *Swimlanes* represent process roles and are used to assign tasks to people. This is realized by first mapping the users to the swimlanes (see lines 3–14 in Fig. 16.14) and then mapping the swimlanes to the tasks (as shown on line 36 in Fig. 16.14). The conceptual difference between allocating a task to a swimlane and assigning a task to a group of users is that, during runtime, a user who selects and executes a work item for a task belonging to a swimlane will automatically get all subsequent work items of the same case for tasks belonging to that swimlane assigned to them. In YAWL, this corresponds to the chained execution routing strategy (see Sect. 2.6.4, page 81).

Currently, the assignment expression used both for tasks and swimlanes supports only the specification of actors. This imposes the limitation that work cannot be offered to a group. It is indeed possible to specify a set of actors (instead of a single one). This work-around has, however, two drawbacks. First, it does not utilize and benefit from the *group* notion. Second, during runtime, the work is not offered to the specified actors, but put into a Pool that can be accessed only by managers. Hence any role-based distribution of work, which one would expect from this work-around, does not occur.

The lifecycle of a work item in jBPM is depicted in Fig. 16.17. When a work item is *created*, it is either directly *allocated* to a user (see the *Assigned To* column in Fig. 16.11) or *offered* to multiple resources (displayed in the *Pooled Actors* column).

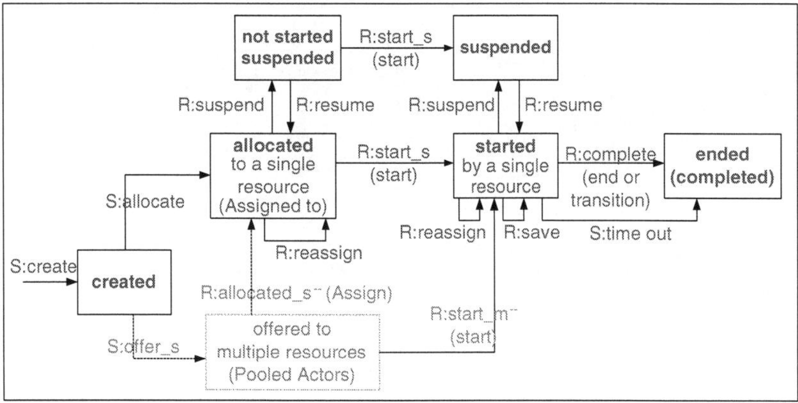

Fig. 16.17 The lifecycle of a work item in jBPM. The *dotted lines* indicate semantics differences from the corresponding YAWL model shown in Fig. 2.34

Currently, work items displayed in the *Pooled Actors* column cannot be checked out and executed by any of the users in the pool. Instead, a manager must first explicitly assign them to a user from the pool. To indicate these subtleties, we have used dotted lines in Fig. 16.17 and added a minus sign to the transition R:allocated_s to show that only certain resources (i.e., managers) are allowed to perform the allocation. Managers can also start and execute work items offered to a pool, which is captured through the R:start_m-transition. After being allocated, a work item can be started and then completed. Work items, both started and not started, can be suspended or reassigned to a different user.

16.3.5 Example

Let us now model the Order Fulfillment process and the Ordering subprocess of Fig. A.1 (page 600) and A.3 (page 603) in jBPM. A draft of the solutions is shown in Fig. 16.18 and 16.19. To save space in the main process, we have used only one swimlane associated with the user Don Vito. As many details are hidden from the graphical models, the corresponding XML solutions are included.

Similar to YAWL, in the Ordering subprocess, the tasks *Ordering*, *CarrierAppointment*, *FreightInTransit*, *Payment*, and *FreightDelivered* are introduced. Each of these tasks is wrapped in a Task Node carrying the same name. To capture the parallel behavior of the tasks *FreightInTransit* and *Payment*, a Fork and a Join Node are introduced to split and synchronize the flow. The time constraint for the *Ordering* task is captured through a timer (see line 16 in Fig. 16.18). To capture the flow of orders that have not been approved, a decision node called *proceedDecision* is introduced (see lines 43–52 in Fig. 16.18). The decision is based on the value of the variable *Approved* (see line 13 in Fig. 16.18), which is required as input from the user during the execution of the task *Ordering*.

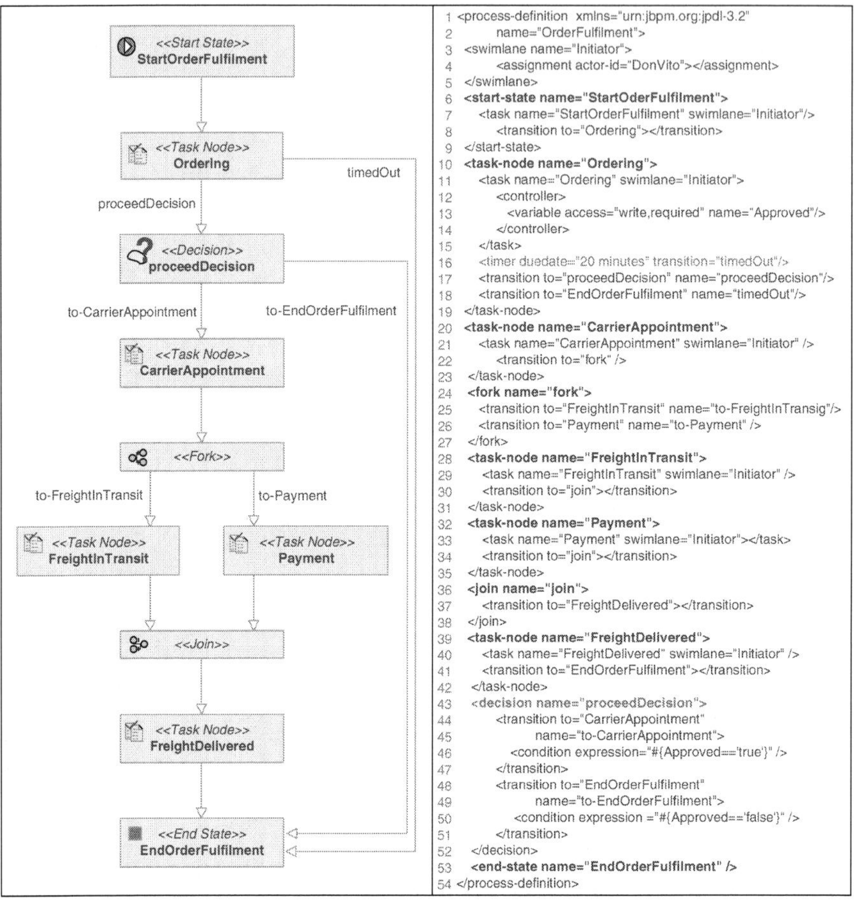

Fig. 16.18 The Order Fulfillment process in jBPM

As discussed in Sect. 16.2.6, the Ordering subprocess contains a sequence, an exclusive choice, a deadline, a loop, and a deferred choice. These are captured in jBPM as follows (see Fig. 16.19):

- *Sequence.* By connecting the corresponding task nodes, that is, *CreatePO*, *ApprovePO*, and *ConfirmPO*, with transitions.
- *Exclusive choice.* Through a decision node called *?OrderApproved* where, according to the value of the variable *Approved*, the transition that satisfies the specified condition is taken (see lines 21–28 in Fig. 16.19).
- *Loop.* Through a transition from task node *ModifyPO* back to task node *ApprovePO* (see line 31 in Fig. 16.19).
- *Deferred choice.* The behavior of the deferred choice is simulated through a task node (with a corresponding task) called *Wait for confirmation*, which contains two outgoing transitions: *modify* and *confirm* (see lines 37–45 in Fig. 16.19).

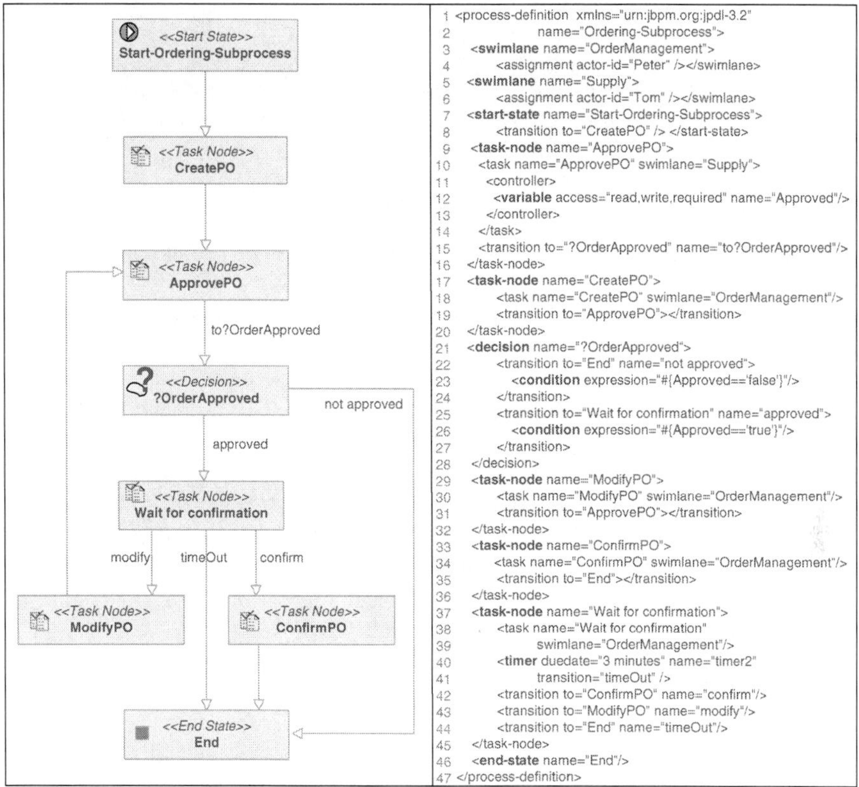


```
1 <process-definition xmlns="urn:jbpm.org:jpdl-3.2"
2             name="Ordering-Subprocess">
3   <swimlane name="OrderManagement">
4       <assignment actor-id="Peter" /></swimlane>
5   <swimlane name="Supply">
6       <assignment actor-id="Tom" /></swimlane>
7   <start-state name="Start-Ordering-Subprocess">
8       <transition to="CreatePO" /> </start-state>
9   <task-node name="ApprovePO">
10    <task name="ApprovePO" swimlane="Supply">
11     <controller>
12      <variable access="read,write,required" name="Approved"/>
13     </controller>
14    </task>
15    <transition to="?OrderApproved" name="to?OrderApproved"/>
16  </task-node>
17  <task-node name="CreatePO">
18      <task name="CreatePO" swimlane="OrderManagement"/>
19      <transition to="ApprovePO"></transition>
20  </task-node>
21  <decision name="?OrderApproved">
22      <transition to="End" name="not approved">
23        <condition expression="#{Approved=='false'}"/>
24      </transition>
25      <transition to="Wait for confirmation" name="approved">
26        <condition expression="#{Approved=='true'}"/>
27      </transition>
28  </decision>
29  <task-node name="ModifyPO">
30      <task name="ModifyPO" swimlane="OrderManagement"/>
31      <transition to="ApprovePO"></transition>
32  </task-node>
33  <task-node name="ConfirmPO">
34      <task name="ConfirmPO" swimlane="OrderManagement"/>
35      <transition to="End"></transition>
36  </task-node>
37  <task-node name="Wait for confirmation">
38      <task name="Wait for confirmation"
39            swimlane="OrderManagement"/>
40      <timer duedate="3 minutes" name="timer2"
41            transition="timeOut" />
42      <transition to="ConfirmPO" name="confirm"/>
43      <transition to="ModifyPO" name="modify"/>
44      <transition to="End" name="timeOut"/>
45  </task-node>
46  <end-state name="End"/>
47 </process-definition>
```

Fig. 16.19 The Ordering subprocess in jBPM

During runtime, after the task *Wait for confirmation* has been started, the user can select one of these alternative transitions and proceed with the flow. Hence the choice is deferred as long as possible. A subtle difference with YAWL is that in jBPM this special task node, or more precisely its task, *Wait for confirmation* has to be started in order to enable the choice. This adds an additional step to the process.

- *Deadline.* Through a timer specified for the *Wait for confirmation* task node (see lines 40–41 in Fig. 16.19). When a timer is defined, the transition to be taken after timer expiry has to be specified. In the example, this is the *timeOut* transition (see line 41) and it is one of the transitions from *Wait for confirmation* task node (see line 44). A side-effect of this definition is that when the *Wait for confirmation* task is started, the transition *timeOut* will be presented, along with the transitions *modify* and *confirm*, as an alternative choice available for selection by the end user. Hence, this solution contains a significant conceptual difference from the solution for YAWL.

16.3.6 Discussion

Similar to YAWL, jBPM offers a graphical notation. A major distinction is, however, that in jBPM the atomic piece of work, i.e., a *Task*, does not have a visual representation. Instead, only *Task Nodes*, which may contain a number of different tasks, are displayed graphically. This raises the question of the purpose of the task node construct. It may be argued that task nodes provide a mechanism for structuring tasks related to each other. This also seems to be the intention with the *Super State* construct. Indeed, a super state can, in addition to tasks, also include other nodes (hence also states) and can be nested. The question of what the task node concept brings to the language in addition to the super state construct still remains, however.

In this vein, it can be noted that in comparison with YAWL's cancelation region construct, the super state construct in jBPM appears to be less general. While in YAWL a cancelation region can contain both tasks that are connected, that is, consecutive to each other or belonging to the same thread of control, and nonconnected tasks (i.e., tasks belonging to segments of parallel or alternative threads), the tasks grouped through a super state are always connected to each other. This means that a behavior defined through a cancelation region in YAWL containing nonconnected tasks may not have a simple representation in jBPM.

Finally, another major difference with YAWL is the lack of automatic form generation support in jBPM. While it may be argued that in general automatically generated forms provide a clumsy interface that is often not the one ultimately used by the end users, they are convenient in the early implementation and test phases of a process. Automatic form generation functionality in a WfMS provides a means to achieve a separation of concerns and process logic design can be considered independent of user interface design.

16.4 Enhydra Shark

Enhydra Shark[4] is a Java workflow engine offering from Together Teamlösungen and ObjectWeb. While Enhydra Shark is an open source offering, its architecture allows for the use of closed-source or proprietary plug-ins to extend it. In contrast with the other open source offerings described in this book, one version of Enhydra Shark is also distributed as a closed-source product. The open source, community version of Enhydra Shark is licensed according to the LGPL. When discussing the capabilities of the system below, we will illustrate them by using the TWE Community edition version 2.2-1 of the editor. We run the case in the Shark engine and administer the work via the TWS Admin application version 2.0-1.

---

[4] enhydra.org

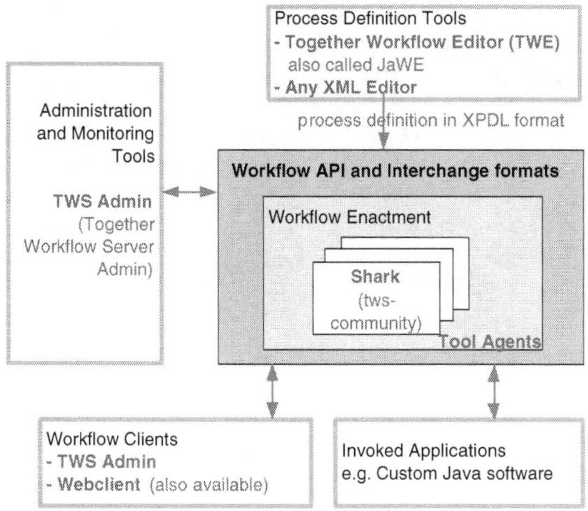

Fig. 16.20 Enhydra Shark – architecture

16.4.1 Architecture

The workflow engine is one of the several products under constant development. The system of products contains the following components (its architecture, based on the WfMC reference model, is shown in Fig. 16.20):

- A workflow engine called Shark
- A Process Definition Tool called Together Workflow Editor (TWE) or JaWE
- A Management Console, TWS Admin, which is a Workflow Client as well as an Administrative and Monitoring Tool
- A set of ToolAgents (e.g., MailToolAgent, SOAPToolAgent, or JavaScriptToolAgent), which provide interfaces for different software, for example, a Javascript interpreter

The Together Workflow Editor (TWE) uses XPDL 1.0 as a process language. The ideas and constructs of XPDL are reflected in Enhydra. The main unit of organization in XPDL is the package. A *package* contains one or several process definitions and can contain global data entities common to process definitions within the package. A process consists of activities and transitions between them. Activities are assigned to be performed by participants. An activity may be performed manually or automatically and the participant may be either a human or an application.

In addition to the textual XML representation, the suppliers of Enhydra Shark provide an in-house developed graphical notation used in the Together Workflow Editor (TWE). In it, a process is visualized as work assignments to participants represented by swimlanes. Screenshots of TWE are shown in Figs. 16.24 and 16.26. As the models are constructed in TWE, a corresponding XML file is updated accordingly.

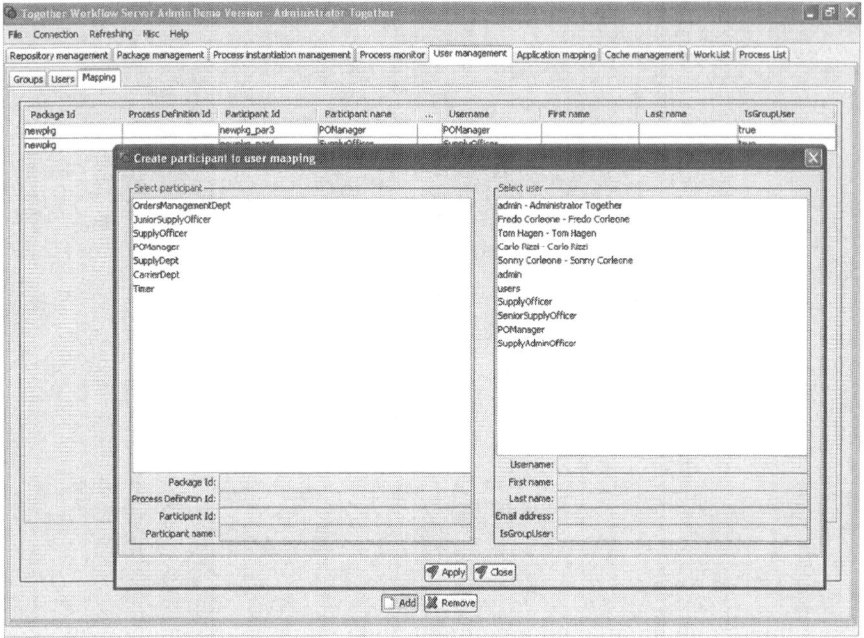

Fig. 16.21 Users–participants mapping in Enhydra Shark

Users and user groups are defined in the engine. For example, one can state that there is a user called Fredo Corleone and that users Tom Hagen and Carlo Rizzi both belong to the user group *SupplyDept*. There are no other simple ways of modeling dependencies between users apart from this mechanism. Other, more complicated kinds of rules must be provided as free text (and are, as a consequence, hard to enforce) or be written directly in, for example, Java code.

When users and user groups have been defined, they are mapped to model participants. The model participant *POManager* can, for example, be mapped to the user Fredo Corleone. The model participant *POManager* can also be mapped to a user group *POManager* in the engine (see Fig. 16.21). Mapping of IT applications is done in a similar way. For example, a model participant "e-mailer" can be mapped to the default email application in the system.

All users have their own individual worklist (see Fig. 16.22). A user may select a work item from their list, possibly perform some actions, for example, update a variable or send an e-mail, and then continue with the next item. If a work item is allocated to a group of users (as a consequence of mapping a participant from the process model to a group of users in the engine), then, when one of the group members selects it, it will disappear from the other group members' lists. Should the user change their mind, then deselecting it will make it appear again on the others' lists. No user (not even a group manager) can select a work item on behalf of another user.

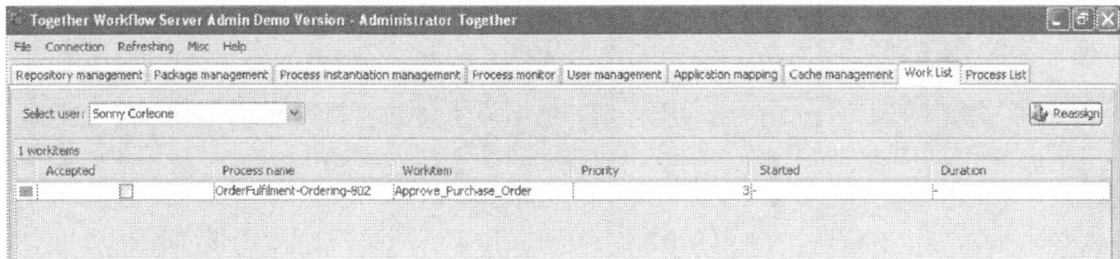

Fig. 16.22 A worklist in Enhydra Shark

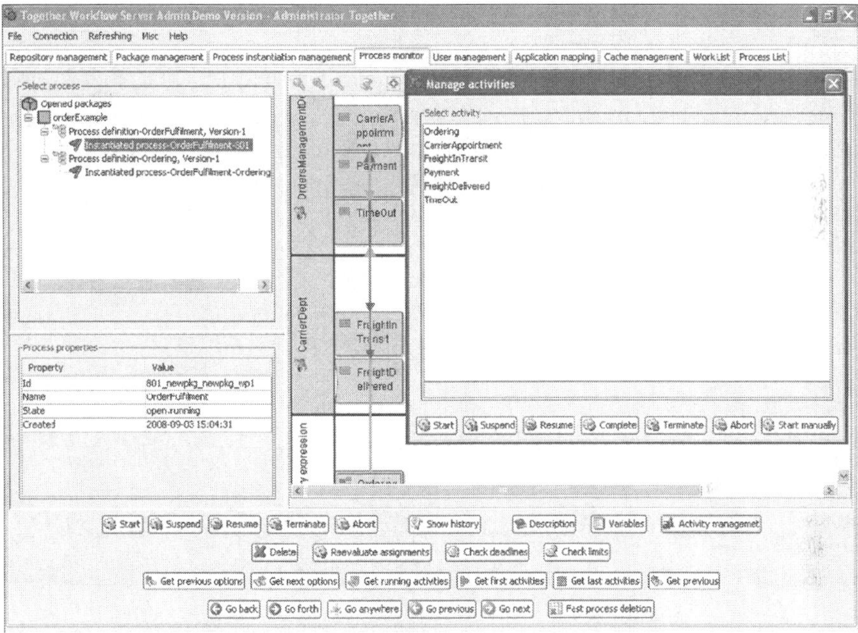

Fig. 16.23 The process monitor tab

Workflow monitoring and administration functionality are provided through the TWS Admin component (see Fig. 16.23). Here, processes or individual activities may be, for example, suspended and resumed, variable values can be read, and historical traces may be inspected. Some limitations may be compensated for by using the administration functionality. For instance, consider the case where there are two instances of the same workflow running concurrently and there is a need to show that one instance has higher priority than the other. It is not possible to express this in the model. It is, however, possible to suspend the enactment of the workflow instance with lower priority in the administration module.

16.4.2 Control-Flow

Enhydra Shark supports all basic control-flow concepts. The basic building blocks are *activities* and *transitions*. A transition between two activities captures a sequence; when the first activity is completed, the point of control in the process thread advances to the next activity. Multiple transitions flow from or into an activity. Conditions that constrain the transitions from one activity to the next, called *transition restrictions*, can be specified. For instance, when multiple threads are entering/exiting an activity, a transition restriction can specify whether it is an AND or an XOR Join/Split activity. For an XOR Split, *transition conditions* are specified for the transitions leaving the XOR activity node.

An example of the basic control-flow constructs is given in Fig. 16.24. The *Ordering* activity has a split transition restriction associated with it. The split is of type XOR, which is visualized in TWE by a regular rectangle. The fact that there are conditions specified for the transitions leaving and XOR activity is visualized by thicker arrows or different color of the transitions (for simplicity, we have decided not to discuss the syntax of the conditions here). The *CarrierAppointment* activity is an AND-split activity, which is visualized by a rectangle with a convex edge. An AND-join activity is visualized by a rectangle with a concave edge (see, e.g., the *FreightDelivered* AND-join activity).

A process can be hierarchically structured through the use of block activities. A *block activity* contains an activity set which is a subprocess. Both YAWL and OpenWFE provide similar decomposition mechanisms. Another way to structure a process hierarchically is by letting an activity in a process call other processes.

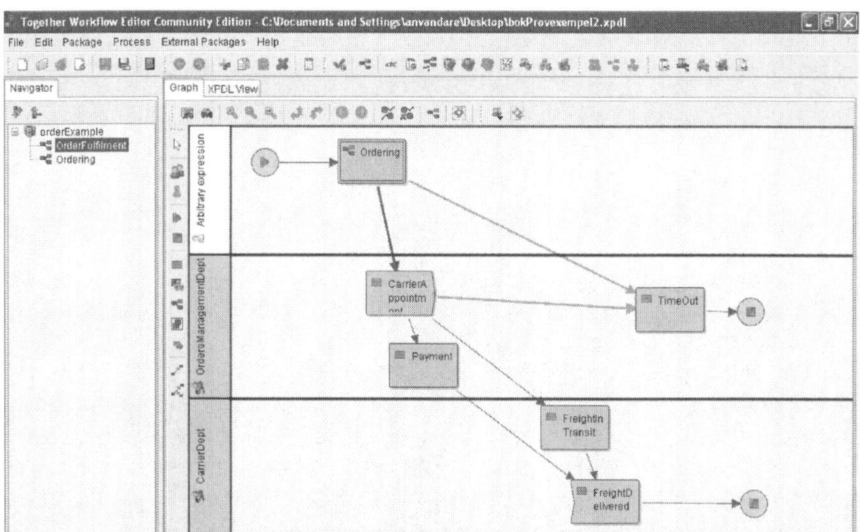

Fig. 16.24 The Order Fulfillment process

In other words, an activity can invoke a process that is specified independently. Such functionality was discussed before for OpenWFE and jBPM. In YAWL, this corresponds to the functionality provided by worklets (see Chaps. 4, 5, and 11).

One form of support for triggers offered by Enhydra Shark is the specification of deadlines on activities. To receive a trigger, such as a deadline expiry or a mail reception, a process needs to execute a specific activity that can handle that trigger. A deadline is in itself not graphically visualized, but the alternative path taken at a deadline expiry is explicitly modeled. In Fig. 16.24, the transitions leading to the *TimeOut* activity specify such alternative paths to be taken at expiry of the deadlines specified for the *Ordering* and the *CarrierAppointment* activities.

To indicate the start and end of a process, Shark TWE uses start and end symbols, which are graphically denoted by circles. An activity without an in-going transition (other than one from the optional start symbol) is a start activity, while an activity without an outgoing transition (other than one to the optional end symbol) is an end activity. End activities are allowed to have transitions going back to themselves. A process model may contain more than one end activity, which differs from YAWL, where a process model has only one output condition.

A special kind of activity in Enhydra Shark is the *routing activity*, which is a "dummy" activity used only for routing the control-flow within a process. An example of such a "dummy" activity is the activity *TimeOut* in Fig. 16.24. Recall that routing activities are also available in YAWL. Note, however, that the *TimeOut* routing activity in Enhydra Shark's solution is an additional activity, which does not have a directly corresponding routing activity in the YAWL solution (see Fig. A.1 on page 600).

16.4.3 Data

Much of the data handling in Enhydra Shark can be understood by the way workflow variables are treated. In Enhydra Shark, a package aggregates one or several workflows. A variable may be defined at the process level or at the package level with the Datafield tag. A datafield positioned at the process level defines a variable as accessible by all instances of the particular process, whereas a datafield positioned at the package level defines a variable as accessible by all process instances in the package. A consequence of this is that a variable defined at the package level may be used and reused by several processes. A package level variable is, however, not a global variable in the usual sense. When a process is instantiated, a copy of the variable is made accessible to the process instance. This copy resides entirely within the process instance and is not visible outside it. A package level variable cannot be used as a means of communication between process instances.

At design time, the definition of an activity can involve reading or modifying variable values. For example, a transition condition associated with an activity may depend on a certain variable having a certain value. A user executing the activity at runtime will then be confronted with an input box to change the value and enable the transition. Variables can also be defined for manipulation during an automatic procedure. For instance, a variable may be incremented to keep a count of iterations in a loop. The administration console allows the inspection of variables at any time during process execution.

Enhydra Shark supports communication of data to and from the environment through the use of *tool agents*. One use of tool agents is to write data from one process instance into an external database. This data may subsequently be read from the database by another process instance. Communication, when a process calls a subprocess, is done through process parameters. In a process call, a formal process parameter is replaced by an actual variable value.

Communication between activities within and across workflows is accomplished by use of variables and parameter passing. For example, in Fig. 16.24 and 16.26, a variable *Approved* is declared at the package level. The subprocess is defined to be called with one formal parameter. The actual parameter in a call is the value contained in the *Approved* variable. This is an In–Out type of parameter, that is, its value can be modified by both the calling and the called process.

16.4.4 Resources

Figure 16.25 shows a state transition diagram for the life-cycle of a work item in Enhydra Shark. A work item is first created by the system, which means that the preconditions for the enablement of the associated activity have been met. This work item is offered either to a single resource or to a group of resources. When it is offered, the offer is nonbinding, that is, a resource can choose to accept or reject it, and the work item remains on offer until it is finally selected by one of the resources to which it is offered. Once selected by a resource, the work item is allocated to that resource on a binding basis, that is, the resource is expected to complete the work

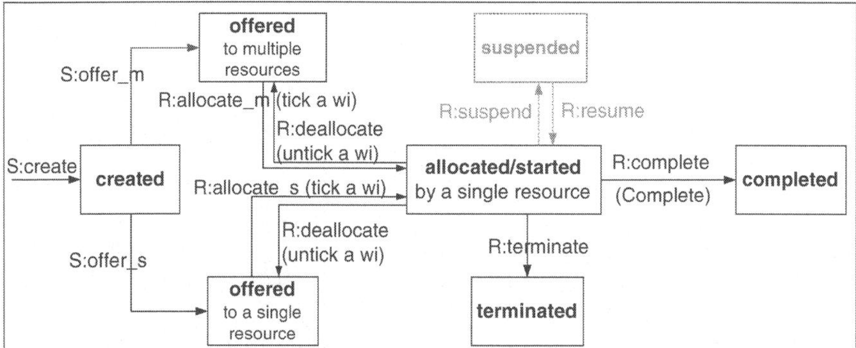

Fig. 16.25 The life-cycle of a work item (wi) in Enhydra Shark. The prefixes S and R in the names of the transitions are abbreviations for system and resource and indicate the initiator of a specific transition

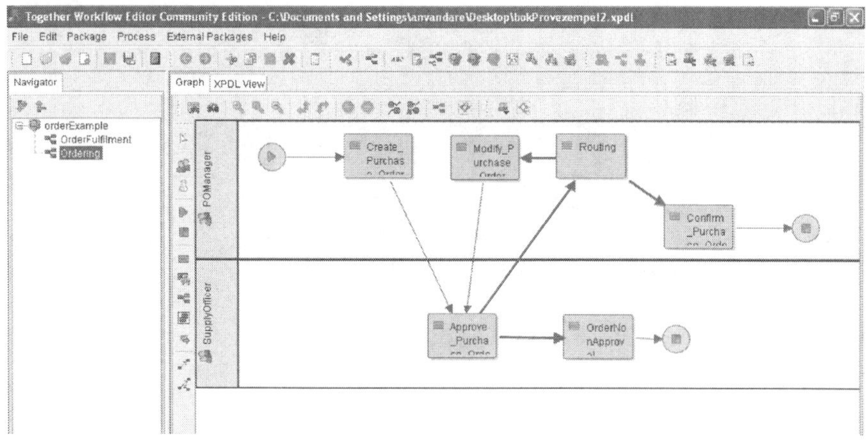

Fig. 16.26 The Ordering subprocess

item at some point in the future. During execution, the work item can be suspended and resumed until finally it is completed or terminated. The latter is often an indication of a failure. In contrast to YAWL, no difference is made between the allocated and started states.

16.4.5 Example

Let us now further elaborate the definitions of the Order Fulfillment process and the Ordering subprocess in Enhydra Shark and discuss how they compare with the corresponding solutions in YAWL. To capture the processes, five activities are introduced: *Ordering*, *CarrierAppointment*, *FreightInTransit*, *Payment*, and *FreightDelivered* (Fig. 16.24). The *Ordering* activity is detailed in the subprocesses shown in Fig. 16.26 and is soon explained. The definitions of the other four activities are left as an exercise for the reader.

The initial activity in the workflow is *Ordering*. Transitioning from *Ordering* can be done in one of the two ways: either through the successful completion of the *Ordering* activity and continuing with the *CarrierAppointment* activity or if a time-out expires. In the case of a time-out, the whole workflow is closed. If the workflow is not timed out and the order is approved, then the workflow continues with the *CarrierAppointment* and subsequent activities.

As discussed in Sect. 16.2.6, the Ordering subprocess contains a sequence, an exclusive choice, a deadline, a loop, and a deferred choice. These are captured in Enhydra Shark as follows (see Fig. 16.26):

- *Sequence.* By connecting the corresponding activities, i.e., *Create_Purchase_-Order*, *Approve_Purchase_Order*, *Routing*, and *Confirm_Purchase_Order* with transitions.
- *Exclusive choice.* Through specifying the *Approve_Purchase_Order* activity as an XOR activity and defining conditions for its outgoing transitions specifying that if the purchase order is approved, the process will continue to the *Routing* activity and if it is not approved, it will continue to the *OrderNonApproved* activity. The *OrderNonApproved* activity is a routing activity that does not have any counterpart in the corresponding process model in YAWL (shown in Fig. A.3 on page 603).
- *Deferred choice.* Similar to the solution for Ruote (discussed in Sect. 16.2.6), the behavior of the deferred choice is replaced by an exclusive choice, where the choice is made by the system. It is based on data entered into the system before the moment of choice. In the example, a special activity called *Routing* is introduced. In this case, *Routing* is not a "dummy" activity. During its execution, the user is asked to provide input on whether the purchase order needs to be modified. To capture the two alternative branches of execution, that is, the activity *Modify_Purchase_Order* or *Confirm_Purchase_Order*, the activity *Routing* is specified as an XOR activity and conditions are imposed on its outgoing transitions used for directing the flow based on the data value entered by the user. Note that the replacement of the deferred choice construct with an exclusive choice is a fundamental difference between the solutions in Enhydra Shark and YAWL.
- *Loop.* Through a transition from *Modify_Purchase_Order* back to the *Approve-_Purchase_Order* activity.
- *Deadline.* The deadline in the *Order Timeout* task in the YAWL model has not been precisely captured in the process model in Enhydra Shark. The reason for this is that timers can be set on activities or block activities but not on process segments, as is done in the original YAWL model. Therefore, instead of limiting the time for modifying and confirming a purchase order, we have introduced a timer for limiting the execution time of the entire Ordering subprocess. This is a deviation from the semantics in the original model and reveals another difference between Enhydra Shark and YAWL.

16.4.6 Discussion

Similar to YAWL, Enhydra Shark offers a graphical notation. A major difference is, however, that the graphical notation in Enhydra Shark is based on XPDL 1.0, which is a standard promoted by the WfMC. Although it can be argued that the use of a standard can be an advantage, for instance, when considering interoperability with other systems, it should also be noted that the use of a standard may be a constraint. Limitations imposed by a chosen standard are reflected in the systems implementing the standard. Such limitations can be worked around in a specific implementation, but this is at the expense of standard compliance. Because of some inherent limitations present in XPDL, a conscious choice was made for YAWL not to build on it, but to develop its own graphical language, which is more expressive than the existing languages and standards. While Enhydra Shark relies on extendability of XPDL to implement complex behavior, YAWL offers a powerful base set of constructs to achieve expressiveness so that additional coding is not necessary.

In addition, and this is a second major difference, YAWL as a language has formally defined semantics, whereas no such formalization is available for XPDL. While the need for and advantages of formal foundations for WfMSs in particular and information systems in general are commonly recognized, only a few WfMSs actually rely on solid formal foundations. A consequence of working with a language/standard like XPDL, which is only described in natural language and lacks precise and formal semantics, is that implementations of it can vary depending on the interpretations of the constructs made by those developing WfMSs.

A third difference is that Enhydra Shark, in addition to its open source version, also offers a closed source version. Some of the functionality available for the closed source version is not available in the open source version, but it is still visible in the interface. A side-effect of this is that users of the open source version see more functions than they are able to access. In contrast, when working with YAWL "what you see is what you get."

16.5 Epilogue

In this chapter, we briefly introduced three major open source WfMSs: OpenWFE, jBPM, and Enhydra Shark. The presentation was structured along the control-flow, data, and resource perspectives. Some major differences of each system with YAWL were outlined. Here we summarize the differences across the three systems and YAWL as follows:

- *Graphical notation of the control-flow.* From the presented systems, OpenWFE is the only one not offering a graphical editor. The reasoning behind the omission of a graphical editor is that when processes are implemented, programming skills are necessary. While it can be argued that deep process modeling knowledge is indeed needed for understanding some advanced process models, the standpoint of YAWL has been that a simple graphical notation facilitates the communication with business users.
- *Distinction between offered, allocated, and started states.* YAWL is the only system to distinguish between these three different states. It does so in order to capture actual "real life" behavior closely. It can be argued that in cases when a work item is offered to a sole resource, a distinction between offered and allocated is not needed and can hence be ignored (as done, for instance, in jBPM). It should be noted, however, that this approach does not cover the scenarios where work is offered to multiple resources, and in some contexts, a distinction between offered and allocated states may be critical from a legal point of view. Furthermore, a distinction between offered and allocated states enables a

proper workflow analysis studying waiting times and bottlenecks in processes and systems, which is lost in systems like OpenWFE and Enhydra Shark.

- *Support for different resource allocation strategies.* All the tools presented in this chapter support a single resource allocation strategy, that is, work items are directly distributed to a resource or offered to a group of resources. Mechanisms such as Shortest Queue and Round-Robin (described in Sect. 2.6.4), where the system selects a user from a group, are not supported.
- *Automatic form generation.* Similar to YAWL, OpenWFE and Enhydra Shark provide automatic form generation functionality. jBPM had this functionality in its previous release, but not in the current one. Instead, functionality for manual report generation is provided. The advantage of manual form creation is that the layout of forms can be enhanced, which is often required by end users. Hence, support for manual form creation is often appreciated. However, this hardly reduces the advantages of having automated form generation functionality, which is convenient during process implementation and testing phases. Providing automated form generations enables separation of concerns for process developers: first, process logic is addressed, then interface design can take place.
- *Support for standards.* Enhydra Shark is the only tool presented here to rely on a standard, namely XPDL 1.0. While, in general, it may be argued that the use of standards is advantageous as it facilitates interoperability, in some cases, the use of standards may pose limitations and hamper the development. For instance, XPDL lacks a graphical notation, support for some frequently appearing control-flow and resource patterns, and a proper formal foundation.
- *Formal foundation.* YAWL is the only open source tool covered here that is based on a formal foundation. This both removes all ambiguities present in most tools lacking such a foundation and facilitates automated model reasoning and verification.
- *Closed source version in addition to the open source version.* Enhydra Shark is the only one of the examined tools to provide a closed source version in addition to the open source version. Both versions use similar administration and monitoring tool, but for the open source version some of the functionality is stripped, that is, some of the displayed buttons are not active. This means that at a first glance Enhydra Shark looks more mature than the other tools presented in this book. A more thorough evaluation is needed, however, to confirm this impression.
- *Extendability.* While Enhydra Shark relies on the extendability of XPDL to implement complex behavior and jBPM requires additional Java coding for this purpose, YAWL offers a powerful base set of process modeling constructs in order to reduce the need for coding.

Two topics we did not discuss when we introduced OpenWFE, jBPM, and Enhydra Shark are verification functionality and support for dynamic workflows. Verification functionality refers to automated analysis techniques for discovering errors in process models, such as deadlocks, which are only possible to apply in process models with a formal semantics. None of the systems introduced in this chapter provide such functionality. As demonstrated in Chap. 20, YAWL provides support for two different verification techniques: verification using reduction rules and verification using structural invariant properties. When it comes to dynamic workflows, that is, the dynamic change of a process definition during the execution of one of its cases, in addition to YAWL, only OpenWFE provides support for this. Chapter 4 explains the mechanisms in YAWL for dealing with dynamic workflows. For information on how OpenWFE deals with this, we refer to the documentation of the tool available on its web site.

Exercises

Exercise 1. Outline three main differences between YAWL and the following:
(a) OpenWFE
(b) jBPM
(c) Enhydra Shark

Exercise 2. Consider the Payment subprocess shown in Fig. A.5 on page 607.
(a) Model it in Ruote's process definition language.
(b) Implement the solution in Ruote and run it through Densha.

Exercise 3. Consider the Payment subprocess shown in Fig. A.5 on page 607.
(a) Model it in jPDL.
(b) Implement the solution in jBPM and run it.

Exercise 4. Consider the Payment subprocess shown in Fig. A.5 on page 607.
(a) Model it in Enhydra Shark's process definition language.
(b) Implement the solution in Enhydra Shark and run it.

Exercise 5. Implement the Freight in Transit subprocess in Fig. A.6 on page 609 in the following:
(a) OpenWFE
(b) jBPM
(c) Enhydra Shark

Guideline: The introduction provided to the tools in this chapter may not be sufficient to solve all the difficulties contained in the example. For extensive introductions to the tools, we refer to the corresponding tool manuals. As a first step, you can implement a simplified solution. Afterwards, you can gradually add complexity to your implementations, studying the tool documentation provided by the developers.

Chapter Notes

More information about the tools presented in this chapter can be found in the corresponding specifications, which are available online from the tools' sites. Additional material concerning jBPM is published in [67]. A detailed analysis and comparison of the systems based on the control-flow pattern (from Sect. 2.2) is available in [266]. A workflow patterns analysis of XPDL 1.0 (which is the language in Enhydra Shark) is presented in [4]. Furthermore, in the article "Fundamentals of Control-flow in Workflows" [131], it is shown that the textual "definitions" of the various control-flow concepts of XPDL 1.0 are open to fundamentally different interpretations. Hence, interoperability cannot be guaranteed through the use on standards containing solely natural language definitions.

Part VII
Advanced Topics

Chapter 17
Process Mining and Simulation

Moe Wynn, Anne Rozinat, Wil van der Aalst, Arthur ter Hofstede, and Colin Fidge

17.1 Introduction

So far we have seen how workflow models can be defined in YAWL and how the YAWL workflow engine allows a model to drive business processes. Once such a model becomes operational, the Engine automatically logs activity, keeping track of task start and completion events, their time of occurrence, the resources involved, and so on (cf. Chap. 9, Sect. 9.7). These logs are a valuable source of information about the way a business process actually performs in practice, and can be used as a basis for operational decision making.

Process mining is a technology that uses event logs (i.e., recorded actual behaviors) to analyze workflows. This is a valuable outcome in its own right, because such dynamically captured information can alert us to practical problems with the workflow model, such as "hotspots" or bottlenecks, that cannot be identified by mere inspection of the static model alone.

Further, the information extracted through process mining can be used to *calibrate simulations* of the workflow's potential future behaviors. In effect, this gives us a "fast forward" capability, which allows future activity to be predicted from the current system state and explored for various anticipated scenarios.

In this chapter, we explain how YAWL's logs can be mined, and how the information extracted from them can be used to calibrate simulations of expected future behavior. This is done using two additional tools: the process mining tool *ProM* and the modeling and simulation package *CPN Tools*.

17.2 Payment Process

In this chapter, the *Payment* subprocess from the running example will be used to showcase the mining and simulation capabilities of YAWL and ProM (see Fig. 17.1). The payment process is concerned with the processing and approval of payment

M. Wynn (✉)
Queensland University of Technology, Brisbane, Australia,
e-mail: m.wynn@qut.edu.au

A.H.M. ter Hofstede et al. (eds.), *Modern Business Process Automation*,
DOI 10.1007/978-3-642-03121-2_17,

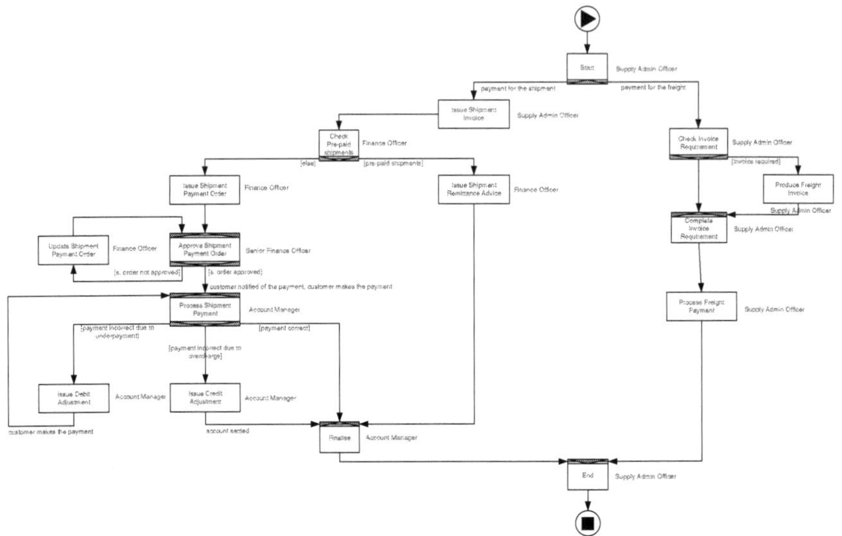

Fig. 17.1 Payment subprocess

invoices and is handled by a number of staff members with varying roles such as supply admin officer, finance officer, account manager, etc. For the purposes of mining and simulation, we assume that (1) all the tasks are being carried out by a staff member (i.e., a human resource, and not, for example, a web service) and (2) tasks are assigned to staff using role-based allocations (e.g., the *Issue Shipment Invoice* task is to be carried out by a staff member with the *Supply Admin Officer* role). We also assume that there are 13 members of staff available in the organization, which includes five capable of acting as "supply admin officers," three capable of acting as "finance officers," two capable of acting as "senior finance officers," and three capable of acting as "account managers." Note that these assumptions are made to create an organizational context that can be used to illustrate the concepts presented in this chapter.

17.3 Process Mining and YAWL

Through the application of Process Mining,[1] workflow execution logs can be retrieved and subsequently subjected to analysis using the process-related information found in those logs. Process mining techniques provide valuable insight into control-flow dependencies, data usage, resource involvement, and various performance related statistics (e.g., case arrival rates or average times that resources take to perform instances of a certain task).

[1] See also www.processmining.org for further information.

One of the leading software tools for process mining is the ProM framework. This framework is extensible and provides more than 230 plug-ins that support various process mining techniques. Furthermore, ProM provides import and conversion capabilities, where process models specified in a range of formalisms (Petri nets, EPCs, BPMN, BPEL, YAWL, etc.) can be imported and mapped to corresponding models in other formalisms. This section describes how YAWL's logs can be imported into ProM, and how these logs can be subsequently analyzed to discover interesting information about a running process supported by YAWL.

17.3.1 Exporting YAWL Execution Logs

In the YAWL workflow system, log entries are created whenever an activity is enabled, started, completed, or canceled. The log entries include the time when this event occurred and the actor who was involved. The logs also contain information about data values that have been entered and used during the execution of workflows. Therefore, historical data about completed process instances are available and can be retrieved. For this purpose, a function has been created that extracts historical data related to a particular specification from the workflow engine and exports the corresponding event logs in the so-called *M*ining *XML* (MXML) log format. This function is available as YAWLDB ProM*Import*. The data in MXML format can then be used for post-execution analysis in the ProM environment. Some sample data for the payment example is shown in Listing 17.1.

```
        <ProcessInstance id="3f9dfc70-5420-40e7-b9f7-329b5c6f0ded">
            <AuditTrailEntry>
                <WorkflowModelElement>
                    Check_PrePaid_Shipments_10
                </WorkflowModelElement>
                <EventType>start</EventType>
                <Timestamp>2008-07-08T10:11:18.104+01:00</Timestamp>
                <Originator>JohnsI</Originator>
            </AuditTrailEntry>
            <AuditTrailEntry>
                <Data>
                    <Attribute name="PrePaidShipment">true</Attribute>
                </Data>
                <WorkflowModelElement>
                    Check_PrePaid_Shipments_10
                </WorkflowModelElement>
                <EventType>complete</EventType>
                <Timestamp>2008-07-08T10:11:28.167+01:00</Timestamp>
                <Originator>JohnsI</Originator>
            </AuditTrailEntry>
        ...
        </ProcessInstance>
        ...
</Process>
```

Listing 17.1 Log entries for the start and completion of task *check prepaid shipments* carried out by resource *JohnsI* with the data attribute *prepaid shipment* set to "true"

17.3.2 Post-execution Analysis with ProM

The correctness, effectiveness, and efficiency of business processes are vital to an organization. A process definition that contains errors may lead to angry customers, backlog, damage claims, and loss of goodwill. Traditionally, significant emphasis has been placed on model-based analysis, that is, reality is captured in a model and this model is used to analyze the process. Typical examples are *verification*, that is, establishing the correctness of a process definition, and *performance analysis*, that is, evaluating its ability to meet requirements with respect to throughput times, service levels, and resource utilization. Less emphasis has been placed on the derivation or calibration of such models, that is, how to "capture reality in a model." An exception is the field of *data mining*, where large data sets (e.g., customer data) are analyzed to provide new insights (e.g., customers who buy X also buy Y). Data mining techniques typically focus on the construction of relatively simple models. Moreover, there is no emphasis on the underlying processes, for example, process models hardly play a role in mainstream data mining techniques.

Unlike classical data mining techniques, *process mining* specifically aims at the analysis of a business process. The goal of process mining is to extract information (e.g., process models) from execution logs, that is, process mining describes a family of a-posteriori analysis techniques exploiting the information recorded in the event logs. Typically, these approaches assume that it is possible to sequentially record events such that each event refers to an activity (i.e., a well-defined step in the process) and is related to a particular case (i.e., a process instance). Furthermore, some mining techniques use additional information such as the performer or originator of the event (i.e., the person/resource initiating or executing the activity), the timestamp of the event, or data elements recorded with the event (e.g., the size of an order).

Process mining aims to address the problem that most "process/system owners" have limited information about what is actually happening. In practice, there is often a significant gap between what is prescribed or supposed to happen and what *actually* happens. Only a concise assessment of reality, which process mining strives to deliver, can help in verifying process models, and ultimately be used in system or process redesign efforts.

Process mining techniques cover a wide range of analysis approaches. Process mining can be used to discover models, for example, process models or organizational models. Based on a log, one can construct a Petri net or social network describing what the typical mode of operation is. Other process mining techniques focus on conformance checking, that is, monitoring, locating, and measuring deviations.

It is impossible to give a complete overview of the many types of analysis possible. Therefore, this chapter focuses on three examples of possible approaches to post-execution analysis of log data. These examples fulfill the following criteria: (a) they are relatively easy to explain; (b) they expose information not already known (e.g., we are not aiming at rediscovering the existing process model); (c) they cover different perspectives of the process (timing, resources, and data); and (d) they require the presence of historical logs.

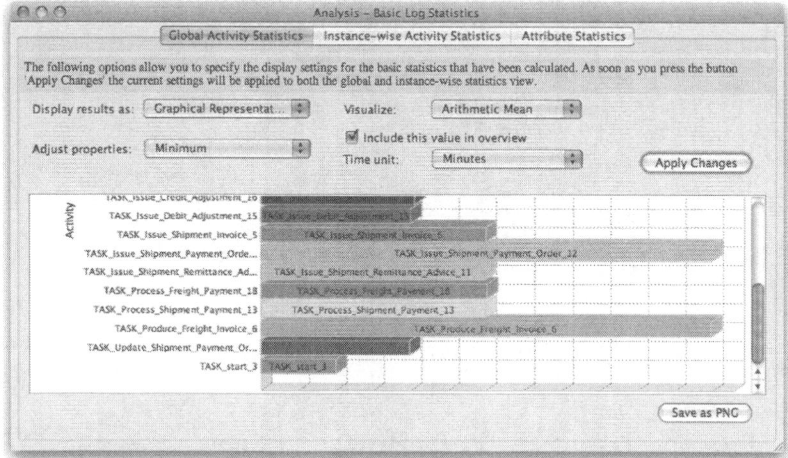

Fig. 17.2 Timing information such as execution times and arrival rates can be extracted from the event log. Here the average execution times for tasks in the payment process are shown

17.3.3 Basic Performance Analysis

Performance analysis provides valuable information about case arrival rates, values of data attributes, and durations of tasks, such as throughput times and waiting times. This type of analysis aims at identifying potential process improvements through examination of the runtime behavior of processes, and can also be exploited for the purpose of obtaining realistic data as a starting point for simulation experiments. The performance analysis metrics are calculated on the basis of events in the life-cycle of work items and the timestamps of these events as found in the execution log. For instance, the execution time of a particular work item is determined by taking the time difference between the start and complete events of this work item. ProM supports the calculation of performance metrics such as execution times, waiting times, and case arrival rates through the Basic Log Statistics plug-in as shown in Fig. 17.2.

Although the example shown is rather simple, more complex forms of performance analysis are supported as well. For example, other plug-ins can be used to identify behavioral patterns that relate to the throughput time of a process, or to highlight bottlenecks directly in the process model by visualizing where the cases spent the most time waiting to be processed further.

17.3.4 Resource Analysis

Another application area for process mining concerns the resource perspective of process models. Process mining techniques can be used to unearth the behavior of resources during the execution of processes. As an example, consider the Dotted

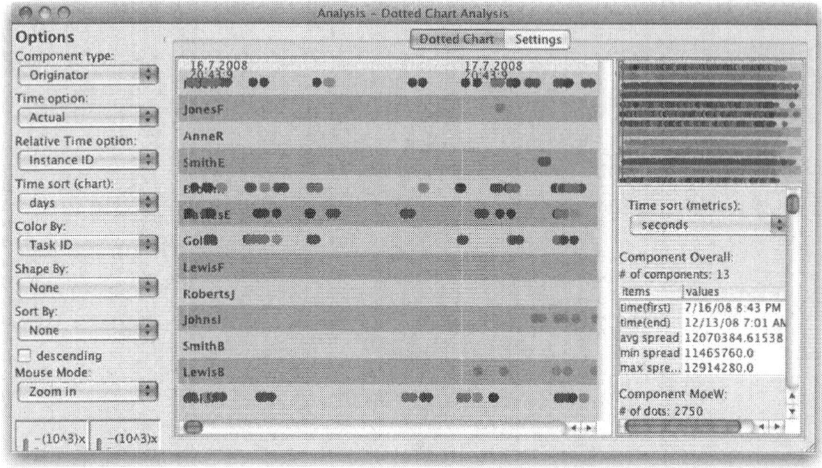

Fig. 17.3 Originator tasks and communication flows can be analyzed based on the log. Here the activities of the resources in the payment process are visualized over time

Chart Analysis provided in ProM, which shows the activities carried out by various resources in a Gantt chart-like manner. Figure 17.3 shows how task-originator information derived from execution logs can be used to visualize the workload performance of resources over a period of time.

More sophisticated forms of analysis involving the resource perspective are also supported by ProM. For example, social network mining can be used to find interaction patterns, and to identify individuals who have a central social role in an organization. Furthermore, role hierarchy mining can be used to identify specialists and generalists for a given process on the basis of the variety of tasks an individual has performed in the past. Finally, it is possible to discover groups of people who work together based on the similarity of tasks they perform, or to discover staff assignment rules that provide insights into organizational practices.

17.3.5 Analysis of Cases with Interesting Properties

The examples presented thus far have focused on *all* instances of a process model in order to derive meaningful information about the behavior of that process. In some situations it may also be interesting to focus on a subset of cases of a particular process. This subset is determined through the application of a selection criterion. For example, one may wish to identify all cases where the customer pays less than the required amount (underpayment). Underpayment can result in the *Process Shipment Payment* task having to be performed multiple times. ProM's LTL (Linear Temporal Logic) Checker plug-in can be used to separate process instances that fulfill a particular property from those that do not. In Fig. 17.4, one can see that

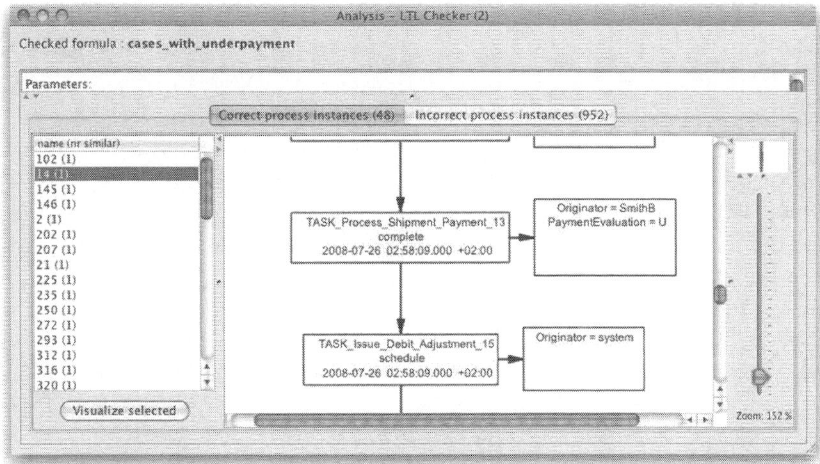

Fig. 17.4 A subset of cases that possess some interesting property can be extracted and further analyzed in isolation. Here the LTL Checker detected 48 cases of underpayment

this plug-in has detected 48 (out of 1,000) process instances where underpayment has occurred. These cases can now be analyzed further to find out more about the impact and the root causes of the underpayment. Another example is the checking for violations of the four-eyes principle, where two different tasks should not be performed by the same person (e.g., a manager should not approve his/her own travel request).

In this section, it has been shown that environments like YAWL provide detailed execution data about the processes they support. These data can be utilized to apply a wide range of analysis techniques that yield additional insight about the runtime characteristics of a process. In the next section, however, historical information is not only used to learn about past executions, but also to predict the future performance of a process.

17.4 Process Simulation and YAWL

Business process simulation is a powerful tool for process analysis and improvement. One of the main challenges is to create simulation models that *accurately* reflect the real-world process of interest. Moreover, we do not want to use simulation just for answering strategic questions but also for tactical and even operational decision making. To achieve this, different sources of simulation-relevant information need to be leveraged. This section presents a new way of creating a simulation model for a business process. In this approach, we integrate design, historical, and state information extracted from a workflow management system such as YAWL.

17.4.1 Approach

Figure 17.5 illustrates our approach. The considered scenario involves a *workflow system* that supports some *real-world process* based on a *workflow and organizational model*. Note that the workflow and organizational models have been designed before enactment and are used for the configuration of the workflow system. During the enactment of the process, the activities performed are recorded in *event logs*. An event log records events related to the offer, start, and completion of work items, for example, an event may be "Mary completed the approval activity for insurance claim XY160598 at 16:05 on Monday 21-1-2008."

The right-hand side of Fig. 17.5 is concerned with enactment using a workflow system, while the left-hand side focuses on analysis using simulation. To link enactment and simulation, three types of information readily available in workflow systems are used to create and initialize the simulation model.

- *Design information.* The workflow system has been configured based on an explicit process model describing control and data flows. Moreover, the workflow system uses organizational data, for example, information about users, roles, groups, etc.
- *Historical information.* The workflow system records all of the events that take place in "event logs" from which the complete history of the process can be reconstructed. By analyzing historical data, probability distributions for workflow events and their timing can be extracted.
- *State information.* At any point in time, the workflow process is in a particular state. The current state of each process instance is known and can be used to initialize the simulation model. Note that this current state information includes the control-flow state (i.e., "tokens" in the process model), case data, and resource data (e.g., resource availability).

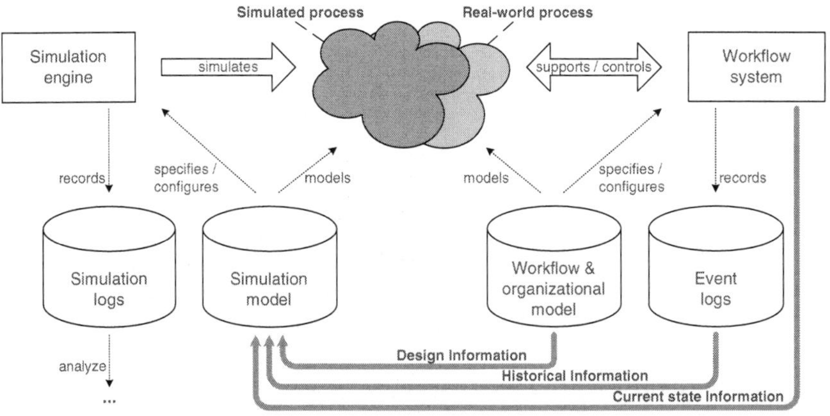

Fig. 17.5 Overview of our integrated workflow management (*right*) and simulation (*left*) system

By merging the above three sources of information into a simulation model, it is possible to construct an *accurate model based on observed behavior* rather than a manually constructed model, which approximates the workflow's anticipated behavior. Moreover, the current state information supports a "fast forward" capability, in which simulation can be used to explore different scenarios with respect to their *effect in the near future*. In this way, simulation can be used for *operational decision making*.

Based on this approach, the system design in Fig. 17.5 allows different simulation experiments to be conducted. For the "as-is" situation, the simulated and real-world processes should overlap as much as possible, that is, the two process "clouds" in Fig. 17.5 should coincide. For the "to-be" situation, the observed differences between the simulated and real-world processes can be explored and quantified. In our implementation, it is ensured that the simulation logs have the same format as the event logs recorded by the workflow system. In this way, the *same tools* can be used to analyze both simulated and real-world processes.

A crucial element of the approach in Fig. 17.5 is that the *design*, *historical*, and *state* information provided by the workflow system are used as the basis for simulation. Table 17.1 describes this information in more detail.

The design information is static, that is, this is the specification of the process and supporting organization that is provided at design time. This information is used to create the structure of the simulation model. The historical and state information are dynamic, that is, each event adds to the history of the process and changes the current state. Historical information is aggregated and is used to set parameters in the simulation model. For instance, the arrival rate and processing times are derived by aggregating historical data, for example, the (weighted) average over the last 100 cases is used to fit a probability distribution. Typically, these simulation parameters are not very sensitive to individual changes. For example, the average processing time typically changes only gradually over a longer period. The current state, however, is highly sensitive to change. Individual events directly influence the current state and must be directly incorporated into the initial state of the simulation. Therefore, design information can be treated as static, while historical information evolves gradually, and state information is highly dynamic.

Table 17.1 Process characteristics and the data sources from which they are obtained

Design information *(obtained from the workflow and organizational model used to configure the workflow system)*	Historical information *(extracted from event logs containing information on the actual execution of cases)*	State information *(based on information about cases currently being enacted using the workflow system)*
• Control and data flow (activities and causalities)	• Data value range distributions	• Progress state of cases (state markers)
• Organizational model (roles, resources, etc.)	• Execution time distributions	• Data values for running cases
• Initial data values	• Case arrival rate	• Busy resources
• Roles for each task	• Availability patterns of resources	• Runtimes for cases

To perform simulation experiments for operational decision making, state-of-the-art *process mining* techniques are needed to analyze the simulation and event logs and to generate the simulation model. To demonstrate the applicability of our approach, the system shown in Fig. 17.5 has been implemented using YAWL and ProM. YAWL is used as the workflow management system and has been extended to provide high-quality design, historical, and state information. The process mining framework ProM has been extended to merge the three types of information into a single simulation model. These extensions are discussed in the following section.

17.4.2 Architecture

Figure 17.6 depicts the landscape of tools and information sources that are involved in the creation of the simulation model for a YAWL process. In the top left corner is the process designer who initially creates a YAWL process specification that is to be enacted on a *YAWL Engine*. The *YAWL Editor* has access to a *Resource DB*, which contains information about the relationships between roles and resources in an organization. This is needed to support resource distribution mechanisms in the Editor, and to associate tasks with roles (i.e., to specify which people are allowed to perform which tasks in the process). This organizational model is also used by the Engine during enactment when newly created work items are offered to users of the workflow system. Finally, the Engine records log data (*YAWL Logs*) during the execution of each case.

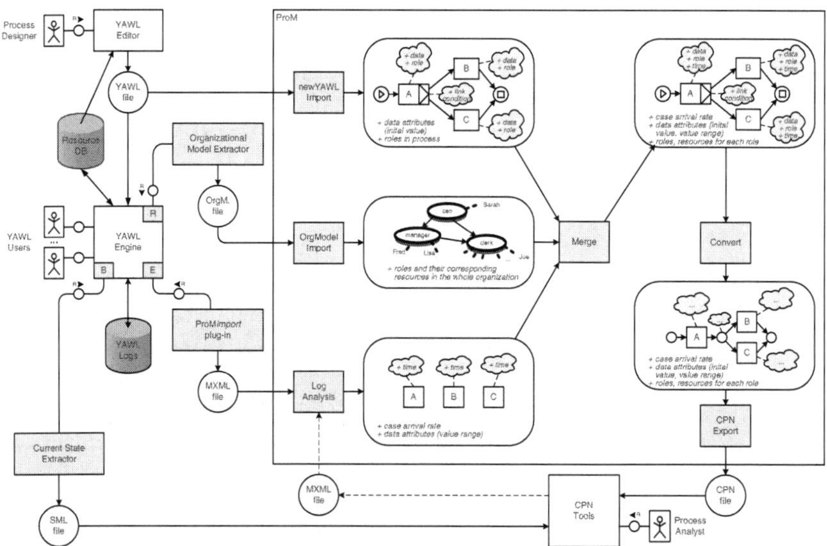

Fig. 17.6 Overview of the tools and information sources used for process mining and simulation in the context of YAWL

All these data sources (the process specification, the organizational model, and the execution log containing historical data) are relevant for the creation of a simulation model that reflects reality as closely as possible. The *ProM* framework allows for the creation of a Colored Petri Net model based on these data sources. Furthermore, knowledge about the current state (operational characteristics) of the process, which is available from the Engine, is needed for short-term simulation tasks. By making this current state information available through an SML file, it can be loaded dynamically into the *CPN Tools* simulation package to set a marking that reflects the current state of the process. Finally, information about the simulation itself can be used to generate simulation logs that in turn can be analyzed using the ProM framework.

The next section provides a scenario for the payment process on which simulation experiments will be performed (Sect. 17.4.3). Then, it will be shown how simulation-relevant information can be extracted from the YAWL System in the form of a *YAWL file*, an *OrgModel file*, an *MXML log file*, and an *SML current state file* (Sect. 17.4.4). Afterwards, it is illustrated how these files are used to create a simulation model in ProM (Sect. 17.4.5) and how the current state is loaded into the generated CPN model (Sect. 17.4.6). Finally, a number of simulation experiments for the payment process scenario are performed (Sect. 17.4.7).

17.4.3 A Simulation Scenario

To highlight the effect of using real case data in simulation scenarios, we assume here that the Payment process has been running in the YAWL workflow system for some time. Process mining techniques are then used to extract the following statistics from the YAWL logs.

- Case arrival rate: 50 payments per week
- Throughput time: 5 working days on average
- 30% of shipments are prepaid
- 50% of orders are approved first-time
- 20% of payments are underpaid
- 10% of payments are overpaid
- The remaining 70% of payments are correct
- 80% of orders require invoices
- 20% of orders do not require invoices

In Sect. 17.3, we explained how event logs can be exploited to extract even further characteristics of the process. For example, using the LTL Checker, it is possible to find out that cases involving underpayment take significantly more time to be completed than other cases. So, we are interested in how much the case load would be reduced if there is a way to avoid underpayments (e.g., by requiring an automatic debit transfer).

Consider the following scenario. The end of the financial year (30 June) is only 4 weeks away and we would like all payments to be processed before then. Currently, there is a backlog of 30 payments in various stages of processing, some of which have been in the system for more than a week. New payments are expected to come in at a rate of 50 payments per week. Since it is essential that most of these applications are processed before the end of the financial year (the "time horizon" of interest), management would like to perform simulation experiments based on the current state to determine whether or not the backlog can be cleared in time using the current number of available resources. If necessary, more resources could be allocated. Also the effect of avoiding underpayments should be evaluated.

17.4.4 Extracting Simulation Information from YAWL

Table 17.1 described the three types of information available in a workflow system: design information, historical information, and state information. Now, let us look at how this information can be obtained from the YAWL workflow system.

The following four types of data can be extracted from YAWL as shown in Fig. 17.6:

- An MXML log file provides historical information about a process in the form of event logs (see Listing 17.1). As mentioned in Sect. 17.3.1, it is possible to generate YAWL logs from specifications that have been deployed in the Engine. Here we assume that the simulation experiments are being carried out on "as-is" process models for which historical data is available. This historical data is used for mining information about case arrival rates and distribution functions for the data values used in future simulation experiments.
- A YAWL file contains the (static) design information for a particular process model, including the control and data flow, initial data values, and resource assignment strategies. This YAWL file is created in the Editor and is exported as an XML file. Listing 17.2 shows an excerpt of the YAWL file for the payment process. This file is read by the Engine to execute the process. For simulation purposes, this YAWL file is imported into ProM to generate the simulation model.
- An OrgModel file, on the other hand, provides an up-to-date organizational model in the standard format required by ProM. The YAWL workflow system provides access to the organizational database (through its interface R). The Organizational Model Extractor function is used to extract all available role and resource data in the YAWL System. This information is used to identify available roles and resources that are relevant for a given specification and is used in setting up the resource availability information in the simulation experiments. Some sample data with the roles of *supply admin officer* and *senior finance officer* are shown in Listing 17.3.
- An SML current state file provides data values, timestamps, and resource information for currently running cases of a given specification in the workflow system. The YAWL workflow system provides access to current state information

```
<specificationSet xmlns="http://www.yawlfoundation.org/yawlschema" xmlns:xsi="
   http://www.w3.org/ 2001/XMLSchema-instance" version="2.0"
   xsi:schemaLocation="http://www.yawlfoundation.org/yawlschema c:/temp/
   YAWL_Schema2.0.xsd">
  <specification uri="Payment_subprocess.ywl">
     <schema xmlns="http://www.w3.org/2001/XMLSchema" />
     <decomposition id="Payment" isRootNet="true" xsi:type="NetFactsType">
        <localVariable>
                <name>InvoiceRequired</name>
                <type>boolean</type>
                <namespace>http://www.w3.org/2001/XMLSchema</namespace>
                <initialValue>false</initialValue>
        </localVariable>
                ...
    <processControlElements>
      <inputCondition id="InputCondition_1">
        <flowsInto><nextElementRef id="start_3"/></flowsInto>
      </inputCondition>
      <task id="start_3">
        <name>start</name>
        <flowsInto><nextElementRef id="Issue_Shipment_Invoice_5"/></flowsInto>
        <flowsInto><nextElementRef id="Check_Invoice_Requirements_7"/><
           flowsInto>
        <join code="xor" />
        <split code="and" />
        <resourcing>
            <offer initiator="system">
                <distributionSet><initialSet>
                <role>RO-202fd4df-b130-4512-9bc2-badc8f5dab7d</role>
                </initialSet></distributionSet>
          </offer>
          <allocate initiator="user"/>
          <start initiator="user"/>
        </resourcing>
        <decomposesTo id="start"/>
      </task>
        ...
      <outputCondition id="OutputCondition_2"/>
        </processControlElements>
    </decomposition>
    ...
    </specification>
</specificationSet>
```

Listing 17.2 An excerpt from a YAWL model for the *Payment subprocess*

for all running workflow specifications through interface B. The Current State Extractor function is created to extract current state information for a running specification from the Engine and to export this information as a CPN Tools input file as shown in Listing 17.4.

17.4.5 Creating a Simulation Model in ProM

There are three data sources from which the simulation model is built: the YAWL Engine file, the MXML execution log, and the OrgModel file. These sources need to be imported and merged. Note that we need process information beyond that for pure control-flow, that is, additional information like data and time that can be

```
<OrgModel>
    <OrgEntity>
        <EntityID>RO-202fd4df-b130-4512-9bc2-badc8f5dab7d</EntityID>
        <EntityName>supply admin officer</EntityName>
        <EntityType>Role</EntityType>
    </OrgEntity>
    <OrgEntity>
        <EntityID>RO-5299e7bc-8556-4db1-82c8-45af3a6f73d4</EntityID>
        <EntityName>finance officer</EntityName>
        <EntityType>Role</EntityType>
    </OrgEntity>
   ...
    <Resource>
        <ResourceID>PA-773916d2-415b-99ec-2ad019b9e945</ResourceID>
        <ResourceName>JohnsI</ResourceName>
        <HasEntity>RO-5299e7bc-8556-4db1-82c8-45af3a6f73d4</HasEntity>
    </Resource>
    ...
</OrgModel>
```

Listing 17.3 An excerpt from an organizational model with roles and resources, where resource Johnsl has the role *finance officer*

```
fun getInitialCaseData()=[
(21, {PrePaidShipment = false,PaymentEvaluation =C,InvoiceRequired =
false, OrderApproval = false}),
(18, {PrePaidShipment = false,PaymentEvaluation = C,InvoiceRequired =
false,OrderApproval = false}),
(33, {PrePaidShipment = false,PaymentEvaluation = C,InvoiceRequired =
true,OrderApproval = false}),
(29, {PrePaidShipment = false,PaymentEvaluation = C,InvoiceRequired =
true,OrderApproval = false}),
(30, {PrePaidShipment = false,PaymentEvaluation = C,InvoiceRequired =
true,OrderApproval = false}),
(31, {PrePaidShipment = true,PaymentEvaluation = C,InvoiceRequired =
true,OrderApproval = true}), ... ];
fun getNextCaseID() =33;
fun getInitialTokensExePlace(pname:STRING) = case pname of
"TASK_Complete_Invoice_Requirement_8`E"=>[(16,"-48153","MoeW")] |
"TASK_Finalise_17`E"=>[(15,"-48170","AnneR"),(33,"-48175","SmithB")] |
... | _ => empty;
fun getInitialTokens(pname:STRING) = case pname of "Process
`COND_c_Complete_Invoice_Requirement_8_Process_Freight_Payment_18_"=>
[(14,"48233"),(31,"-48000")] |
"Process`COND_c_Issue_Shipment_Remittance_Advice_11_Finalise_17_"=>
[(18,"-48117"),(26,"-34000")] | ... | _ => empty;
fun getBusyResources() = ["MoeW","AnneR","LewisB","JohnsI","JonesF","SmithB"];
fun getCurrentTimeStamp() = "20257972";
fun getTimeUnit() = "Min";
```

Listing 17.4 An excerpt from a current state CPN Tools input file for the *Payment* process

attached either to the process as a whole or to certain elements within the process (see, e.g., the items in the clouds in the overview picture in Fig. 17.6, which reflect extra information attached to activities and choices). The merged model is then converted into a Petri net-based process, which can be exported to the CPN Tools simulator.

As illustrated in Fig. 17.6, four basic steps need to be performed within ProM to generate the simulation model:

Step 1: The workflow, the organizational model, and the event log are imported from the YAWL workflow system and analyzed.

- The information that can be obtained from the Engine file encompasses the complete YAWL process model, including the roles for individual tasks, data flows, and link conditions at choice points in the process.
- From the MXML execution log information about the case arrival rate, value range distributions for data attributes and observed execution times for tasks in the process can be obtained.
- The OrgModel file provides information about the relationship between all roles and resources in the organization.

Step 2: Simulation-relevant information from the organizational model and log analysis are integrated into the YAWL model.

Step 3: The YAWL model is converted into a Petri net model (because our simulation tool is based on Colored Petri Nets), in which all of the extra information (e.g., time and data) that is relevant for the simulation model is preserved.

Step 4: Finally, the integrated and converted Petri net model is exported as a CPN model.

By using information extracted from the YAWL file, the OrgModel file, and the MXML log, the parameters of the generated CPN model are automatically configured to reflect the behavior of the payment process. The CPN Tools system can then be used to simulate the generated model. However, to produce useful results, we do not want to start from an empty initial state. Instead, the current state of the actual YAWL System is loaded into the CPN model as the initial state for the simulation.

17.4.6 Loading the Current State

To carry out simulation experiments for operational decision making purposes (the "fast forward" approach), it is essential to include the current state of the workflow system. This allows us to make use of the data values for current cases as well as the status of the work items for current cases within the simulation experiments.

The following information is obtained about the current state and is introduced as the initial state of a simulation run:

- All the running cases of a given specification and their marking
- All the data values associated with each case
- Information about enabled work items
- Information about executing work items and the resources in use
- The date and time at which the current state file is generated

When the empty initial state file of the generated simulation model is replaced with the file depicted in Listing 17.4, tokens are created in the CPN model that reflect the current system status (see Fig. 17.7). For example, there are 30 cases in the system (represented as tokens in the *Case Data* place) and 7 of the 13 resources are currently available (in the *Resources* place).

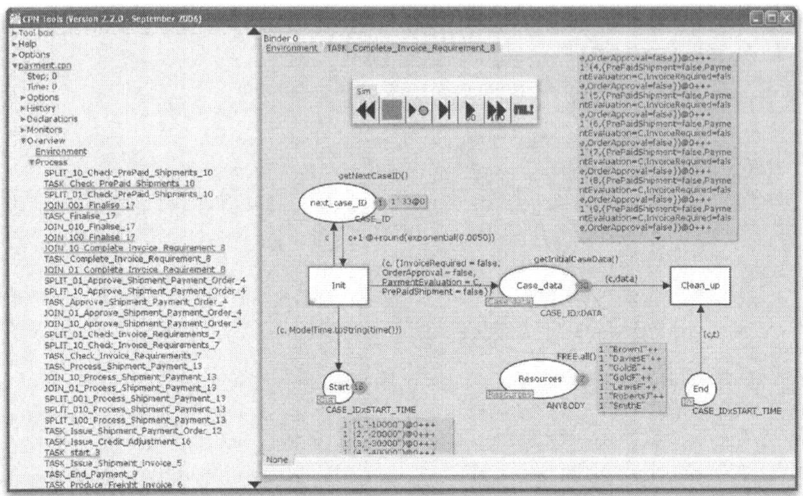

Fig. 17.7 The generated simulation model with the loaded current state file

We now continue the scenario described in Sect. 17.4.3, that is, simulation experiments for the coming 4 weeks are performed to investigate possible future behaviors extrapolated from the current situation.

17.4.7 Analyzing the Simulation Logs

The process from the generated CPN model is simulated for four different scenarios, as shown in Fig. 17.8:

1. An empty initial state (The “empty” trajectory in Fig. 17.8)
2. After loading the current state file with the 30 cases that are currently in the system and with no modifications to the model, that is, the “as is” situation (The “as is” trajectory in Fig. 17.8)
3. After loading the current state file but adding 13 extra resources (doubling the amount of resources for each role), that is, a possible “to be” situation to help clear the backlog more quickly (The “to be A” trajectory in Fig. 17.8)
4. After loading the current state file and changing the model so that underpayments are no longer possible (The “to be B” trajectory in Fig. 17.8)

The figure compares the number of cases in the workflow system for an example simulation run of each scenario over the coming 4 weeks. In case of the “empty” scenario, the simulation starts with zero payment process instances in the system. This does not reflect the normal situation nor does it capture our current backlog of cases. Only after the processing “pipeline” fills does this simulation represent the normal behavior of the payment process (i.e., with around 50 applications arriving

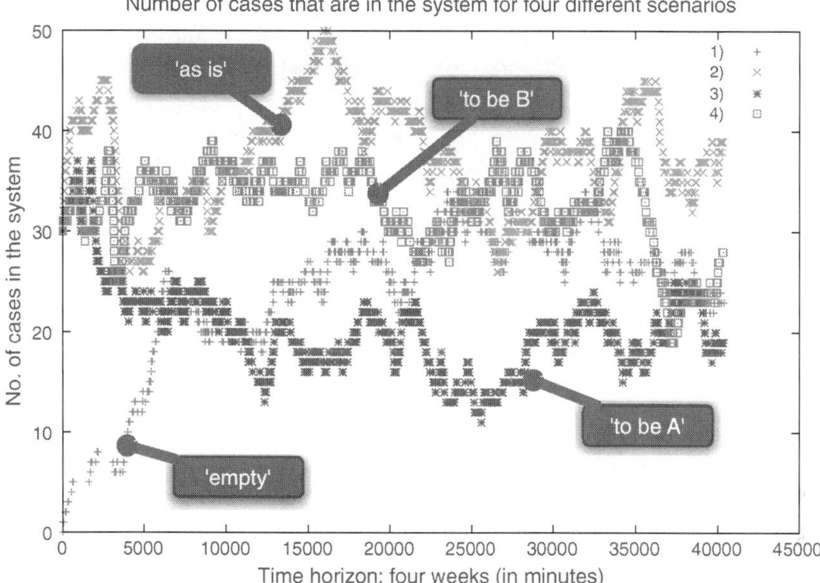

Fig. 17.8 The number of cases in the simulated process for the different scenarios, up until the time horizon. While the scenario with the empty state initially has zero applications, the other scenarios are initialized by loading 30 applications from the current state file

per week). This highlights the value of starting the simulation with the actual current state of the system.

The other three scenarios load a defined initial state, which contains the 30 cases that are assumed to be currently in the system. Furthermore, one can observe that in the scenarios where extra resources are added to the process or underpayments are avoided (the "to be A" and "to be B" trajectories), the case load decreases more quickly to an acceptable level than without further intervention (the "as is" trajectory). However, the "to be B" scenario, which simulates the process under the assumption that there will be no further underpayments, does not seem to perform as well as we might have thought. In this way, the effect of possible measures to address the problem at hand can be assessed, that is, we can compare different "what if" scenarios in terms of their estimated effects.

Note that Fig. 17.8 merely shows individual scenarios. Of course, it is vital not to do an analysis just based on example scenarios. CPN Tools supports the notion of replication to do many experiments. This can be used to compute confidence intervals. By doing so it is possible to indicate how reliable the results are, for example, the statement "after one week the number of cases in the system will be less than 30" can be made with 95% confidence.

As just explained, CPN Tools have powerful simulation capabilities, which can be leveraged to support statistical analyses of specific process characteristics. Furthermore, while CPN Tools provide built-in logging facilities and even generate gnuplot scripts that can be used to plot certain properties of the simulated process, MXML event log fragments are also generated during simulation, similar to the one shown in Listing 17.1 for the workflow log. These fragments can then be combined using the CPN Tools filter of the ProM*import* framework.

The ability to use the same toolset for analyzing simulation logs as well as the actual workflow logs constitutes a significant advantage because the simulation analysis results can be more easily related to the initial properties of the process. In particular, as we support the loading of current cases into the initial state at the beginning of the simulation, *we can easily combine the real process execution log ("up to now") and the simulation log (which simulates the future "from now on") so as to look at the process execution in a unified manner* (with the possibility of tracking both the history and the future of particular cases that are in the system at this point in time).

17.5 Conclusion

Process mining and simulation technologies can exploit the log data produced by the YAWL Engine. Process mining allows us to analyze the actual behavior of the system, rather than the behavior the workflow designer assumed it would have, and thus supports calibration of the system. Simulation allows us to predict future behaviors, and thus supports operational decision making. Both of these capabilities greatly increase the value of our YAWL models.

Exercises

These exercises aim to provide you with hands-on experience in performing process mining analysis and simulation experiments using ProM and YAWL. You may need to download ProM,[2] ProM*Import*,[3] and CPN Tools.[4]

Exercise 1. Open the MXML log from the payment process in ProM. Analyze the activity durations using the Basic Log Statistics plug-in to find out which task takes the most time. Is the task that is *on average* the most time-consuming also the one that *accumulates* the most time in total? If not, can you imagine why this is the case?

Exercise 2. Consider again the running example in this chapter. Assume that we now wish to ensure that the *Approve Shipment Payment Order* task is carried out by the *same* senior finance officer if the order is not approved in the first instance (i.e.,

[2] prom.sourceforge.net/

[3] promimport.sourceforge.net/

[4] wiki.daimi.au.dk/cpntools/

the retain familiar resource pattern). Perform LTL checks on the given MXML log to determine how many cases satisfy this criterion.

Exercise 3. Download the repair example log from the process mining tutorial page[5] and open it in ProM. Read the "Process Mining in Action" tutorial to learn how to clean the log and discover a process model. Use the Conformance Checker to make sure your discovered model fits all the completed cases in the log. Then, convert your model into a YAWL model, export it as a YAWL file, and import the Engine file into the Editor.

Exercise 4. Download the example files from the simulation tutorial page.[6] Follow the steps in the provided tutorial from Sect. 4.2 on to create a CPN model. Open the generated CPN model in CPN Tools. Go to the "Environment" page and change the case arrival rate from `exponential(0.01)` to `exponential(0.1)`. This will increase the simulated case load as the average inter-arrival rate of new cases is reduced from 100 to 10 min. Load the provided current state file into your modified model and simulate it for at least 1,000 cases.[7] Open the generated MXML log from your simulation in ProM and use the Performance Analysis with Petri net plug-in to find out where the biggest bottlenecks are. Where is the most waiting time accumulating and more resources are needed? Make a new model in which you add resources to cope with the increased case load. Compare the simulation results of your new model with the previous results.

Chapter Notes

The advantages of adding simulation capabilities to a workflow environment have been recognized for some time, including the abilities to monitor activities, predict future behaviors, and implement real-time optimizations [31]. Although statistical theories are the traditional way of forecasting average load or throughput in network-like systems, such mathematical models are rarely available for a particular business process and may be difficult to apply. Although several authors show the applicability of analytical queuing models [53], in reality simulation is the only widely used approach for predicting future behaviors from business process models [137]. In general, this capability is known as *business process simulation* [248]. Business process simulation offers the ability to perform "as is" and "what if" analyses that reflect the actual, rather than hypothetical, system state [110]. Accurate simulation of business processes is also an essential foundation for predicting the effectiveness of *business process reengineering*, that is, making changes to a workflow model to overcome performance problems identified during its operation [119].

[5] prom.win.tue.nl/research/wiki/prom/tutorials

[6] prom.win.tue.nl/research/wiki/prom/yawltutorial

[7] Hint: Add a guard such as `[c<=1,000]` to the "Init" transition to make sure your simulation terminates.

The idea to focus on the transient behavior of a workflow by taking the real current state as the initialization of the simulation model was first presented in [207].

Apart from the approach described in this chapter, a variety of tools are potentially helpful for business process simulation. These include specific business process modeling tools such as Protos and ARIS, business process management tools such as FLOWer and FileNet, and general purpose simulation tools such as Arena [127]. Many commercial tools (Protos, ARIS, iBPS, FileNet, etc.) include some simulation features [122].

The YAWL capabilities described in this chapter rely on three technologies: data logging, process mining, and simulation.

Logging. Accurately and comprehensively logging data is a basic prerequisite for workflow-based simulation [38]. Indeed, most workflow tools implement some form of data logging for disaster recovery and to support auditing, and many have monitoring and measurement features for real-time display of system statistics as a "management dashboard." Thus the Engine's ability to log data is not unusual; however, YAWL has been customized to record additional log data specifically for the purposes of driving simulations.

In particular, logging the start and completion of a task separately allows for parallelism in the resulting simulation model because time-consuming activities can overlap [198]. Also, our integration of three existing tools, the YAWL Engine, ProM, and CPN Tools, offers a lightweight and highly flexible implementation compared to fully integrated tools.

Mining. Process mining involves extracting dynamic attributes of a system from observations (logs) of its actual behavior [21] and has been recognized for some time as a useful adjunct to workflow modeling [30]. A wide variety of mathematical techniques exist for driving simulations, but all rely on the availability of appropriate statistical data concerning the "typical" behavior of the system under study. The approach described in this chapter extends previous work on process mining as a starting point for CPN-based simulations [216].

Although business process simulation models can be constructed by manually entering *assumed* behaviors of tasks and resources [245], such approaches are inevitably less accurate than extracting this data automatically. To support a comprehensive workflow simulation, multiple "views" of the system's behavior need to be extracted from the logs, including control-flow, and performance and resource usage perspectives, among others [215].

Simulation. Simulation typically requires the (costly) development of a model for the system of interest and experimentation with its behavior under different assumptions. As we already have a workflow model, it is natural to want to reuse this for the simulation model itself, as we have done. Developing a simulation model separately from the original workflow model is both time-consuming and error prone [125]. Another approach is to develop a library of reusable simulation components (themselves expressed in a workflow language) as building blocks for constructing simulation models [246].

The results of simulation must, of course, be interpreted with care – the outcomes produced always contain a significant degree of uncertainty, and users of business process simulation tools need to understand their limitations [59]. Indeed, recent research is focusing on ways of validating the "correctness" of such simulation models, although this remains a significant challenge as each simulation model is unique [228]. Furthermore, the modeling of human resources is often especially problematic [18].

Overall, YAWL's approach to process mining and simulation provides the advantages of previous approaches to business process simulation and does so with a lightweight and flexible combination of existing tools. More information on our work can be found in [217, 218].

Chapter 18
Process Configuration

Florian Gottschalk and Marcello La Rosa

18.1 Introduction

The Order Fulfillment process of Genko Oil was set up based on recommendations from the Voluntary Inter-industry Commerce Solution Association (VICS).[1] These recommendations are best practices derived from the experience of hundreds of organizations in the area of logistics and supply chain management. Although these companies offer products varying in many aspects, they all have similar order fulfillment processes in place. For this process, an efficient collaboration with many trading partners is very important. Thus, companies prefer to use standards and suggestions, such as the VICS recommendations, for setting up their workflows instead of trying to innovate in executing these processes.

Of course, not all companies have the same requirements on order fulfillments. For example, Zula Exquisite might only provide *Full Truck-Load* and *Single Package Shipments* and neglect *Partial Truck-Loads*. Contrary to Genko Oil, they therefore do not need to implement those parts of the *Carrier Appointment* subprocess dealing with *Partial Truck-Loads*. Still, there will be many commonalities with the order fulfillment process deployed at Genko Oil. Thus, instead of reimplementing this process from scratch, it is far more attractive for Zula Exquisite to amend the YAWL workflow model deployed at Genko Oil to their needs.

Process configuration can prove beneficial when process adaptations are needed to suit specific requirements, for example, if a process model that depicts a certain business process in a general manner should be adapted to the requirements of an individual organization. It restricts the possible behavior with respect to a previously built process model in a controlled way before the process' execution. The configuration of a process model therefore takes place in an adaptation phase between build time and runtime of a process model. Hence, at build time process, designers can integrate several variants of how a particular process can be executed into a single

[1] www.vics.org

F. Gottschalk (✉)
Eindhoven University of Technology, Eindhoven, the Netherlands
e-mail: f.gottschalk@tm.tue.nl

A.H.M. ter Hofstede et al. (eds.), *Modern Business Process Automation*,
DOI 10.1007/978-3-642-03121-2_18,

process model. Afterwards, the model user selects those parts relevant to the specific setting (e.g., an organization or project). All irrelevant model parts can then be automatically removed before the process is executed at runtime. This approach not only allows organizations like Zula Exquisite to adapt an existing model for the same process to individual needs, but it also enables organizations like VICS to provide templates for workflow models that are easily adaptable to user requirements while still conforming to their recommendations. These templates are called *configurable reference process models*. Furthermore, process configuration decisions can be mapped to domain related, natural language questions. In this way, users who are not very proficient in workflow modeling can adapt a process model through simply answering a questionnaire.

This chapter is organized as follows. Section 18.2 discusses in more detail what process configuration means. Afterwards, Sect. 18.3 introduces a configurable YAWL (C-YAWL) notion by showing how YAWL processes can be restricted through process configuration. How such a configuration can be steered through a questionnaire with domain-related questions is depicted in Sect. 18.4. Section 18.5 finally shows how the configuration decisions of a C-YAWL model lead to a new, executable YAWL model. The corresponding software tools are introduced in Sect. 18.6. The chapter concludes with a summary and suggestions for further readings.

18.2 How Does Process Configuration Work?

When developing a configurable workflow model, the aim is a model that can be adapted without any manual modeling efforts. Thus, during such an adaptation phase, it is not possible to add behavior that has not been modeled beforehand. Instead, all the possible behavior must already be contained in the configurable model. The configuration of the model to individual needs is then based on restricting this behavior.

To understand how the process behavior of a YAWL model can be restricted, theoretic results from analyzing inheritance relations between process models must be explored first. For this, let us abstract for a moment from YAWL and instead use the very simple notation of labeled transition systems (LTSs) to depict the process flow. This will help in understanding the essence of process configuration without being influenced by a specific process modeling language like BPMN, EPCs, BPEL or YAWL. Afterwards, we will discuss how these theoretical results can be applied to the behavior represented in a YAWL workflow model.

An LTS is a directed graph where transitions depict the possibilities to switch between states. The states are represented as the nodes of the graph, while the transitions are represented as the arcs in between the states. For an example, let us have a look at the LTS labeled A on the left-hand side of Fig. 18.1. Initially that process is in state s_1. In this state there is a choice to do either a, b, or c leading to states s_2, s_3 or s_4. In s_3 there is the choice to do either d or e. Executing d leads to state s_2,

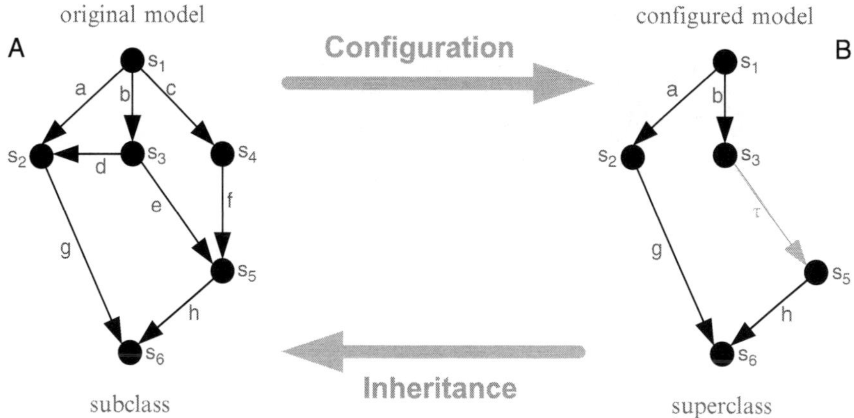

Fig. 18.1 The configuration of an LTS can be represented through inheritance relations

that is, the same state that is reached by executing a initially and so on. The process completes when the final state s_6 is reached either from s_2 through executing g or from s_5 by executing h. Thus, *an LTS can always be only in one state at a time* and the various transitions leaving a state depict the options for state changes.

When configuring a model, the goal is to execute a subset of this behavior only. For example, in LTS B on the right-hand side of Fig. 18.1, no transition other than a, b, g and h can be executed. More explicitly, the behavior should always be that either the execution of a is followed by g or the execution of b is followed by h. That means, compared to the original model, the execution of e between b and h must be skipped, which is indicated through the grey arc labeled τ.

Inheritance relations between two models can be used to identify two operators necessary to configure such models. The basic idea of inheritance is to construct different subclasses that, on the one hand, inherit all behavior and features of a common superclass, and on the other hand, extend this behavior with additional behavior or features. *The inheritance relation between two models is thus the inverse of the configuration relation between the two models*, where the original model extends the behavior of the configured model (see Fig. 18.1).

The first operator completely inhibits functionality of the original model (i.e., the subclass) in the configured model (i.e., the superclass). The corresponding inheritance relation says that a model B is a superclass of another model A if it is not possible to distinguish the behavior represented by A from the behavior represented by B when only those transitions of A are executed which also exist in B. In Fig. 18.1, for example, the only transitions leaving s_1 are a and b. Thus, when comparing the behavior of A and B, we are not allowed to execute transition c that exists in A, but not in B. We say that all transitions of A that are not present in B are *blocked* in B.

The second operator also compares the behavior of the two models, but now the execution of all transitions in both models is allowed. During the comparison,

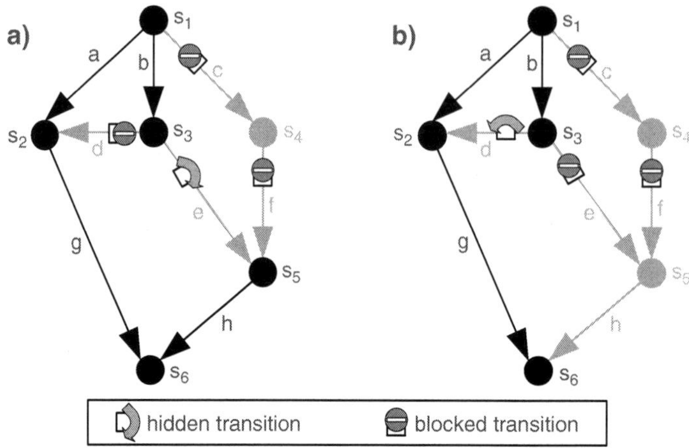

Fig. 18.2 Configurations of the LTS

however, only those transition executions of the subclass (A) are taken into consideration that also exist in the superclass (B). For example, we can execute from s_1 in A first the transition b, then e, and finally h. In model B, we execute first b and afterwards directly h as we skip e. If we now consider only the transitions of B, that is, transitions a, g, b, and h, these two traces are identical. We say, the transitions of A that do not exist in B, such as transition e in Fig. 18.1, are *hidden* in B.

Using these two operators in Fig. 18.1, B can thus be derived from A by blocking c, d, and f and hiding e as can be seen in Fig. 18.2a. The difference between both operators becomes clear if, for example, transition d would be hidden instead of blocked (see Fig. 18.2b). In this case, a trace of executing b and afterwards g would become possible in the configured model. This is not possible when transition d is blocked. On the other hand, when e is blocked instead of hidden, then it is no longer possible to execute h after b. Hence, both operators are needed to restrict the possible behavior of an LTS.

In the following section, these results are transferred to the configuration of YAWL models.

18.3 Configuring YAWL Models

The blocking and hiding operators that were introduced in the previous section apply to the transitions of an LTS. Transitions are the active elements of an LTS as they lead to changes of state in the model. In YAWL, the execution of a task leads to a change of the overall state of a process instance. Thus, the active elements of a YAWL model are the tasks. When we want to configure a YAWL model, we therefore have to configure its tasks.

Compared to transitions in LTSs, YAWL tasks comprise far more behavior as they allow different split and join patterns, the start of multiple instances, as well as the cancelation of other tasks through cancelation regions. Therefore, simply saying that we configure a whole task as blocked or hidden would not be fine-grained enough. Instead we have to identify when and how each task can change the overall state of the process. That means, a state change that is represented by a single transition in an LTS corresponds in YAWL to the combination of one of the alternatives how a task can be triggered with one of the alternatives how a task can be completed.

18.3.1 Input Port Configuration

Let us first analyze when a task can be started. For a YAWL task, this is defined through its pre-conditions and the specified join behavior. A task with an XOR-join, for example, has several incoming arcs from (potentially implicit[2]) pre-conditions. The task, however, requires only one of these pre-conditions to be marked by a token to be executed because it consumes only a single token from one of its marked pre-conditions (see Fig. 18.3a). Each of the incoming arcs therefore represents a

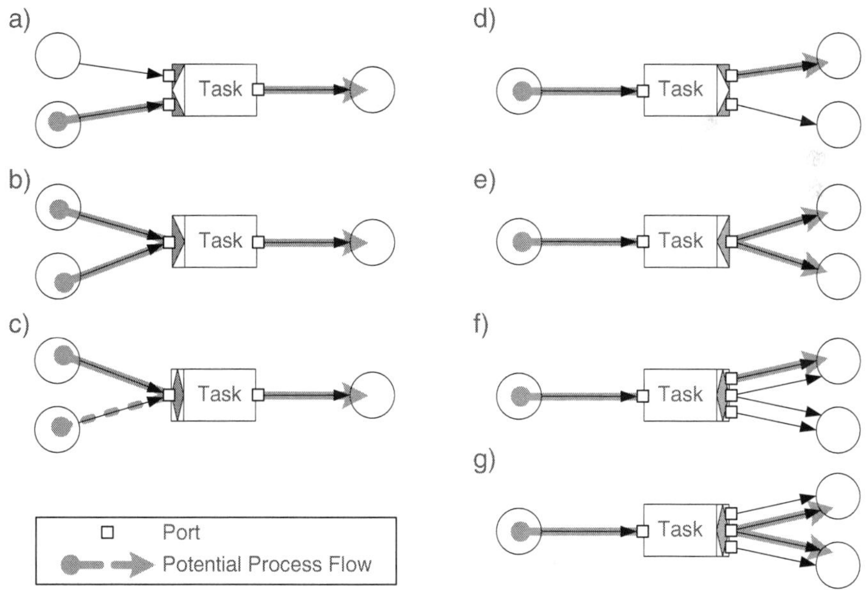

Fig. 18.3 Ports of a YAWL task

[2] see Chap. 2, Sect. 2.4.1 for more information on implicit conditions

different alternative to change the state of the process. For that reason, each of these alternatives must be configurable in the way shown for transitions in LTSs, that is, through blocking and hiding. In the following, each of these alternatives of how a task can be started is called an *input port* of the task.

The number of input ports is different for tasks with an AND-join. Here, the control-flows of all the incoming branches are synchronized (see Fig. 18.3b). That means, for its execution the task needs to consume tokens from all the pre-conditions, which implies that it can change the state of the process only if all these pre-conditions are marked with a token. *Only this single alternative to execute the task must thus be configurable.* In a similar way, there is also only one alternative when a task with an OR-join can be executed and thus change the state of the process. This is when there is no longer a chance that any further token can arrive in any of the task's pre-conditions[3] (see Fig. 18.3c).

As a transition in an LTS, an input port can be blocked or hidden by configuration. If it is blocked, the task cannot be executed through that port. Instead, the token must be consumed from the particular pre-condition by another task. If the input port is configured as hidden, this also implies that the task itself is not executed. But then the task is still able to consume and forward the token, which means the task can still change the state of the process. Only the task's execution is skipped.

18.3.2 Output Port Configuration

How the state of the process is changed depends on the outcome of the task's execution, that is, it depends on which conditions are marked after the task's execution. This is determined through the split behavior of a task. An XOR-split marks only one post-condition (see Fig. 18.3d). Thus, each post-condition corresponds to a different change of the process state. A task with an OR-split cannot only mark a single post-condition, but also any combination of the task's post-conditions (see Fig. 18.3f, g). As the overall state of a YAWL model depends on which conditions are marked, each of these marking combination conforms to a different overall state change. A task with an AND-split simply marks all its post-conditions after its execution (see Fig. 18.3e). Thus in this case only a single alternative to change the state of the model exists.

Not all these state changes might be desired. Thus, it should be possible to inhibit undesired state changes by configuration. For this, each combination in which post-conditions can be marked after the execution of a task is called an *output port* of the task. To generally inhibit the corresponding state change, this output port can then be explicitly blocked. In this way, the marking of the corresponding set of post-conditions is no longer possible.

[3] see Chap. 3 for more information on the semantics of the OR-join

An output port cannot be hidden because, considered on its own, it does not correspond to the full state change of a transition in an LTS. The state change depicted by a transition in an LTS rather corresponds to a complete task execution in YAWL, that is, the consumption of tokens from pre-conditions and the production of tokens in the post-conditions. Each LTS transition corresponds to such a possible combination. Its configuration therefore corresponds to a combination of an input port configuration with an output port configuration. If one of them is blocked, the execution of this combination is inhibited, which corresponds to the blocking of the particular transition in the LTS. If the input port is hidden and the output port is enabled, then this corresponds to hiding the particular transition as the task will not be executed but the target state will be reached. If neither the input port nor the output port is blocked or hidden, then the task can be executed as usual. We then say that the ports remain *enabled*.

18.3.3 Example

Figure 18.4 provides some examples of task configurations, where conditions have been tagged with a unique identifier. The model depicts the configuration of Genko Oil's *Carrier Appointment* subprocess such that it is applicable to Zula Exquisite. As mentioned in the introduction, Zula Exquisite ships only full *Truck-Loads* or *Single Packages*, but no partial *Truck-Loads*. To reflect this situation in the YAWL model, the output port from task *Prepare Transportation Quote* to the branch handling *Partial Truck-Loads* is blocked, indicated by the *Do not enter* symbol on the corresponding arc. To ship single packages, Zula Exquisite uses common express mail services, which do not require delivery appointments as these service providers have fixed delivery routes. For that reason, a delivery appointment does not need to be arranged and the particular process step can be skipped. To reflect this in the configuration, the input port of the corresponding task is configured as hidden (indicated by the "jumping" arrow).

The shipment information document that Zula Exquisite uses to ship full *Truck-Loads* requires both a prearranged pickup appointment and a prearranged delivery appointment. To enforce the availability of this information, we have blocked all input ports of task *Create Shipment Information Document* that do not enforce this information, that is, the ports from conditions c_6, c_7, and c_8. Only the two input ports, which require arranging both appointments beforehand, remain enabled. As all appointments are arranged before the shipment information document is created, there is no need to arrange any of those appointments afterwards. Thus, also all output ports of this task leading to such an arrangement are blocked. Task *Create Shipment Information Document* has an OR-split leading to seven output ports: one port to each of the two subsequent tasks *Arrange Pickup Appointment* and *Arrange Delivery Appointment*, one port to condition c_{10}, one port triggering the paths to both tasks at the same time, one port triggering the paths to task *Arrange Pickup Appointment* and to condition c_{10}, one port triggering both the paths to task *Arrange*

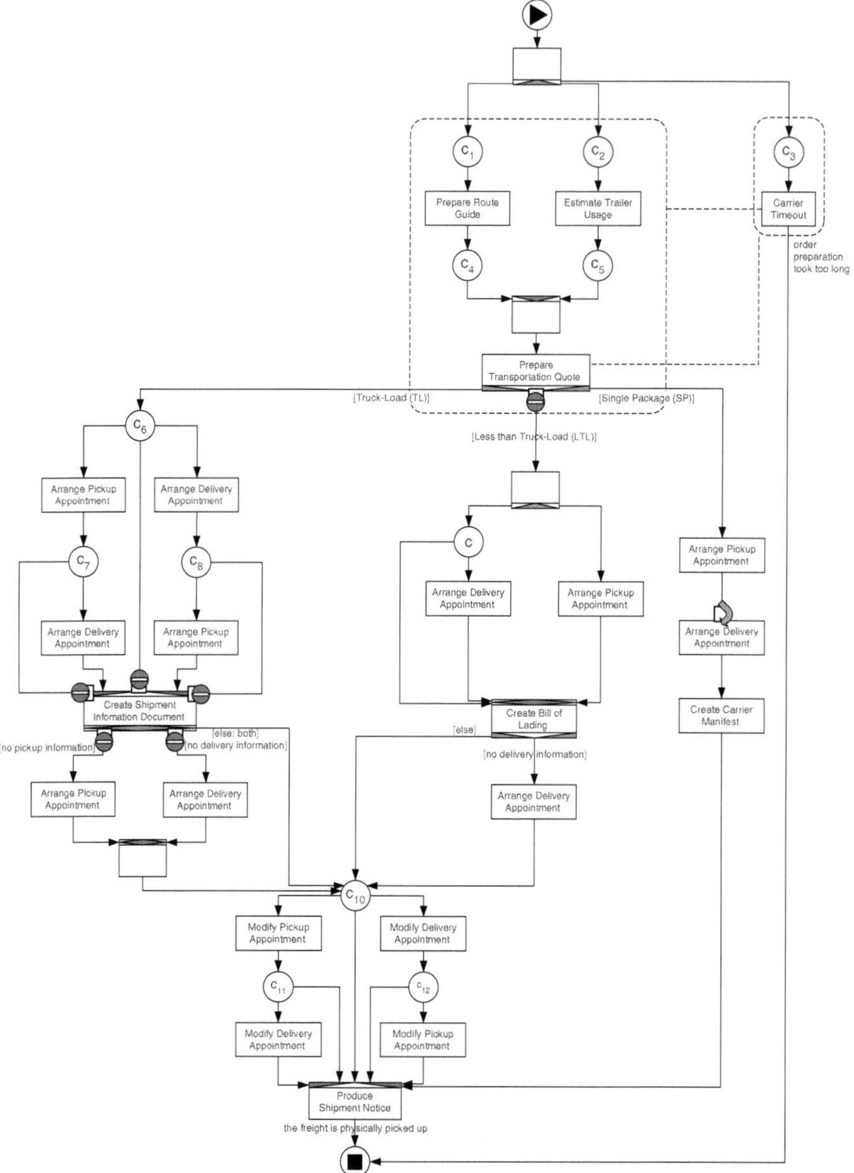

Fig. 18.4 The *Carrier Appointment* process configured according to the needs of Zula Exquisite

Delivery Appointment and to condition c_{10}, and one port triggering the paths to all three of these elements. By directly blocking an arc, it is implied that all ports involving this arc are blocked (if only a subset of these ports should be blocked, a dedicated list of the blocked ports will be provided as an annotation of the task). Indeed, by evaluating the result of task *Shipment Information Document*, the outgoing branch tagged as [else] would be automatically chosen anyway. However, blocking these ports explicitly, allows skipping the evaluation of the results at runtime, and thus simplifies the process's runtime execution.

18.3.4 Further Configuration Options

Besides the state changes along the control-flow arcs, a YAWL task can also change the state of a process through consuming all tokens from its cancelation region. While this might be desirable in many cases to preserve correct process execution, there are also situations when the task's influence on other process parts should be restricted. For example, a configurable YAWL model might contain two tasks for doing the same work at different stages of a process. One of these two tasks is in the cancelation region of a third task. If this third task is executed, the first occurrence of the task in question is canceled, that is, it is no longer guaranteed to be executed. In this case, the second occurrence of the task will guarantee that the work is still executed. However, if the execution of the second occurrence of the task is inhibited by process configuration, this guarantee is lost. Hence, in this case it might be necessary to block the cancelation of the task's first occurrence. Thus, a *cancelation port* is assigned to every cancelation region. This cancelation port can then either remain enabled, in which case the triggering of an enabled or hidden cancelation task would still cause the cancelation of all behavior in the cancelation region, or it can be configured as blocked, which then inhibits that the execution of tasks in the cancelation region is canceled while executing the task connected to the cancelation region.

YAWL's multiple instance tasks depict multiple executions of a single task as one task.[4] As such a task represents more behavior than a simple state change, this behavior can also be restricted by process configuration. On the one hand, it is possible to restrict the freedom of how many task instances can be started maximally by decreasing the maximal number. For example, if an original value would allow a maximum of ten instances to be started, this can be restricted in a configuration to a maximum of five instances to be started. On the other hand, the minimal number of started instances can be increased. For example, a multiple instance task might allow multiple checks while requiring at least one execution of the task, that is, at least one check. By increasing this minimal number of task instances to two through process configuration, the behavior can be changed such that the check is performed at least twice, that is, a four-eyes principle is enforced. Thus, both these configuration options limit the freedom of how many process instances can be started, increasing the lower bound of task instances that are started or decreasing the upper bound. Furthermore, the threshold value of a multiple instance task determines how many of these started instances need to be completed. Also this freedom

[4] see Chap. 2, Sect. 2.4.1 for more information on multiple instance tasks.

Table 18.1 Configuration options for a task with n incoming/outgoing process branches

Configurable element	No. of ports	Configuration options
AND-join	1	Enabled, blocked, hidden
XOR-join	n	Enabled, blocked, hidden
OR-join	1	Enabled, blocked, hidden
AND-split	1	Enabled, blocked
XOR-split	n	Enabled, blocked
OR-split	$2^n - 1$	Enabled, blocked
Cancelation region	1	Enabled, blocked
Multiple instances		Reduce maximal number of instances, increase minimal number of instances, increase threshold, forbid dynamic creation of instances

can be restricted by increasing the threshold. Finally, by restricting a dynamic creation of task instances to a static creation, it is no longer possible to dynamically start additional task instances while the multiple instance task is running.

To summarize, in C-YAWL the behavior of a YAWL model can be restricted by configuration in four ways: by hiding and blocking input ports, by blocking output ports, by blocking a task's cancelation region, and by restricting the parameters of a multiple instance task (see Table 18.1).

As configuration operators are applied at the granularity of single tasks, the configuration of a YAWL workflow would typically require a deep understanding of the process model and its notation. Furthermore, each task's configuration would very likely affect (directly or indirectly) other parts of the process model. For example, the blocking of one port can also indirectly deny the execution of the flow downstream. Therefore, an approach that can streamline the configuration process by abstracting from the intricacies of YAWL (or any other specific modeling notation) would prove beneficial, especially if it could also improve understanding the impact of configuration decisions throughout the workflow model.

In the next section, an approach to process configuration will be presented that is based on the use of questionnaires. These questionnaires allow process users, or more generally, subject matter experts, to reason in terms of the domain concepts involved.

18.4 Steering Process Configuration Through Questionnaires

So far we have seen how the behavior of a YAWL model can be restricted by means of, for example, the blocking and hiding operators. The reason for hiding or blocking a certain port should derive from practical needs for the corresponding functionality, that is, it should be driven by the requirements of the organization or project for which we want to configure our workflow. That means, before we can answer a technical question like "Do we need to hide the input port of the task *Arrange Delivery Appointment*?," we first have to ask ourselves "Do we need to arrange delivery appointments when shipping *Single Packages*?." Only the answers to such domain-related questions can lead to the necessary configuration decisions on the process model level.

Moreover, configuring a YAWL model to the requirements of a specific setting requires a user to be proficient in both the YAWL notation and the application domain. This assumption might be unrealistic in application domains where users are unfamiliar with modeling notations. This is the case of the supply chain management domain in which the Order Fulfillment process model has been constructed: users of this process would very likely be logistics experts only.

In this section, it will be shown how process configuration can be facilitated via a questionnaire-driven approach, where variations are presented in terms of domain-related questions expressed in natural language, which are mapped to configuration decisions on a YAWL model. First, the features of questionnaire models will be presented, which allow a systematical encoding of questions and their answers. Second, it will be shown how such models can be linked with a YAWL model for the automatic derivation of a configured workflow thereof.

18.4.1 Questionnaire Models

A questionnaire model is a structured representation of domain choices and their dependencies. Key components of a questionnaire model are *domain facts* and *questions*. Domain facts capture domain choices, that is, features that may or may not exist in a given setting of the domain. For example, three possible features of a logistics domain are "Truck-Load," "Less than Truck-Load," and "Single Package," which denote three shipment types (or products) a logistics department of a company may provide. A domain fact (fact for short) is modeled as a boolean variable that the user can set to *true* if they are interested in having the respective feature, or *false* otherwise. For example, at Genko Oil, all the above facts are set to true, while for Zula Exquisite "Less than Truck-Load" is false.

Facts can be grouped into questions according to their content, so that facts belonging to the same question can be set at once. In our example, a possible question to group the above three facts is "Which shipment types have to be provided?" Questions are expressed in natural language. Therefore, they can be answered by domain experts without extensive knowledge of the underlying process model.

Each fact is given a default value, which is used to suggest the most recurring value for that fact. In the Order Fulfillment example, fact "Single Package" is assigned a default value of true as the majority of logistics organizations, including Genko Oil, typically provide such a product because it is the easiest of the three to be handled.

Moreover, a fact can be marked as "mandatory" if it needs to be explicitly set by the user. This is used to prevent users from overlooking important aspects of

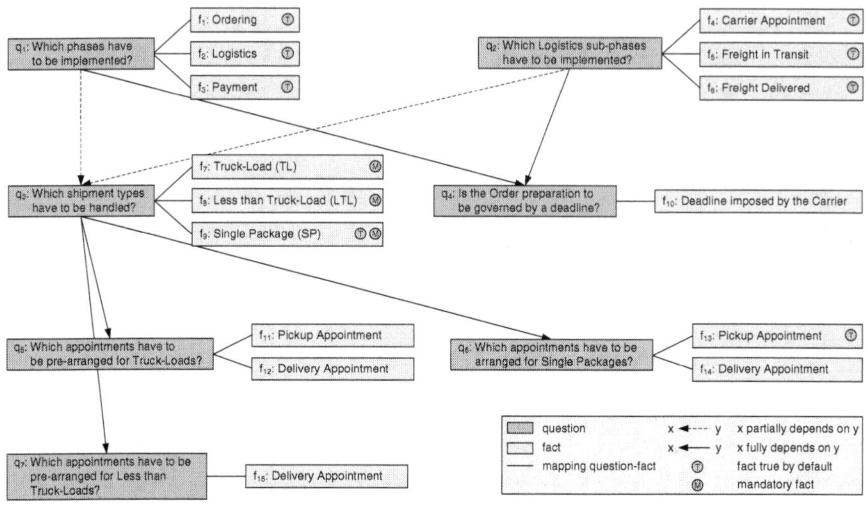

Fig. 18.5 The Questionnaire model for the Supply Chain Management domain (excerpt)

a domain. In fact, a question that has at least a mandatory fact must be explicitly answered by the user, while a question with no mandatory facts can be skipped. In this case the question is automatically answered by using the default values of its facts.

Figure 18.5 shows an excerpt of a possible questionnaire model for the supply chain management domain, where all questions and facts are assigned a unique identifier and a description.

In this model, two questions refer to high-level decisions for this domain. Question q_1, "Which phases have to be implemented," inquires after the business functions that should be supported, among *Ordering* (f_1), *Logistics* (f_2), and *Payment* (f_3). Meanwhile, question q_2 refers to the subphases within Logistics, which can be *Carrier Appointment* (f_4), *Freight in Transit* (f_5) and *Freight Delivered* (f_6). Other questions instead refer to more specific aspects of the domain, such as the shipment types to be provided (q_3), the possibility of enabling a deadline for the order preparation (q_4), and the (pre-)arrangement of pickup and delivery appointments for each shipment type (q_5, q_6, q_7).

As we can see, there exist interdependencies among the above questions. For example, an implication of setting to true any fact of q_2 (the *Logistics* subphases) is to set f_2 (*Logistics*) to true in q_1. This is because only if Logistics is implemented, any of its subphases can also be implemented. Similarly, the answer given to the remaining questions (q_3 to q_7) is affected by the decision taken in q_2 whether to enable Logistics or not. Indeed, it would make no sense to inquire after the shipment types to be handled, the deadline on the order preparation, or the arrangement of pickup and delivery appointments, if the organization will not support *Logistics*. In particular, which appointments can be agreed upon depends on the shipment type that is selected to be handled: the facts of q_5 depend on f_3, those of q_6 on f_5, and so on.

As facts are boolean variables, their interactions are captured in a questionnaire model by means of boolean expressions over their values, which are called *domain constraints*. For example, the interaction between q_1 and q_2 can be modeled by the constraint $(f_4 \ or \ f_5 \ or \ f_6) \ implies \ f_2$, which essentially means that if any fact of q_2 is selected, then f_2 must be set to true as well. On the other hand, if f_2 is deselected, all the facts of q_2 must be set to false too. Similar constraints can be defined to capture the interactions among the other questions.

Questionnaire models also allow the definition of an order relation for posing questions to users. There are two types of *order dependencies* that can be defined among questions: *partial* and *full* dependencies. A partial dependency (represented by a dashed arrow in Fig. 18.5) is used to model an optional precedence between two questions: for example, q_3 will be posed after either q_1 or q_2 have been answered. A full dependency (depicted by a full arrow) captures a mandatory precedence: for example, q_4 will be posed only after both q_1 and q_2 have been answered. These dependencies can be arbitrary so long as undesired cycles are avoided.

The combined use of constraints and dependencies can be exploited by a supporting tool (cf. Sect. 18.6) to provide an interactive interface to the user when answering a questionnaire. Specifically, according to the constraints, the answers given to some questions can affect the answers of subsequent questions. For instance, questions q_1 and q_2 correspond to high-level decisions: if Logistics is disabled in q_1, all the subsequent questions become irrelevant (with all their facts being set to false). A similar result is obtained by disabling *Carrier Appointment* in q_2. Therefore, if these most discriminating questions are asked first, it is possible to (partly) infer the answer to subsequent questions automatically. In our example, q_1 and q_2 have not been assigned any order dependencies so as to be posed first.

In conclusion, a domain configuration is the result of completing a questionnaire where questions are posed in an order consistent with the order dependencies, and each fact is set to a boolean value complying with the domain constraints. This experience is interactive as the user is aided in the completion of the questionnaire according to the choices they take.

18.4.2 From Answers to Configured YAWL Process Models

Questionnaire models are meant to be constructed and linked to configurable YAWL process models by experienced modelers. This should be done in collaboration with subject matter experts, who possess expertise in the application domain. By answering questions, domain experts assign values to domain facts and once the questionnaire has been completed, the resulting facts' valuation is used to configure the respective YAWL process model through an individualization algorithm. An overview of this approach is depicted in Fig. 18.6.

The idea is that each configuration decision for a port, a task's multiple instance parameter, or a cancelation region in a YAWL model is captured by a boolean variable, namely *process fact*. Only if the process fact is true, the specific configuration

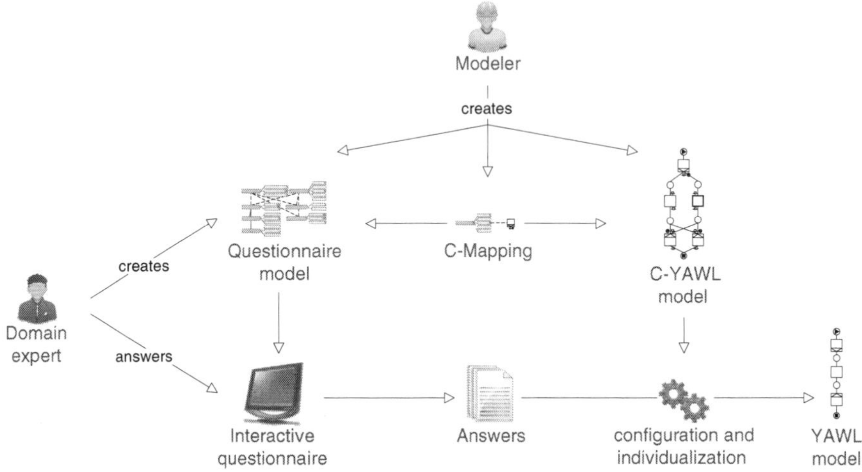

Fig. 18.6 The questionnaire-based framework for process models configuration

is applied, for example, an input port is hidden. Each process fact is then linked to the domain facts of a questionnaire model through a boolean expression. Such expressions embody the constraints of the domain. Thus, a process fact is set to true whenever the corresponding boolean expression evaluates to true.

Let us have a look at some mapping examples. Question q_6 asks whether to arrange pickup and delivery appointments for *Single Packages*. Here domain fact f_{14} is set to true if delivery appointments have to be arranged, and to false otherwise. This fact can be mapped to the input port of task *Arrange Delivery Appointment* (in the process branch handling *Single Packages*), such that when f_{14} is false, the port is set to hidden. Therefore, if the configuration *hidden* of this port is encoded with a process fact, say p_1, the boolean expression linking the latter with f_{14} would be: *not* f_{14} *implies* p_1. Likewise, f_{13} is mapped to the value *hidden* of the input port of task *Arrange Pickup Appointment* in the same process branch, so as to skip this task if the pickup does not need to be arranged.

In a similar way, the three facts of question q_3 can be mapped to the value *blocked* of the three output ports of task *Prepare Transportation quote*, such that when any of these facts is false, the respective port is blocked.

A more complex mapping exists between the facts of q_5 and the process elements in the branch handling *Truck-Loads*. Depending on the values of f_{11} and f_{12}, a number of ports can be blocked at the same time. In case both facts are true, that is, both appointments need to be prearranged, the input ports of task *Create Shipment Information Document* from conditions c_6, c_7, and c_8 will be blocked, as well as its output ports to the [no pickup information] and [no delivery information] branches. The blocking of the input ports allows both the pickup and delivery appointments to be executed (in any order), while the blocking of the output ports avoids the rescheduling of the appointments. This situation is depicted in Fig. 18.7a. Similarly, if only one of the two appointments can be prearranged, or none of them

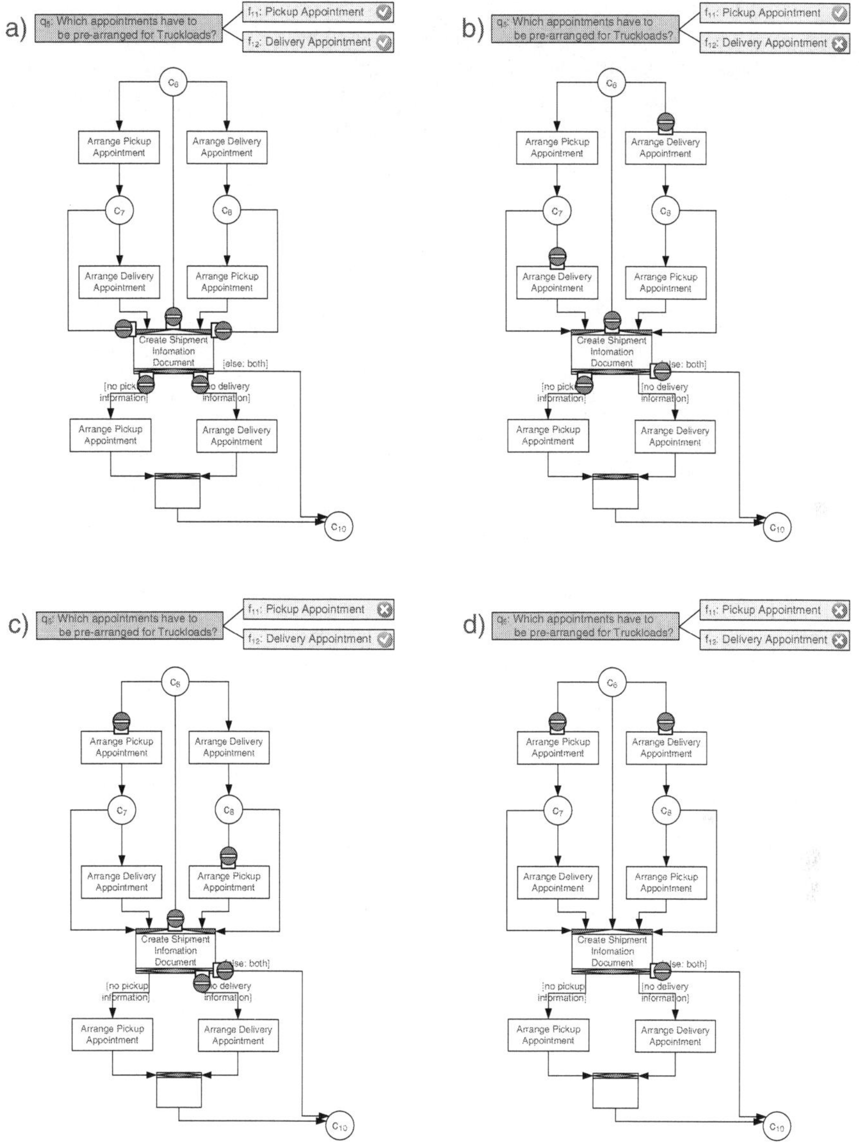

Fig. 18.7 Illustration of how the *Truck-Load* fragment of the *Carrier Appointment* process changes on the basis of the answer given to question q_5: (**a**) if both the appointments are to be prearranged, (**b**) if only the *Pickup* needs to be prearranged, (**c**) if only the *Delivery* needs to be prearranged, (**d**) if there is no prearrangement

can be, a different set of ports needs to be blocked. These situations are depicted in Figure 18.7b–d. Whether a domain fact should be mapped to a hidden port or to a blocked port depends on what model transformations one expects to achieve by answering the questionnaire.

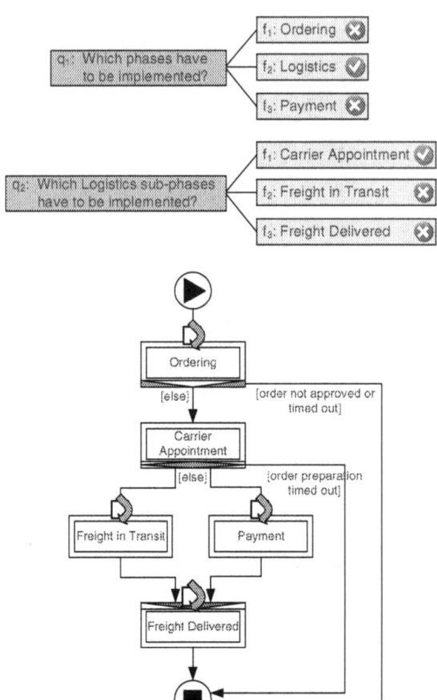

Fig. 18.8 The configuration of the Order Fulfillment process when only *Carrier Appointment* is selected in the questionnaire

The advantage of using questionnaires to configure more than one port at a time is twofold: on the one hand, the user can save time during configuration (especially in huge process models), and on the other hand, it is easier to visualize the impact of a domain decision (e.g., prearrange *Truck-Load* appointments) on the process model. Let us consider the mapping for questions q_1 and q_2. Their domain facts are mapped to the input ports of the first net in the decomposition hierarchy of the Order Fulfillment process, such that when any of these facts is set to false, the corresponding port is configured as *hidden*. Figure 18.8 shows the configuration of the Order Fulfillment process when only the *Carrier Appointment* phase in Logistics is implemented, which corresponds to facts f_2 and f_4 being true and all other facts being false. In this process model, composite task *Carrier Appointment* is the only one not being skipped. The complexity of the initial model has been dramatically reduced as only one of the five composite tasks will be executed. This has been achieved by answering only two questions.

18.5 Applying Configuration Decisions to YAWL Models

The C-YAWL models to which the configuration decisions are applied, that is, the models which are used as the basis for configuration, integrate all the process execution variants. By configuration, the intention is to eliminate all the behavior from the model, which is irrelevant in a specific context. Until now, we have discussed how such intentions can be expressed through configuration decisions. However, we have not yet derived any YAWL models that only contain the desired behavior.

For this, obviously, all the undesired parts of the model must be removed. Based on the configuration decisions, this removal can be performed completely automatically in two phases. First, all the direct implications of configuration decisions on tasks and arcs are implemented in the YAWL model. Second, the implications of these configuration decisions on other elements in the workflow are resolved in a clean-up phase.

18.5.1 Applying Configuration Decisions to Tasks

The first phase starts with adjusting the parameters of multiple instance tasks, as well as implementing blocked cancelation ports and blocked output ports as these require local changes in the model. The implementation of the input port configurations should happen only afterwards as this might require some copying of the other configuration decisions.

To adjust the parameters of multiple instance tasks as described in Sect. 18.3.4, the values for the minimal number of instances that need to be started and the threshold for the number of instances that need to complete for a successful completion of the overall task are increased according to the corresponding configuration values. Also, the maximal number of instances that can be started is decreased according to the corresponding configuration value. Furthermore, the instance creation is changed to a static creation if necessary. In the example from Sect. 18.3.4, the task parameters could change from `[1,10,1,dynamic]` to `[2,5,2,static]` in this way.

Implementing blocked cancelation regions in the configured model is similarly straightforward. For all tasks with such a blocked cancelation region, all references to the elements contained in the cancelation region are simply deleted. For the enabled cancelation regions the references are preserved.

The configuration of output ports influences how threads of control can leave the task. If a port is blocked, there should be no way that the corresponding subsequent conditions are marked. In case of an XOR-split, each port refers exactly to one outgoing arc. Thus, if this port is blocked, the corresponding arc can simply be removed.[5] In this way, any process flow through this port that would mark the particular, subsequent condition is inhibited. Inhibiting the process flow is not that simple in case of an OR-split. Here several ports can refer to the same arc. In this case, the arc can only be removed if all the ports referring to the arc are blocked. For example, let us have a look at the first configuration in Fig. 18.9. Here we can remove the arc to $c4$ as all ports referring to it are blocked. The only ports enabled

[5] If the arc that is chosen as default, is removed, YAWL automatically defines the last arc as the new default arc and uses it, even if the corresponding predicate evaluates to false.

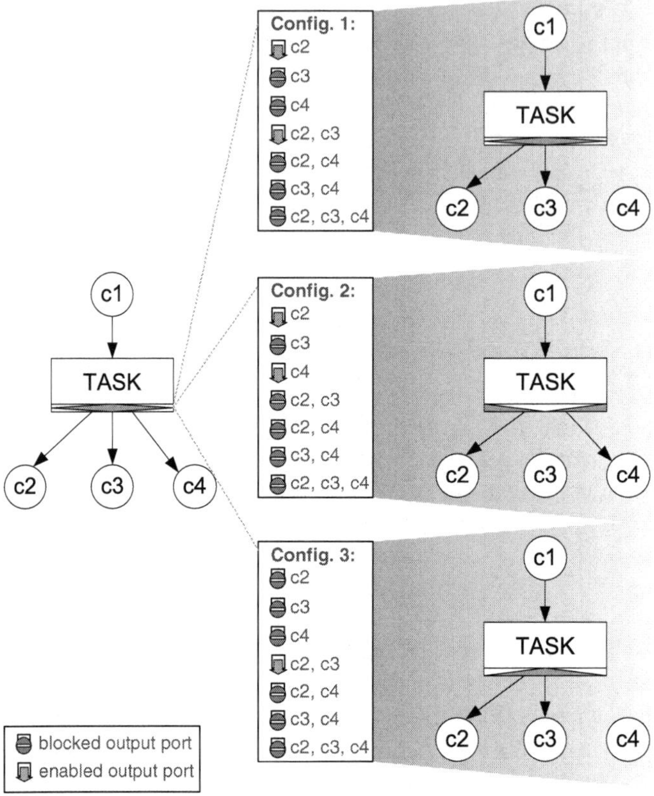

Fig. 18.9 Three configurations of a task with an OR-split transformed into new models

are the ones leading to $c2$ and to both $c2$ and $c3$. Although the port to $c3$ is blocked, the arc to $c3$ must be kept in the net because of the enabled port leading to both $c2$ and $c3$. The blocking of such ports where the corresponding arc has to remain in the net is then realized through adapting the arc's predicate.[6] By concatenating the predicate from the arc to $c2$ with the original predicate from the arc to $c3$ as the new predicate for the arc to $c3$, the blocking of the port to (only) $c3$ can, for example, be implemented for the first configuration of Fig. 18.9. That means that if the predicate for the arc leading to $c2$ would be $x > 25$ and the original predicate for the arc leading to $c3$ was $y < 100$, then the new predicate for the arc to $c3$ would be $(x > 25) \wedge (y < 100)$. The condition $c3$ can then be marked only if the arc's original predicate is true (i.e., $y < 100$), and in addition the predicate to $c2$ is true (i.e., $x > 25$), which thus always implies the marking of $c2$ in addition (for which $x > 25$ is the only condition).

[6] see Chap. 8, Sect. 8.3 for more information on how to set up predicates

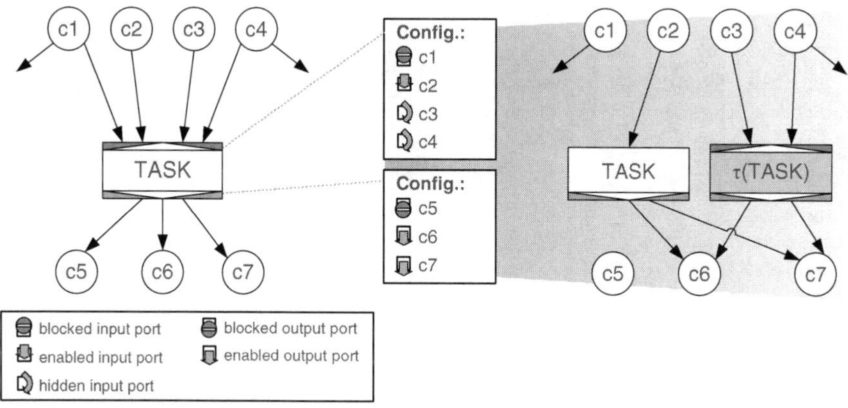

Fig. 18.10 The configurations of a task's input ports transformed into a new model

In two special cases, this complex changing of predicates can be avoided. First, if all the output ports referring to a combination of arcs are blocked, then it is possible to change the split behavior of the task from an OR-split to an XOR-split and keep the predicates as they were before. For example, in *Configuration 2* of Fig. 18.9, the port leading to $c2$ only as well as the port leading to $c4$ only are enabled, while all other ports are blocked. Thus, a simple XOR-split between these two process flows is sufficient. If, on the other hand, all ports but one are blocked as in *Configuration 3* of Fig. 18.9, the behavior is exactly captured by an AND-split. All predicates of the resulting task are simply set to true.

After having done these three transformations, the input port configurations of tasks can be implemented. Task with an AND-join or OR-join have just a single input port. In case this port is blocked, the corresponding task is not allowed to be executed. Therefore, all arcs leading to the task are simply removed. In case this port is hidden, the incoming arcs cannot be simply removed – the process flow has to continue in this case. However, the task should not be executed either. Thus, in these cases, any task decomposition[7] must be removed and the task has to be transformed into a so-called silent task (or τ task). That means, in this way, the task no longer leads to any work or to any changes of data, corresponding to the behavior of τ transitions in LTSs.

Tasks with an XOR-join have several input ports. The configuration of these ports can therefore vary. Each port refers, however, to exactly one arc. Hence, to inhibit the flow through a blocked input port, the corresponding arc can simply be removed. This is, for example, the case in Fig. 18.10, where the input port corresponding to the arc from $c1$ to the task is blocked. In the resulting model on the right, the arc is therefore removed. Besides being blocked, an input port can also be hidden, such

[7] see Chap. 8, Sect. 8.2 for more information on how to assign decompositions like codelets, external applications, or subnets to tasks

as the ports from $c3$ and $c4$ in Fig. 18.10. The task's decomposition must then be skipped. However, in case of an XOR-join, the decomposition of the task cannot be generally skipped for all process instances as the decomposition might still be needed by instances arriving through other enabled ports of the same task (such as the port from $c2$ in Fig. 18.10). For that reason it is necessary to duplicate the current task (i.e., with multiple instance, cancelation port, and output port configurations already applied). While one copy captures the original task, the task's decomposition is eliminated from the other copy, that is, this copy becomes the silent τ-version of the task without any behavior. Any multiple instance parameters or cancelation regions remain the same for both copies. Also the outgoing arcs of both tasks target the same subsequent conditions. In case of a direct connection to the subsequent tasks in the original model, an intermediate condition will be introduced for this (i.e., the implicit conditions between the two tasks will be made explicit). Then, the arcs of all enabled ports are directed to the original task while the arcs of hidden input ports are redirected to the silent copy (compare the arc from $c2$ with the arcs from $c3$ and $c4$ on the right of Fig. 18.10).

Figure 18.11 shows the resulting YAWL model after applying all the configuration decisions depicted in Fig. 18.4 as described. All undesired routings of the process are no longer possible and the *Arrange Delivery Appointment* task for single packages is replaced with a silent τ task such that it can be skipped. However, while the arcs incorporating undesired behavior are removed, the undesired tasks themselves are still in the net. Just their reachability is inhibited. The second phase thus has to rid the model of such nonexecutable elements.

18.5.2 Cleaning-up the Configured YAWL Model

To be reachable, a model element needs to be on a path from the input condition. Additionally, a YAWL model requires that the final condition indicating the workflow's completion can always be reached, that is, there also needs to be a path from every element to the final condition. To identify all the elements that are on such a path, we search for all paths from the initial condition to the final condition and mark all elements that are such a path. All the nonmarked elements will be removed from the net.

When removing all these elements, it is important to take into consideration that a task with an AND-join needs tokens in all its preceding conditions to be executed. If a condition that precedes an AND-join is removed, this condition can no longer be marked and hence the task should not be executed any longer either. For that reason, tasks subsequent to a removed condition have to be removed whenever they have an AND-join behavior. As the removal of a task might break an existing path between the input condition and the output condition (because the task was on such a path), the check for paths between the input condition and the output condition has to be repeated afterwards. The removal of the elements not on such a path must be continued until a search confirms that all elements are on such a path (or the

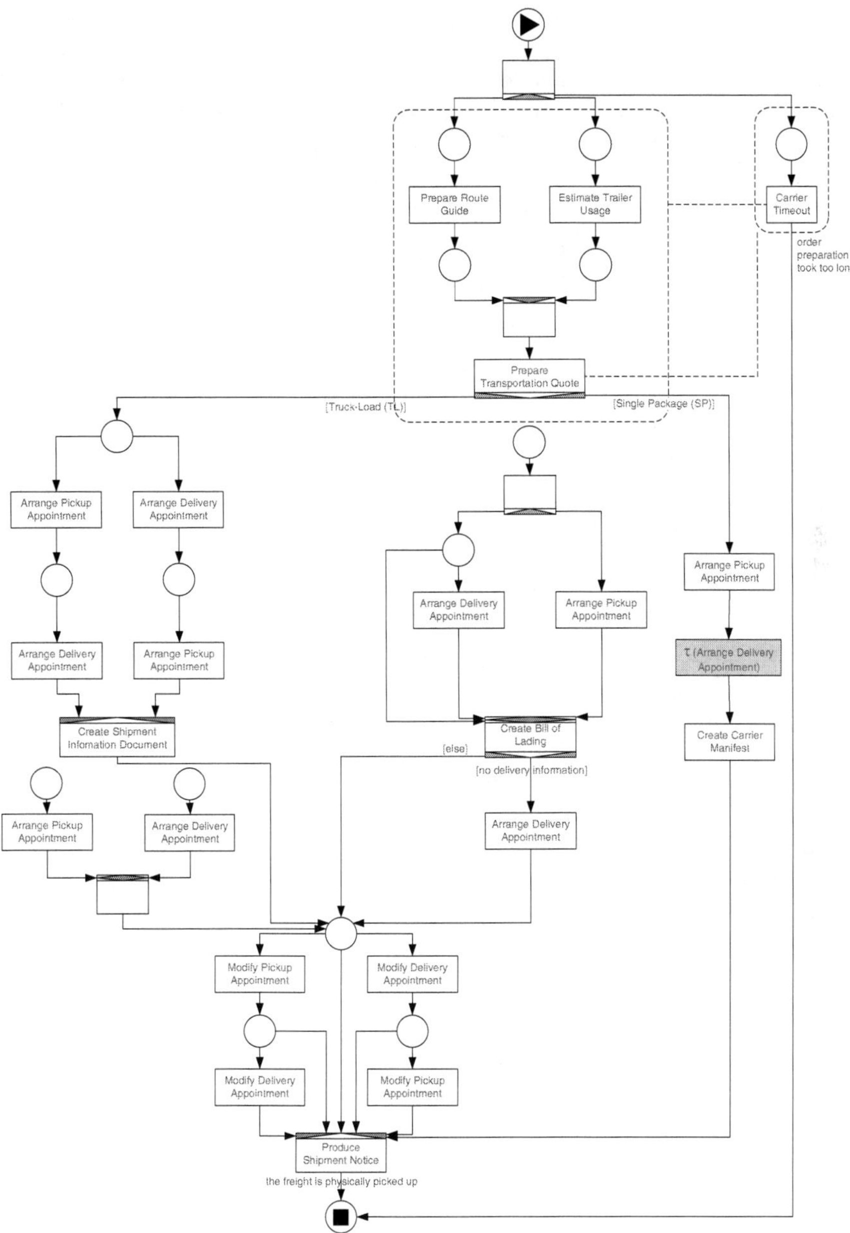

Fig. 18.11 The *Carrier Appointment* process after applying the configuration decisions, but without a clean-up of the process

only remaining elements are the input and output conditions, which then indicates a misconfiguration).

If all elements are on a path between input condition and output condition, the clean-up is complete. The resulting model, such as the one in Fig. 18.12 for the *Carrier Appointment* subprocess of Zula Exquisite, can be directly loaded as a workflow specification in the Engine. It can also be imported into the Editor to make further manual changes, for example, to update the necessary resources for certain tasks.

18.6 Tool Support

YAWL workflow models can be configured through the *Synergia* toolset. The purpose of this toolset is to foster synergism between domain models and process models via the use of questionnaires. *Synergia* assists domain experts and process modelers with creating questionnaire models, mapping questionnaires to process models, answering questionnaires and applying the result of a questionnaire to a process model for the configuration of the latter.

Synergia consists of a set of rich client applications, with each application taking care of a specific task. The tools composing *Synergia* are shown in Fig. 18.13. Currently, the toolset supports the configuration of process models defined in C-YAWL and in C-EPC - the latter being the configurable extension to the Event-driven Process Chains (EPC) language.

The first tool is the *Questionnaire Designer*, which allows users to visually create questionnaire models. Besides the definition of questions, domain facts, order dependencies, and domain constraints, it is possible to assign textual guidelines to questions and facts, which will be used to provide advice to users while the questionnaire is being answered. Moreover, the tool can detect cyclic dependencies in the order of questions and inconsistencies in the domain constraints. This prevents the user from producing invalid questionnaire models that may deadlock while being answered due to an order dependency that cannot be resolved. A screenshot of the tool with its validation feature is shown in Fig. 18.14.

A serialization of a questionnaire model (.qml) is generated by the *Questionnaire Designer* tool. This file can be imported into *Quaestio*, which is the second tool of *Synergia*. *Quaestio* prompts questions to users in an order consistent with the order dependencies defined in the questionnaire model. The tool is interactive as it prevents users from entering conflicting answers to subsequent questions by dynamically enforcing the domain constraints according to the answers given at any time. In this way, questions are posed only if they are relevant to the context. Questions can be explicitly answered or can be skipped if they do not contain mandatory facts. In this case, they are automatically answered by using the default values of their facts. Furthermore, questions already answered can be rolled back should a decision need to be reconsidered. A screenshot of *Quaestio* showing the questionnaire for the Order Fulfillment process model is depicted in Fig. 18.15. It represents the state in which question q_1 ("Which phases have to be implemented?") has already

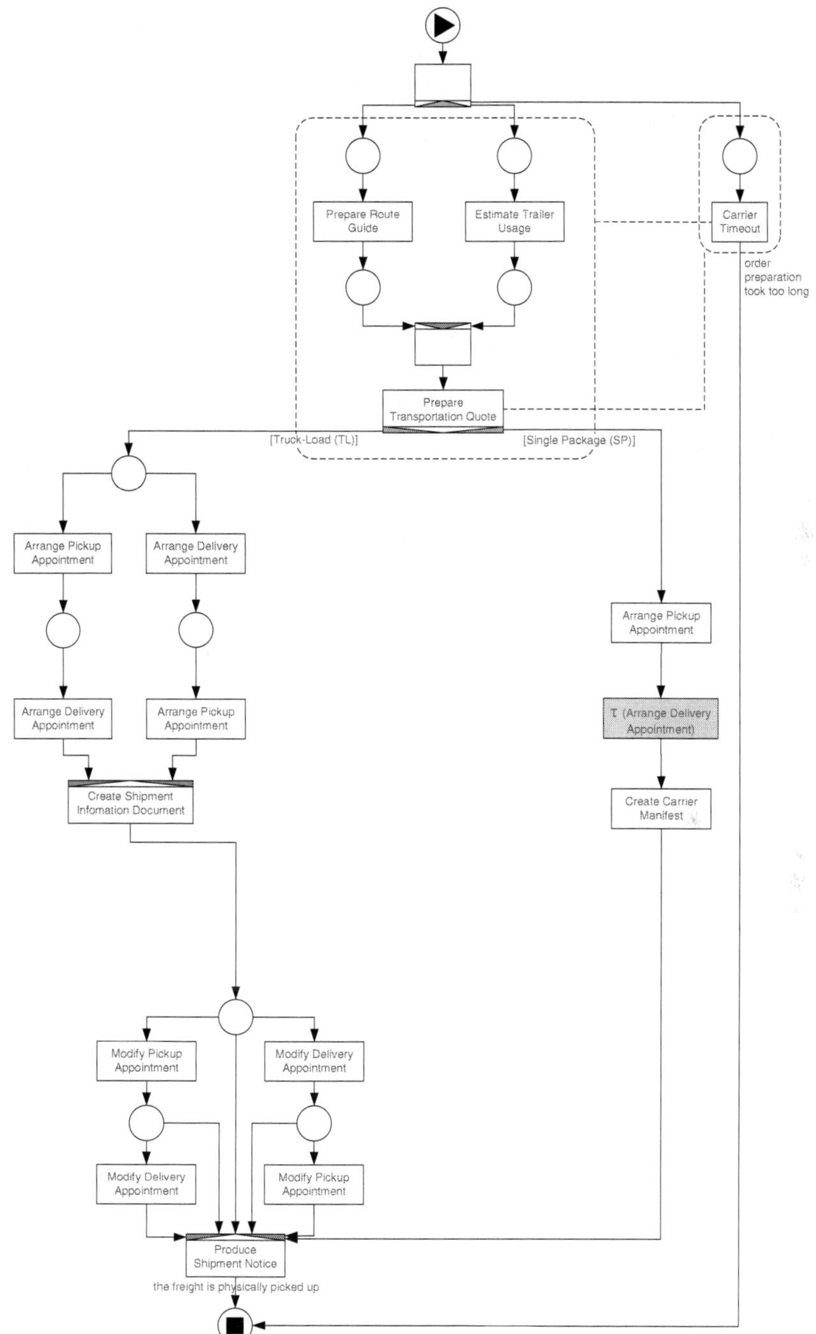

Fig. 18.12 The cleaned up YAWL model for the *Carrier Appointment* process of Zula Exquisite

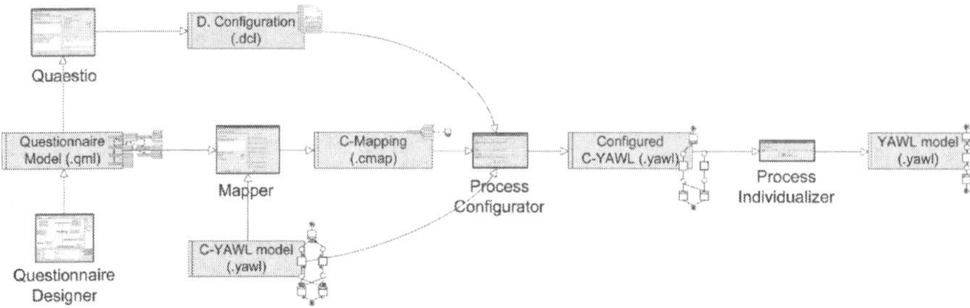

Fig. 18.13 *Synergia* configuration toolset – overview of the software tools

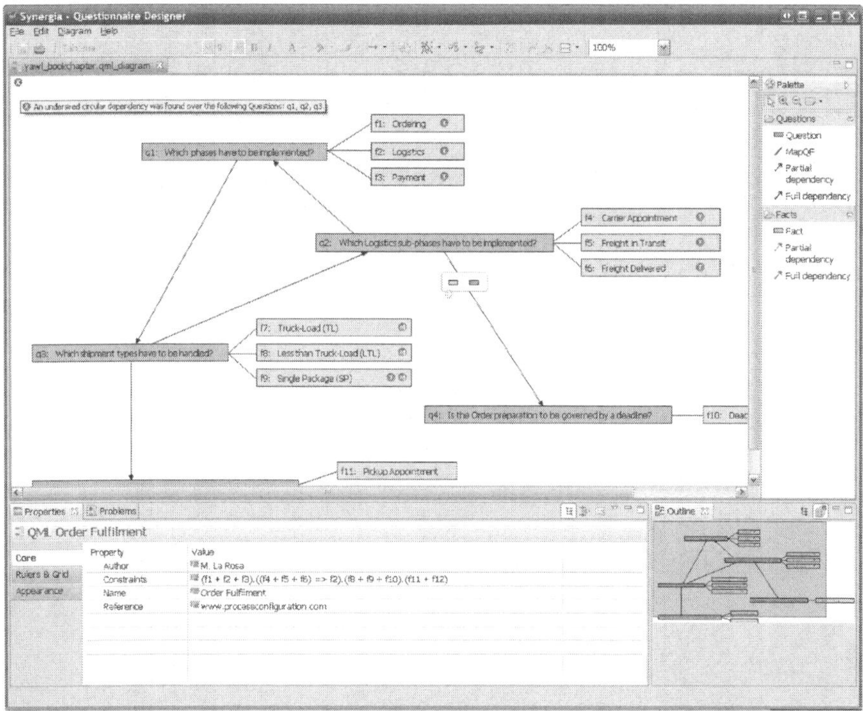

Fig. 18.14 *Questionnaire Designer* – detection of a cyclic dependency among three questions

been answered and the user is presented questions q_2 ("Which Logistics subphases have to be implemented?") and q_3 ("Which shipment types have to be handled?"), with the latter question having become available after answering q_1. The facts and guidelines for q_2 are also shown.

Once a questionnaire has been completed, it can be exported as a domain configuration (.dcl). This file can be used by the *Process Configurator* tool to configure a process model, that is, to apply the configuration decisions to the process model as shown in Sect. 18.4.2. The *Process Configurator* needs a serialization of a process model (e.g., .yawl for YAWL) and a configuration mapping (.cmap) between

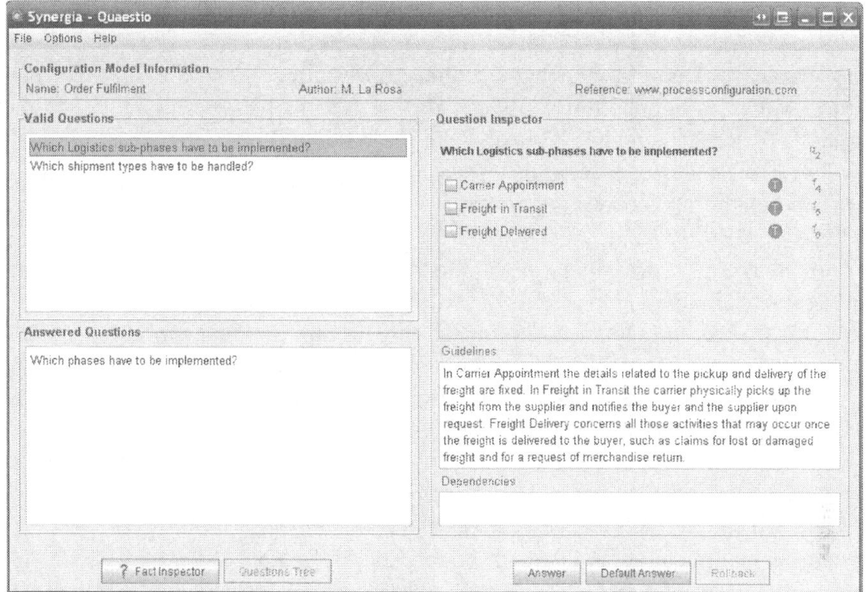

Fig. 18.15 *Quaestio* – the interactive questionnaire tool

the process model and the questionnaire model. Mappings can be generated by the *Mapper* tool, which allows the connection of process facts to domain facts by means of boolean expressions.

Finally, the *Process Individualizer* tool performs the individualization algorithm as discussed in Sect. 18.5. Given a configured process model as input, this tool takes care of removing all those process fragments that are no longer required, and generates a new process model as output. The algorithm preserves the syntactic correctness of the input model: the resulting model will be well-formed, provided the input model is well-formed. The tool currently generates YAWL models from configured C-YAWL models and EPC models from configured C-EPC models.

18.7 Summary

Process configuration enables domain experts, who are usually not proficient in modeling notations, to adapt process models to specific settings, for example, a new organization or project. While process configuration does not extend the behavior represented by a process model, it allows users to restrict the already modeled behavior in such a way that only desired steps of the process will be executed. Thus, a configurable process model is meant to incorporate execution variants which uses can select from.

To provide configuration opportunities to YAWL models, input and output ports are defined for each task of a YAWL model. By blocking such ports, the process flow through the port is inhibited and thus tasks subsequent to this port will not be executed. An input port of a task can also be hidden. This means, the execution of the task itself should be skipped, while the process flow continues afterwards. Moreover, the behavior of multiple instance tasks and tasks' cancelation regions can be restricted through process configuration. This extension to the YAWL notation is called C-YAWL.

Modeling and configuring the variations of a process model in terms of input and output ports being blocked or hidden requires a thorough knowledge of both the YAWL modeling notation and of the domain in which the process has been constructed. An approach based on the use of questionnaires can streamline the configuration of process models. A questionnaire model defines a set of questions grouping the features a domain can support. The possible values these features, called domain facts, can take are restricted by a set of domain constraints, which model the interdependencies of the domain. Also, a questionnaire model defines the order in which questions can be posed to users.

Questionnaire models are then linked to the ports of a YAWL process model by means of boolean expressions. Whenever such an expression evaluates to true on the basis of an answer to the questionnaire, the given port is configured accordingly. In this way, it is possible to configure a number of YAWL ports simultaneously by answering a single question. This also applies to the configuration of multiple instance tasks and cancelation regions.

The fact that questionnaire models are expressed in terms of domain concepts enables subject matter experts, who are usually not proficient in modeling notations, to benefit from process configuration. Moreover, it allows them to better estimate the impact of a domain decision (e.g., disabling *Truck-Loads* shipments) throughout the process model.

Once a C-YAWL process model has been configured, it can automatically be transformed into a well-formed YAWL model, including only those process variants that have been selected in the configuration. To load this new process specification into the Engine, the workflow developer only needs to adjust the resource assignment and the necessary data parameters.

In this chapter, the benefits of using process configuration for configurable reference process models were demonstrated. However, the concepts of process configuration and questionnaires might have a wider applicability beyond the scope of reference models. For example, it can be useful to apply process configuration when implementing seasonal switches between different process variants. Questionnaires could also be used to steer decisions independent of a process model's control-flow. In general, the various elements of process configuration can be beneficial whenever a selection from a set of predefined variants needs to be made.

Exercises

Exercise 1. Via the use of input and output ports, configure the Payment subprocess of Fig. A.5 for the following:

- To disable the payment for the freight (only the payment for the shipment is to be performed)
- To disable the approval of *Shipment Payment Orders*
- To disable the possibility of having *Debit* adjustments.

What will the resulting process look like after applying this configuration?

Exercise 2. Can the cancelation region of task *Carrier Timeout* be blocked without violating the soundness of the net?

Exercise 3. Complete the questionnaire model of Fig. 18.5 to incorporate the variations of the *Payment* subprocess defined in Exercise 1. Also, create a mapping between the new facts and the new ports in terms of boolean expressions.

Exercise 4. Using the Worklet Service (cf. Chap. 11), several subprocesses can be assigned to a single atomic task in YAWL, that is, there is a runtime choice among the execution of the various subprocesses. How could such a choice be restricted by configuration? Define the required ports and their possible configuration values in the configuration of worklets.

Exercise 5. The following *Ordering* subprocess (Fig. 18.16) is used at Zula Exquisite. Combine it with the *Ordering* subprocess at Genko Oil (Fig. A.3), so as to obtain a single YAWL model. Afterwards, determine the necessary configurations to derive the two original models again.

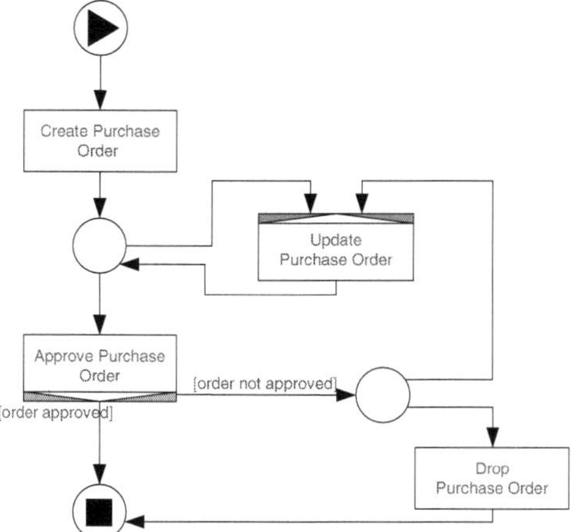

Fig. 18.16 The ordering subprocess at Zula Exquisite

Chapter Notes

This chapter introduced the topic of process configuration through the use of LTSs and the application of the concept of inheritance of workflow behavior. The interested reader can find more information on the configuration of process models defined as LTSs in [106] and on the inheritance of workflow behavior in [8]. After this introduction, the chapter focused on the configurable YAWL notation and presented a questionnaire-based approach to steer the configuration of YAWL models. The technical details of the C-YAWL notation are provided in [107], while a complete description of questionnaire models can be found in [141]. The application of questionnaire models to the configuration of process models defined in C-YAWL and C-EPC is discussed in [140, 143], whereas in [108, 144] the reader can find insights on the application of this approach in practice.

Furthermore, this chapter presented an individualization algorithm that transforms a configured YAWL process model while preserving the syntactical correctness of the model. The formal algorithm is available in [107], while an in-depth discussion on the topic of correctness in process configuration can be found in [10]. Specifically, here the authors define a technique for incrementally preserving the correctness of a process model during its configuration, given some assumptions on the structure of the configurable model. This technique is first defined in the context of Petri nets and then extended to the specificities of C-EPCs.

More information on process configuration can also be found on the *Process Configuration* web-site,[8] where the *Synergia* toolset described in this chapter is available for download along with several examples.

The possibility of reusing process models as best-practice templates has led to a number of commercial reference models for various industry sectors. For example, the Order Fulfillment model is inspired by the Voluntary Inter-industry Commerce Solutions (VICS) model.[9] Other well-known examples are the Information Technology Infrastructure Library (ITIL),[10] the Supply Chain Operations Reference Model (SCOR),[11] and the SAP R/3 reference model [130], which was utilized to capture the processes supported by the SAP's ERP system. An overview and classification of reference models can be found in [93].

Although existing reference models as the above ones depict recommendations for the execution of specific processes, they hardly provide insights on how these recommendations can be adapted to the individual needs of an organization. In this chapter we presented process configuration for YAWL models as one such adaptation mechanism. A similar mechanism is suggested in [214] for the configuration of process models defined in EPCs (via the C-EPC extension). The approach in [42] is based upon the principle of model projection. As a *reference* process model

[8] www.processconfiguration.com

[9] www.vics.org

[10] www.itil.co.uk

[11] www.supply-chain.org

typically contains information for multiple application scenarios, it is possible to create a projection of an EPC model for a specific scenario (e.g., a class of users) by fading out those process branches that are not relevant to the scenario in question. Another proposal is defined in [199], which relies on stereotype annotations to accommodate variability in a so-called *variant-rich process model*. Although stereotypes are an extensibility mechanism of UML, in this proposal they are applied to both UML Activity Diagrams and BPMN models. A subset of these stereotypes also appears in [201] while the general idea of annotating model elements to represent variability has also been investigated in [71].

The above approaches, including the one presented in this chapter, focus on the configuration of the process control-flow. For the application of configuration mechanisms to other process perspectives such as the resources and the business objects participating in a process, the reader is referred to [142]. Here the authors define a process meta-model to capture complex task-resource and task-object associations and configurations thereof. This meta-model is embodied in an extension of the C-EPC notation, namely integrated C-EPCs (C-iEPCs).

Process configuration allows users to adapt process models by restricting their behavior. However, other adaptation mechanisms are available in literature, such as modification, abstraction, generalization, and specialization. A comparison of different adaptation mechanisms is provided in [42]. This might help the reader to identify if process configuration is the required method for adapting a process model to their specific scenario.

Chapter 19
Process Integration

Lachlan Aldred

19.1 Introduction

Distributed architectures depending on their nature require different forms of coupling. Web-solutions with a database back-end are traditionally tightly coupled with one another. Meanwhile, the transaction processing systems running in banks rely on loose couplings with the systems in other banks. This can be observed in the tendency for inter-bank transactions to be reconciled overnight. The nature of any business problem that brings about integrating distributed resources dictates the format and style of their coupling. The same goes for integrating processes. Business processes depending on the business problem they solve will require the whole range of coupling styles usually observable in distributed applications. Furthermore, the fact that there are usually many instances of any business process running in parallel means that a whole range of challenges that commonly do not occur in distributed applications always occur. Terms such as "synchronous," "asynchronous," "blocking," "nonblocking," "directed," and "nondirected" are often used to refer to the degree of coupling required by an architecture or provided by a middleware. However, these terms are used with various connotations. Although various informal definitions have been provided, there is a lack of an overarching formal framework to unambiguously communicate architectural requirements with respect to (de-)coupling.

This chapter does not propose a solution to process integration. It does not discuss the range of modeling elements that would allow us to model an integrated process. Nor, does it discuss an implementation that solves process integration. What it does is define and frame the problems of process integration. This way we can understand what a solution must be able to do in order to support the modeling and execution of integrated process models. This way we can measure and evaluate some of the "solutions" based on their real merits rather than the claims of the creators of those solutions.

L. Aldred
Queensland University of Technology, Brisbane, Australia

A.H.M. ter Hofstede et al. (eds.), *Modern Business Process Automation*,
DOI 10.1007/978-3-642-03121-2_19,

In this chapter, we are going to introduce some of the technical and conceptual challenges and hurdles to the problems of process integration. These challenges are presented as patterns or requirements for integrated processes. In this presentation, we have attempted to not only deal with requirements that are unique to process integration, but also tie these notions into commonly understood requirements for Enterprise Application Integration (EAI).

19.1.1 What is Process Integration?

The first workflows were able to coordinate the actions of people in an organization. Workflow systems matured to schedule complex collaborations of both people and applications. To achieve this, workflow systems provided interfaces allowing interactions with invoked applications and interfaces for interoperation with other workflow enactment services. The desire for common interfaces, enabling workflow systems to invoke applications and other workflow systems, inspired core aspects of the Workflow Reference Model (WRM), authored by the Workflow Management Coalition. The WRM defined five standard interfaces for a workflow system back in 1995. Three of the five interfaces (1, 2, and 5) dealt with other issues not related to process integration, but two of the five interfaces (3 and 4) focused on discrete aspects of process integration. One enabled local processes to invoke remote applications, while the other addressed the ability to interact with other workflow systems. This is basically an Interface-Based Process Integration Architecture, which basically means that the process engine provides interfaces for integration and someone has to design/develop/deploy software to call those interface APIs to achieve process integration.

Later, BPM systems became available and they offered "turn-key" integration support. This was their distinguishing feature. They typically came with a box of plug-ins for integration. Someone had already created and *commoditized* reuseable software components that would connect, for instance, a Staffware process to a Java Message Service (JMS) server. This integration style was basically a incremental improvement over *interface-based integration* – in that it significantly reduced the development effort, but did not change the process integration paradigm in any significant way.

BPEL4WS 1.0 burst into the scene in 2002 and it proposed a Service Oriented Architecture (SOA) approach to process integration. In BPEL4WS, *Invoked Applications* (Interface 3 – WRM) became invocations of remote services. And the process itself was able to be presented to the enterprise as a service; rolling *Other Workflow Enactment Services* (Interface 4 – WRM) into the SOA paradigm. This service-oriented approach was, for its time, a bit of a revelation in which invoked applications and processes became just calls made to a Web-service. And the process itself could be exposed as a service to other remote processes and application through a commonly understood SOA architecture; but correlation sets were the real revelation introduced by BPEL4WS. This is because the most significant challenge in process integration is the fact that there are many cases running at the same time and incoming messages must be routed to the right context or case. BPEL's correlation sets solve this age-old problem in an interesting way. The interesting thing about correlation sets is that they allow us to extract correlation values from message payloads (such as a *Purchase Order ID*) as opposed to having each endpoint create/read/match arbitrary unique IDs in message headers and have the receiver send them back in response headers. Thus, in BPEL4WS, and its successor WS-BPEL, the addition of correlation sets and an SOA approach to process integration has been a giant leap towards a more abstract, more model-driven approach to process integration. *And that is our goal.*

An example of a process integration, in the classical sense, is two processes, in separate organizations, actively managing a supply chain. Each process is running in a BPM system or workflow management system. However, network and software heterogeneity and/or manual processes add an incredible amount of complexity. How can we model this heterogeneity? We are going to extend this notion of process integration to include the most challenging case – where one process is running in any BPM/workflow system and the other is totally manual.[1] This notion of process integration will have to embrace network and messaging heterogeneity rather than hide behind assumptions about both parties using SOAP/JMS/HTTP, etc.

19.1.2 Major Issues with Process Integration at the Conceptual Level

Spolsky is well known in the technical community for blogging that "all nontrivial abstractions, to some degree, are leaky" [239]. The rationale behind his assertion is that there is no perfect implementation of an abstraction. One example he uses is the reliability of the Transmission Control Protocol (TCP). TCP is a reliable abstraction over the unreliable Internet Protocol (IP). According to Spolsky, if the pet snake chews through the cable leading to your computer, TCP is not going to deliver its payload – to an extent the implementation (sending signals over the cable) has leaked up to TCP. The conceptual nature of an abstraction allows us to hide a great deal of complexity that exists underneath. That is why, as users of technology, we love to use abstractions. It could be assumed that the real success of an abstraction can be measured by the degree to which it can remain true to its abstract semantics, that is, by the degree to which it does not leak.

At process design time, the tasks of a business process are abstractions of changes in remote applications or changes to the physical world. However, it should be pointed out that, as far as abstractions go, they actually do not leak *too much*. At

[1] Be mindful that similar challenges exist to integrating any process – regardless of whether the process is manual or automated and irrespective of the brand of process automation software.

runtime, the process engine "knows" exactly what task needs to be performed next and under what conditions. The states (conditions) are precisely known at any stage of the process, and in formal workflow languages, such as YAWL, their current states can be fully understood and their future states even predicted. The design of a task is precise and has its own state-transition model; everything is precise and predictable.

A workflow engine can be precise and predictable in an unpredictable world because it delegates all of its work. Naturally, it also delegates the problem of having to abstract the unpredictable real world as well. If the process has told a person to approve a form and that person has decided to go on strike, the workflow/BPM system does not need to know that. It only needs to follow its designed process model, which may wait 10 days and then delegate a compensatory action. If the real-world delegate's abstraction fails (or leaks), the process engine, being a delegator, is naturally going to delegate the problem of fixing the leak. Dealing with failures to abstract is the responsibility of an application or a piece of integration middleware, etc. When a process engine delegates process integration to a plug-in, it also delegates making sure that leakages of process integration are minimal, and it delegates transforming those inevitable leakages (e.g., a network exception, a lost message, a duplicate message, a receive timeout) and their handling. Put another way, it is safe to assume that any design-time modeling of process integration will leak, and when it does, the process integration plug-in needs to handle it gracefully – avoiding scripting and code.

Abstractions of bridging components between the process engine and remote applications are never clean. The state of the remote process is probably unknown. Collaborative distributed processes are constantly in a state of uncertainty, as to the state of the remote process. For instance, a message arrives from a remote process asking this process to perform an action. Then the same message arrives again. Do we do the same action again or ignore the second message? If this incoming message contains a request to pay $9,000.00 to a beneficiary, we will want to know whether this is meant to be executed once or twice.

The "Two Generals Paradox" demonstrates that we cannot know if the remote process is acting in collaboration with us. For example, if we send a message to the other general to attack the city at 2 pm, can we be certain that he got the message? If he sends a message back to us can he be certain that we got his response? If we send a response to his response confirming that we got his response, we know that he will hesitate unless our response got through. How do we know that our response got through? And on goes the problem. Adding to uncertainty between remote processes is an innate lack of trust, or uncertainty that the remote process owner is truly acting in our best interest. Indeed remote processes are plagued by uncertainty about each other's state. To minimize this uncertainty, messaging technologies introduced feedback mechanisms such as remote exceptions, event-handlers, and request-response. Naturally we would want to have all of these techniques, and more, at our disposal during process design time.

19.1.3 Why is it Relevant at the Conceptual Level?

In a context of uncertainty and technological heterogeneity, it is essential that any conceptual solution to process integration is precise in its semantics, consistent, and dependable, but most importantly it should elegantly abstract messaging and communication in the process model. It should also elegantly handle any leakage. Interprocess communication is potentially fraught with all manner of problems and is far from trivial.

YAWL has always possessed an extensibility mechanism, allowing users to create a task template specifically suited to a particular type of work. These are called Custom Services. Such a task template can be dragged into the process model, allowing us to build a useful process. Therefore, intuitively, we could create a YAWL Custom Service able to connect to an email server and send a message, and we could create another YAWL Custom Service to send/receive SMS messages, etc. However, such an approach to integration inevitably brings details about the technology into the process layer. Furthermore, there is no opportunity to declare the interaction patterns in an abstract manner. Two different technologies may interact using the same pattern, but if they use different custom services (i.e., one for email, and the another for SMS), the similarity of their interaction patterns[2] would not be the slightest bit apparent in the process integration model.

Perhaps the most challenging aspect of process integration is correlation of multiple messages along the same email address/fax number/FTP location, etc. with the right instance of a process. With many instances of a process this can be tricky. For example, if a fax arrives concerning one of the many concurrent purchase orders, we need to ensure that the fax gets correlated with the relevant purchase order. Obviously, if there is one YAWL Custom Service for faxes and an entirely different one (perhaps built by someone else) for handling JMS messages, then each must handle correlation in its own way. JMS messages have a field that can be set with a correlation ID. Faxes have source phone numbers, etc. Without any conceptual framework for correlation, we will quickly become embroiled in JMS headers and fax numbers; and we will have failed to capture the model cleanly – subverting our architectural goals. Capturing correlation in a process model, in an abstract manner, is extremely challenging when the conceptual framework for integration is missing or is piecemeal.

Without a correlation foundation, our process models will inevitably groan under the weight of all this low-level, technology-specific code. It is technically challenging and error-prone to pull correlation IDs out of JMS messages and compare them with values of process variables, and then find the matching process and push the message to it. Clearly a conceptually clean way of performing correlation, which is not tied to a particular technology, will reduce our modeling effort and reduce the amount of time we spend testing and debugging the process-based solutions that we create.

---

[2] An SMS message sent over the phone network is unidirectional and asynchronous, as is an email sent over SMTP.

As mentioned earlier, a YAWL process does not perform its tasks; it simply schedules these tasks and requests the appropriate YAWL Services to do the tasks it delegates. Likewise, the Engine does not transport data to remote processes. It delegates this problem to any capable YAWL Service. This is beneficial – in that we can bind YAWL processes to any form of middleware through custom services. But this is potentially problematic if we delegate different middleware technologies to disparate YAWL Services. Consider that each middleware-connected YAWL service may be fundamentally different, having its middleware work differently, support different interaction patterns, and be invoked differently. For example, a buyer enters his credit card details, the details are sent to a synchronous e-payment gateway, but the fact that there is a synchronous response from the gateway is not visible in the model. It is only when you drill into the inputs and outputs of the task and its corresponding YAWL service that such details become apparent. Then when the follow-up email, containing the invoice, is sent to that buyer, it is pushed onto the local POP server and the Internet takes care of everything else. Obviously, these interactions are completely different. For starters, one is elementary and "asynchronous," while the other is a synchronous request-response. Yet, these differences would not be visible in the process model.

In the context of uncertainty about the state of remote collaborators, the *way* messages are sent and received is almost as important as *what* we send and receive. In other words, we definitely do not want the way we invoke these remote processes to be fully abstracted away from our process model to the degree that we cannot see any difference between a synchronous and asynchronous message exchange. To make these appear the same would ultimately lead to ambiguity and even greater uncertainty about our trading/collaborating partners. This is one of the major reasons why it is so important to expose the differences between interaction pattern(s) of the different middleware technologies being used when we model integrated processes.

When processes *talk* to other processes, there are vast numbers of potential complications that can arise. For instance, two copies of the same message may arrive, or the message may not arrive at all. Certainty of the present-time state of the partner process is not possible. Messages, or events, can arrive unexpectedly. Some events need to be passed between processes instantaneously or they become stale; for example, a snapshot of the current trading prices of shares in a set of company stocks, because the information being shared is timely and has a limited *shelf life*. Other events are not timely, but must make it through no matter what – for instance, bank transfers.

19.1.4 Summary

With large numbers of parallel processes that also happen to swap messages, the problem of correlating any given message to its *right* process instance can become a real burden for process modelers if not handled elegantly.

This chapter introduces fundamental concepts in an interaction that play a major role in realizing solutions to these problems.

Therefore, we can conclude the following:

- One solid test of a process integration is that it should be conceptually able to model processes communicating over fax/email using the same frame of reference as process integrations over SOAP/JMS/HTTP, etc.
- Naturally, it follows that another solid test of a process integration is its ability to master conversation/correlation modeling.

There are of course many other factors to consider, but these are probably the most likely to be sure sign of a weak solution if they are not handled cleanly.

19.1.5 About This Chapter

Having discussed the issues related to process integration, the remainder of this chapter presents fundamental requirements for a solution to process integration. They are presented in the form of patterns. These patterns should be regarded as an expression of what a process integration should be capable of supporting. Section 19.2 introduces the many different ways processes can be coupled together and their impacts. Sections 19.3–19.5 discuss batch messaging, request-response, and conversations, respectively. Sections 19.6–19.8 discusses some advanced requirements unique to process integration. Section 19.9 concludes the chapter.

19.2 Coupling Dimensions

Next we will present these fundamental concepts as patterns. The Workflow Patterns (see Chap. 2) provided a measure by which we can assess the modeling power of any workflow language, and they were used as a benchmark during the creation of YAWL. Likewise, the patterns in this chapter are intended to help us understand and assess languages for process integration. The first set of patterns denote the structure and design of an interaction using three dimensions. The first dimension addresses *thread-coupling*. Thread-coupling is a major factor in integration design. It encapsulates whether the process threads will be waiting or not for the presence of communication. The second dimension, *time*, is related to whether two participants need to both be participating in an interaction at the exact same moment for the interaction to occur. The last dimension, *space*, deals with whether the processes interacting with one-another address each other directly or using aliases, and whether a message sent is duplicated for all listeners or sent to only one. Having introduced these dimensions, we will proceed to explain how they relate to alternative forms of middleware.

For now, we will apply these dimensions to unidirectional interactions only. We will address bi-directional interactions (e.g., request–response) in Sect. 19.4.

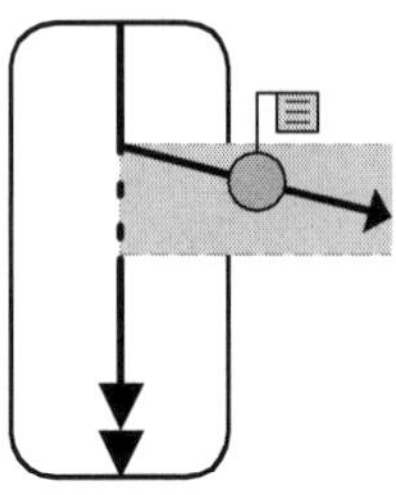

Fig. 19.1 Notation, blocking send

19.2.1 The Thread Engagement Dimension

Thread-decoupling enables "nonblocking communication." Nonblocking communication allows the sender to fire off a message and then forget it. It allows the receiver to not have to wait for messages, but essentially *be woken up* when one has arrived. Thread-coupling, by contrast, allows a sender to get live feedback about the success or failure of the interaction. If a process tries to contact another process about something timely, or important, it may be helpful to get immediate, feedback about the outcome. This is a lot easier to do using a time-coupled approach. Next, we will see that thread coupling and decoupling can be broken down into four basic patterns.

Pattern 1: Blocking send. A *blocking send* is the basic pattern describing when the sending process waits while the message is being sent. A blocking send is characterized by a task in a process that waits until the message has left the process execution environment (Fig. 19.1).[3] In a blocking send situation, the sending process will know when the message has successfully left the process. Once we know the message has left, the process will be able to continue, comfortable in this knowledge.

Pattern 2: Nonblocking send. A *nonblocking send* is the alternative to a blocking send. It describes the pattern where the sending process is able to continue immediately – regardless of whether the message has left or not. A nonblocking send means that the task of transmitting the message is decoupled from the overall flow of the process. Figure 19.2 presents a notation for nonblocking send.

Support for the nonblocking send pattern is fairly uncommon in middleware. Websphere MQ and Microsoft Message Queues (MSMQ) claim to support it; however, their support is only partial. Nonblocking send is uncommon in RPC-based technologies.

Pattern 3: Blocking receive. A *blocking receive* pattern occurs when a process wishing to receive a message waits for the message to arrive while halting the flow of

[3] If the interaction was bidirectional, it would block for longer (Sect. 19.4).

Fig. 19.2 Notation, nonblocking send

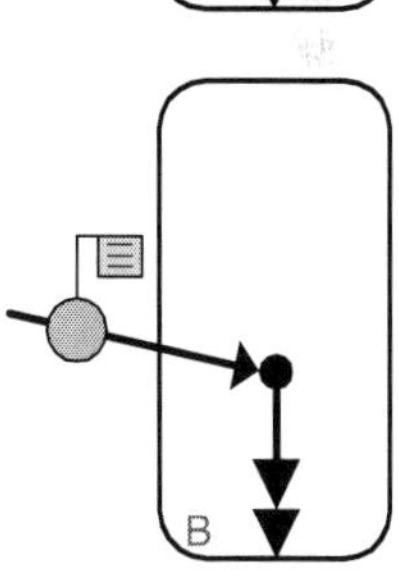

Fig. 19.3 Notation, blocking receive

Fig. 19.4 Notation, nonblocking receive

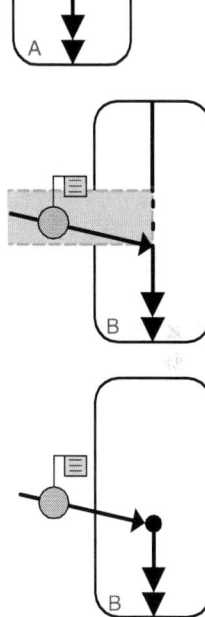

the business process. The process resumes only when the message has arrived. This means that the flow of the process is highly coupled to the arrival of message.

If non-arrival of a message is as important (and needs to be handled) as the message arriving, then it may be wise to consider making the process wait for the message to arrive. An example of where this construct is useful is in waiting for a customer to pay before shipping the goods. By making the process wait for a message, it is easy to extend the process to do something else (i.e., a compensation strategy) when the message does not arrive. Figure 19.3 presents a notation for blocking receive.

Pattern 4: Nonblocking receive. A *nonblocking receive* pattern occurs when the receiving process is not waiting for the message to arrive. In many situations, we do not care if a message fails to arrive. However, what if something needs to be done *if* a message arrives. An incoming message affects the process, but its absence is not going to impact the processes success or failure. In other words, if something comes up we handle it; otherwise the process proceeds *business-as-usual.* A nonblocking-receive pattern is the right pattern to use in this situation. It is illustrated in Fig. 19.4.

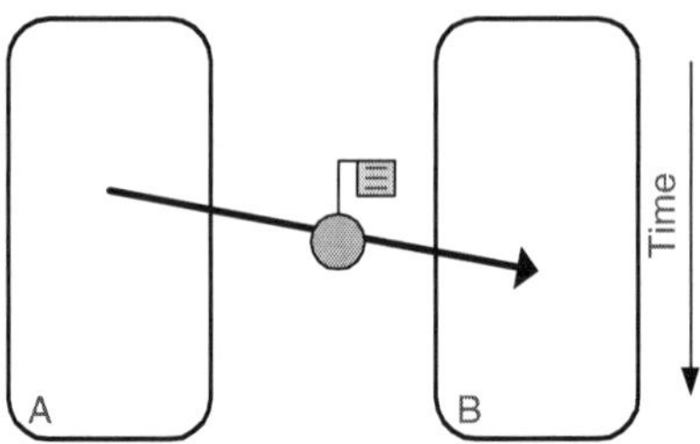

Fig. 19.5 Notation, time-coupled

Nonblocking receive is an extremely suitable pattern to capture the process scenario where an external event may be triggered. This pattern forms the foundation of most asynchronous architectures. However, nonblocking or asynchronous techniques when applied to scenarios where our process depends on the arrival of a message force us to check and recheck whether the message has arrived yet, and if it does not arrive in time we have to take additional action. These scenarios are more easily modeled using blocking-receive techniques, where a timeout can be used to take remedial action if the message does not arrive.

A nonblocking send is a necessary condition, but not a sufficient condition to achieve total thread-decoupling, which is to say that the receive action must also be nonblocking. Rephrased, if both the send and receive are blocking (nonblocking) then a total thread coupling (decoupling) occurs. A partial thread decoupling occurs when the send is blocking and the receive nonblocking or vice-versa.

19.2.2 The Time Dimension

Time, or timing, is a fundamental aspect of an interaction. Like the thread-engagement dimension, the time dimension has a large impact on the amount of feedback (or lack thereof) possible between two integrated processes. In an interaction, time is either coupled or decoupled.

Pattern 5: Time-coupled interactions require that both processes are participating in an interaction at the same moment. In time-coupled arrangements, both processes begin the interaction and complete it at the same moments. A depiction of time-coupled interactions is presented in Fig. 19.5.

Because the sender and receiver are involved in a time-coupled interaction concurrently, it is possible to know the success/failure immediately. When thread-coupling is used alongside time-coupling, it is even easier to recognize success/failure. Middleware such as RMI, RPC, and SOAP/HTTP make good use of this fact.

Payment gateways frequently use time-coupling to give feedback to the customer. Wouldn't it be a poor service if you entered your credit card details online only to get an email the next day letting you know that the bank systems were unavailable at the time of payment? With time-coupling, however, there is a trade-off between ease-of-feedback and just getting on with the job. With time-coupling, both processes

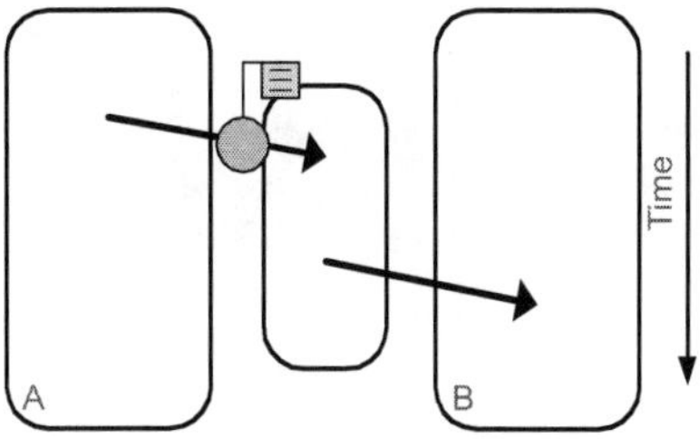

Fig. 19.6 Notation, time-decoupled

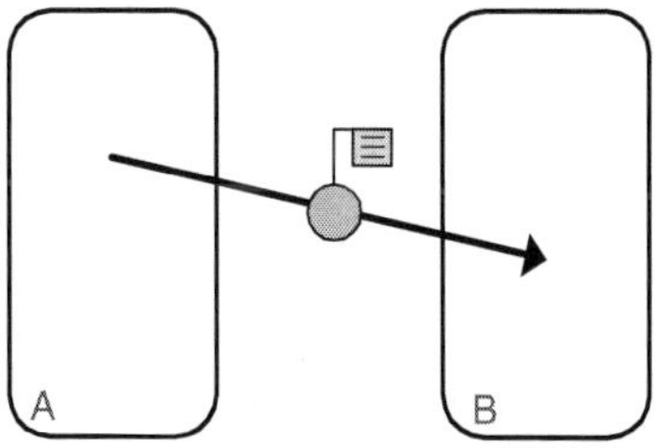

Fig. 19.7 Notation, space-coupled

have to be online and available. If any link in the chain is broken, then integration cannot occur.

Pattern 6: Time-decoupled interactions, by converse, do not support feedback about the success/failure of the interaction; however, the popularity of time-decoupled integration technologies such as JMS, email, SMS, etc. is testament to the fact that feedback does not always matter. Figure 19.6 is a notional depiction of time-decoupled integration.

If a process wants to send a message and then get on with other things, then this pattern is optimal. For example, a weather station is regularly sending its readings to the weather bureau's central systems. Another example is commercial banks reconciling account transfers overnight.

The pattern basically assumes that the messages are stored somewhere enroute and implementations are built this way. Middleware solutions such as Websphere MQ and MSMQ are examples.

19.2.3 The Space Dimension

The dimension of space encapsulates where, or to whom, the message is being sent. We have identified three patterns of integration residing in the space dimension. All are extremely common and have their own unique strengths.

Pattern 7: Space coupled. A *space-coupled* interaction is essentially one where the sending process knows about the receiving process. It is also known as direct addressing and is common in many integration techniques and technologies. It is depicted in Fig. 19.7.

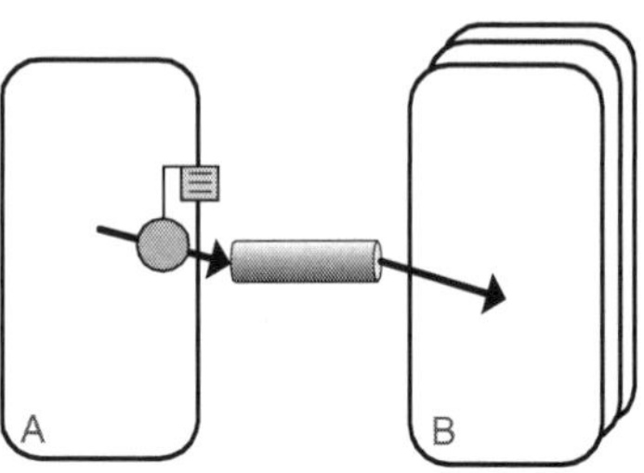

Fig. 19.8 Notation, space-decoupled

Space-coupling is simple and direct. There are no proxies and layers of indirection that the true destination of a message is hiding behind. Hence it is observable in many of the clean and simple integration techniques available (e.g., these include SOAP/HTTP, fax, email, FTP, etc.).

A purchase order process, where incoming orders are received over fax, is an example of space-coupling, as is a cosmetics company exposing an email address that starts a complaints handling process.

Pattern 8: Space-decoupled. A space-decoupled interaction (akin to indirect addressing) occurs where the sender process interacts with the receiver process through an intermediate address. Such a technique is slightly more elaborate than direct addressing. The sender only knows about the indirect address, and the receiver registers itself with the indirect address as the receiver for it. Hence the two processes are decoupled in space, that is, from each other in the addressing domain. Space-decoupling is depicted in Fig. 19.8.

Typically, the more sophisticated middleware technologies offer space-decoupling. These include, for example, JMS, Websphere MQ, and CORBA/IDL. Uses of space-decoupling include clustering and load balancing applications (because many receivers can share the workload of incoming messages). Additionally, space-decoupling enables hot-swapping of an old process with a new one. This is achieved by de-registering the old process to receive messages while registering the new one in its place.

Pattern 9: Space-decoupled topic (Publish-Subscribe). Space-decoupling can take on a slightly different flavor called space-decoupled topic (Publish-Subscribe). Space-decoupling, as presented, allows a message to be passed indirectly from the sender to a receiver; however, publish-subscribe allows a message to be passed from a sending process, with the difference being that every registered receiver gets a copy of the sender's message. This pattern is related to the *observer pattern* used in software engineering. In the observer pattern, an object known as the subject maintains a list of its observers, and notifies them of any state changes to the subject.

Publish-Subscribe is depicted graphically in Fig. 19.9. Integration technologies such as MSMQ, Websphere MQ, and JMS enable this pattern of interaction. A process that is configured to broadcast the fluctuating trade price of a stock to various day trading applications is one application of Publish-Subscribe. Another example of its use is where a flight booking process is monitoring changes to the conditions concerning a flight booking.

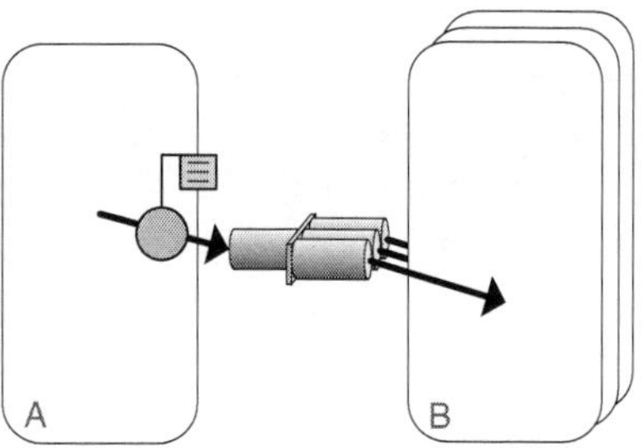

Fig. 19.9 Notation, space-decoupled – topic

In summary: Space-coupling allows any sender to direct the message to *one* known receiver. Space-decoupling allows the sender to interact with any receiver process that happens to be registered with the indirect channel at that time. Note that there can be many such receiving processes sharing the channel, in which case we have load-balancing or clustering – as the interaction will reach only one of the receivers. Space-decoupling (topic), on the other hand, allows the sender to interact with *all* receiver processes that happen to be registered with the indirect channel at that time. Each one will get its own copy of the message.

19.3 Batch Messaging

Many instances of the same process integration typically share the same communication resources (i.e., they share the same channel(s)). Sometimes, copies of the same message need to be sent to many partner processes. Either way, situations constantly arise in process integrations where we need to handle and process many messages.

Pattern 10: Multicast involves sending a message to many receivers, and so it in effect is an alternative solution to the space-decoupled (topic) pattern. However, there are some subtle differences. The first one is that the sender is burdened with interaction with each receiver directly – making it a more cumbersome technique. However, it is absolutely necessary as we cannot guarantee that each of our intended recipients are subscribed to a common topic. Multicast becomes essential where each recipient process uses a different technology set to receive messages. It is also useful if the list of recipient processes is not known until runtime; at which point, the list of recipients can be stored in a variable and then communicated with.

Pattern 11: Message aggregation. The complement to multicast is the multiple message aggregation or aggregate receipt. Essentially many messages are received as an aggregated batch. They may be aggregated according to some common property of data. An example of this is correlating two incoming line-item messages for the same purchase order.

Aggregated message receiving may occur for a group of unrelated messages that a process needs to handle en-masse. For example, a banking process may consume many account transactions from remote banks and process them en-masse, overnight, while reconciling its accounts.

Pattern 12: Filtering. Message aggregation indiscriminately consumes a large set of messages together. But what if some of those messages are irrelevant to our process? How do we keep them out? Message filtering removes the burden on the process of having to check each message in a batch and only keep the ones we want. It also removes the need to worry about putting the messages back if another process actually *is* interested in our unwanted messages – because our process would not consume them in the first place.

As an example, an auction house only processes bids that are greater than the current highest bid. Message filtering over the contents of the message header is supported by JMS.

19.4 Seeking Feedback: Bidirectional Interactions

Many integrated processes need immediate feedback in response to the messages they send. Bidirectional interactions are an enabler of this feedback. A *request-response* is the archetypal bidirectional interaction.[4] The term *Bidirectional interaction* is preferred because *request-response* implies that the interaction is both time-coupled and thread-coupled.

As mentioned earlier, one-way interactions that are both time-coupled and thread-coupled dramatically increase the certainty of the sender about the success or failure of the interaction. However, in such one-way interactions, there is no way to immediately tell (1) whether the receiving process understood the message, and (2) what the receiver has done with the message. So if the aim is to increase certainty of state, the best approach is to use bidirectional interaction styles that are both time- and thread-coupled.

This is fairly intuitive, but it is important to know that if we choose time-decoupled and thread-decoupled techniques like email, fax, or FTP, we will not easily be able to achieve the same certainty. And of course building an aggregation of email interactions to mimic a request-response is nontrivial. The range of integration technologies on offer provide varying degrees of certainty about the remote state of a process integration. Certainty about remote process state is difficult with *Fire-and-forget* techniques such as email, SMS, and FTP. Conversely, time-coupled, two-way technologies (such as CORBA) offer strong certainty about remote process state. Message-Oriented Middleware technologies (such as JMS) contain rich APIs allowing alternative interaction styles. These can offer extreme uncertainty about the interaction or total certainty depending on the interaction style chosen.

[4] Middleware technologies such as CORBA and SOAP/HTTP support request-response.

Bidirectional messaging adds additional possibilities – for instance, delivery receipt, and the reporting of receiver-side faults. A delivery receipt (i.e., system acknowledgement) is an event returned to the requester that its message has been successfully received. A delivery receipt (as distinguished from a response) does not imply that the targeted endpoint has processed the message – just that it has received it. Finally, a receiver side fault being propagated back to the requester indicates that an error occurred during the processing of the request message. In our surveys of various forms of middleware, we have observed five main patterns of response:

Pattern 13: Acknowledgement. a signal gets returned to the requester, indicating that the message was successfully processed. This is useful in any application where we need to be certain that the message has been processed, for example, in military communications.

Pattern 14: Response is a standard response, containing meaningful information. This is useful in any application where the requester needs information back from the responder, for example, the response to a book price query.

Pattern 15: Fault. A problem occurs in the responder. Rather than just hide the fault message inside the standard response, the fault is packaged in a purpose-built container and sent back. This is useful in applications where the sender wants to know about, and handle, a problematic request. For example, a form is being rendered and several database queries need to be used to inject data into the form. If the database queries fail on the server, the client needs to load a plain, unpopulated form in order to keep the customer engaged.

Pattern 16: Receive response: blocking. the requester blocks for the response. This is useful in applications where the response comes back quickly, for example, in a credit-card payment gateway. It is supported by JMS, for example.

Pattern 17: Receive response: nonblocking the requestor thread uses a nonblocking technique to receive the response. This is useful in applications where the response is likely to come back days or maybe weeks later. It is usually performed by composing two interactions together – one in each direction. Even though this is a little *clumsy*, it can be done with virtually any integration technology.

In the business world, we need to be pragmatic about our choice of integration middleware. This is necessary because we cannot force our trading partners to adopt *our* integration middleware. If the trading partner uses email only, then email is what we use. But that should not stop us from designing certainty into our process models. It just forces us, as designers of these solutions, to raise our game. The ideas presented thus far describe the various *styles* of communication commonly performed by different forms of middleware. Next we will look higher up the abstraction layer at ideas strongly aligned with process integration models.

19.5 Composed Interactions

As mentioned already, *correlation*, (i.e., *the ability to detect two or more related messages from a much larger set of messages – as being related to one another*), is a fundamental problem in business processes. *Conversations*, on the other hand, refer to a moment in time where we have observed two or more *end-points*,[5] exchanging related messages for a given purpose. Note the distinction: *Correlation* is an ability, while *conversation* is an actual event that can be observed between many endpoints.

When hundreds, or perhaps, thousands of process instances are each *having their own conversations* with remote processes, we as designers need to grapple with the fact that there is a real risk of conversations getting their wires crossed, or more accurately, of messages leaking between different conversations.

Naturally, a process integration language needs to be able to model both correlation and conversations. There are many middleware technologies and techniques that already solve this problem. Sometimes it is enough just to leverage them. However, sometimes that is not enough, and we need to model something more sophisticated. For instance, we need something more when using middleware technologies that are not capable of supporting correlation, or when modeling "chatty" conversations involving many end-points. This section presents options for modeling correlations and conversations in processes, starting from techniques that leverage correlation-capable middleware, through to highly abstract modeling techniques that leverage sophisticated features of process engines.

19.5.1 Correlation Patterns

Pattern 18: Instance channels allow each process instance to be given its own channel. This solves the problem of correlation by effectively distinguishing sets of related messages at their source by sending related messages over discrete pathways. If a channel is dedicated to one process instance, this removes the possibility of many processes getting their *wires crossed*. The *REpresentational State Transfer* (REST) architectural style advocates such an approach, claiming that this approach more closely follows the architecture of the Web. The REST approach proposes applying these ideas to Web-service development. Naturally these same ideas transfer over to the process modeling, given the opportunity. Such an approach seems intuitive to many designers.

Pattern 19: Request-response correlation is basically the correlation of two messages, a request and response, by leveraging a request-response middleware technology. The technology does the correlation, significantly reducing the amount of modeling effort. Like *Instance Channels*, this pattern does not require any additional

[5] An end-point is a process, resource, etc. able to send and/or receive messages.

effort to achieve the aims of correlation – merely taking advantage of circumstances. The request is naturally correlated with the response. Many forms of middleware offer request-response functionality, for example, HTTP request-response or CORBA-RPC. These forms of middleware set up a dedicated connection between the requester and the responder. Once this connection is created, the requester and the responder do not need to know about what message is related to what, nor do they need to know about locations or addressing. They only need to know about the connection; even though, in practical terms, that connection is short-lived.

Pattern 20: Token-based correlation works by using "correlation tokens" inside the headers of messages. These correlation tokens are compared and a match indicates that two messages correlate. Some business process solutions do this automatically, though it can be done by manually writing to and reading from message headers. If done automatically, this technique is clean and workable.

19.5.2 Conversation Patterns

While the *Correlation Patterns* generally leverage opportunity, or features built into middleware, the *Conversation Patterns* depend on a match-making service that is able to match tasks that wish to receive a message with a set of potentially receivable messages. This match-making service can be part of the process engine or part of an integration layer available to, but not part of, the process engine – the actual architecture does not matter. A distinguishing feature of such offerings is that this match-making layer effectively hides many of the technical complexities of correlating messages, allowing process modelers to capture the intent with a minimal amount of modeling effort, and is perhaps able to be represented graphically.

Pattern 21: Property-based conversations should be modeled by expressing a function (often called a property)[6] over one messaging task, and connecting it with a related function over another messaging task in the same net. This can be done declaratively. The match-making service then applies these functions to incoming and outgoing messages. When messages pass through the match-making service, they produce a value. Matching values indicate correlated messages. The properties being applied to messages are part of the process design. This technique is conceptually clean, even if it is a little counter-intuitive at first. The WS-BPEL `correlation-set` works this way, and many adopters of WS-BPEL struggle with this concept.

Pattern 22: Compound property-based conversations can be modeled by expressing two or more functions (or properties) over one messaging task and connecting it

[6] Properties are basically functions that, when applied to a message, produce a simple string, integer, or date value. They are in practice, often, just an XPath expression.

with the same number of functions (or properties) over another messaging task in the same net. For example, the task *Send Line Item Price Query* of an order fulfillment process may apply `/PurchaseOrder/LineItem[i]` as one property and `/PurchaseOrder/@id` as the other property. The combination of these values will be used to determine matching messages for the task *Receive Line Item Price*.

Pattern 23: Conversation nesting/overlap is necessary, on those rare instances, where a messaging task is taking part in more than one conversation. For example, conversation nesting would be needed where (in an order fulfillment process) the task *Send Line Item Price Query* started a conversation called `Line Item Price Query`, and is also participating in pre-existing conversation called `Negotiate Purchase Order`. Task *Send Line Item Price Query* being in a loop is executed once for each line item. On each iteration, it participates in the one `Negotiate Purchase Order` conversation and a `Line Item Price Query` conversation dedicated to that iteration.

19.6 Event-based Process Patterns

Loosely speaking *messages* and *events* are interchangeable terms. However, events are often used to refer to specific types of messages that signify a *state change* in a remote process or application. There is a set of patterns that need to be supported in a process integration that are specifically concerned with changing the intended execution plan when remote events are received. Henceforth, they are being referred to as event-based process patterns.

Pattern 24: Event-based process instance creation. there needs to be a way of creating a process instance when an event is received by a process engine. For example, an email to a shipper may need to trigger the shipment stage of the order fulfillment process, or a JMS message may need to be used to trigger a debt adjustment process in finance. It is probably true that most processes begin with a form; however, the problem remains: *How do we begin one without a form?*

Pattern 25: Event-based task enablement is useful where a task should only be started if a certain precondition, announced by an event, is met. For example, the task *Check Customer Credit History* in an order fulfillment process should not be enabled unless the incoming order is over $5,000 (US) and there are some unpaid bills. The *Begin Debt Recovery Proceedings* task of the same process should not be enabled (allowed to be started) unless there are three unpaid bill reminders sent from finance in the last month and the credit check failed. If and only if these events occur should these tasks be offered to a worklist for someone to action.

Pattern 26: Event-handling is necessary because not all incoming events are part of the core-flow of a process. There are plenty of situations in the business world where we should only take a certain action if something *unusual* happens. For example, the subnet *Cancel Order Fulfillment Shipment* should only be started if we receive a message from a trusted financial organization containing a credit card *stop order* for a certain customer. This subnet should obviously not be part of most process instances, and thus does not form part of the core process. However, should certain events occur, the core-flow of the process will need to be drastically changed – hence the need for supporting event-handling.

Pattern 27: Event-based task interruption helps solve the problem where an incoming event should interrupt a running task. This is obviously useful in many situations. For example, the task *Receive Payment* in an order fulfillment process should be interrupted from its wait-state if a timeout occurs and we need to send a reminder. Another use of event-based task interruption is where a request-response goes awry. If a request-response causes the remote trading partner to throw an exception, then the response is replaced with the exception. At process design time, we will need a way to model that the incoming exception interrupts the normal behavior of waiting for a response.

19.7 Transformations

Pattern 28: Process message vs. wire message. Messages over-the-wire differ, in terms of format, depending on the integration middleware we are using. It would indeed be extremely taxing at process design time if we had to worry about these differences. Imagine having to worry about *big-endian* or *little-endian* byte ordering for integers; or having to worry about whether the SOAP envelope in our message conforms to the SOAP standard. We as designers of integrated business processes do not and should not care about such low-level issues. Naturally, these problems should be taken care of by the process modeling/execution technology.

Pattern 29: Data transformations In the distributed world, we cannot control the data formats used at remote sites; consequently, the processes we create inevitably need to convert external data formats to internal ones. For example, a request for a tender is sent to four trading partners as an email. One trading partner replies with an MS Word attachment sent over email. Another trading partner replies to the quote request with a PDF document attachment. The third partner replies by fax. These three formats all need to be presented to the review panel electronically in some cohesive manner. We are not proposing that we have built or will build an MS-Word/email/fax transformation service, but what is required is a conceptual framework that allows the opportunity to plug such a service into our processes easily.

To ensure consistency of data format in the processes we build, it makes a lot of sense to do this transformation *before* the data is passed into the process we create.

19.8 Process Discovery

Process discovery is akin to *service discovery*. It is the ability of two or more automatic processes to begin coordinating actions where they were not fully known at the time of process design. This translates to a process being able to "learn" about new trading opportunities at runtime. Given that integrating with remote processes – with a great degree of *certainty* about the state of your partner(s) – requires effort, the problem of dynamic process discovery remains elusive for all but trivial problems. Additionally, dynamic discovery requires automated matching of the semantics, which possibly lifts the problem of automated process integration into the realm of artificial intelligence. Nevertheless, we know of three patterns concerning dynamic discovery.

Pattern 30: Service registries are well known places where services can be advertised on the Web. They are essentially a Yellow Pages for Web Services. They are an opportunity for a service provider to announce a service to the world and for service users to look up and find the service they require. Standards technologies such as Universal Description Discovery and Integration (UDDI) enable a structured and parsable way for services to be dynamically shared.

Describing a service is nontrivial. There are many subtle aspects to any service. Notions of quality, standards compliance, longevity, voluntary/mandatory warranties, and the legal framework of the country where the service is hosted are just a few. Indeed, the problem of capturing all this information is nontrivial, but perhaps the greater challenge is automatically deciding which is the best fit to our needs. For that reason we would consider this pattern, while interesting, to be a noncore pattern – meaning that it only need to be supported once sufficient progress has been made in the fields of artificial intelligence, service discovery, and ontologies.

Pattern 31: Channel passing is where a process forwards contact details it *knows* about to another process. This process, in turn, starts collaborating with the newly learned contact.

An example of channel passing is the supply chain scenario. A retailer forwards the contact details of a customer to a shipping process. The shipping process then uses this contact to conduct the shipment process to the customer. Theoretical frameworks such as π-calculus are well known for their ability to conceptualize channel passing. The JMS API supports sharing of JMS channels.

Pattern 32: Process mobility, like channel mobility, involves moving resources during process runtime, but in this case, a process instance is moved from one process engine to another. The idea is that a process is started in one site, and then during process runtime it gets moved to another site. This could be helpful in patient care scenarios, for example, where a process gets started in one hospital, but due to capacity constraints the patient is moved to another hospital. The process instance for that patient gets moved to the new hospital with the patient. Process mobility adds a raft of new technical challenges to core process integration, including migration of process instance, process model, any format translations necessary between sites, peripheral data migration (data that is relevant to a process instance but not strictly part of it, e.g., a patient history). These challenges make process mobility a noncore problem. Consequently, it is mentioned here for completeness but is not essential in a process integration language.

19.9 Conclusion

In software engineering circles patterns have helped the process of object-oriented analysis and design. Commonly encountered problems can be solved with patterns that express the essence of a design.

The process integration patterns, on the other hand, are not intended to guide the *process of process design* to such a great degree. Consequently, they do not have long detailed sections in them about how to implement them in YAWL, BPEL, etc. Their primary purpose is to shape our thinking about the sorts of problems a process integration needs to solve.

Second, selection of a process execution platform is a precarious and difficult task. If there were some indication of the likely problems of process integration, then we would at least have some hope of being able to judge alternative process integration platforms based on their technical merits. This set of patterns is essentially inspired by an analysis of different forms of integration middleware; an analysis of middleware aggregation technologies such as enterprise message bus technologies; and an analysis of process integration technologies. No process integration technology is likely to solve all of these requirements. However, the requirements for your process integration project may have many common elements with this list. And if so those patterns could be used to select the right process integration platform for your needs.

Finally, these process integration patterns provide a concise set of requirements for *process integration-oriented* extensions to YAWL.

Exercises

Exercise 1. Provide a scenario where blocking receive constructs are of great help.

Exercise 2. What are the three dimensions of decoupling systems?

Exercise 3. Wave-monitoring systems and seismic monitoring systems are integrated across several continents. Various business rules need to be defined as to what happens when different combinations of alarms get triggered. Also different processes are triggered based on the degree of the wave-size or the size of the seismic event. What patterns are likely to help to capture these differing rules and why?

Exercise 4. A payment gateway process and a purchase order process need to coordinate with one another. There are a number of different actions taken by the purchase order process – depending on factors such as credit card validity and account information of the purchaser. List the patterns that would help integrate these processes and explain why.

Exercise 5. A remote partner site uses manual processes, and they rely heavily on email and fax in order to flow documents through that process. What are the advantages/disadvantages of creating technology-specific process integration bridges and custom YAWL services (i.e., one for fax and another for email)?

Chapter Notes

It is extremely difficult to truly know the state of remote processes, despite the frequent exchange of messages. This is one of the greatest challenges in process integration. The Two Generals Paradox, defined by Gray [109], illustrates this idea. It shows that, when there is the possibility of a message not arriving, there is no sequence of messages that can guarantee that each process *knows* the state of the other.

Another major challenge in integration is conceptualizing integration processes. This challenge can be understood by looking at the history of techniques for integrating processes. In the mid 1990s, the Workflow Reference Model [269] was proposed by the Workflow Management Coalition (WfMC) as a set of standards encompassing the various interfaces that a workflow management system should support. It groups related functionality into five distinct interfaces. The WfMC Reference Model includes Interface 1 – *Process Definition Import/Export*, Interface 2 – *Client Apps/Worklist Handlers*, Interface 3 – *Invoked Applications*, Interface 4 – *Other Workflow Enactment Services*, and Interface 5 – addressed *Administration and Monitoring Tools*.[7] BPEL4WS [69] represented a major change in thinking about process integration. First, it enabled process integration models to leverage SOA-related technologies such as WSDL, SOAP, etc. Furthermore, it provided a conceptual handle for exposing the process as a service using the same Web-service technologies. It also provided a conceptually clean mechanism for ensuring that incoming Web-service requests get routed to the *right* process instance. This is particularly important to process integrations because large numbers of parallel processes may swap messages and we need to ensure that any message reaches the *right* process instance [32]. But more importantly this should not be too hard or clumsy to model.

Middleware is an enabler of integrated processes. The various types of middleware systems in use today are fundamentally different, conceptually, and in terms of the *interaction patterns* they support. They look and feel completely different and

[7] www.wfmc.org/reference-model.html#workflow_reference_model.

this can lead to poor choices of a solution for a given integration problem. Works such as [33,120] attempt to tease out the fundamental aspects of integration through a patterns-based approach. Intuitively there are those that started as *thread-coupled* solutions (e.g., CORBA-RPC [184]) and those that started as *thread-decoupled* solutions (e.g., JMS [116]); however, as implementations matured they became more powerful and flexible, leading to those old lines becoming frequently blurred.

For more information about the patterns of coupling, which exists at the heart of process integration, see [33]. This article is a detailed analysis of middleware technologies and how they support fundamental interaction patterns. A formal semantics is also provided.

Nonblocking, or asynchronous, techniques are less appropriate for use in scenarios where our process must receive a message in order to complete. These scenarios are more easily modeled using blocking-receive techniques.

Most of the interaction patterns are supported; however, for the *nonblocking send pattern* (see Snir and Otto [238] or Eugster [91]) support is rare.

Time-decoupled interactions basically assumes that the messages are stored somewhere enroute and implementations are built this way. For example, middleware solutions such as Websphere MQ [259] and MSMQ [172] servers deployed in a "hub and spoke" architecture offer time-decoupled messaging.

The publish-subscribe pattern mentioned in the text is akin to the *observer pattern*. The observer pattern is frequently used in software development. More information on it can be found in [100].

The REST architectural style proposed a clean conceptual solution for giving any Web Resource its own URL. This was in keeping with *the spirit of the Web*. The use of REST to support business process correlation has been discussed by zur Muehlen et al. in [175]. More about REST can be found in [94].

To have consistency of data format in the processes we build, it makes a lot of sense to do transformation as soon as possible. This approach is proposed by Bussler in his book on B2B Integration [52].

To discover processes dynamically, we need ways to capture a great deal of information about them. Not just the inputs, outputs, and what they do, but also many soft aspects, such as delivery mode, payment terms and conditions, refunds policy, etc. [186]. Obviously such detailed aspects only add to the burden for machine-to-machine negotiations. Nevertheless, dynamic process/service discovery and process mobility continue to be an active area for IT researchers. π-calculus is a conceptual proposal to support process mobility, which is still in its infancy. More about π-calculus can be found in [173].

Chapter 20
Verification

Eric Verbeek and Moe Wynn

20.1 Introduction

Chapter 2 introduced the soundness property on a special class of Petri nets called WF-nets (WorkFlow nets). To reiterate, a WF-net is sound if and only if the following requirements are met:

- Any executing instance of the WF-net must eventually terminate
- At the moment of termination, there must be precisely one token in the end place and all other places are empty
- No dead tasks

Van der Aalst has shown that soundness of a WF-net corresponds to boundedness and liveness of an extension of that WF-net. As boundedness and liveness of Petri nets are both decidable, soundness of WF-nets is also decidable. Based on this observation, the Woflan tool was built, which uses standard Petri-net techniques to decide soundness.

However, as Chap. 1 has already mentioned, the class of WF-nets is not sufficient to reason about YAWL nets. For this, we need to extend the WF-nets with the concept of reset arcs, yielding RWF-nets (Reset WorkFlow nets). Using reset arcs, the cancelation features of YAWL can be captured in a natural way: if some task or condition is canceled by some task, then the corresponding place is reset by the corresponding transition.

Unfortunately, soundness of an RWF-net does not correspond to boundedness and liveness of an extension of this net, as the example RWF-net in Fig. 20.1 shows. Although this RWF-net is sound, it is unbounded as the place p may contain an arbitrary number of tokens. Furthermore, the reachability problem is undecidable for reset nets. As the liveness property corresponds to a reachability problem, the liveness property cannot be decided for arbitrary reset nets. On top of this, YAWL also includes the OR-join construct, which has been neglected so far in this chapter.

E. Verbeek (✉)
Eindhoven University of Technology, Eindhoven, the Netherlands,
e-mail: h.m.w.verbeek@tue.nl

A.H.M. ter Hofstede et al. (eds.), *Modern Business Process Automation*,
DOI 10.1007/978-3-642-03121-2_20,

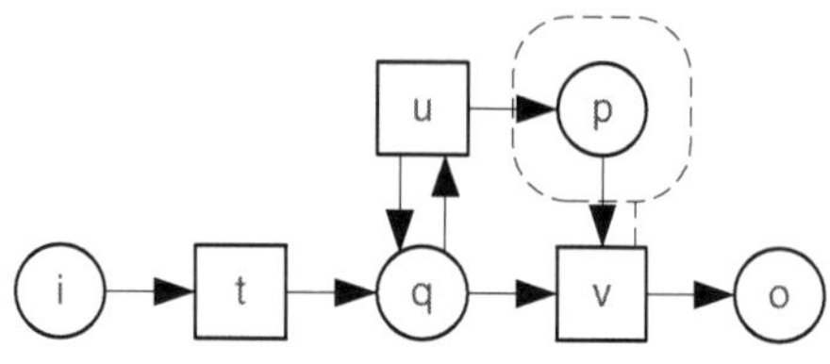

Fig. 20.1 Soundness of an RWF-net does not require boundedness

Based on these observations, answering the question whether some YAWL net is sound is a challenging problem. This chapter tries to tackle this problem using advanced techniques. As mentioned, the question is undecidable in general. Therefore, we allow ourselves to simplify the question a bit whenever we see no other way. For this reason, this chapter:

- Introduces a number of soundness-preserving reduction rules for YAWL nets and RWF-nets. Applying these rules simplifies the net to verify, which simplifies the soundness problem
- Introduces some weaker notions of soundness that might help the YAWL developer in cases where soundness itself cannot be decided

The remainder of this chapter is organized as follows. First, we answer the question of when we consider a YAWL net to be sound. Second, we show how soundness can be decided for many nets by constructing their state spaces. Third, we introduce a number of reduction rules (both for RWF-nets and for YAWL nets) that simplify the soundness problem. Fourth, for those cases where the simplified problem is still too complex, we propose the use of structural invariant properties. Although these structural properties are necessary, they are not sufficient. As a result, using them, we can only show that an RWF-net is not sound, but not that it is sound.

This chapter does not provide any formal definition, formal theorem, or formal proof. Rather than going into these details, this chapter keeps at a high level. The reader interested in these details is referred to the Chapter Notes.

20.2 Preliminaries

As mentioned in Chap. 2, YAWL has been inspired by a class of Petri nets called WF-nets. For these WF-nets, a soundness criterion has been defined (recall from Chap. 2), which can be transferred to YAWL nets with ease: a YAWL net is sound if (and only if) it satisfies the following three requirements:

- *Option to complete:* Any executing instance of the YAWL net must eventually terminate
- *Proper completion:* At the moment of termination there must be precisely one token in the end condition and all other conditions are empty
- *No dead tasks:* Any task can be executed for some case

However, possibly, these requirements are too strong. As an example, consider a YAWL net that includes two decision points (say, two XOR splits) that are controlled by the same case attribute. As a result, if we go left at the first decision point, we also go left at the second, and vice versa. However, as we abstract from the data perspective when considering soundness, the fact that both decisions are synchronized is lost, and the possibility that we go left at one and right at the other is also taken into account. As a result, the YAWL net may be considered not sound, while in fact it is.

The *relaxed soundness* property takes these synchronized decision points into account. A YAWL net is relaxed sound if (and only if) all its tasks are relaxed sound, and a task is relaxed sound if some proper completing case exists for which the task is executed. Obviously, the intent of a YAWL net is to forward a running case towards its completion. If some task cannot help in completing any case, then there has to be an error in the model. This is what the relaxed soundness property is about.

Another advantage of the relaxed soundness property is that it allows for a simplistic view on the OR-join. When trying to decide soundness, we need to check for every OR-join over and over again whether it is enabled or not, which makes it extremely hard, if possible at all, to capture this by a mere reset net. Relaxed soundness, however, will give us the input combinations that are covered by some proper completing execution paths. Therefore, we can initially allow all possible input combinations, and the relaxed soundness property will filter out those combinations that are hopeless, and for which the OR-join should be disabled.

Still, the relaxed soundness property requires the entire state space to be constructed, which is known to be impossible in general. For many YAWL nets, it will be possible to construct the entire state space, but YAWL nets exist for which this is not possible. Therefore, our approach so far (soundness and relaxed soundness) is not complete. To make the approach more complete, we introduce a third notion of soundness, called *weak soundness*. A YAWL net is weak sound if (and only if) the following three requirements are satisfied:

- *Weak option to complete:* It is possible to complete a case
- *Proper completion:* Once a case has been completed, no references to that case are left behind
- *No dead tasks:* Any task can be executed for some case

Note that only the first requirement differs from the soundness requirements.

As Chaps. 1 and 2 have explained, YAWL is grounded in reset nets to provide support for cancelation regions. Recall that a reset net is a Petri net that allows transitions to reset places, that is, to remove all tokens from these places. YAWL's cancelation regions can be modeled by reset nets in a natural way: If some task cancels a region, then the corresponding transition resets all places in the corresponding region. For example, Fig. 20.2 shows a first version[1] of the *Carrier Appointment* decomposition, where the part between the *Prepare Transportation Quote* and the

[1] This version predates the version throughout the rest of this book, and contains a subtle error that was overlooked by the YAWL experts who designed the entire model.

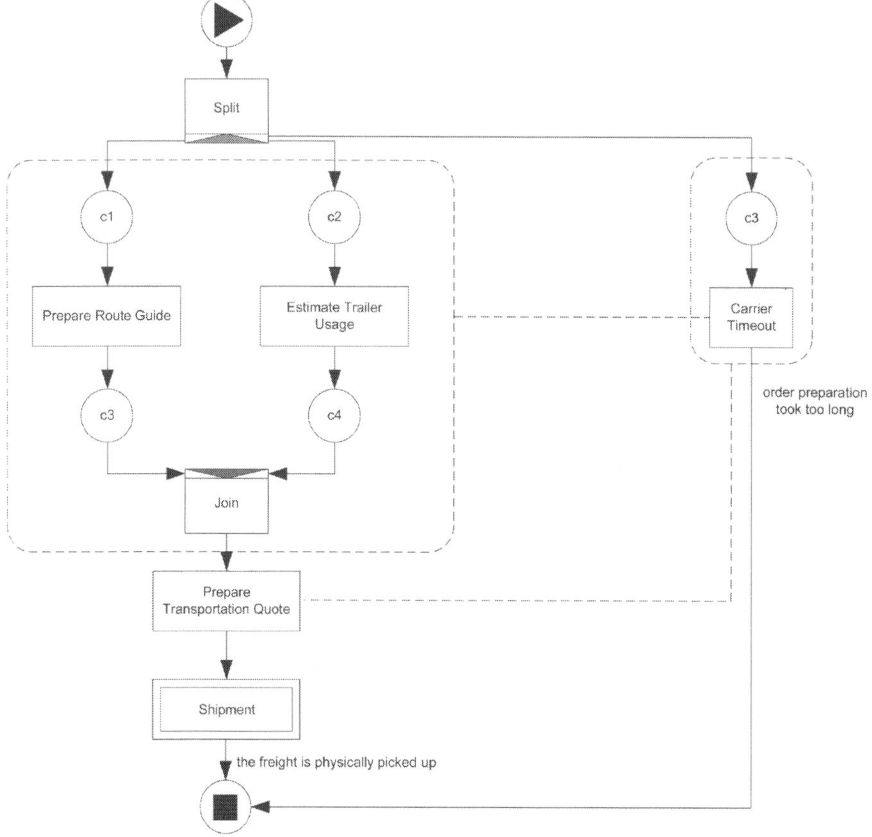

Fig. 20.2 The simplified carrier appointment decomposition

output condition has been replaced by a single composition task named *Shipment*. For ease of reference, we have labeled the other conditions as well. This decomposition contains two cancelation regions, and Fig. 20.3 shows how this fragment can be captured by an RWF-net. Conceptually, every task is captured by a *Busy* place, a *Start* transition for every possible input combination, and a *Complete* transition for every possible output combination, and conditions (including implicit conditions) are captured by places.

Furthermore, YAWL's OR-joins also benefit from these reset nets, as their semantics use reset nets to determine when they are enabled. Fortunately, the question whether some state is *coverable* from a given state is decidable for reset nets, which is sufficient for this purpose. A state is coverable if some state is reachable that contains the given state (see also Chap. 2). This also explains why the weak soundness property is decidable, as the weak soundness property can be expressed in terms of coverability:

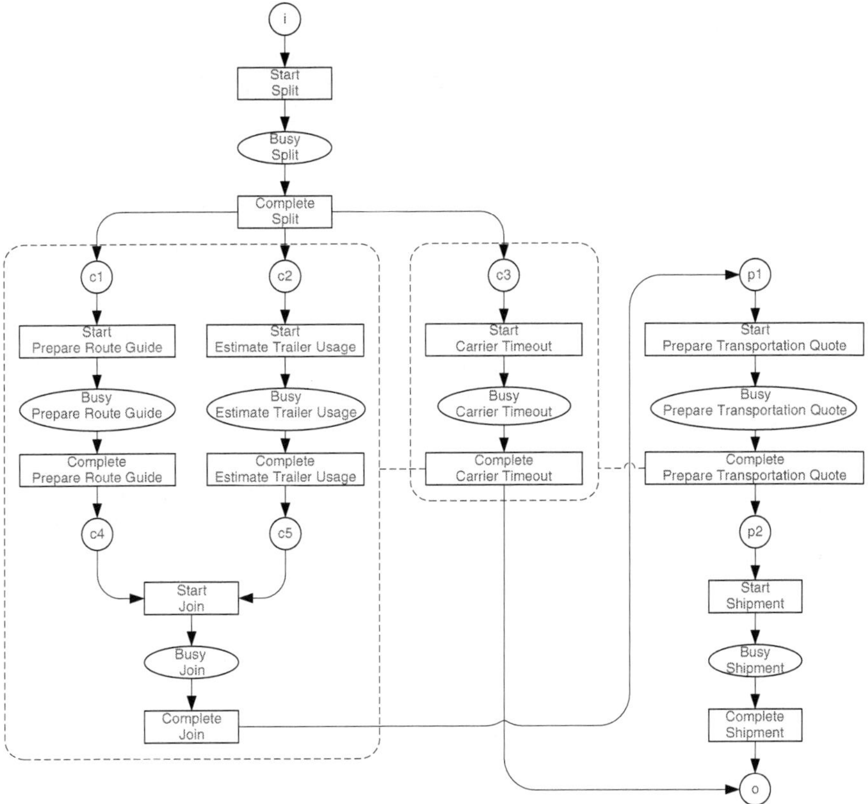

Fig. 20.3 The decomposition of Fig. 20.2 as an RWF-net

- *Weak option to complete:* The completion state is coverable from the initial state
- *Proper completion:* No state exceeding the proper completion state is coverable from the initial state
- *No dead tasks:* For every task a state is coverable that enables the task

In contrast, the first requirement of the soundness property requires reachability, as it requires all reachable states.

20.3 Soundness of YAWL Models

Unfortunately, the question whether some state is reachable from a given state is not decidable. In other words, it is not possible to construct an algorithm that can tell for any YAWL net whether completion is always reachable from the initial state. Nevertheless, for specific YAWL nets, this may still be possible. As an example, we take the RWF-net from Fig. 20.3.

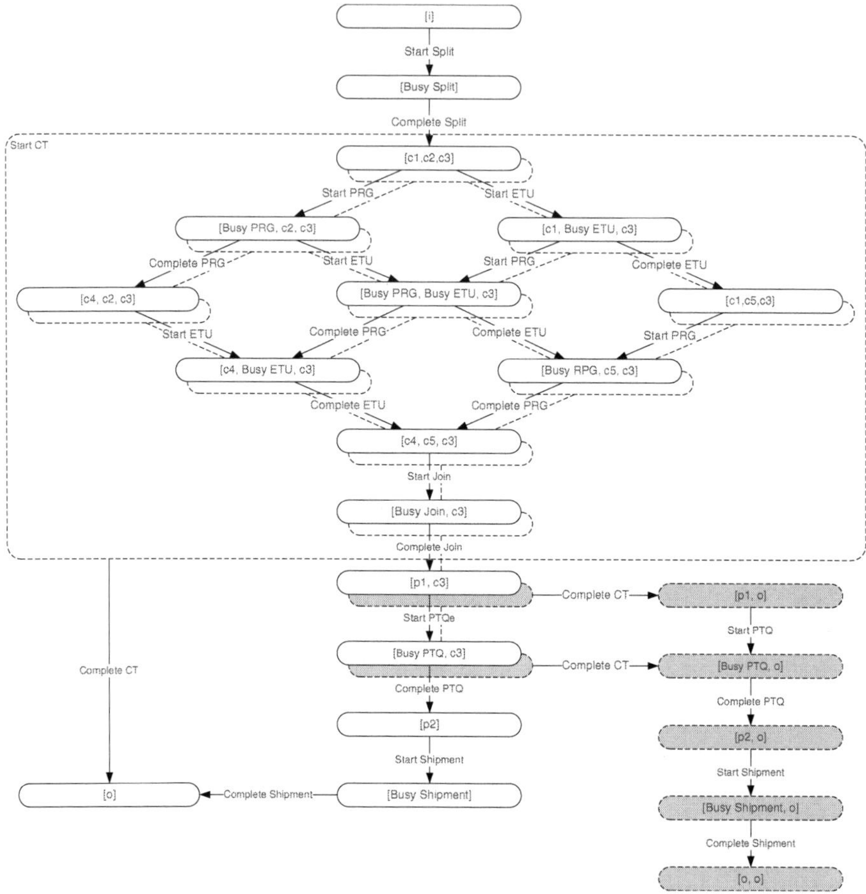

Fig. 20.4 The state space of the RWF-net shown in Fig. 20.3

First, we construct the state space of the RWF-net, which contains 34 states and is shown (in a compact way) in Fig. 20.4. For sake of completeness, we mention that we have used acronyms instead of long task names, and that the dotted (shadow) states can be reached by executing the *Start Carrier Timeout* (*Start CT*) transition. The reason why some of the shadow states are filled will become clear further on; please ignore these fills for the time being. Second, we check the three requirements for soundness on this state space.

- *Option to complete:* It is always possible to mark the sink place o, hence it is always possible to complete a case
- *Proper completion:* Several states exist that exceed the state $[o]$. Examples are $[p1, o]$ and $[o, o]$. As a result, completion may be improper
- *No dead tasks:* For any transition, an edge exists in the state space. Hence, no task is dead

We conclude that the simplified *Carrier Appointment* decomposition is not sound. It is straightforward to check that this erroneous behavior is also present in the original *Carrier Appointment* decomposition, hence the entire Order Fulfillment model is not sound.

This error is caused by the fact that the *Carrier Timeout* task does not cancel the *Prepare Transportation Quote* task nor its immediately preceding (implicit) condition. As a result, the latter task may be enabled or in progress, which cannot be canceled by the former.

As mentioned before, perhaps this error is due to some case attribute. Perhaps some attribute exists that prevents the completion of the *Carrier Timeout* task as soon as the *Prepare Transportation Quote* task has been enabled. As we take only the control-flow of the YAWL net into account for verification, we abstracted from all such attributes. Hence, it might be interesting to check whether the decomposition is relaxed sound.

To check relaxed soundness, we restrict the state space to that part that is covered by proper completing execution paths. In other words, we strip all states from the state space from which proper completion is not possible anymore. In Fig. 20.4, all filled states are to be stripped. For relaxed soundness, we now need to check whether all tasks are covered by this state space, that is, do all tasks occur on some edge in this state space? It is straightforward to check that this is indeed the case. Hence, the decomposition is relaxed sound, and provided that all other decompositions in the entire model are also relaxed sound, the entire model is also relaxed sound.

Until now, we have a kind of overlooked the possible problem that a YAWL task is not a transition and vice versa. In the example above, every task was captured by a single *Start* transition and a single *Complete* transition, but this need not be the case. A second fragment from the *Carrier Appointment* decomposition clarifies this. Figure 20.5 shows this fragment, which includes the *Create Bill of Lading* task that has an OR-join semantics. When converting such an OR-join to a reset net, there are no alternatives but to introduce a transition for every possible input combination, which is shown by Fig. 20.6. However, from the seven possible input combinations, only two are covered by proper completing execution paths (the other five have fills in Fig. 20.6).

Now, for verification it is important to note that all inputs are covered by the relaxed sound transitions. As a result, all inputs to this task are viable. If some inputs (outputs) of some tasks are not covered by relaxed sound transitions, then a warning can be issued that some incoming (outgoing) edge of the corresponding task cannot be successfully traversed; if such an edge would be traversed, proper completion would be impossible. When taking relaxed soundness into account, we can disregard possible input combinations that cannot lead to proper completion, which enhances the effectiveness (more warnings can be issued).

In a similar way, we can use relaxed soundness and the state space to detect whether some tasks or conditions can be removed from a certain cancelation region. If a certain task cannot cancel a given task or condition, or if canceling this given task or condition prohibits proper completion, then a warning can be issued that the cancelation region can be simplified by removing from it the given task or condition. However, these warnings can also be achieved by using the coverability property:

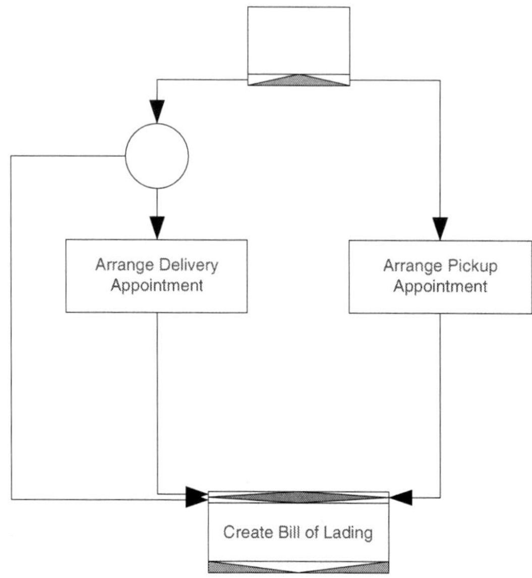

Fig. 20.5 OR-join construct in the Carrier Appointment decomposition

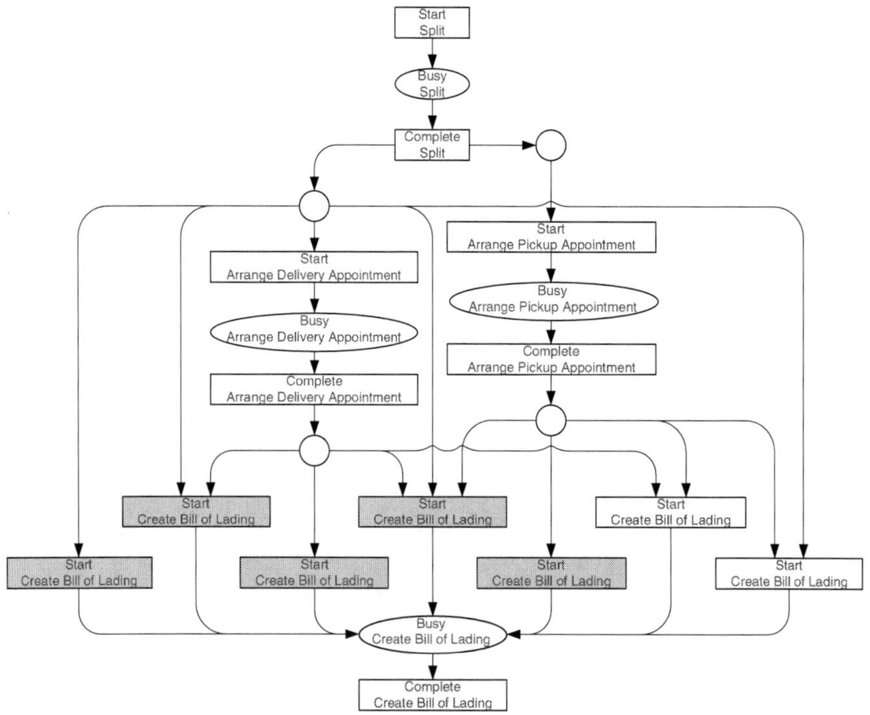

Fig. 20.6 RWF-net for OR-join construct

- A condition can be removed from a task's cancelation region if the state where
  - the condition is satisfied and
  - the task is enabled

  cannot be covered from the initial state.
- A task can be removed from another task's cancelation region if the state where both tasks are enabled is not coverable from the initial state.

Thus, in contrast to the OR-join (OR-split) warnings, the cancelation region warning can be complete.

Although for relaxed soundness it is possible to model an OR-join by a transition for every possible input combination, for soundness this is, in general, not possible. Take, for example, the RWF-net fragment as shown by Fig. 20.6. Clearly, the two places in the *Delivery* branch are mutually exclusive. However, the two top filled transitions require both places to be marked simultaneously. Hence, these transitions are dead, which violates the soundness property. Furthermore, both the *Delivery* and the *Pickup* branch will contain a token. However, the three bottom-filled transitions only remove one of these tokens, which violates the completion from being proper.

Because of the non-local semantics of an OR-join, a YAWL net with one or more OR-joins cannot be converted to an RWF-net *without some approximation*. As a result, we perform the soundness analysis for YAWL nets with OR-joins directly on the YAWL level and not on the RWF-net level. That is, a YAWL net is *not* converted to an RWF-net before performing the state space analysis to determine the soundness property. If the state space of a given YAWL net is finite, then it is possible to decide whether a particular YAWL net is sound.

If the soundness property of a YAWL net with OR-join cannot be determined, we can also attempt to use coverability analysis on the RWF-net level to decide whether the net is weak sound. In this case, we transform all OR-joins in the YAWL net into XOR-joins first, and then translate this resulting YAWL net into an RWF-net. Because of this potential semantic loss, it is only possible to determine whether a YAWL net with OR-joins is weak sound in some cases.

Similarly, because of the expensive nature of OR-joins, we propose to provide users with warnings when unnecessary OR-joins are present in the net. This is made possible by looking at the entire state space of the YAWL net with OR-joins, if available. In case we have an OR-join task that is enabled only if all inputs are available, then a warning can be issued that this task can be an AND-join, and if we have an OR-join task that is enabled only if exactly one input is available, then a warning can be issued that this task can be an XOR-join.

20.4 Soundness-Preserving Reduction Rules

The success of our verification approach hinges on the ability to generate a state space. If we can construct a state space, then we can decide soundness, and we can provide a complete diagnosis report. Otherwise, we have to revert to using either relaxed soundness or weak soundness (or structural invariant properties), in which cases the results might be incomplete. Although it is known that construction of a state space is infeasible for some YAWL nets, we want to improve our chances of success in case it is feasible. This means that we try to avoid a state space explosion as much as possible. A popular way to do so is to reduce the size of the model beforehand, that is before generating its state space.

Of course, if we reduce the model, we change it in some way. This makes diagnosing the possible errors in the model a very complex task, as we need to be aware of this change and we need to be able to transfer errors in the reduced model to errors in the original model. A fine point to note is that the reduction should be such that any errors present in the original model are also present in the reduced model: The reduction should preserve the sought-after qualities, which in our case is the soundness property.

Reductions can be applied on either the YAWL net or on the underlying RWF-net. For the latter, it is important to note that (ideally) there should be no OR-joins present in the YAWL net, as these do not transfer well onto reset net constructs. Therefore, we first assume that no OR-joins are present, and introduce a number of reset net-based reduction rules. Later on, we will transfer these reduction rules to the level of the YAWL net and allow for OR-joins as well.

20.4.1 Reduction Rules for Reset Nets

We propose a number of reduction rules for reset nets to keep the size of the corresponding state space at bay. These rules are extensions of a number of well-known reduction rules for Petri nets. As the presence of reset arcs could invalidate these rules, we extend the rules with additional requirements on these reset arcs.

For the rules proposed, we do assume a YAWL context. As a result, only the place that corresponds to the input condition (also known as the source place) is marked initially. The other places are empty, which allows us to abstract from the initial marking. As a side-effect, some rules are by definition not applicable. As an example, we mention the *elimination of self-loop place* rule. This rule allows to reduce some place, provided that it is marked initially and provided that it satisfies some additional requirements. As only the source place is marked initially, and as this place does not satisfy these additional requirements, there is no reason to extend this rule.

Figure 20.7 wraps up the remainder of this section on reduction rules for reset nets by giving a visual representation for all rules. Every rule potentially removes (or replaces by an arc) the filled places and transitions.

20.4.1.1 Fusion of Series Places (FSP)

Recall that our aim is to reduce the size of the state space, that is, the number of reachable states. Typically, the state explosion problem is caused by a number of

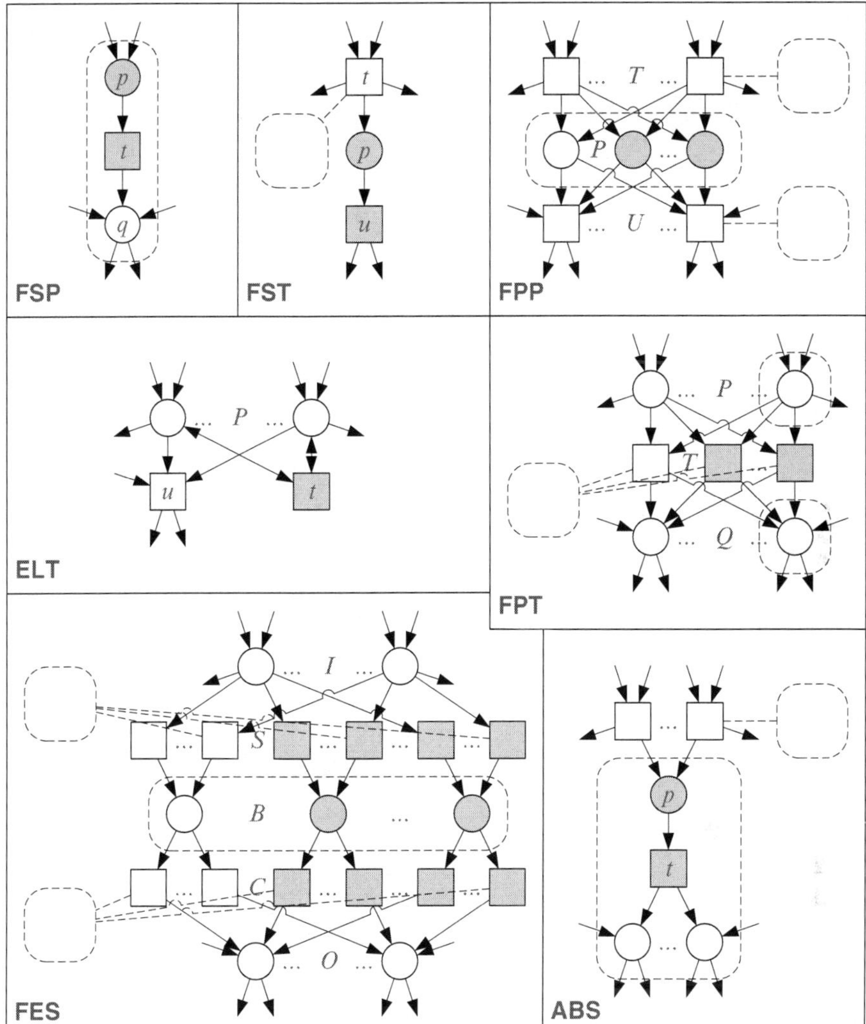

Fig. 20.7 A visual representation of all reduction rules for reset nets

parallel paths, which each contains a number of places. If these paths are truly parallel, then the number of reachable states would equal the product of the number of places of each path. Therefore, if we would be able to reduce the number of places on such paths, and this could have a profound effect on the state space size.

The first rule, called the *fusion of series places (FSP)* rule, attempts to reduce two places, say p and q, which are connected by a single transition, say t. Provided that:

- p is the only input and q the only output of t and
- t is the only output of p,

then we can either:

- Transfer all input arcs from p to q (we have to introduce arc weights here if such an arc to q already exists) and
- Remove both p and t together with their (incoming or outgoing) arcs

or

- Transfer all input and output arcs from q to p (we have to introduce arc weights here if such an arc from p already exists) and
- Remove both q and t together with their (incoming or outgoing) arcs.

The question now is, what requirements do we need to pose on reset arcs such that the rule is still applicable?

Transition t should not reset any place

As t has only p as input and only q as output, the remainder of the net cannot influence the moment t is being executed. In other words, transition t is some kind of loose cannon. However, the moment t executes should not matter, as after it has been removed it cannot matter anymore. Therefore, we cannot allow t to reset any place, *unless* we can guarantee that the effect of this reset is always identical. In general, this requires a state space search, which is not a good idea as we are still trying to reduce the net prior to constructing a state space.

Places p and q should be reset by the same set of transitions

To see this, it helps to consider the tokens in place p to be present in some way in place q as well. After all, any token from p can be transferred to q by executing transition t at any moment in time. Therefore, any transition that resets p should reset q as well, as in the mean time the token it was about to reset may have been transferred by t to q.

Together, these requirements are sufficient to ensure that the result of this rule preserves the soundness property. Figure 20.8 shows the result of applying this rule to the reset net introduced earlier: filled transitions and places have been removed by the rule. The state space of the resulting net contains only eight (was: 34) states (three states if transition *Complete Carrier Timeout* is executed in an improper way and five if it is not executed or executed in a proper way).

20.4.1.2 Fusion of Series Transitions (FST)

A second rule that allows for removal of places is the *fusion of series transitions (FST)* rule. While the previous rule looked for a series containing two places and one transition, this rule looks for a series containing two transitions and one place. Furthermore, while the previous rule allowed the second place to have additional input, this rule allows for the first transition to have additional output.

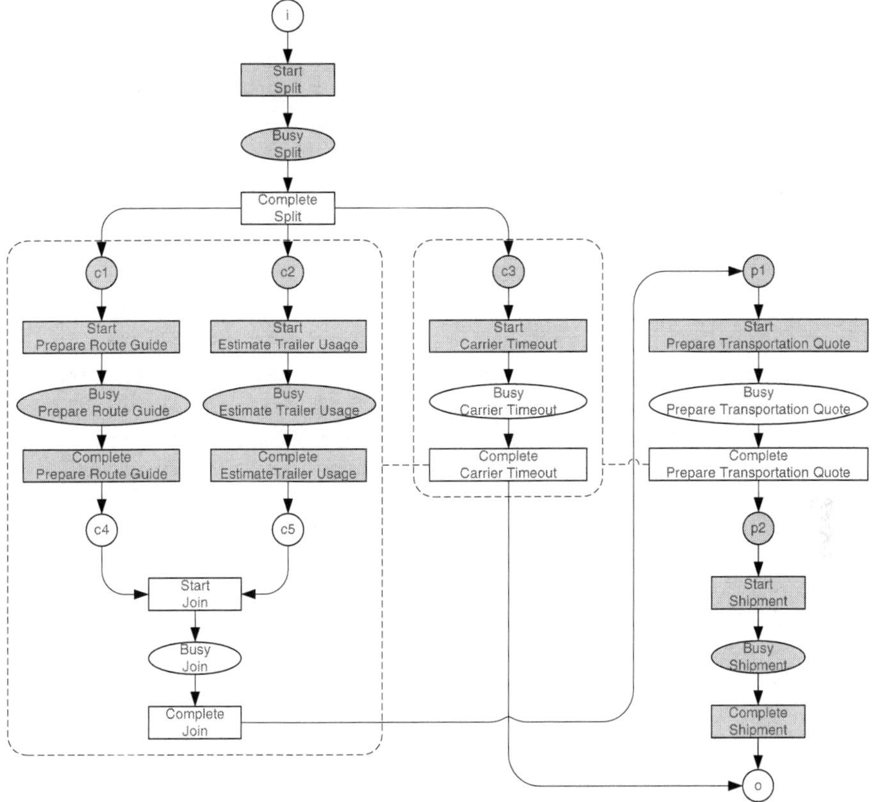

Fig. 20.8 The result of applying the *fusion of series places* rule to the RWF-net of Fig. 20.3

Given two transitions, say t and u, and one place, say p, this rule can remove a place and a transition. Provided that:

- t is the only input and u the only output of p and
- p is the only input of u,

then we can either:

- Transfer all output arcs from u to t and
- Remove both p and u together with their incident arcs

or

- Transfer all inputs and outputs from t to u and
- Remove both p and t together with their incident arcs

Again, the question is what additional requirements we should pose for possible reset arcs.

Place p is reset by the same set of transitions as any output of transition u is

Analogous to the previous rule, we can argue that the execution of transition u cannot be influenced. Hence, the effect of the execution of any transition should not be influenced by an intermediate execution of u. All outputs of u should agree on the set of transitions that reset them, and that p should be reset by the same set of transitions as well.

Transition u does not reset

If u resets some place, then any transition that requires that place as input is influenced by the execution of u, which is undesirable. As with the previous rule, we could construct a state space to determine whether u actually resets a place (resetting an empty place is excluded from this), in which case the reset arc could be allowed, but that constructing a state space at this point defeats the purpose of the rule.

Together, these requirements are sufficient to ensure that this rule preserves soundness.

20.4.1.3 Fusion of Parallel Places (FPP)

The two previous rules reduce the size of the state space by removing places from the net prior to constructing the state space. The *fusion of parallel places (FPP)* rule also removes a place, but this does not lead to a direct reduction in the size of the state space. However, after this rule has been applied, one of the other rules could then be applied, which might reduce the state space size. As such, the aim of this rule is not to reduce the state space per se, but to enable other rules to do so.

Given a set of places, say P, and two sets of transitions, say T and U, this rule can remove a number of places. Provided that:

- Every place from P has the first set of transitions (T) as inputs and
- Every place from P has the second set of transition (U) has outputs,

then we can remove all-but-one places from P.

This rule has only one additional requirement for reset arcs.

All places in P are being reset by the same set of transitions

If none of the places are reset places, then it is obvious that the rule holds. Assume that one of the places from P can be reset by some transition. The only way to guarantee that soundness is preserved by the reduction is to have the remaining place being reset by the same transition as well, which, in turn, requires the other places from P to be reset by this transition.

As a result of this rule, place *p5* can be filled as well in Fig. 20.8, which allows either the *fusion of series places* rule or the *fusion of series transitions* rule to remove both *c4* and *Start Join*. Note that *Busy Join* cannot be removed by either of these

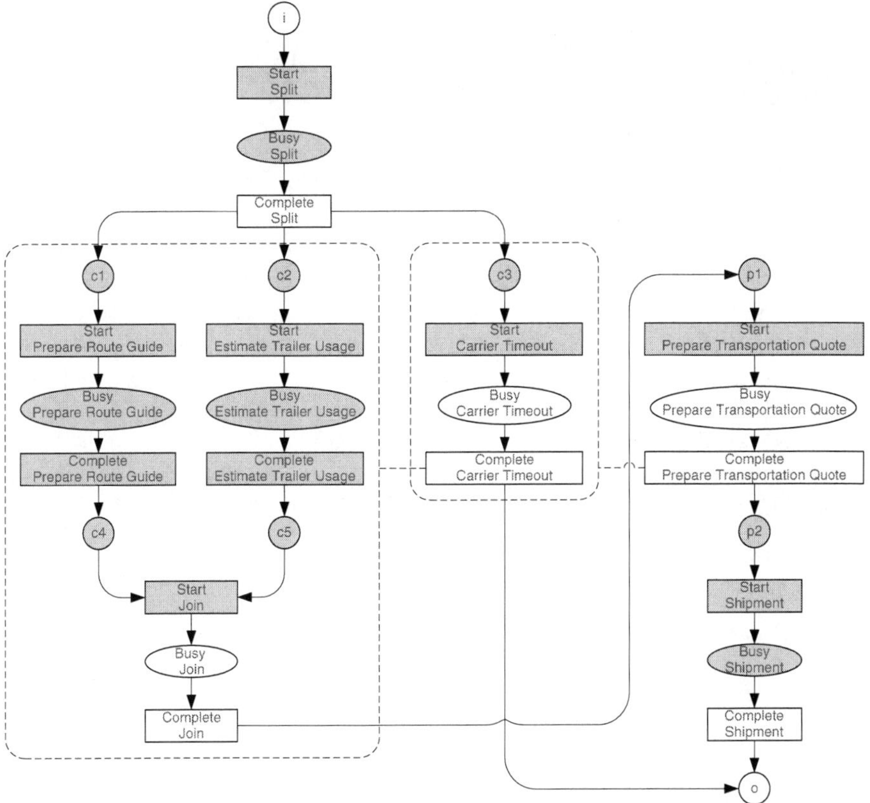

Fig. 20.9 The result of applying the three rules known so far to the RWF-net of Fig. 20.3

rules, as it is being reset by *Complete Carrier Timeout*, while place *Busy Prepare Transportation Quote* is not. Figure 20.9 shows the result.

20.4.1.4 Fusion of Parallel Transitions (FPT)

Similar to the previous rule, the *fusion of parallel transitions (FPT)* rule aims to enable any other rule, but while the previous rule attempts to do so by removing parallel places, this rule attempts to do so by removing parallel transitions.

Given a set of transitions, say T, and two sets of places, say P and Q, this rule can remove all-but-one transitions from T. Provided that:

- Every transition from T has the first set of places (P) as inputs, and
- Every transition from T has the second set of places (Q) as outputs,

then we can remove all-but-one transitions from T.

All transitions in T reset the same set of places

The argument for this is similar to the argument for the *fusion of parallel places* rule: all transitions should have an identical effect.

20.4.1.5 Eliminate Self-Loop Transitions (ELT)

A third rule that aims to enable other rules is the rule to remove an elementary cycle. If executing a transition does not change the current state, then we can ignore (or remove) that transition, provided a state exists that enables the transition. If the transition is never enabled, then removing it is not allowed as this might not preserve the third soundness requirement. Therefore, we restrict ourselves to transitions for which it is known that another transition is enabled as soon as it is enabled, that is, its inputs should be a subset of that other transition.

Given two transitions, say t and u, and a set of places, say P, this rule can remove the transition t. Provided that:

- Transition t has the set of places P as inputs and as outputs, and
- Transition u has at least the set of places P as inputs,

then we can remove transition t.

Transition t resets no place

Transition t may only reset empty places, as the effect of executing t should be identical to the effect of not executing it. As mentioned before, constructing a state space to determine whether t can only reset empty places defeats the purpose of the rule. Therefore, we require that t does not reset any place.

20.4.1.6 Abstraction (ABS)

The next rule is based on the *abstraction* rule from Desel and Esparza. As for the first two rules, this rule allows for the removal of a place and a transition. As such, it aims to reduce the size of the state space in a direct way.

Given a transition, say t, and a place, say p, this rule can remove both t and p. Provided that:

- t is the only output of p and
- p is the only input of t,

then we can:

- Add arcs from any input of p to any output of t and
- Remove both t and p.

As for both *fusion of series* rules, we may have to introduce arc weights if arcs already exist from some inputs of p to some outputs of t. Also, because place p

may have multiple inputs, the *fusion of series transitions* rule may not be enabled, and that because transition t may have multiple outputs, the *fusion of series places* rule may not be enabled.

The additional requirements of this rule match the requirements of both *fusion of series* rules and for similar reasons.

Place p is reset by the same set of transitions as any output of transition t is

Transition t does not reset any place

20.4.1.7 Fusion of Equivalent Subnets (FES)

The last rule for reducing reset nets is a rule specifically tailored towards the reduction of alternative XOR-join-XOR-split tasks in YAWL nets. Recall that in a reset net a YAWL task is captured by a *busy* place, a set of *Start* transitions, and a set of *Complete* transitions. Now assume that we have several tasks that meet the following requirements:

- The inputs of all tasks are identical
- The outputs of all tasks are identical
- The cancelation regions of all tasks are identical
- All tasks have XOR-join and XOR-split decorators

then, we should be able to remove all-but-one of these tasks. Unfortunately, none of the rules presented so far is enabled on the resulting reset net fragment. Therefore, we include a specific rule for this construct. This rule *allows to remove* the subnets (or fragments) corresponding to all-but-one of the corresponding YAWL tasks. As this rule is directly aimed at reset nets, we present the additional requirements for reset arcs right away.

Given a set of *Busy* places B, a set of *input* places I, a set of *output* places O, a set of *Start* transitions S, and a set of *Complete* transitions C, and provided that:

- For every combination of an input place and a *Busy* place, there exists a *Start* transition such that the input place is its only input and the *Busy place* is its only output
- For every combination of a *Busy* place and an output place, there exists a *Complete* transition such that the *Busy* place is its only input and the output place is its only output
- All transitions are covered by the previous two items
- All transitions sharing the same input place as input (i.e., which correspond to the XOR-join of the same YAWL task) share the same set of reset places
- All transitions sharing the same output place as output (i.e., which correspond to the XOR-split of the same YAWL task) share the same set of reset places
- All busy places are reset by the same set of transitions,

then we can remove:

- All-but-one *Busy* places and
- All *Start* and *Complete* transitions not connected to the remaining *Busy* place.

20.4.2 Reduction Rules for YAWL Nets

On the basis of the reduction rules for reset nets, we can construct a number of reduction rules for YAWL nets, as long as no OR-joins are present. As we discussed the rules for reset nets in some length, we only briefly discuss the rules for YAWL nets. Figure 20.10 visualizes these rules:

- The *fusion of series conditions (FSC)* rule *allows to remove* a condition and a task, provided that the task does not reset and the place is reset by the same set of transitions as the output condition of the transition is.
- The *fusion of series tasks (FST)* rule *allows to remove* a condition and a task, provided that the task does not reset and the place is not being reset.
- The *fusion of parallel tasks (FPT)* rule *allows to remove* all-but-one parallel AND-join–AND-split tasks, while the *fusion of alternative tasks (FAT)* rule *allows to remove* alternative XOR-join–XOR-split tasks.
- The *fusion of parallel conditions (FPC)* rule *allows to remove* all-but-one parallel conditions, provided that all input tasks are XOR-splits and all input transitions are XOR-joins, while the *fusion of alternative conditions (FAC)* rule *allows to remove* all-but-one alternative conditions, provided that all input tasks are XOR-splits and all input transitions are XOR-joins.
- The *elimination of self-loop tasks (EST)* rule *allows to remove* an AND-join–AND-split task for which the set of input conditions is identical to the set of output conditions, provided that some other AND-join task contains at least the set of output conditions.
- The *fusion of AND-join–AND-split tasks (FAAT)* rule *allows to merge* an AND-split task and an AND-join task, provided that the output conditions of the former match the input conditions of the latter, while the *fusion of XOR-join-XOR-split tasks (FXXT)* rule *allows this* for an XOR-split task and a XOR-join task.

In some rules, some join-split decorators are kept, while in others the decorators are removed as well. The *fusion of series tasks* rule is an example of the former, while the *fusion of AND-join–AND-split task* is an example of the latter. The latter is visualized by filling the join-split constructs, while for the former these constructs are left as they were.

Also, some tasks have multiple inputs (outputs) while they do not have a join (split) decorator. This should be read as that for such a task the join (split) decorator is of no importance, it can either be an AND decorator, an XOR decorator, an OR decorator, or simply no decorator at all.

Apart from these rules, there are also two YAWL reduction rules that do take the OR-join into account. As one of these rules is a specialization of the other, we start with the more general rule: the *fusion of incoming edges to an OR-join (FOR1)* rule. This rule is quite similar to the *fusion of parallel conditions* rule and the *fusion of alternative conditions* rule, but

- Requires only one split task and one join task, while the other two rules allow for multiple splits and joins, and

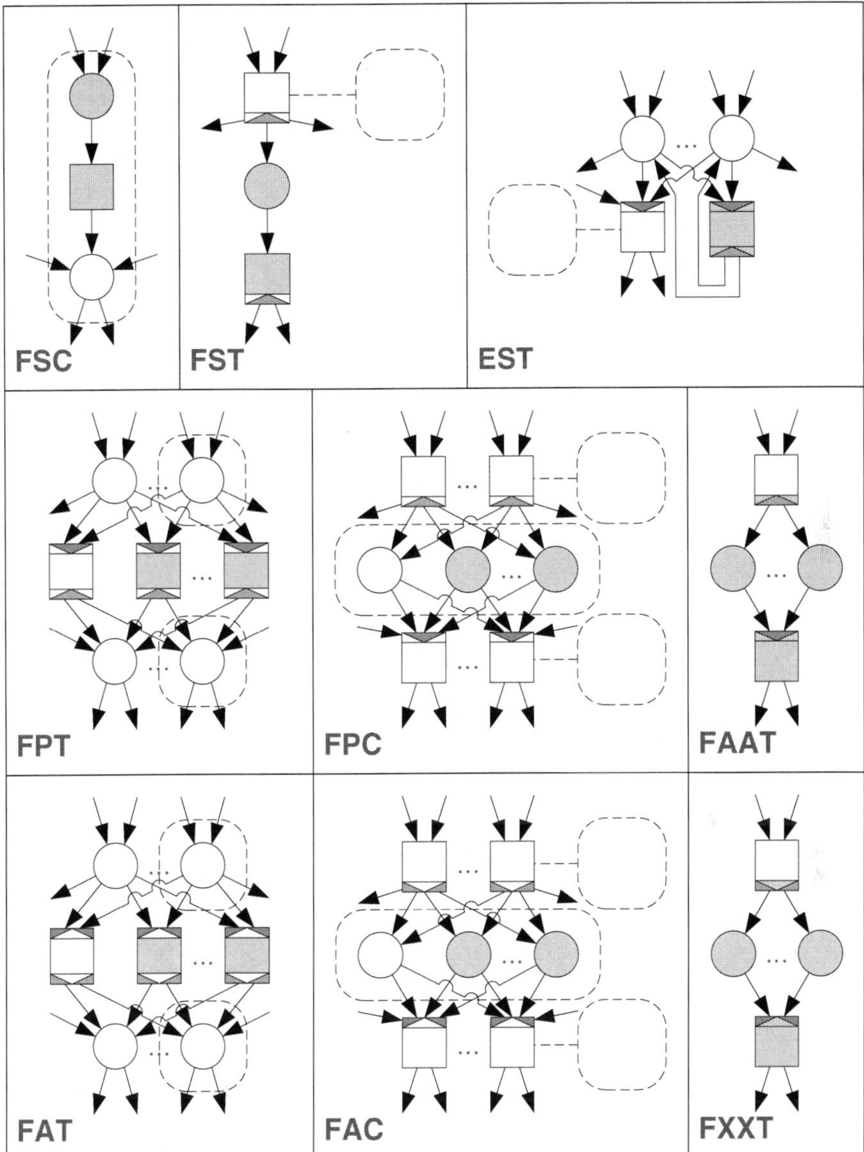

Fig. 20.10 A visual representation of all reduction rules for YAWL without OR-joins

- Allows for the join construct to be an OR-join, while the *parallel* rule requires an AND-join and the *alternative* rule an XOR-join.

Given a set of conditions, which contains at least two conditions, an OR-join task, and an arbitrary split task, this rule can remove all-but-one of the conditions. Provided that:

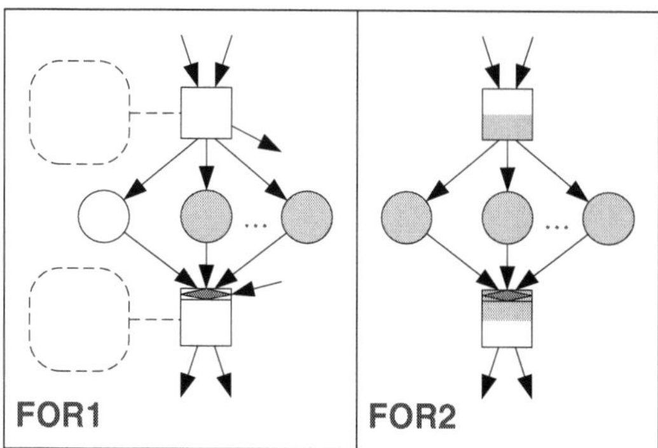

Fig. 20.11 A visual representation of both OR-join rules

- The conditions are being canceled by the same set of tasks
- The OR-join task is the output of every condition from the set
- The other task is the input of every condition from the set
- The OR-join task is not on some cycle with any other OR-join task,

then we can remove all-but-one of these conditions. The OR-join task is allowed to have additional inputs, while the other task is allowed to have additional outputs.

The OR-join is indifferent to the number of conditions in-between any preceding task and itself, as long as there is at least one. Therefore, we can simply remove all-but-one of them.

The second rule is a specialization of this rule, as it:

- Forbids the OR-join task to have additional inputs,
- Forbids the other task to have additional outputs,
- Forbids both tasks to have a (nonempty) cancelation region, and
- Forbids the OR-join to be on some cycle with some other OR-join task.

Using the *fusion of incoming edges to an OR-join (FOR2)* rule, we can now first remove all-but-one conditions in-between both tasks, which leaves only one input place for the OR-join. As a result, this OR-join decorator can be removed, and the *fusion of series tasks* rule can then be applied to reduce the condition and one of the tasks. However, because of the middle step (removing an OR-join decorator), this rule is not just a combination of both other rules. Figure 20.11 visualizes both rules.

After having applied all these soundness-preserving reduction rules over and over again, the resulting state space could be considerably smaller than the original state space. This could have a positive effect on the verification of the soundness properties, which all are based on this state space. Nevertheless, there is no guarantee that the resulting state-space is even finite in size, let alone be small enough to make the verification feasible. In such situations, we can use structural transition invariants that can be used to *approximate* the relaxed soundness property and diagnose the YAWL net based on this approximation.

20.5 Structural Invariant Properties

Recall that relaxed soundness uses the proper completing execution paths in a YAWL net, that is, those execution paths that start with a case in the input condition and end with the case in the output condition. As an example, take the YAWL net shown by Fig. 20.12, and ignore the dotted task t for now. Possible execution paths in this net include the following:

- $n\ c\ s$
- $n\ c\ e\ ce\ s$
- $nw\ cw\ w\ cw\ sw$

Now, if we would add the additional task t that transfers a case from the output condition to the input condition, then all these proper execution paths correspond to cyclic execution paths that include this additional task. As a result, all relaxed sound tasks would be covered by cyclic execution paths.

It is a known fact that in a Petri net any cyclic execution path corresponds to a semi-positive transition invariant. Basically, a transition invariant is a weighted sum of the transitions such that the current state would not change if all these transitions would be executed simultaneously according to their weight (a negative weight for some transition means that it is executed in the backward direction: it produces tokens for its input places and consumes tokens from its output places),

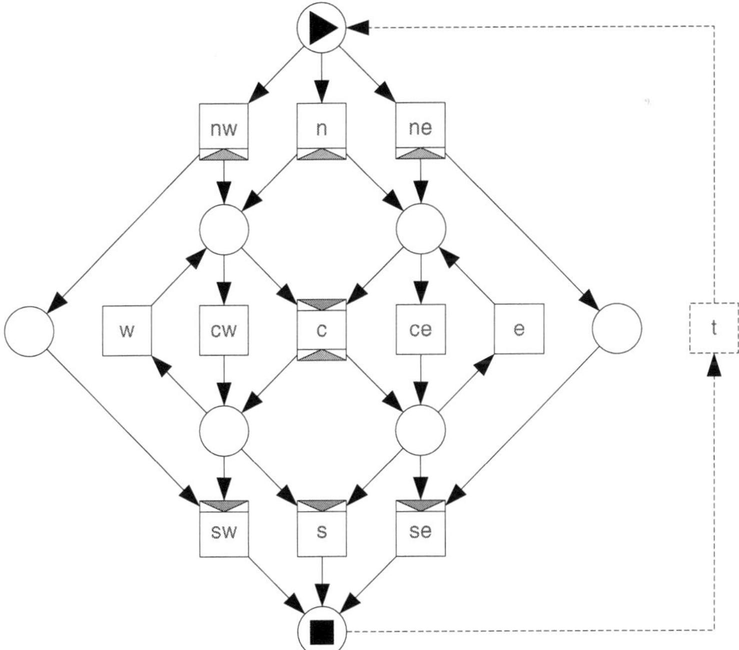

Fig. 20.12 The semi-positive transition invariant $c + e + nw + sw + t$ is not realizable

while a semi-positive transition invariant is a transition invariant where none of the weights are negative. As any cyclic execution paths lead back to the same state, the weighted set of executed transitions has to be a semi-positive transition invariant. In the example as shown by Fig. 20.12, possible invariants include the following:

- $c + n + s + t$
- $ce + e$
- $c + e + nw + sw + t$
- $c - ce - cw$

Note that only the first three invariants are semi-positive invariants, and that only the first and third include the additional task t.

So far, we have obtained the following:

1. Any relaxed sound transition is on some proper execution path.
2. Any proper execution path corresponds to a cyclic execution path after having added task t.
3. Any corresponding cyclic execution path corresponds to a semi-positive transition invariant that covers t.

Combining these three results, we obtain the fact that any relaxed sound transition is covered by some semi-positive transition invariant that covers t. Thus, if some transition is not covered by these invariants, then it cannot be relaxed sound.

Unfortunately, as Fig. 20.12 shows, this does not work the other way: examples exist for which some invariants do not correspond to some proper completion path. Take, for example, the third invariant mentioned above: $c+e+nw+sw+t$. Although this invariant satisfies the requirements (it is semi-positive and it covers t), it does not correspond to a proper execution path as the path blocks after having executed nw. The root cause of this is that not every invariant may be *realizable* in the model. For an invariant to be realizable, we need to be able to order the transitions in such a way that the transitions could be executed by the model in that order. For some invariants, this may not be possible. As a result, we may not be able to locate all relaxed transitions by this technique, as some not-relaxed-sound transitions may be covered by invariants that are not realizable.

Transition invariants are structural properties, which means that they can be decided using only the structure (and not the behavior, i.e., the state space) of the net. Thus, if constructing the state space is no option, we could still construct transition invariants and use these invariants for diagnostic purposes.

20.6 Tools

The techniques mentioned in this chapter have been implemented in two tools: The YAWL Editor and WofYAWL. While the Editor is part of the entire YAWL suite, the WofYAWL tool is a stand-alone tool that works on YAWL Engine files.

The Editor contains algorithms to construct a state space, if possible. To help in constructing a state space, it can reduce the YAWL net as much as possible while preserving the soundness property. As mentioned before, if a YAWL net contains OR-joins, the YAWL semantics are used to construct a state space. Otherwise, a YAWL net is converted to an RWF-net and the reset net semantics are used to construct the state space. Given that a state space can be constructed, the Editor can decide soundness and decide which OR-joins can be simplified. If the construction of the state space is not feasible, the Editor can still decide weak soundness for YAWL nets without OR-joins and for some YAWL nets with OR-joins and decide which tasks and conditions could be removed from certain cancelation regions.

Figure 20.13 shows the verification result on the unsound *Carrier Appointment* decomposition described earlier. The first warning states that this example may complete while the task *Prepare Transportation Quote* and its immediately preceding implicit condition may still be enabled. For sake of readability, we include the first four messages on this decomposition here:

1. ResetNet Analysis Warning: Tokens could be left in the following condition(s) when the net has completed: [[Atomic Task:Prepare_Transportation_Quote_4, Condition:c{null_41_Prepare_Transportation_Quote_4}]]
2. ResetNet Analysis Warning: The net Carrier_Appointment does not satisfy the weak soundness property.
3. ResetNet Analysis Warning: The net Carrier_Appointment does not have proper completion. A marking 2OutputCondition_2 larger than the final marking is reachable.
4. ResetNet Analysis Warning: The net Carrier_Appointment does not satisfy the soundness property.

As a result (see also the fourth warning), this model is not sound.

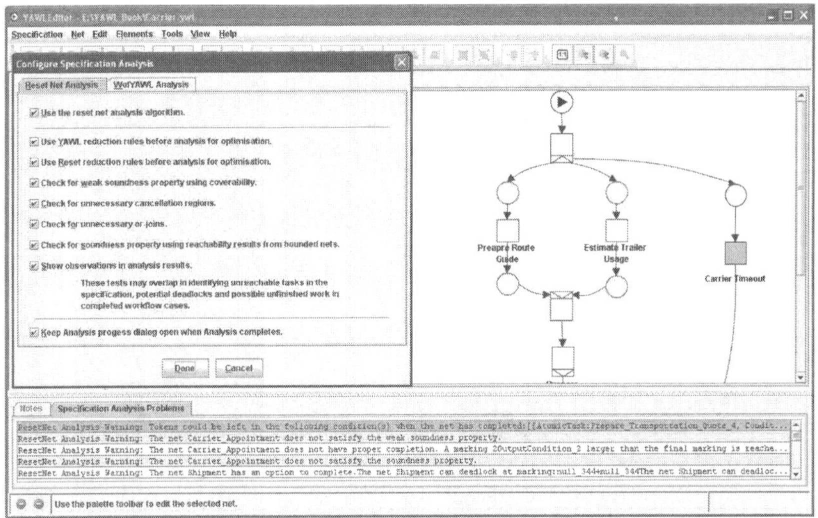

Fig. 20.13 The Carrier Appointment example in the YAWL editor

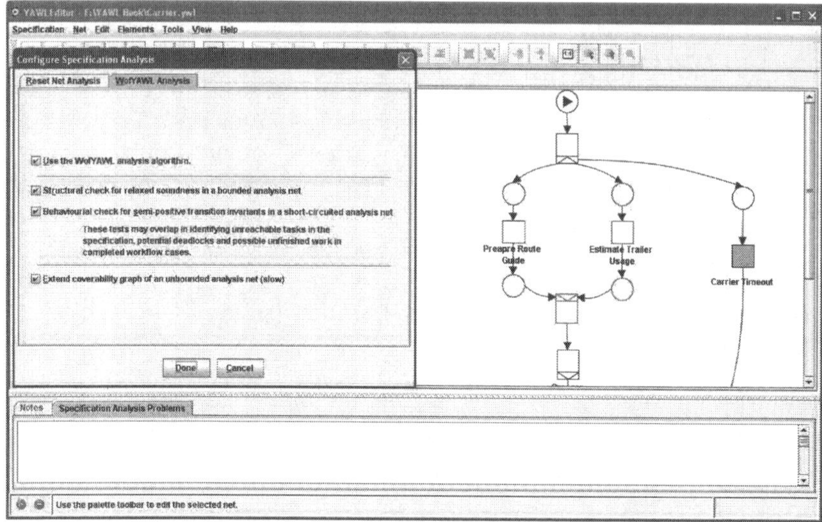

Fig. 20.14 The Carrier Appointment example in the YAWL editor, using WofYAWL

WofYAWL is a spin-off of the workflow verification tool Woflan that is especially tailored towards verification of YAWL nets. Based on a YAWL Engine file, it can also try to construct the entire state space, but, unlike the Editor, it uses only the reset net-based reduction rules when reducing the model before constructing the necessary state space. Given a state space, WofYAWL can decide relaxed soundness or use transition invariants to approximate relaxed soundness.

The Editor and WofYAWL have been loosely integrated in the sense that the Editor can start WofYAWL on the YAWL net at hand and can display WofYAWL's messages in the Editor pane. Figure 20.14 shows the results of WofYAWL's diagnosis of the unsound *Carrier Appointment* decomposition. As this decomposition is relaxed sound, WofYAWL reports that no problems were discovered.

20.7 Concluding Remarks

This chapter has introduced a number of desirable properties for YAWL nets, which include the following:

- *Soundness:* Can every case complete in a proper way and is every task viable?
- *Weak soundness:* Can a new case complete, is completion always proper, and is every task viable?
- *Relaxed soundness:* Can any task contribute to the proper completion of a case?

The weak soundness property is decidable, but the soundness property and the relaxed soundness property are not. This means that for certain example of YAWL nets, a definitive answer on whether the YAWL net is (relaxed) sound can not be given. Nevertheless, many YAWL nets may exist for which this answer can be given.

To help in the computation of these properties, two sets of reduction rules have been introduced. By first applying these reduction rules, the YAWL net becomes smaller, which might alleviate the problem of computing the property itself. The first set of reduction rules operates on the reset net that underlies a YAWL net, while the second set operates on the YAWL net itself. As the OR-join construct cannot be captured successfully by reset nets, the first set of rules can be used only in the absence of this construct.

In case the computation of the properties is too hard to compute even after the reductions have been applied, this chapter has shown how the relaxed soundness property can be approximated using only structural properties of the underlying reset net. These properties are structural as they do not require the current state, and hence do not require the entire state space to be constructed.

The YAWL Editor has been extended to enable a YAWL process designer to check the above mentioned properties for a YAWL net at hand. For the non-structural properties, the Editor provides native support, while it relies on the WofYAWL tool for the structural properties. Based on these techniques, the Editor can also check whether:

- Some OR-joins can be simplified to either an XOR-join or an AND-join
- Some tasks or conditions can be removed from some cancelation region

These advanced verification techniques help the developer of a YAWL net to correct possible errors in the YAWL net at hand, which is shown nicely by the *Carrier Appointment* decomposition. The original *Carrier Appointment* decomposition was developed by two YAWL experts. Nevertheless, it contained a flaw that allowed for improper completion, which shows the value of verification support for all developers.

Exercises

Exercise 1. Determine whether the *Carrier Appointment* decomposition (shown in Fig. 20.2) is sound, relaxed sound, and/or weak sound.

Exercise 2. Consider the OR-join construct shown by Fig. 20.5. Assume that the input condition directly precedes the anonymous AND-split task, and that the output condition directly succeeds the *Create Bill of Lading* task. Construct a state space (use the YAWL OR-join semantics) for this YAWL net, and determine whether it is sound, relaxed sound, and/or weak sound. Also determine whether the OR-join can be simplified to either an AND-join or a XOR-join.

Exercise 3. As Fig. 20.3 shows, the *fusion of series places* cannot reduce the *Busy Join* place and the *Complete Join* transition. Determine why not.

Exercise 4 (Advanced). The three soundness properties all abstract from the ordering between tasks. Show that the second option of the *fusion of series transitions* rule (where inputs and outputs are transferred from t to u) does not necessarily preserve the ordering between tasks.

Exercise 5. Determine which places and transition could be reduced from the net of Fig. 20.3 if it would only use the *fusion of series transitions* rule, and to what extent this differs from the result as shown by Fig. 20.8.

Exercise 6. Show that the rules as shown by Fig. 20.10 are correct using the rules as shown by Fig. 20.7. To do so, first convert a YAWL fragment onto a reset net fragment, apply reduction rules on the reset net level, and convert the resulting reset net fragment back to a YAWL net fragment.

Exercise 7 (Advanced). Show how transition invariants can be computed in the presence of reset arcs, which, by nature, disturb these invariants. Hint: transition invariants are trivial to compute if every reset arc could be replaced by an *inhibitor* arc. An inhibitor arc from a place to a transition prevents execution of the transition as long as the place is marked.

Chapter Notes

Petri Nets

In his overview paper on Petri nets, Murata [179] has provided a number of reduction rules that preserve both liveness and boundedness. In their book on free-choice Petri nets, Desel and Esparza [77] have given three different reduction rules that also preserve liveness and boundedness. In contrast with the rules provided by Murata, the Desel and Esparza rules are complete: any live and bounded nets can be reduced to a simple net containing only one place, one transition, and two arcs. However, the decision whether some rule is applicable is easy for the Murata rules, while it is complex for two of the three Desel and Esparza rules.

Workflow Nets

Workflow nets have been introduced by van der Aalst [2] as a subclass of Petri nets to capture the essential behavior of workflow processes.

Soundness

In the same paper, van der Aalst [2] has introduced the soundness property on workflow nets, which is considered as the minimal requirement any workflow process should satisfy. Based on this soundness property, the Woflan tool has been build by Verbeek and others [250]. The WofYAWL [249] tool is a spin-off of the Woflan tool. As this soundness property abstracts from data, which might influence the flow of control, the notion of relaxed soundness has been introduced by Dehnert [75]. A workflow net that is sound is also relaxed sound, but not vice versa. Both soundness and relaxed soundness are not decidable in general when taking reset nets into account, as reachability is undecidable [12, 13, 80, 272]. Therefore, Wynn [272, 274] has introduced the notion of weak soundness,[2] which is decidable even for reset nets.

Reset Nets

More information on reset nets can be found in [72, 80, 81, 95, 96]. The fact that coverability is decidable for reset nets has been shown by Dufourd and others [80], which was used by Wynn [272, 274] to show that weak soundness is decidable for reset nets. The soundness preserving reduction rules have been introduced by Wynn and others [251, 272, 277] and implemented in the YAWL Editor.

YAWL Nets

The OR-join semantics of YAWL nets has been discussed by Wynn and others [273]. The simplistic view on the OR-join semantics which is allowed by the relaxed soundness property has been introduced by Verbeek and others [249]. Finally, the set of YAWL net reduction rules has been introduced by Wynn and others [272, 276].

[2] The term "weak soundness" used here has a different meaning to the one introduced by Martens [158].

Part VIII
Case Studies

Chapter 21
YAWL4Healthcare

Ronny Mans, Wil van der Aalst, Nick Russell, Arnold Moleman, Piet Bakker, and Monique Jaspers

21.1 Introduction

Hospitals face increasing pressure to both improve the quality of the services delivered to patients and to reduce costs. As a consequence of the open-market approach adopted in healthcare provision, hospitals must compete with each other on the basis of performance indicators such as total treatment costs for a patient, waiting time before treatment can start, etc. Patients visiting the hospital will no longer accept long waiting times, and increasingly have specific demands with respect to the planning of their appointments and quality of services they will receive.

The aforementioned issues place significant demands on hospitals in regard to how the organization, execution, and monitoring of work processes is performed. Workflow Management Systems (WfMSs) offer a potential solution as they support processes by managing the flow of work, such that individual work items are done at the right time by the proper person. The main benefit of utilizing these kinds of systems is that processes can be executed faster and more efficiently. In addition, these processes can be closely monitored, potentially enhancing patient safety.

It is generally agreed that WfMSs are mature enough to support administrative processes that are relatively stable and fixed in form. However, hospital processes are typically *diverse*, require *flexibility*, and often involve *multiple medical departments* in diagnostic and treatment processes. Even for patients who have the same condition, the actual diagnostic and treatment process that they undergo may vary considerably. Furthermore, it is often necessary to change the course of the treatment process based on the results of tests performed and the way in which a patient responds to individual treatments.

These kinds of issues seriously hamper the introduction of WfMSs into the healthcare domain and require a complex range of requirements to be addressed. The different kinds of flexibility that are required during various stages of the care process further complicate the application of process technology. Current workflow

R. Mans (✉)
Eindhoven University of Technology, Eindhoven, the Netherlands,
e-mail: r.s.mans@tue.nl

A.H.M. ter Hofstede et al. (eds.), *Modern Business Process Automation*,
DOI 10.1007/978-3-642-03121-2_21,

systems fall short in their support for flexibility, a fact which is recognized in the literature.

Therefore, an interesting and challenging question is: *What are the considerations in regard to coping with process flexibility when applying workflow technology to the healthcare domain?* In other words, what kind of flexibility need to be delivered and under which conditions?

To give an answer to this question, the following approach is pursued. First, healthcare processes are examined. In the healthcare domain, there are many processes with varying characteristics. Consequently, these processes require different workflow system flexibility. For example, imagine what the process would look like for cataract surgery or what happens when a patient, having suffered a heart attack, is brought into the emergency department. Nowadays, cataract surgery is a highly standardized process in which it is completely clear what needs to be done from the start of the process until the end. In contrast to this, for the patient who is brought into the emergency department, conditions can change rapidly and it needs to be decided on the spot what step(s) will be done next. Clearly, these two cases differ in the requirements related to system flexibility.

Second, a complex real-life healthcare process at the Academic Medical Center (AMC) hospital in the Netherlands is considered. The AMC is a large university hospital affiliated with the University of Amsterdam (UvA). The process under consideration is the diagnostic process of patients visiting the gynecological oncology outpatient clinic. This process involves several disciplines and is considered to be illustrative for other healthcare processes. To fully support the healthcare process, its process flexibility requirements are identified.

To investigate the extent of process flexibility that can be provided by current WfMS offerings, the care process has been implemented in YAWL and also in three other WfMSs (FLOWer, ADEPT1, and Declare), which demonstrate various kinds of flexibility. YAWL was chosen because it is a powerful open source system providing comprehensive support for the Workflow Patterns and provides execution flexibility through its Worklet capability. FLOWer is considered to be the most successful commercial system providing flexible process execution based on the so-called case-handling approach, which focuses on the data rather than on the control-flow aspects of process execution. ADEPT1 is a workflow system, developed at the University of Ulm, which provides flexibility by supporting dynamic change meaning that the process model for an individual case can be adapted at runtime. Finally, Declare is a workflow system, developed at Eindhoven University of Technology, which provides flexibility through its declarative process definition language, which specifies what should be done instead of how it should be done. Declare can act as a service of YAWL and is discussed in more detail in Chap. 12 (The Declare Service).

In this chapter, the focus is on the implementation in the YAWL workflow system. The main emphasis is on how process flexibility can be provided to support the process and its ongoing operation in an environment that is subject to various kinds of change. However, as the chapter is about an application of YAWL, other issues that need to be tackled when configuring YAWL to support the healthcare process are also considered, that is, the control-flow, data, and resource perspectives. Finally, at the end of this chapter, the YAWL implementation is compared with those in FLOWer, ADEPT1, and Declare.

Before elaborating on its implementation, healthcare processes and their characteristics are explored in Sect. 21.2. Then, the gynecological oncology healthcare process is introduced in Sect. 21.3 together with an examination of the kinds of flexibility that need to be provided by a workflow system. In Sect. 21.4, the implementation of the candidate healthcare process in YAWL is discussed. Finally, Sect. 21.5 concludes the chapter.

21.2 Healthcare Processes

In this section, the focus is on healthcare processes themselves. First, different care concepts that exist in healthcare are introduced to illustrate how care processes could be organized. Then, the different process characteristics of these care concepts are considered to distinguish care processes from each other. Based on these different process characteristics, different kinds of healthcare processes are derived.

In healthcare, many different (innovative) concepts can be found relating to the delivery of care. Some of these concepts focus on the patient and related needs while others focus on the care providers. In the following section, three widely used concepts are introduced, which help in illustrating how care can be provided to different patient groups and identify different process characteristics.

The first concept that is discussed is *managed care*. Managed care is defined as a process in which continuous improvement of the measurable results of the treatment of a specific disease is encouraged. The focus is on effectiveness, quality, price, and the extent of care that is to be delivered. To apply this concept, the care processes need to be highly structured. One of the hospitals using the managed care concept is the Shouldice hospital in Canada. Only one kind of surgery is performed in this hospital, namely abdominal hernia repair, which is a relatively low tech procedure. The entire surgical procedure is meticulously planned, keeping the costs as low as possible. On day one, the patient is examined; on day two, surgery takes place; and on day four, the patient is discharged. Where a patient does not exactly fit the profile for standard surgery, there are risks of complications and the patient is not admitted.

The second concept is *stepped care*. Using this concept, care is gradually offered. The goal of stepped care is on delivering the right intensity of care at the right moment and it is specifically focused on people with chronic diseases. Typically, the treatment of a patient suffering from a chronic disease takes years and consists mainly of medical tests or treatments performed at regular intervals. An example of stepped care is the treatment of rheumatoid arthritis patients. In the beginning, an intake and some medical tests are performed in order to diagnose the patient. Once the patient is diagnosed and found to suffer from rheumatoid arthritis, the process typically continues with regular sequences of lab tests and visits to the outpatient clinic. Results of tests and outcomes of patient visits are examined to decide on the next treatment phase.

The last concept that is discussed is *integrated care*. Integrated care is defined as the bringing together of inputs, delivery, management and organization of services related to diagnosis, treatment, care, rehabilitation, and health promotion. The focus of integrated care is to combine parts into one coherent entity. Different medical departments that are required for the treatment of specific patient groups are brought together. This allows for the multidisciplinary treatment of a patient. An example of integrated care is the treatment of a patient with breast cancer. The treatment of patients with this condition requires the collaboration of several medical departments, such as radiology, radiotherapy, internal medicine, surgery, and hematology. Typically, the process consists of an intensive period of care that is delivered to diagnose and treat a patient.

From the above discussion of the different care concepts, it becomes obvious that care can be delivered in many different ways. Indeed, many patients' care processes differ substantially. Even though patients are in the same patient group, the accompanying process may exhibit many different execution outcomes. Next, the main characteristics necessary to arrive at a good understanding of coexisting healthcare processes are considered. Figure 21.1 shows the main kinds of healthcare processes that are distinguished in this section and how they relate to each other. For each of the care concepts presented, the kind of healthcare process that they belong to is identified.

In healthcare, a distinction can be made between *medical treatment* patient-specific processes and *organizational* processes. Organizational processes consist of organizational tasks in which cooperation between people from different departments is a vital characteristic. An organizational process is repetitive but nontrivial. Examples of organizational tasks include the planning of medical tests, reports that need to be written and evaluated, and preparations preceding a medical test. In contrast, medical treatment processes can be seen as a diagnostic–therapeutic cycle. This cycle consists of patient observation, medical reasoning, and decision making. Decision making requires various actions: diagnostic tests that need to be performed and/or therapeutic treatments that need to be undertaken. Unlike medical treatment processes, organizational processes do not provide any support for medical decision making by means of medical guidelines or medical pathways. In the remainder of this chapter, only organizational processes are considered.

One of the most important characteristics that influences the course of a care process is called *predictability*. The predictability of a process can range from high

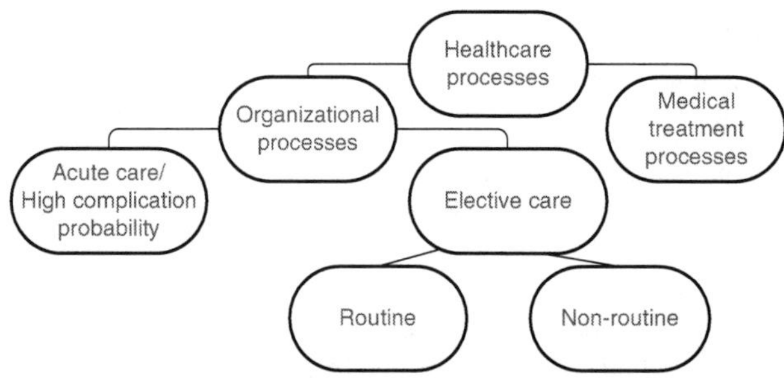

Fig. 21.1 Overview of the main kinds of healthcare processes

to low, indicating to what extent the course of an illness and the corresponding treatment can be predicted. The predictability is low for *acute care* processes, because when treating critically ill patients, conditions change rapidly. Furthermore, the predictability is also low for care processes that deal with patients for whom the probability of complications is high. These kinds of processes require dramatic changes in particular parts, once a complication occurs, which consequently impacts the overall process.

In contrast to acute care, there is *elective care* for which it is medically sound to postpone treatment for days or weeks. In principle, this type of care can be planned in advance, but there are also many possible variations here. Elective care can be distinguished in care processes that can be completely planned from beginning to end, care processes for which parts of the process can be planned (process fragments), and processes which need to be planned step by step.

Care processes that can be completely planned are called *routine processes*, which are typically defined for a comprehensive treatment of diseases for which an evidence knowledge base concerning the diagnostic course and outcome of therapy exists. For these diseases, a diagnostic-treatment path can be defined, outlining the different diagnostic and therapeutic procedures and their specific timing. In other words, routine processes are possible for well-defined complaints with almost 100% certainty about the process required and the outcome. For routine processes, a treatment path can be defined which defines the different steps in the process and their timing. It is vital that well understood knowledge exists about the course of the process. For routine care processes, the managed care concept can be utilized.

Contrary to routine processes, there are also *nonroutine processes* in which the clinician proceeds in a step-by-step way, checking the patient's reaction to a treatment and deciding about the steps to be taken next. Throughout the process, this can mean that some parts of the process can be planned (process fragments) while others cannot. The ability to plan or not plan a process may also depend on other factors besides existing evidence concerning the course of a disease. For example, it may depend on the patient's conditions. Second, it may also depend on the kind of complaints the patient is suffering from. Third, a clear understanding of the complaints from which the patient is suffering also influences whether the process can be planned.

Often, the decision about the next steps to be taken is not made by just one clinician. When complex care needs to be delivered, cooperation between various clinicians across different medical specialties and departments is needed to decide on and execute parts of the individual patient's care plan, which can be realized through an integrated care concept.

21.3 Gynecological Oncology

In this section, the selected healthcare process is discussed in more detail. The healthcare process under consideration is the diagnostic course of patients visiting the gynecological oncology outpatient clinic at the AMC. First, the healthcare process is introduced in Sect. 21.3.1. Then, in Sect. 21.3.2, the characteristics of this process are considered in more detail in order to identify which types of flexibility are needed to accommodate this process using workflow technology.

21.3.1 Introduction

The gynecological oncology healthcare process is an extensive process (comprising 230 activities) concerned with the process of diagnosing a patient once they are admitted to the AMC hospital for treatment. One of the main reasons to choose this particular process is that it involves many medical (sub)specialties, such as radiology, anesthetics, and pathology. The care process can be considered to be nontrivial and illustrative for other healthcare processes, both in the AMC and in other hospitals.

The healthcare process under consideration actually consists of two distinct processes. For the first of these, Fig. 21.2 shows the top page of the YAWL model. This process deals with all the steps that may be taken when a patient is referred to the AMC hospital for treatment up to the point where they are diagnosed. The second process deals with meetings, which are organized on a weekly basis, to discuss the medical status of current patients and to decide about their final diagnosis before the treatment phase is started. In this chapter, only the first process is discussed in detail.

The discussion of the respective process is commenced by elaborating on the top part of the process shown in Fig. 21.2. The process starts with the *referral patient and preparations for first visit* composite activity. This subprocess deals with the steps that need to be taken when a patient is referred to the AMC hospital and the preparations that are needed for the first visit of the patient to the outpatient clinic. The next step in the process is the *visit outpatient clinic* composite activity, where the patient visits the outpatient clinic for a consultation with a clinician. Such a consultation can also be done by telephone (*consultation by telephone*). During a visit or consultation, the patient discusses his/her medical status with the clinician and it is decided whether any diagnostic tests or treatments need to be (re)scheduled or canceled.

The actual tests that may be needed and the number of them can vary greatly depending on the conditions of the patient. So, for one patient, a Magnetic Resonance Imaging (MRI) may be sufficient, while for another patient, various tests may be needed, such as a Computed Tomography scan (CT-scan) or an electrocardiography (ECG). For each of these tests, different medical departments can be involved. The execution of these tests is modeled by the *examinations* composite task, which allows for the concurrent instantiation of a number of different tests for a patient. This node has been modeled in the YAWL model in Fig. 21.2. In this way, when a patient is visiting the outpatient clinic (shown in the top part of the figure), it is possible for a series of subprocesses to run concurrently for each test that has been requested. As the execution of individual tests associated with each of these

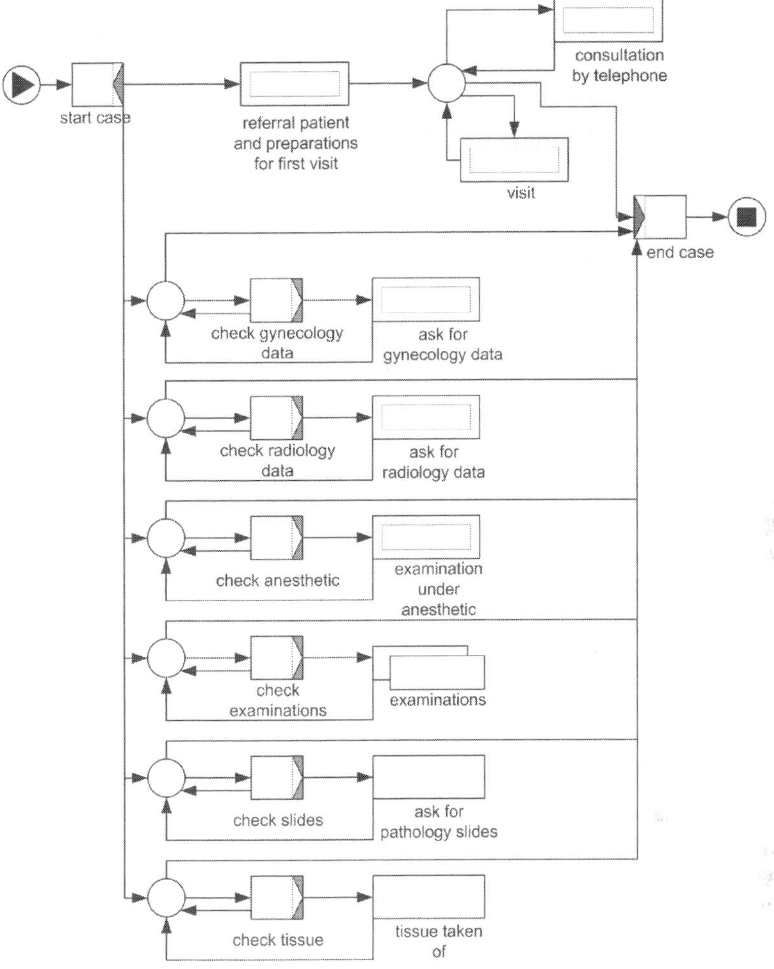

Fig. 21.2 General overview of the gynecological oncology healthcare process

subprocesses can be complex and time consuming, there is no guarantee that all of them will be finished before the start of a next patient consultation. Furthermore, it is also necessary to cater for the fact that new medical tests may become available in the future and that they might be requested by a clinician. Clearly, this imposes specific demands on process flexibility, which need to be managed properly in order to support the process.

However, for each patient there is also another series of subprocesses that may run concurrently. As for the medical tests, the completion of these tasks cannot be determined beforehand. For example, in the *ask for gynecology data* composite task, a telephone call is made to the referring hospital with the request to send a patient's gynecological data to the AMC hospital so that it can be processed. However, it cannot be determined beforehand when this process will be finished, hence the completion of this subprocess cannot be modeled as a requirement for the start of another activity, such as a visit to the gynecologist.

As the implementation of the healthcare process in the context of the healthcare scenario is demonstrated in Sect. 21.4, it is necessary to discuss the *referral patient and preparations for first visit* and *visit outpatient clinic* composite tasks in more detail. It is also worthwhile giving some more background on the actual operation of the healthcare process.

In Fig. 21.3, the subprocess underlying the *referral patient and preparations for first visit* is shown, which describes the initial stages of the healthcare process. Here, a clinician or nurse at the AMC is called by a clinician of a referring hospital and a decision is made about the timing of the first visit of the patient (*plan first visit*) and about which medical tests may need to be scheduled. The process also shows some administrative tasks that might need to be executed. For example, it is possible to make appointments for an "MRI" or a "preassessment" test or a fax may need to be sent to the pathology department (*send fax to pathology*) requesting the pathology slides to be sent to the AMC. It is important to note that the process shown in Fig. 21.3 is considered to be a standardized procedure for these patients at the AMC. From the figure, it can be inferred that there are several courses of actions that may be taken.

Finally, the *visit outpatient clinic* process is discussed in more detail. Figure 21.4 provides an overview of this subprocess. During such a consultation, the medical status of the patient is discussed and a decision is made whether to (re)schedule or cancel diagnostic tests (task *make conclusion*). For any diagnostic tests that need to be scheduled, application forms need to be filled in by the clinician before the actual execution of these tests (task *fill in application form*). Moreover, at different stages during the process, several administrative tasks, such as handing out brochures (task *additional information with brochures*) and producing a patient card (task *make card patient*), may be necessary. These tasks are generally performed by a nurse.

21.3.2 The Need for Process Fragments

In the previous section, insights have been given into the complexity of the candidate healthcare process. As the complete process consists of around 230 activities, only a small part of the overall process model can be discussed. In this section, its characteristics are considered and linked with the different kinds of healthcare processes and their characteristics that have been explored in Sect. 21.2. This aids in characterizing the process flexibility that is necessary. Finally, it is shown how this can be provided by the YAWL System.

The gynecological process, of which an overview is shown in Fig. 21.2, is an organizational process. In this process, there are organizational tasks such as the registration of a patient (*register patient* in Fig. 21.4), planning of appointments (*plan CT* and *plan MRI* in Fig. 21.3). Moreover, a clinician can choose any

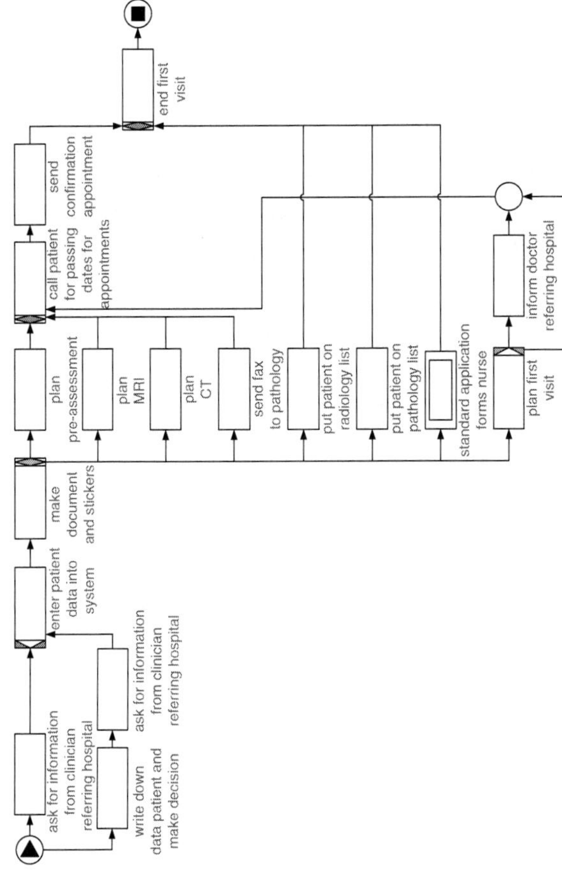

Fig. 21.3 Initial stages of the gynecological oncology healthcare process

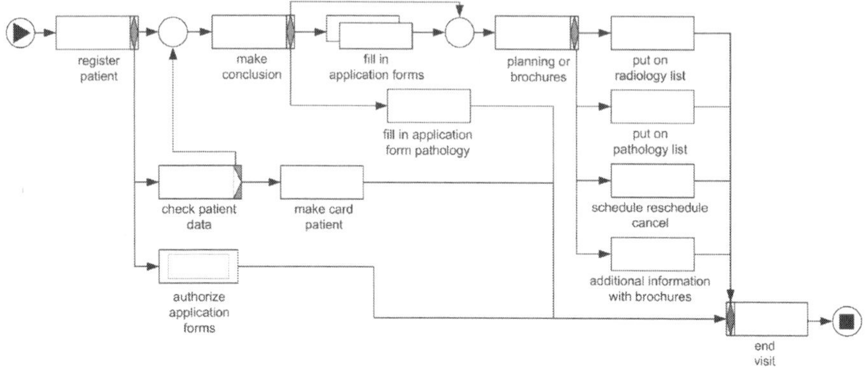

Fig. 21.4 Visit of the patient to the outpatient clinic

diagnostic test that he/she considers relevant, but does not receive any automated support in selecting the most appropriate diagnostic test.

For this group of patients, it is still medically sound to postpone treatment for some days. This means that the healthcare process under consideration can be typified as an elective care process. Generally speaking, the series of diagnostic tests to be conducted is known. However, the actual tests that are ultimately performed, and their number, are based on the actual condition of the patient. In addition, the selection of the tests also depends on the outcome and the particular quality of tests performed earlier. As a consequence, at the beginning of the process it is not known which diagnostic tests are to be performed for a given patient. This only becomes clear during the course of the diagnostic process. This is illustrated in Fig. 21.5, which shows a multiple instance task that allows multiple diagnostic tests to run concurrently. However, some tests are typically performed in the diagnostic process of this patient group. So, they can be put into a repertoire of possible tests. For example, for this patient group, typically an MRI or CT-scan is performed, which is illustrated by the process fragments having a continuous line. Moreover, in the future, new tests may become available or a diagnostic test may be needed which has never been ordered before and therefore does not belong to the repertoire yet. This requires that the repertoire can be extended with new tests in the future. In Fig. 21.5, this is illustrated by the process fragment with a question mark in it.

Earlier, it was indicated that the gynecological oncology healthcare process is an elective care processes. However, for the ultimate selection of diagnostic tests, a group of clinicians proceeds in a step-by-step way, deciding about the next actions to be taken. So, the process can be planned and some tasks in the process can be clearly defined. However, the ultimate realization of some tasks in the process (e.g., the diagnostic tests) needs to be deferred until runtime. In addition, the decision about the next steps to be taken is done in close cooperation with specialists from different medical specialties. Linking this back to the different kinds of healthcare processes and their characteristics defined in Sect. 21.2, it can be concluded that the gynecological oncology healthcare process is an elective, nonroutine, healthcare process in which *complex care* needs to be delivered.

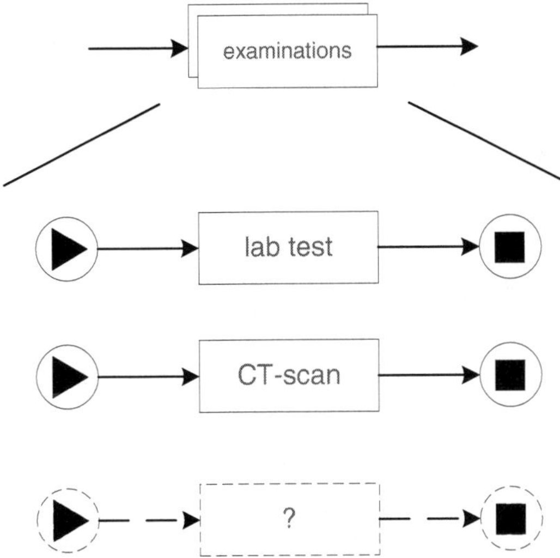

Fig. 21.5 Illustration of the diagnostic tests that can be chosen – in the future, new diagnostic tests may become available or tests may be chosen, which have never been selected before

21.4 Realization

In this section, the realization of the process in the YAWL System is considered. The whole process implemented in the YAWL System consists of around 230 tasks and took over 120 hours to implement. This illustrates that the implementation is far from trivial. For the control-flow, data, and resource perspective, it is discussed how they were realized and which services of the YAWL environment have been used. Moreover, during the discussion it is shown how the required flexibility to accommodate the gynecological oncology healthcare process can be provided in YAWL. Afterwards, the system is demonstrated in the context of a healthcare scenario. Finally, at the end of the chapter, the corresponding implementations in FLOWer, ADEPT1, and Declare are considered.

21.4.1 Control-Flow Perspective

The YAWL models shown in Figs. 21.2–21.4 form only a small part of the total process. Note that these models have been used for the implementation in YAWL to support the healthcare process. The most significant challenge that arose during this development activity was that, for multiple diagnostic tests, processes can run concurrently. In addition, there is also the possibility for a process to be initiated for a completely new diagnostic test. For such a new diagnostic test, it may even be necessary to define the accompanying process fragment from scratch.

In YAWL, this problem can easily be solved using the Worklet Service. The Worklet Service allows certain process steps to be linked to a repertoire of possible actions, which in this case are process fragments, known as worklets (cf. Chap. 11). Based on the properties of the case and other context information, the right worklet can be chosen at runtime. Also, during process execution, it is possible to add new worklets to the repertoire. In the example in Fig. 21.5, the *examinations* multiple instance task is linked to the Worklet Service. At runtime, multiple process fragments can be chosen from a repository of possible process fragments as the basis for executing this particular task. In total, it took around 30 hours to implement the control-flow perspective.

21.4.2 Resource Perspective

During the execution of the healthcare process, each task is performed by a single person having a specific role. For example, in Fig. 21.4, the *register patient* task is performed by a nurse and the *make conclusion* task is performed by a clinician. YAWL allows for role-based distribution of tasks to human resources, where a human resource is referred to as a participant. A participant may have one or more roles. In the healthcare process, for each task, a single role is specified, indicating which participants are allowed to perform a task. At runtime, a participant uses the worklist handler of the Resource Service to perform work items that he/she is allowed to undertake.

In total, it took around 5 hours to define the organizational perspective. This was due to the fact that the distribution of work for the tasks in the healthcare process is done in a simple and straightforward way.

When implementing the resource perspective, one main drawback was identified. For all of the tasks in a process it is assumed that each of them are performed at an arbitrary future point in time when a participant becomes available. This allows work items to be presented in a worklist, where any suitable participant (clinician, administrator) can select and start working on them. However, in the healthcare domain, participants (e.g., clinicians) are expensive, limited in number, and have a scarce availability. Moreover, many tasks require the involvement of several participants. Typically, for these kind of tasks and participants, an appointment-based approach is used to guarantee the most effective use of the available participants. An example of such a task is patient John who is scheduled to have a physical examination at 10 o'clock with assistant Rose and clinician Nick.

Today's WfMSs offer work items to users through specific worklists in which participants select the work items they will perform. However, no calendar-based scheduling is offered for these work items. In other words, no appointments can be scheduled for such tasks so that they are scheduled in the calendars of the participants involved in the actual performance of the task in order to ensure that they occur at a precise preagreed time suitable for all participants involved.

Currently, YAWL offers no support for the scheduling of work items and the allocation of a single work item to multiple participants. A related concept is the team concept, in which a participant (e.g., a manager) can see the work queues of the members in a team.

21.4.3 Data Perspective

The primary goal of implementing the healthcare process in YAWL is to investigate the flexibility that needs to be offered by the WfMS to support the execution of the healthcare process. Therefore, the data support that has been implemented in YAWL is only intended for routing purposes. Nevertheless, some complex data types still needed to be defined in XML schema (cf. Chap. 8 (The Design Environment)) in order to ensure that work is performed in the right way. For example, a complex data type has been defined, allowing a clinician to indicate which examinations need to be (re)scheduled or canceled. Furthermore, for each (re)scheduled or canceled examination, a remark can be added if needed. The automatically generated form for this data type can be seen in Fig. 21.7. In total, it took around 30 hours to define the data perspective, indicating that it was not a trivial task.

21.4.4 Scenario

In this section, the operation of the healthcare process in the YAWL System in the context of a healthcare scenario is discussed. Imagine that patient Sue is referred to the AMC hospital for treatment. At the *ask for information from clinician referring hospital* step of Fig. 21.3, clinician Nick decides that she needs to undergo an "MRI," "CT-scan," and a preoperative screening (task *plan preassessment*). These events must be planned by a nurse. In addition, the referring hospital needs to be requested to send their pathological slides to the AMC (task *send fax to pathology*). All the work items that must be performed as a result of this decision have to be undertaken after the execution of the *make document and stickers* task. Figure 21.6 shows the worklist of a nurse for a given case after the completion of this task where the previously mentioned decisions have been made.

After all necessary steps shown in Fig. 21.3 have been performed, patient Sue can visit the hospital. During this task (*make conclusion*), the status of the patient is discussed and a decision is made about the next steps to be undertaken in the process. Figure 21.7 shows part of the corresponding form that is presented to the clinician, allowing the (re)scheduling or cancelation of diagnostic tests and consultations.

In the form in Fig. 21.7, the clinician has indicated that a "lab test" and an "MRI" are needed. Consequently, at the end of the visit, two separate worklets need to be started to accomplish these tests. This is initiated by triggering the *examinations*

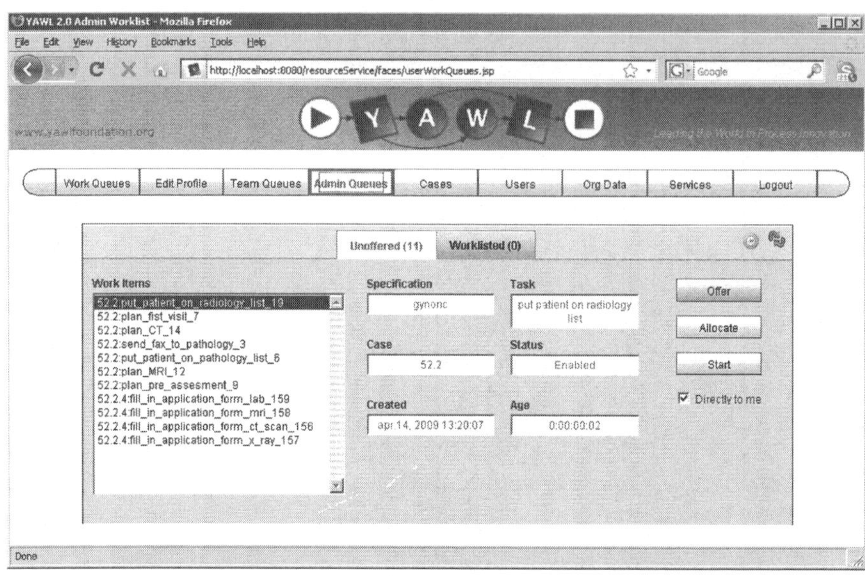

Fig. 21.6 The work items that need to be performed for a case after the *make documents and stickers* task has been completed

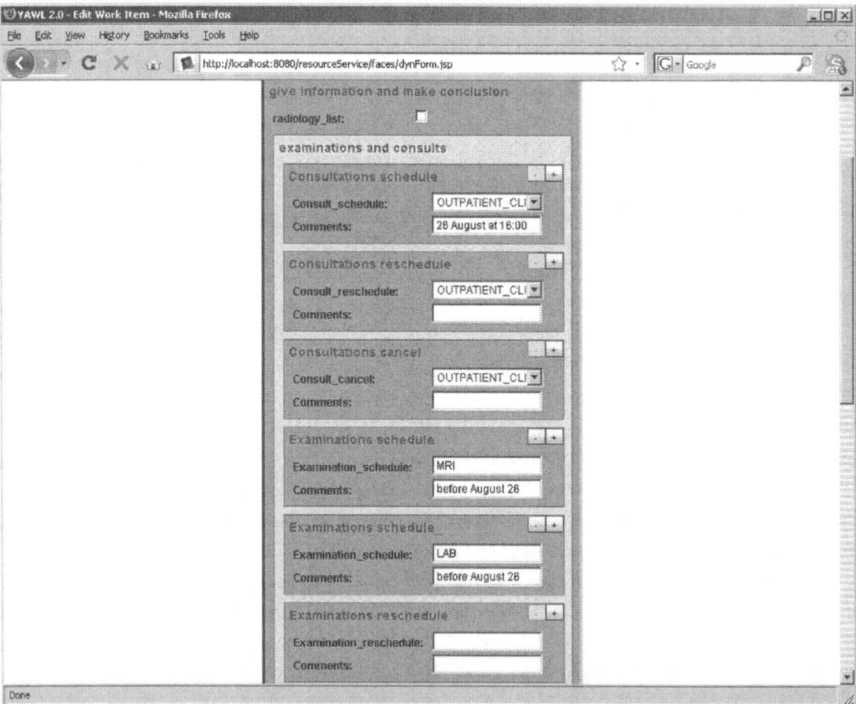

Fig. 21.7 Form in which the clinician requests a lab test and an MRI

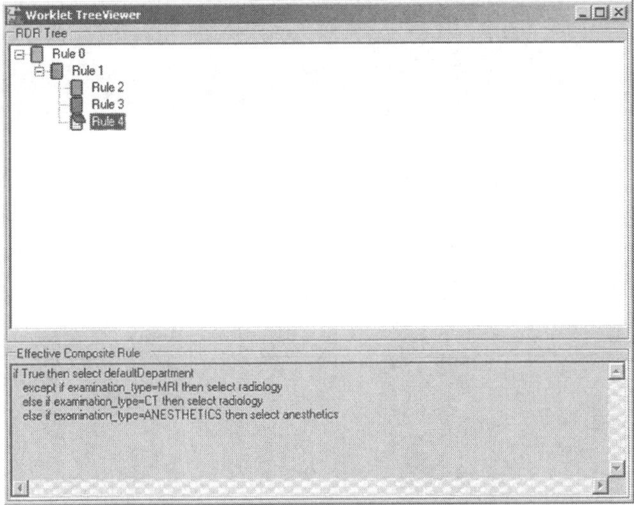

Fig. 21.8 Ripple-Down Rule Tree for the *examinations* multiple instance task

task. As already indicated, this task is connected to the Worklet Service. Figure 21.8 depicts a screenshot of the Rules Editor showing the Ripple-Down Rule Tree for the *examinations* task. This Ripple-Down Rule Tree defines under which conditions a worklet is selected. In this case, based on the decision of the clinician, a worklet needs to be started for the lab test and also for the MRI.

From the Ripple-Down Rule Tree in Fig. 21.8, it can be inferred that a radiology worklet will be started for an MRI or CT scan and that an anesthetics worklet will be started if an anesthetics test needs to be performed. However, if none of these rules apply, then a default worklet will be started. In this case, this would mean that a radiology worklet will be started for the MRI test. However, the clinician has decided that a lab test also needs to be performed for patient Sue, for which there is currently no specific rule in the Ripple-Down Rule Tree. As a consequence, the default worklet will be started for the lab test. Figure 21.9 shows which work items are offered because of the selection of these two worklets. Here the default department work item is offered because the default worklet is selected for the lab test.

Unfortunately, the selection of the default department worklet triggers the performance of a lab test, which is not the most appropriate one. Instead, the dynamic addition of a more suitable worklet needs to be initiated. To cater for this, the Worklet Service allows for the dynamic addition of new rules to the Ripple-Down Rule Tree. Each rule indicates the context conditions under which an individual worklet needs to be chosen. In this case, it is necessary to define that the lab worklet is to be chosen when a lab test has been requested. Note that the worklet can be chosen from the repository, if it is already present, or it can be defined from scratch.

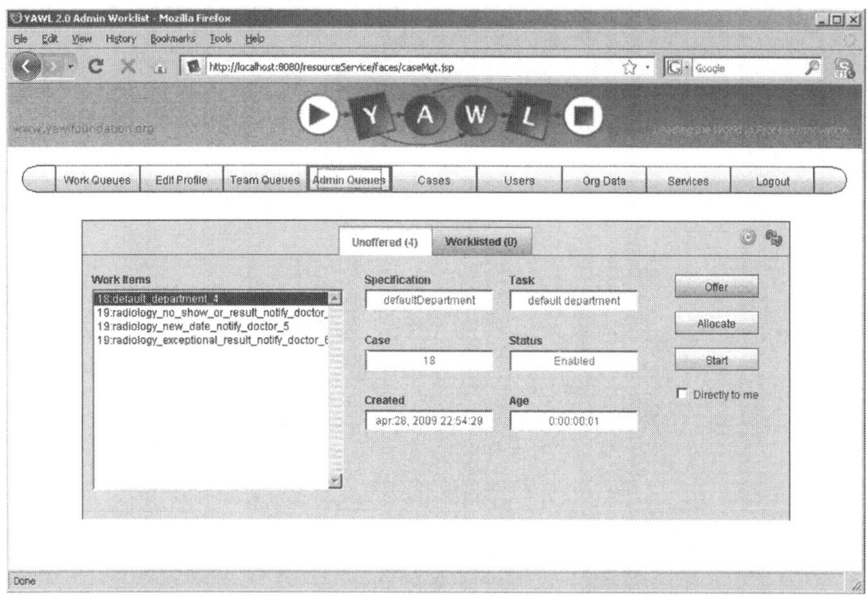

Fig. 21.9 Work items offered because of the selection of the radiology and default worklet

Figure 21.10 shows the addition of a rule to the tree set in order to achieve the above mentioned requirement. Moreover, it has also been decided to replace the default worklet with the lab worklet. The consequence of this action can be seen in Fig. 21.11, which shows the new work items that are offered after this replacement. Here it can be seen that the default department work item has disappeared and is replaced by other work items that are offered due to the selection of the lab worklet.

For the successful application of workflow technology in healthcare, the sort of process flexibility that was demonstrated above, and the extent to which this flexibility can be delivered, is of prime importance. For every diagnostic test which is requested, a worklet needs to be started. Based on a Ripple-Down Rule Tree, the Worklet Service automatically selects a worklet. Nevertheless, it can be the case that the most appropriate worklet is not selected. In this case, the Worklet Service allows for the replacement of a worklet by a more appropriate one in a controlled way. It would not be acceptable to the medical professionals who are participating in the process (e.g., a clinician) that a less suitable worklet is chosen for a diagnostic test and that it cannot be replaced by another one. Offering such a facility is therefore of the utmost importance.

21.4.5 Implementation in FLOWer, ADEPT1, and Declare

In this section, the alternate implementations of the gynecological oncology healthcare process in FLOWer, ADEPT1, and Declare are discussed. These systems were

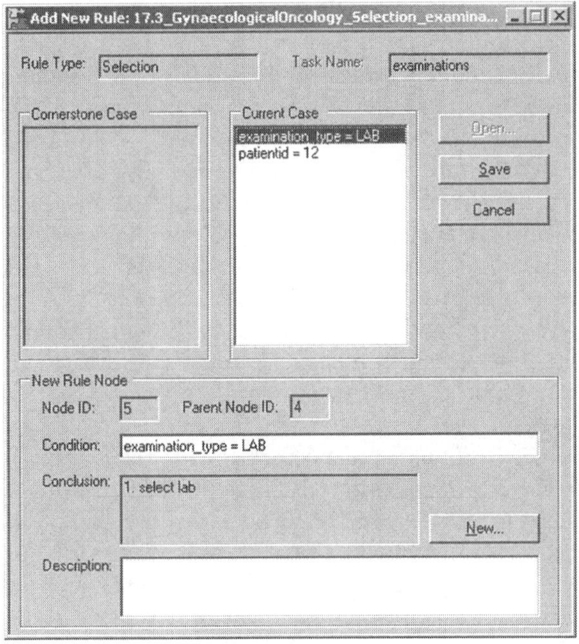

Fig. 21.10 Replacement of the default worklet by the lab worklet

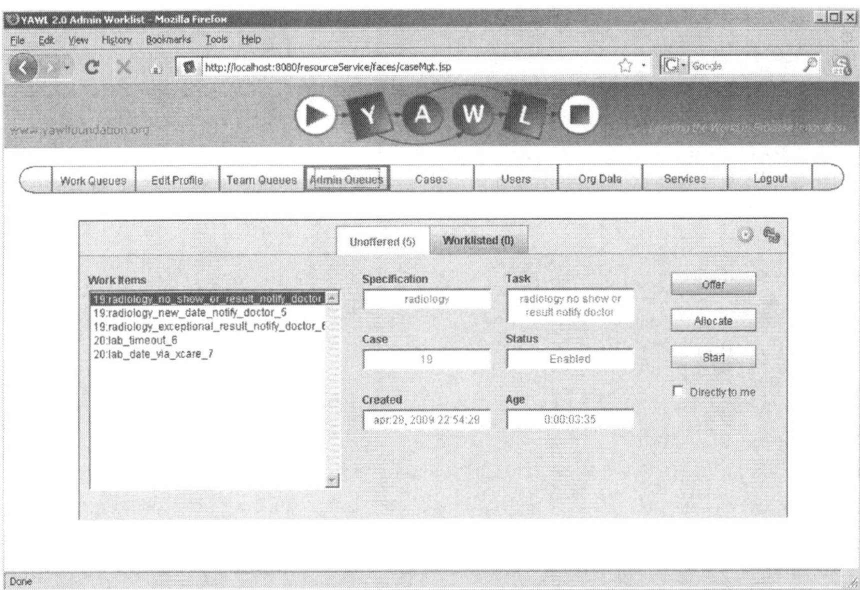

Fig. 21.11 The work items for the default worklet are replaced by the work items for the lab worklet

selected as they demonstrate various approaches to facilitating process flexibility. Moreover, for the flexibility provided by each system, it is indicated for which kind of healthcare processes, as defined in Sect. 21.2, it can provide optimal support.

FLOWer adds flexibility by allowing for deviation from an existing process by skipping or redoing steps in the process. Moreover, it is possible to view/add/modify data elements before and after the corresponding activities have been executed. An example is the skipping of the *plan MRI* step of Fig. 21.3. Likewise, the *enter patient data* step can be repeated after the performance of the *call patient for passing dates for appointment* step. In summary, it appears that FLOWer can provide optimal support for routine processes. Currently, YAWL does not provide any support for the kind of flexibility offered by the FLOWer case-handling paradigm. This does not mean that YAWL cannot support routine processes. However, it can only provide support for routine processes in which all possible scenarios can be encoded in the design time model.

ADEPT1 provides flexibility by supporting *dynamic change* that allows the process model for one individual case to be changed. In doing so, it is possible to deviate from the premodeled process template in a secure and safe way. For example, ADEPT1 allows for the insertion of steps or the deletion of steps. So, when realizing the process in Fig. 21.3 in ADEPT1, it is possible, for a running case, to dynamically add the activity *order drug* after the activity *make document and stickers* and before the activity *plan first visit*, which allows for the ordering of a drug in-between the activities *make document and stickers* and *plan first visit*. This means that ADEPT1 allows dramatic changes to be made to the process. By doing so, ADEPT1 can provide optimal support for acute care processes and care processes that deal with patients for whom the probability that complications might arise is high. YAWL provides partial support for these kinds of processes, as through the Exlets approach it is possible to remove a task from the process model for one individual case. Furthermore, a case can be terminated prematurely when a desired point in the process for the case is reached.

Declare provides flexibility through the language used for specifying processes, called ConDec. ConDec is a declarative process modeling language. This means that it specifies what should be done instead of specifying how it should be done, as is the case with imperative languages (e.g., YAWL, FLOWer). Users can execute activities in any order and as often as they want, but they are bound by certain specified rules, called constraints. For example, when implementing Fig. 21.3, in Declare it can be defined that activities *enter patient data into system* and *make document and stickers* need to be executed at least once, while leaving the order in which they need to be executed unspecified. Moreover, as is the case for ADEPT1, Declare supports dynamic change. In summary, Declare can provide optimal support for acute care processes and care processes that deal with patients for whom the probability that complications might arise is high. In addition, Declare can also provide support for routine processes. If the individual Declare system is compared with YAWL, it can be seen that YAWL provides optimal support for elective, nonroutine, healthcare processes as it allows for the ultimate realization of some tasks to be deferred until runtime. Declare does not provide any support for this kind of flexibility and as a result its consequence is that it can provide limited support for elective, nonroutine, healthcare processes.

In comparison to FLOWer, Declare allows for a less strict execution of the process, as the language used by Declare is a declarative process modeling language instead of an imperative one. On the other hand, FLOWer allows for deviation from an existing process. If all four systems are compared that the healthcare process has been implemented in, it can be seen that individual systems tend to exhibit a degree of specialization in their approach to process flexibility. This has the consequence that different systems need to be used in conjunction with each other in order to fully support all types of care processes that might be encountered. An interesting achievement in this context is that in Chap. 9 it is demonstrated that YAWL, Declare, and the Worklet Service can all interoperate, thereby combining the powers of the various approaches.

21.5 Conclusions

In this chapter, it has been discussed how the gynecological oncology healthcare process can be supported by the YAWL workflow system. The process is an elective, nonroutine, healthcare process in which complex care needs to be delivered. For these kinds of healthcare processes, process flexibility is vital in the sense that it can never be determined beforehand which diagnostic tests are ultimately needed during the course of a process and which new diagnostic tests may become available in the future. While some tasks in the process can be clearly defined, this is not possible for other tasks and their ultimate realization needs to be deferred until runtime.

In YAWL, process flexibility can be achieved using the Worklet Service. For the tasks in the process that are not fully specified, the most suitable worklet can be selected prior to the enablement of the task. In addition to this, it is possible to deviate from the chosen worklet by selecting an alternative one or by redefining it completely. To the best of our knowledge, YAWL is the only system providing full support for this kind of flexibility. The other systems, FLOWer, ADEPT1, and Declare, do not offer this kind of process flexibility, which means that optimal support for the gynecological oncology healthcare process cannot be provided by these systems. However, some WfMSs provide pieces of functionality that is comparable to the functionality that is provided by the Worklet Service though full support for the flexibility mentioned cannot be provided.

The implementation of the healthcare process in four workflow systems revealed two main issues. As already indicated, one issue is the lack of calendar-based scheduling support. Healthcare is a prime example of a domain where effective work item execution is often tied to the availability of scarce and busy resources. Therefore, it is of the utmost importance that the scheduling of appointments for these resources is done in an efficient way, that is suitable both for the medical staff and also for the patients being treated.

Another major shortcoming of WfMSs is also applicable to the healthcare domain. For a given patient, many small processes often execute concurrently. These processes can start at any point in time and can also be terminated at any point in time. For example, imagine that it has been modeled that a blood test needs to be finished before the next visit of a patient to the hospital. If the completion of the blood test is delayed for one reason or another, a consequence may be that the next scheduled visit of the patient cannot take place and needs to be canceled and a new appointment made.

Clearly, what is missing in current WfMSs is that often a certain magnetism exists between the processes that are running in parallel for a patient, which is not dealt with properly. This can also be demonstrated by the blood test example. If the patient dies, then not only the blood test process needs to be terminated but also the one containing the visit to the hospital and vice versa.

Exercises

Exercise 1. Please consider the Ripple-Down Rule Tree shown in Fig. 21.12, which shows the original Ripple-Down Rule Tree for the *examinations* multiple instance task shown in Fig. 21.8. In Sect. 21.4, it can be seen that a default worklet is started for a lab test. Rather, a lab worklet should be started for the lab test. Please change the tree of Fig. 21.12 in such a way that a lab worklet is started for a lab test instead of the default worklet. As condition, you may use "examination_type = LAB," and as conclusion "lab" may be used.

Exercise 2. Please consider Fig. 21.3. How many work items can at most be enabled concurrently for this figure, given that for subnet *standard application forms nurse* at most four work items can be concurrently enabled. You may assume that for this figure only one case is running and no other tasks are enabled in other subprocesses in the gynecological oncology healthcare process.

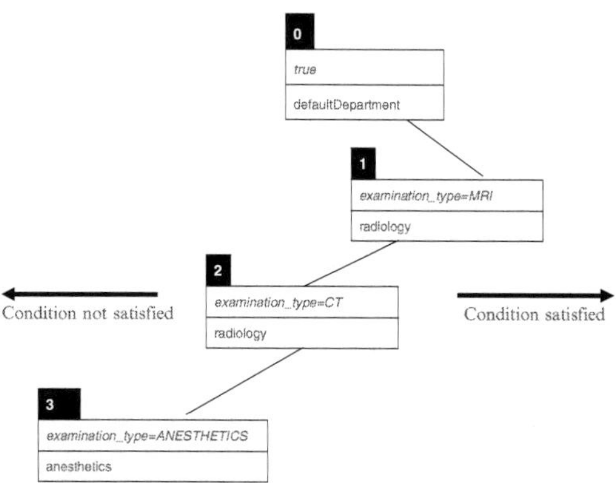

Fig. 21.12 The Ripple-Down Rule Tree belonging to Exercise 1

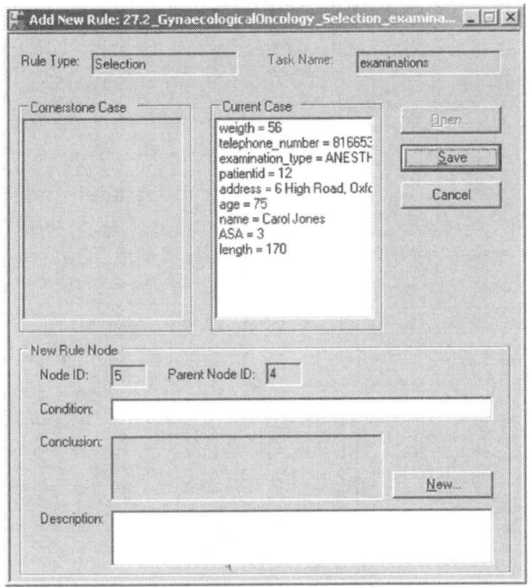

Fig. 21.13 Addition of a rule in the Rules Editor (Exercise 4a)

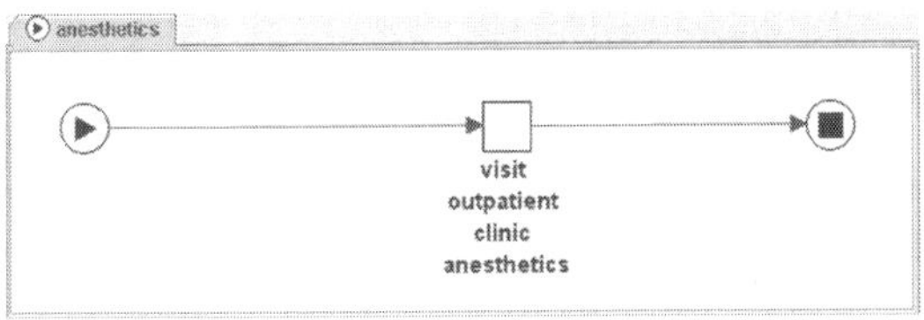

Fig. 21.14 Standard anesthetics worklet (Exercise 4b)

Exercise 3. The gynecological oncology healthcare process has been defined as a nonroutine healthcare process according to the characterization of healthcare processes defined in Sect. 21.2.

(a) For now, assume that this care process is a routine healthcare process instead of a nonroutine healthcare process. Describe which kind of process flexibility needs to be provided to accommodate the process with workflow technology (if any!).
(b) For the flexibility that is described in question (a), how should the models shown in Figs. 21.2–21.4 be changed?

Exercise 4. In the Ripple-Down Rule Tree of Fig. 21.12, it can be seen that a standard anesthetics worklet is chosen if an anesthetics test needs to be performed. The anesthetics worklet instance can be seen in Fig. 21.14. As you can see, only a visit to the outpatient clinic of anesthetics is defined. Assume that Carol Jones needs to visit the anesthetics outpatient clinic. However, as she is an old lady and her ASA classification is set to 3,[1] a single visit to the outpatient clinic is not enough. Rather, the anesthetics worklet needs to be replaced by a more dedicated worklet for dealing with patients like Carol Jones.

(a) In Fig. 21.13, you can see the panel of the Rules Editor for adding a rule to the Ripple-Down Rule Tree Set. In the "Current Case" panel, you see the data attributes that exist for the running case together with their current values. Based on these values, define a condition such that a more dedicated worklet can be selected for patients older than 70 years and having an ASA classification of more than 2.
(b) In Fig. 21.14, you can see a standard anesthetics worklet Instance. Extend the process definition of this worklet in such a way that it satisfies the following process description. For patients older than 70 years and having an ASA classification of more than 2, several diagnostic tests are required. But, first of all, the patient should pay a visit to the outpatient clinic of anesthetics so that a clinician can check the condition of the patient. During this check, the clinician decides whether additional diagnostic tests are needed for the patient. For these tests, the clinician can make a selection out of the following list of tests: ECG, consultation with a cardiologist, consultation with an internist, and a lung function test. If additional tests are needed, the clinician again needs to see the patient to discuss the results of the tests after which the process ends. However, it can also be the case that no additional tests are needed. In this case, the process immediately ends after the first visit to the outpatient clinic. Furthermore, you may assume that the tests selected are executed only once.

Chapter Notes

The advantages of successfully applying workflow technology are that processes are executed more rapidly and more efficiently [208]. However, notwithstanding these advantages, the widespread adoption and dissemination of WfMSs in the healthcare domain is the exception rather than the rule [180]. One of the problems that has to be dealt with to support healthcare processes by WfMSs is that flexibility needs to be provided by the system [241]. Unfortunately, current workflow systems fall short in this area, a fact that is recognized in the literature [17, 22, 89, 135]. Furthermore, additional requirements, that need to be satisfied by WfMSs in order to be applied in the healthcare domain, can be found in [17, 89, 135, 212]. The work presented in [154] specifically focusses on the process flexibility requirements that need to be fulfilled by workflow systems, in order to be successfully applied in a hospital

[1] ASA is a frequently used scoring scale used by anesthesia providers for indicating the patient's overall physical health or sickness preoperatively.

environment. Related to this is [177], which identifies a set of requirements addressing the aspect of process flexibility. These set of requirements are presented in the form of Process Flexibility Patterns.

As WfMSs YAWL, FLOWer [22], ADEPT1 [203], and Declare (see Chap. 12 (The Declare Service)) have been selected to implement (parts of) the healthcare process that has been studied in this chapter. The selected systems cover various flexibility paradigms: adaptive workflow (ADEPT1), case handling (FLOWer), and declarative workflow (Declare). The identified Process Flexibility Patterns in [177] are used to evaluate the process flexibility of the previously mentioned systems.

Many different (innovative) concepts can be found related to the delivery of care. In [253–255], respectively, the integrated care, stepped care, and managed care concepts are considered. These concepts differ in the way whether the focus is on the patient and his/her needs or whether the focus is on the care providers. From these care concepts, it becomes obvious that care can be delivered in many ways and with varying characteristics. Varying characteristics of healthcare processes can be found in [49, 147, 154, 171, 252].

In this chapter, two main shortcomings of WfMSs have been identified. One issue is the lack of calendar-based scheduling support. In [155], it has been investigated how WfMSs can be augmented with scheduling facilities to address this shortcoming.

Chapter 22
YAWL4Film

Chun Ouyang

22.1 Introduction

In recent years, business process management (BPM) and workflow technology has reached a certain level of maturity, with a great potential to deliver benefits in a wide range of application areas. Nevertheless, it is typically applied by companies with a high adoption level of information technology. As part of the Australian Research Council Centre of Excellence for Creative Industries and Innovation, the BPM Group at Queensland University of Technology decided to explore the potential of applying BPM and workflow technology to the field of *screen business*.

The field of screen business is characterized by business processes with high demands for creativity and flexibility. These processes span a value chain consisting of four major phases: *development*, *preproduction*, *production*, and *postproduction*. Out of these four phases, *production* is the most expensive phase in screen business as it is when the movie is actually created and shot. Also, it is during production that the majority of cast and crew are contracted and the majority of equipment and resources utilized. At present, a film production is a highly manual process that requires handling large amounts of heterogeneous data on a daily basis and coordinating many geographically distributed stakeholders. Not surprisingly, such a process is time-consuming and error-prone, and can easily increase the risk of delays in the schedule. Applying a workflow system can provide benefits to the film production process by automating, streamlining, and optimizing the process. YAWL offers such capabilities and moreover its service-oriented architecture facilitates the development of sophisticated extensions to tailor workflow systems to the needs of screen business.

Within the above context, YAWL was applied to the automation of film production processes. The term "film production process" hereby focuses on the chain of information processing accompanied with the shooting activities during a film production. A case study was carried out in close collaboration with the Australian

C. Ouyang
Queensland University of Technology, Brisbane, Australia,
e-mail: c.ouyang@qut.edu.au

A.H.M. ter Hofstede et al. (eds.), *Modern Business Process Automation*,
DOI 10.1007/978-3-642-03121-2_22,

Film Television and Radio School (AFTRS). A major outcome of this collaboration was the development of an application platform called YAWL4Film. YAWL4Film exploits the principles of BPM in order to coordinate work distribution within production teams, to automate the daily document processing and report generation, to ensure data synchronization across distributed nodes, to archive and manage all shooting related documents systematically, and to document experiences gained in a film production project for reuse in the future. The case study can be seen as a proof of concept to demonstrate how film production can benefit from the application of BPM technology.

This chapter describes the development of YAWL4Film from design, implementation, to deployment. The rest of the chapter is organized as follows. Section 22.2 introduces background knowledge of film production processes and the current status of software support for film production. Section 22.3 describes the design and implementation of YAWL4Film, which consists of a film production process model in YAWL and a customized user interface with a professional look-and-feel. The process deployment and system deployment within the context of film production are discussed in Sect. 22.4. Evaluations, feedback, and reported perceptions of using the YAWL4Film system from two pilot projects completed at the AFTRS in 2007 are in Sect. 22.5. Finally, Sect. 22.6 concludes the chapter.

22.2 Overview of Film Production Processes

A standard film production process covers the entire period of the actual shooting of a motion picture, which can be as short as a day or last for several years (e.g., a very large production). The shooting procedure is carried out on a daily basis. During each day, a number of activities such as acting, camera, and sound recording are performed in studio or on location. While shooting is taking place, designated onset crew collect the information associated with each of these activities via corresponding production forms. Let us consider some examples. The *continuity* person (also known as the *script supervisor*) is in charge of the *continuity log*, which records details of each scene shot such as the timing, the lighting condition, the takes of choice of the director. After the shooting is wrapped up, the continuity person fills in the *continuity daily*, which summarizes of all the scenes shot on the day. The *camera assistant* is responsible for filling in the *camera sheet*(s)[1] with the photographic details of each shot (e.g., the counter reading and the film length of each take). The *sound recordist* completes the *sound sheet*(s) of all dialogue or effects recorded on set. The *second assistant director* (2nd AD) maintains a time sheet-like document called the *2nd AD report* for logging the data related to the working hours of the crew and cast, exceptions like delays and accidents, etc. All the above information is later gathered and collated to generate the *daily progress report* (DPR) at

[1] Note that individual camera (or sound) sheet is created for each camera (or sound) roll used on a day.

Table 22.1 List of documents created/used during film production

Place	Document name	Person in charge	Input document
On set	Continuity log	Continuity	–
	Continuity daily	Continuity	Continuity log
	Camera sheet	Camera assistant	–
	Sound sheet	Sound recordist	–
	2nd AD report	2nd AD	Call sheet
Production office	DPR	Production coordinator (edit) Production manager (review)	Continuity daily Camera sheet Sound sheet 2nd AD report Previous day's DPR
	Call sheet	Production coordinator (edit) Production manager (review)	Shooting schedule
(From preproduction)	Shooting schedule	1st AD	Cast list Crew list Location notes

the *production office*. A DPR is a summary of the shooting information for each day and serves the purpose of keeping track of a production's progress and expenses.

In addition to the generation of DPR, the production office monitors requirements for the subsequent shooting days and communicates frequently with onset crew to work out and finalize the *call sheet* for the next day. A call sheet is a one stop form for all necessities and logistics of a particular shooting day, and is usually issued to all cast and crew one day in advance. It takes as input the *shooting schedule*, which is a project plan for the entire production period, and elaborates the daily schedule of scene shooting by adding detailed timings for each production team member, specific costume, location, make-up, prop, transport requirements, and so on. As schedule of scenes often changes, a daily call sheet usually needs to be updated several times before it can be finalized.

Table 22.1 summarizes the list of production documents mentioned earlier. All these documents are created during production except the shooting schedule, which is created in preproduction but is often revised and updated during production. A detailed analysis of these documents shows a large amount of information overlap. Consider the DPR as an example. Almost all the information on the DPR can be obtained from four production forms (i.e., the continuity daily, camera sheet, sound sheet, and 2nd AD report) and the previous day's DPR (if present). Another example of information overlap is that around 60% of the information on a call sheet can be retrieved from the shooting schedule.

After a close study on the current status of tool support in film industry, it is surprising to see that so far there exists little tool support for information processing during the production of a film. Every document is created and handled manually. In particular, all the production forms and reports are paper-based (sometimes with carbon copies). A DPR or call sheet, which may be a Word or Excel document, is manually completed. As a consequence, delays in generating the production documents become frequent. For example, in principle, a DPR should be generated at the end of the current shooting day, but very often it is delayed to the day after to finish the previous day's DPR. The lack of tool support also causes delays in data synchronization between various production documents. An obvious example is the delay in updating the production schedule. When changes to the shooting schedule occur, in principle, the schedule should be revised first and then the call sheet updated accordingly, while in reality, after the shooting starts, any revision to the schedule is made only to the daily call sheet. The entire shooting schedule will not be updated until it becomes out-of-date and is hard to follow for composing call sheets.

In summary, there is an opportunity to optimize and automate film production processes to avoid errors and delays that happen often in current film production processes, which finally leads to the reduction of production costs. Moreover, by saving the time otherwise spent in costly and tedious activities, the production team can invest more in creative activities, such as rescheduling of scenes due to unexpected changes, thus increasing the team's work efficiency and responsiveness to changes.

22.3 YAWL4Film Design and Implementation

YAWL4Film was developed to support the automation of film production processes. It consists of a YAWL model capturing the control-flow, data, and resource perspectives of a film production process. Also, it extends the general YAWL System with a customized user interface to support templates used in professional filmmaking.

22.3.1 The Film Production Process Model

Figure 22.1 depicts a film production process model in YAWL. Tasks are represented as rectangles that may have an icon indicating whether they are manual or automatic. A task without an icon is an "empty" task that appears only for control-flow routing purposes. An instance of the process model starts with the collection of specific documents (i.e., cast list, crew list, location notes, and shooting schedule) generated during the preproduction phase. Next, the shooting starts, which is carried out on a daily basis.

22.3.1.1 Control-Flow Perspective

The tasks for each day are performed along two main parallel subprocesses. One subprocess begins with task *Begin Call Sheet* and ends with task *Finish Call Sheet* as in Fig. 22.1. It focuses on the production of a call sheet for the next day. The

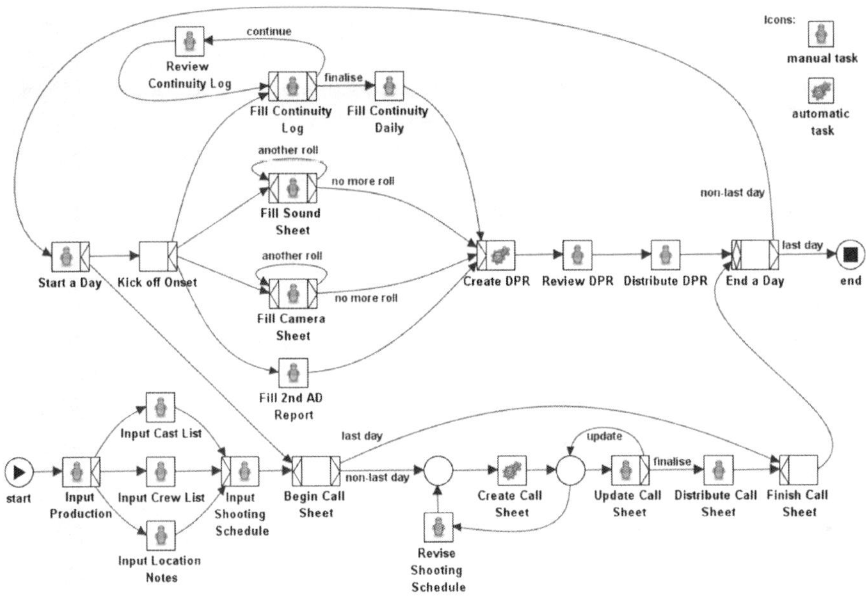

Fig. 22.1 A film production process model in YAWL

call sheet is created automatically from the shooting schedule and then updated with certain information that is not included in the schedule, such as individual cast and crew call times. The schedule of scenes for the subsequent shooting days often changes due to various reasons such as delays in the shooting of scheduled scenes, unexpected weather change, absence of designated cast and crew members, breakdown of onset equipment, etc. Once the schedule has been revised, the call sheet can be recreated accordingly. In addition, the call sheet information may also need updating during the production. In the call-sheet production subprocess, task *Update Call Sheet* has an XOR-split, which determines whether to continue updating or to finalize the call sheet based on the user input collected by the task. The use of a deferred choice between task *Update Call Sheet* and task *Revise Shooting Schedule* then makes it possible to revise the shooting schedule at any time before the call sheet is finalized. Task *Create Call Sheet* is an automatic task that upon execution generates a draft of the call sheet that contains the information available from the most recent version of the shooting schedule. The call sheet may go through any number of revisions and once finalized will be distributed to every cast and crew member. Given that a call sheet is always produced one day in advance, no call sheet is generated on the last shooting day. This is captured by the conditional outgoing branch of task *Begin Call Sheet* that is directly connected with task *Finish Call Sheet*, thus bypassing the actual production of a call sheet on the last day.

The other subprocess begins with task *Kick off Onset* and ends with task *Distribute DPR* as in Fig. 22.1. It specifies the sequence of onset shooting activities and supports the production of a DPR. Task *Kick off Onset* is an AND-split whose incoming flow diverges into four parallel outgoing branches to capture individual shooting activities. On each branch tasks are executed to record information of each shooting activity into the corresponding onset documents (i.e., continuity log and continuity daily, sound sheet, camera sheet, and 2nd AD report). Usually, a *production manager* or *1st AD* needs to check the status of the shooting continuity at a certain point of the day, for example, during a meal break. In the process model, the loop containing a sequence of task *Fill Continuity Log* followed by task *Review Continuity Log* allows the *continuity* person to submit a work-in-progress version of the continuity log for review, and then to resume the work after the review. In parallel, there may be multiple sound or camera sheets to be filled in during a day, which is captured by the self-loop of task *Fill Sound Sheet* or that of task *Fill Camera Sheet*. Next, upon completion of the above onset documents, a DPR is automatically generated and passed on to the *production manager* for review. After the review is finished, the DPR is circulated to certain crew members such as *producer* and *executive producer*.

It is interesting to see how the OR-join associated with task *End a Day* behaves. Before the first shooting day starts, an instance of the call sheet branch is executed for producing the first day's call sheet. As it is the only active incoming branch to task *End a Day*, the task will be performed once the call sheet has completed, without waiting for the completion of a DPR. In this case, the OR-join behaves like an XOR-join. On the other hand, if both call sheet and DPR branches are active (which is the case for the rest of the shooting days), the OR-join behaves like an AND-join. Finally, task *End a Day* has an XOR-split that determines whether it is the last shooting day. If so, the entire process ends. Otherwise, the flow loops back to task *Start a Day*, which initiates the shooting on the next day.

22.3.1.2 Data Perspective

A film production is a data-intensive process. The YAWL model in Fig. 22.1 handles all the 11 production and preproduction documents listed in Table 22.1. To capture the information on these documents, a number of complex data types are defined using XML schema. For example, Fig. 22.2 sketches the complex data types used to specify the information on a call sheet and the way they are built upon each other in an informal manner. Each data type is presented in a text box labeled by the name of the data type at the top of the box. It consists of one or more elements. Each element has a name and a data type, which can again be a complex data type or a simple data type (e.g., Boolean, Date, Integer, String, Time, Integer). In case of a complex data type, it needs to be defined. In Fig. 22.2, an arc with an arrow head is used to refer such a complex data type to its definition. In a data type, an element may appear exactly once (indicated by 1 enclosed in the pair of brackets following the element name), may or may not appear (0:1), or may appear one or multiple times (1:m).

In Fig. 22.2, the data type *GeneralInfoType* captures the common information that is shown on every production document. The *CallSheetInfoType* is the main data type that offers a structure for capturing call sheet information. It is built upon

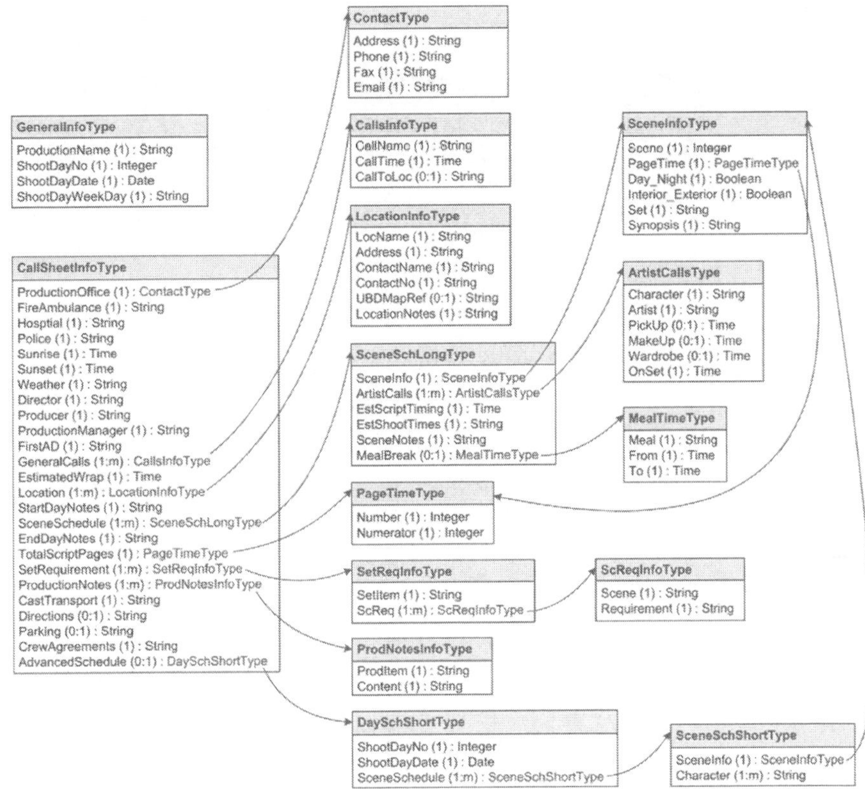

Fig. 22.2 Schematic representation of the data types for specifying a call sheet

a number of smaller data types. All these data types are valid and usable within the entire process model.

The data manipulation mechanism in YAWL, as discussed in Chap. 2, supports document processing and report generation by YAWL4Film. Consider task *Update Call Sheet*. Figure 22.3 depicts how this task handles data at runtime. There are three task variables: *GeneralInfo* (input only), *CallSheetInfo* (input and output), and *Finalize* (output only). When a work item of type *Update Call Sheet* is checked out, the values of its input variables are determined from the contents of the net variables (*General* and *CallSheet*). An example of the data extracted by this inbound mapping is shown inside the task symbol of *Update Call Sheet*. Later, when the work item is checked in, the output variables of the task are used to update the corresponding net variables (*CallSheet* and *Finalize*).

The data that the Engine supplies to the above work item of type *Update Call Sheet* is used to populate a Web form for the call sheet (see Fig. 22.7). Using this form, the user may perform updates to the call sheet, such as inserting "start-of-day notes" and he/she may indicate whether to finalize the call sheet (final submission) or to keep updating it (partial submission). This decision is captured in the output

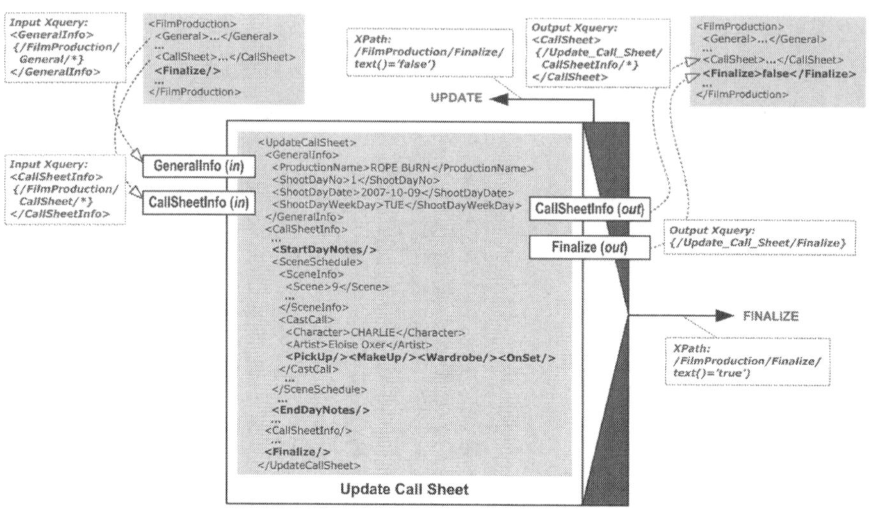

Fig. 22.3 Data manipulation by task *Update Call Sheet*

parameter *Finalize*. When the work item is checked in later, the updated call sheet and the value of *Finalize* are stored in the net variables. The value of *Finalize* is then used to determine which outgoing flow of the XOR-split is taken.

22.3.1.3 Resource Perspective

During a film production, each task is performed by someone who plays a certain role. For example, a camera sheet is filled in by the *camera assistant*, while a sound sheet is filled in by the *sound recordist*. This allows one to specify at design time one or multiple roles to which instances of the task will be distributed at runtime, that is, *role-based task distribution* to human resources (cf. Chap. 2). In YAWL, a human resource is referred to as a participant. Each participant may play one or more roles, hold one or more positions (each of which belongs to an Org group), and possess a number of capabilities. YAWL supports organization data management, maintenance, and monitoring functionalities. Below are two examples of the administrative forms used in YAWL4Film, which are provided by the YAWL runtime environment. One in Fig. 22.4 is for managing film production roles such as *production manager*, the other in Fig. 22.5 is for maintaining information about human resources such as an individual crew member who plays the role of *production manager*.

The resource allocation for each of the manual tasks in the film production process is specified using the resource manager wizard provided by the Editor (cf. Chap. 8). Figure 22.6 shows a screenshot of Step 2 of the resource manager wizard for specifying the resource requirements of task *Update Call Sheet*. It provides a list of crew members (see in "Participants") and a list of production roles (see in "Roles") available from the existing organization data. A selection of two roles,

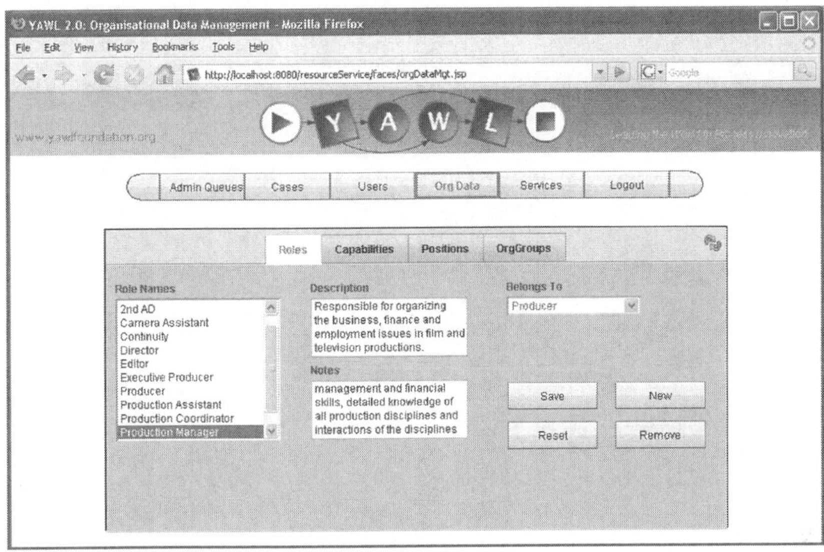

Fig. 22.4 Admin screen for managing film production roles

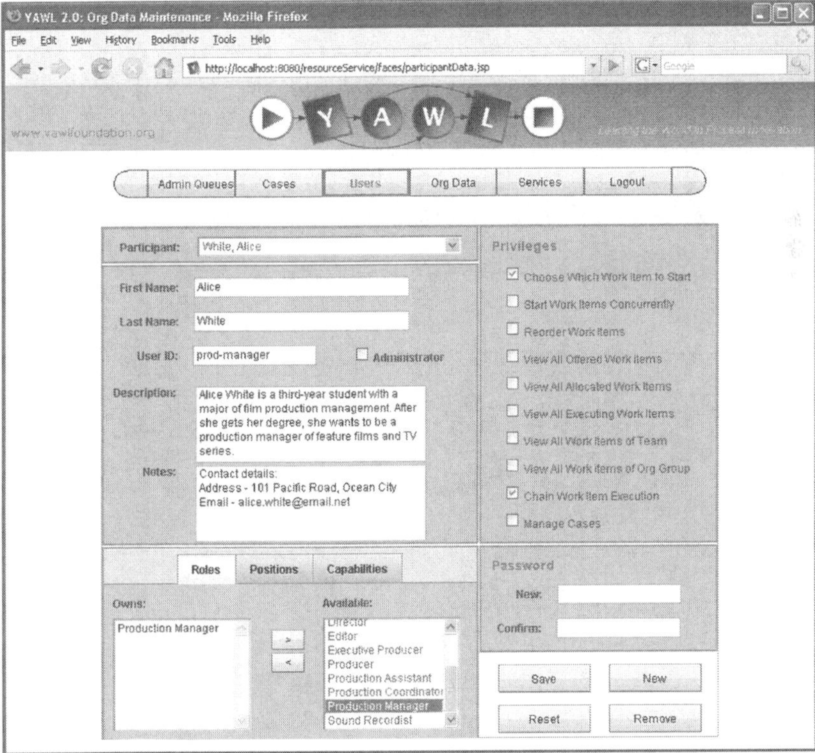

Fig. 22.5 Admin screen for maintaining details about individual crew members

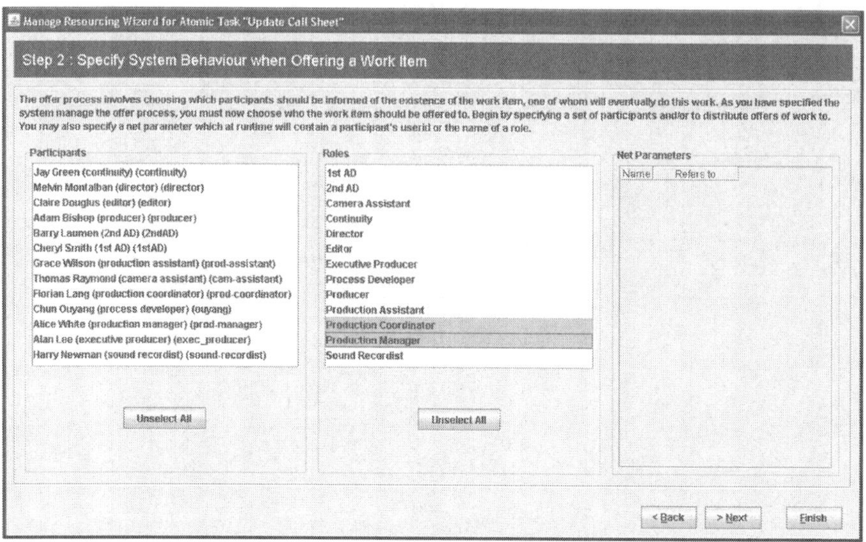

Fig. 22.6 Specifying at design time that a work item *Update Call Sheet* would be offered to crew members who play a role of production coordinator or production manager

production coordinator and production manager, indicates that the crew members who play either of these two roles will be offered a work item of type *Update Call Sheet* at runtime.

22.3.2 Customized Forms of Film Production Documents

Most of the manual tasks in the film production process are performed to record and gather information for specific film production documents. They require input from the user by means of forms. YAWL supports automatic form generation for each work item. These forms are dynamically generated and follow a standard (default) template. This default template, however, cannot be used to capture a production form with a professional look and feel. To support templates used in professional filmmaking, custom Web forms need to be created and then linked to the work-list handler. YAWL supports the use of custom forms. As mentioned in Chap. 8, at design time one can link a Web form to the corresponding task by supplying the URL of this form as part of the task's context. It should be mentioned that there is no specific way in which a custom form should be created, and it is entirely up to system/form developers to decide how to design and implement the form. In the following, how custom forms were implemented in YAWL4Film is discussed as an example of YAWL custom form application.

The custom Web forms of film production documents were developed using mainly HTML and standard Java technology. HTML was used to create a template capturing as much as possible the format of a standard paper form used at the

ROPE BURN

DIRECTOR: MELVIN MONTALBAN | PRODUCER: ADAM BISHOP

TUES, 9-10-2007 Shoot Day 1 of 3

Production Managers: ALICE WHITE
1st AD: CHERYL SMITH
Police: Eastwood Police Station ph (02) 9858 5944 **Hospital:** Ryde Hospital 1 Denistone Road Eastwood NSW 2212 ph (02) 9874 0199
Fire/Ambulance: 000

Production Office

Address Australian Film Television and Radio School: Corner Epping and Balaclava Roads, North Ryde, NSW

Phone +61.2.9805 6676 **Fax** +61.2.9887 1030 **Email** ropeburnproduction@gmail.com

Weather

Sunrise: 05:24:00 **Sunset:** 18:02:00
Forecast:
Partly Cloudy Min 14 Max 21

Call Times

Calls	Time	Location
Crew	08:00:00	AFTRS
Location	08:00:00	AFTRS

...

Shooting Schedule

Start of Day Notes ABSOLUTELY NO FOOD OR DRINK (EXCEPT FOR WATER BOTTLES) IN STUDIO

Sc: 9 **Pages:** 4/8 **Timing:** 00:00:25 Night / INT **Set:** DRESSING ROOM
Synposis: Charlie's not going to Europe with them

Character	Artist	PU	M/UP	WR	On Set
CHARLIE	Eloise Oxer	0630	0745	0715	0815
SIMONE	Amelia Best	0620	0715	0845	0815

Insert Row | Delete Row
Shoot Times: 09:00-11:15
Scene Notes: BLOCK-THROUGH 0815-0830 THEN LIGHT/COMPLETE M/UP AND WR 0830-0900

Sc: 2 **Pages:** 1 2/8 **Timing:** 00:01:07 Night / INT **Set:** DRESSING ROOM
Synposis: Simone and Charlie get it on but are interrupted.
Character Artist PU M/UP WR On Set

...

Partial Submission ◉ Final Submission ○

Print Preview | Print | Save | Submission | Upload

Fig. 22.7 An example of a customized user form – call sheet

AFTRS, and Java was used to render the logic of data information in the form. For example, Fig. 22.7 depicts the Web form for task *Update Call Sheet* (see Fig. 22.3), as seen by the *production coordinator* or the *production manager*.

On each form, a user can *load* a data file into the form, *save* his/her input into a local data file, *print* a copy of the form, and *submit* the form back to the worklist handler once it has been completed. Moreover, each form provides data validation upon save and submission to prevent the generation of invalid data documents. An example of an invalid data input could be that a field of *Integer* type is supplied with

ROPE BURN PRODUCTION OFFICE
Australian Film Television and Radio School: Corner Epping and Balaclava Roads, North Ryde, NSW
Telephone: +61.2.9805 6676 Facsimile: +61.2.9887 1030 Email: ropeburnproduction@gmail.com

POLICE: Eastwood Police Station ph (02) 9858 5944
FIRE/AMBULANCE: 000
HOSPITAL: Ryde Hospital 1 Denistone Road Eastwood NSW 2212 ph (02) 9874 0199

ROPE BURN

DIRECTOR: MELVIN MONTALBAN | **PRODUCER:** ADAM BISHOP

TUES 9 October 2007 | Shoot Day 1 of 3

PRODUCTION MANAGER(S): ALICE WHITE
1ST AD: CHERYL SMITH

AFTRS STUDIO 1
Cnr Epping and Balaclava Roads North Ryde NSW
UBD: 374 D12
Be sure to wear your security pass at all times.
Location Contact:Production Coord - Emma

Crew Call:	08:00:00	AFTRS
Location Call:	08:00:00	AFTRS
Makeup/Hair Call:	07:00:00	AFTRS
Wardrobe Call:	07:00:00	AFTRS
Unit Call:	07:00:00	AFTRS
Breakfast:	08:00:00 - 08:15:00	AFTRS
Est Wrap:	18:45:00	

Sunrise: 05:24:00
Sunset: 18:02:00
Weather: Partly Cloudy Min 14 Max 21

ABSOLUTELY NO FOOD OR DRINK (EXCEPT FOR WATER BOTTLES) IN STUDIO

SC	PAGE TIMING	D/N INT/EXT	SET/LOCATION	CHARACTER	ARTIST	PU	WR	M/UP	ON SET
BLOCK-THROUGH 0815-0830 THEN LIGHT/COMPLETE M/UP AND WR 0830-0900									
9	0 4/8 00:00:25	Night INT	DRESSING ROOM @ AFTRS Studio 1 *Charlie's not going to Europe with them*	CHARLIE	Eloise Oxer	0630	0745	0715	0815
				SIMONE	Amelia Best	0620	0715	0845	0815
09:00-11:15									
Eloise (CHARLIE) to M/UP to REMOVE BLOODIED FIGURES SFX MAKE-UP									
2	1 2/8 00:01:07	Night INT	DRESSING ROOM @ AFTRS Studio 1 *Simone and Charlie get it on but are*	CHARLIE	Eloise Oxer	CLD	CLD	CLD	CLD
				SIMONE	Amelia Best	CLD	CLD	CLD	CLD

. . .

Fig. 22.8 A printer-ready version of the custom form of the call sheet shown in Fig. 22.7

a value of *String* type. This validation is interactive: any field of the form which has been filled out with invalid data is reported to the user with suggestions for correction. This function is particularly useful when the forms are very complex and thus error-prone. Apart from the above, a *print-preview* function allows the user to generate a printer-ready document from the Web form that resembles the hard copy format used in practice at the AFTRS. The print-out version can then be distributed to each cast and crew member in the film production team, as is the case of the call sheet shown in Fig. 22.8, which has been generated from the Web form in Fig. 22.7.

Finally, it is worth mentioning that the development of custom user interface is in general not an easy task. Custom user interface is not the core of a system but usually plays an important role in determining the final acceptation of the system by the end users. In YAWL4Film, the development of custom Web forms had gone through extensive discussions with people from AFTRS. The initial design of the forms were based on a few sets of sample film production documents used in practice at the AFTRS. The forms were then refined for many times through intensive interactions with the end users at the AFTRS. In total, 14 custom forms were developed in YAWL4Film, and it took approximately 500 h of a research assistant and 50 h of a professional programmer to develop these forms.

22.4 YAWL4Film Deployment

The deployment of YAWL4Film involves process deployment and system deployment. First, the film production process model was deployed in the YAWL runtime environment for enactment. Second, the system was deployed in a typical film production setting, which involves a central production office and a mobile unit for film shooting. It should be noted that YAWL4Film was originally developed in YAWL beta 8.2 and has been recently migrated to YAWL 2.0.

22.4.1 Process Deployment

The film production process model described in the previous section can be directly loaded into the Engine for execution. The Engine supports worklist handling where each work item is handled based on the three interaction points, that is, offered, allocated, and started, within a work item life-cycle. The user interface in the runtime environment is presented by means of Web pages and forms within a browser. An administrator can upload/unload a process specification as well as launch and cancel process instances through the case management administration form.

Figure 22.9 depicts an admin screen, as viewed by an administrator, for managing and monitoring work queues. The screen displays the list of all enabled work items at the beginning of a shooting day (i.e., after a work item *Start a Day* has been completed). The information about each work item includes the process specification, the identifier of the process instance (i.e., the case number), the task that the work item belongs to, its creation time and age, the human resources that the work item is assigned to, and the status of the work item within its life-cycle. For example, the work item of type *Update Call Sheet* is enabled and offered to two

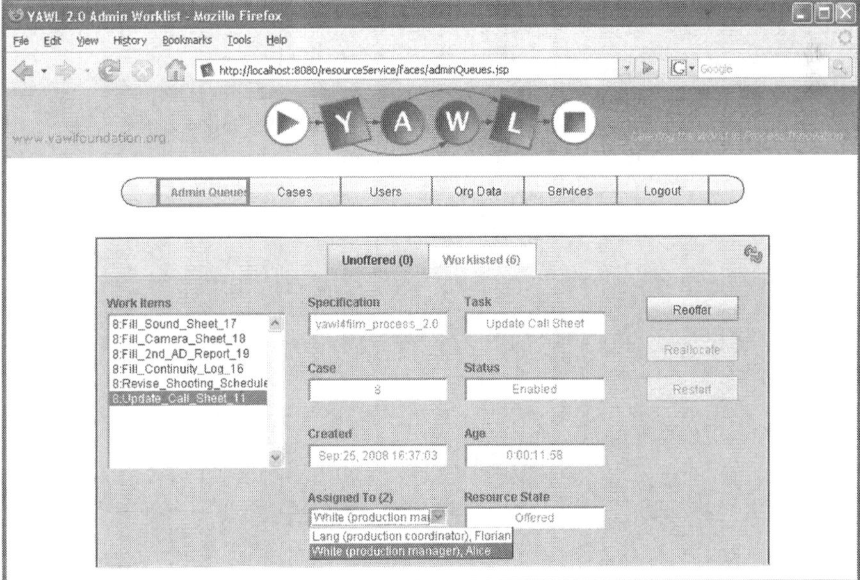

Fig. 22.9 Admin screen for management of work queues when a shooting day starts

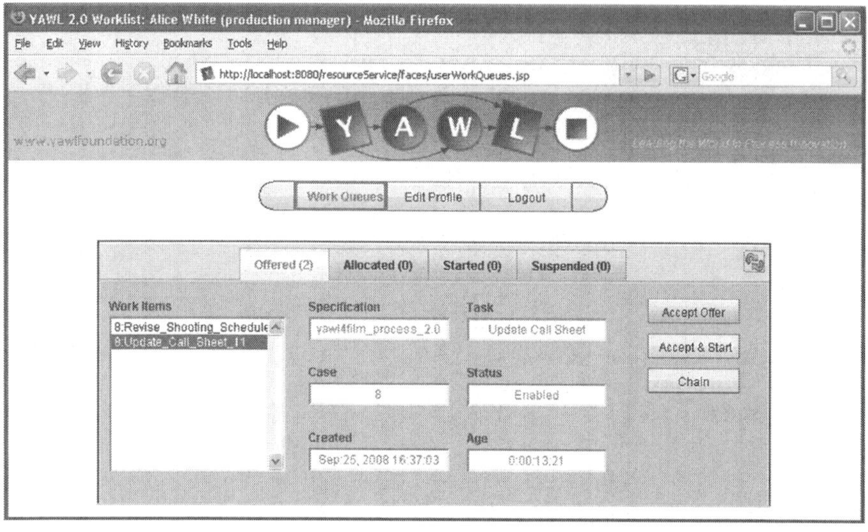

Fig. 22.10 Screen of an offered work queue of the production manager when a shooting day starts

crew members – one acting as the production manager and the other acting as the production coordinator.

The screen of an offered work queue for the production manager when a shooting day starts is shown in Fig. 22.10. There are two enabled work items: *Revise Shooting Schedule* and *Update Call Sheet*. Because of a deferred choice between these two work items (see the process model in Fig. 22.1), as soon as one of them is accepted, the other will be disabled and therefore disappear from the work queues. Note that in this screen, the "Chain" functionality is enabled for the *production manager* as he/she is granted a privilege to chain work item execution in the profile (see in Fig. 22.5). If the production manager decides to accept and start the work item *Update Call Sheet*, this work item will move to the started work queue as shown in Fig. 22.11. The "View/Edit" function[2] then links to the custom Web form of the call sheet (see in Fig. 22.7) associated with the work item.

22.4.2 System Deployment

A film production process usually involves a central production office and a film shooting unit. As with traditional workflow systems, YAWL4Film can be deployed as illustrated in Fig. 22.12. A server that runs the Engine is set up in the production office. The connection for communication between the production office and

[2] By default this links to an automatically generated Web form for user input.

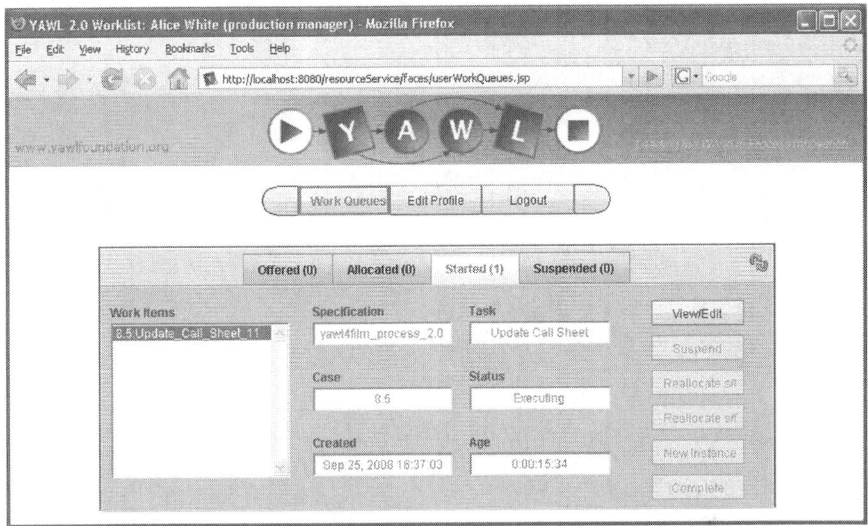

Fig. 22.11 Screen of a started queue of the production manager containing the work item *Update Call Sheet* by the production manager

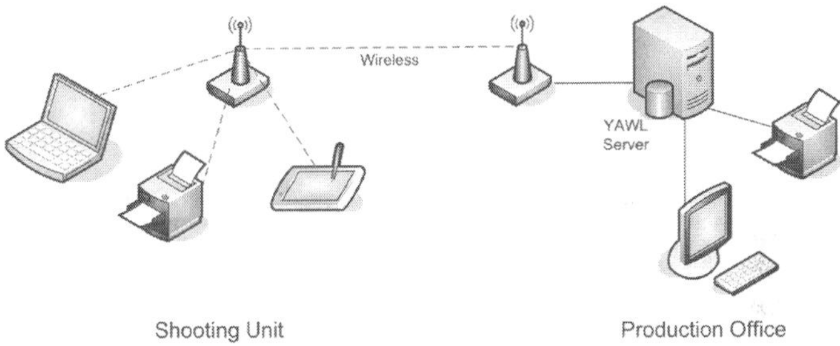

Fig. 22.12 Deployment scenario I (with connectivity) – traditional set-up

the shooting unit is available all the time via wired/wireless networks. The crew members can access the YAWL server via laptops, tablet PCs, etc.

However, unlike traditional workflow applications, the connection for communication between distributed stakeholders during a film production is not always guaranteed. For example, the designated location for film shooting may have no standard Internet or phone coverage to facilitate communication between the production office and the shooting unit(s). Although a dedicated wireless connection may be set up to cover the whole area between the production office and the unit, it is usually not feasible due to the budget constraints in many film production projects. In such cases, it is not possible to rely on a single centralized YAWL System, and the infrastructure in Fig. 22.12 must be revised. So far two alternative approaches have been considered.

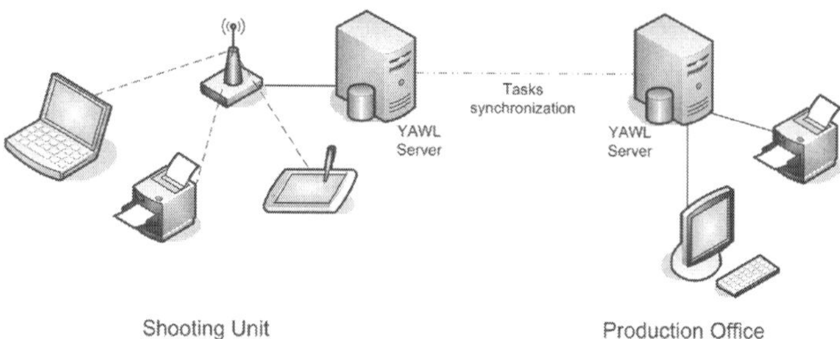

Fig. 22.13 Deployment scenario II (no connectivity) – decentralization and off-line task synchronization using execution logs

One approach is to split the centralized YAWL System into two systems: one runs at the production office, the other at the shooting location. This is illustrated in Fig. 22.13. Deploying two YAWL systems implies executing two instances of the production process model (one in each system). Every shooting day, the instance running at the production office is directly responsible for the production of the call sheet and the review and distribution of the DPR, while the instance running at the shooting unit is responsible for coordinating the completion of the shooting documents and generating the DPR. These two process instances are dependent on each other, as the former requires the DPR for revision and distribution, while the latter requires the call sheet for preparation of the onset shooting documents for the unit crew. Hence, daily synchronization between the two process instances need to occur at tasks *Kick off Onset* (where the unit needs the previous day's call sheet) and *Review DPR* (where the production office needs the current day's DPR). Specifically, all tasks between *Kick off Onset* and *Review DPR* will be executed off-line at the shooting site, and their execution logs will be replayed back at the production office after the shooting ends for the day. A number of tasks will then be performed in the production office in the evening and the execution logs of the tasks performed during the evening will be replayed the next morning in the YAWL System running at the shooting site. These operations are achieved by means of the "log replay" functionality of the Engine, which allows one to bring the execution state of the Engine to a given state by replaying logs.

The above revised infrastructure for deployment can overcome the issue of the lack of connectivity and at the same time help to maintain the integrity of workflow execution. However, the set-up of an additional YAWL server at the shooting location also requires the necessary technical support on set, and to initiate the "log replay" in the Engine is so far a manual task that requires knowledge about YAWL. As a result, the production team, especially the mobile production unit, would expect the introduction of extra amount of work, device, and even personnel. Hence, the solution may be hard to adopt.

The aforementioned considerations lead us to a second approach, which is to run "standalone" custom forms for the unit crew to fill in on set and to later synchronize the data from these forms to generate the DPR at the production office. The original custom forms, as introduced in Sect. 22.3.2, are linked to the worklist handler. In this case, data is communicated between the forms and the Engine, and hence the forms require a running Engine, that is, they cannot work on their own. The standalone forms are obtained by disabling the data communication between the forms and the Engine. All the functionalities on the original forms are provided except the *submission* function (see Fig. 22.7), which requires communication with the engine. Instead of reading data from the Engine, the standalone forms are initially supplied with default data. After the user fills out the forms, the data will not be checked back to the Engine (as the *submission* function is disabled on the forms), but it can be (locally) saved to a data file (as the *save* function is still available on the forms). In this way, standalone forms do not rely on a running Engine, and are also independent of network connection. During the system deployment at the AFTRS, the end users called the standalone forms the "off-line" forms, and the original forms the "online" forms.

The standalone forms are implemented for continuity log, continuity daily, sound sheet, camera sheet, and 2nd AD report. Each of these forms is installed in the computer device (e.g., tablet PC) of the designated crew member. The data synchronization between the production office and the shooting unit is bidirectional. On the one hand, at the end of each shooting day, the saved files containing the data from each of the standalone onset forms will be physically brought to the production office (e.g., by a *production runner*). The files will then be loaded into the respective online forms in order to generate the DPR. On the other hand, when the process starts the flow of another shooting day, the initial data for the onset forms will be generated and saved to files at the production office. The saved files will then be brought back to the shooting unit in the morning of the day after, before physically starting the new shooting session. This scenario is illustrated in Fig. 22.14. Although the flow of the onset shooting tasks is executed off-line (i.e., at the production office) later in a day, it does ensure the integrity of the user input data throughout the production process, which seems to be the most important issue for the end users. Also, deployment of the system is easy and does not introduce much extra work nor require YAWL-specific knowledge.

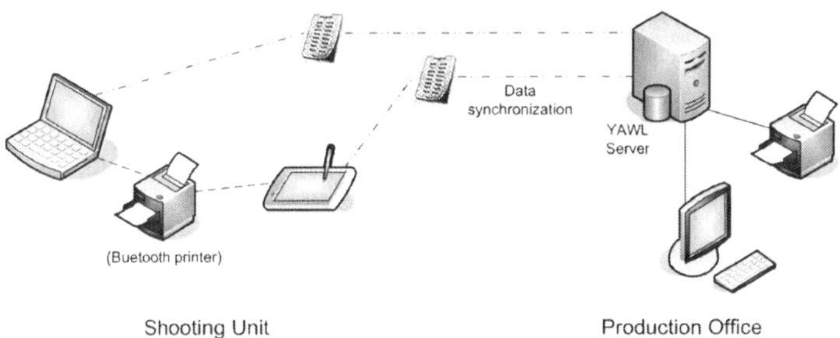

Fig. 22.14 System deployment scenario III (no connectivity) – standalone custom forms application and off-line data synchronization

22.5 Pilot Applications: *Rope Burn* and *Family Man*

In Australia, the AFTRS is the national training and research facility for Graduate Diploma, Masters courses, and short courses in film and TV production. YAWL4Film was deployed on two film production projects in the AFTRS in 2007.

Project 1, *Rope Burn*, was a three-day shoot in studio with 30 onset crew, 6 cast, and 6 production office crew. The office was run by a professional production manager, and supervised by a student producer. Project 2, *Family Man*, was a three-day shoot on location and in studio with 35 crew, 5 cast, and 4 production office crew. A semi-professional production manager was contracted and supervised by a student producer.

In Project 1, the connection for communication between the production office and the shooting unit (in studio) was available via wireless networks all the time, so that the system was set up in a traditional way as illustrated in Fig. 22.12. In Project 2, the connection for communication between the production office and the shooting location was not guaranteed. In this case, the system was deployed as illustrated in Fig. 22.14, which applied standalone custom forms on location and then performed the off-line data synchronization at the end of each day. For hardware set up in the two projects, both laptops and tablet PCs (with stylus-enabled user input) were used by onset crew members – continuity and 2nd AD.[3]

In both productions, the YAWL System shadowed the process of call sheet generation, DPR generation, and cast and crew database updates. For *Rope Burn*, the system was used on set alongside the traditional paper method of data capture for continuity and 2nd AD, and later for *Family Man*, the system totally replaced paper method for continuity and 2nd AD.

From the feedback from both projects, it was clear that the system has the potential to save time, and create more precise documentation:

> *I have managed over a dozen productions offices, and the amount of time this device[4] could save is incredible. Seeing the system up and running makes me realize how manual and laborious many of the activities are in any production office.*
>
> (Production Manager in *Rope Burn*)

> *I found the electronic form simple and easy to fill in. It was really just the same as using a paper form, but much cleaner and neater, for example, no messy handwriting, smudges, or crumpled paper.*
>
> (2nd AD in *Family Man*)

> *I so often make errors when calculating DPR or even the Call Sheet, it is much easier to use the tool to double check figures and ratios.*
>
> (Production Manager in *Family Man*)

---

[3] In both projects, students from the Camera and Sound Departments were not part of the testing and the system supervisor and technical assistant entered their data manually into the system.

[4] In both projects, users often employed terms like "device" and "tool" to refer to YAWL4Film.

The feedback also indicated that, once users became familiar with the tablet PC, the data input was significantly streamlined:

> *There is a bit of a knack to filling in the details using an electronic tablet and pen, but with a small amount of practice I found a way to do it that I was most comfortable with.*
>
> (2nd AD in *Rope Burn*)

> *Writing on the machine should be as fast as handwriting. The system in itself is pretty easy to use.*
>
> (Continuity in *Family Man*)

Finally, the crew members in both projects indicated that the more information one could store, such as scripts and schedule, the more useful the tool could become. Such feedback suggests that the YAWL system should be used right from the preproduction phase, e.g. during schedule planning and editing, so that information gathered during the preproduction phase can be exploited to better coordinate the production phase.

22.6 Epilogue

In this case study, YAWL was applied to support automation of film production processes, known as YAWL4Film. During the development of YAWL4Film, the generic YAWL System was extended with a number of additional modules tailored to the needs of the film industry. These include customized renderers, user form generators, report generators and data synchronization modules. YAWL4Film was also evaluated from its trial application in two film productions at the AFTRS. The positive feedback and comments received from these two projects bode well for the deployment of YAWL4Film in an educational setting and increased our confidence in the potential of this system for use in real-world film productions.

As is clear from the feedback of crew members of the pilot projects, YAWL4Film would benefit from providing support for activities in other phases of the screen business value chain, particularly preproduction. Also, we envisage deploying the YAWL4Film platform in the context of very large production projects, for example, those involving multiple production teams spread across multiple locations and employing several shooting units.

Exercises

Exercise 1. List two characteristics of processes within the screen business. Draw briefly the screen business value chain. Position the film production processes within this business value chain, and explain how it would benefit from application of workflow automation.

Exercise 2. In the film production process model shown in Fig. 22.1, task *End a Day* has an OR-join for synchronizing/merging two parallel branches – one concerned with the DPR and the other with the call sheet. Is it possible, without changing the process behavior, to replace this OR-join with other non-OR-join constructs or a combination of these? If so, construct the resulting process model. If not, explain why the model cannot be changed.

Exercise 3. The process model in Fig. 22.1 captures the entire film production process life-cycle from the first shooting day to the last shooting day. This can be referred to as Approach 1. Consider another approach, referred to as Approach 2, where a smaller process model is constructed to capture daily shooting process and is deployed on each shooting day, that is, one process instance is carried out for each day. In this approach, one needs to ensure that data dependencies between the process instances of two consecutive shooting days are handled properly. Discuss and compare these two approaches.

Exercise 4. The process model in Fig. 22.1 captures a film production process that involves one production office and one shooting unit. Extend this process model to support a film production process that employs more than one shooting units. (Hint: consider the use of multiple instance tasks)

Exercise 5. For the process deployment in Sect. 22.4.1, consider the offered work queue shown in Fig. 22.10 and the started work queue shown in Fig. 22.11 (both are work queues of the production manager). The offered work queue contains 2 work items (see "Offered (2)" in Fig. 22.10): *Revise Shooting Schedule* and *Update Call Sheet*. Once the work item *Update Call Sheet* is started, it is moved to the started work queue. At the same time, the offered queue becomes empty (as indicated by "Offered (0)" in Fig. 22.11). Discuss why the work item *Revise Shooting Schedule* is also removed from the offered queue once the work item *Update Call Sheet* is started.

Chapter Notes

During the development of YAWL4Film, the information about a standard film production process was collected from a number of sources: books [61, 121], Internet (see, e.g., the *filmmaking* entry on Wikipedia), and interviews with domain experts from AFTRS and from Porchlight, an independent film production company in Sydney.

YAWL4Film has been reported on in a few publications [188–190]. In [189], YAWL4Film is introduced as an example of a Web-scale workflow system for film production. The paper discussed workflow technology and Web technology used in developing YAWL4Film. It targeted an audience with general IT background. The publications [188, 190] both faced a different community of audience with background in art, social science, cultural science, and/or creative industries.

Thereby YAWL4Film was introduced as a process innovation to the domain of film production.

The development of YAWL4Film continued after its pilot applications at the AFTRS. In May 2008, a revised prototype focusing on call sheet generation of YAWL4Film was successfully deployed in the production of a full feature film by Porchlight. More details can be found in [190]. Having this success, the YAWL4Film team started designing and working on the development of Genie Workbench[5]. Genie Workbench is a suite of film and television production software that assist filmmakers in many production tasks. The first version of Genie Workbench consists of three software modules, namely Genie Schedule, Genie Cast, and Genie Crew, and was launched on 27 March 2009.

[5] www.genieworkbench.com

Part IX
Epilogue

Chapter 23
Epilogue

Wil van der Aalst, Michael Adams, Arthur ter Hofstede, and Nick Russell

23.1 Overview

This book presents both the YAWL language and the supporting toolset. YAWL fits well in the transition from workflow management systems focusing on automation of structured processes to *Business Process Management* (BPM) systems aiming at a more comprehensive support of a wider variety of business processes. The development of YAWL was triggered by limitations of existing software for process automation. Although much has been written about business processes, there is a lack of consensus about how they are best described for the purposes of analysis and subsequent enactment. This has resulted in a plethora of approaches for capturing business processes. Despite an abundance of standardization initiatives, there is no common agreement on the essential concepts. Standardization processes are mainly driven by vested business interests. The large and influential software companies in the BPM area are involved in multiple standardization processes (to keep all options open), but do not wholeheartedly commit to any of them.

The inherent complexity of business processes and the question of what fundamental concepts are necessary for business process modeling, enactment, and analysis gave rise to the development of a *collection of workflow patterns*. These patterns describe process modeling requirements in a language-independent manner. The *Workflow Patterns Initiative* started in the late nineties when the first set of 20 *control-flow patterns* were identified. These patterns where inspired by the design patterns for object-oriented design by Gamma et al. and focused on the ordering of tasks in business processes (e.g., parallelism, choice, synchronization, etc). Later this set of patterns was extended to more than 40 patterns. Moreover, driven by the success of the control-flow patterns, also other perspectives were analyzed using a patterns-based approach. This resulted in *data patterns* (dealing with the passing of information, scoping of variables, etc.), *resource patterns* (dealing with roles, task allocation, work distribution, delegation, etc.), *exception handling*

W. van der Aalst
Eindhoven University of Technology, Eindhoven, the Netherlands,
e-mail: w.m.p.v.d.aalst@tue.nl

A.H.M. ter Hofstede et al. (eds.), *Modern Business Process Automation*,
DOI 10.1007/978-3-642-03121-2_23,

patterns (dealing with exceptional situations), *flexibility patterns* (to make processes more adaptive and adaptable), *interaction patterns* (for modeling the "glue" between processes), and so on. All of these patterns resulted in a systematic overview of the possible, and maybe also expected, functionality of an ideal BPM solution.

The various sets of workflow patterns (especially the control-flow, data, and resource patterns) were used for the analysis of modeling languages (UML Activity Diagrams, BPMN, EPCs, etc.), proposed standards (BPEL, XPDL, etc.), and workflow management systems (Staffware, FLOWer, WebSphere, FileNet, jBPM, OpenWFE, etc.). These patterns-based evaluations revealed the weaknesses of contemporary offerings. The patterns demonstrated to be an effective tool to expose the Achilles heel of a language, standard, or system. The Workflow Patterns Initiative directly influenced the development of systems and standards. Moreover, the patterns are frequently used in selection processes. Vendors of BPM systems were typically annoyed by such an analysis as it rigorously exposed weaknesses of their products. One of the typical responses would be that indeed functionality is missing, but that it is "impossible to provide a useable language that would cover a substantial set of patterns". Their argument would be that, to support more patterns, the language would become unnecessarily complex. The criticism of BPM vendors led to the development of *YAWL* (Yet Another Workflow Language), that is, *the YAWL initiative was motivated by the desire to show developers of workflow technology that it is possible to have a simple yet powerful language, covering a wide range of patterns.*

YAWL started as a "paper exercise" showing that the initial collection of control-flow patterns can be supported by a language simpler than those used by existing commercial tools. It turned out that just a few powerful constructs are needed to support a wide variety of desired behaviors. The starting point of YAWL was a deep understanding of Petri nets and the inherit limitations of existing languages. Although Petri nets are surprisingly expressive and allow for all kinds of analysis, it is obvious that such a simple language does not support particular patterns dealing with cancelation, synchronization of active branches only, and multiple concurrently executing instances of the same task. Hence constructs were added to YAWL to deal with these patterns. YAWL was given formal semantics in terms of state transition systems, and several dedicated analysis techniques were developed. This convincingly showed that *it is possible to provide a simple and clear language able to deal with many control-flow patterns.*

After a while it turned out that YAWL was more than just a paper exercise. There was a substantial interest in both academia and industry in further developing YAWL. This resulted in the realization of the first YAWL System in 2003. The first version of the system offered comprehensive support for the original control-flow patterns (all patterns except Implicit Termination), but provided only limited functionality in terms of the data and resource patterns. Driven by the identification of more patterns in all of the perspectives mentioned before (control-flow, data, resource, exception, flexibility, interaction, etc.), the functionality of YAWL was gradually extended to provide a complete BPM solution.

This book describes the YAWL language and related software. The YAWL System is set up using a *Service Oriented Architecture* (SOA). The core system handles the control-flow and data perspectives. The other perspectives are predominantly supported by dedicated YAWL Services. For example, flexibility is supported by the Worklet and Declare services. The YAWL System, and associated open-source systems such as ProM,[1] also provide a wide range of highly innovative analysis techniques. These techniques are described in detail in this book. Moreover, YAWL is compared to existing languages, standards, and systems, and two application domains are discussed (YAWL4Healthcare and YAWL4Film).

23.2 Positioning of YAWL

YAWL, and also this book, serve two purposes. On the one hand, YAWL and the associated software provide a comprehensive BPM solution that can be used by academics and practitioners. On the other hand, the YAWL initiative aims to influence vendors and standardization bodies to provide more functionality. Below we elaborate on both purposes.

YAWL as a Comprehensive BPM Solution

In part IV, we described the core system of YAWL. This part shows that YAWL provides all the functionality that may be expected from a classical workflow management system. The fact that it is based on the patterns makes the system more expressive. As a result, it is easier to realize suitable support for complex processes and typically fewer work-arounds are needed. The various services provide additional support for, for example, flexible processes. Moreover, unlike most existing BPM systems, there is support for verification and innovative forms of analysis, such as process mining and short-term simulation.

The fact that YAWL is open-source makes it an interesting platform for both academics and practitioners. Organizations that want to use workflow technology can quickly get started without an elaborate and costly selection process. YAWL can be installed in a few minutes and the software is for free. Commercial BPM systems tend to be expensive and difficult to install. YAWL is also an ideal platform for BPM scientists. It is easy to extend YAWL, because it is open-source, uses standard XML technology, and has clearly defined interfaces.

YAWL as Tool to Influence Vendors and Standards

As indicated before, YAWL started as a paper exercise to show that it is possible to provide a simple, yet very powerful, workflow language. Hence, YAWL is not only positioned as a comprehensive BPM solution but also aims at advancing the maturity of (other) languages, standards, and systems. In this sense, YAWL can be seen as an extension of the Workflow Patterns Initiative influencing workflow developers, software vendors, standardization bodies, and so on. Here the role of end users is

[1] www.processmining.org

vital. If organizations do not ask for more functionality, commercial products are not likely to evolve into more mature solutions. YAWL can help end-users to gain a better understanding of the required BPM functionality and use this in selection processes.

23.3 Analysis

While the traditional focus of workflow technology is on automation, *more mature BPM approaches emphasize on process analysis*. Therefore, analysis techniques play(ed) an important role in the development of YAWL. Figure 23.1 shows an overview of YAWL, highlighting its analysis capabilities. YAWL models are used to model real-life processes and serve as input for the Engine and related services. Note that the term "YAWL models" should be interpreted in a broad sense, that is, process models, organizational models, Ripple-Down Rules, and so on. The Engine and related services use these models to support the corresponding processes in some context denoted as the "world" in Fig. 23.1. While doing this, events are recorded. For analysis, two artifacts can be used: (a) YAWL models and (b) event logs. Figure 23.1 shows four examples of analysis activities supported by YAWL: (1) verify, (2) check conformance, (3) discover, and (4) simulate.

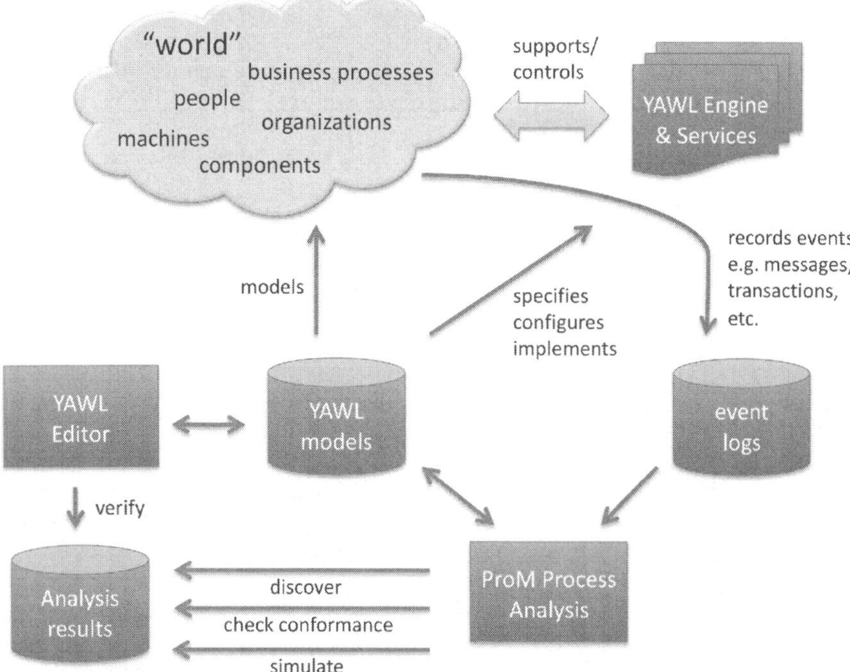

Fig. 23.1 Overview of analysis in the context of YAWL

Verification aims to determine whether the model has some obvious flaws (e.g., deadlocks, livelocks, etc.). An important concept is the notion of *soundness*, which states that (a) it should always be possible to be complete, (b) when completing a case no dangling references should be left behind, and (c) there are no dead tasks. In the context of Workflow nets, several powerful analysis techniques have been developed. For example, the tool Woflan is using all kinds of Petri-net-based analysis techniques to discover errors and to assess correctness. The functionality of Woflan has been incorporated in YAWL (and ProM). However, the YAWL language is much more expressive than workflow nets or Petri nets. The ability to have cancelation regions and the presence of OR-joins make analysis more difficult. Therefore, dedicated analysis techniques have been developed and implemented in the YAWL Editor. Given the expressiveness of YAWL, this functionality can easily be adapted for languages such as BPMN, UML Activity Diagrams, EPCs, and so on.

Traditionally, most of BPM research focussed on *model-based analysis*, that is, only a model is used as input for analysis. However, more and more information about (business) processes is recorded by information systems in the form of so-called "event logs." IT systems are becoming more and more intertwined with these processes, resulting in an "explosion" of available data that can be used for analysis purposes. The goal of *process mining* is to extract process-related information from event logs, for example, to automatically *discover* a process model by observing events recorded by some information system. However, process mining is not limited to discovery and also includes *conformance checking* (investigating whether reality conforms to a given model and vice versa). Moreover, the availability of event logs enables *innovative simulation approaches*. A particular challenge is the discovery of *complete* simulation models with the control-flow, resources, allocation rules, timing information, and so on, in place. This requires the ability to merge and integrate the different perspectives, for example, the simulation model is derived for the YAWL model while timing information and information about the availability of resources are derived from the log. The tight coupling of simulation with process mining enables new forms of analysis. An example is "short-term simulation", where the simulation experiment can start in the *current state* based on logs or based on the state of the information system. Unlike traditional simulation that typically starts in some "empty" state, this form of simulation uses the real current situation to initialize the model. By using a complete model covering all perspectives and the information about the current state, it is possible to construct an accurate model based on observed behavior rather than a manually constructed model, which approximates the workflow's anticipated behavior. Moreover, the state information supports a "fast forward" capability, in which simulation can be used to explore different scenarios with respect to their effect in the near future. In this way, simulation can be used for *operational decision making*.

The tight coupling between YAWL and the process mining tool ProM enables innovative forms of analysis such as process discovery, conformance checking, and short-term simulation. It is expected that, in the future, also predictions and recommendations are supported through the combination of both systems.

As BPM technology matures, more information will become available in the form of (executable) process models and event logs. This will make the analysis techniques supported by YAWL and related tools even more relevant.

23.4 Next Steps

To conclude, we would like to encourage the reader to actively use the information provided in this book and to think about next steps.

If you are a *developer of workflow technology*, this book provides a rich source of innovative ideas and a check-list for missing functionality. When developing or extending a BPM solution, the patterns could be used as an objective means to assess the desired and available workflow support. Moreover, if there is a mismatch between desired and available workflow support, this book provides hints on how to realize missing functionality in a generic manner.

For *software developers using workflow technology*, the goal is not to develop a new BPM product but to select and deploy an existing one. As YAWL is open-source and is easy to install, it is relatively easy to develop an initial prototype for a pilot project. Moreover, this book will assist in a better understanding of the basic BPM concepts.

For *BPM researchers* YAWL provides an ideal platform for experimenting with new ideas. YAWL is very open and has clearly defined interfaces. Therefore, it is easy to extend and add experimental functionality (e.g., new analysis techniques, new flexibility mechanisms, innovative services, etc.).

If you are a *workflow consultant or analyst*, this book provides a good overview of the state-of-the-art in BPM. Unlike many other books in this space, this book is more concrete as it is centered around a particular system. However, the concepts addressed are not system specific and the reader is not bothered with technical details. As a result, the book can be used as a reference for the selection and evaluation of BPM technology.

If you are an *end user*, it is very important to come up with the right requirements. Many BPM projects fail *because end users are unable to clearly express requirements* during the selection and development process. Only at the end, it becomes clear that much functionality is missing. This book can help to bridge the classical gap between business and IT. Moreover, the nature of YAWL allows for pilot projects where a running prototype is realized in a short time-span and with minimal costs. Experience shows that playing with a prototype is much more effective than having endless discussions about process models and notations.

Part X
Appendices

Appendix A
The Order Fulfillment Process Model

Marcello La Rosa, Stephan Clemens, and Arthur ter Hofstede

A.1 Introduction

This appendix describes the Order Fulfillment process followed by a fictitious company named Genko Oil. The process is freely inspired by the VICS (Voluntary Inter-industry Commerce Solutions) reference model[1] and provides a demonstration of YAWL's capabilities in modeling complex control-flow, data, and resourcing requirements.

A.2 Overall Process

The Genko Oil company features four departments: the *Order Management Department* (OD), the *Supply Department* (SD) including the *Warehouse*, the *Carrier Department* (CD), and the *Finance Department* (FD). The Order Fulfillment process model is divided into the following phases:

- Ordering
- Logistics, which includes
  - Carrier Appointment
  - Freight in Transit
  - Freight Delivered
- Payment

The Order Fulfillment process model is shown in Fig. A.1, where each of the above phases is captured by a composite task. The orders remitted by customers

[1] www.vics.org

M. La Rosa
Queensland University of Technology, Brisbane, Australia,
e-mail: m.larosa@qut.edu.au

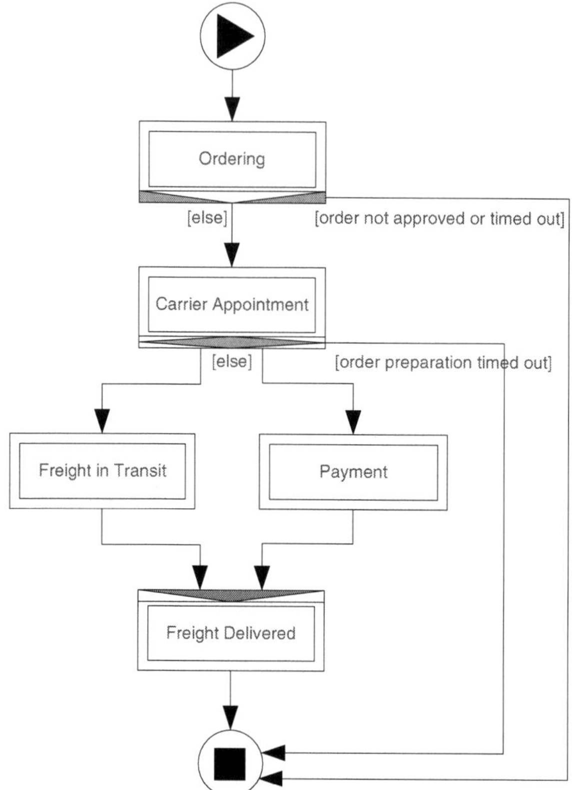

Fig. A.1 The Order Fulfillment process model in YAWL

are processed by the Orders Management Department. To keep the example manageable, a number of simplifying assumptions were made. For example, we assume that an order does not lead to more than one shipment. On the other hand, orders from different clients may be combined into a single shipment. A single package occupies only a fraction of a truck and can correspond to one of a fixed number of sizes.

The process starts with the *Ordering* task, where purchase orders can be created, modified, approved, or rejected. If an order is rejected or not confirmed in time, the process terminates, otherwise the logistical arrangements start in the task *Carrier Appointment*. This task deals with the preparation of the shipment quote, with delivery and pick-up arrangements and culminates in the actual pickup of the freight from the *Carrier Department*. If the quote is not prepared in time, the whole process terminates and the order is canceled. If the freight is picked up, tasks *Payment* and *Freight in Transit* are started in parallel. The former deals with the processing of the payment for the freight and for the shipment. The latter allows the Order department to issue inquiries after the status of the freight in transit, and handles notifications of the shipment's progress from the *Carrier Department*. This task terminates with the actual delivery of the freight to the customer. After completion of payment and

Table A.1 Participants of the Order Fulfillment process – name, position and role

User id	First name	Last name	Position	Role
ao	Arturo	de Ofstede	FD clerk	Finance Officer
bva	Billy	Van Arsdale	SD clerk	Junior Supply Officer
cc	Carmine	Cuneo	CD clerk	Shipment Planner
ccr	Connie	Corleone Rizzi	Head of warehouse	Warehouse Admin Officer
cm	Carmine	Marino	OD clerk	PO Manager, Client Liaison
cmc	Captain	McCluskey	FD clerk	Finance Officer
cr	Carlo	Rizzi	SD clerk	Junior Supply Officer
dcc	Don Carmine	Cuneo	FD clerk	Account Manager
dvc	Don Vito	Corleone	CEO	Order Fulfillment Manager
eb	Emilio	Barzini	FD clerk	Finance Officer
fc	Fredo	Corleone	Head of OD	PO Manager
jf	Johnny	Fontaine	CD clerk	Shipment Planner
jj	Jaggy	Jovino	SD clerk	Senior Supply Officer
jl	Joe	Lucadello	SD clerk, Warehouse clerk	Senior Supply Officer
jw	Jack	Woltz	Warehouse clerk	Warehouse Officer
ka	Kay	Adams	Head of CD	Carrier Admin Officer
mac	Mama	Corleone	CD clerk	Shipment Planner
mb	Momo	Barone	CD clerk	Courier
mc	Michael	Corleone	OD clerk	PO Manager
mlr	Marcello	La Rosa	CD clerk	Courier
pc	Peter	Clemenza	CD clerk	Courier
sc	Sonny	Corleone	OD clerk	PO Manager
sca	Stefano	Clemenza	CD clerk	Shipment Planner, Courier
st	Sal	Tessio	OD clerk	Client Liaison
th	Tom	Hagen	Head of SD, Assistant head of OD	Senior Supply Officer, Supply Admin Officer
vmc	Vincent 'Vinnie'	Mancini-Corleone	OD clerk	PO Manager
vs	Virgil 'The Turk'	Sollozzo	Head of FD	Senior Finance Officer

delivery, task *Freight Delivered* handles loss or damage claims and requests for return of merchandise. If no claim or request is lodged within a certain time frame, the process terminates.

Table A.1 lists all participants of the Order Fulfillment process with their user identifier, name, positions and roles. Each participant has default password "apple", whereas the YAWL administrator has user identifier "admin" and password "YAWL". Figure A.2 depicts the organizational chart of Genko Oil, where each participant belongs to one or more positions within a department.

All participants except *Don Vito Corleone* have the privilege to choose which work item to start, to start work items concurrently, and to reorder work items. *Don Carmine Cuneo* can also chain work item execution, and *Tom Hagen* and *Virgil "The Turk" Sollozzo* can also view all work items of their organizational group. *Don Vito Corleone* can only manage cases in his role of Manager of the Order Fulfillment process.

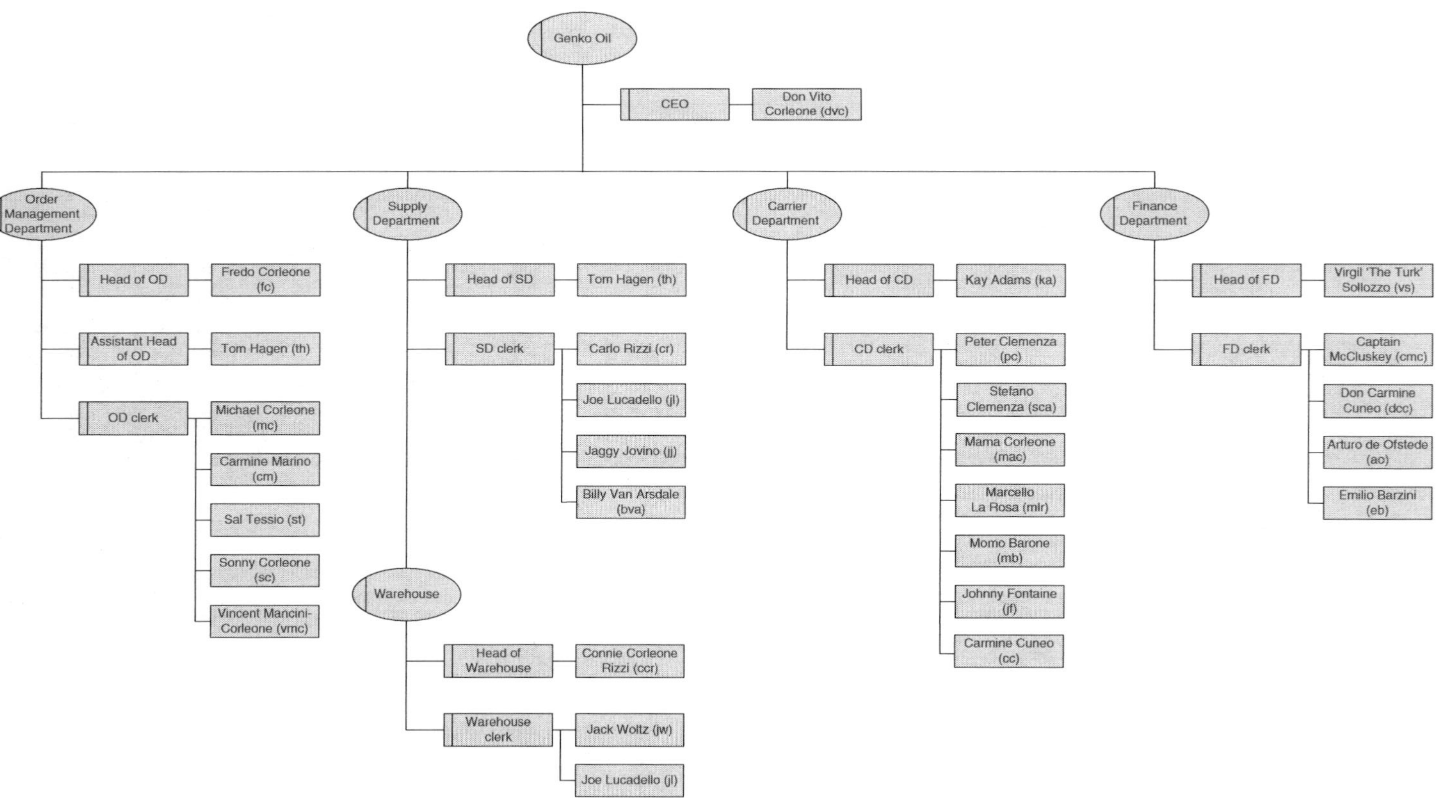

Fig. A.2 Organizational chart of the Genko Oil company

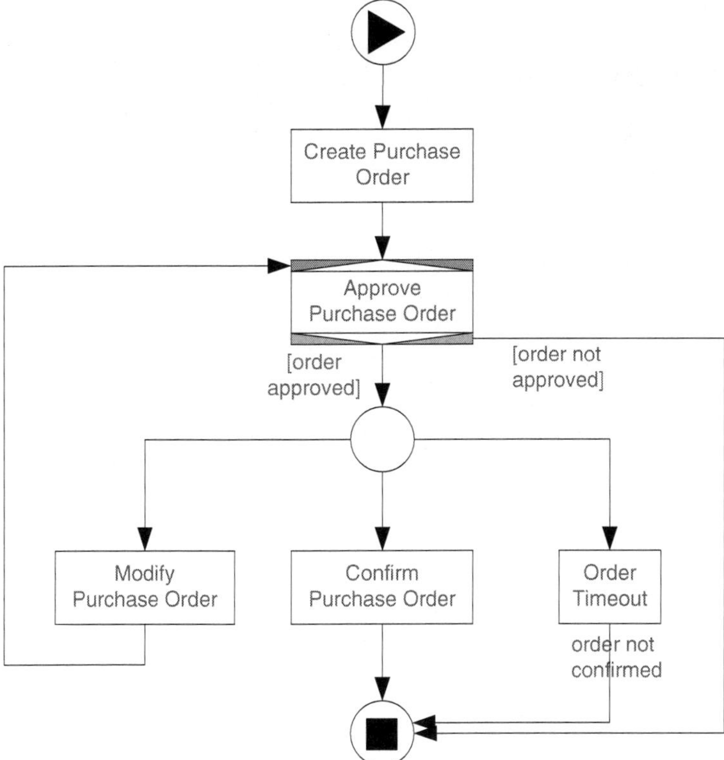

Fig. A.3 The *Ordering* subprocess

A.3 Ordering

The *Ordering* subprocess starts with the creation of a Purchase Order (PO) by the Order Management Department (see Fig. A.3). A PO needs to be approved by the *Supply Department* and may then be subject to a number of modifications, though it requires confirmation within a certain time frame.

The creation of a PO is handled by an *OD clerk*, who may choose to reallocate the task to another *PO Manager* with or without the work performed on it thus far. A *PO Manager* may also choose to relinquish working on the task and have the system offer it again to the available *PO Managers*. Moreover, a *PO Manager* may suspend working on the creation of a PO and choose to resume working on it at some later stage. Finally, a *PO Manager* may volunteer to be the main entry point for processing POs during a certain period of time. When a *PO Manager* is offered a task to create a PO, and they volunteer for it, the system will initiate the task automatically. Upon completing the PO, the *PO Manager* needs to decide which *PO Manager* will work on modification requests as they may eventuate at a later stage. The default *PO Manager* for PO modifications and confirmations is *Carmine Marino* (user id "cm").

The completed PO is passed on to a *Supply Officer* who needs to approve it. If the *Supply Officer* who allocates this task to themselves is the *Head of SD*, they may choose to delegate this task to an *SD clerk* who reports to them. *Supply Officers* choose which approval tasks they will work on, and once they have chosen such a task, they may decide when to actually start working on it. This interaction pattern with the system is the default one for the various tasks that need to be performed in the Genko Oil company.

A PO contains information about the client's company (e.g., name, address and business number), information about the order (e.g., order number and date, currency, order terms, line items), the freight cost, and the delivery location. Moreover, it is possible to specify whether the order needs to be invoiced and whether it is part of a prepayment agreement between the client and Genko Oil.

Once a PO has been approved, repeated modifications may be requested. These are tracked by a revision number attached to the PO, which is increased at each change. Each of these changes again need to be approved. If the original PO or any modification is rejected, the order process ends. Moreover, the PO needs to be confirmed within 3 days, otherwise it is discarded and the process terminates.

A.4 Carrier Appointment

After confirmation of a PO, a route guide needs to be prepared and the trailer usage needs to be estimated. The route guide is prepared by determining the trackpoints that are going to be visited during the shipment. The trailer usage is determined by estimating the number of packages for the shipment, where each package has an identifier and a fixed volume of 25, 50, 100, or 200 lbs. These operations are performed in parallel by the two tasks *Prepare Route Guide* and *Estimate Trailer Usage* (see Fig. A.4). The former task is allocated to the Shipment Planner with the shortest work queue, while the latter task is allocated to the Shipment Planner who was allocated an instance of this task the longest time ago. If either task takes too long, a time out is triggered, which leads to the cancelation of the PO and the termination of the overall process. This timer is set to 5 days for a PO with one line item and is increased by 1 day for each additional line item. This calculation is performed by the automated task *Calculate Carrier Timeout*, which is assigned to a codelet to perform the required additions.

If both tasks *Prepare Route Guide* and *Estimate Trailer Usage* are completed in time, a *Supply Officer* can perform the task *Prepare Transportation Quote* by establishing the shipment cost based on the number of packages and on the total volume of the freight, and by assigning a shipment number to the order number. Once a *Supply Officer* has chosen to perform this task, they have the privilege to reallocate it to someone else without loss of the work performed thus far, and to suspend and resume working on the task. In addition, the system automatically starts the task for a *Supply Officer* once they have committed themselves to performing it.

After the task *Prepare Transportation Quote*, based on the total volume of the freight and on the number of packages, a distinction is made among shipments that

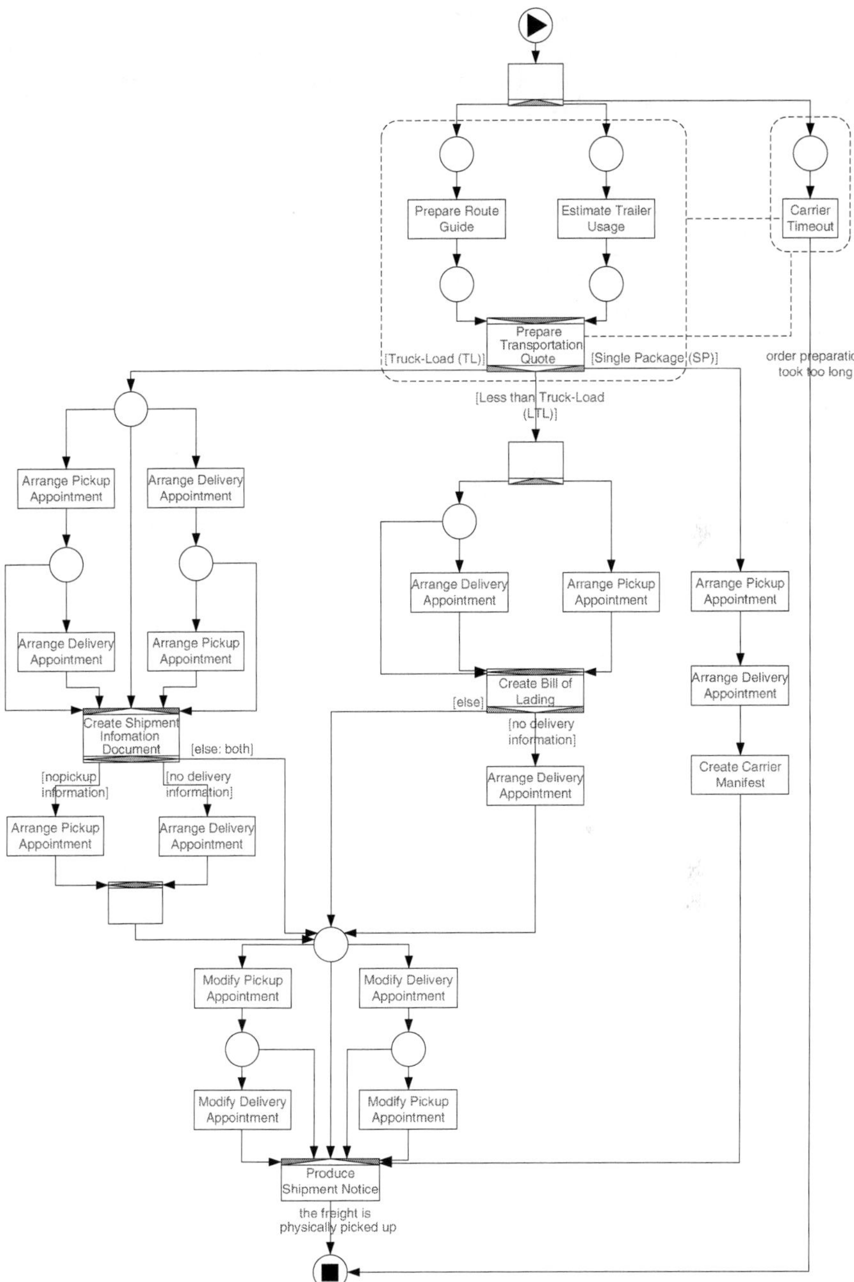

Fig. A.4 The *Carrier Appointment* subprocess

require a full truck load (total volume greater than or equal to 10.000 lbs), those that require less than a truck load (total volume less than 10.000lbs and more than one package), and those that simply concern a single package (total volume less than 10.000 lbs). For shipments that require a full truck load, *Client Liaisons* from the *OD* try to arrange a Pickup appointment and a Delivery appointment by specifying the location for pickup/delivery and any specific instructions. The *Client Liaisons* associated with these two tasks should be different. It is possible that only one of these or even none of these appointments is made before a *Senior Supply Officer* holding a *Master's in Supply Chain and Logistics Management* decides to create a Shipment Information document.

The Shipment Information document is used by the *Senior Supply Officer* to specify an authorization code and a consignee number for the shipment number. After the creation of this document, any missing appointments are made, though at this time a *Warehouse Officer* takes charge of arranging a Pickup appointment and a *Supply Officer* takes care of arranging a Delivery appointment, and there are subsequent opportunities to modify them until a *Warehouse Admin Officer* produces a Shipment Notice after which the freight can actually be picked up from the *Warehouse*. Modifications of Pickup appointments are handled by a *Warehouse Officer*, while modifications of Delivery appointments are taken care of by a *Supply Officer*.

When the shipment consists of more than one package but a dedicated truck is not required, then a *Warehouse Officer* arranges a Pickup appointment and a Client Liaison tries to arrange a Delivery appointment. Afterwards, a *Senior Supply Officer*, who holds a *Bachelor's in Supply Chain and Logistics Management*, creates a Bill of Lading, which, similar to the Shipment Information document, requires the specification of an authorization code and a consignee number. If no Delivery appointment was made prior, a *Supply Officer* takes care of this and the remainder of the process is the same as for a shipment that requires a dedicated truck.

For shipments consisting of a single package, the process is straightforward. All that needs to be done is by a *Supply Officer* – one who has the most experience in performing this particular task (identified using the "Round Robin by Experience" allocation strategy) – to create a Motor Carrier Pickup manifest. This is done by specifying only an authorization code. Afterwards, a Shipment Notice is produced by a *Warehouse Admin Officer* and the freight is ready for pickup.

A Shipment Notice provides a summary of the shipment, which includes the shipment number, the order number, the number of packages, the pickup and delivery appointments, and a variable indicating whether the shipment is a full truck load. The *Warehouse Admin Officer* has to indicate when the freight loading on the carrier's truck started and completed and provide the details of the driver (number and name), the delivery instructions, and a deadline for the claims, which will be used later on in the subprocess *Freight Delivered*. Delivery instructions contain textual instructions, the delivery date, and the delivery location (the latter being retrieved from the Route Guide).

A.5 Payment

After a freight has been picked up, the *Payment* subprocess can start (see Fig. A.5). This process has two components, one which is concerned with payment for the shipment and one which is concerned with payment for the freight.

The first task that needs to be performed in dealing with the payment for a shipment is the production of a Shipment Invoice containing the shipment costs related to the order number, the shipment number, and the company to be invoiced. This task is handled by a *Supply Admin Officer*. During the creation of the PO in the *Ordering* subprocess, a *PO Manager* specified if shipments were paid in advance. In this case, all that is required is for a *Finance Officer* to issue a Shipment Remittance Advice, where he/she specifies the amount being debited to the client. Otherwise, a *Finance Officer* issues a Shipment Payment Order. This document includes a shipment payment order number that refers to both the order and the shipment numbers, the shipment costs, and the details of the beneficiary, which includes the company, the name and code of the beneficiary bank, and the name and number of the bank account.

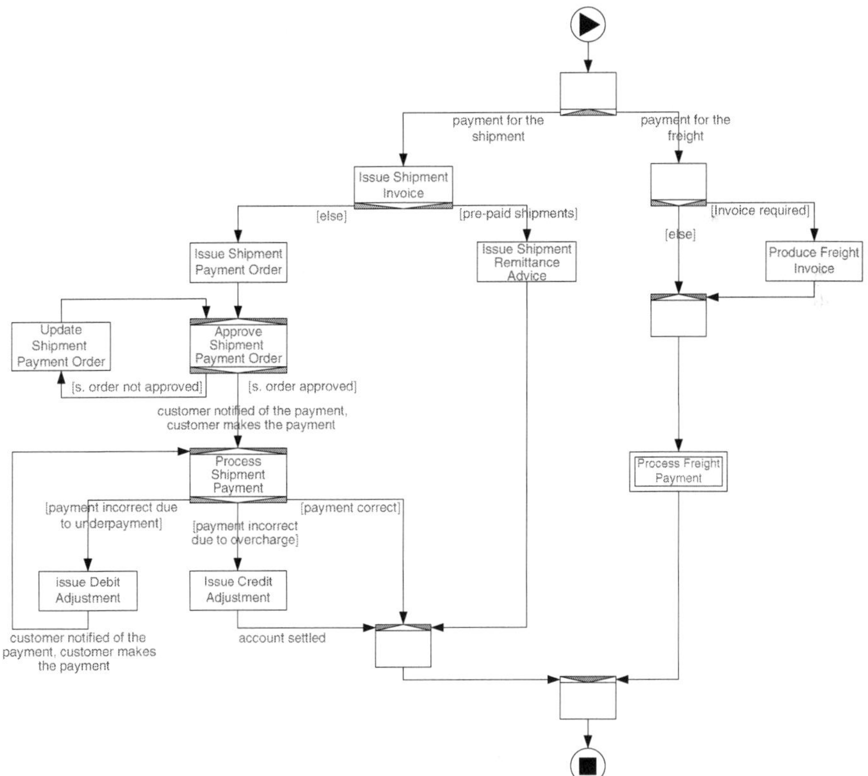

Fig. A.5 The *Payment* subprocess

A *Senior Finance Officer* who is the superior of this *Finance Officer* has to approve this document. A number of updates may be made to this document by the *Finance Officer* who issued it, but each of these need to be approved by the *Senior Finance Officer*. After the document is finalized, an *Account Manager* can process the payment for the shipment once the client has made the payment, by specifying the balance. It is possible that the client has underpaid. This case requires a debit adjustment (the amount still due is specified), the Account Manager needs to notify the client and the payment needs to be reprocessed. A client can also overpay and then the *Account Manager* needs to perform a credit adjustment (the reimbursement amount and the beneficiary details are specified). In the latter case and in case of correct payment, the shipment payment process is completed. An *Account Manager* can choose to have adjustment tasks to be started immediately for them upon completion of the processing of a Shipment Payment in order to expedite matters.

As regards the payment for a freight, a *PO Manager* specifies during creation of a PO whether an invoice is required or not. Should an invoice be required, a Freight Invoice is created by a *Supply Admin Officer*, containing the freight costs related to the order number and the company to be invoiced. If a Product Invoice is not required, then processing of the freight payment can start right away, if not, it can begin after the creation of the Freight Invoice. The processing of a freight payment is an involved subprocess, which we will not consider further.

A.6 Freight in Transit

The *Freight in Transit* subprocess is performed in parallel with the *Payment* subprocess and is concerned with tracking progress of the delivery of an order and with handling client inquiries (see Fig. A.6).

A delivery truck may visit multiple destinations and some of these may be designated trackpoints that may assist in finding out where a certain shipment is. When a trackpoint is visited, a Trackpoint Notice is issued by the *Courier*, registering the truck's arrival and departure time, plus additional notes. The *Courier* may choose to skip this task as it may lead to undesirable delay in some cases.

Once all trackpoints have been visited, *Carrier Admin Officers* need to log a Trackpoint Order Entry for each trackpoint, which contains a report for the specific trackpoint so that this information can be audited in the future. Trackpoint Order Entries can be logged in parallel so long as there are Carrier Admin Officers available to work on the entries. While entries are being logged, customer enquiries need to be addressed by a *Client Liaison*. Once completed, new enquiries can be ignored and a *PO Manager* creates an Acceptance Certificate to register that the freight has been physically delivered. An Acceptance Certificate refers to an order number and a shipment number and contains the acceptance date and delivery notes.

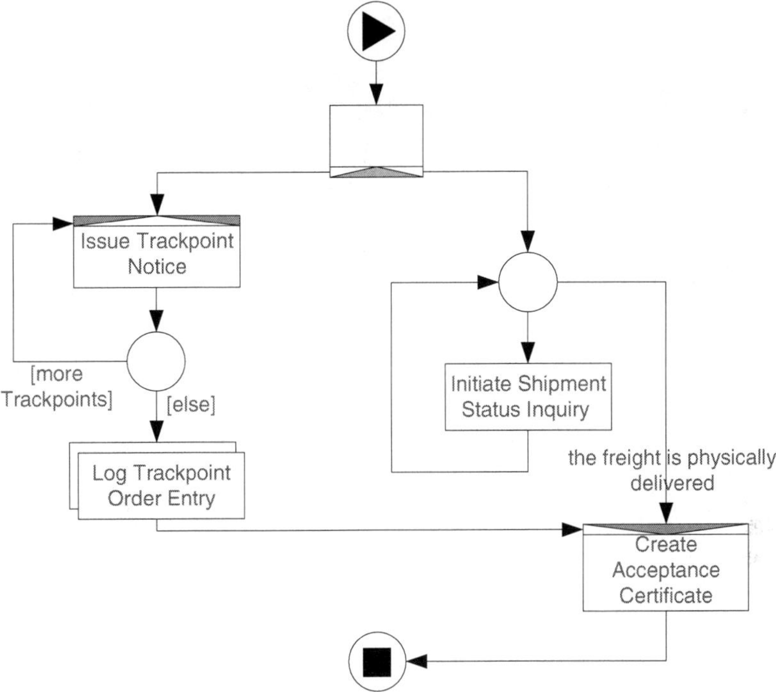

Fig. A.6 The *Freight in Transit* subprocess

A.7 Freight Delivered

After the delivery of the freight, clients may request a Return of Merchandise or may want to lodge a Loss or Damage Claim (see Fig. A.7). A Return of Merchandise document contains the reason for return and the line items to be returned, while a Loss or Damage Claim indicates the reason for claim. Both these documents are lodged into the system by a *Client Liaison* upon the client's request. If no client communications are received within a certain period after the delivery – the length of which is specified in the Shipment Notice – the Order Fulfillment process is considered to be successfully completed.

The request for a Return of Merchandise needs to be approved by a *Senior Supply Officer*. If approved, an involved Return Management process starts, which will not be considered further. Similarly, a Loss or Damage Claim also needs to be approved by a *Senior Supply Officer*, and if approved, will involve the process of actually managing this Loss or Damage Claim. Again, this is something we will not consider any further. In either case, if approval is not granted, the process is considered to be completed.

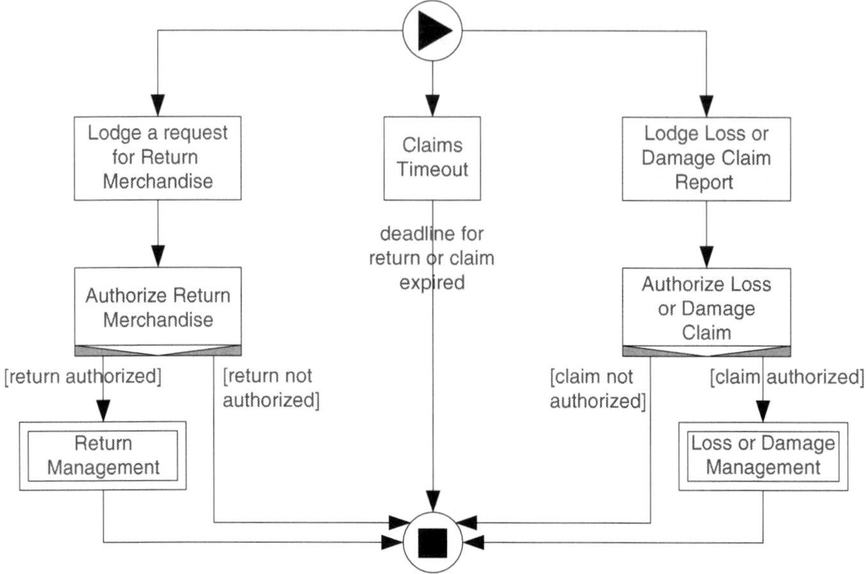

Fig. A.7 The *Freight Delivered* subprocess

A.8 Showcased YAWL features

The Order Fulfillment process demonstrates a wide range of language features. This section illustrates some advanced ones.

Cancelation Region: The subprocess *Carrier Appointment* contains two cancelation sets. The first one is triggered by the timer task *CarrierTimeout*, which, once completed, leads to the termination of the Order Fulfillment process by consuming tokens in tasks *Prepare Route Guide*, *Estimate Trailer Usage*, *Prepare Transportation Quote*, and in the conditions in-between. Similarly, the second cancelation set is triggered after completion of task *Prepare Transportation Quote*, leading to the consumption of tokens in the task *CarrierTimeout* and in its preceding condition.

Multiple Instance Task: In subprocess *Freight in Transit*, multiple instance task *Log Trackpoint Order Entry* reads all Trackpoint Notices created in task *Issue Trackpoint Notice* and assigns each notice to a *Carrier Admin Officer* who can attach a report to it. Multiple *Carrier Admin Officers* can work in parallel on this task.
OR-join: Subprocess Carrier Appointment features two OR-join constructs. The first one is in the truck load branch after tasks *Arrange Pickup Appointment* and *Arrange Delivery Appointment*. The second OR-join is attached to task *Create Bill of Lading* in the less than truck load branch.

Timer Task: Subprocess Carrier Appointment features a timer task that fires upon work item starting (task *Carrier Timeout*), while subprocess *Freight Delivered* features a timer task that fires on work item enablement (task *Claims Timeout*). The deadline associated with both these timer tasks is dynamically determined via a net variable.

Codelet: Automated task *Calculate Carrier Timeout* (subprocess *Carrier Appointment*) determines the timeout for the preparation of the Transportation Quote based on the number of Order Lines in the PO. This is done by using the *ExampleCodelet* in YAWL, which allows one to make simple additions.

Extended attributes: In task *Prepare Route Guide* of subprocess *Carrier Appointment*, task variables `OrderNumber` and `DeliveryLocation` are both Input and Output because their values need to be shown to the user and then used to compose the content of net variable `Prepare_Route_Guide`. However, to avoid that these values are modified by the user, the two variables have been assigned extended attribute `readOnly` with value `true`.

Dynamic Task Allocation: In subprocess *Ordering*, task *Modify Purchase Order* is allocated to the participant whose user id is stored in net variable `PO_Manager`. This variable is set by task *Create Purchase Order* before executing *Modify Purchase Order*.

Retain Familiar: When a single package is shipped, the participant who arranges the pickup appointment (task *Arrange Pickup Appointment* of subprocess Carrier_Appointment) has also to arrange the Delivery Appointment (task *Arrange Delivery Appointment*).

Four Eyes Principle: In full truck load shipments, tasks *Arrange Delivery Appointment* and *Arrange Pickup Appointment* of subprocess *Carrier Appointment* are never offered to the same participant during the same process instance.

Distribution Set Filter: The initial distribution set assigned to task *Create Purchase Order* is filtered such that this task can be piled to a single participant (runtime filtering). Similarly, a filter on capabilities is applied to the distribution set assigned to tasks *Create Shipment Information Document* (which is offered only to a *Senior Supply Officer* holding a *Master's in Supply Chain and Logistics Management*) and *Create Bill of Lading* (which is only offered to a *Senior Supply Officer* holding a *Bachelor's in Supply Chain and Logistics Management*).

Allocation Strategy: In subprocess *Carrier Appointment*, task *Estimate Trailer Usage* is allocated to the Shipment Planner who was allocated this task the longest time ago (allocation by time). Similarly, in case of single package shipments, task *Arrange Pickup Appointment* is allocated to the *Supply Officer* who executed this task the most (allocation by experience).

Privileges: Any participant with role of *Account Manager* has the privilege to chain the execution of work items assigned to them. In this way an *Account Manager* can have tasks *Issue Debit Adjustment* and *Issue Credit Adjustment* of subprocess *Payment* immediately started for them upon completion of task *Process Shipment Payment*.

A.9 Setup

The Order Fulfillment process has been tested to be executed by any YAWL 2.0 engine release (YAWL4Study, YAWL4Enterprise, and YAWLive). This section provides instructions on how to successfully launch a case of the Order Fulfillment process.

First, import the organizational data associated with the Order Fulfillment example into the yawl database:

1. Start the YAWL Engine
2. Open the Control Centre and log in as administrator (id "admin", password "YAWL")
3. Click on *Org Data*
4. If the YAWL database has already been populated with resources, backup that data by clicking on the icon "Export Org Data to file" on the top-right corner
5. Click on the icon "Import Org Data from file" on the top-right corner and select the file *orderfulfillment.ybkp* from the directory *[YAWL Installation folder]/misc/examples/orderfulfillment/*

A message box at the bottom should appear and indicate that the operation has succeeded by listing the number of resources that have been imported in terms of participants, groups, capabilities, and positions.

Now, a case of the Order Fulfillment process may be launched:

1. Click on *Cases* in the Control Centre
2. Upload the file *orderfulfillment.yawl* from the directory *[YAWL Installation folder]/misc/examples/orderfulfillment/*
3. Select the loaded specification *OrderFulfillment*
4. Click on *Launch Case*

Log out as administrator and log in as an employee of Genko Oil to view and edit the first work item. For example, log in as Fredo Corleone (user id "fg," password "apple") and start the creation of a PO.

Appendix B
Mathematical Notation

Nick Russell

This appendix outlines mathematical notations used in this book, which are not in general use and hence merit some further explanation.

Given two sets X and Y, a function $f : X \rightarrow Y$ can be defined, indicating that for $x \in X$ there exists *exactly one* element $y \in Y$ such that $f(x) = y$. If f applies to *all* members of X, then f is considered to be a *total* function (generally denoted as $f : X \rightarrow Y$). If the function does not apply to all members of X, then f is considered to be a *partial* function and is denoted $f : X \nrightarrow Y$.

A function $f : X \rightarrow Y$ can be considered to be comprised of a set of ordered pairs $(x, y) \in f$, where $x \in X, y \in Y$, and $f(x) = y$. The domain restriction operator (dom) when applied to f yields the set of elements $x \in X$ that participate in the function, that is, $dom(f) = \{x \mid (x, y) \in f\}$.

In the context of a function $f : X \rightarrow Y$, range restriction of f over a set $R \subseteq X$ is defined by $f \rhd R = \{(x, y) \in f \mid y \in R\}$.

The difference between two sets A and B can be defined as $A \backslash B = \{x \in A \mid x \notin B\}$.

$\mathbb{P}(X)$ denotes the power set of X, where $Y \in \mathbb{P}(X) \Leftrightarrow Y \subseteq X$.

$\mathbb{P}^+(X)$ denotes the power set of X without the empty set, that is, $\mathbb{P}^+(X) = \mathbb{P}(X) \backslash \{\varnothing\}$.

Let $V = \{v_1, \ldots v_n\}$ be a (nonempty) set and $<$ a strict total order over V, then $[V]^<$ denotes the sequence $[v_1, \ldots v_n]$ such that $\forall_{1 \leq i \leq j < n}[v_i < v_j]$ and every element of V occurs precisely once in the sequence. $[V]$ denotes the sequence in arbitrary order. Sequence comprehension is defined as $[E(x) \mid x \leftarrow [V]^<]$, yielding a sequence $[E(v_1) \ldots E(v_n)]$.

N. Russell
Eindhoven University of Technology, Eindhoven, the Netherlands,
e-mail: n.c.russell@tue.nl

A bag is a finite multi-set of elements from a given alphabet A and can be considered to take the form of a function from A to the set of natural numbers $\mathbb{N}$. A bag only contains a finite number of elements from A, each of which is mapped to a non-zero value. For a given bag X comprised of elements from alphabet A where $a \in A$, $X(a)$ denotes the number of occurrences of a in A, a notion also known as the cardinality of a in X.

The explicit enumeration of a bag uses a similar notation to that for sets except that square brackets are used instead of curly brackets and superscripts identify the cardinality of individual elements. For example $[a^2, b^3, c]$ denotes the bag with two elements a, three elements b and one c. The bag $[a^2 \mid P(a)]$ contains two elements a such that for every element a function $P(a)$ holds where P is some predicate on the symbols of the alphabet A under consideration. To denote individual elements of a bag, the "$\in$" symbol is used as for sets. For any bag X over alphabet A and element $a \in A$, $a \in X$ if and only if $X(a) > 0$.

The sum of two bags X and Y, denoted $X + Y$, is defined as $[a^n \mid a \in A \wedge n = X(a) + Y(a)]$. The difference of X and Y, denoted $X - Y$, is defined as $[a^n \mid a \in A \wedge n = max((X(a) - Y(a)), 0)]$.

Appendix C
The Original Workflow Patterns

Nick Russell

Since the first group of 20 control-flow patterns were identified and documented in 1999–2000, the Workflow Patterns have been subject to ongoing scrutiny and enhancement. In 2006, a comprehensive review was conducted of the original control-flow patterns, which resulted in the collection being augmented with a further 23 new patterns. Some of these were specializations of existing patterns, while others recognized earlier gaps in the scope of the pattern set.

Table C.1 identifies the original control-flow patterns that were specialized into two or more patterns that more precisely reflect specific problem/solution scenarios that were previously amalgamated in the original pattern definitions.

See www.workflowpatterns.com for an up-to-date and detailed description of all workflow patterns.

Table C.1 Refinements of original control-flow patterns

Original patterns	Revised patterns
Synchronization	– Synchronization – Generalized AND-Join
Synchronizing Merge	– Structured Synchronizing Merge – Local Synchronizing Merge – General Synchronizing Merge
Discriminator	– Structured Partial Join – Local Partial Join – General Partial Join
Multiple Instances with a priori Design Time Knowledge	– Multiple Instances with a priori Design Time Knowledge – Static Partial Join for Multiple Instances – Canceling Partial Join for Multiple Instances

(continued)

N. Russell
Eindhoven University of Technology, Eindhoven, the Netherlands,
e-mail: n.c.russell@tue.nl

Table C.1 (continued)

Original patterns	Revised patterns
Multiple Instances without a priori Design Time Knowledge	– Multiple Instances without a priori Design Time Knowledge – Dynamic Partial Join for Multiple Instances
Interleaved Parallel Routing	– Interleaved Parallel Routing – Interleaved Routing – Critical Section
Cancel Activity	– Cancel Task – Cancel Region – Cancel Multiple Instance Task – Complete Multiple Instance Task

Table C.2 Retained and new control-flow patterns

Original patterns	Revised patterns
Retained patterns	
Sequence	Sequence
Parallel Split	Parallel Split
Exclusive Choice	Exclusive Choice
Simple Merge	Simple Merge
Multiple Choice	Multiple Choice
Multiple Merge	Multiple Merge
Arbitrary Cycles	Arbitrary Cycles
Implicit Termination	Implicit Termination
Multiple Instances without Synchronization	Multiple Instances without Synchronization
Multiple Instances with a priori Runtime Knowledge	Multiple Instances with a priori Runtime Knowledge
Deferred Choice	Deferred Choice
Milestone	Milestone
Cancel Case	Cancel Case
New Patterns	
	Thread Merge
	Thread Split
	Recursion
	Explicit Termination
	Structured Loop
	Transient Trigger
	Persistent Trigger

Table C.2 identifies those patterns that were essentially retained in the same format in the revised pattern set and also shows the new patterns that were introduced as a consequence of the review.

References

1. W.M.P. van der Aalst, Three good reasons for using a Petri-net-based workflow management system, ed. by S. Navathe, T. Wakayama. In *Proceedings of the International Working Conference on Information and Process Integration in Enterprises (IPIC'96)*, Cambridge, MA, USA, November 1996, pp. 179–201
2. W.M.P. van der Aalst. The application of Petri nets to workflow management. J. Circ. Syst. Comput. **8**(1), 21–66 (1998)
3. W.M.P. van der Aalst, Formalization and verification of Event-driven Process Chains. Inform. Software Tech. **41**(10), 639–650 (1999)
4. W.M.P. van der Aalst, Patterns and XPDL: A critical evaluation of the XML process definition language, BPM Center Report BPM-03-09, BPMcenter.org (2003)
5. W.M.P. van der Aalst, M. Adams, A.H.M. ter Hofstede, M. Pesic, M.H. Schonenberg, ed. by L. Chen, C. Liu, Q. Liu, Ke Deng. In Database Systems for Advanced Applications, DASFAA 2009 International Workshops: BenchmarX, MCIS, WDPP, PPDA, MBC, PhD, Brisbane, Australia, April 20–23, 2009, Lecture Notes in Computer Science, vol. 5667, Brisbane, Australia, April 2009 (Springer, Berlin, 2009), pp. 319–333
6. W.M.P. van der Aalst, L. Aldred, M. Dumas, A.H.M. ter Hofstede, Design and implementation of the YAWL system, ed. by A. Persson, J. Stirna. In *Proceedings of the 16th International Conference on Advanced Information Systems Engineering (CAiSE'04)*, Lecture Notes in Computer Science, vol. 3084, Riga, Latvia, June 2004 (Springer, Berlin, 2004), pp. 142–159
7. W.M.P. van der Aalst, A.P. Barros, A.H.M. ter Hofstede, B. Kiepuszewski, Advanced workflow patterns, ed. by O. Etzion, P. Scheuermann. In *Proceedings of the 5th IFCIS International Conference on Cooperative Information Systems (CoopIS'2000)*, Lecture Notes in Computer Science, vol. 1901, Eilat, Israel, September 2000 (Springer, Berlin, 2000), pp. 18–29
8. W.M.P. van der Aalst, T. Basten, Inheritance of workflows: An approach to tackling problems related to change. Theor. Comput. Sci. **270**(1–2), 125–203 (2002)
9. W.M.P. van der Aalst, J. Desel, E. Kindler, On the semantics of EPCs: A vicious circle, ed. by M. Rump, F.J. Nüttgens. In *Proceedings of the EPK 2002: Business Process Management using EPCs*, Trier, Germany, November 2002 (Gesellschaft für Informatik, Germany, 2002), pp. 71–80
10. W.M.P. van der Aalst, M. Dumas, F. Gottschalk, A.H.M. ter Hofstede, M. La Rosa, J. Mendling, Preserving correctness during business process model configuration. Formal Aspect. Comput. (2009). doi: 10.1007/s00165-009-0112-0
11. W.M.P. van der Aalst, K.M. van Hee, *Workflow Management: Models, Methods and Systems*, new edn. (MIT, Cambridge, MA, 2004)
12. W.M.P. van der Aalst, K.M. van Hee, A.H.M. ter Hofstede, N. Sidorova, H.M.W. Verbeek, M. Voorhoeve, M.T. Wynn, Soundness of workflow nets: Classification, decidability, and analysis, BPM Center Report BPM-08-02, BPMCenter.org (2008)
13. W.M.P. van der Aalst, K.M. van Hee, A.H.M. ter Hofstede, N. Sidorova, H.M.W. Verbeek, M. Voorhoeve, M.T. Wynn, Soundness of workflow nets with reset arcs is undecidable! ed. by J. Kleijn, M. Koutny. In *Proceedings of the International Workshop on Concurrency Methods*

Issues and Applications (CHINA'08), Xi'an, China, June 2008 (Xidian University, China, 2008), pp. 57–72

14. W.M.P. van der Aalst, A.H.M. ter Hofstede, Workflow patterns: On the expressive power of Petri-net-based workflow languages, ed. by K. Jensen. In *Proceedings of the 4th International Workshop on Practical Use of Coloured Petri Nets and the CPN Tools*, Aarhus, Denmark, August 2002, pp. 1–20, Technical Report DAIMI PB-560 (2002)
15. W.M.P. van der Aalst, A.H.M. ter Hofstede, YAWL: Yet Another Workflow Language. Inform. Syst. **30**(4), 245–275 (2005)
16. W.M.P. van der Aalst, A.H.M. ter Hofstede, B. Kiepuszewski, A.P. Barros, Workflow patterns. Distr. Parallel Database **14**(3), 5–51 (2003)
17. W.M.P. van der Aalst, S. Jablonski, Dealing with workflow change: Identification of issues and solutions. Int. J. Comput. Syst. Sci. Eng. **15**(5), 267–276 (2000)
18. W.M.P. van der Aalst, J. Nakatumba, A. Rozinat, N. Russell, Business process simulation: How to get it right? BPM Center Report BPM-08-07, BPMcenter.org (2008)
19. W.M.P. van der Aalst, M. Pesic, DecSerFlow: Towards a truly declarative service flow language, ed. by M. Bravetti, M. Nunez, G. Zavattaro. In *Proceedings of the International Conference on Web Services and Formal Methods (WS-FM'06)*, Lecture Notes in Computer Science, vol. 4184, Vienna, Austria, 2006 (Springer, Berlin, 2006), p. 1–23
20. W.M.P. van der Aalst, M. Pesic, in *Specifying and Monitoring Service Flows: Making Web Services Process-Aware*, ed. by Baresi, Di Nitto. Test and Analysis of Web Services (Springer, Berlin, 2007), pp. 11–56
21. W.M.P. van der Aalst, A.J.M.M. Weijters, L. Maruster, Workflow mining: Discovering process models from event logs. IEEE Trans. Knowl. Data Eng. **16**(9), 1128–1142 (2004)
22. W.M.P. van der Aalst, M. Weske, D. Grünbauer, Case handling: A new paradigm for business process support. Data Knowl. Eng. **53**(2), 129–162 (2005)
23. M. Adams, *Facilitating Dynamic Flexibility and Exception Handling for Workflows*, PhD thesis, Queensland University of Technology, Brisbane, Australia, 2007
24. M. Adams, D. Edmond, A.H.M. ter Hofstede, The application of Activity Theory to dynamic workflow adaptation issues, ed. by J. Hanisch, D. Falconer, S. Horrocks, M. Hillier. In *Proceedings of the 2003 Pacific Asia Conference on Information Systems (PACIS 2003)*, Adelaide, Australia, July 2003 (University of South Australia, Australia, 2003), pp. 1836–1852
25. M. Adams, A.H.M. ter Hofstede, W.M.P. van der Aalst, D. Edmond, Dynamic, extensible and context-aware exception handling for workflows, ed. by R. Meersman, Z. Tari. In *Proceedings of the 15th International Conference on Cooperative Information Systems (CoopIS'07)*, Lecture Notes in Computer Science, vol. 4803, Vilamoura, Portugal, November 2007 (Springer, Berlin, 2007), pp. 95–112
26. M. Adams, A.H.M. ter Hofstede, D. Edmond, W.M.P. van der Aalst, Facilitating flexibility and dynamic exception handling in workflows through worklets, ed. by O. Bello, J. Eder, O. Pastor, J. Falcão e Cunha. In *Proceedings of the CAiSE'05 Forum*, FEUP edições, Porto, Portugal, June 2005, pp. 45–50
27. M. Adams, A.H.M. ter Hofstede, D. Edmond, W.M.P. van der Aalst, Worklets: A service-oriented implementation of dynamic flexibility in workflows, ed. by R. Meersman, Z. Tari et al. In *Proceedings of the 14th International Conference on Cooperative Information Systems (CoopIS'06)*, Lecture Notes in Computer Science, vol. 4275, Montpellier, France, November 2006 (Springer, Berlin, 2006), pp. 291–308
28. A. Agrawal, M. Amend, M. Das, M. Ford, M. Keller, M. Kloppmann, D. König, F. Leymann, R. Müller, G. Pfau, K. Plösser, R. Rangaswamy, A. Rickayzen, M. Rowley, P. Schmidt, I. Trickovic, A. Yiu, M. Zeller, Web services human task (ws-humantask). version 1.0, http://download.boulder.ibm.com/ibmdl/pub/software/dw/specs/ws-bpel4people/WS-HumanTask_v1.pdf (sighted 1 May 2009), June 2007
29. A. Agrawal, M. Amend, M. Das, M. Ford, M. Keller, M. Kloppmann, D. König, F. Leymann, R. Müller, G. Pfau, K. Plösser, R. Rangaswamy, A. Rickayzen, M. Rowley, P. Schmidt, I. Trickovic, A. Yiu, M. Zeller. Ws-bpel extension for people (bpel4people). version 1.0, http://download.boulder.ibm.com/ibmdl/pub/software/dw/specs/ws-bpel4people/BPEL4People_v1.pdf (sighted 1 May 2009), 2007

30. R. Agrawal, D. Gunopulos, F. Leymann, Mining process models from workflow logs, ed. by H-J. Schek, F. Saltor, I. Ramos, G. Alonso. In *Proceedings of the 6th International Conference on Extending Database Technology (EDBT'98)*, Lecture Notes in Computer Science, vol. 1377, Valencia, Spain, March 1998 (Springer, Berlin, 1998), pp. 469–484
31. M. Aguilar, A.J.G. Pater, Business process simulation: A fundamental step supporting process centered management, ed. by P. Farrington, H. Nembhard, D. Sturrock, G. Evans. In *Proceedings of the 1999 Winter Simulation Conference (WSC'99)*, Phoenix, AZ, USA, December 1999 (ACM, NY, 1999), pp. 1383–1392
32. L. Aldred, W.M.P. van der Aalst, M. Dumas, A.H.M. ter Hofstede, Communication abstractions for distributed business processes, ed. by J. Krogstie, A. Opdahl, G. Sindre. In *Proceedings of the 19th International Conference on Advanced Information Systems Engineering (CAiSE'07)*, Lecture Notes in Computer Science, vol. 4495, Trondheim, Norway, June 2007 (Springer, Berlin, 2007), pp. 409–423
33. L. Aldred, W.M.P. van der Aalst, M. Dumas, A.H.M. ter Hofstede, Dimensions of coupling in middleware. Concurrency Comput. Pract. Ex. (2009). doi: 10.1002/cpe.1414
34. G. Alonso, D. Agrawal, A. El Abbadi, M. Kamath, G. Gunthor, C. Mohan, Advanced transaction models in workflow contexts, ed. by S.Y.W. Su. In *Proceedings of the 12th International Conference on Data Engineering (ICDE'96)*, New Orleans, LA, USA, February 1996 (IEEE Computer Society, MD, 1996), pp. 574–581
35. A. Alves, A. Arkin, S. Askary, C. Barreto, B. Bloch, F. Curbera, M. Ford, Y. Goland, A. Guízar, N. Kartha, C.K. Liu, R. Khalaf, D. König, M. Marin, V. Mehta, S. Thatte, D. van der Rijn, P. Yendluri, A. Yiu, Web services business process execution language. version 2.0, OASIS standard, http://www.oasis-open.org/specs/#wsbpelv2.0 (sighted 1 May 2009), 2007
36. T. Andrews, F. Curbera, H. Dholakia, Y. Goland, J. Klein, F. Leymann, K. Liu, D. Roller, D. Smith, S. Thatte, I. Trickovic, S. Weerawarana, Business process execution language for web services. version 1.1, http://xml.coverpages.org/BPELv11-May052003Final.pdf (sighted 27 April 2009), 2003
37. E. Badouel, P. Darondeau, Theory of regions, ed. by W. Reisig, G. Rozenberg. Lectures on Petri Nets I: Basic Models, Advances in Petri Nets, Lecture Notes in Computer Science, vol. 1491 (Springer, Berlin, 1998), pp. 529–586
38. A. Bahrami, D. Sadowski, S. Bahrami, Enterprise architecture for business process simulation, ed. by D. Medeiros, E. Watson, J. Carson, M. Manivannan. In *Proceedings of the 1998 Winter Simulation Conference (WSC'98)*, Washington, DC, USA, December 1998 (ACM, NY, 1998), pp. 1409–1413
39. J.E. Bardram, I love the system – I just don't use it! ed. by S.C. Payne, W. Prinz In *Proceedings of the International ACM SIGGROUP Conference on Supporting Group Work (GROUP'97)*, Phoenix, AZ, USA, November 1997 (ACM, NY, 1997), pp. 251–260
40. L. Baresi, E. Di Nitto (ed.), *Test and Analysis of Web Services* (Springer, Berlin, 2007)
41. P. Barthelmess, J. Wainer, Workflow systems: a few definitions and a few suggestions, ed. by N. Comstock, C. Ellis. In *Proceedings of the ACM Conference on Organizational Computing Systems (COOCS'95)*, Milpitas, CA, USA, 1995 (ACM, NY, 1995), pp. 138–147
42. J. Becker, P. Delfmann, R. Knackstedt, Adaptive reference modeling: Integrating configurative and generic adaptation techniques for information models, ed. by J. Becker, P. Delfmann. In *Reference Modeling – Efficient Information Systems Design Through Reuse of Information Models, Proceedings of the Reference Modeling Conference (RM'06)*, Passau, Germany, February 2007 (Springer, Berlin, 2007), pp. 27–58
43. J. Becker, M. Kugeler, M. Rosemann (ed.), *Process Management. A Guide for the Design of Business Processes* (Springer, Berlin, 2003)
44. J. Becker, R. Schütte, *Handelsinformationssysteme*, (in German) 2nd edn. (Redline Wirtschaft, Frankfurt am Main, Germany, 2004)
45. P. Berens, in *The FLOWer Case Handling Approach: Beyond Workflow Management*, chapter 15, ed. by M. Dumas et al. Process-Aware Information Systems: Bridging People and Software Through Process Technology (Wiley-Interscience, NY, 2005), pp. 363–395

46. P. Bichler, G. Preuner, M. Schrefl, Workflow transparency, ed. by A. Olivé, J.A. Pastor. In *Proceedings of the 9th International Conference on Advanced Information Systems Engineering (CAiSE'97)*, Lecture Notes in Computer Science, vol. 1250, Barcelona, Spain, June 1997 (Springer, Berlin, 1997), pp. 423–436
47. A. Borgida, T. Murata, Tolerating exceptions in workflows: A unified framework for data and processes, ed. by D. Georgakopoulos, W. Prinz, A.L. Wolf. In *Proceedings of the International Joint Conference on Work Activities Coordination and Collaboration (WACC'99)*, San Francisco, CA, USA, February 1999 (ACM, NY, 1999), pp. 59–68
48. M. Brambilla, S. Ceri, S. Comai, C. Tziviskou, Exception handling in workflow-driven web applications, ed. by A. Ellis, T. Hagino. In *Proceedings of the 14th International Conference on World Wide Web (WWW2005)*, Chiba, Japan, May 2005 (ACM, NY, 2005), pp. 170–179
49. E. Bredenhoff, R. Schuring, R. Caljouw, Zorg in focus, op zoek naar routine in het zorgproces. (in Dutch). Medisch Contact **59**(34), 1304–1308 (2004)
50. F. van Breugel, M. Koshkina, Models and verification of BPEL, http://www.cse.yorku.ca/~franck/research/drafts/tutorial.pdf (sighted 1 May 2009), 2006
51. A. Brogi, R. Popescu, From BPEL processes to YAWL workflows, ed. by M. Bravetti, M. Núñez, G. Zavattaro. In *Proceedings of the 3rd International Workshop on Web Services and Formal Methods (WS-FM'06)*, Lecture Notes in Computer Science, vol. 4148, Vienna, Austria, 2006 (Springer, Berlin, 2006), pp. 107–122
52. C. Bussler, *B2B Integration: Concepts and Architecture* (Springer, Berlin, 2003)
53. J.A. Buzacott, Commonalities in reengineered business processes: Models and issues. Manag. Sci. **42**(5), 768–782 (1996)
54. F. Casati, A discussion on approaches to handling exceptions in workflows. ACM SIGGROUP Bull. **20**(3), 3–4 (1999)
55. F. Casati, S. Ceri, S. Paraboschi, G. Pozzi, Specification and implementation of exceptions in workflow management systems. ACM Trans. Database Syst. **24**(3), 405–451 (1999)
56. F. Casati, S. Ilnicki, L. Jin, V. Krishnamoorthy, M-C. Shan, Adaptive and dynamic composition in eFlow, ed. by B. Wangler, L. Bergman. In *Proceedings of the 12th International Conference on Advanced Information Systems Engineering (CAiSE 2000)*, Lecture Notes in Computer Science, vol. 1789, Stockholm, Sweden, 2000 (Springer, Berlin, 2000), pp. 13–31
57. D. Chiu, Q. Li, K. Karlapalem, A logical framework for exception handling in ADOME workflow management system, ed. by B. Wangler, L. Bergman. In *Proceedings of the 12th International Conference on Advanced Information Systems Engineering (CAiSE 2000)*, Lecture Notes in Computer Science, vol. 1789, Stockholm, Sweden, 2000 (Springer, Berlin, 2000), pp. 110–125
58. D.K.W. Chiu, Q. Li, K. Karlapalem, ADOME-WFMS: Towards cooperative handling of workflow exceptions, ed. by A. Romanovsky, C. Dony, J. Lindskov Knudsen, A. Tripathi. In *Advances in Exception Handling Techniques*, Lecture Notes in Computer Science, vol. 2022 (Springer, Berlin, 2001), pp. 271–288
59. R. Chokshi, Decision support for call center management using simulation, ed. by P. Farrington, H. Nembhard, D. Sturrock, G. Evans. In *Proceedings of the 1999 Winter Simulation Conference (WSC'99)*, Phoenix, AZ, USA, December 1999 (ACM, NY, 1999), pp. 1634–1639
60. E.M. Clarke, O. Grumberg, D.A. Peled, *Model Checking* (MIT, Cambridge, MA, 1999)
61. B. Clevé, *Film Production Management*, 3rd edn. (Focal Press, Burlington, MA, 2006)
62. P. Compton, B. Jansen, Knowledge in context: A strategy for expert system maintenance, ed. by J. Siekmann. In *Proceedings of the 2nd Australian Joint Artificial Intelligence Conference (AI'88)*, Lecture Notes in Artificial Intelligence, vol. 406, Adelaide, Australia, November 1988 (Springer, Berlin, 1988), pp. 292–306
63. P. Compton, P. Preston, B. Kang, The use of simulated experts in evaluating knowledge acquisition. In *Proceedings of the 9th Banff Knowledge Acquisition for Knowledge Based Systems Workshop*, Banff, Canada, February 1995
64. J. Cortadella, *Petrify: a tutorial for the designer of asychronous circuits*, Universitat Politécnica de Catalunya, Barcelona, Spain, http://www.lsi.upc.es/petrify (sighted 20 May 2007), December 1998

65. J. Cortadella, M. Kishinevsky, L. Lavagno, A. Yakovlev, Deriving Petri nets from finite transition systems. IEEE Trans. Comput. **47**(8), 859–882 (1998)
66. COSA BPM product description, http://www.cosa-bpm.com/project/docs/COSA_BPM_5_Productdescription_eng.pdf (sighted 13 March 2007), 2005
67. M. Cumberlidge, *Business Process Management with JBoss jBPM: A Practical Guide for Business Analysts* (PACKT Publishing, Birmingham, UK, 2007)
68. F. Curbera, M.J. Duftler, R. Khalaf, W. Nagy, N. Mukhi, S. Weerawarana, Unraveling the web services web: An introduction to SOAP, WSDL, and UDDI. IEEE Internet Comput. **6**(2), 86–93 (2002)
69. F. Curbera, Y. Goland, J. Klein, F. Leymann, D. Roller, S. Thatte, S. Weerawarana, Business process execution language for web services, version 1.0. Standards proposal by BEA Systems, International Business Machines Corporation, and Microsoft Corporation (2002)
70. T. Curran, G. Keller, A. Ladd, SAP R/3 Business Blueprint: Understanding the Business Process Reference Model, in *Enterprise Resource Planning Series* (Prentice Hall PTR, NJ, 1997)
71. K. Czarnecki, M. Antkiewicz, Mapping features to models: A template approach based on superimposed variants, ed. by R. Glück, M.R. Lowry. In *Proceedings of the 4th International Conference on Generative Programming and Component Engineering (GPCE'05)*, Lecture Notes in Computer Science, vol. 3676, Tallinn, Estonia, September 2005 (Springer, Berlin, 2005), pp. 422–437
72. P. Darondeau, Unbounded Petri net synthesis, ed. by J. Desel, W. Reisig, G. Rozenberg. In *Lectures on Concurrency and Petri Nets, Advances in Petri Nets*, Lecture Notes in Computer Science, vol. 3098, Eichstätt, Germany, 2003 (Springer, Berlin, 2003), pp. 413–428
73. G. Decker, A. Grosskopf, A. Barros, A graphical notation for modeling complex events in business processes, in *Proceedings of the 11th IEEE International Enterprise Distributed Object Computing Conference (EDOC'07)*, Annapolis, MD, USA, October 2007 (IEEE Computer Society, MD, 2007), pp. 27–36
74. G. Decker, J. Mendling, Instantiation semantics for process models, ed. by M. Dumas, M. Reichert, M.-C. Shan. In *Proceedings of the 6th International Conference on Business Process Management (BPM'08)*, Lecture Notes in Computer Science, vol. 5240, Milan, Italy, September 2008 (Springer, Berlin, 2008), pp. 164–179
75. J. Dehnert, *A Methodology for Workflow Modelling: from Business Process Modelling towards Sound Workflow Specification*, PhD thesis, Technische Universität Berlin, Berlin, Germany, 2003
76. W. Derks, J. Dehnert, P. Grefen, W. Jonker, Customized atomicity specification for transactional workflows, ed. by H. Lu, S. Spaccapietra. In *Proceedings of the 3rd International Symposium on Cooperative Database Systems and Applications (CODAS'2001)*, Beijing, China, April 2001 (IEEE Computer Society, MD, 2001), pp. 155–164
77. J. Desel, J. Esparza, *Free Choice Petri Nets*. Cambridge Tracts in Theoretical Computer Science, vol. 40 (Cambridge University Press, Cambridge, UK, 1995)
78. R. Dijkman, M. Dumas, C. Ouyang, Formal semantics and analysis of BPMN process models. Inform. Software Tech. **50**(12), 1281–1294 (2008)
79. B.F. van Dongen, W.M.P. van der Aalst, A meta model for process mining data, ed. by M. Missikoff, A. De Nicola. In *Proceedings of the CAiSE'05 Workshops (EMOI-INTEROP Workshop)*, CEUR Workshop Proceedings, vol. 160, Porto, Portugal, June 2005, pp. 309–320, CEUR-WS.org
80. C. Dufourd, A. Finkel, P. Schnoebelen, Reset nets between decidability and undecidability, ed. by K. Larsen, S. Skyum, G. Winskel. In *Proceedings of the 25th International Colloquium on Automata, Languages and Programming (ICALP'98)*, Lecture Notes in Computer Science, vol. 1443, Aalborg, Denmark, July 1998 (Springer, Berlin, 1998), pp. 103–115
81. C. Dufourd, P. Jančar, P. Schnoebelen, Boundedness of reset P/T nets, ed. by J. Wiedermann, P. van Emde Boas, M. Nielsen. In *Proceedings of the 26th International Colloquium on Automata, Languages and Programming (ICALP'99)*, Lecture Notes in Computer Science, vol. 1644, Prague, Czech Republic, July 1999 (Springer, Berlin, 1999), pp. 301–310

82. M. Dumas, W.M.P. van der Aalst, A.H.M. ter Hofstede (ed.), *Process-Aware Information Systems: Bridging People and Software through Process Technology* (Wiley-Interscience, NY, 2005)
83. M. Dumas, A. Grosskopf, T. Hettel, M.T. Wynn, Semantics of standard process models with OR-joins, ed. by R. Meersman, Z. Tari. In *Proceedings of the 15th International Conference on Cooperative Information Systems (CoopIS'07)*, Lecture Notes in Computer Science, vol. 4803, Vilamoura, Portugal, November 2007 (Springer, Berlin, 2007), pp. 41–58
84. M. Dumas, A.H.M. ter Hofstede, UML activity diagrams as a workflow specification language, ed. by M. Gogolla, C. Kobryn. In *Proceedings of the 4th International Conference on the Unified Modeling Language (UML 2001)*, Lecture Notes in Computer Science, vol. 2185, Toronto, Canada, October 2001 (Springer, Berlin, 2001), pp. 76–90
85. Eastman Software, *RouteBuilder Tool User's Guide* (Eastman Software Inc, Billerica, MA, 1998)
86. J. Eder, W. Liebhart, The workflow activity model (WAMO), ed. by S. Laufmann, S. Spaccapietra, T. Yokoi. In *Proceedings of the 3rd International Conference on Cooperative Information Systems (CoopIS'95)*, Vienna, Austria, May 1995, pp. 87–98
87. J. Eder, W. Liebhart, Workflow recovery, in *Proceedings of the 1st IFCIS International Conference on Cooperative Information Systems (CoopIS'96)*, Brussels, Belgium, June 1996 (IEEE Computer Society, MD, 1996), pp. 124–134
88. A. Ehrenfeucht, G. Rozenberg, Partial (set) 2-structures - part 1 and part 2. Acta Inform. **27**(4), 315–368 (1989)
89. C.A. Ellis, K. Keddara, G. Rozenberg, Dynamic change within workflow systems, ed. by N. Comstock, C. Ellis, R. Kling, J. Mylopoulos, S. Kaplan. In *Proceedings of the Conference on Organizational Computing Systems*, Milpitas, CA, USA, August 1995 (ACM, NY, 1995), pp. 10–21
90. A. Elmagarmid (ed.), *Database Transaction Models for Advanced Applications* (Morgan Kaufmann, CA, 1992)
91. P. Eugster, P. Felber, R. Guerraoui, A. Kermarrec, The many faces of publish/subscribe. ACM Comput. Surv. **35**(2), 114–131 (2003)
92. P. Fettke, P. Loos, Classification of reference models - a methodology and its application. Inform. Syst. E. Bus. Manag. **1**(1), 35–53 (2003)
93. P. Fettke, P. Loos, J. Zwicker, Business process reference models: Survey and classification, ed. by C. Bussler, A. Haller et al. In *Business Process Management Workshops – BPM 2005 International Workshops, BPI, BPD, ENEI, BPRM, WSCOBPM, BPS. Revised Selected Papers*, Lecture Notes in Computer Science, vol. 3812, Nancy, France, September 2005 (Springer, Berlin, 2005), pp. 469–483
94. R.T. Fielding, *Architectural Styles and the Design of Network-based Software Architectures*. PhD thesis, University of California, Irvine, CA, USA, 2000
95. A. Finkel, J.-F. Raskin, M. Samuelides, L. Van Begin, Monotonic extensions of petri nets: Forward and backward search revisited. Electron. Notes Theor. Comput. Sci. **68**(6), 1–22 (2002)
96. A. Finkel, P. Schnoebelen, Well-structured transition systems everywhere! Theor. Comput. Sci. **256**(1–2), 63–92 (2001)
97. L. Fischer (ed.), *Workflow Handbook 2004* (Future Strategies Inc, Lighthouse Point, FL, 2004)
98. M. Fowler, *Patterns of Enterprise Application Architecture* (Addison-Wesley, Boston, MA, 2003)
99. B.R. Gaines, Induction and visualisation of rules with exceptions, ed. by J. Boose, B. Gaines. In *Proceedings of the 6th Banff AAAI Knowledge Acqusition of Knowledge Based Systens Workshop*, Banff, Canada, October 1991, pp. 7.1–7.17
100. E. Gamma, R. Helm, R. Johnson, J. Vlissides, In *Design Patterns: Elements of Reusable Object-Oriented Software*. Professional Computing Series (Addison-Wesley, Boston, MA, 1995)
101. D. Georgakopoulos, M.F. Hornick, A.P. Sheth, An overview of workflow management: From process modeling to workflow automation infrastructure. Distr. Parallel Database **3**(2), 119–153 (1995)

102. M. Georgeff, J. Pyke, *Dynamic process orchestration* (White paper, Staffware PLC, Maidenhead, 2003)
103. R. Gerth, D. Peled, M.Y. Vardi, P. Wolper, Simple on-the-fly automatic verification of linear temporal logic, ed. by R. Gotzhein, J. Bredereke. In *Proceedings of the 15th IFIP WG6.1 International Symposium on Protocol Specification, Testing and Verification*, Kaiserslautern, Germany, October 1996 (Chapman and Hall, London, 1996), pp. 3–18
104. D. Giannakopoulou, K. Havelund, Automata-based verification of temporal properties on running programs. In *Proceedings of the 16th IEEE International Conference on Automated Software Engineering (ASE'01)*, San Diego, CA, USA, November 2001 (IEEE Computer Society, MD, 2001), pp. 412–416
105. D. Giannakopoulou, F. Lerda, From states to transitions: Improving translation of LTL formulae to Büchi automata, ed. by D.A. Peled, M.Y. Vardi. In *Proceedings of the 22nd IFIP WG 6.1 International Conference on Formal Techniques for Networked and Distributed Systems (FORTE'02)*, Lecture Notes in Computer Science, vol. 2529, Houston, TX, USA, November 2002 (Springer, Berlin, 2002), pp. 308–326
106. F. Gottschalk, W.M.P. van der Aalst, M.H. Jansen-Vullers, Configurable process models – a foundational approach, ed. by J. Becker, P. Delfmann. In *Reference Modeling – Efficient Information Systems Design Through Reuse of Information Models, Proceedings of the Reference Modeling Conference (RM'06)*, Passau, Germany, February 2007 (Springer, Berlin, 2007), pp. 59–78
107. F. Gottschalk, W.M.P. van der Aalst, M.H. Jansen-Vullers, M. La Rosa, Configurable workflow models. Int. J. Cooper. Inform. Syst. **17**(2), 177–221 (2008)
108. F. Gottschalk, T.A.C. Wagemakers, M.H. Jansen-Vullers, W.M.P. van der Aalst, M. La Rosa, Configurable process models – experiences from a municipality case study, ed. by P. van Eck, J. Gordijn, R. Wieringa. In *Proceedings of the 27th International Conference on Advanced Information Systems Engineering (CAiSE'09)*, Lecture Notes in Computer Science, vol. 5565, Amsterdam, The Netherlands, June 2009 (Springer, Berlin, 2009), pp. 486–500
109. J. Gray, Notes data base operating systems In *Operating Systems, An Advanced Course* (Springer, London, 1978), pp. 393–481
110. A. Greasley, Effective uses of business process simulation, ed. by J. Joines, R. Barton, K. Kang, P. Fishwick. In *Proceedings of the 2000 Winter Simulation Conference (WSC 2000)*, Orlando, FL, USA, December 2000 (ACM, NY, 2000), pp. 2004–2009
111. U. Greiner, J. Ramsch, B. Heller, M. Löffler, R. Müller, E. Rahm, Adaptive guideline-based treatment workflows with AdaptFlow, ed. by K. Kaiser, S. Miksch, S.W. Tu. In *Proceedings of the Symposium on Computerized Guidelines and Protocols (CGP 2004)*, Prague, Czech Republic, April 2004 (IOS Press, Amsterdam, 2004), pp. 113–117
112. D. Grigori, F. Casati, U. Dayal, M.-C. Shan, Improving business process quality through exception understanding, prediction, and prevention, ed. by P. Apers, P. Atzeni, S. Ceri, S. Paraboschi, K. Ramamohanarao, R. Snodgrass. In *Proceedings of the 27th International Conference on Very Large Data Bases (VLDB'01)*, Rome, Italy, September 2001 (Morgan Kaufmann, San Francisco, CA, 2001), pp. 159–168
113. C. Hagen, G. Alonso, Flexible exception handling in the OPERA process support system, In *Proceedings of the 18th International Conference on Distributed Computing Systems (ICDCS'98)*, Amsterdam, The Netherlands, May 1998 (IEEE Computer Society, MD, 1998), pp. 526–533
114. C. Hagen, G. Alonso, Exception handling in workflow management systems. IEEE Trans. Software Eng. **26**(10), 943–958 (2000)
115. T.A. Halpin, *Information Modeling and Relational Databases: From Conceptual Analysis to Logical Design* (Morgan Kaufmann, San Francisco, CA, 2001)
116. M. Hapner, R. Burridge, R. Sharma, J. Fialli, K. Stout, Java Message Service, version 1.1. Technical Specification – Final Release, April 2002, http://java.sun.com/products/jms/docs.html (sighted 1 May 2009)
117. P. Heinl, S. Horn, S. Jablonski, J. Neeb, K. Stein, M. Teschke, A comprehensive approach to flexibility in workflow management systems, ed. by G. Georgakopoulos, W. Prinz, A.L. Wolf.

In *Proceedings of International Joint Conference on Work Activities Coordination and Collaboration (WACC'99)*, San Francisco, CA, USA, February 1999 (ACM, NY, 1999), pp. 79–88

118. C. Hensinger, M. Reichert, T. Bauer, T. Strzeletz, P. Dadam, ADEPT workflow – advanced workflow technology for the efficient support of adaptive, enterprise-wide processes, in *the Software Demonstration Track of the Conference on Extending Database Technology*, Konstanz, Germany, March 2000, pp. 29–30
119. V. Hlupic, S. Robinson, Business process modelling and analysis using discrete-event simulation, ed. by D. Medeiros, E. Watson, J. Carson, M. Manivannan. In *Proceedings of the 1998 Winter Simulation Conference (WSC'98)*, Washington, DC, USA, December 1998 (ACM, NY, 1998), pp. 1363–1369
120. G. Hohpe, B. Woolf, *Enterprise Integration Patterns: Designing, Building, and Deploying Messaging Solutions* (Addison-Wesley, Boston, MA, 2004)
121. E.L. Honthaner, *The Complete Film Production Handbook* (Focal Press, Woburn, MA, 2001)
122. G. Hook, *Simulation-enabled business process management, that "extra ingredient" for intelligent BPM* (White paper, Lanner Group, TX, 2003)
123. S.Y. Hwang, J. Tang Consulting past exceptions to facilitate workflow exception handling. Decis. Support Syst. **37**(1), 49–69 (2004)
124. S. Jablonski, C. Bussler, *Workflow Management: Modeling Concepts, Architecture and Implementation* (Thomson Computer Press, London, UK, 1996)
125. S. Jain, L.M. Collins, R.W. Workman, E.C. Ervin, Development of a high-level supply chain simulation model, ed. by B. Peters, J. Smith, D. Medeiros, M. Rohrer. In *Proceedings of the 2001 Winter Simulation Conference (WSC'01)*, Arlington, VA, USA, December 2001 (ACM, NY, 2001), pp. 1129–1137
126. S. Jajodia, L. Kerschberg (ed.), *Advanced Transaction Models and Architectures* (Kluwer, Norwell, MA, 1997)
127. M. Jansen-Vullers, M. Netjes, Business process simulation, ed. by K. Jensen. In *Proceedings of the 7th Workshop and Tutorial on Practical Use of Coloured Petri Nets and the CPN Tools*, Aarhus, Denmark, October 2006 (University of Aarhus, Denmark, 2006), http://www.daimi.au.dk/CPnets/workshop06/cpn/papers/Paper03.pdf (sighted 30 April 2007)
128. K. Jensen, Coloured Petri Nets. Basic Concepts, Analysis Methods and Practical Use, *Volume 1*. Monographs in Theoretical Computer Science (Springer, Berlin, 1997)
129. G. Keller, M. Nüttgens, A.-W. Scheer, Semantische prozessmodellierung auf der grundlage "ereignisgesteuerter prozessketten (EPK)" in Germany. Heft 89 (Institut für Wirtschaftsinformatik, Saarbrücken, Germany, 1992)
130. G. Keller, T. Teufel, *SAP R/3 Process Oriented Implementation: Iterative Process Prototyping* (Addison-Wesley, Boston, MA, 1998)
131. B. Kiepuszewski, A.H.M. ter Hofstede, W.M.P. van der Aalst, Fundamentals of control flow in workflows. Acta Inform. **39**(3), 143–209 (2003)
132. E. Kindler, On the semantics of EPCs: A framework for resolving the vicious circle, ed. by J. Desel, B. Pernici, M. Weske. In *Proceedings of 2nd International Conference on Business Process Management (BPM'04)*, Lecture Notes in Computer Science, vol. 3080, Potsdam, Germany, June 2004 (Springer, Berlin, 2004), pp. 82–97
133. E. Kindler. On the semantics of EPCS: Resolving the vicious circle. Data Knowl. Eng. **56**(1), 23–40 (2006)
134. M. Klein, C. Dellarocas, A knowledge-based approach to handling exceptions in workflow systems. J. Comput. Support. Collaborative Work **9**(3–4), 399–412 (2000)
135. M. Klein, C. Dellarocas, A. Bernstein, Adaptive workflow systems. Special Issue of Computer Supported Cooperative Work **9**(3–4), 265–267 (2002)
136. M. Kloppmann, D. Koenig, F. Leymann, G. Pfau, A. Rickayzen, C. von Riegen, P. Schmidt, I. Trickovic, WS-BPEL extension for sub-processes – BPEL-SPE, Joint white paper, IBM and SAP, 2005, http://download.boulder.ibm.com/ibmdl/pub/software/dw/webservices/ws-bpelsubproc/ws-bpelsubproc.pdf (sighted 1 May 2009)

137. R. Klungle, Simulation of a claims call center: A success and a failure, ed. by P. Farrington, H. Nembhard, D. Sturrock, G. Evans. In *Proceedings of the 1999 Winter Simulation Conference (WSC'99)*, Phoenix, AZ, USA, December 1999 (ACM, NY, 1999), pp. 1648–1653
138. C. Kruse, *Referenzmodellgestütztes Geschäftsprozeßmanagement: Ein Ansatz zur prozeßorientierten Gestaltung betriebslogistischer Systeme* in German (Gabler, Wiesbaden, Germany, 1996)
139. K. Kuutti, Activity Theory as a Potential Framework for Human-Computer Interaction Research, ed. by Nardi. In *Context and Consciousness: Activity Theory and Human-Computer Interaction* (MIT, Cambridge, MA, 1996), pp. 17–44
140. M. La Rosa, *Managing Variability in Process-Aware Information Systems*, PhD thesis, Queensland University of Technology, Brisbane, Australia, 2009
141. M. La Rosa, W.M.P. van der Aalst, M. Dumas, A.H.M. ter Hofstede, Questionnairebased variability modeling for system configuration. Software Syst. Model. **8**(2), 251–274 (2009)
142. M. La Rosa, M. Dumas, A.H.M. ter Hofstede, J. Mendling, F. Gottschalk, Beyond control-flow: Extending business process configuration to roles and objects, ed. by Q. Li, S. Spaccapietra, E. Yu, A. Olivé. In *Proceedings of the 27th International Conference on Conceptual Modeling (ER'08)*, Lecture Notes in Computer Science, vol. 5231, Barcelona, Spain, October 2008 (Springer, Berlin, 2008), pp. 199–215
143. M. La Rosa, F. Gottschalk, M. Dumas, W.M.P. van der Aalst, in *Linking domain models and process models for reference model configuration*, ed. by A.H.M. ter Hofstede, B. Benatallah, H.-Y. Paik. Business Process Management 2007 Workshops, Lecture Notes in Computer Science, vol. 4928, Brisbane, Australia, September 2008 (Springer, Berlin, 2008), pp. 417–430
144. M. La Rosa, A.H.M. ter Hofstede, M. Rosemann, K. Shortland, Bringing process to post production. In *Proceedings of the CCI International Conference "Creating Value: Between Commerce and Commons"*, Brisbane, Australia, June 2008 (Queensland University of Technology, Australia, 2008)
145. K.R. Lang, M. Schmidt, Workflow-supported organizational memory systems: An industrial application. In *Proceedings of the 35th Hawaii International International Conference on Systems Science (HICSS-35 2002)*, Big Island, HI, USA, January 2002 (IEEE Computer Society, MD, 2002), p. 208
146. P. Lawrence (ed.), *Workflow Handbook 1997* (Wiley, NY, 1997)
147. R. Lenz, M. Reichert, IT support for healthcare processes - premises, challenges, perspectives. Data Knowl. Eng. **61**(1), 49–58 (2007)
148. M. de Leoni, W.M.P. van der Aalst, A.H.M. ter Hofstede, Visual support for work assignment in process-aware information systems, ed. by M. Dumas, M. Reichert, M.-C. Shan. In *Proceedings of the 6th International Conference on Business Process Management (BPM'08)*, Lecture Notes in Computer Science, vol. 5240, Milan, Italy, September 2008 (Springer, Berlin, 2008), pp. 68–84
149. F. Leymann, D. Roller, Workflow-based applications. IBM Syst. J. **36**(1), 102–123 (1997)
150. F. Leymann, D. Roller, *Production Workflow: Concepts and Techniques* (Prentice Hall, NJ, 2000)
151. L. Liu, C. Pu, ActivityFlow: Towards incremental specification and flexible coordination of workflow activities, ed. by D.W. Embley, R.C. Goldstein. In *Proceedings of the 16th International Conference on Conceptual Modeling (ER'97)*, Lecture Notes in Computer Science, vol. 1331, Los Angeles, CA, USA, November 1997 (Springer, Berlin, 1997), pp. 169–182
152. R. Lu, S. Sadiq, V. Padmanabhan, G. Governatori, Using a temporal constraint network for business process execution, ed. by G. Dobbie, J. Bailey. In *Proceedings of the 17th Australasian Database Conference (ADC'06)*, Hobart, Australia, January 2006 (Australian Computer Society, Australia, 2006), pp. 157–166
153. Z. Luo, A. Sheth, K. Kochut, J. Miller, Exception handling in workflow systems. Appl. Intell. **13**(2), 125–147 (2000)
154. R.S. Mans, W.M.P. van der Aalst, N.C. Russell, P.J.M. Bakker, Flexibility schemes for workflow management systems, ed. by R. Lenz, M. Peleg, M. Reichert. In *Pre-Proceedings of*

the 2nd International Workshop on Process-Oriented Information Systems in Healthcare (ProHealth'08), Milan, Italy, 2008, pp. 50–61

155. R.S. Mans, N.C. Russell, W.M.P. van der Aalst, A.J. Moleman, P.J.M. Bakker, Augmenting a workflow management system with planning facilities using Colored Petri Nets, ed. by K. Jensen. In *Proceedings of the 9th Workshop on the Practical Use of Coloured Petri Nets and CPN Tools*, Aarhus, Denmark, 2008 (University of Aarhus, Denmark, 2008), pp. 143–162, http://www.daimi.au.dk/CPnets/workshop08/cpn/papers/Paper09.pdf (sighted 1 May 2009)
156. Y. Mansuri, J.G. Kim, P. Compton, C. Sammut, A comparison of a manual knowledge acquisition method and an inductive learning method. In *Proceedings of the 1st Australian Workshop on Knowledge Acquisition for Knowledge Based Systems (AKAW'91)*, Pokolbin, Australia, August 1991 (Pokolbin, 1991), pp. 114–132
157. D.C. Marinescu, Internet-Based Workflow Management: Towards a Semantic Web, Wiley Series on Parallel and Distributed Computing, vol. 40. (Wiley-Interscience, NY, 2002)
158. A. Martens, On compatibility of web services. Petri Net Newsletter **65**, 12–20 (2003)
159. S. Mehrotra, R. Rastogi, H.F. Korth, A. Silberschatz, A transaction model for multidatabase systems. In *Proceedings of the 12th International Conference on Distributed Computing Systems (ICDCS'92)*, Yokohama, Japan, June 1992 (IEEE Computer Society, MD, 1992), pp. 56–63
160. J. Mendling, *Detection and Prediction of Errors in EPC Business Process Models*, PhD Thesis, Vienna University of Economics and Business Administration,Vienna, Austria, 2007
161. J. Mendling, *Metrics for Process Models: Empirical Foundations of Verification, Error Prediction and Guidelines for Correctness*, Lecture Notes in Business Information Processing, vol. 6 (Springer, Berlin, 2008)
162. J. Mendling, W.M.P. van der Aalst, Formalization and verification of EPCs with OR-Joins based on state and context, ed. by J. Krogstie, A.L. Opdahl, G. Sindre. In *Proceedings of the the 19th International Conference on Advanced Information Systems Engineering (CAiSE'07)*, Lecture Notes In Computer Science, vol. 4495, Trondheim, Norway, June 2005 (Springer, Berlin, 2005), pp. 439–453
163. J. Mendling, B.F. van Dongen, W.M.P. van der Aalst, Getting rid of the OR-Join in business process models. In *Proceedings of the 11th IEEE International Enterprise Distributed Object Computing Conference (EDOC'07)*, Annapolis, MD, USA, October 2007 (IEEE Computer Society, MD, 2007), pp. 3–14
164. J. Mendling, B.F. van Dongen, W.M.P. van der Aalst, Getting rid of OR-Joins and multiple start events in business process models. Enterprise Inform. Syst. **2**(4), 403–419 (2008)
165. J. Mendling, M. Moser, G. Neumann, Transformation of yEPC business process models to YAWL, ed. by H. Haddad. In *Proceedings of the 21st Annual ACM Symposium on Applied Computing*, vol. 2, Dijon, France, April 2006 (ACM, NY, 2006), pp. 1262–1267
166. J. Mendling, G. Neumann, M. Nüttgens, Towards workflow pattern support of event-driven process chains (EPC), ed. by M. Nüttgens, J. Mendling. In *Proceedings of the 2nd GI Workshop – XML Interchange Formats for Business Process Management (XML4BPM'05)*, Karlsruhe, Germany, March 2005, pp. 23–38, http://wi.wu-wien.ac.at/home/mendling/XML4BPM2005/xml4bpm-2005-proceedings-mendling.pdf (sighted 1 May 2009)
167. J. Mendling, G. Neumann, M. Nüttgens, Yet Another Event-Driven Process Chain, ed. by W.M.P. van der Aalst, B. Benatallah, F. Casati, F. Curbera. In *Proceedings of the 3rd International Conference on Business Process Management 2005 (BPM'05)*, Lecture Notes in Computer Science, vol. 3649, Nancy, France, September 2005 (Springer, Berlin, 2005), pp. 428–433
168. J. Mendling, G. Neumann, M. Nüttgens, Yet Another Event-driven Process Chain - modeling workflow patterns with yEPCs. Enterprise Model. Inform. Syst. Architect. Int. J. **1**(1), 3–13 (2005)
169. J. Mendling, M. Nüttgens, EPC Markup Language (EPML) - an XML-based interchange format for event-driven process chains (EPC). Inform. Syst. E. Bus. Manag. **4**(3), 245–263 (2006)

170. J. Mendling, H.M.W. Verbeek, B.F. van Dongen, W.M.P. van der Aalst, G. Neumann, Detection and prediction of errors in EPCs of the SAP reference model. Data Knowl. Eng. **64**(1), 312–329 (2008)
171. G.G. van Merode, S. Groothuis, A. Hasman, Enterprise resource planning for hospitals. Int. J. Med. Informat. **73**(6), 493–501 (2004)
172. Microsoft Corporation, Microsoft Message Queuing homepage, http://www.microsoft.com/windowsserver2003/technologies/msmq/default.mspx (sighted 1 December 2008)
173. R. Milner, *Communicating and Mobile Systems: The π Calculus* (Cambridge University Press, Cambridge, 1999)
174. M. zur Muehlen, in *Workflow-based Process Controlling. Foundation, Design, and Implementation of Workflow-driven Process Information Systems*. Advances in Information Systems and Management Science, vol. 6 (Logos, Berlin, 2004)
175. M. zur Muehlen, J. Nickerson, K. Swenson, Developing web services choreography standards - the case of REST vs. SOAP. Decis. Support Syst. **40**(1), 9–29 (2005)
176. R. Muller, U. Greiner, E. Rahm, AgentWork: a workflow system supporting rule-based workflow adaptation. Data Knowl. Eng. **51**(2), 223–256 (2004)
177. N. Mulyar, W.M.P. van der Aalst, N.C. Russell, Process flexibility patterns, BETA Working Paper Series WP 251 (Eindhoven University of Technology, Eindhoven, The Netherlands, 2008)
178. N.A. Mulyar, *Patterns for Process-Aware Information Systems: An Approach Based on Colored Petri Nets*, PhD thesis, Eindhoven University of Technology, Eindhoven, The Netherlands, 2009
179. T. Murata, Petri nets: Properties, analysis and applications, Proc. IEEE **77**(4), 541–580 (1989)
180. M. Murray, Strategies for the successful implementation of workflow systems within healthcare: A cross case comparison. In *Proceedings of the 36th Annual Hawaii International Conference on System Sciences*, Big Island, HI, USA, January 2003 (IEEE Computer Society, MD, 2003), pp. 166–175
181. B.A. Nardi (ed.), *Context and Consciousness: Activity Theory and Human-Computer Interaction* (MIT, Cambridge, MA, 1996)
182. M. Nüttgens, F.J. Rump, Syntax und semantik ereignisgesteuerter prozessketten (EPK), (in German) ed. by J. Desel, M. Weske. In *Proceedings of Promise 2002*, Lecture Notes in Informatics, vol. 21, Potsdam, Germany, 2002 (Springer, Berlin, 2002), pp. 64–77
183. Object Management Group, Unified Modeling Language (UML) superstructure, OMG Final Adopted Specification ptc/04-10-02 (Object Management Group, Needham, MA, 2002)
184. Object Management Group, Common Object Request Broker Architecture: Core specification, 3.0.3. (Object Management Group, Needham, MA, 2004), http://www.omg.org/docs/formal/04-03-01.pdf (sighted 1 December 2008)
185. Object Management Group, Business Process Modeling Notation (BPMN) version 1.0, OMG Final Adopted Specification dtc/06-02-01 (Object Management Group, Needham, MA, 2006)
186. J. O'Sullivan, D. Edmond, A.H.M. ter Hofstede, What's in a service? Towards accurate description of non-functional service properties. Distr. Parallel Database J. – Special Issue on E-Services **12**(2–3), 117–133 (2002)
187. C. Ouyang, M. Dumas, W.M.P. van der Aalst, A.H.M. ter Hofstede, J. Mendling. From business process models to process-oriented software systems. ACM Trans. on Software Eng. and Meth. 19(1):article 2 (37 pages), August 2009
188. C. Ouyang, A.H.M. ter Hofstede, M. La Rosa, M. Rosemann, K. Shortland, D. Court, Camera, set, action: Automating film production via business process management. In *Proceedings of the CCI International Conference "Creating Value: Between Commerce and Commons"*, Brisbane, Australia, June 2008 (Queensland University of Technology, Australia, 2008), http://www.cci.edu.au/publications/camera-set-action (sighted 1 May 2009)
189. C. Ouyang, M. La Rosa, A.H.M. ter Hofstede, M. Dumas, K. Shortland, Towards webscale workflows for film production. IEEE Internet Comput. **12**(5), 53–61 (2008)

190. C. Ouyang, K. Wang, A.H.M. ter Hofstede, M. La Rosa, M. Rosemann, K. Shortland, D. Court, Camera, Set, Action: Process Innovation for Film and TV Production. Cult. Sci. **1**(2) (2008); http://cultural-science.org/journal/index.php/culturalscience/article/view/17/60 (sighted 1 May 2009)
191. Pallas Athena, Case handling with FLOWer: Beyond workflow, Positioning paper, Pallas Athena, Apeldoorn, The Netherlands, 2000, http://www.pallas-athena.com/downloads/eng_flower/flowerup.pdf (sighted 13 May 2007)
192. M. Pesic, *Constraint-Based Workflow Management Systems: Shifting Control to Users*, PhD Thesis, Eindhoven University of Technology, Eindhoven, The Netherlands, 2008
193. M. Pesic, W.M.P. van der Aalst, A declarative approach for flexible business processes, ed. by J. Eder, S. Dustdar. In *Business Process Management Workshops, Workshop on Dynamic Process Management (DPM'06)*, Lecture Notes in Computer Science, vol. 4103, Vienna, Austria, September 2006 (Springer, Berlin, 2006), pp. 169–180
194. M. Pesic, M.H. Schonenberg, W.M.P. van der Aalst, DECLARE: Full support for loosely-structured processes. In *Proceedings of the 11th IEEE International Enterprise Distributed Object Computing Conference (EDOC'07)*, Annapolis, MD, USA, October 2007 (IEEE Computer Society, MD, 2007), pp. 287–298
195. M. Pesic, M.H. Schonenberg, N. Sidorova, W.M.P. van der Aalst, Constraint-based workflow models: Change made easy, ed. by R. Meersman, Z. Tari. In *Proceedings of the 15th International Conference on Cooperative Information Systems (CoopIS'07)*, Lecture Notes in Computer Science, vol. 4803, Vilamoura, Portugal, November 2007 (Springer, Berlin, 2007), pp. 77–94
196. J.L. Peterson, *Petri Net Theory and the Modelling of Systems* (Prentice-Hall, NJ, 1981)
197. C.A. Petri, *Kommunikation mit Automaten*, (in German) PhD thesis, Institut für instrumentelle Mathematik, Bonn, Germany, 1962
198. S.S. Pinter, M. Golani, Discovering workflow models from activities lifespans. Comput. Ind. 53(3), 283–296 (2004)
199. F. Puhlmann, A. Schnieders, J. Weiland, M. Weske, Variability mechanisms for process models, PESOA-Report TR 17/2005, Process Family Engineering in Service-Oriented Applications (PESOA), BMBF-Project (Hasso Plattner Institut, Postdam, Germany, 2005)
200. A.B. Raposo, H. Fuks, Defining task interdependencies and coordination mechanisms for collaborative systems, ed. by M. Blay-Fornarino, A.M. Pinna-Dery, K. Schmidt, P. Zaraté. In *Cooperative Systems Design*, Frontiers in Artificial Intelligence and Applications, vol. 74, Saint-Raphaël, France, 2002 (IOS Press, Amsterdam, 2002), pp. 88–103
201. M. Razavian, R. Khosravi, Modeling variability in business process models using UML, ed. by S. Latifi. In *Proceedings of the 5th International Conference on Information Technology: New Generations (ITNG'08)*, Las Vegas, NV, USA, April 2008 (IEEE Computer Society, MD, 2008), pp. 82–87
202. J. Recker, M. Indulska, M. Rosemann, P. Green, Do process modelling techniques get better? A comparative ontological analysis of BPMN. In *Proceedings of the 16th Australasian Conference on Information Systems (ACIS'05)*, Sydney, Australia, November 2005
203. M. Reichert, P. Dadam, ADEPTflex - supporting dynamic changes of workflows without losing control. J. Intell. Inform. Syst. **10**(2), 93–129 (1998)
204. M. Reichert, P. Dadam, T. Bauer, Dealing with forward and backward jumps in workflow management systems. Software Syst. Model. **2**(1), 37–58 (2003)
205. M. Reichert, S. Rinderle, U. Kreher, H. Acker, M. Lauer, P. Dadam, ADEPT next generation process management technology-tool demonstration, ed. by N. Boudjlida, D. Cheng, N. Guelfi. In *Forum Proceedings of the 18th Conference on Advanced Information Systems Engineering (CAiSE'06 Forum)*, CEUR Workshop Proceedings, vol. 231, Luxembourg, June 2006, pp. 11–14; CEUR-WS.org.
206. M. Reichert, S. Rinderle, U. Kreher, P. Dadam, Adaptive process management with ADEPT2. In *Proceedings of the 21st International Conference on Data Engineering (ICDE'05)*, Tokyo, Japan, April 2005 (IEEE Computer Society Press, MD, 2005), pp. 1113–1114
207. H.A. Reijers, W.M.P. van der Aalst, Short-term simulation: Bridging the gap between operational control and strategic decision making, ed. by M.H. Hamza. In *Proceedings of the*

IASTED International Conference on Modelling and Simulation, Philadelphia, PA, USA, May 1999 (IASTED/Acta Press, Canada, 1999), pp. 417–421
208. H.A. Reijers, W.M.P. van der Aalst, The effectiveness of workflow management systems: Predictions and lessons learned. Int. J. Inform. Manag. **56**(5), 457–471 (2005)
209. W. Reisig, in *Petri Nets, An Introduction*. EATCS, Monographs on Theoretical Computer Science (Springer, Berlin, 1985)
210. A. Reuter, F. Schwenkreis, ConTracts a low-level mechanism for building generalpurpose workflow management-systems. Data Eng. Bull. **18**(1), 4–10 (1995)
211. D. Richards, V. Chellen, P. Compton, The reuse of Ripple Down Rule knowledge bases: Using machine learning to remove repetition, ed. by P. Compton, R. Mizoguchi, H. Motoda, T. Menzies. In *Proceedings of the 2nd Pacific Knowledge Acquisition Workshop (PKAW'96)*, Coogee, Australia, October 1996 (UNSW Press, Australia, 1996), pp. 293–312
212. S. Rinderle, M. Reichert, P. Dadam, Correctness criteria for dynamic changes in workflow systems: a survey. Data Knowl. Eng. **50**(1), 9–34 (2004)
213. S. Rinderle, B. Weber, M. Reichert, W. Wild, Integrating process learning and process evolution: A semantics based approach, ed. by W.M.P. van der Aalst, B. Benatallah, F. Casati, F. Curbera. In *Proceedings of the 3rd International Conference on Business Process Management (BPM'05)*, Lecture Notes in Computer Science, vol. 3649, Nancy, France, September 2005 (Springer, Berlin, 2005), pp. 252–267
214. M. Rosemann, W.M.P. van der Aalst, A configurable reference modelling language. Inform. Syst. **32**(1), 1–23 (2007)
215. A. Rozinat, R.S. Mans, M. Song, W.M.P. van der Aalst, Discovering Colored Petri Nets from event logs. Int. J. Software Tool Tech. Tran. **10**(1), 57–74 (2008)
216. A. Rozinat, R.S. Mans, M. Song, W.M.P. van der Aalst, Discovering simulation models. Inf. Syst. **34**(3), 305–327 (2009)
217. A. Rozinat, M.T. Wynn, W.M.P. van der Aalst, A.H.M. ter Hofstede, C.J. Fidge, Workflow simulation for operational decision support using design, historic and state information, ed. by M. Dumas, M. Reichert, M-C. Shan. In *Proceedings of the 6th International Conference on Business Process Management (BPM'08)*, Lecture Notes in Computer Science, vol. 5240, Milan, Italy, September 2008 (Springer, Berlin, 2008), pp. 196–211
218. A. Rozinat, M.T. Wynn, W.M.P. van der Aalst, A.H.M. ter Hofstede, C.J. Fidge, Workflow simulation for operational decision support. Data Knowl. Eng. **68**(9), 834–850 (2009)
219. N. Russell, Foundations of Process-Aware Information Systems, PhD thesis, Queensland University of Technology, Brisbane, Australia, 2007
220. N. Russell, W.M.P. van der Aalst, A.H.M. ter Hofstede, Exception handling patterns in process-aware information systems. BPM Center Report BPM-06-04, BPMcenter.org (2006)
221. N. Russell, W.M.P. van der Aalst, A.H.M. ter Hofstede, Workflow exception patterns, ed. by E. Dubois, K. Pohl. In *Proceedings of the 18th International Conference on Advanced Information Systems Engineering (CAiSE'06)*, Lecture Notes in Computer Science, vol. 4001, Luxembourg, June 2006 (Springer, Berlin, 2006), pp. 288–302
222. N. Russell, W.M.P. van der Aalst, A.H.M. ter Hofstede, D. Edmond, Workflow resource patterns: Identification, representation and tool support, ed. by O. Pastor, J. Falcão e Cunha. In *Proceedings of the 17th Conference on Advanced Information Systems Engineering (CAiSE'05)*, Lecture Notes in Computer Science, vol. 3520, Porto, Portugal, June 2005 (Springer, Berlin, 2005), pp. 216–232
223. N. Russell, W.M.P van der Aalst, A.H.M. ter Hofstede, P. Wohed, On the suitability of UML 2.0 activity diagrams for business process modelling, ed. by M. Stumptner, S. Hartmann, Y. Kiyoki. In *Proceedings of the 3rd Asia-Pacific Conference on Conceptual Modelling (APCCM2006)*, CRPIT, vol. 53, Hobart, Australia, 2006 (Australian Computer Society, Australia, 2006), pp. 95–104
224. N. Russell, A.H.M. ter Hofstede, W.M.P. van der Aalst, N. Mulyar, Workflow control-flow patterns: A revised view. BPM Center Report BPM-06-22, BPMCenter.org (2006)
225. N. Russell, A.H.M. ter Hofstede, D. Edmond, W.M.P. van der Aalst, Workflow data patterns: Identification, representation and tool support, ed. by L. Delcambre, C. Kop, H.C. Mayr,

J. Mylopoulos, O. Pastor. In *Proceedings of the 24th International Conference on Conceptual Modeling (ER'05)*, Lecture Notes in Computer Science, vol. 3716, Klagenfurt, Austria, October 2005 (Springer, Berlin, 2005), pp. 353–368

226. H. Saastamoinen, G.M. White, On handling exceptions, ed. by N. Comstock, C. Ellis. In *Proceedings of the ACM Conference on Organizational Computing Systems (COCS'95)*, Milpitas, CA, USA, August 1995 (ACM, NY, 1995), pp. 302–310
227. SAP, *SAP Advanced Workflow Techniques*, SAP AG, 2006, https://www.sdn.sap.com/irj/servlet/prt/portal/prtroot/docs/library/uuid/82d03e23-0a01-0010-b482-dccfe1c877c4 (sighted 17 March 2007)
228. R.G. Sargent, Validation and verification of simulation models, ed. by R. Ingalls, M. Rossetti, J. Smith, B. Peters. In *Proceedings of the 2004 Winter Simulation Conference (WSC'04)*, Washington, DC, USA, December 2004 (ACM, NY, 2004), pp. 17–28
229. A.-W. Scheer, *CIM – Computer integrated manufacturing: der computergesteuerte Industriebetrieb* (in German) (Springer, Berlin, 1987)
230. A.-W. Scheer, *Business Process Engineering: Reference Models for Industrial Enterprises* (Springer, Berlin, 1994)
231. A.-W. Scheer, *ARIS – Business Process Frameworks*, 2nd edn. (Springer, Berlin, 1998)
232. A.-W. Scheer. *ARIS – Business Process Modeling*, 3rd edn. (Springer, Berlin, 2000)
233. A.-W. Scheer, O. Thomas, O. Adam, in *Process Modeling Using Event-Driven Process Chains*, ed. by M. Dumas et al. Process-Aware Information Systems: Bridging People and Software Through Process Technology (Wiley-Interscience, NY, 2005), pp. 119–146
234. T. Scheffer, Algebraic foundation and improved methods of induction of Ripple Down Rules, ed. by P. Compton, R. Mizoguchi, H. Motoda, T. Menzies. In *Proceedings of the 2nd Pacific Rim Workshop on Knowledge Acquisition (PKAW'96)*, Sydney, Australia, October 1996 (UNSW Press, Australia, 1996), pp. 279–292
235. B. Scholz-Reiter, E. Stickel (ed.), *Business Process Modelling* (Springer, Berlin, 1996)
236. M.H. Schonenberg, R.S. Mans, N.C. Russell, N. Mulyar, W.M.P. van der Aalst, Process flexibility: a survey of contemporary approaches, ed. by J.L.G. Dietz, A. Albani, J. Barjis. In *Advances in Enterprise Engineering I (4th International Workshop CIAO! and 4th International Workshop EOMAS, held at CAiSE'08)*, Lecture Notes in Business Information Processing, vol. 10, Montpellier, France, June 2008 (Springer, Berlin, 2008), pp. 16–30
237. S. Seidel, M. Adams, A. ter Hofstede, M. Rosemann, Modelling and supporting processes in creative environments. In *Proceedings of the 15th European Conference on Information Systems (ECIS'07)*, St. Gallen, Switzerland, June 2007
238. M. Snir, S. Otto, S. Huss-Lederman, D. Walker, J. Dongarra, *MPI – The Complete Reference: The MPI Core*, 2nd edn. (MIT, Cambridge, MA, 1998)
239. J. Spolsky, The law of leaky abstractions, November 2002, http://www.joelonsoftware.com/articles/LeakyAbstractions.html (sighted September 2008)
240. J.L. Staud, *Geschäftsprozessanalyse: Ereignisgesteuerte Prozessketten und Objektorientierte Geschäftsprozessmodellierung für Betriebswirtschaftliche Standardsoftware* (in German), 3rd edn. (Springer, Berlin, 2006)
241. M. Stefanelli, Knowledge and process management in health care organizations. Meth. Inform. Med. **43**(5), 525–535 (2004)
242. L.A. Stein, Challenging the computational metaphor: Implications for how we think. Cybern. Syst. **30**(6), 473–507 (1999)
243. R. van Stiphout, T. Dirk Meijler, A. Aerts, D. Hammer, R. le Comte, TREX: Workflow transaction by means of exceptions, ed. by H.-J. Schek, F. Saltor, I. Ramos, G. Alonso. In *Proceedings of the 6th International Conference on Extending Database Technology (EDBT'98)*, Lecture Notes In Computer Science, vol. 1377, Valencia, Spain, March 1998 (Springer, Berlin, 1998), pp. 21–26
244. D.M. Strong, S.M. Miller, Exceptions and exception handling in computerized information processes. ACM Trans. Inform. Syst. **13**(2), 206–233 (1995)
245. H. Tarumi, T. Matsuyama, Y. Kambayashi, Evolution of business processes and a process simulation tool. In *Proceedings of the 6th Asia-Pacific Software Engineering Conference (APSEC'99)*, Takamatsu, Japan, December 1999 (IEEE, MD, 1999), pp. 180–187

246. T.W. Tewoldeberhan, A. Verbraeck, S.S. Msanjila, Simulating process orchestrations in business networks: A case using BPEL4WS, ed. by Q. Li, T.-P. Liang. In *Proceedings of the 7th International Conference on Electronic Commerce (ICEC'05)*, Xi'an, China, August 2005 (ACM, NY, 2005), pp. 471–477
247. O. Thomas, A.-W. Scheer, Tool support for the collaborative design of reference models – a business engineering perspective. In *Proceedings of the 39th Hawaii International Conference on Systems Science (HICSS-39 2006)*, Kauai, HI, USA, January 2006 (IEEE Computer Society, MD, 2006).
248. K. Tumay, Business process simulation, ed. by J. Charnes, D. Morrice, D. Brunner, J. Swain. In *Proceedings of the 1996 Winter Simulation Conference (WSC'96)*, Coronado, CA, USA, December 1996 (ACM, NY, 1996), pp. 93–98
249. H.M.W. Verbeek, W.M.P. van der Aalst, A.H.M. ter Hofstede, Verifying workflows with cancellation regions and OR-joins: An approach based on relaxed soundness and invariants, Comput. J. **50**(3), 294–314 (2007)
250. H.M.W. Verbeek, T. Basten, W.M.P. van der Aalst, Diagnosing workflow processes using Woflan, Comput. J. **44**(4), 246–279 (2001)
251. H.M.W. Verbeek, M.T. Wynn, W.M.P. van der Aalst, A.H.M. ter Hofstede, Reduction rules for Reset/Inhibitor nets. BPM Center Report BPM-07-13, BPMcenter.org (2007)
252. G. de Vries, J.W.M. Bertrand, J.M.H. Vissers, Design requirements for health care production control systems. Prod. Plann. Contr. **10**(6), 559–569 (1999)
253. H.J. Vrijhoef, L.M. Steuten, Innovatieve zorgconcepten op een rij: integrated care (4) (in Dutch). Tijdschrift voor Gezondheidswetenschappen **83**(8), 513–514 (2005)
254. H.J. Vrijhoef, L.M. Steuten, Innovatieve zorgconcepten op een rij: managed care (2) (in Dutch). Tijdschrift voor Gezondheidswetenschappen **83**(6), 384–385 (2005)
255. H.J. Vrijhoef, L.M. Steuten, Innovatieve zorgconcepten op een rij: stepped care (5) (in Dutch). Tijdschrift voor Gezondheidswetenschappen **84**(1), 55–56 (2005)
256. J. Wainer, F. de Lima Bezerra, Constraint-based flexible workflows, ed. by J. Favela, D. Decouchant. In *Proceedings of the 9th International Workshop on Groupware: Design, Implementation, and Use (CRIWG 2003)*, Lecture Notes In Computer Science, vol. 2806, Autrans, France, September 2003 (Springer, Berlin, 2003), pp. 151–158
257. M. Wang, H. Wang, Intelligent agent supported flexible workflow monitoring system, ed. by A. Banks Pidduck, J. Mylopoulos, C.C. Woo, M. Tamer Ozsu. In *Proceedings of the 14th International Conference on Advanced Information Systems Engineering (CAiSE'02)*, Lecture Notes in Computer Science, vol. 2348, Toronto, Canada, May 2002 (Springer, Berlin, 2002), pp. 787–791
258. B. Weber, W. Wild, R. Breu, CBRFlow: Enabling adaptive workflow management through conversational case-based reasoning, ed. by P. Funk, P.A. González Calero. In *Proceedings of the 7th European Conference for Advances in Case Based Reasoning (ECCBR'04)*, Lecture Notes in Computer Science, vol. 3155, Madrid, Spain, August 2004 (Springer, Berlin, 2004), pp. 434–448
259. WebSphere, WebSphere MQ family homepage, http://www-01.ibm.com/software/integration/wmq/ (sighted 1 December 2008)
260. M. Weske, *Business Process Management: Concepts, Languages, Architectures* (Springer, Berlin, 2007)
261. S. White, in *Process Modeling Notations and Workflow Patterns*, ed. by L. Fischer. Workflow Handbook 2004 (Future Strategies Inc, Lighthouse Point, FL, 2004), pp. 265–294
262. P. Wohed, W.M.P. van der Aalst, M. Dumas, A.H.M. ter Hofstede, Analysis of web services composition languages: The case of BPEL4WS, ed. by I.Y. Song, S.W. Liddle, T.W. Ling, P. Scheuermann. In *Proceedings of the 22nd International Conference on Conceptual Modeling (ER'03)*, Lecture Notes in Computer Science, vol. 2813, Chicago, IL, USA, October 2003 (Springer, Berlin, 2003), pp. 200–215
263. P. Wohed, W.M.P. van der Aalst, M. Dumas, A.H.M. ter Hofstede, N. Russell, Pattern-based analysis of the control-flow perspective of UML activity diagrams, ed. by L. Delcambre, C. Kop, H. Mayr, J. Mylopoulos, O. Pastor. In *Proceedings of the 24th International Conference on Conceptual Modeling (ER'05)*, Lecture Notes in Computer Science, vol. 3716, Klagenfurt, Austria, October 2005 (Springer, Berlin, 2005), pp. 63–78

264. P. Wohed, W.M.P. van der Aalst, M. Dumas, A.H.M. ter Hofstede, N. Russell, On the suitability of BPMN for business process modelling, ed. by S. Dustdar, J.L. Fiadeiro, A. Sheth. In *Proceedings of the 4th International Conference on Business Process Management (BPM'06)*, Lecture Notes in Computer Science, vol. 4102, Vienna, Austria, September 2006 (Springer, Berlin, 2006), pp. 161–176
265. P. Wohed, E. Perjons, M. Dumas, A.H.M. ter Hofstede, Pattern based analysis of EAI languages – the case of the business modeling language, ed. by O. Camp, M. Piattini. In *Proceedings of the 5th International Conference on Enterprise Information Systems (ICEIS 2003)*, vol. 3, Angers, France, April 2003 (Escola Superior de Tecnologia do Instituto Politecnico de Setubal, Portugal, 2003), pp. 174–184
266. P. Wohed, N. Russell, A.H.M. ter Hofstede, B. Andersson, W.M.P. van der Aalst, Patterns-based evaluation of open source BPM systems: The cases of jBPM, OpenWFE, and Enhydra Shark. Inform. Software Tech. **51**(8), 1187–1216 (2009)
267. P.Y.H. Wong, J. Gibbons, A process semantics for BPMN, ed. by S. Liu, T.S.E. Maibaum, K. Araki. In *Proceedings of the 10th International Conference on Formal Engineering Methods (ICFEM'08)*, Lecture Notes in Computer Science, vol. 5256, Kitakyushu-City, Japan, October 2008 (Springer, Berlin, 2008), pp. 355–374
268. D. Worah, A.P. Sheth, Transactions in Transactional Workflows, ed. by S. Jajodia, L. Kerschberg. In *Advanced Transaction Models and Architectures* (Kluwer, Norwell, MA, 1997), pp. 3–34
269. Workflow Management Coalition, Workflow reference model, (Workflow Management Coalition, Winchester, UK, January 1995), http://www.wfmc.org/standards/docs/tc003v11.pdf (sighted 1 April 2009)
270. Workflow Management Coalition, Terminology and glossary, Reference WFMC-TC-1011, Issue 3.0, (Workflow Management Coalition, Winchester, UK, 1999), http://www.wfmc.org/standards/docs/TC-1011_term_glossary_v3.pdf (sighted 1 April 2009)
271. Workflow Management Coalition, Introduction to workflow, (Workflow Management Coalition, Winchester, UK, 2002) http://www.wfmc.org/introduction_to_workflow.pdf (sighted 14 April 2009)
272. M.T. Wynn, *Semantics, Verification, and Implementation of Workflows with Cancellation Regions and OR-joins*, PhD Thesis, Queensland University of Technology, Brisbane, Australia, 2006
273. M.T. Wynn, D. Edmond, W.M.P. van der Aalst, A.H.M. ter Hofstede, Achieving a general, formal and decidable approach to the OR-Join in workflow using reset nets, ed. by G. Ciardo, P. Darondeau. In *Proceedings of the 26th International Conference on Application and Theory of Petri nets and Other Models of Concurrency*, Lecture Notes in Computer Science, vol. 3536, Miami, FL, USA, June 2005 (Springer, Berlin, 2005), pp. 423–443
274. M.T. Wynn, W.M.P. van der Aalst, A.H.M. ter Hofstede, D. Edmond, Verifying workflows with cancellation regions and OR-Joins: An approach based on reset nets and reachability analysis, ed. by S. Dustdar, J. Fiadeiro, A. Sheth. In *Proceedings of the 4th International Conference on Business Process Management (BPM'06)*, Lecture Notes in Computer Science, vol. 4102, Vienna, Austria, September 2006 (Springer, Berlin, 2006), pp. 389–394
275. M.T. Wynn, W.M.P. van der Aalst, A.H.M. ter Hofstede, D. Edmond, Synchronization and cancelation in workflows based on reset nets, Int. J. Cooper. Inform. Syst. **18**(1), 63–114 (2009)
276. M.T. Wynn, H.M.W. Verbeek, W.M.P. van der Aalst, A.H.M. ter Hofstede, D. Edmond, Reduction rules for YAWL workflows with cancellation regions and OR-joins. Inform. Software Tech. **51**(6), 1010–1020 (2009)
277. M.T. Wynn, H.M.W. Verbeek, W.M.P. van der Aalst, A.H.M. ter Hofstede, D. Edmond, Soundness-preserving reduction rules for reset workflow nets, Inform. Sci. **179**(6), 769–790 (2009)
278. J. Yan, Y. Yang, G.K. Raikundalla, Towards incompletely specified process support in SwinDeW - a peer-to-peer based workflow system, ed. by W. Shen, Z. Lin, J-P.A. Barthès,

T. Li. In *Proceedings of the 8th International Conference on Computer Supported Cooperative Work in Design (CSCWD 2004)*, Lecture Notes in Computer Science, vol. 3168, Xiamen, China, May 2004 (Springer, Berlin, 2004), pp. 328–338

279. L. Zeng, J-J. Jeng, S. Kumaran, J. Kalagnanam, Reliable execution planning and exception handling for business process, ed. by B. Benatallah, M.-C. Shan. In *Proceedings of the 4th International Workshop for Technologies for E-Services (TES 2003)*, Lecture Notes in Computer Science, vol. 2819, Berlin, Germany, September 2003 (Springer, Berlin, 2003), pp. 119–130
280. O. Zukunft, F.J. Rump, From Business Process Modelling to Workflow Management: An Integrated Approach, ed. by B. Scholz-Reiter, E. Stickel. In *Business Process Modelling* (Springer, Berlin, 1996), pp. 3–22

Index

Symbols

A

B

C

D

E

F

G

H

I

J

L

M

Q

R

S

T

U

V

W

Z

Cited Author Index

Acronyms

AFTRS Australian Film Television and Radio School

AMC Academic Medical Center

BPA Business Process Automation

BPD Business Process Diagram

BPEL Business Process Execution Language

BPM Business Process Management

BPML Business Process Modeling Language

BPMN Business Process Modeling Notation

BPMS Business Process Management System

C-EPC Configurable Event-driven Process Chain

C-iEPC Configurable Integrated Event-driven Process Chain

C-YAWL Configurable YAWL

CORBA Common Object Request Broker Architecture

CPN Coloured Petri nets

CRM Customer Relationship Management

CTL Computational Tree Logic

EAI Enterprise Application Integration

EPC Event-driven Process Chain

EPML EPC Markup Language

ERP Enterprise Resource Planning

ITIL Information Technology Infrastructure Library

JMS Java Message Service

JSF Java Server Faces

LTL Linear Temporal Logic

LTS Labeled Transition System

MSMQ Microsoft Message Queues

OASIS Organization for the Advancement of Structured Information Standards

OMG Object Management Group

ORM Object Role Modeling

PDM Product Data Management

POX Plain Old XML

ProM Process Mining Framework

RDR Ripple-Down Rules

REST Representational State Transfer

RMI Remote Method Invocation

RPC Remote Procedure Call

SCOR Supply Chain Operations Reference Model

SMTP Simple Mail Transfer Protocol

SOA Service Oriented Architecture

SOAP Simple Object Access Protocol

SSADM Structured Systems Analysis and Design Method

TCP Transmission Control Protocol

TWE Together Workflow Editor

UDDI Universal Description Discovery and Integration

UML Unified Modeling Language

VDM Vienna Development Method

VICS Voluntary Inter-industry Commerce Solutions

WfMC Workflow Management Coalition

WfMS Workflow Management System

Woflan Workflow Analyzer

WoPeD Workflow Petri Net Designer

WRM Workflow Reference Model

WSDL Web Services Description Language

WSFL Web Services Flow Language

XML Extensible Markup Language

XPDL XML Process Definition Language

YAWL Yet Another Workflow Language

Useful Websites

The following is a list of web sites related to this book that readers may find useful.

`www.yawlbook.com`

This site is an extension to this book, and contains errata, links and resources, and other material related to the book's topics.

`www.yawlfoundation.org`

The "official" YAWL site, the central site for all things YAWL. From here, extensive information on YAWL publications, case studies, research, workflow patterns, teaching resources, manuals, videos, and links to most of the other sites listed. Also, those interested in contributing to the YAWL Environment can find information and download the *Contributor's Agreement* from this site.

`sourceforge.net/projects/yawl/`

The host site for the open source YAWL system. From this site, offical YAWL releases, source code, and associated tools can be downloaded. This site also hosts a discussion forum and a documentation repository, containing user guides and technical manuals for the various YAWL system versions.

`www.workflowpatterns.com`

On this web site you will find detailed descriptions of patterns for the various perspectives relevant for process-aware information systems: control-flow, data, resource, and exception handling – many with explanatory animations. In addition, you will find detailed evaluations of various process languages, (proposed) standards for web service compositions, and workflow systems in terms of these patterns.

www.bpmcenter.org

The BPM Center is a collaborative virtual research centre in the area of Business Process Management (BPM), and aims to address aspects concerning all phases of the lifecycle of business processes, including modeling, verification, enactment, monitoring, mining, and diagnosis.

www.processmining.org

The Process Mining web site, containing research publications, presentations, and videos pertaining to process mining (extracting information from event logs). Also, the ProM Environment, its associated tools and plug-ins, and documentation can be sourced here.

www.processconfiguration.com

A site dedicated to documenting research and experiences in the area of business process configuration. The Synergia toolset for process configuration and associated documentation can be downloaded from this site.

www.yawl4film.com

This site contains resources for YAWL4Film, which extends YAWL with a suite of tools specifically designed for the film industry. Publications, instructional videos, and other information about the YAWL4Film initiative can be found here.

declare.sf.net

The Declare Service and associated documentation and publications can be downloaded from this site.